Photoelectrochemical Solar Fuel Production

Sixto Giménez • Juan Bisquert
Editors

# Photoelectrochemical Solar Fuel Production

## From Basic Principles to Advanced Devices

Springer

*Editors*
Sixto Giménez
Institute of Advanced Materials (INAM)
Universitat Jaume I
Castelló de la Plana, Spain

Juan Bisquert
Institute of Advanced Materials (INAM)
Universitat Jaume I
Castelló de la Plana, Spain

ISBN 978-3-319-29639-5 ISBN 978-3-319-29641-8 (eBook)
DOI 10.1007/978-3-319-29641-8

Library of Congress Control Number: 2016936880



Printed on acid-free paper

This Springer imprint is published by Springer Nature
The registered company is Springer International Publishing AG Switzerland

*To Professor Pedro Salvador, a pioneer of photoelectrochemistry*

# Preface

The supply of water, food, and energy on a global scale constitutes the most relevant challenge to sustain humankind development in the twenty-first century. Focusing on energy, the wide availability of resources is essential to meet the needs of a growing world population with increasing living standards and hence high energy demand. In this context, chemical fuels constitute a key energy vehicle in the global scheme, since energy is available in a form that can be easily stored and used upon demand. Fossil fuels are excellent for transportation and generation of electricity and heat, but they arise from finite reserves that will be depleted in a few decades. In addition, burning oil and natural gas produces heavy pollution and greenhouse effect, leading to progressive warming of the earth crust with the concomitant risks of climate change. Consequently, we urgently need chemical fuels that can be derived from a widely available source, and that yield a benign residue when converted into other energy vectors. The production of a green and renewable chemical fuel is a major challenge for science and technology.

Water and sunlight are abundant and decentralized natural resources, which constitute ideal candidates for fuel production. Sunlight is the largest readily available source of energy in the world, and water can be split into $O_2$ and $H_2$, the latter being one of the most valuable energy vectors due to its high energy density and its clean combustion in fuel cells. Turning solar photons into a convenient fuel that stores energy as chemical bonds, either $H_2$ gas or a hydrocarbon derived by further dark reactions, has been a longstanding goal of research, particularly at those times when conventional energy sources experienced trouble. Inspired by natural photosynthesis, the process satisfying the energy needs of all living matter on earth, we can prepare inorganic or organic semiconductor materials that absorb photons and convert them into energetic electronic carriers. These carriers endowed with excess free energy can launch desired chemical and electrochemical reactions, provided that their forward pathway probability is larger than recombination rates.

Semiconductor materials have demonstrated their ability to split water into separate $O_2$ and $H_2$ gases, or to reduce $CO_2$, and have been investigated for over

40 years in the scientific field of photoelectrochemistry. The oil scarcity during the supply crisis in the 1970s was a major driving force to search for a semiconductor material capable of extensive energy generation based on the photoelectrochemical properties of the semiconductor/electrolyte interface. In principle one could employ abundant materials processed by low-cost synthetic procedures (as titanium dioxide or iron oxide) for the large-scale production of solar fuels. However, in practice, these semiconductors face important limitations, rendering the conversion efficiency from solar photons to hydrogen very low. It is well recognized that these materials are useful model systems but far from a technologically viable solution.

There exist a number of stringent requirements for a semiconductor to transform water into hydrogen at high performance, assisted only by the solar radiation. A principal factor impeding the viability of such device is the difficult balance between two opposing properties. The first is the necessity of a narrow semiconductor bandgap to realize the optical absorption of a large portion of the solar spectrum, as photons with energy less than the bandgap cannot be absorbed. On the other hand, in order to drive water oxidation and reduction reactions involving significant overpotential, a separation of valence and conduction band in the semiconductor quite in excess of the water splitting free energy of 1.23 eV is needed. Furthermore, in order to efficiently realize all the functions of charge separation, collection, and charge transfer across a semiconductor/electrolyte interface, excellent semiconductor properties are needed, including high conductivity, low recombination rates, and excellent surface catalytic properties. Another important drawback is related to the fact that some materials with the required properties lack the necessary stability under photoelectrochemical operation due to decomposition caused by the photogenerated carriers. All in all, the "artificial photosynthesis" based on a single successful semiconductor photoelectrode could not be realized.

Nevertheless, the demand for renewable primary energy vectors such as solar-based $H_2$ or hydrocarbon fuel constitutes an urgent priority in the near future global energy context. Recently, the research community has started a sustained effort towards the realization of the photoelectrochemical solar fuel production, adopting new avenues of research and shaping a dynamic and rapidly growing field. The current activity has benefited from the basic "classical" knowledge on semiconductor photoelectrochemistry established in the last century, combined with an expanded array of innovative tools and scientific directions, and together with the technological research needed for the demonstration of viable device engineering methods. The present research on solar fuel production applies and develops new methods of nanomaterials production, incorporates advanced photovoltaic devices, designs original catalytic materials and coatings, and introduces new semiconductor materials, as well as new nanoscale, electrochemical, optoelectronic, computational, and surface characterization techniques. This development of the field adopts a broad perspective with respect to the candidate materials and device configurations to target a low-cost high-conversion-efficiency and durable device. A combination of mixed metal oxides, nonoxide conventional semiconductors, and molecular systems is exploited in order to find the most suitable materials and interfaces. The use of combinatorial methods for high-throughput studies involves the investigation of ternary or quaternary compositions, maximizing the possibilities to optimize optical, electronic, and catalytic properties contributing to efficient solar fuel production.

The combination of photoelectrochemical and photovoltaic concepts, catalysis, advanced materials and nanostructuring strategies, facilitates a variety of routes, which may produce important progress towards effective applications. Indeed, different approaches towards reliable technological devices coexist in the research arena, which show potential advantages as well as drawbacks. As mentioned earlier, it is well established that capturing sunlight with a single junction material is a too demanding task for a high voltage application as water splitting. Consequently, one of the most promising approaches involves the combination of different light absorbing materials with complementary spectral properties adding up the voltages in *tandem* connection. The strategy to separate device functions in different materials and interfaces relaxes the stringent conditions that a single semiconductor must satisfy to perform all water splitting steps. Therefore light active semiconductors can be coupled with suitable catalysis layers. However, when different junctions, parts, and connections are required, there is a price to pay in terms of increased complexity and decreased integration. Therefore charge separation at semiconductor/electrolyte junctions, where electrochemical reactions also occur, continues to be a favorite research topic. On the other extreme, one may couple effective PV cells with dark catalysis. As a compromise, a photovoltaic cell can supply additional voltage to a photoelectrochemical cell.

In summary, a wide variety of approaches exists towards the established goal and often involves device configurations that use different types of fundamental properties. However, a strong connection occurs between the different methods. The innovations and discovery obtained in the massive research effort towards the solar fuel production eventually may become of broader interest and significance. Advances in one particular topic of research are likely to fertilize others. This is why it appears useful for students and researchers to present a general perspective of the field on a unified basis, starting from the fundamental knowledge set of established photoelectrochemical and catalysis concepts, and moving to a variety of materials and devices that constitute the vanguard of investigation in this field. This is the general purpose of this book, aiming at introducing the main trends of current research in photoelectrochemical solar fuel production from the standpoint of a firm scientific basis.

The book is organized in three topical parts, starting with fundamentals, followed by the experimental protocols and characterization techniques. The final and larger part explains the main types of materials and devices applications. The introductory chapter by L. M. Peter is a summary of the thermodynamics and kinetics properties of the semiconductor/electrolyte interface, which provides an understanding of how band bending combined with electron transfer model explains the photocurrent. This analysis shows the central role of surface catalysis in all fuel production cells; hence two chapters examine the central reactions that are considered today for solar fuel production. Doyle and Lyons explain the properties of the oxygen evolution reaction, which slow kinetics is a major barrier for the expansion of solar fuel production. Sudhagar, Nakata, and coworkers discuss the materials and properties for catalytic photoreduction to form hydrogen gas or to reduce $CO_2$.

The second part of the book addresses the experimental methods that are used to understand and guide the development of solar fuels materials, interfaces, and devices. First W. Smith provides a broad and detailed view of the main types of devices that are currently investigated, and he describes their parts, principles of operation, as well as the methods of measurement for the characterization of operational properties. Jaegermann, Kaiser, and coworkers critically discuss the different alternatives of solar fuel production approaches. The second part of the chapter gives fundamental insights into the structure of the critical interfaces, like the formation of semiconductor/passivation layer/co-catalyst/electrolyte interfaces. A detailed view of such properties is obtained by extensive application of photoelectron spectroscopies. Bisquert, Giménez, and coworkers describe the recent advances in the application of impedance spectroscopy technique in photoelectrochemical system, aiming to develop a profound characterization of kinetic steps that occur in the formation of photocurrent, with particular attention to the role of surface states in the competition between charge transfer and recombination. These previous two chapters recognize the complexity of the catalyzed interface and the many interactions and effects that determine the dynamics of the photoelectrochemical behavior. Subsequently Lin and Boettcher address this central problem and describe advanced characterization methods able to examine separate components of the interface during operation.

The last part of the book is devoted to the description of relevant materials and devices for photoelectrochemical solar fuel production. The first four chapters provide a detailed panorama of the most advanced materials developed in the last years. Abdi, Berglund, and Van de Krol describe the current research status of multinary metal oxides and discuss the main challenges of using these materials as photoelectrode materials along with future outlook for their application. Fujii provides an extensive description of nonoxide materials, particularly nitrides, chalcogenides, and arsenides, emphasizing the controllability of their electronic properties and the strategies needed to enhance the stability of these materials under solar fuel production conditions. Skorupska and Parkinson review the main advances on combinatorial synthesis and screening of oxide-based materials providing a detailed overview of the experimental methods employed for high-throughput combinatorial studies, focusing on oxygen evolution photoanodes, hydrogen evolution photocathodes, and electrocatalysts for both oxygen and hydrogen evolution. Thorne, He, and Wang discuss how the nanoscale synthesis of materials entails both advantages and disadvantages for the performance of semiconductor photoelectrodes. They describe the morphologies of nanostructured materials as well as some of the most relevant synthetic methods and critically discuss the performance of nanostructured photoelectrodes compared with their bulk counterparts. The last two chapters of this part illustrate how optimized materials can be integrated in advanced devices, based on different principles, targeting technological exploitation of photoelectrochemical solar fuel production. Sivula extensively describes how the use of tandem devices, combining two or more semiconductor junctions, can lead to efficient devices with up to 30 % conversion efficiency even assuming large losses. Particularly, key examples of tandem cells are discussed, taking into account the choice of semiconductor material, the design geometry of the device with respect to resistance losses due to membranes and ionic conduction, and techno-economic considerations. Finally, Coggins and Meyer provide an overview of the dye-sensitized photoelectrosynthesis cell devices, covering the fundamentals of the device concept, as well as the different strategies employed to achieve competitive efficiencies in prototype device.

Castelló de la Plana, Spain

Sixto Giménez
Juan Bisquert

# Contents

# List of Acronyms

| | |
|---|---|
| AACVD | Aerosol-assisted chemical vapor deposition |
| AAO | Anodized aluminum oxide |
| ABPE | Applied bias photon to current conversion efficiency |
| AEM | Anion exchange membrane |
| AIST | (National Institute of)Advance Industrial Science and Technology |
| ALD | Atomic layer deposition |
| AM | Air mass |
| APCE | Absorbed photon to current conversion efficiency |
| APCVD | Atmospheric pressure chemical vapor deposition |
| APT | Atom-proton transfer |
| ARUPS | Angle-resolved UPS |
| AZO | Aluminum doped zinc oxide |
| bb | Band bending |
| BPM | Bipolar membrane |
| BSCF | $Ba_{0.5}Sr_{0.5}Co_{0.8}Fe_{0.2}O_{3-d}$ |
| CB | Conduction band |
| CBD | Chemical bath deposition |
| CE | Counter electrode |
| CEM | Cation exchange membrane |
| CHT | Combinatorial high throughput |
| CNL | Charge neutrality level |
| CoBi | Cobalt-based catalysts deposited from borate electrolytes |
| Co-OEC | Co oxygen evolved complex |
| CoPi | Cobalt-based catalysts deposited from phosphate electrolytes |
| CUS | Coordinately unsaturated |
| CV | Cyclic voltammetry |
| CVD | Chemical vapor deposition |
| DC | Direct current |
| DFT | Density functional theory |
| DI | Distillate water |

| | |
|---|---|
| DOE | U.S. Department of Energy |
| DOS | Density of states |
| DSPEC | Dye-sensitized photoelectrosynthesis cells |
| DSSC | Dye-sensitized solar cell |
| DWE | Dual-working-electrode |
| EATP | Electron-atom-proton transfer |
| EC | Electrochemically prepared amorphous oxide |
| EC | Electrochemical converter |
| EC | Equivalent circuit |
| EC | Electrocatalyst |
| ECR | Electron cyclotron resonance |
| EDX | Energy dispersive X-ray spectroscopy |
| EFRC | Energy Frontier Research Center |
| EPR | Electron paramagnetic resonance |
| EQE | External quantum efficiency |
| ESCA | Electron spectroscopy for chemical analysis |
| eV | Electron volts |
| FF | Fill factor |
| FTO | Fluorine doped tin oxide |
| FVD | Flame vapor deposition |
| FWHM | Full width at half-maximum |
| g | Gaseous |
| HArPoON | Heterogeneous anodes rapidly perused for oxygen overpotential neutralization |
| HBE | Hydrogen binding energy |
| HEC | Hydrogen evolution catalyst |
| HER | Hydrogen evolution reaction |
| HF | Hydrofluoric acid |
| HOMO | Highest occupied molecular orbital |
| HTE | High-throughput experimentation |
| IMPS | Intensity-modulated photocurrent spectroscopy |
| IPCE | Incident photon to current conversion efficiency |
| IQE | Internal quantum efficiency |
| IS | Impedance spectroscopy |
| ITO | Indium doped tin oxide |
| JCAP | Joint Center for Artificial Photosynthesis |
| l | Liquid |
| LCAO | Linear combination of atomic orbitals |
| LED | Light-emitting diode |
| LMMR | Light-modulated microwave reflectance |
| MBE | Molecular beam epitaxy |
| MCC | Molecular cobaltate clusters |
| MD | Molecular dynamics |
| MeCN | Acetonitrile |

| | |
|---|---|
| MIGS | Metal-induced gap states |
| MIS | Metal insulator semiconductor |
| MLCT | Metal to ligand charge transfer |
| MOCVD | Metal-organic chemical vapor deposition |
| MOS | Metal oxide semiconductor |
| MPP | Maximum power point |
| NHE | Normal hydrogen electrode |
| oc | Open circuit |
| occ | Occupied |
| OCP | Open circuit potential |
| OEC | Oxygen evolution catalyst |
| OEC | Oxygen evolving complex |
| OER | Oxygen evolution reaction |
| OHP | Outer Helmholtz plane |
| op | Operational |
| ORR | Oxygen reduction reaction |
| OSDC | Optical scanning droplet cell |
| ox | Oxidized |
| PC | Propylene carbonate |
| PCET | Proton-coupled electron transfer |
| PDS | Potential determining step |
| PEC | Photoelectrochemical/photoelectrochemistry |
| PECVD | Plasma-enhanced chemical vapor deposition |
| PEM | Proton exchange membrane |
| PES | Photoelectron spectroscopy |
| ph | Photo |
| PL | Photoluminescence |
| PLD | Pulsed laser deposition |
| PMMA | Poly(methyl methacrylate) |
| PS | Polystyrene |
| PSII | Photosystem II |
| PTFE | Polytetrafluoroethylene |
| PV | Photovoltaic |
| QD | Quantum dot |
| QDSSC | Quantum dot sensitized solar cell |
| RDS | Rate determining step |
| RE | Reference electrode |
| red | Reduced |
| RHE | Reversible hydrogen electrode |
| SC | Semiconductor |
| SCE | Standard calomel electrode |
| SCLI | Semiconductor electrolyte interface |
| SCR | Space charge region |
| SEAL | Solar Energy Activity Laboratory |

| | |
|---|---|
| SECM | Scanning electrochemical microscopy |
| SEM | Scanning electron microscopy |
| SETI | Search for Extraterrestrial Intelligence |
| SHArK | Solar hydrogen activity research kit |
| SHE | Standard hydrogen electrode |
| SLJ | Semiconductor liquid junction |
| SoLiAS | Solid/liquid analysis system |
| SPM | Scanning probe microscope |
| STFCE | Solar-to-fuel conversion efficiency |
| STH | Solar-to-hydrogen conversion efficiency |
| STHCE | Solar-to-hydrogen conversion efficiency |
| SWE | Single-working-electrode |
| SXPS | Soft X-ray photoelectron spectroscopy |
| TC | Thermally prepared crystalline oxide |
| TCO | Transparent conducting oxide |
| TEM | Transmission electron microscopy |
| UHV | Ultra-high vacuum |
| UNC | University of North Carolina at Chapel Hill |
| UNC EFRC | University of North Carolina at Chapel Hill Energy Frontier Research Center |
| unocc | Unoccupied |
| UPS | Ultraviolet photoelectron spectroscopy |
| UV | Ultraviolet |
| UV–vis | Ultraviolet–visible (spectroscopy) |
| vac | Vacuum level |
| VB | Valence band |
| VLS | Vapor-liquid-solid |
| WE | Working electrode |
| XANES | X-ray absorption near edge spectroscopy |
| XAS | X-ray absorption spectroscopy |
| XPS | X-ray photoelectron spectroscopy |
| XRC | X-ray rocking curve |
| XRD | X-ray diffraction |
| YAG | Yttrium aluminum garnet |

# Contributors

**Fatwa F. Abdi** Helmholtz-Zentrum Berlin für Materialien und Energie GmbH, Institute for Solar Fuels, Berlin, Germany

**Sean P. Berglund** Helmholtz-Zentrum Berlin für Materialien und Energie GmbH, Institute for Solar Fuels, Berlin, Germany

**Luca Bertoluzzi** Institute of Advanced Materials (INAM), Universitat Jaume I, Castelló, Spain

**Juan Bisquert** Institute of Advanced Materials (INAM), Universitat Jaume I, Castelló, Spain

**Shannon W. Boettcher** Department of Chemistry and Biochemistry, University of Oregon, Eugene, OR, USA

**Michael K. Coggins** Department of Chemistry, University of North Carolina at Chapel Hill, Chapel Hill, NC, USA

Eastman Chemical Company, Kingsport, TN, USA

**Anitha Devadoss** Photocatalysis International Research Center (PIRC), Tokyo University of Science, Noda, Chiba, Japan

**Richard L. Doyle** Electrochemical Materials and Energy Group, Tyndall National Institute, University College Cork, Cork, Ireland

**Katsushi Fujii** Global Solar Plus Initiative, The University of Tokyo, Tokyo, Japan

**Akira Fujishima** Photocatalysis International Research Center (PIRC), Tokyo University of Science, Noda, Chiba, Japan

Research Institute for Science & Technology, Tokyo University of Science, Noda, Chiba, Japan

**Sixto Giménez** Institute of Advanced Materials (INAM), Universitat Jaume I, Castelló, Spain

**Yumin He** Department of Chemistry, Boston College, Chestnut Hill, MA, USA

**Isaac Herraiz-Cardona** Institute of Advanced Materials (INAM), Universitat Jaume I, Castelló, Spain

**Wolfram Jaegermann** Institute of Materials Science and Excellency Graduate School for Energy Science and Engineering, Jovanka-Bontschits-Straße 2, Technical University Darmstadt, Darmstadt, Germany

**Bernhard Kaiser** Institute of Materials Science and Excellency Graduate School for Energy Science and Engineering, Jovanka-Bontschits-Straße 2, Technical University Darmstadt, Darmstadt, Germany

**Joachim Klett** Institute of Materials Science and Excellency Graduate School for Energy Science and Engineering, Jovanka-Bontschits-Straße 2, Technical University Darmstadt, Darmstadt, Germany

**Roel van de Krol** Helmholtz-Zentrum Berlin für Materialien und Energie GmbH, Institute for Solar Fuels, Berlin, Germany

**Fuding Lin** Department of Chemistry and Biochemistry, University of Oregon, Eugene, OR, USA

**Michael E.G. Lyons** Trinity Electrochemical Energy Conversion and Electrocatalysis Group, School of Chemistry & AMBER National Centre, CRANN Research Institute, Trinity College Dublin, Dublin 2, Ireland

**Thomas J. Meyer** Department of Chemistry, University of North Carolina at Chapel Hill, Chapel Hill, NC, USA

**Kazuya Nakata** Photocatalysis International Research Center (PIRC), Tokyo University of Science, Noda, Chiba, Japan

Research Institute for Science & Technology, Tokyo University of Science, Noda, Chiba, Japan

**Bruce A. Parkinson** Center for Photoconversion & Catalysis, University of Wyoming, Laramie, WY, USA

**Laurence M. Peter** Department of Chemistry, University of Bath, Bath, UK

**Nitish Roy** Photocatalysis International Research Center (PIRC), Tokyo University of Science, Noda, Chiba, Japan

**Kevin Sivula** Institute of Chemical Sciences and Engineering, École Polytechnique Fédérale de Lausanne, Lausanne, Switzerland

**Katarzyna Skorupska** Department of Chemistry, University of Wyoming, Laramie, WY, USA

Center for Photoconversion & Catalysis, University of Wyoming, Laramie, WY, USA

**Wilson A. Smith** Materials for Energy Conversion and Storage (MECS), Department of Chemical Engineering, Faculty of Applied Sciences, Delft University of Technology, Delft, The Netherlands

**Pitchaimuthu Sudhagar** Photocatalysis International Research Center (PIRC), Tokyo University of Science, Noda, Chiba, Japan

**Chiaki Terashima** Photocatalysis International Research Center (PIRC), Tokyo University of Science, Noda, Chiba, Japan

Research Institute for Science & Technology, Tokyo University of Science, Noda, Chiba, Japan

**James E. Thorne** Department of Chemistry, Boston College, Chestnut Hill, MA, USA

**Raman Vedarajan** School of Materials Science, Japan Advanced Institute of Science and Technology (JAIST), Nomi, Ishikawa, Japan

**Dunwei Wang** Department of Chemistry, Boston College, Chestnut Hill, MA, USA

**Jürgen Ziegler** Institute of Materials Science and Excellency Graduate School for Energy Science and Engineering, Jovanka-Bontschits-Straße 2, Technical University Darmstadt, Darmstadt, Germany

# Part I
# Fundamentals

# Chapter 1
# Semiconductor Electrochemistry

Laurence M. Peter

## 1.1 Introduction

The advent of semiconductor electrochemistry can be traced back at least as far as the mid-1950s to Brattain and Garrett's experiments on the germanium/electrolyte interface (Brattain and Garrett 1955) and Letaw and Bardeen's paper (Letaw and Bardeen 1954) on the "electrolytic analog transistor", which described the characteristics of a system in which a germanium electrode was contacted by a variety of electrolytes, including redox couples. The fundamental principles of semiconductor electrochemistry were established during the 1970s in work on monocrystalline semiconductor electrodes by Gerischer (1961), Pleskov (Pleskov and Gurevich 1986), Memming (1978), Bard (1980), and others. An excellent account of basic concepts as well as a review of work before 1980 is provided by Morrison's book *Electrochemistry at Semiconductor and Oxidized Metal Electrodes* (Morrison 1980), which includes a comprehensive bibliography. A more up-to-date book is Memming's *Semiconductor Electrochemistry* (Memming 2015), which will be released as a second edition in early 2015. Other useful books include Sato's *Electrochemistry at Metal and Semiconductor Electrodes* (Sato 1998). In addition, volume 6 of the *Encyclopedia of Electrochemistry* (Licht et al. 2002) deals exclusively with semiconductor electrodes and photoelectrochemistry. A lucid discussion of many fundamental aspects of semiconductor junctions can be found in Würfel's recently updated book *Physics of Solar Cells* (Würfel and Würfel 2009). The reader is directed to these sources for more detailed treatments of some of the topics covered in this chapter.

L.M. Peter (✉)
Department of Chemistry, University of Bath, Bath BA2 7AY, UK
e-mail: l.m.peter@bath.ac.uk



S. Giménez, J. Bisquert (eds.), *Photoelectrochemical Solar Fuel Production*,
DOI 10.1007/978-3-319-29641-8_1

## 1.2 A Brief Summary of Semiconductor Physics

### 1.2.1 Energy Bands and the Fermi-Dirac Distribution

Semiconductors are materials in which the electron energy levels are grouped into valence and conduction bands that are separated by a forbidden energy gap. The density of states distributions corresponding to the valence and conduction bands are generally rather complicated functions of energy (see Fig. 1.1a), but near the band edges they often approach the simple parabolic forms shown in Fig. 1.1b. In this chapter, we shall further simplify the band diagram for semiconductors to the form shown in Fig. 1.1c. The critical energies, as far as we are concerned, are the energies of the conduction and valence band edges, $E_C$ and $E_V$, which also define the band gap energy $E_{gap} = E_C - E_V$.

In general, for a distribution of energy levels, the electron occupation probability is defined by the Fermi–Dirac function $f(E)$.

$$f(E) = \frac{1}{1 + \exp\left(\frac{E - E_F}{k_B T}\right)} \tag{1.1}$$

where $E_F$ is the Fermi energy. It can be seen that for $E > E_F$, the occupation probability falls rapidly towards zero over an energy range of a few $k_B T$, whereas for $E < E_F$ it rises rapidly to 1. For $E = E_F$, the occupation probability is 0.5.

Let us consider an ultrapure semiconductor that has no doping impurities. Electrons can be thermally excited across the forbidden gap, leaving electron vacancies (holes) in the valence band. The concentrations of electrons ($n$) and holes ($p$), which must be equal under these conditions, can be expressed in terms of the Fermi–Dirac probability function for the presence ($f$) or absence ($1 - f$) of electrons, respectively.

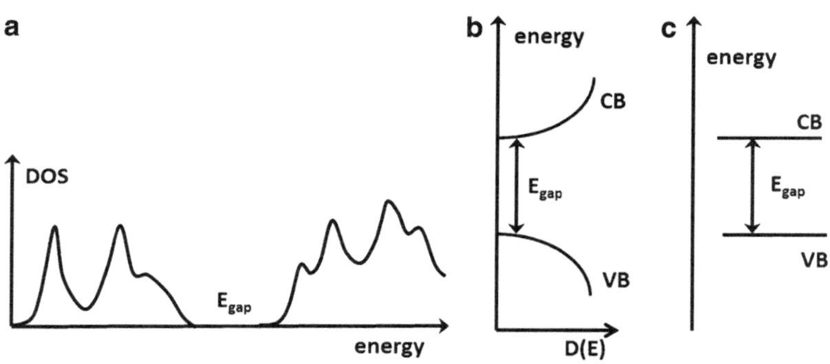


**Fig. 1.1** (**a**) Density of states (DOS) for a typical semiconductor; (**b**) parabolic approximation near band edges; (**c**) simplified band diagram used in this chapter

$$n = \int_{E_C}^{\infty} D_C(E) f(E) \mathrm{d}E$$
$$p = \int_{-\infty}^{E_V} D_V(E)[(1 - f(E))] \mathrm{d}E \tag{1.2}$$

where $D_C$ and $D_V$ are the densities of states distributions, and $N_C$ and $N_V$ are the *effective densities of states* in the conduction and valence band. The equilibrium concentrations of electrons and holes are related by the mass balance equation.

$$np = N_C N_V \exp\left(-\frac{E_{gap}}{k_B T}\right) \tag{1.3}$$

If $N_C$ and $N_V$ are equal, it follows that the Fermi energy $E_F$ for an intrinsic (undoped) semiconductor is situated at the mid-gap position because $n = p$. It can be seen from the exponential term in Eq. (1.3) that the concentrations of electrons and holes in an intrinsic semiconductor increase rapidly with temperature.

## 1.2.2 Doped Semiconductors and the Fermi Energy

In most cases, we are dealing with doped semiconductors. These contain donor (n-type doping) or acceptor (p-type doping) atoms (or vacancies) in the lattice that can produce electrons or holes according to the equilibria

$$\mathrm{D} \rightarrow \mathrm{D}^+ + \mathrm{e_{CB}} \qquad \mathrm{A} \rightarrow \mathrm{A}^- + \mathrm{h_{VB}} \tag{1.4}$$

The ionization energies for these reactions are generally of the order of $k_B T$, so that the free electron or hole concentrations at room temperature are determined by the doping level rather than by excitation of electrons across the bandgap. In this case, the Fermi energy is also determined by $N_D$ or $N_A$, and if $n \ll N_C$ or $p \ll N_V$, the Fermi Dirac function reduces to the Boltzmann function.

$$n = N_C \exp\left(-\frac{E_C - E_F}{k_B T}\right) \qquad p = N_V \exp\left(\frac{E_V - E_F}{k_B T}\right) \tag{1.5}$$

$N_C$ and $N_V$ for silicon are of the order of $10^{19}$ cm$^{-3}$, so that an n-type doping concentration of $10^{16}$ cm$^{-3}$ corresponds to $E_C - E_F \approx 180$ meV. Similarly, p-type doping at the same level gives $E_F - E_V \approx 180$ meV. Figure 1.2 contrasts an intrinsic semiconductor with n- and p-type doped materials with the same bandgap.

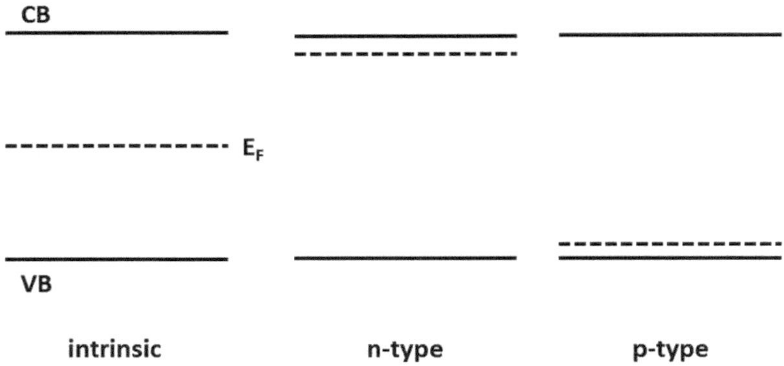


**Fig. 1.2** Band diagrams intrinsic and doped semiconductors, showing the position of the Fermi energy $E_F$

### *1.2.3 Fermi Energy and Electrochemical Potential*

The *electrochemical potential* of a charged species i is a partial molar quantity. It measures the way that the Gibbs energy $G$ of a system changes when i is added to a phase that has an inner potential $\varphi$.

$$\overline{\mu}_i = \left(\frac{\partial \overline{G}}{\partial n_i}\right)_{T,p,n_j,\varphi} \tag{1.6}$$

Conceptually, the electrochemical potential can be split into components associated with the chemical species and its charge.

$$\overline{\mu}_i = \mu_i + z_i F \varphi \tag{1.7}$$

The first term is the chemical potential and the second term represents the electrical work done in transferring one mole of charge $z_i F$, where $z_i q$ is the total charge on the species i. The absolute value of the chemical potential of electrons, $\mu_n$, is the *work function*, $\Phi$. It represents the energy required to excite an electron from a bound state in the metal to a state corresponding to a free electron in vacuum. The *chemical potential* of electrons and holes depends on their concentration. Here we change from molar concentrations to number densities, so that the chemical potential is per electron or hole.

$$\mu_n = \mu_n^0 + k_B T \ln\frac{n}{N_C} \qquad \mu_p = \mu_p^0 + k_B T \ln\frac{p}{N_V} \tag{1.8}$$

where $\mu_n^0$ and $\mu_p^0$ are the standard chemical potentials of holes and electrons. It can be seen from Eq. (1.8) that $N_C$ and $N_V$ correspond to the standard states for electrons and holes, respectively.

It can be shown that the electrochemical potentials of electrons and holes are related to the Fermi energy by (Würfel and Würfel 2009)

$$\overline{\mu}_{\mathrm{n}} = E_{\mathrm{F}} \quad \text{and} \quad \overline{\mu}_{\mathrm{p}} = -E_{\mathrm{F}} \tag{1.9}$$

As we shall see in the next section, these thermodynamic relationships can also be applied under non-equilibrium conditions for an illuminated semiconductor.

### *1.2.4 The Illuminated Semiconductor and Quasi-Fermi Energies*

The absorption of photons with energy greater than the band gap of a semiconductor leads to excitation of electrons from the valence band to the conduction band. Under photostationary conditions of steady illumination at open circuit (no current flow), the excitation of electron hole pairs is balanced by their recombination. Recombination may involve band to band transfer, Auger processes or transfer via defect or impurity states, either in the bulk or at the surfaces of the semiconductor (Landsberg 1991). The concentrations of excess carriers, $\Delta n$ and $\Delta p$, must of course be equal. If we ignore surface recombination for the moment, recombination in a doped semiconductor is a pseudo first-order process if the concentration of photogenerated majority carriers (i.e. electrons for an n-type semiconductor) is lower than the equilibrium dark concentration, so that the majority carrier concentration is essentially constant. The recombination process determines the *minority carrier lifetime*, $\tau_{\mathrm{min}}$. This is an important bulk property of a semiconductor, and values range from $10^{-3}$ s for ultrapure silicon to $10^{-9}$ s or less for many compound semiconductors. If the bulk lifetime is long enough, carriers can diffuse to the surface of the semiconductor, where they may recombine via surface states associated with defects or impurities. This will occur if the minority carrier diffusion length

$$L_{\mathrm{min}} = \sqrt{D_{\mathrm{min}}\tau_{\mathrm{min}}} = \sqrt{\frac{k_{\mathrm{B}}T}{q}u_{\mathrm{min}}\tau_{\mathrm{min}}} \tag{1.10}$$

is comparable with the sample thickness. Here $D_{\mathrm{min}}$ and $u_{\mathrm{min}}$ are the diffusion coefficient and mobility of minority carriers, respectively. In highly pure silicon, $L_{\mathrm{min}}$ is greater than the typical wafer thicknesses, so that recombination occurs predominantly via the surface unless steps are taken to remove surface states, for example, by hydrogen termination. *Surface recombination* is characterized by a surface recombination velocity, $\upsilon_{\mathrm{sr}}$, that depends on the thermal velocity, $\upsilon_{\mathrm{th}}$, of electrons and holes, and the concentration, $N_{\mathrm{ss}}$, and capture cross-section, $\sigma$, of surface states.

$$v_{sr} = v_{th}\sigma N_{ss} \tag{1.11}$$

Surface recombination is an important loss mechanism in many devices, such as solar cells and photoelectrodes. In the latter case, surface recombination competes with interfacial electron transfer, lowering the quantum efficiency of processes such as light-driven water splitting.

The generation of electron hole pairs in a semiconductor perturbs the thermal equilibrium that is described in terms of the Fermi energy, and the concentrations of holes and electrons now depend on the rates at which electron–hole pairs are created and removed. In the absence of surface recombination, the excess electron and hole concentrations under open circuit conditions are determined only by the generation rate $G$ and the bulk minority carrier lifetime. For an n-type semiconductor under steady state illumination, the excess hole concentration $\Delta p$ is therefore defined by

$$\frac{d\Delta p}{dt} = G - \frac{\Delta p}{\tau_p} = 0 \tag{1.12a}$$

so that

$$\Delta p = \Delta n = G\tau_p \tag{1.12b}$$

The photoexcited electrons and holes equilibrate rapidly with lattice phonons (in less than $10^{-12}$ s), but recombination takes longer. For this reason, the distribution of thermally relaxed electrons and holes can be described by quasi-Fermi energies $E_{Fn} = \overline{\mu}_n$ and $E_{Fp} = -\overline{\mu}_p$. Consider the specific example of an n-type semiconductor with a bandgap of 1 eV, $N_C = N_V = 10^{20}$ cm$^{-3}$, a doping concentration of $10^{17}$ cm$^{-3}$, an excitation rate of $G = 10^{20}$ cm$^{-3}$, and a hole lifetime of $10^{-6}$ s. From Eq. (1.5), we find that the dark Fermi energy $E_F$ is situated ca. 0.2 eV below the conduction band, and from Eq. (1.3), we find that the product $np \approx 10^{23}$ cm$^{-6}$ in the dark. Since $n = 10^{17}$ cm$^{-3}$ for our n-doped semiconductor, the equilibrium hole concentration given by the mass balance condition is $p = 10^6$ cm$^{-3}$. The excess hole and electron concentrations under steady state illumination can be calculated from Eq. (1.12b): the values are $\Delta n = \Delta p = 10^{11}$ cm$^{-3}$. It can be seen that the change in electron concentration brought about by illumination is negligible, whereas the hole concentration increases by five orders of magnitude under illumination. The steady state concentrations of electrons and holes are described by the quasi-Fermi levels (QFLs)

$$n + \Delta n = N_C \exp\left(-\frac{E_C - E_{Fn}}{k_B T}\right) \text{ and } p + \Delta p = N_V \exp\left(\frac{E_V - E_{Fp}}{k_B T}\right) \tag{1.13}$$

Since $n \gg \Delta n$, $E_{Fn} \approx E_F$, whereas $E_{Fp}$ is close to mid gap in this example, as illustrated in Fig. 1.3.

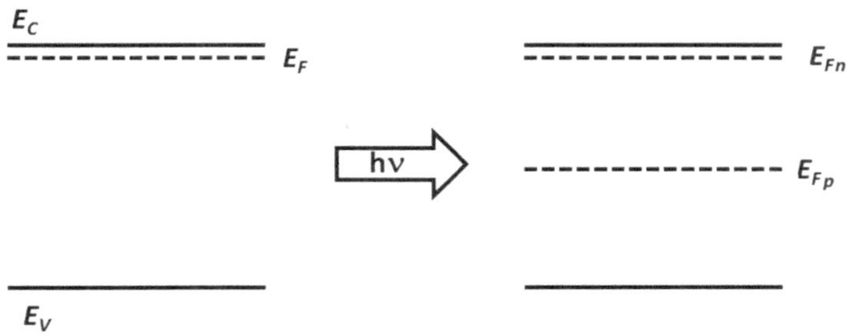


**Fig. 1.3** Splitting of the Fermi energies brought about by illumination. See text for values used in the calculation for a specific example. Note that the Fermi energy of the minority carriers (holes) is strongly affected by illumination, whereas the Fermi energy of majority carriers remains essentially the same

The free energy stored by electron hole pairs is represented by the sum of their electrochemical potentials. Since the charges on holes and electrons are of opposite sign, the $q\varphi$ terms in their electrochemical potentials cancel, so that the energy of electron hole pairs is given by the sum of their *chemical potentials* and therefore by the *difference* between their quasi Fermi energies.

$$\mu_{\mathrm{eh}} = \mu_{\mathrm{n}} + \mu_{\mathrm{p}} = E_{\mathrm{Fn}} - E_{\mathrm{Fp}} \tag{1.14}$$

The objective of any practical energy conversion device based on the absorption of light by a semiconductor is to extract the largest possible fraction of this chemical energy, either as electrical power in a solar cell or as a chemical product (a fuel) such as hydrogen in a photoelectrolysis cell.

### 1.2.5 Semiconductor Junctions and Energy Scales

Semiconductor junctions of different kinds are important in many types of electronic devices, including solar cells. We can illustrate the process of junction formation by considering a metal–semiconductor contact as an example. If an n-type semiconductor is contacted by a metal with a higher work function, electrons will initially move from the semiconductor to the metal, generating a *contact potential difference* that establishes electronic equilibrium. The removal of electrons from the semiconductor leaves behind a region of positive *space charge* near the semiconductor surface, consisting of ionized donors. If we make the simplifying assumption that we can neglect the charge due to electrons in the space charge region (SCR) of width $W_{\mathrm{SC}}$ (the *abrupt depletion approximation*), the electric field and potential distribution in this space charge region can be obtained by integrating Poisson's equation in one dimension.

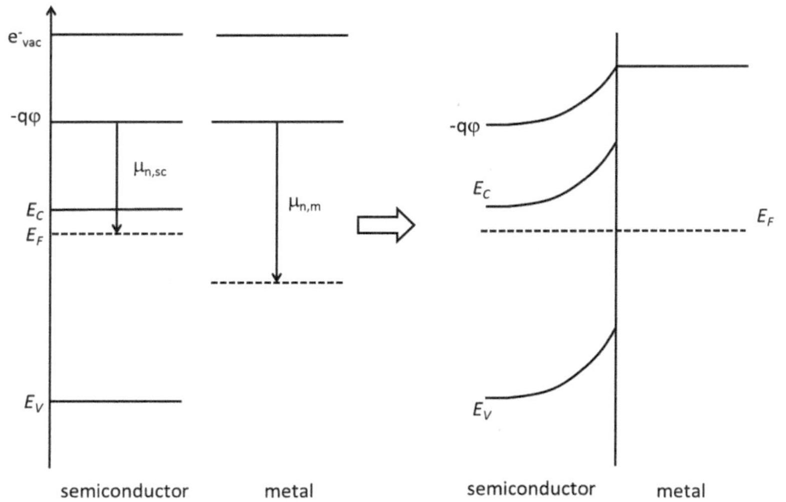


**Fig. 1.4** Formation of a Schottky barrier between an n-type semiconductor and a metal with a higher work function. Note that the Fermi energy is constant across the junction at equilibrium

$$\frac{\partial^2\varphi}{\partial x^2} = -\frac{\rho(x)}{\varepsilon\varepsilon_0} = -\frac{N_\mathrm{d}}{\varepsilon\varepsilon_0} \tag{1.15}$$

Integration with the boundary conditions that the electric field and the electrical potential are zero at $x = W_\mathrm{SC}$ shows that the electric field varies linearly across the space charge region and the electrical potential varies with the square of distance. The parabolic distance dependence of $-q\varphi$ that is apparent in Fig. 1.4 corresponds to *band bending* in the space charge region. It is important to note that although the electrical potential varies with distance, the electrochemical potential of electrons is constant across the junction at equilibrium, as indicated by the horizontal line representing the Fermi energy. This means that in the dark at equilibrium, there is *no driving force* for the separation of thermally generated electrons and holes. The variations of equilibrium electron density and electrical potential with distance are such that $\overline{\mu}_\mathrm{n}$ is constant (cf. Eq. (1.7)).

Similar considerations apply to the formation of semiconductor p–n junctions, see e.g. Sze (1981). However, the central focus of this chapter is the semiconductor/electrolyte junction. The difference here is that the electrolyte is not an electronically conducting phase—there are no holes and electrons to consider. Instead, electrons are localized on energy levels in the solvent and ions. At first sight, it might appear that the definition of equilibrium in terms of Fermi energies would be difficult or impossible in this case. However, if the electrolyte contains a *redox couple*, electronic equilibrium can be achieved by electron transfer across the semiconductor/electrolyte interface. For a one-electron reaction

$$\mathrm{O} + \mathrm{e}^- \rightarrow \mathrm{R} \tag{1.16}$$

the equilibrium condition can be expressed in terms of the electrochemical potentials of the reactants and products.

$$\overline{\mu}_{\mathrm{O}} + \overline{\mu}_{\mathrm{n}} = \overline{\mu}_{\mathrm{R}} \text{ or } \overline{\mu}_{\mathrm{n}} = \overline{\mu}_{\mathrm{R}} - \overline{\mu}_{\mathrm{O}} \tag{1.17}$$

We see that at equilibrium the electrochemical potential of electrons is equal to the *difference* between the electrochemical potentials of R and O. Bearing in mind the definition of electrochemical potential (Eq. (1.7)), we can expand Eq. (1.17) into chemical and electrical components.

$$\overline{\mu}_{\mathrm{n}} + \overline{\mu}_{\mathrm{n}} - q\varphi_{\mathrm{sc}} = \overline{\mu}_{\mathrm{O}} - \overline{\mu}_{\mathrm{R}} = \mu_{\mathrm{O}} - \mu_{\mathrm{R}} + q(z_{\mathrm{R}} - z_{\mathrm{O}})\varphi_{\mathrm{sol}} \tag{1.18}$$

Noting that $(z_{\mathrm{O}} - z_{\mathrm{R}})$ is equivalent to the number of electrons transferred (1 in our example), we can now obtain the difference in inner potentials (the *contact potential difference*) between the semiconductor and the solution as

$$\Delta\varphi = \varphi_{\mathrm{sc}} - \varphi_{\mathrm{sol}} = \frac{1}{q}(\mu_{\mathrm{n}} + \mu_{\mathrm{R}} - \mu_{\mathrm{O}}) \tag{1.19}$$

Comparison with the metal semiconductor junction in Fig. 1.4 shows that the band bending in the semiconductor is now determined by the chemical potential (work function) of the electrons in the semiconductor and the difference between the chemical potentials of the redox species R and O.

Equation (1.6) also allows us to define the *redox Fermi energy* or Fermi level ($E_{\mathrm{F,\ redox}}$) in terms of the standard chemical potentials of O and R and their number densities as

$$E_{F,redox} = \overline{\mu}_{\mathrm{R}} - \overline{\mu}_{\mathrm{O}} = \left( \overline{\mu}_{\mathrm{R}}^{0} - \overline{\mu}_{\mathrm{O}}^{0} \right) + k_{\mathrm{B}}T \ln\frac{N_{\mathrm{R}}}{N_{\mathrm{O}}} \tag{1.20}$$

The redox Fermi level therefore plays the same role as the Fermi level of any other contact, such as a metal or semiconductor, allowing construction of band diagrams like the one in Fig. 1.4.

Equation (1.20) corresponds to the well-known *Nernst equation* that describes the equilibrium electrode potential of a redox couple (for the general case where $n \neq 1$).

$$U_{\mathrm{eq}} = U^0 - \frac{RT}{nF}\ln\frac{C_{\mathrm{R}}}{C_{\mathrm{O}}} = U^0 - \frac{k_{\mathrm{B}}T}{nq}\ln\frac{C_{\mathrm{R}}}{C_{\mathrm{O}}} = U^0 - \frac{k_{\mathrm{B}}T}{nq}\ln\frac{N_{\mathrm{R}}}{N_{\mathrm{O}}} \tag{1.21}$$

where $U$ has been used to distinguish the electrode potential from energy $E$.

It is not possible to measure the interfacial potential difference defined by Eq. (1.19). Instead, the measurable quantity $U_{\mathrm{eq}}$ is defined as the potential difference (i.e. voltage) between a metal electrode in contact with the redox system and a second metal electrode in contact with a solution containing protons (pH = 0) and hydrogen (in equilibrium with $H_2$ in the gas phase at a pressure of 1 bar). This second electrode is the *standard hydrogen electrode* (SHE). The SHE is chosen as an arbitrary zero of potential for the definition of *standard reduction potentials*. By contrast, electrochemical potentials of electrons (and hence Fermi energies) in semiconductors are usually defined as shown in Fig. 1.4. In order to convert from the standard hydrogen voltage scale to the vacuum scale, we need to know the redox Fermi energy, $E_{F,SHE}$, corresponding to the standard reaction

$$H^+ + e^- \rightarrow \tfrac{1}{2} H_2 \tag{1.22}$$

$E_{F,SHE}$ has been obtained by a range of non-thermodynamic arguments (Morrison 1980; Trasatti 1986; Sato 1998) and the generally accepted value is −4.5 eV on the vacuum scale. This allows us to relate redox Fermi energies on the vacuum scale to the standard reduction potentials on the SHE scale.

$$E^0_{F,\mathrm{redox}} = -4.5\,\mathrm{eV} - qU^0_{\mathrm{redox}} \tag{1.23}$$

It can be seen that redox couples with $U^0$ values that are positive on the hydrogen scale have redox Fermi levels that are lower than $E_{F,SHE}$ on the vacuum scale, whereas redox couples with negative potentials with respect to the SHE have Fermi energies lying above $E_{F,SHE}$. So, for example, the $Fe^{3+}/Fe^{2+}$ redox couple ($U^0$ = +0.77 V) will oxidize hydrogen to $H^+$ under standard conditions, whereas the $Cr^{3+}/Cr^{2+}$ couple ($U^0$ = −0.42 V) will reduce protons to hydrogen as illustrated in Fig. 1.5.

### *1.2.6 Electron Transfer at the Semiconductor/Electrolyte Interface*

Figure 1.5 illustrates the calculation of the net free energy changes that are involved in electron transfer. Thus, the reaction

$$Cr^{2+} + Fe^{3+} \rightarrow Cr^{3+} + Fe^{2+} \tag{1.24}$$

is exoenergetic with a free energy change per ion equal to $-q$[0.77 V − (−0.42 V)] = −1.19 eV. The actual process of electron transfer is more complex because the occupied (reduced) and vacant (oxidized) energy levels of the ions of a redox couple like $Cr^{3+}/Cr^{2+}$ in solution differ, since the interactions with the solvent depend on the *charge* on the ions. As a consequence, the energy levels are split into two, with the occupied (R) level lying below the vacant (O) level. In addition, the fluctuations of solvent dipoles arising from the thermal energy $k_BT$ lead to large

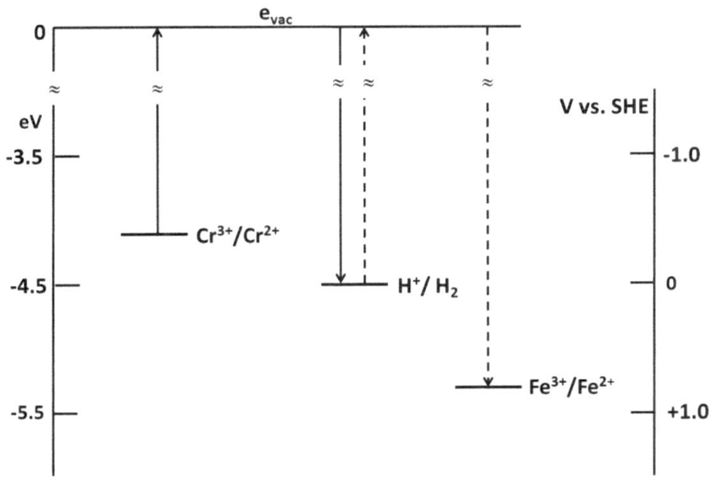


**Fig. 1.5** Scheme illustrating the redox Fermi levels for the $Cr^{3+}/Cr^{2+}$ ($U^0 = -0.42$ V) and $Fe^{3+}/Fe^{2+}$ ($U^0 = +0.77$ V) redox couples on the vacuum scale relative to $E_{F,SHE}$. The *arrows* show energy changes corresponding to the *spontaneous* electron processes (i.e. negative change in free energy) in the two cases. $Cr^{2+}$ reduces $H^+$ to hydrogen and $Fe^{3+}$ oxidizes $H_2$ to $H^+$. Note that the left hand *y* axis is interrupted to allow the vaccuum level to be shown

temporal variations of the electron affinity and ionization energy. The electron energies can be described by Gaussian probability distributions of the form

$$W(E) = \frac{1}{(4\pi\lambda k_B T)^{1/2}} \exp\left[-\frac{(E - E^0)^2}{4\lambda k_B T}\right] \tag{1.25}$$

Here $E^0$ is the most probable electronic energy level and $\lambda$ is the *reorganization energy*. The distributions $W(E)$ shown in Fig. 1.6 are the basis of the *Marcus theory* of electron transfer (Marcus 1964; Morrison 1980) and of the description of electron transfer processes at semiconductors introduced by Gerischer (1961) It should be noted that the probability distributions $W(E)$ are not electronic density of states functions of the type encountered in the case of solid lattices. Instead, they represent the probability of the electronic energy level fluctuating to a particular value as a consequence of dynamic changes in the ionic environment, see e.g. Morrison (1980).

It is clear that the addition of an electron to the O level or removal of an electron from the R level must involve not only electron transfer, but also additional rearrangement ("reorganization") of the solvent around the ions in order for the initial and final states to correspond to the most likely energies of O and R shown in Fig. 1.6. *Adiabatic* electron transfer is only possible if the O and R energies are the same. The most probable process therefore involves the O and R states in the region where the two $W(E)$ distributions overlap. Since the probabilities of the redox

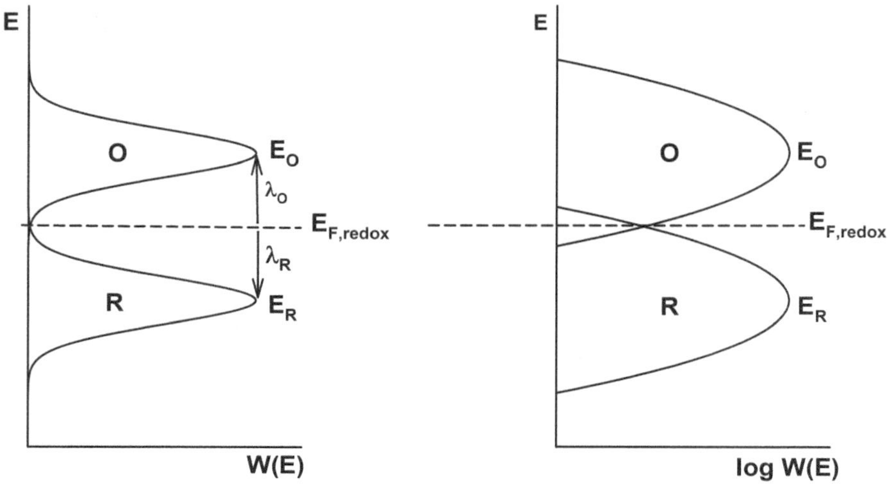


**Fig. 1.6** Gaussian probability distributions for a redox couple (equal concentrations of O and R). The logarithmic plot on the right illustrates the intersection point corresponding to the redox Fermi energy $E_{F,redox}$. The width of the distributions is determined by the reorganization energies $\lambda_O$ and $\lambda_R$ (equal in this example). Values of $\lambda$ lie in the range 1–2 eV

electron level being occupied or unoccupied are equal at this energy, it corresponds to the standard redox Fermi energy, $E_{F,redox}$.

The Marcus theory of electron transfer can be applied to *outer sphere* reactions, that is to say reactions in which no bonds are broken or formed. In the case of metal ligand complexes, such as $Fe(CN)_6^{3-/4-}$, the reorganization energy is split into two parts corresponding to changes in the metal ligand bonding and to changes in ion solvation. Many reactions of interest, for example those involved in photoelectrolysis, are not outer sphere reactions because bonds are formed or broken and often the reaction involves several steps and may involve surface-bound species.

Electron transfer reactions at metal and semiconductor electrodes involve overlap of the energy distributions with filled and vacant electron states in the electrode, as shown in Fig. 1.7. The Fermi energy in metals marks the boundary between occupied and unoccupied states and, as Fig. 1.7 shows, electron exchange with the redox couple will take place close to $E_F$ (i.e. within a few $k_BT$ of $E_F$) and involve the exponential tails of the O and R distributions. The situation is different for a semiconductor, since overlap of the O and R states can only occur for energies above $E_C$ and below $E_V$ because there are no electron states in the forbidden gap. The oxidation of R can therefore occur by two distinct mechanisms—injection of an electron into the conduction band or reaction with a hole in the valence band. Similarly, reduction of O can involve either transfer of an electron from the conduction band or injection of a hole into the valence band. Which of these mechanisms (i.e. conduction or valence band electron exchange) predominates will depend on the overlap integrals for the conduction and valence bands with the distribution functions for O and R.

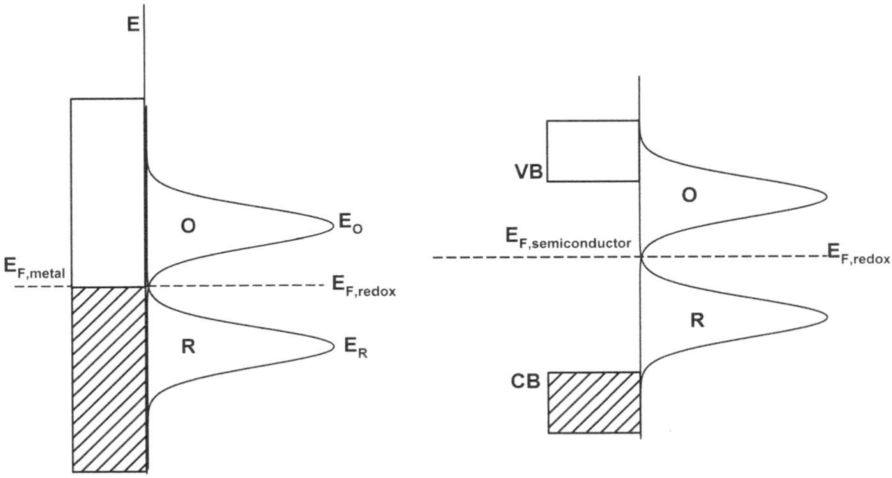


**Fig. 1.7** Comparison of the redox equilibrium at metal and semiconductor electrodes. Electron exchange with the metal electrode involves states close to the Fermi level. Electron exchange at the semiconductor electrode involves valence and conduction band states, since there are no states available at the Fermi energy, which lies in the forbidden gap. We can therefore distinguish between valence and conduction band electron exchange

### *1.2.7 Potential Distribution Across the Semiconductor/ Electrolyte Junction*

The preceding discussion shows that—as far as constructing band diagrams is concerned—the redox Fermi energy plays the same role as the electron Fermi level in a metal contact. It follows that if a junction is formed between an n-type semiconductor and an electrolyte that has a lower Fermi energy, a depletion layer will be formed equivalent to the one shown in Fig. 1.4 for a metal semiconductor contact. Similarly, a depletion layer will be formed in a p-type semiconductor if the redox Fermi energy is higher than the Fermi energy of the semiconductor. Figure 1.8 illustrates the distributions of charge, electrical field, and electrical potential across the semiconductor/electrolyte junction. Integration of Poisson's equation (Eq. (1.15)) using the abrupt depletion approximation gives the potential gradient at the surface of the semiconductor (Eq. (1.26a)). A second integration gives the potential drop across the space charge region (Eq. (1.26b)), which allows us to define $W_{SC}$ as a function of potential drop $\Delta\phi_{SC}$ and the doping density $N$ (Eq. (1.26c)), giving Eq. (1.26d) for the electric field.

$$\left.\frac{d\phi}{dx}\right|_{x=0} = \frac{qNW_{SC}}{\varepsilon_{SC}\varepsilon_0} \tag{1.26a}$$

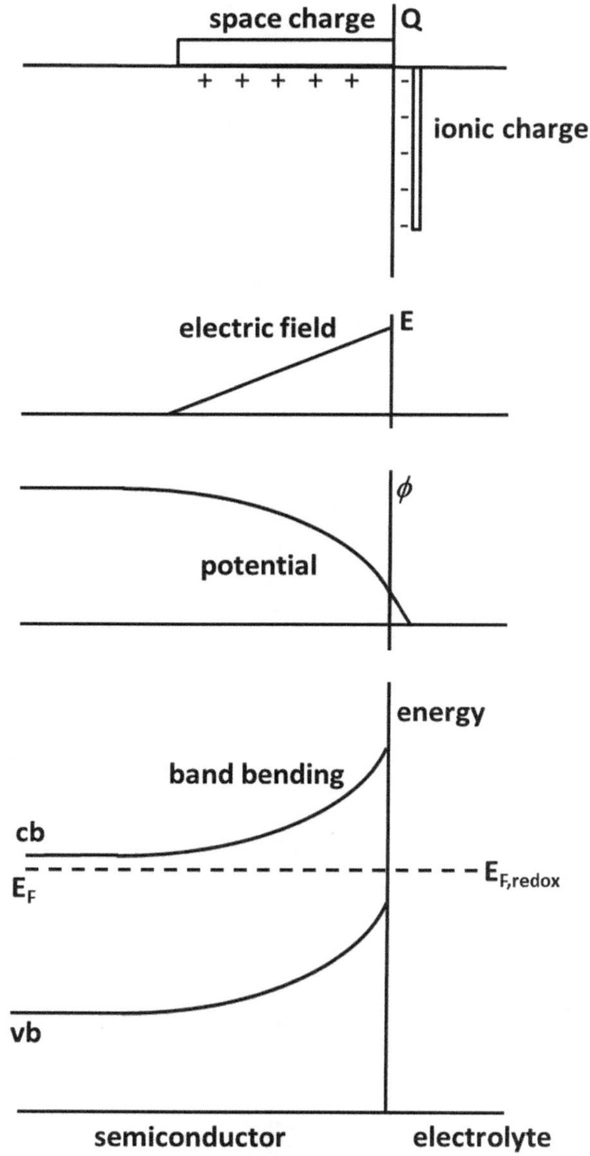


**Fig. 1.8** Distribution of charge, electrical field, electrical potential, and band energies showing the depletion condition that is developed when an n-type semiconductor is brought into contact with a redox electrolyte with a lower Fermi level. It can be seen that the majority of the potential drop across the junction occurs in the space charge region. Note that the width of the space charge region is generally much greater than the width of the Helmholtz layer

$$\Delta\phi_{SC} = \frac{qNW_{SC}^2}{2\varepsilon_{SC}\varepsilon_0} \tag{1.26b}$$

$$W_{SC} = \left(\frac{2\Delta\phi_{SC}\varepsilon_{SC}\varepsilon_0}{qN}\right)^{1/2} \tag{1.26c}$$

$$\left.\frac{d\phi}{dx}\right|_{x=0} = \left(\frac{2qN\Delta\phi_{SC}}{\varepsilon_{SC}\varepsilon_0}\right)^{1/2} \tag{1.26d}$$

If we use a redox electrolyte with a redox Fermi level that is equal to $E_F$ in the semiconductor, no charge is removed from the semiconductor and there is no space charge region. This is the flat band condition. In practice, this situation can be obtained in the absence of a redox couple in the electrolyte by controlling the electrode potential. The potential at which the space charge region vanishes ($W_{SC} = 0$) is the *flat band potential*, $U_{fb}$.

The fixed space charge in the semiconductor is balanced by a net charge of opposite sign in the electrolyte that is due to a rearrangement of the mobile ions which form an *electrical double layer*. In concentrated electrolytes, the ionic counter charge is located very close to the surface of the semiconductor, and we can use the simple *Helmholtz model* of the double layer, where the ionic charge is assumed to be situated on the *outer Helmholtz plane*, see e.g. Bard and Falkner (2001). The region of solvent dipoles located between the layer of solvated ions and the solid surface—the Helmholtz layer—gives rise to the Helmholtz capacitance, $C_H$, which according to the simple Helmholtz model of the electrical double layer is given by

$$C_H = \frac{\varepsilon_H\varepsilon_0}{\delta_H} \tag{1.27}$$

where $\delta_H$ is the width of the Helmholtz layer and $\varepsilon_H$ is the relative permittivity of the solvent, which may be lower than the bulk value due to dielectric saturation at high fields. The potential drop across the Helmholtz layer can be obtained using dielectric continuity ($\varepsilon_{sc}E_{sc} = \varepsilon_H E_H$) as the boundary condition.

$$\Delta V_H = \frac{\varepsilon_{SC}}{\varepsilon_H}\left(\frac{2\Delta\phi_{SC}qN}{\varepsilon\varepsilon_0}\right)\delta_H \tag{1.28}$$

For $\Delta\phi_{SC} = 1$ V, $\varepsilon_H = \varepsilon_{SC} = 10$, $\delta_H = 0.2$ nm, $N = 10^{17}$ cm$^{-3}$, we obtain $\Delta V_H =$ 12 mV. The potential drop across the Helmholtz layer is therefore negligible compared with $\Delta\phi_{SC}$. However, if the doping density is $10^{19}$ cm$^{-3}$, the potential drop across the Helmholtz layer (120 mV) is no longer negligible (note that the electric field in the Helmholtz layer depends on the square root of the doping density (Eq. (1.26d)).

Since the charge stored in the depletion layer is proportional to the $W_{SC}$, we can obtain the *space charge capacitance* per unit area $C_{SC} = dQ_{SC}/d\Delta\phi_{SC}$.

The exact solution gives the *Mott Schottky equation*, written here for an n-type semiconductor considering that $\Delta\phi_{SC} = V + V_{fb}$

$$C_{SC}^{-2} = \frac{2}{q\varepsilon N_D}\left(V + V_{fb} - \frac{k_B T}{q}\right) \tag{1.29}$$

The capacitance is normally measured as a function of the applied potential in an inert electrolyte with a controlled pH rather than in a redox electrolyte. If the doping density is sufficiently low so that the potential drop across the Helmholtz layer can be neglected (see above), $\Delta\phi_{SC}$ in Eq. (1.29) can be replaced by the difference $U - U_{fb}$, where $U$ is the applied potential. It follows that the doping density and flat band potential can be obtained by plotting $C^{-2}$ vs. electrode potential. Normally, $C_{SC}$ is much smaller than $C_H$, so that the measured capacitance is close to $C_{SC}$. However, if the doping density is high, it is necessary to take into account the fact that the space charge capacitance is connected in series with the Helmholtz capacitance (Degryse et al. 1975). This does not affect the slopes of Mott Schottky plots, but displaces the intercepts, since $C^{-2}$ at $U = U_{fb}$ is equal to ${C_H}^{-2}$, rather than zero. Unfortunately, reliable values of $C_H$ are not available for semiconductor electrodes. $C_H$ values for metals generally lie in the range 20–100 μF cm$^{-2}$ depending on factors, such as surface roughness. Figure 1.9 illustrates Mott Schottky plots for p and n-type semiconductor electrodes, showing how the flat band potential is related to the Fermi energies and the valence and conduction band energies. It can be seen that the flat band potential for a p-type electrode is expected to be more positive than that of its n-type counterpart by an amount that is close to $E_{gap}/q$.

The voltage drop across the Helmholtz double layer is also affected by the *ionic* charge on the semiconductor surface, which depends on the solution composition. In the case of oxide semiconductors for example, the acid/base properties of surface groups mean that the ionic charge varies with pH. Experimentally, it is found that the flat band potential of oxide semiconductors such as ZnO and $TiO_2$ follows a Nernstian dependence of the form

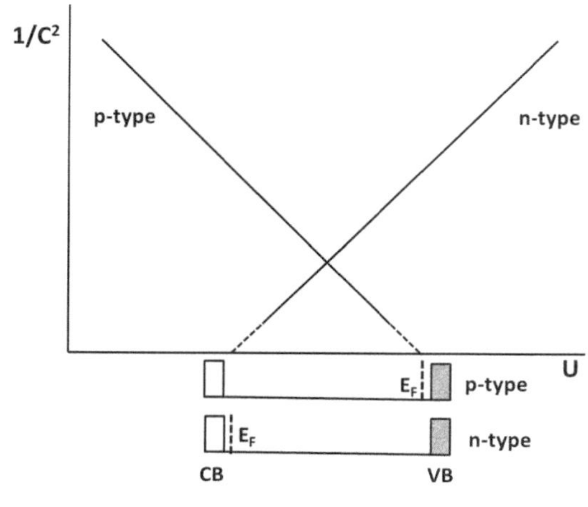


**Fig. 1.9** Schematic Mott Schottky plots for the same material doped p-type and n-type. The figure shows how the flat band potentials are related to the Fermi energies and the band energies. Note that more positive potentials correspond to lower lying electron energies

$$U_{\mathrm{fb}} = U_{\mathrm{fb}}(\mathrm{pH0}) - 0.059\mathrm{pH} \tag{1.30}$$

The pH dependence of the ionic charge on oxide surfaces is also reflected in the electrophoretic behaviour of colloidal particles, which is used to determine the *point of zero charge* (pzc). It is important to note that other ions apart from $H^+$ and $OH^-$ may play a role in determining surface charge if they can be adsorbed at surface sites. Xu and Schoonen (2000) have compiled a comprehensive list of flat band potentials and pzc values for a range of semiconducting minerals in order to calculate the absolute positions of the energy bands on the vacuum scale. Tables of pzc values for oxides can also be found in Kosmulski (2009). In the case of metal chalcogenide semiconductors, the corresponding chalcogenide ions (e.g. $HS^-$ for sulfides) are also potential determining, giving rise to a non-Nernstian pH dependence of $U_{\mathrm{fb}}$. Gerischer (1989) has given a lucid discussion of a number of issues regarding the pH dependence of $U_{\mathrm{fb}}$ that lie outside the scope of this chapter.

### 1.2.8 Surface States and Fermi Level Pinning

The semiconductor interface may also be able to store electronic charge in energy levels that lie in the forbidden gap. For a pristine solid surface in vacuum, intrinsic surface states arise from the symmetry-breaking discontinuity in the crystal lattice, and they can be described either in terms of the nearly free electron model (Shockley states) or in terms of the tight binding approximation (Tamm states). Extrinsic surface states, by contrast, arise from crystal defects or from bonding interactions with a solid or liquid contacting phase. In practice, it is very difficult to prepare semiconductor/liquid junctions without a significant concentration of surface states. Rare exceptions are monocrystalline silicon in fluoride solutions, where etching leads to the formation of a hydrogen terminated surface with an exceptionally low concentration of surface states (Yablonovitch et al. 1986), and ZnO, which can be photo-etched in mildly acidic electrolytes to give an almost ideal junction (Dewald 1960).

Since the electron occupation of surface states varies with potential, the potential drop across the Helmholtz region will vary as a function of the electronic surface charge.

$$\Delta V_{\mathrm{H,ss}} = \frac{\Delta Q_{\mathrm{SS}}}{C_{\mathrm{H}}} = \frac{q}{C_{\mathrm{H}}} \int_{E_{\mathrm{V}}}^{E_{\mathrm{C}}} f(E - E_{\mathrm{F}}) g_{\mathrm{ss}}(E) \mathrm{d}E \tag{1.31}$$

Here $f(E - E_{\mathrm{F}})$ is the Fermi Dirac function and $g_{\mathrm{ss}}(E)$ is the energy distribution of surface states. For simplicity, we can use the zero Kelvin approximation that all surface states below $E_{\mathrm{F}}$ are filled and those above $E_{\mathrm{F}}$ are empty as illustrated schematically in Fig. 1.10.

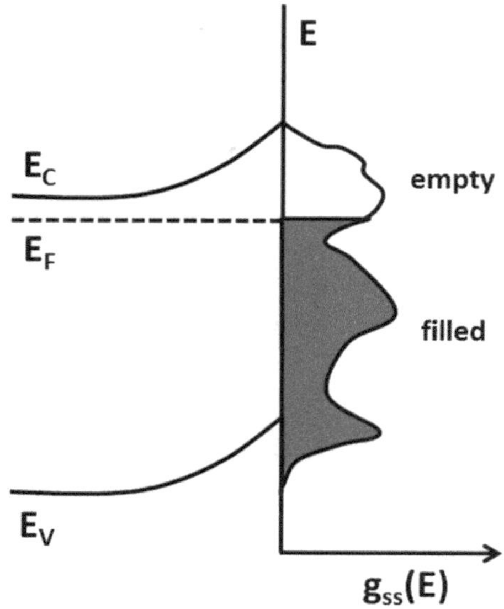


**Fig. 1.10** Occupation of electronic surface states showing filled states below $E_F$ and empty states above $E_F$. Charging of the surface modifies the potential drop across the Helmholtz layer

In the presence of surface states, the relationship between applied potential $(U - U_{fb})$ and the potential drop across the space charge region $(\Delta\phi_{SC})$ is no longer linear. An example is illustrated in Fig. 1.11 for a Gaussian distribution of surface states centred on an energy level 1 eV below the conduction band of an n-type semiconductor. As the voltage is made more positive, the Fermi level moves down, emptying electrons from the surface state distribution and altering the charge on the surface. It can be seen that in the region where the Fermi level moves through the peak of the distribution, $\Delta\phi_{SC}$ increases more slowly because the change in surface charge (as electrons are removed from the defect states) alters the potential drop across the Helmholtz layer. The Fermi level appears to become 'stuck' in the surface state distribution; a phenomenon referred to as *Fermi level pinning*. In the context of semiconductor electrochemistry, the term *band edge unpinning* is perhaps more appropriate. In the example shown in Fig. 1.11, the conduction and valence band energies move down by over 300 meV as the Fermi level passes through the surface state distribution. If the distribution of surface states is very broad, then the slope of the plot of $\Delta\phi_{SC}$ vs. applied potential may be linear, but with a slope less than 1. This means that doping densities derived from Mott Schottky plots may be incorrect.

The charging of surface states corresponds to the charging of a surface state capacitance $dQ_{ss}/dU$. This capacitance can be large if the density of surface states is high, as in the example shown in Fig. 1.11. Figure 1.12 illustrates the capacitance peak that corresponds to the Gaussian surface state distribution shown in Fig. 1.11. Since Mott Schottky plots are measured using ac impedance techniques, surface states will have an impact on the kind of response that is measured experimentally. In principle, the surface state capacitance appears in parallel with the space charge

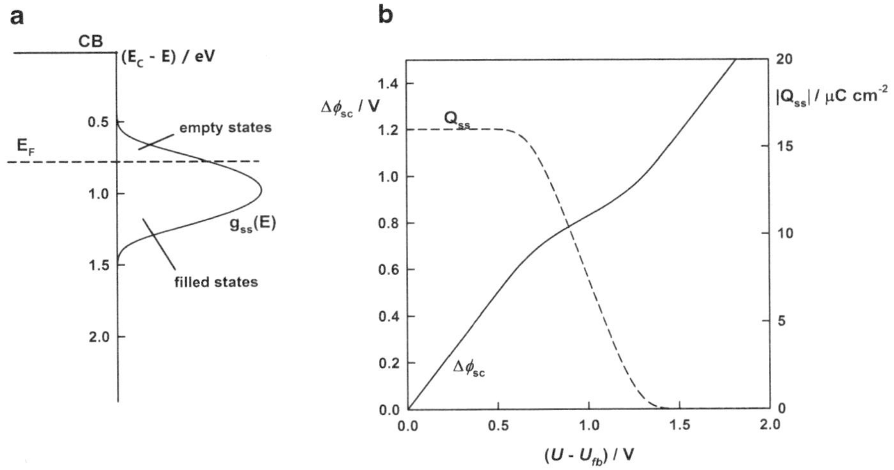


**Fig. 1.11** Fermi level pinning (or band edge unpinning) caused by charging of surface states. (**a**) Gaussian distribution $g_{ss}(E)$ of surface states centred 1 eV below the conduction band. (**b**) Potential dependence of surface state charge ($Q_{SS}$) and bend potential drop across the SCR ($\Delta\phi_{SC}$). Total surface state density $= 10^{14}$ cm$^{-2}$ and $C_H = 50$ μF cm$^{-2}$

**Fig. 1.12** Surface state capacitance corresponding to the Gaussian surface distribution used to calculate Fig. 1.11. Total surface state density $= 10^{14}$ cm$^{-2}$

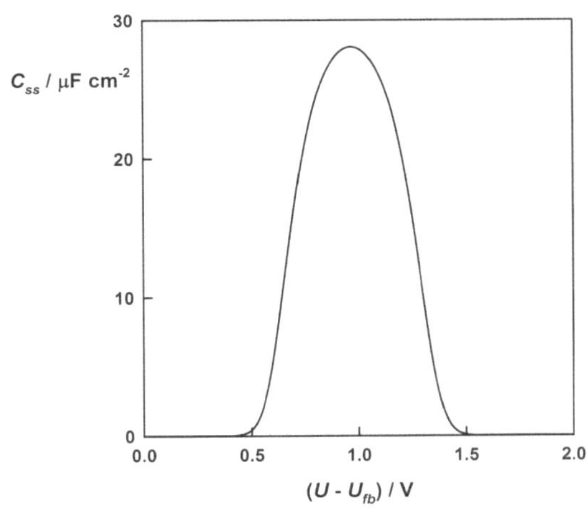


capacitance, so that $C_{ss}$ and $C_H$ are added. However, equilibration of surface states following changes in the applied potential is not instantaneous. This means that when the potential, and hence the bulk Fermi level, is modulated, the surface state occupancy will lag behind as a consequence of the need to equilibrate electrons in surface energy levels with electrons in the conduction band—a thermally activated process. The relaxation time for surface states can be represented by a series combination of a resistance $R_{ss}$ with the surface state capacitance, see e.g. Allongue and Cachet (1985). The impedance of the semiconductor/electrolyte interface can therefore be modelled using the equivalent circuit shown in Fig. 1.13.

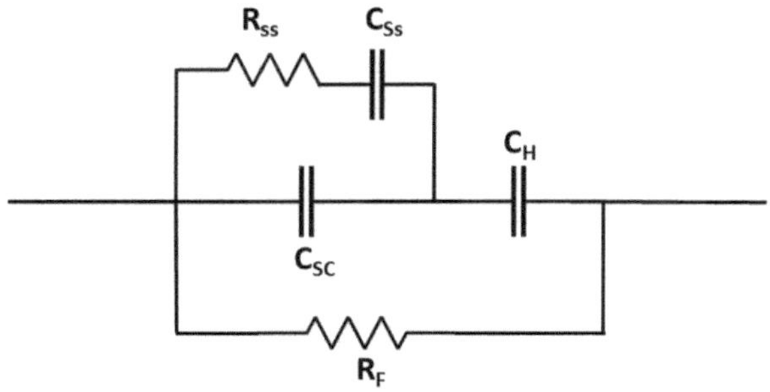


**Fig. 1.13** Equivalent circuit of the semiconductor electrolyte interface. $R_{ss}C_{ss}$ is the time constant for the charging of the surface state capacitance $C_{ss}$. $C_{SC}$—space charge capacitance; $C_H$—Helmholtz capacitance; $R_F$—charge transfer resistance

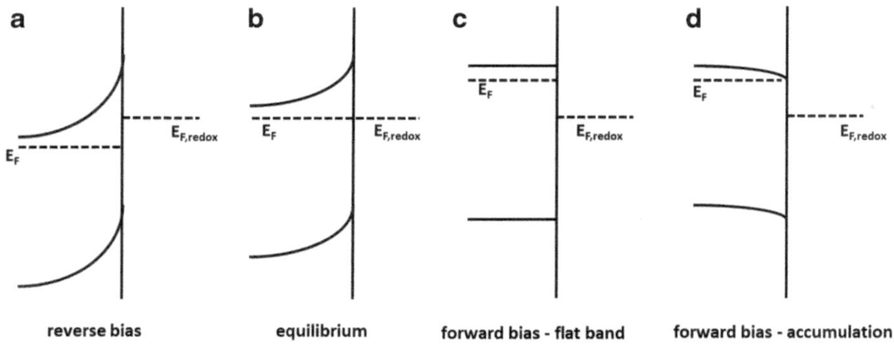


**Fig. 1.14** Diode behaviour of the (n-type) semiconductor/redox electrolyte junction in the dark. The junction blocks electron flow from right to left under reverse bias, but passes an electron current from left to right as forward bias reduces band bending. For more negative potentials, an accumulation region forms and the electrode behaves like a metal

## *1.2.9 Current–Voltage Characteristics of the Semiconductor/ Electrolyte Junction in the Dark*

So far we have focussed our attention on the behaviour of the semiconductor/ electrolyte junction under depletion conditions. If an n-type semiconductor electrode is in equilibrium with a redox system with a lower Fermi energy (i.e. at open circuit), the bands are bent after equilibration of the Fermi levels, as shown in Fig. 1.14a. If a positive potential is applied (reverse bias), the Fermi level in the semiconductor moves downward relative to the solution redox level and electron flow from the redox system to the semiconductor becomes thermodynamically possible (Fig. 1.14b). However, the rate of electron injection from the redox species R into the conduction band of the semiconductor is determined by the energy difference between $E_{F,redox}$ and $E_C$. This energy barrier is the same as the *Schottky barrier* that is formed with high work function metals, and it restricts current flow to a small constant value known as the *reverse saturation current*. Now if a negative potential is applied (forward bias—Fig. 1.14c), the band bending decreases and electron flow from the semiconductor to the species O in the electrolyte becomes thermodynamically possible. The current will increase as the potential is made more negative and the concentration of electrons at the surface rises. Beyond the flat band potential, electrons accumulate near the surface of the semiconductor, which now behaves more like a metal electrode (Fig. 1.14d). The current voltage characteristic of the junction in the dark therefore corresponds to that of a diode, blocking under reverse bias and passing current under forward bias.

## 1.3 Photoelectrochemical Processes

### *1.3.1 The Gärtner Equation*

So far we have only considered the behaviour of the semiconductor/electrolyte junction in the dark. When photons with energy $h\nu > E_{gap}$ are absorbed by a semiconductor, electrons are excited across the forbidden gap, creating *electron hole pairs*. For a junction initially held at the equilibrium potential as shown in Fig. 1.14, the local concentrations of electrons and holes will be perturbed from their equilibrium (dark) values giving rise to a splitting of the dark Fermi level into quasi-Fermi levels for electrons and holes (see Fig. 1.3). Since the holes generated by illumination can accept electrons from the species R, electron flow from right to left now becomes possible and a photocurrent is produced that depends on the generation, collection, and reaction of holes in the junction region. This generation collection problem was originally solved by Gärtner for solid state junctions (Gärtner 1959) and adopted by Butler (1977) for the semiconductor/electrolyte junction. Consider a flat bulk semiconductor illuminated through an electrolyte solution. The important length scales involved in formulation of the problem are illustrated in Fig. 1.15.

We begin by assuming that the reaction of holes at the surface is fast because it involves electron transfer from a fast outer sphere redox species, i.e. no bond breaking or formation is involved. Secondly, we assume that recombination at the surface or in the space charge region can be neglected. These assumptions are equivalent to stating that all electron hole pairs generated in the space charge region are separated by the gradients of the corresponding quasi-Fermi levels. It should be noted here that, in general, the driving force for the transport of electrons and holes is the gradient of the appropriate quasi-Fermi level. The current density is therefore given by (Würfel and Würfel 2009):

$$j = j_n + j_p = nu_n\left(\frac{\partial E_{Fn}}{\partial x}\right) + pu_p\left(\frac{\partial E_{Fp}}{\partial x}\right) = nu_n\left(\frac{\partial \overline{\mu}_n}{\partial x}\right) - pu_p\left(\frac{\partial \overline{\mu}_p}{\partial x}\right) \tag{1.32}$$

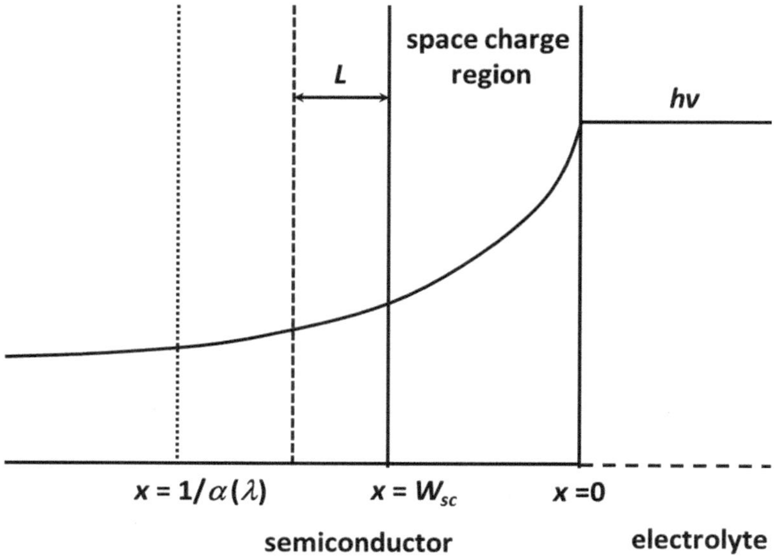


**Fig. 1.15** Characteristic length scales considered in the derivation of the Gärtner equation. $W_{SC}$ is the width of the space charge region and $1/\alpha$ is the wavelength-dependent penetration depth of the incident light

where $n$ and $p$ are the concentrations and $u_n$ and $u_p$ are the mobilities of electrons and holes. Remembering that the electrochemical potential can be separated *conceptually* into chemical and electrical parts, we can define currents associated with the gradient of chemical potential (*diffusion current*) and with the gradient of electrical potential (*drift current*). However, these currents cannot be separated experimentally. Nevertheless, models of the type discussed here are often referred to as "drift diffusion" models.

So far we have assumed that all holes generated in the space charge region reach the surface, where they accept electrons from the species R. We must also consider holes generated beyond the edge of the space charge region. These have a finite diffusion length, and if they reach $W_{SC}$, we assume that they will be transported to the surface to accept electrons from R. The solution of this problem is the *Gärtner equation*, which defines the ratio of the electron flux measured in the external circuit to the incident photon flux. This ratio is referred to as the *external quantum efficiency* (EQE) or the *incident photon to current conversion efficiency* (IPCE).

$$\mathrm{EQE} = \frac{j_{\mathrm{photo}}}{qI_0} = 1 - \frac{\exp(-\alpha W_{\mathrm{sc}})}{1 + \alpha L_{\mathrm{min}}} \tag{1.33}$$

Here $I_0$ is the incident photon flux corrected for reflection losses, $\alpha$ is the absorption coefficient at the photon energy concerned, and $L_{\mathrm{min}}$ is the minority carrier diffusion length.

The assumptions made in the derivation of the Gärtner equation are often not valid for real systems. Firstly, surface recombination is difficult to eliminate, and secondly, recombination in the space charge region may be important, particularly if interfacial electron transfer is so slow that the concentration of minority carriers builds up at the interface. The reactions involved in photoelectrolysis (hydrogen or oxygen evolution) are slow multistep processes involving adsorbed intermediates, and Eq. (1.33) no longer describes the current voltage behaviour because recombination is dominant, particularly for low band bending. This means that the EQE is much lower than predicted by the Gärtner equation.

The net current across the illuminated semiconductor/electrolyte interface is made up of the current due to the reactions involving photogenerated minority carriers as well as majority carriers, which becomes more important as the band bending is reduced under forward bias. The current voltage characteristics of an n-type semiconductor/redox electrolyte junction are illustrated in Fig. 1.16. In this particular case, the penetration depth of light is similar to the width of the space charge region and $\alpha L_p < 1$, so that the photocurrent increases with increased band bending. The forward bias current increases exponentially in the same way as it would for a metal electrode, and the net current drops to zero close to the flatband

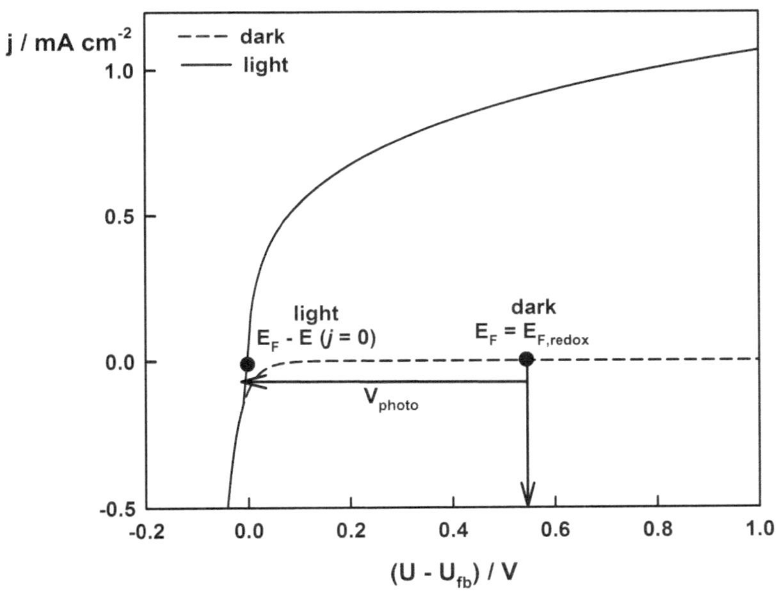


**Fig. 1.16** Current voltage curves for an n-type semiconductor in the dark and under illumination. The photocurrent for the illuminated electrode was calculated using the Gärtner equation. $I_0 = 10^{16}$ cm$^{-2}$ s$^{-1}$, $W_{SC}$ $(U - U_{fb} = 1\ \text{V}) = 10^{-5}$ cm, $\alpha = 10^5$ cm$^{-1}$. The dark current increases exponentially with forward bias. At open circuit in the dark, the electrode potential is determined by the equilibrium potential of the redox system. Under illumination, the open circuit potential is determined by the condition $j = 0$. The change in potential brought about by illumination corresponds to the photovoltage, $V_{photo}$, that would be measured in a two electrode cell

potential in this ideal case. If the electrode is illuminated at open circuit, its potential will change from the equilibrium value determined by the redox couple to a new value where the net current is zero. This change is the *photovoltage*, $V_{photo}$, and for an ideal junction, it is close to the difference between the redox potential and the flat band potential.

### 1.3.2 Recombination in the Space Charge Region and at the Surface

If electron–hole recombination occurs in the space charge region or at the surface, holes are lost before they can accept electrons from solution species. In this case, the steady state photocurrent voltage plots will deviate from the ideal form predicted by the Gärtner equation. A typical example showing the delayed onset of the photocurrent is sketched in Fig. 1.17.

The effects of space charge (SCR) recombination were considered over 30 years ago by Reichman (1980), and around the same time El Guibaly and Colbow considered SCR recombination, surface recombination, and the rate of hole transfer at the interface. Both types of recombination are expected to result in a delayed onset of the photocurrent, as illustrated in Fig. 1.17. The onset is predicted to move towards the flat band potential as the light intensity is increased.

Surface recombination was also considered by Peter et al. in a series of papers (Li et al. 1984; Peter et al. 1984; Li and Peter 1985, 1986) published in the 1980s. The most important aspect of this work was the explanation of the transient effects

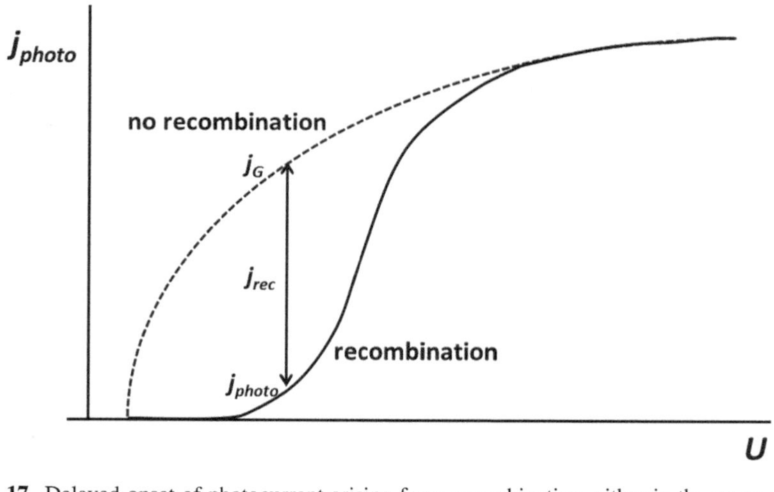


**Fig. 1.17** Delayed onset of photocurrent arising from recombination, either in the space charge region or at the surface. $j_G$ is the current predicted by the Gärtner equation, $j_{photo}$ is the experimental photocurrent, and $j_{rec}$ is the recombination current

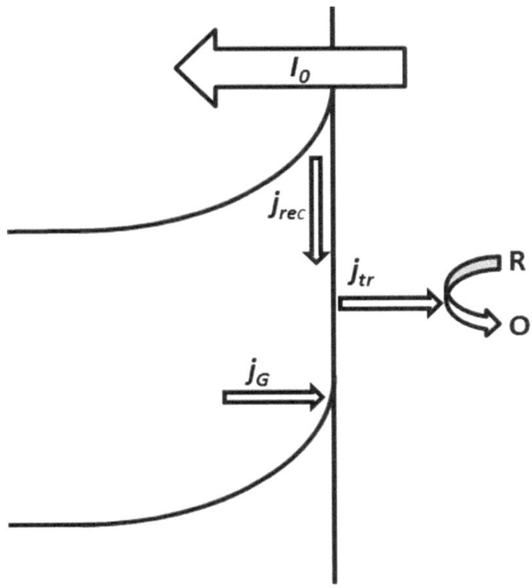


**Fig. 1.18** Simple kinetic scheme describing the competition between charge transfer and surface recombination (see text for definition of fluxes)

seen when chopped light is used to record voltammograms. When the light is switched on, photogenerated minority carriers move towards the interface giving rise to an almost instantaneous displacement current corresponding to a change in the charge stored in the space charge capacitance (see Fig. 1.18). If interfacial electron transfer is slow, the concentration of minority carriers will build up near the interface. The minority carriers may be free or they may be trapped, for example, in surface states. Since the surface recombination current is proportional to the surface hole concentration, $j_{\mathrm{rec}}$ increases from zero when the light is switched on to a steady state value for longer times. The simplest kinetic scheme to describe these processes is shown in Fig. 1.18. We assume that the current $j_{\mathrm{G}}$ due to holes moving to the surface is given by the Gärtner equation. The current associated with surface recombination is due to electrons moving in the same direction as the holes, so the current has opposite sign. At this point, we make a simplifying assumption that the kinetics of interfacial charge transfer and of surface recombination can be described in terms of surface rather than volume concentrations. Thus, the currents due to hole transfer and recombination in Fig. 1.18 are given by

$$j_{\mathrm{tr}} = qk_{\mathrm{tr}}p_{\mathrm{surf}} \quad j_{\mathrm{rec}} = qk_{\mathrm{rec}}p_{\mathrm{surf}} = q\sigma\upsilon_{\mathrm{th}}n_{x=0}p_{\mathrm{surf}} \tag{1.34}$$

Here $k_{\mathrm{tr}}$ and $k_{\mathrm{rec}}$ are first-order rate constants ($\mathrm{s}^{-1}$), $p_{\mathrm{surf}}$ is the surface concentration ($\mathrm{cm}^{-2}$) of holes, $\sigma$ is the capture cross-section for electron hole recombination, and $\upsilon_{\mathrm{th}}$ is the thermal velocity.

It can be seen from Fig. 1.18 that the time-dependent hole concentration is determined by the difference between the rate of arrival of holes at the surface ($j_{\mathrm{G}}$) and the rate of their removal by recombination ($j_{\mathrm{rec}}$) and transfer ($j_{\mathrm{tr}}$).

$$q\frac{\mathrm{d}p_{\mathrm{surf}}(t)}{\mathrm{d}t} = j_{\mathrm{G}}(t) - j_{\mathrm{tr}}(t) - j_{\mathrm{rec}}(t) = j_{\mathrm{G}}(t) - qk_{\mathrm{tr}}p_{\mathrm{surf}}(t) - qk_{\mathrm{rec}}p_{\mathrm{surf}}(t) \quad (1.35)$$

When the light is switched on, the initial (excess) concentration of holes is zero, so $j_{\mathrm{rec}} = 0$ and we measure the 'charging' current $j_{\mathrm{G}}$. As $p_{\mathrm{surf}}$ increases with time, $j_{\mathrm{rec}}$ increases proportionately and the measured photocurrent falls if interfacial hole transfer is sufficiently slow that recombination can begin to compete for holes. Finally, in the steady state, the hole concentration becomes constant and the measured photocurrent decays to a plateau. The time-dependent hole concentration is obtained by solving Eq. (1.35) for the case of a square light pulse of duration $\tau$.

$$\begin{aligned} p_{\mathrm{surf}}(t<\tau) &= \frac{j_{\mathrm{G}}}{q(k_{\mathrm{tr}}+k_{\mathrm{rec}})}(1-\exp[-(k_{\mathrm{tr}}+k_{\mathrm{rec}})t]) \\ p_{\mathrm{surf}}(t>\tau) &= p_{\mathrm{surf}}(\tau)\exp[-(k_{\mathrm{tr}}+k_{\mathrm{rec}})t] \end{aligned} \quad (1.36)$$

The exponential decay of the photocurrent seen after switching the light on is given by

$$\frac{j_{\mathrm{photo}}(t) - j_{\mathrm{photo}}(\infty)}{j_{\mathrm{photo}}(0) - j_{\mathrm{photo}}(\infty)} = \exp[(k_{\mathrm{tr}}+k_{\mathrm{rec}})t] \quad (1.37)$$

and the steady state current is given by

$$\frac{j_{\mathrm{photo}}(\infty)}{j_{\mathrm{photo}}(0)} = \frac{k_{\mathrm{tr}}}{(k_{\mathrm{tr}}+k_{\mathrm{rec}})} \quad (1.38)$$

When the light is switched off after some time $\tau$, $j_{\mathrm{G}}$ falls almost instantaneously to zero, and the recombination of the residual holes at the surface gives rise to a transient current of *opposite sign* to the photocurrent as electrons move to the surface to recombine with the holes. This 'overshoot' current also decays exponentially, in this case to zero.

$$\frac{j(t>\tau)}{j(0)-j(\tau)} = \exp[-(k_{\mathrm{tr}}+k_{\mathrm{rec}})t] \quad (1.39)$$

The time dependence of the surface hole concentration and photocurrents is shown in normalized form in Fig. 1.19. Since $k_{\mathrm{tr}} = k_{\mathrm{rec}}$ in this example, only half of the holes arriving at the surface are transferred to the solution under steady state conditions. This simple analysis shows that the values of $k_{\mathrm{tr}}$ and $k_{\mathrm{rec}}$ can be obtained from the decay lifetime and the ratio of steady state to instantaneous current.

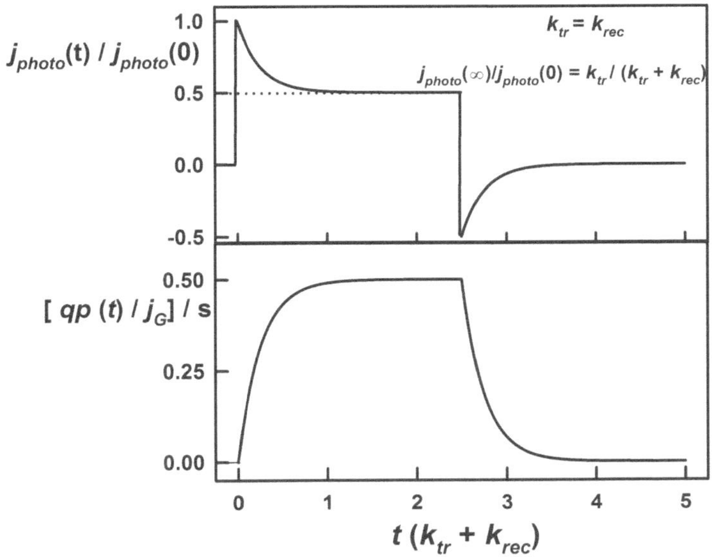


**Fig. 1.19** Normalised plots of transient photocurrent and corresponding surface hole concentration for $k_{tr} = k_{rec}$. The exponential decay time constant $= (k_{tr} + k_{rec})^{-1}$ and the ratio of the steady state current to the 'spike' $= k_{tr}/(k_{tr} + k_{rec})$, allowing $k_{tr}$ and $k_{rec}$ to be determined

### *1.3.3 Rates and Rate Constants of Photoelectrochemical Processes*

The rate ($cm^{-2}\ s^{-1}$) of a heterogeneous photoelectrochemical process involving minority carriers can be expressed in terms of the concentrations or number densities ($cm^{-3}$) of the reacting species p and R. So, for example, the rate of reaction of holes with a species R is given by the rate equation

$$v = k_{het} pR \tag{1.40}$$

It follows that the heterogeneous rate constant $k_{het}$ has units $cm^4\ s^{-1}$. Royea et al. (1997) have calculated the maximum rate constant for outer sphere reactions at semiconductor electrodes to lie in the range $10^{-17}$–$10^{-16}$ $cm^4\ s^{-1}$. This maximum value is obtained when the distribution $W(E)$ of the redox species overlaps optimally with the relevant semiconductor band. In general, of course, the rate constant will be lower by an amount that depends exponentially on the activation energy. If the concentration of redox reactant is much higher than the hole concentration, one can also use the pseudo first-order heterogeneous rate constant $k_{het}R$ ($cm\ s^{-1}$) to describe the kinetics. Finally, if we consider that electron transfer involves holes located within a tunnelling distance $\delta$ of the interface, we can recover the first-order rate constant used in the preceding section.

$$k_{tr} = \frac{k_{het}R}{\delta} \tag{1.41}$$

Inserting order of magnitude values $R = 6 \times 10^{18}$ cm$^{-3}$ (10 mM), $\delta = 10^{-8}$ cm, we find the corresponding maximum value of $k_{tr} = 6 \times 10^{9}$ s$^{-1}$. It is interesting to compare this with the value of $k_{rec}$ close to flat band potential, when $n_{x=0}$ in Eq. (1.33) approaches the bulk value. Taking $n_{x=0} = 10^{17}$ cm$^{-3}$, $\sigma = 10^{-16}$ cm$^2$, and $\upsilon_{th} = 10^5$ cm s$^{-1}$ gives $k_{rec} = 10^4$ s$^{-1}$, which is more than five orders of magnitude lower than the maximum $k_{tr}$. This explains why using fast outer sphere redox electrolytes completely suppresses surface recombination, even close to the flat band potential.

Remarkably slow photocurrent transients are commonly seen in photoelectrochemical water splitting experiments (photoelectrolysis) (Peter 2013). An example is shown in Fig. 1.20 for a polycrystalline hematite ($\alpha$-$Fe_2O_3$) electrode. The relaxation of the photocurrent to its steady state value takes several seconds, indicating slow charge transfer and recombination.

The reason for such a slow transient response is that here we are not dealing with a fast outer sphere redox reaction. The water oxidation reaction involves four holes and four protons, and it involves bond breaking and formation of high energy surface bound intermediates. The activation energy for the rate-determining step in the reaction is evidently very high. This poses a problem in terms of achievable efficiency because recombination can compete effectively with charge transfer, even at positive potentials, where the increase in band bending lowers $n_{x=0}$.

The theoretical treatment of the transient photocurrent response outlined above has been extended to interpret experiments in which the light intensity is modulated by a small amount, rather than switched on and off. This approach, which is known as *intensity modulated photocurrent spectroscopy* (IMPS) (Li and Peter 1986; Peat and Peter 1987; Ponomarev and Peter 1995b; Peter et al. 1997; Fermin et al. 1999), has a number of advantages since it is a small amplitude method that allows

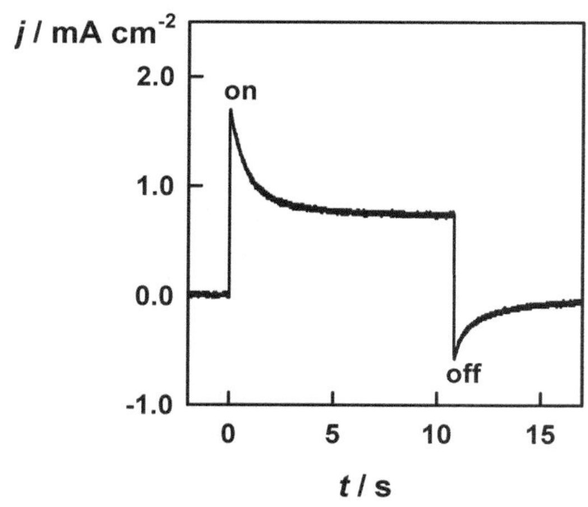


**Fig. 1.20** Transient photocurrent response of $\alpha$-$Fe_2O_3$ electrode at 0 V vs. Ag|AgCl, showing the decay and overshoot characteristic of surface electron–hole recombination. Electrolyte 1.0 M NaOH

linearization in more complex systems. This method has also been applied extensively to study electron transport in dye-sensitized solar cells (Dloczik et al. 1997; Fisher et al. 2000; Dunn and Peter 2009). For an illustration of the power of the IMPS technique as a way of accessing rate constants for water splitting, see e.g. Peter et al. (2012), Peter (2013).

The kinetic scheme shown in Fig. 1.18 has also been used to interpret measurements using *Impedance spectroscopy* (IS), and Ponomarev et al. (Ponomarev and Peter 1995a) as well as Fermin et al. (1999) were able to demonstrate that PEIS and IMPS give consistent results when interpreted within the relatively simple theoretical framework outlined here. The theory has even been used to interpret the results of *light-modulated microwave reflectance* (LMMR) measurements, and here again it has been shown that the IMPS and LMMR give consistent results in terms of the kinetics (Schlichthörl et al. 1995; Cass et al. 2003a, b). These aspects of the transient and periodic photocurrent responses of semiconductor electrodes have been reviewed in detail elsewhere (Peter and Vanmaekelbergh 1999; Peter and Tributsch 2008), and the interested reader is directed to these sources and the original papers.

### *1.3.4 Quasi Fermi Levels and the Concept of Overpotential*

The concept of *overpotential* is well-established in electrochemistry, where it is used to describe the difference between the applied potential and the equilibrium potential in the case of metal electrodes. Changes in applied potential at the metal electrode interface appear primarily across the Helmholtz layer, changing the relative electrochemical potentials of the electrons and the reacting redox couple O/R. Some fraction of this change in energy alters the activation energy for electron transfer, leading to the well-known potential dependence of the heterogeneous rate constants for reduction and oxidation.

$$\begin{aligned}\vec{k} &= k^0 \exp\left[\frac{-\beta nF\left(U - U^0\right)}{RT}\right] \\ \overleftarrow{k} &= k^0 \exp\left[\frac{(1-\beta)nF\left(U - U^0\right)}{RT}\right]\end{aligned} \tag{1.42}$$

where $\beta$ is the cathodic transfer coefficient and $k^0$ is the standard heterogeneous rate constant (cm $s^{-1}$). In essence, the $nF(U - U^0)$ term here represents the difference in Fermi energies of the metal and the redox system. The key point here is that it is the *internal energies* of the reactants that are altered by the changes in potential.

The term overpotential is often applied to semiconductor photoelectrodes, but as we have seen above, in the ideal case very little of any change in potential appears across the Helmholtz layer under depletion conditions. So does the term overpotential make any sense? The answer to this is yes, because we can still define it in terms of the difference between the quasi Fermi level (QFL) of minority carriers and the redox Fermi level. We can illustrate this with an order of magnitude calculation for an n-type semiconductor. Suppose we have a fast redox system with $k_{het} = 10^{-20}$ cm$^4$ s$^{-1}$ and a concentration of $6 \times 10^{18}$ cm$^{-3}$ (10 mM). If the photon flux is $10^{16}$ cm$^{-2}$ and the EQE is 1.0 for simplicity, then the steady state hole concentration given by Eq. (1.40) will be $1.7 \times 10^{17}$ cm$^{-3}$. If we assume that the effective density of valence band states is $N_V = 10^{20}$ cm$^{-3}$, the quasi-Fermi level at the surface $E_{Fp}$ $(x = 0)$ will be 160 meV above the valence band. The overpotential is then given by

$$q\eta = E_{F,\mathrm{redox}} - E_{Fp}(x = 0) \tag{1.43}$$

Equation (1.13) shows that $E_{Fp}$ will move 59 meV closer to the valence band for each factor of 10 increase in light intensity. For slower reactions, the hole QFL will be so close to the valence band that essentially $q\eta$ becomes $E_{F,\mathrm{redox}} - E_V$. This overpotential is clearly related to the concentration of holes and not to the potential drop in the Helmholtz layer; in other words, it arises from the *entropic term* in the free energy and not from the internal (potential-dependent) term.

It is worth noting here that the applicability of the quasi Fermi-level concept (Reineke and Memming 1992a, b) to reactions at semiconductor photoelectrodes was the subject of considerable discussion in the early years of photoelectrochemistry, when Nozik (1978) suggested that a stochastic model of charge transfer was more appropriate than the statistical model inherent in the quasi-Fermi level approach. This point of view was contested by Lewis and co-workers (Kumar et al. 1992), who presented experimental and theoretical evidence that the quasi-Fermi level approach and conventional chemical kinetics are adequate for the description of electron transfer at semiconductor electrodes. For an enlightening discussion of these differing points of view, the reader is referred to an excellent but sadly neglected paper by Salvador (2001).

In order to calculate the quasi-Fermi levels in the space charge region of the junction, it is necessary to use numerical methods to solve the 'drift diffusion problem' using appropriate boundary conditions, one of which is the rate of interfacial minority carrier transfer and the other is the rate of surface recombination. Cendula et al. (2014) have recently obtained solutions for thin film electrodes by assuming that only the minority carrier QFL changed by illumination. The majority carrier concentration in the space charge region is therefore fixed by the local band bending, greatly simplifying the calculations. However, more work is needed to obtain solutions that are relevant for slow multistep reactions such as those involved in water splitting.

Under short-circuit conditions (or under potentiostatic conditions where the potential $U = U_{\mathrm{redox}}$), the Fermi level splitting in the bulk of the semiconductor (the neutral region) is negligible, i.e. the hole and electron QFLs converge, so that the QFL profile at the junction varies as shown schematically in Fig. 1.21. Here the value of the hole QFL is fixed as discussed above. The rate constant for the reaction of holes depends on the overlap of the hole QFL with the distribution $W(E)$ for the species R, which in the example shown in Fig. 1.21 is very favourable. The situation is more complex in the case of water splitting since two redox systems are involved, $H^+/H_2$ and $O_2/H_2O$, neither of which can be represented by the simple one-electron outer sphere model shown in Fig. 1.21.

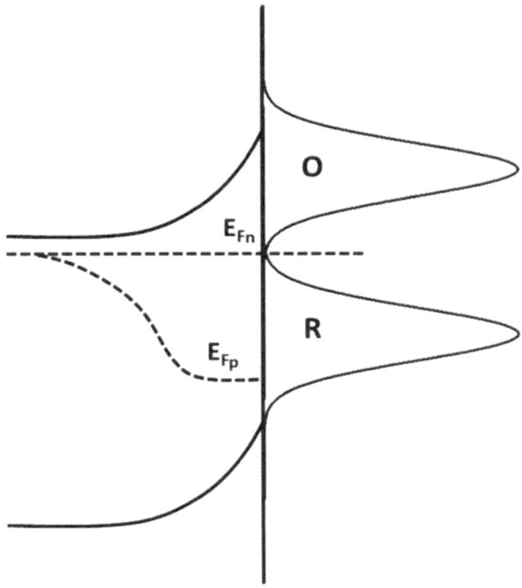


**Fig. 1.21** Quasi Fermi levels for an n-type semiconductor photoelectrode held at the redox potential (equivalent to short circuit in a two-electrode cell). Holes at the surface accept electrons from occupied levels of the reduced redox species R. The rate constant for the reaction of holes depends on the overlap of the hole QFL with the distribution $W(E)$ for the species R

## 1.4 Nanostructured Semiconductor Electrodes

### *1.4.1 Photocatalysis and Photosynthesis at Dispersed Semiconductor Particles*

So far discussion has been limited to well-defined bulk semiconductor electrodes with planar interfaces. However, the important topic of *photocatalysis* is largely based on using dispersed nanoparticles or colloidal semiconductor particles. Interest here has focussed on oxidation and reduction reactions taking place at the surface of the illuminated particles. For example, oxygenated dispersions of $TiO_2$ are used to destroy organic pollutants such as dyes or chlorinated organic species. In these cases, the oxidation reactions are driven by the holes generated by illumination, and the oxygen is reduced by the photogenerated electrons. The oxidizing power of holes in $TiO_2$ is sufficiently high that the final product of oxidation can be $CO_2$ and, for example, chloride ions. This process is therefore referred to as mineralization. The oxidation of organic species by oxygen is generally exoenergetic, but proceeds at a negligible rate in the absence of a catalyst. The process of exoenergetic oxidation involving illuminated oxides such as $TiO_2$ is therefore referred to as *photocatalysis*. By contrast, light-driven reactions that are endoenergetic, such as water splitting, should be referred to as *photosynthetic*

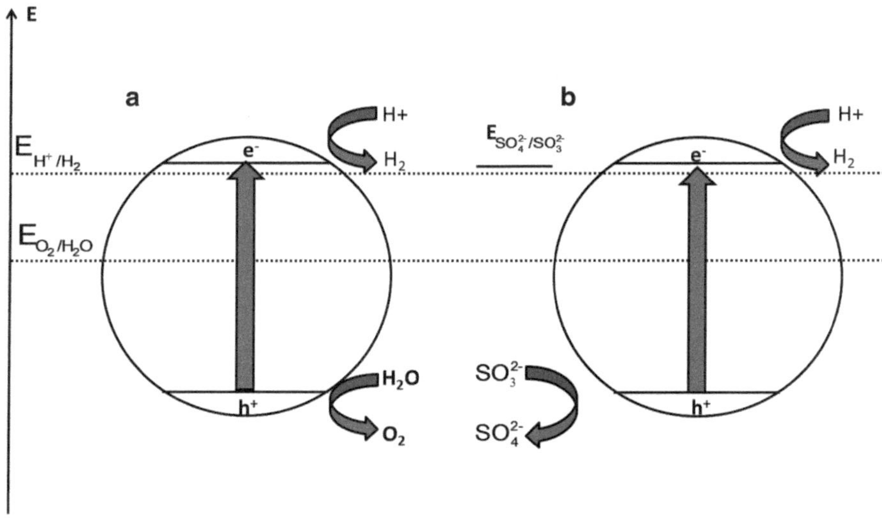


**Fig. 1.22** Contrast between (**a**) a photosynthetic (endoenergetic) reaction and (**b**) photocatalytic (exoenergetic) reaction. (**a**), the light energy is used to generate hydrogen and oxygen (an endoenergetic process). (**b**), the light is used to 'catalyze' the reduction of protons to hydrogen by the strongly reducing electron donor species $SO_3^{2-}$, a reaction that is already thermodynamically feasible in the dark but takes place at a negligible rate

*reactions*, although the term photocatalysis is widely misused in this context. The confusion is often compounded when a so-called *sacrificial agent* is used as an electron donor in order to generate hydrogen from illuminated colloids. In this case, the generation of hydrogen may be exoenergetic. In other words, the electron donor is thermodynamically capable of reducing protons to hydrogen. In this case, we are again dealing with photocatalysis and not with photosynthesis. These two different situations are illustrated schematically in Fig. 1.22 using the example of sulfite as a sacrifical electron donor. The $SO_4^{2-}/SO_3^{2-}$ redox potential is −0.11 V vs. the reversible hydrogen potential, so that the reduction of protons by $SO_3^{2-}$ is exo-energetic (negative change in Gibbs energy). Similarly, if persulfate or Ce (IV) are used as electron acceptors, the generation of oxygen is exoenergetic since their standard reduction potentials are more positive than the standard oxygen reduction potential.

### *1.4.2 Potential Distribution for Nanospheres and Nanorods in an Electrolyte*

In Sect. 1.2.7, we showed that the width of the space charge regions for a planar semiconductor electrode is given by

$$W_{SC} = \left(\frac{2\Delta\phi_{SC}\varepsilon\varepsilon_0}{qN}\right)^{1/2} \quad (1.44)$$

If we consider a bulk anatase ($TiO_2$, $\varepsilon_r \approx 30$) electrode doped n-type by oxygen vacancies with a concentration of $10^{17}$ $cm^{-3}$, $W_{SC}$ is around 130 nm for $\Delta\phi_{SC} = 0.5$ V. Typical anatase particle sizes are in the range 20–50 nm, so we clearly run into the problem that there is no room to accommodate the space charge region. In other words, the space charge region must occupy the entire particle. The radial distribution of electrical potential in a spherical particle was derived by Albery and Bartlett (1984).

$$\phi(r) = \frac{qN}{6\varepsilon\varepsilon_0}\left(r - r_W^2\right)\left(1 + \frac{2r_W}{r}\right) \quad (1.45)$$

Let us consider a spherical anatase particle with a diameter of 50 nm and $N_d = 10^{17}$ $cm^{-3}$. The Fermi level can be lowered by varying the redox potential of the electrolyte, inducing a space charge in the particle. This process reaches a limit when the entire particle becomes depleted. As Fig. 1.23 shows, the maximum band bending that can be achieved is only around 6 meV because at this point the space charge extends to the centre of the particle ($r_W = 0$ in Eq. (1.45)). If the redox potential is lowered further, the band bending remains constant, but the electron concentration still continues to fall as the Fermi level moves downwards. If the doping density is $10^{18}$ $cm^{-3}$, saturation occurs at a band bending ten times higher

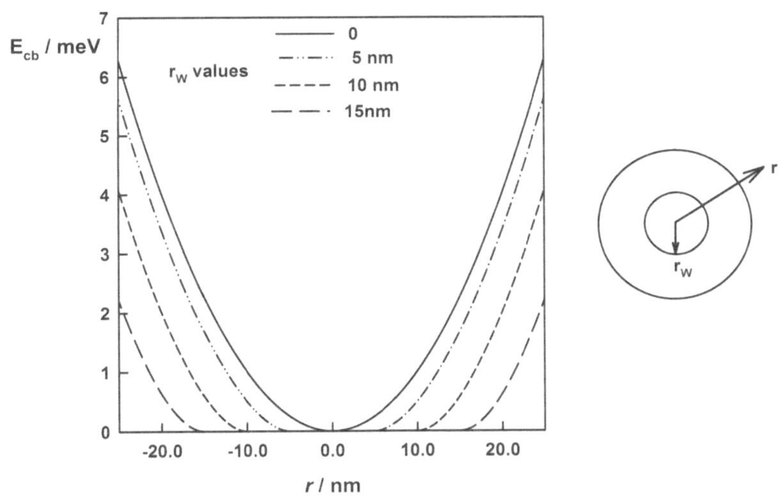


**Fig. 1.23** Band bending for a spherical anatase particle ($N_d = 10^{17}$ $cm^{-3}$, $r = 25$ nm, $\varepsilon = 30$) as a function of $r_W$ (cf. Eq. (1.44)). The radius $r_W$ defines the edge of the space charge region. The maximum band bending of ca. 6 meV is reached when $r_W$ is reduced to zero, i.e. the space charge region extends to the centre of the particle

(ca. 60 meV). For the lower doping density, the band bending is smaller than the average thermal energy $k_BT$, so the effects of the band bending on charge carrier distributions can be neglected. For the higher doping density, the band bending is large enough to affect the distribution of charge carriers in the particle. A more detailed discussion of the implications of the equilibration of particles with a redox electrolyte can be found in Bisquert et al. (1999).

Solutions for the development of space charge in cylindrical nanorods are also available in the literature (Tena-Zaera et al. 2008). In terms of the cross-sectional geometry defined in Fig. 1.23, the voltage drop across the space charge region in a cylinder with radius $R$ is given by

$$\Delta\phi_{SC} = \frac{qN}{2\varepsilon\varepsilon_0}\left[\frac{1}{2}\left(R^2 - r_W^2\right) + R^2 \ln\left(\frac{r_W}{R}\right)\right] \tag{1.46}$$

This solution has been used to describe the capacitance voltage behaviour of arrays of highly-doped ZnO nanowires. Equation (1.46) predicts that the space charge region in 100 nm diameter ZnO nanorods ($\varepsilon_r = 10$, $N_d = 10^{19}$ cm$^{-3}$) would only be 10 nm for a band bending of 1 eV. This means that there is a large quasi-neutral 'core' in the nanorods so that excess electrons can be extracted efficiently when the electrode is illuminated. By contrast, the space charge region will extend almost to the centre of the rod if the doping density is lowered to $10^{17}$ cm$^{-3}$, and the rods will become depleted at higher voltages.

### *1.4.3 Mesoporous Semiconductor Electrodes*

Interest in nanostructured semiconductor electrodes increased sharply following the discovery that mesoporous anatase electrodes coated with a ruthenium complex could be used to fabricate dye-sensitized solar cells (DSC) (Oregan and Grätzel 1991). The porous electrodes can be made by sintering a colloidal paste of anatase nanoparticles with sizes in the range 20–40 nm that has been spread on the surface of a conducting glass sheet by doctor blading. The resulting structure has a porosity of around 50 % allowing electrolyte to penetrate into the film. The physics and chemistry of the DSC lie outside the scope of this chapter, and details can be found elsewhere, for example in references Peter (2007a, b) and (2011). The potential distribution in an array of interconnected nanoparticles contacted by a redox electrolyte such as the $I_3^-/I^-$ couple used in DSC is relevant to the performance of the cell. Since the $I_3^-/I^-$ redox Fermi level lies around 1 eV below the conduction band of $TiO_2$, the equilibrium concentration of holes is very small. When the dye-coated electrode is illuminated, electrons are injected into the conduction band of the $TiO_2$, and the quasi Fermi level moves up towards the conduction band, generating a photovoltage of 0.6 V or more, corresponding to an increase in free electron concentration of at least a factor of $10^{10}$. In addition to photogenerated free electrons in the conduction band, there is a very high concentration of electrons trapped in an exponential distribution of energy levels that spans the forbidden gap. The negative charge in the interconnected nanoparticles is balanced by a net positive ionic charge that is primarily located at the outer Helmholtz plane. This counter charge—as well as the orientation of the solvent dipoles—results in very effective *screening* of the electronic charge, so that electron transport occurs predominantly by diffusion.

In the absence of a redox electrolyte, the Fermi level in mesoporous electrodes such as $TiO_2$ and ZnO can be controlled potentiostatically. As a consequence of electrolyte screening, very high electron concentrations can be produced, changing the optical absorbance of the transparent electrodes (Boschloo and Fitzmaurice 1997, 1999). Other porous semiconductors such as porous silicon (Feng and Tsu 1997; Sailor 2011) and porous GaP (Vanmaekelbergh et al. 1996) have also been studied extensively, and here too shielding plays an important role in controlling the electronic properties of these materials when they are permeated by solvents or electrolytes. The behaviour of all of these porous electrodes differs substantially from their bulk counterparts. For an in-depth discussion of the physical electrochemistry of nanostructured electrodes, the reader is referred to a comprehensive review by Bisquert (2008).

While electron transport in the DSC is quite well-understood, less is known about the behaviour of photogenerated minority carriers. Nanostructured electrodes of materials such as iron oxide (hematite) appear to offer significant advantages for light-driven water splitting (Cesar et al. 2009; Brillet et al. 2010) and this is usually interpreted in terms of the hole diffusion length. However, as shown in Sect. 1.2.4, the minority carrier diffusion length is a property of the bulk semiconductor, where the majority carriers are in large excess. The situation in a mesoporous semiconductor is likely to be quite different, depending on whether or not the width of the space charge region $W_{SC}$ is smaller than the particle size $d$. If $W_{SC} \ll d$, then the situation is similar to the one described in the previous section for nanorods, with well-defined bulk and space charge regions. However, if $W_{SC}$ is comparable with d, complete depletion can occur as in the case of mesoporous anatase electrodes. There is then no bulk for which the minority carrier lifetime can be defined. In fact, since hole reactions in water splitting are very slow (see Sect. 1.3.3), hole diffusion is not the relevant property. What matters is the competition between charge transfer and recombination, either in the space charge region or at the surface. At present, models for these processes still need to be developed.

## 1.5 Conclusions

This chapter has placed its main emphasis on fundamental aspects of semiconductor electrochemistry in order to lay the foundations for a clear understanding of the behaviour of at least more ideal systems. However, it is important to bear in mind that many of the assumptions and approximations made for ideal systems may not be appropriate for real semiconductor electrodes. The trend in recent years away from bulk electrodes to various types of nanostructured electrodes opens up new perspectives, but many of the descriptions of these more complex systems are limited at present to a qualitative level. In order to make further progress in our understanding of nanostructured electrodes, we shall need better theory and efforts to achieve more quantitative measurements and interpretation. The next few years are likely to see substantial steps forward involving the application of new experimental methods accompanied by realistic 3-D modelling of nanostructured electrodes.

**Acknowledgment** The author thanks Gabriela Kissling (Bath), David Tilley (Zurich), and Sixto Giménez (Castelló) for reading of the manuscript carefully and for useful suggestions.

## References

Albery WJ, Bartlett PN (1984) The transport and kinetics of photogenerated carriers in colloidal semiconductor electrode particles. J Electrochem Soc 131:315–325

Allongue P, Cachet H (1985) Band-edge shift and surface-charges at illuminated n-GaAs aqueous electrolyte junctions. Surface state analysis and simulation of their occupation rate. J Electrochem Soc 132:45–52

Bard AJ (1980) Photoelectrochemistry. Science 207:139–144

Bard AJ, Faulkner LR (2001) Electrochemical methods: fundamentals and applications. John Wiley and Sons Inc, New York

Bisquert J (2008) Physical electrochemistry of nanostructured devices. Phys Chem Chem Phys 10:49–72

Bisquert J, Garcia-Belmonte G, Fabregat-Santiago F (1999) Modelling the electric potential distribution in the dark in nanoporous semiconductor electrodes. J Solid State Electrochem 3:337–347

Boschloo G, Fitzmaurice D (1997) Electron accumulation in nanoporous-nanocrystalline $TiO_2$ electrodes. Proc Symp Photoelectrochem 97:84–87

Boschloo G, Fitzmaurice D (1999) Electron accumulation in nanostructured $TiO_2$ (anatase) electrodes. J Phys Chem B 103:7860–7868

Brattain WH, Garrett CGB (1955) Experiments on the interface between Germanium and an electrolyte. Bell Syst Tech J 34:129–176

Brillet J, Grätzel M, Sivula K (2010) Decoupling feature size and functionality in solution-processed, porous hematite electrodes for solar water splitting. Nano Lett 10:4155–4160

Butler MA (1977) Photoelectrolysis and physical properties of semiconducting electrode WO3. J Appl Phys 48:1914–1920

Cass MJ, Duffy NW, Peter LM, Pennock SR, Ushiroda S, Walker AB (2003a) Microwave reflectance studies of photoelectrochemical kinetics at semiconductor electrodes. 1. Steady-state, transient, and periodic responses. J Phys Chem B 107:5857–5863

Cass MJ, Duffy NW, Peter LM, Pennock SR, Ushiroda S, Walker AB (2003b) Microwave reflectance studies of photoelectrochemical kinetics at semiconductor electrodes. 2. Hydrogen evolution at p-Si in ammonium fluoride solution. J Phys Chem B 107:5864–5870

Cendula P, Tilley SD, Gimenez S, Bisquert J, Schmid M, Grätzel M, Schumacher JO (2014) Calculation of the energy band diagram of a photoelectrochemical water splitting cell. J Phys Chem C 118:29599–29607

Cesar I, Sivula K, Kay A, Zboril R, Graetzel M (2009) Influence of feature size, film thickness, and silicon doping on the performance of nanostructured hematite photoanodes for solar water splitting. J Phys Chem C 113:772–782

Degryse R, Gomes WP, Cardon F, Vennik J (1975) Interpretation of Mott-Schottky plots determined at semiconductor-electrolyte systems. J Electrochem Soc 122:711–712
Dewald JF (1960) The charge distribution at the Zinc Oxide-electrlolite interface. J Phys Chem Solids 14:155–161
Dloczik L, Ileperuma O, Lauermann I, Peter LM, Ponomarev EA, Redmond G, Shaw NJ, Uhlendorf I (1997) Dynamic response of dye-sensitized nanocrystalline solar cells: characterization by intensity-modulated photocurrent spectroscopy. J Phys Chem B 101:10281–10289
Dunn HK, Peter LM (2009) How efficient is electron collection in dye-sensitized solar cells? Comparison of different dynamic methods for the determination of the electron diffusion length. J Phys Chem C 113:4726–4731
Feng Z, Tsu R (1997) Porous silicon. World Scientific, Singapore
Fermin DJ, Ponomarev EA, Peter LM (1999) A kinetic study of CdS photocorrosion by intensity modulated photocurrent and photoelectrochemical impedance spectroscopy. J Electroanal Chem 473:192–203
Fisher AC, Peter LM, Ponomarev EA, Walker AB, Wijayantha KGU (2000) Intensity dependence of the back reaction and transport of electrons in dye-sensitized nanacrystalline $TiO_2$ solar cells. J Phys Chem B 104:949–958
Gärtner WW (1959) Depletion layer photoeffects in semiconductors. Phys Rev 116:84–87
Gerischer H (1961) Advances in electrochemistry and electrochemical engineering, vol 1. Interscience, New York
Gerischer H (1989) Neglected problems in the pH-dependence of the flatband potential of semiconducting oxides and semiconductors covered with oxide layers. Electrochim Acta 34:1005–1009
Kosmulski M (2009) Compilation of PZC and IEP of sparingly soluble metal oxides and hydroxides from literature. Adv Colloid Interf Sci 152:14–25
Kumar A, Santangelo PG, Lewis NS (1992) Electrolysis of water at $SrTiO_3$ PHOTOELECTRODES - Distinguishing between statistical and stochastic formalisms for electro transfer processes in fuel forming photoelectrochemical systems. J Phys Chem 96:834–842
Landsberg PT (1991) Recombination in semiconductors. Cambridge University Press, Cambridge
Letaw H, Bardeen J (1954) Electrolytic analog transistor. J Appl Phys 25:600–606
Li J, Peat R, Peter LM (1984) Surface recombination at semiconductor electrodes. 2. Photoinduced near-surface recombination centers in p-GaP. J Electroanal Chem 165:41–59
Li J, Peter LM (1985) Surface recombination at semiconductor electrodes. 3. Steady-State and Intensity modulated photocurrent response. J Electroanal Chem 193:27–47
Li J, Peter LM (1986) Surface recombination at semiconductor electrodes. 4. Steady-state and intensity modulated photocurrents at normal GaAs electrodes. J Electroanal Chem 199:1–26
Licht S, Bard AJ, Stratmann M (2002) Encyclopedia of electrochemistry: semiconductor electrodes and photoeletrochemistry, vol 6. Wiley-VCH, Weinheim
Marcus RA (1964) Chemical + electrochemical electron transfer theory. Annu Rev Phys Chem 15:155–196
Memming R (1978) Top current chemistry. Springer, New York
Memming R (2015) Semiconductor electrochemistry. Wiley VCH, Weinheim
Morrison SR (1980) Electrochemistry of semiconductor and metal electrodes. Plenum Press, New York
Nozik AJ (1978) Photoelectrochemistry. Applications to solar energy conversion. Annu Rev Phys Chem 29:189–222
Oregan B, Grätzel M (1991) A low-cost, higher efficiency solar cell based on dye-sensitized colloidal $TiO_2$ films. Nature 353:737–740
Peat R, Peter LM (1987) Intensity modulated photocurrent spectroscopy of n-GaAs. Berichte Der Bunsen-Gesellschaft Phys Chem Chem Phys 91:381–386
Peter LM (2007a) Characterization and modeling of dye-sensitized solar cells. J Phys Chem C 111:6601–6612

Peter LM (2007b) Dye-sensitized nanocrystalline solar cells. Phys Chem Chem Phys 9:2630–2642
Peter LM (2011) The Grätzel cell: where next? J Phys Chem Lett 2:1861–1867
Peter LM (2013) Energetics and kinetics of light-driven oxygen evolution at semiconductor electrodes: the example of hematite. J Solid State Electrochem 17:315–326
Peter LM, Li J, Peat R (1984) Surface recombination at semiconductor electrodes. 1. Transient and steady-state photocurrents. J Electroanal Chem 165:29–40
Peter LM, Ponomarev EA, Fermin DJ (1997) Intensity-modulated photocurrent spectroscopy: reconciliation of phenomenological analysis with multistep electron transfer mechanisms. J Electroanal Chem 427:79–96
Peter LM, Tributsch H (2008) Nanostructured and photoelectrochemical systems for solar photon conversion, vol 3. Imperial College Press, London
Peter LM, Vanmaekelbergh D (1999) Time and frequency resolved studies of photoelectrochemical kinetics. In: Alkire RC, Kolb DM (eds) Advances in electrochemical science and engineering, vol 6. Wiley-VCH, Weinheim, pp 77–164
Peter LM, Wijayantha KGU, Tahir AA (2012) Kinetics of light-driven oxygen evolution at alpha-$Fe_2O_3$ electrodes. Faraday Discuss 155:309–322
Pleskov IV, Gurevich YY (1986) Semiconductor electrochemistry. Consultants Bureau, New York
Ponomarev EA, Peter LM (1995a) A comparison of intensity modulated photocurrent spectroscopy and photoelectrochemical impedance spectroscopy in a study of photoelectrochemical hydrogen evolution at p-InP. J Electroanal Chem 397:45–52
Ponomarev EA, Peter LM (1995b) A generalized theory of intensity modulated photocurrent spectroscopy (IMPS). J Electroanal Chem 396:219–226
Reichman J (1980) The current-voltage characteristics of semiconductor-electrolyte junction photovoltaic cells. Appl Phys Lett 36:574–577
Reineke R, Memming R (1992a) Comparability of redox reactions at n-type and p-type semiconductor electrodes. 1. The quasi Fermi level concept. J Phys Chem 96:1310–1317
Reineke R, Memming R (1992b) Comparability of redox reactions at n-type and p-type semiconductor electrodes. 2. Electrochemical overpotential and recombination in view of the quasi Fermi level concept. J Phys Chem 96:1317–1323
Royea WJ, Fajardo AM, Lewis NS (1997) Fermi golden rule approach to evaluating outer-sphere electron-transfer rate constants at semiconductor/liquid interfaces. J Phys Chem B 101:11152–11159
Sailor M (2011) Porous silicon in practice. Wiley-VCH Verlag GmbH & Co, Weinheim
Salvador P (2001) Semiconductors' photoelectrochemistry: a kinetic and thermodynamic analysis in the light of equilibrium and nonequilibrium models. J Phys Chem B 105:6128–6141
Sato N (1998) Electrochemistry at metal and semiconductor electrodes. Elsevier, Amsterdam
Schlichthörl G, Ponomarev EA, Peter LM (1995) An investigation of hydrogen evolution at p-Si by intensity modulated photocurrent spectroscopy and photomodulated microwave reflectivity. J Electrochem Soc 142:3062–3067
Sze SM (1981) Physics of semiconductor devices. John Wiley & Sons Inc, New York
Tena-Zaera R, Elias J, Levy-Clement C, Bekeny C, Voss T, Mora-Sero I, Bisquert J (2008) Influence of the potassium chloride concentration on the physical properties of electrodeposited ZnO nanowire arrays. J Phys Chem C 112:16318–16323
Trasatti S (1986) The absolute electrode potential. An explanatory note (recommendations 1986). Pure Appl Chem 58:955–966
Vanmaekelbergh D, Marin FI, van De Lagemaat J (1996) Transport of photogenerated charge carriers through crystalline GaP networks investigated by intensity modulated photocurrent spectroscopy. Berichte Der Bunsen-Gesellschaft Phys Chem Chem Phys 100:616–626
Würfel P, Würfel U (2009) Physics of solar cells. From basic principles to advanced concepts. Wiley, Weinheim
Xu Y, Schoonen MAA (2000) The absolute energy positions of conduction and valence bands of selected semiconducting minerals. Am Mineral 85:543–556
Yablonovitch E, Allara DL, Chang CC, Gmitter T, Bright TB (1986) Unsually low surface-recombination velocity on silicon and germanium surfaces. Phys Rev Lett 57:249–252

# Chapter 2
# The Oxygen Evolution Reaction: Mechanistic Concepts and Catalyst Design

**Richard L. Doyle and Michael E.G. Lyons**

## 2.1 Introduction

The splitting of water into hydrogen and oxygen is a vital component of a promising renewable energy infrastructure (Zhou 2005; Lewis and Nocera 2006). Hydrogen gas is regarded by many as the fuel of the future. With the highest mass energy density of any fuel and its clean combustion in air, hydrogen could be considered the ultimate clean energy carrier (Dinga 1985; Schlogl 2010). However, the realisation of a largely hydrogen based renewable energy solution—the hydrogen economy—depends heavily on the development of cost-effective, green production technologies (Crabtree et al. 2004). Hydrogen is readily produced on a large scale by steam reforming of natural gas, yet this non-renewable process is inherently environmentally offensive producing significant quantities of carbon dioxide (Häussinger et al. 2000). The catalytic splitting of water, on the other hand, offers a clean, renewable and potentially cost-effective route to the production of hydrogen gas (Ohi 2005; Tributsch 2008; Zeng and Zhang 2010).

R.L. Doyle (✉)
Electrochemical Materials and Energy Group, Tyndall National Institute, University College Cork, Lee Maltings, Cork, Ireland
e-mail: richard.doyle@tyndall.ie

M.E.G. Lyons
Trinity Electrochemical Energy Conversion and Electrocatalysis Group, School of Chemistry & AMBER National Centre, CRANN Research Institute, Trinity College Dublin, Dublin 2, Ireland
e-mail: melyons@tcd.ie



S. Giménez, J. Bisquert (eds.), *Photoelectrochemical Solar Fuel Production*,
DOI 10.1007/978-3-319-29641-8_2

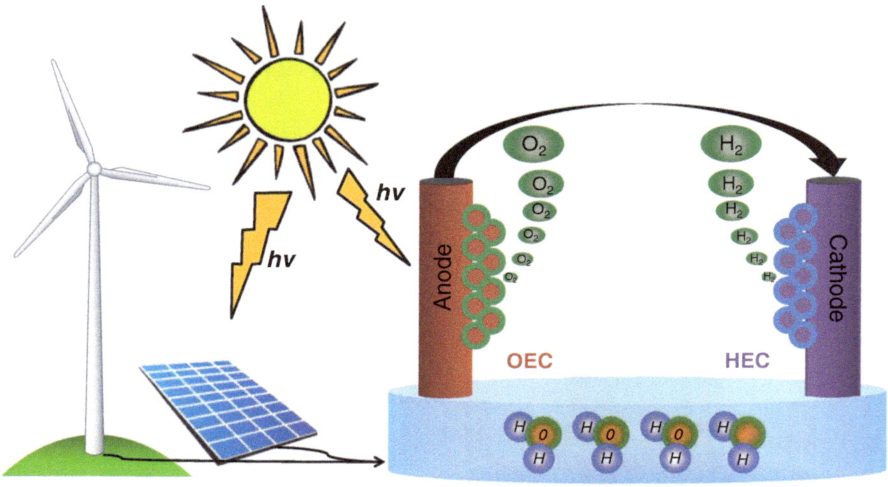


**Fig. 2.1** Schematic representation of the splitting of water via electrolysis, utilising electricity derived from renewable sources such as wind and solar, and photoelectrolysis, where the electrodes directly harvest the solar energy. Oxygen evolution catalysts (OEC) are located on the anode and hydrogen evolution catalysts (HEC) are located on the cathode. Adapted from Joya et al. (2013)

In general, the overall water splitting process can be represented as follows, with molecular hydrogen and oxygen generated individually at the cathode and anode, respectively.

$$2H_2O + \text{Energy} \rightarrow 2H_2 + O_2 \tag{2.1}$$

In principle, the energy required to drive the water splitting reaction can be obtained from any of a number of sources making it a highly versatile energy conversion technology, as illustrated in Fig. 2.1. One of the more attractive options is the coupling of electrochemical water splitting devices—water electrolysers—with grid scale renewable energy harvesting technologies such as wind turbines or photovoltaics (Turner 2004; Grätzel 2005). In this way, water electrolysis could act as a local energy storage system permitting the implementation of these intermittent energy sources on a global scale (Lewis and Nocera 2006). An alternative method, in the context of solar-to-fuel conversion, is photoelectrolysis or light-driven water splitting (Brimblecombe et al. 2009; Nocera 2012; Joya et al. 2013). In this approach, light harvesting mechanisms, typically involving semiconductor materials, are incorporated into the electrode design so that the necessary solar energy is harvested directly by the electrode materials. Although technically more demanding than electrolysis from the point of view of electrode composition, this standalone approach is growing in popularity off the back of substantial progress in the design of artificial photosynthetic cells—the artificial leaf (Nocera 2012; Joya et al. 2013). However, regardless of the route taken, the viability of these systems as sustainable hydrogen production technologies is, in the end, dependent on the electrochemistry of oxygen. The generation of molecular oxygen at the anode is the most energy-intensive step in the overall water splitting process (Hall 1983; Dau et al. 2010; Marinia et al. 2012).

To understand this, consider the fundamental operating requirement of a water splitting system: the equilibrium potential $V_{eq}$ of 1.23 V. In order to generate hydrogen at a specific rate, a voltage at least equal in magnitude to 1.23 V must be supplied to the system. For water electrolysis, this operational voltage depends on the kinetics of the water splitting reactions and the design of the electrolyser unit such that:

$$V_{op} = V_{eq} + \eta_A + |\eta_C| + \eta_\Omega \tag{2.2}$$

where $\eta_A$ and $\eta_C$ are the overpotentials required to overcome the kinetic barriers for the oxygen evolution reaction (OER) at the anode and the hydrogen evolution reaction (HER) at the cathode, respectively, and $\eta_\Omega$ is the additional overpotential required to compensate for resistance losses within the device (Trasatti 1994). The efficiency of the electrolyser unit is therefore reflected in the degree to which $V_{op}$ deviates from $V_{eq}$. In an ideal system, $\eta_A$ and $\eta_C$ would be close to zero and $V_{op}$ would depend only on $\eta_\Omega$, which could be minimised through efficient design of the electrolyser unit. In reality, though, this is never the case and kinetic limitations are significant, accounting for up to 85 % of the total efficiency losses (Greeley and Markovic 2012). In alkaline solution, the OER and HER can be described by the following two electrochemical reactions:

$$4OH^- \rightarrow O_2 + 2H_2O + 4e^- \; (V_{Anode} = 1.23 \text{ V vs. RHE}) \tag{2.3}$$

$$4H_2O + 4e^- \rightarrow 2H_2 + 4OH^- (V_{Cathode} = 0.0 \text{ V vs. RHE}) \tag{2.4}$$

and in acid the corresponding reactions are:

$$2H_2O \rightarrow O_2 + 4H^+ + 4e^- \; (V_{Anode} = 1.23 \text{ V vs. RHE}) \tag{2.5}$$

$$4H^+ + 4e^- \rightarrow 2H_2 \; (V_{Cathode} = 0.0 \text{ V vs. RHE}) \tag{2.6}$$

where $V_{Anode}$ and $V_{Cathode}$ are the equilibrium potentials for the OER and HER, respectively. In practice, the efficiency of water electrolysis is limited by the large anodic overpotential of the OER. Whereas the HER can proceed with appreciable currents at potentials close to its equilibrium potential (Markovic and Ross 2002; Sheng et al. 2010), the OER requires a substantial overpotential to generate measurable current densities. Consequently no working electrolysis device can approach the minimum operational voltage of 1.23 V (Greeley and Markovic 2012). Similarly, the slow kinetics of the OER complicate the design of photoelectrolysis devices. In direct light-driven water splitting it is not sufficient for the band-gap of the semiconductor to just match the equilibrium potential for water splitting, it must also be large enough to easily overcome the OER overpotential if a reasonable rate is to be achieved (Valdés et al. 2012). Thus, understanding and optimising the oxygen evolution process is seen as one of the remaining grand challenges for both physical electrochemistry and energy science.

Bearing that in mind, this chapter focuses on the principles underlying efficient oxygen evolution electrocatalysis. Over the past fifty years, considerable research effort has been devoted to the design, synthesis and characterisation of oxygen evolution anode materials with the aim of achieving useful rates of active oxygen evolution at the lowest possible overpotentials. Owing to their stability under oxygen evolution conditions, metal oxides have emerged as the material of choice for catalysing the OER (Katsounaros et al. 2014). Even in the case of the solid metal electrodes, the anodic OER always occurs at an oxidised surface (Conway 1995). Accordingly, a wide range of oxide-based catalysts have been developed. Frequently, the leading OER anode materials are based on $RuO_2$ and $IrO_2$, since these materials exhibit the lowest overpotentials for the OER at practical current densities (Michas et al. 1992). However, the high cost and low natural abundance of these materials renders their widespread commercial utilisation impractical (Kinoshita 1992). This fact has motivated an extensive search for earth abundant catalytic materials with a particular emphasis on the various oxides of first row transition metals (Fabbri et al. 2014; Galán-Mascarós 2015). Although they are typically unstable in acidic media, their relatively low cost and long term corrosion resistance in alkaline solution makes them attractive, alternative OER materials. Yet despite these efforts, OER catalysts are still predominantly developed by an intuitively trial and error approach. The type and method of preparation of the oxide along with its composition and surface morphology can all have a significant influence on its catalytic performance with the result that knowledge of the intrinsic activity of the material can remain elusive (Surendranath and Nocera 2012; Doyle et al. 2013). The central challenge for oxygen evolution research is, therefore, uncovering the mechanistic details and structural motifs necessary for efficient catalysis. Such a realisation could facilitate a unified theory of oxygen evolution catalysis, greatly accelerating the design of ever more efficient and cost effective catalysts.

Considering the extent of the research on oxide-based oxygen evolution catalysts, this chapter does not attempt an exhaustive review, but rather seeks to emphasise the key concepts common to these materials and to the study of the OER in general. The OER is fundamentally an electrocatalytic reaction and accordingly this chapter begins with a traditional electrochemical approach, outlining the practical and theoretical application of an ensemble of electrochemical techniques. This type of approach provides a valuable kinetic analysis of the OER upon which further mechanistic investigations should be based. In the second section of this chapter, modern computational descriptions of the OER are discussed. Theoretical calculations provide a thermodynamic basis for understanding the electrochemical activity of oxide materials and are a useful complement to a comprehensive experimental study. With the fundamental knowledge gained in these sections, the chapter then progresses to consider OER catalyst design principles. In the third section, the contribution of single-parameter "descriptors" to the understanding of the intrinsic activity of OER catalysts is examined and in the fourth section, several prominent methods for enhancing the activity of OER catalysts are described. Taken together, these approaches form a multi-dimensional view of the requirements for efficient OER catalysis. Finally, this chapter concludes with a molecular level consideration of the nature of the OER active site at metal oxide catalysts that seeks to bridge the fields of heterogeneous electrocatalysis and homogeneous molecular catalysis.

## 2.2 Electrochemical Perspectives

Electrocatalysis can be broadly defined as the ability of an electrode material or surface to accelerate the rate of an electrochemical process. The core of any electrocatalytic study is therefore the elucidation of the kinetic parameters that describe the electron-transfer reactions at the interface. Key parameters of interest include the transfer coefficient and the corresponding Tafel slope, as well as the reaction orders of the mechanistically significant reactants. This section begins with a general overview of these important parameters focusing on practical considerations relating to their application in a kinetic study of the OER. Building on this, a simple electrochemical approach for the construction of a meaningful mechanistic interpretation of the OER is presented.

### 2.2.1 *The Tafel Slope*

In any electrochemical reaction, the fundamental observables are the current ($i$) and the potential ($V$). For a given interfacial process, the current is a manifestation of the rate of the interfacial reaction and can be shown to be dependent on the applied potential (Bard and Faulkner 2000). In essence, varying the potential is equivalent to changing the driving force for the electrode process (Surendranath and Nocera 2012). Thus, the relationship between the current and potential is a primary concern in electrocatalysis.

Steady-state Tafel plot analysis is the most widely applied technique in the study of electrocatalytic reactions. For a multistep reaction such as the OER, the relationship between the steady-state anodic current and the applied potential can be represented in the following general form:

$$i = i_0 \exp\left(\frac{\alpha_a F \eta}{RT}\right) \tag{2.7}$$

where $i_0$ is the exchange current which represents the rate of the forward and reverse reactions at the equilibrium potential, $\alpha_a$ is the transfer coefficient for the anodic reaction, $\eta$ is the overpotential, $F$ is Faraday's constant, $R$ is the gas constant and $T$ is temperature. This expression holds in the absence of mass transport limitations, which would alter the concentration of reactants at the electrode surface relative to their bulk values (Guidelli et al. 2014a, b). That is, the reaction is assumed to be under kinetic control. Now, if the current is expressed in logarithmic form, the Tafel relationship is revealed:

$$\log(i) = \log(i_0) + \eta/b \tag{2.8}$$

or alternatively, Eq. (2.8) can be expressed in the original format proposed by Tafel (1904):

$$\eta = a + b\log(i) \tag{2.9}$$

where $a$ is a constant and $b$ is the Tafel slope given by:

$$b = \frac{\partial \eta}{\partial \log i} = \frac{2.303RT}{\alpha_a F} \tag{2.10}$$

From Eqs. (2.8) and (2.9) it is clear that a linear relationship between $\eta$ and $\log(i)$ is predicted such that the Tafel slope $b$, typically expressed in units of millivolts (mV) per decade of current ($dec^{-1}$), can be readily extracted from plots of $\log(i)$ vs. $\eta$ or $\eta$ vs. $\log(i)$.[1] It is noteworthy that substitution of the applied potential $V$ for $\eta$ has no impact on the measured Tafel slope and is often the preferred protocol for multistep reactions, where a particular reaction step may control the overall kinetic profile making it difficult to define the thermodynamic potentials for the individual steps (Surendranath and Nocera 2012).

The significance of the Tafel slope in electrocatalysis is multifaceted. Depending on the interpretation, it can provide the means for both a quantitative and mechanistic characterisation of an electrocatalytic process. At its most basic level, the Tafel slope is a sensitivity parameter giving a measure of the rate of increase of electrode potential or polarisation with the log of the current density. As such it is considered a useful parameter for the practical evaluation of electrocatalytic materials. Notably, Conway et al. (1987) contend that the Tafel slope is a more important measure of electrocatalytic performance than the commonly utilised $i_0$, which is a measure of the intrinsic kinetic rate of a reaction. This is rationalised by the fact that

[1] Traditionally Tafel plots were recorded using Galvanostatic methods where the current was controlled and the potential was measured, as described by Eq. (2.9). In this way, the Tafel slope could be obtained directly from the experimental plots, hence the convention of reporting the Tafel slope in the form of Eq. (2.10). However, due to the ease with which the potential can be controlled using modern potentiostats, Tafel plots are now routinely recorded in the form of Eq. (2.8) and the corresponding Tafel slope is obtained from the inverse slope of the experimental plot.

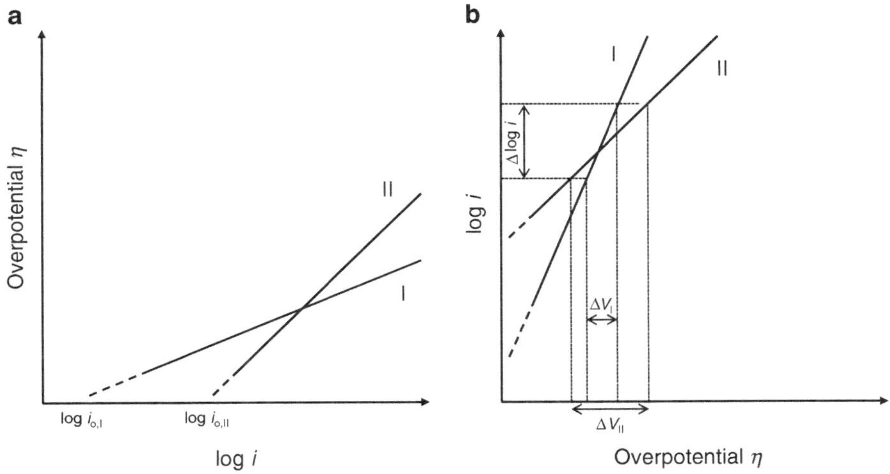


**Fig. 2.2** Schematic diagrams illustrating the significance of Tafel slopes for determining favourable electrocatalytic properties. The quantities $\Delta V_{\rm I}$ and $\Delta V_{\rm II}$ indicate the potential changes necessary to alter the current output of catalyst I and II, respectively, by the value $\Delta \log i$. Adapted from Conway et al. (1987)

$i_0$ only refers to the kinetics at the thermodynamic potential and does not characterise the kinetics of the electrode process at higher, more practical current densities. This point is illustrated in Fig. 2.2 where schematic Tafel plots for catalytic materials with different $i_0$ and Tafel slope values are shown in the form of Eqs. (2.8) and (2.9). In Fig. 2.2a, catalyst I is depicted with a low $i_0$ value relative to catalyst II and so could be considered less active. However, catalyst II clearly has a larger Tafel slope than catalyst I, thereby experiencing greater polarisation with increasing current density. Consequently, catalyst I exhibits a lower overpotential than catalyst II at higher current densities. This distinction is particularly important for systems such as water electrolysers where high current densities at minimum operational voltages are required for economic viability. In this respect, it should also be noted that such systems may require operation over a range of current densities depending on $H_2$ demand (Conway et al. 1987). Therefore, from a practical stand point, it is not just the current density obtainable at a given overpotential that is important but also the rate of change of this current density with overpotential—the inverse Tafel slope $1/b$ given by the slope of Eq. (2.8). As demonstrated in Fig. 2.2b, manipulation of the current density for catalyst I requires much smaller changes in overpotential than for catalyst II, making it a more practical real world material. Hence, low Tafel slopes are widely acknowledged as an indicator of efficient electrocatalytic performance (Conway et al. 1987; Merrill and Dougherty 2008).

Taking a more theoretical perspective, the Tafel slope yields insight into the mechanistic pathway of an electrode process. It can be seen from Eq. (2.10) that the

Tafel slope is defined by the transfer coefficient $\alpha_a$ which is experimentally quantifiable according to (Guidelli et al. 2014a, b):

$$\alpha_a = \frac{1}{b}\left(\frac{2.303RT}{F}\right) \tag{2.11}$$

Moreover, for a multistep reaction consisting of a sequence of elementary steps with a single rate determining step (RDS) $\alpha_a$ can be given by (Parsons 1951; Damjanovic et al. 1966, 1967a, b; Bockris and Khan 1993; Guidelli et al. 2014a, b):

$$\alpha_a = n_f/\nu + n_r\beta \tag{2.12}$$

where $n_f$ is the number of electrons transferred before the RDS, $\nu$ is the stoichiometric number defined as the number of times the RDS occurs for one repetition of the overall reaction, $n_r$ is the number of electrons transferred in each occurrence of the RDS and $\beta$ is the symmetry factor. While there is no general consensus on the physical meaning of $\beta$ it is related to the activation barrier and is generally assumed to be close to 0.5, describing a symmetrical potential energy barrier (Guidelli et al. 2014a, b). In this way, it can be seen that the Tafel slope is a composite parameter, giving information on the stoichiometry and the succession of steps in the overall reaction. Note, for an electron transfer step $n_r$ can be safely assumed to equal one as the transfer of more than one electron at a time is improbable (Guidelli et al. 2014a, b). Therefore, if the first electron transfer step in a sequential reaction is rate-determining then $n_f = 0$ and $\alpha_a = \beta = 0.5$ implying a Tafel slope of 120 mV dec$^{-1}$. Similarly, if the second electron transfer step is rate-determining $n_f = 1$ and $\alpha_a = 1.5$, assuming a typical value of $\nu = 1$, giving a Tafel slope of 40 mV dec$^{-1}$. On the other hand, if the rate-determining step involves a chemical step subsequent to the first electron transfer step $n_r = 0$ and $n_f = 1$. In this case, $\alpha_a = 1$ predicting a Tafel slope of 60 mV dec$^{-1}$. Thus, the elucidation of Tafel slopes can be useful in differentiating between possible reaction mechanisms.

In view of the potential utility of the Tafel slope it is important to discriminate between true kinetic effects and experimental effects. The discussion so far has only considered the possibility of a single linear Tafel region. In practice, however, multiple Tafel regions may be observed. Certainly the presence of two distinct linear regions is a common kinetic feature of the OER Tafel plots presented in the literature (Damjanovic et al. 1966, 1967a; Hrussanova et al. 2004; Guerrini et al. 2007; Doyle and Lyons 2013a, b). Such changes in Tafel slope with increasing potential are most often attributed to either a change in the RDS within a given pathway or to the influence of changing potential on the adsorption of the reaction intermediates (Damjanovic et al. 1966, 1967a), the kinetic basis for which are discussed in Sect. 2.2.3. It is important though to note that increases in the Tafel slope do not necessarily have mechanistic significance and caution should be taken when interpreting multiple Tafel regions in terms of a possible mechanistic pathway. Experimentally, increases in the Tafel slope with applied potential could be the result of a reduction in the effective electrode surface area with increasing gas evolution at the higher applied potentials. Likewise, the onset of mass transport limitations at high overpotentials will reduce the sensitivity of the current response to increases in potential resulting in an increasing Tafel slope. These complications can be readily minimised using a rotating disc electrode which would simultaneously limit the adherence of gas bubbles to the electrode surface while providing a well-defined mass transport regime.

However, the identification of true mechanistically significant changes in Tafel slope is best accomplished using a combination of experimental techniques. Tafel plots can be generated using various steady-state polarisation techniques, Impedance Spectroscopy (IS) and the analysis of open circuit potential decay curves, the specific details of which have been comprehensively reviewed by Doyle et al. (2013). IS offers a useful accessory method for the determination of Tafel slopes. Obtaining equivalent Tafel slopes using IS involves the experimental measurement of the total Faradaic resistance $R_{\text{far}}$ as a function of the applied potential according to:

$$\log\left(\frac{1}{R_{\text{far}}}\right) = \frac{V}{b} + \log\left(\frac{2.303 i_0}{b}\right) \tag{2.13}$$

implying that the inverse slope of a plot of $\log(1/R_{\text{far}})$ against $V$ is equal to the Tafel slope $b$. In this regard, the following caveat should be noted; the parameter $R_{\text{far}}$ is a combination of the charge transfer resistances for all steps in the reaction, regardless of whether they are the RDS, a preceding step or a following step and so mechanistic interpretation of IS Tafel slopes alone is not recommended (Harrington and van den Driessche 2011). That said, a number of studies have highlighted the effectiveness of this type of data treatment when used along side steady-state polarisation techniques (Lyons and Brandon 2009; Doyle and Lyons 2013a, b). A comparison of the steady-state polarisation and IS Tafel plots obtained for a hydrous iron oxide catalyst is presented in Fig. 2.3a. Clear agreement was observed between the two different methods and in this situation, the authors could conclude that the dual Tafel behaviour was in fact mechanistically significant and not simply due to electrode blocking, mass-transport limitations or ohmic effects (Doyle and Lyons 2013a, b).

Steady-state measurements are also complemented by studies of the decay of the open circuit potential (OCP). This approach is frequently adopted in cases where elucidation of the reaction mechanism by steady state polarisation techniques is complicated by continuous alteration of the electrode surface with time. Following an initial polarisation at a potential in the region where OER Tafel behaviour is observed, the variation of the electrode potential at open circuit with time is given by (Bockris et al. 2002):

$$V_{\text{OCP}} = \frac{2.303RT}{\beta F}\log\left(\frac{RT}{\beta F i_0}\right) - \frac{2.303RT}{\beta F}\log(t) \tag{2.14}$$

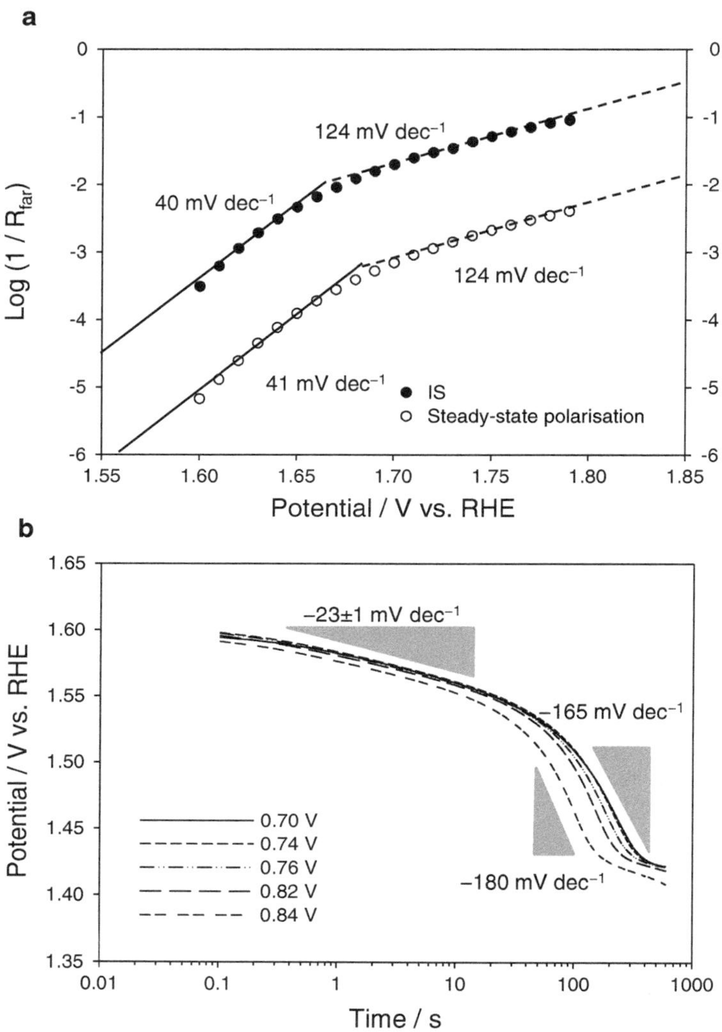


**Fig. 2.3** (**a**) A comparison of the Tafel plots generated using IS and steady-state polarisation data and (**b**) the variation of OCP as a function of log (time) after a 5 min polarisation period at a series of potentials associated with oxygen evolution for a hydrous iron oxide catalyst film in 1.0 M NaOH. Adapted from Doyle et al. (2013)

On this basis, the slope $b_{OCP}$ of a plot of the measured potential or overpotential during the decay as a function of log $t$ is equal to the negative of the Tafel slope $b$ (Bockris et al. 2002). While such simple relationships between the OCP decay slope and the Tafel slope have certainly been observed (Lyons and Floquet 2011), the true power of this approach lies in its ability to describe the potential dependence of the surface capacity and the fractional coverage of surface intermediates. In deriving Eq. (2.14) it was assumed that the electrode surface capacitance is independent of potential (Bockris et al. 2002). In fact, Conway and Bourgault (1959, 1962) and Bourgault and Conway (1960) have shown that this is the only case under which Eq. (2.14) holds. Thus, the observation of numerical identity between the decay slope and the Tafel slope suggests that the surface capacity remains constant during self-discharge. In situations where the surface capacity varies during self-discharge the observed decay slope can differ quite significantly from the steady-state Tafel slope. This is evident in Fig. 2.3b where the corresponding $V$ vs. log $t$ plots for the hydrous iron oxide catalyst are shown. At short decay times a slope $b_{\mathrm{OCP}}$ of ca. $-23$ mV dec$^{-1}$ is obtained, whereas at longer times the slope $b_{\mathrm{OCP}}$ varies between ca. $-160$ and $-180$ mV dec$^{-1}$ depending on the initial polarisation potential. Importantly, these slopes differ considerably from the expected slope of ca. 40 mV dec$^{-1}$ obtained using steady-state polarisation and IS techniques in Fig. 2.3a. To interpret this result two primary limiting cases can be identified from the analysis of Conway and Bourgault (1962). If the surface coverage of intermediates during self-discharge is assumed to be low, it can be shown that the surface capacity decreases with decreasing potential giving the following relationship between the decay slope $b_{\mathrm{OCP}}$ and the true Tafel slope $b$:

$$b_{\mathrm{OCP}} = \frac{b'b}{b - b'} \tag{2.15}$$

where $b'$ is some positive multiple or fraction of $RT/F$. Hence the observed decay slope is predicted to be greater than the Tafel slope. Conversely, if there is high but not full coverage of intermediates the surface capacity increases with decreasing potential and the decay slope is given by:

$$b_{\mathrm{OCP}} = -\frac{b'b}{b' - b} \tag{2.16}$$

indicating that the observed decay slope will be less than the Tafel slope. In this way, the lack of agreement between the decay slopes and Tafel slopes in Fig 2.3 suggests that the surface capacity does not remain constant during self-discharge. This was interpreted in terms of a decrease in the surface concentration of charged *surfaquo* groups, which were considered to be the active sites for oxygen evolution (see Sect. 2.5.4) (Doyle et al. 2013). From an accompanying IS study, these authors found that the surface capacity increased with increasing potential in accordance with the formation of charged surface intermediates (Doyle and Lyons 2013a, b). Therefore, following the reasoning of Conway and Bourgault (1962), at high potentials or short decay times a large proportion of the surfaquo groups are in the higher charged state implying significant surface coverage of intermediates and the observed decay slope is lower than the Tafel slope. On the other hand, for longer times the surface coverage becomes considerably reduced due to increased discharge[2] of the *surfaquo* groups and the decay slope is found to be greater than the Tafel slope. In this way, the relationship between the decay slope and the steady state Tafel slope can provide a useful qualitative characterisation of the kinetics and mechanism of the basic processes involved in an electrocatalytic reaction.

### 2.2.2 Electrochemical Reaction Orders

The relevance of reaction orders in the kinetic study of chemical reactions is well known. Their determination provides the relationship between the reaction rate and the concentration of a particular reactant—a key criterion used in mechanistic interrogations. In the case of electrochemical reactions, the rate of the reaction, expressed as current, is intimately related to the applied potential, as described in Sect. 2.2.1. As a result, the electrochemical reaction order $m_x$ can be expressed in terms of either of these quantities according to the following derivatives (Conway and Salomon 1964):

1.
$$m_{x,i} = \left(\frac{\partial V}{\partial \log a_x}\right)_i$$

2.
$$m_{x,\eta} = \left(\frac{\partial \log i}{\partial \log a_x}\right)_\eta$$

3.
$$m_{x,V} = \left(\frac{\partial \log i}{\partial \log a_x}\right)_V$$

where $a_x$ is the activity of a mechanistically significant reactant $x$ and the subscripts $i$, $\eta$, and $V$ denote conditions of constant current, constant overpotential, and constant applied potential, relative to a pH independent reference electrode, respectively. It should be noted that the parameter $m_{x,V}$ is regarded as the "chemically" or "mechanistically" significant reaction order since it gives the dependence of log (rate/$i$) on log (reactant concentration/$a_x$) in the absence of variations of the double-layer configuration, i.e. constant $V$ (Conway and Salomon 1964). In contrast to the situation outlined in Sect. 2.2.1 for the Tafel slope, substitution of the potential for the overpotential is not innocent with regard to measured reaction orders and the following correction factor is necessary to account for this (Lyons et al. 2014; Gileadi 1993):

[2] Self-discharge is assumed to proceed by an electrochemical mechanism analogous to that of corrosion. That is, the simultaneous occurrence of anodic and cathodic reactions as a mixed potential via a local cell mechanism. In the present case, self-discharge consists of a cathodic oxide or *surfaquo* group reduction process and an anodic oxygen evolution process.

$$m_{x,\eta} = m_{x,V} - \beta \tag{2.17}$$

In addition, the relationship between $m_{x,i}$ and $m_{x,V}$ offers a useful self-consistency check[3] between the experimentally measured Tafel slope $b$ and the mechanistically significant reaction order $m_{x,V}$ as follows (Surendranath and Nocera 2012):

$$m_{x,i} = -b(m_{x,V}) \tag{2.18}$$

However, in subsequent discussions all references to the parameter $m_x$ refer to the mechanistically significant $m_{x,V}$.

In the case of alkaline water electrolysis the pertinent reactant term is the activity of the hydroxide ion $a_{OH^-}$. Thus, a plot of $\log i$ vs. $\log a_{OH^-}$ should be linear with slope equal to the corresponding electrochemical reaction order $m_{OH^-}$. These experimentally determined reaction order values can often be potential dependent and so it is necessary to generate reaction order plots at a range of potentials. This is especially important for systems exhibiting dual Tafel behaviour, as distinct reaction orders can be associated with each Tafel region. A sample reaction order study for a hydrous iron oxide catalyst is shown in Fig. 2.4. An important point which must be considered when interpreting such reaction order data is that, theoretically, reaction orders should be integer values. While the reaction orders in Fig. 2.4 are statistically close to unity, this is not always the case and fractional reaction orders can often be observed. Simply, a reaction order $m_{OH^-} = 1$ suggests that only a single hydroxide ion reacts at each active site for all steps up to and including the RDS. Similarly, if $m_{OH^-} = 2$ then a total of two $OH^-$ equivalents can be assumed to be involved in the overall reaction prior to and including the RDS. However, it can be more difficult to rationalise fractional reaction orders. A number of authors have considered the effect of the diffuse layer potential $\phi^*$ on the observed kinetic parameters (Parsons 1961; Albery 1975; Lyons and Floquet 2011). Here, the diffuse layer corrected electrode potential is represented by $V - \phi^*$ where:

$$\phi^* = \frac{RT}{F} \ln a_x \tag{2.19}$$

and $a_x = a_{OH^-}$ or $a_{H^+}$ depending on the medium. Thus, if the experiments are not performed at a constant ionic strength then the observed reaction order might not correspond to a constant $V - \phi^*$. For example, Lyons and Floquet (2011) suggest that correcting for the diffuse layer potential can account for the fractional reaction order $m_{H^+} = -1.5$ observed for $RuO_2$ layers in acidic media. On the other hand, Carugati et al. (1981) have proposed that fractional reaction orders may arise due to competing or parallel reaction pathways. In such cases, the experimentally

[3] Applying this check to the low Tafel slope data in Fig. 2.4 gives $m_{x,i} = -(0.058)(1.01) = -0.059$, in agreement with the slope of $-0.057$ obtained for a plot of $V$ measured at 1.0 mA $cm^{-2}$ versus $\log a_{OH^-}$.

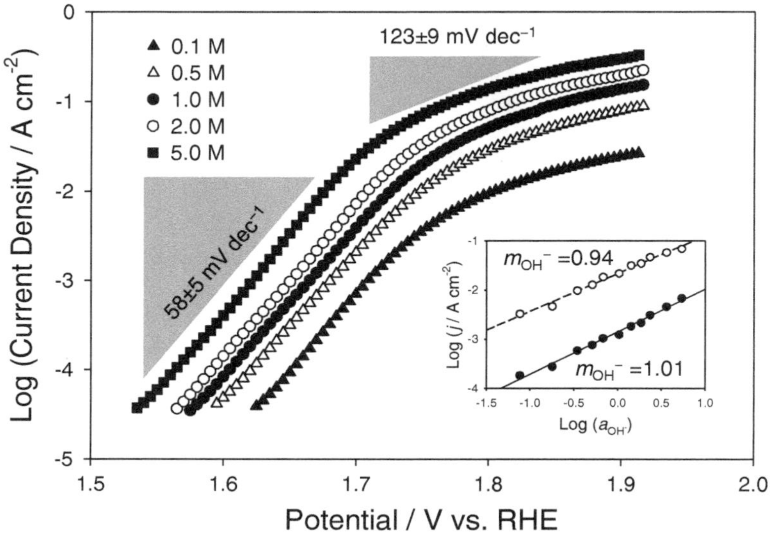


**Fig. 2.4** Steady-state polarisation curves recorded for a hydrous iron oxide catalyst film in a series of aqueous NaOH solutions of varying concentrations. The Tafel regions are indicated by the *triangular features*. Reaction order plots generated at fixed potentials in the low and high Tafel regions are shown in the *inset*. Adapted from Doyle and Lyons (2013)

observed kinetic parameters represent a combination of the parameters expected for the competing pathways, the exact value of which depends on the fraction of catalytic sites following each individual pathway. Then again, fractional reaction orders can also arise depending on the surface coverage of intermediates. Bockris and Ottagawa (1983) observed a reaction order of $m_{OH^-} = 1.5$ for a range of cobalt perovskites whereas Lyons and Brandon (2009) reported the same value for "aged" passive iron oxides. In both cases it was shown that the fractional reaction orders could be rationalised by admitting Temkin rather than Langmuir adsorption conditions.

### 2.2.3 *Mechanistic Analysis*

In terms of a mechanistic analysis of the OER, a major difficulty lies in the fact that the OER is a complex process involving the transfer of four electrons. Since electrons are transferred one at a time the process will, by necessity, be multistep in which distinct intermediates are generated on the electrode surface. Consequently, the OER may follow any of a number of different pathways. In this respect, evaluation of the parameter couple $(b, m_{OH^-})$ will go a long way towards narrowing the possible mechanistic pathways.

Over the years various possible mechanistic schemes have been proposed with notable early studies including the works of Bockris and Otagawa (1983), Krasil'shchikov (1963), Kobussen and Broers (1981), Willems et al. (1984) and O'Grady et al. (1974). In these early mechanistic schemes the OER was usually interpreted in terms of an initial discharge of hydroxide ions at a catalytically active surface site M leading to the formation of discrete adsorbed hydroxide intermediates:

$$M + OH^- \rightarrow MOH + e^- \tag{2.20}$$

Subsequent steps in the reaction were thought to involve the formation of a range of surface adsorbed intermediates such as MO, MOOH or physisorbed peroxide species, which could then react with each other, through various disproportionation or bimolecular decomposition reactions, or undergo nucleophilic attack from the electrolyte to liberate oxygen gas (Dau et al. 2010; Marinia et al. 2012). A summary of the diagnostic criteria for some of these commonly considered pathways is presented in Table 2.1.

In order to derive a rate expression for the OER, and thereby distinguish between reaction pathways, it is useful to view the overall reaction as a sequence of elementary one electron transfer steps and chemical steps. From this perspective, the overall rate of the OER can be expressed in terms of the slowest step or RDS. Depending on which particular elementary step is rate-limiting, the reaction sequence can predict several different Tafel slope and reaction order values. By way of illustration consider the Bockris Electrochemical Oxide Path in Table 2.1. If step 2 (II-2) in this scheme is the RDS then the net reaction flux $f_\Sigma$ for pathway II is given by:

$$f_\Sigma \cong f_2 = i/4FA = k_2^0 a_{OH^-} \theta_{MOH} \exp[\beta F\eta/RT] \tag{2.21}$$

where $f_2$ and $k_2^0$ are the reaction flux and standard rate constant, respectively, for the forward reaction in II-2 and $\theta_{MOH}$ is the fractional surface coverage of the intermediate MOH. That is, $\theta_{MOH}$ represents the proportion of the total number of active sites M that exist in the intermediate state MOH under steady-state conditions. Now, if there is a high surface coverage of intermediates, MOH will be the dominant surface species and $\theta_{MOH}$ can be said to approach unity, $\theta_{MOH} \rightarrow 1$. In this case, $\theta_{MOH}$ is treated as a constant and the net reaction flux (Eq. (2.21)) predicts a reaction order $m_{OH^-} = 1$ and a Tafel slope $b = 2.303(2RT/F)$ or 120 mV dec$^{-1}$, assuming $\beta = 0.5$.

On the other hand, if there is a low surface coverage of intermediates, M will be the dominant surface species and $\theta_{MOH} \rightarrow 0$ as $\theta_M \rightarrow 1$. In this case, it is necessary to express the term $\theta_{MOH}$ in Eq. (2.21) in terms of the constant $\theta_M$. This can be accomplished in one of two ways. The first method, the quasi-equilibrium method, assumes that only the RDS is irreversible and all other steps are in equilibrium (Guidelli et al. 2014a, b). Under these conditions, the rate of the forward and reverse reactions in II-1 should be equal with reaction fluxes given, respectively, by:

**Table 2.1** Kinetic parameters derived for five classical OER mechanistic pathways[a]. Adapted from Bockris and Otagawa (1983)

| Rate-determining step | Langmuir | | | | | Temkin | | | | |
|---|---|---|---|---|---|---|---|---|---|---|
| | | $\frac{\partial V}{\partial \ln i}$ | | $\left[\frac{\partial \ln i}{\partial \ln C_{OH^-}}\right]_{V,\ \zeta^b}$ | | $\frac{\partial V}{\partial \ln i}$ | | | | |
| | $\nu^c$ | $\theta \to 0$ | $\theta \to 1$ | $\theta \to 0$ | $\theta \to 1$ | NA[d] | $A^e$ | NA[d] | $A^e$ | Condition[f] |
| (I) Bockris's Oxide Path (Bockris and Otagawa 1983) | | | | | | | | | | |
| 1. $M + OH^- \to MOH + e^-$ | 4 | 2RT/F | | 1 | | | | | | |
| 2. $2MOH \to MO + M + H_2O$ | 2 | RT/2F | ∞ | 2 | 0 | 2RT/F | | 0.5 | | $r_{OH} \sim r_O$ |
| | | | | | | RT/F | | 1 | | $r_{OH} \gg r_O$ |
| 3. $2MO \to 2\ M + O_2$ | 1 | RT/4F | ∞ | 4 | 0 | RT/2F | RT/F | 2 | 1 | $K_2 \sim 1$ |
| | | | | | | RT/4F | RT/3F | 2 | 1 | $K_2 \ll 1$ |
| (II) Bockris Electrochemical Oxide Path (Bockris and Otagawa 1983) | | | | | | | | | | |
| 1. $M + OH^- \to MOH + e^-$ | 2 | 2RT/F | | 1 | | | | | | |
| 2. $MOH + OH^- \to MO + H_2O + e^-$ | 2 | 2RT/3F | 2RT/F | 2 | 1 | 2RT/F | | 1 | | $r_{OH} \sim r_O$ |
| | | | | | | RT/F | | 1.5 | | $r_{OH} \gg r_O$ |
| 3. $2MO \to 2\ M + O_2$ | 1 | RT/4F | ∞ | 4 | 0 | RT/2F | RT/F | 2 | 1 | $K_2 \sim 1$ |
| | | | | | | RT/4F | RT/3F | 4 | 3 | $K_2 \ll 1$ |
| (III) Krasil'shchikov's Path (Krasil'shchikov 1963) | | | | | | | | | | |
| 1. $M + OH^- \to MOH + e^-$ | 2 | 2RT/F | | 1 | | | | | | |
| 2. $MOH + OH^- \to MO^- + H_2O$ | 2 | RT/F | ∞ | 2 | 1 | ∞ | | 1 | | $r_{OH} \sim r_{O-}$ |
| | | | | | | 2RT/F | | 1.5 | | $r_{OH} \gg r_{O-}$ |
| 3. $MO^- \to MO + e^-$ | 2 | 2RT/3F | 2RT/F | 2 | 0 | 2RT/F | | 0 | | $K_2 \sim 1$ |
| | | | | | | 2RT/F | | 1 | | $K_2 \ll 1$ |
| 4. $2MO \to 2\ M + O_2$ | 1 | RT/4F | ∞ | 4 | 0 | RT/2F | RT/F | 2 | 1 | $K_3 \sim 1$ |
| | | | | | | RT/4F | RT/3F | 2 | 1 | $K_3 \ll 1$ |
| (IV) O'Grady's Path (O'Grady et al. 1974) | | | | | | | | | | |
| 1. $M^z + OH^- \to M^zOH + e^-$ | 2 | 2RT/F | | 1 | | | | | | |
| 2. $M^zOH \to M^{z+1}OH + e^-$ | 2 | 2RT/3F | 2RT/F | 1 | 0 | 2RT/F | | 0 | | $r_1 \sim r_2$[g] |
| | | | | | | RT/F | | 0.5 | | $r_1 \gg r_2$ |

| | | | | | | | | | | |
|---|---|---|---|---|---|---|---|---|---|---|
| 3. $2M^{z+1}OH + 2OH^- \rightarrow M^z + H_2O + O_2$ | 1 | RT/4F | ∞ | 4 | 2 | RT/2F | RT/F | 4 | 3 | $K_2 \sim 1$ |
| | | | | | | RT/4F | RT/3F | 4 | 3 | $K_2 \ll 1$ |
| (V) Kobussen's Path (Kobussen and Broers 1981, Willems et al. 1984) | | | | | | | | | | |
| 1. $M + OH^- \rightarrow MOH + e^-$ | 1 | 2RT/F | | 1 | | | | | | |
| 2. $MOH + OH^- \rightarrow MO + H_2O + e^-$ | 1 | 2RT/3F | 2RT/3F | 2 | 1 | 2RT/F | | 1 | | $r_{OH} \sim r_O$ |
| | | | | | | RT/F | | 1.5 | | $r_{OH} \gg r_O$ |
| 3. $MO + OH^- \rightarrow MO_2H^-$ | 1 | RT/2F | ∞ | 3 | 1 | ∞ | | 1 | | $K_2 \sim 1$ |
| | | | | | | RT/F | | 2 | | $K_2 \ll 1$ |
| 4. $MO_2H^- + OH^- \rightarrow MO_2^- + H_2O + e^-$ | 1 | 2RT/5F | 2RT/F | 4 | 1 | 2RT/F | | 1 | | $K_3 \sim 1$ |
| | | | | | | 2RT/F | | 2 | | $K_3 \ll 1$ |
| 5. $MO_2^- \rightarrow M + O_2 + e^-$ | 1 | 2RT/7F | ∞ | 4 | 0 | RT/F | 2RT/F | 1 | 0.5 | $K_4 \sim 1$ |
| | | | | | | RT/2F | 2RT/3F | 1 | 1.5 | $K_4 \ll 1$ |

[a]Symmetry factors, i.e. $\beta$, $\gamma$ and $\delta$, in all steps, were taken as ½
[b]$\zeta$ is the potential difference between the outer Helmholtz plane and the bulk of the solution
[c]Stoichiometric number
[d]Nonactivated desorption of $O_2$
[e]Activated desorption of $O_2$
[f]$r$ is a coefficient determining the variation of heat of adsorption of a particular species with coverage. Unless stated, $r$ values for each species were taken as equal. $K_i$ is the equilibrium constant of the $i$th step
[g]$r_1$ and $r_2$ refer to $r$ for $M^zOH$ and $r$ for $M^{z+1}OH$, respectively

$$f_1 = k_1^0 a_{OH^-} \theta_M \exp[\beta F\eta/RT] \tag{2.22}$$

and,

$$f_{-1} = k_{-1}^0 \theta_{MOH} \exp[-(1-\beta)F\eta/RT] \tag{2.23}$$

Applying the equilibrium condition $f_1 = f_{-1}$ and rearranging gives the following expression for $\theta_{MOH}$:

$$\theta_{MOH} = K a_{OH^-} \theta_M \exp[F\eta/RT] \tag{2.24}$$

where $K = k_1^0/k_{-1}^0$. Now substituting for $\theta_{MOH}$ in Eq. (2.21), the net reaction flux can be readily estimated as:

$$f_\Sigma \cong k_2^0 (a_{OH})^2 \theta_M \exp[(1+\beta)F\eta/RT] \tag{2.25}$$

Thus, under low surface coverage of intermediates the predicted reaction order and Tafel slope are now $m_{OH^-} = 2$ and $b = 2.303(2RT/3F)$ or 40 mV dec$^{-1}$ respectively.

Alternatively, the steady-state method can be used to provide an expression for $\theta_{MOH}$. According to this method, a short time after initiation of the process, the effective concentrations of all intermediates are assumed to reach values that remain constant in time and therefore, are in a steady state. This implies that the rate of formation and disappearance of any intermediate will be equal, with the result that the rate of change of its concentration will be zero (Guidelli et al. 2014a, b). Applying this condition to $\theta_{MOH}$ gives:

$$\frac{d\theta_{MOH}}{dt} = f_1 - f_{-1} - f_2 \cong 0 \tag{2.26}$$

where the rate of formation of $\theta_{MOH}$ is given by $f_1$ and the rate of its disappearance is $f_{-1} + f_2$. Here, substituting for $f_1, f_{-1}$ and $f_2$ in Eq. (2.26) and again rearranging for $\theta_{MOH}$ gives:

$$\theta_{MOH} = \frac{k_1^0 a_{OH^-} \theta_M \exp[\beta F\eta/RT]}{k_{-1}^0 \exp[-(1-\beta)F\eta/RT] + k_2^0 a_{OH^-} \exp[\beta F\eta/RT]} \tag{2.27}$$

At first glance it appears that the quasi-equilibrium and steady-state methods are inconsistent. However, noting that II-2 is the RDS, it is safe to assume that $k_2^0 \ll k_{-1}^0$. In fact, this assumption is implicit in the quasi-equilibrium method, where the rate constant of the RDS is required to be at least 100 times smaller than those of all steps that precede it (Guidelli et al. 2014a, b). Thus, Eq. (2.27) reduces to Eq. (2.24) and the same expression for the net reaction flux in Eq. (2.25) is obtained.

Hence, distinct kinetic parameters are predicted for low and high surface coverage of intermediates. Further coverage effects can also be considered. Several studies have utilised a Temkin isotherm to describe the adsorption process (Parsons 1958; Thomas 1961; Conway and Gileadi 1962; Bockris and Otagawa 1983; Lyons and Brandon 2009). The limiting coverages $\theta \rightarrow 0$ and $\theta \rightarrow 1$, known as Langmuir adsorption conditions, operate under the assumption that lateral interactions between adsorbed species have no effect on the reaction kinetics. However, for intermediate coverages ($0.2 < \theta < 0.8$) where no single species can be said to be dominant it is reasonable to expect the interactions between adsorbates to be significant. Such cases are best described in terms of a Temkin isotherm. For example, Conway and Gileadi (1962) outlined a model based on the principle that the free energy of adsorption of an intermediate species depends on the total fractional coverage of all adsorbed reaction intermediates as follows:

$$\Delta G_{\theta} = \Delta G^{0} - r_{x}\theta \tag{2.28}$$

where $\Delta G_{\theta}$ and $\Delta G^{0}$ are the adsorption free energies of an intermediate species $x$ in the presence and absence of coverage effects, respectively, and $r_x$ is the rate of change in the free energy of adsorption with the total coverage. According to this model, the free energy of adsorption of an intermediate will decrease with increasing total fractional coverage. For a reaction step consisting of the inter-conversion of two adsorbed intermediates such as II-2, a decrease in the free energy of adsorption of the intermediate species MO with increasing total coverage would result in an increase in the free energy of activation for II-2. In contrast, a decrease in the free energy of adsorption of MOH would cause a decrease in the activation energy. The overall reaction kinetics will, therefore, be influenced by changes in the relative populations of important intermediates. In particular, depending on the relative sensitivities of the free energy of the intermediates to changes in surface coverage, as expressed through their respective $r$ values, further distinct kinetic parameters can be estimated. Sample kinetic analyses under Temkin conditions are outlined by Bockris and Otagawa (1983) and Lyons and Brandon (2009), and the predicted kinetic parameters for II-2 are presented in Table 2.1.

The above discussion is predicated on a slow electron-transfer step determining the overall kinetics of the OER. However, a slow chemical step can equally be rate-limiting. The reaction step in II-2 represents a proton coupled electron transfer. This step could also be decomposed into a sequential proton transfer and electron transfer process as follows:

$$\mathrm{MOH} + \mathrm{OH}^{-} \rightarrow \mathrm{MO}^{-} + \mathrm{H_2O} \tag{2.29}$$

$$\mathrm{MO}^{-} \rightarrow \mathrm{MO} + \mathrm{e}^{-} \tag{2.30}$$

Indeed, this is the only difference between the Bockris Electrochemical Oxide Path and Krasil'shchikov's Path. It can be seen in Table 2.1 that III-2 and III-3 are the equivalent of Eqs. (2.29) and (2.30). In this case, if Eq. (2.29) is the slow step then the reaction flux expression (Eq. (2.21)) can be modified to give:

$$f_{\Sigma} \cong i/4FA = k_2 a_{\mathrm{OH}^-} \theta_{\mathrm{MOH}} \tag{2.31}$$

Note $k_2$ is now a chemical rate constant and there is no potential dependence. Under low coverage Langmuir adsorption conditions $\theta_{\mathrm{MOH}}$ is again replaced by Eq. (2.24) resulting in the following estimate for the net reaction flux:

$$f_{\Sigma} \cong k_2 (a_{\mathrm{OH}^-})^2 \theta_{\mathrm{M}} \exp[F\eta/RT] \tag{2.32}$$

Thus, if a chemical step, such as the reaction outlined in Eq. (2.29), rather than an electron-transfer step is rate-limiting a reaction order $m_{\mathrm{OH}^-} = 2$ and Tafel slope $b = 2.303(RT/F)$ or 60 mV dec$^{-1}$ are predicted.

In this way, the determination of Tafel slopes and electrochemical reaction orders provides valuable insight into the rate-limiting processes at the reaction interface. It is clear that these reaction parameters are strongly influenced by the concentration and interaction of the surface species generated during the OER. This provides a kinetic basis for the rationalisation of multiple experimental Tafel regions. An assumption implicit in the analysis thus far is that a single isotherm can describe the surface coverage of key intermediates. Over short potential ranges such an assumption will likely hold; it has been pointed out by Damjanovic et al. (1966, 1967a) that the range of potentials over which Temkin conditions might be expected to prevail is limited to ca. 150 mV. Over wider potential ranges it is more realistic to expect the surface coverage to vary appreciably. Consider the situation where II-2 is rate determining, the relative proportion of active sites in the MOH state is controlled primarily by the equilibrium in step II-1. Given that II-1 involves the transfer of one electron, one could easily suppose the equilibrium to favour the formation of the MOH species at high overpotentials. Accordingly, the surface coverage of a particular intermediate could be potential dependent, transitioning through states of low, intermediate and high coverage over the course of a polarisation experiment. Examining the simplest case where $\theta_{\mathrm{MOH}}$ changes from low coverage to high coverage with increasing potential, this transition would be characterised by a change in Tafel slope from 40 to 120 mV dec$^{-1}$ with a corresponding change in reaction order from $m_{\mathrm{OH}^-} = 2$ to $m_{\mathrm{OH}^-} = 1$. Interestingly, this dual Tafel behaviour could also be accounted for by a change in RDS. If step II-1 becomes rate-limiting at high overpotentials then the net reaction flux is given simply by Eq. (2.22) which can be readily shown to predict a Tafel slope of 120 mV dec$^{-1}$ and a reaction order $m_{\mathrm{OH}^-} = 1$. In such cases, concurrent IS and OCP decay analysis can be useful in distinguishing between surface coverage effects and a change in the RDS. As noted in Sect. 2.2.1, variations in the oxide surface capacity during active oxygen evolution reflect the changing concentration of surface intermediates (Terezo et al. 2001; Klahr et al. 2012a, b) (see also Chap. 7). Thus, significant increases in the oxide surface capacity with increasing potential, as measured by IS, or large deviations of the OCP decay slope from the

Tafel slope would point to changes in the surface coverage of intermediates rather than a change in the RDS being the reason behind the dual Tafel behaviour.

The foregoing kinetic analysis highlights an inherent complication in attempts at a mechanistic analysis of the OER. That is, the kinetic parameters predicted for a particular pathway are not unique. Hence it is often very difficult to unambiguously identify the pathway operative for a particular system. Combining electrokinetic studies with a more detailed interrogation of the reactive interface is essential for the development of a complete mechanistic picture. Cyclic voltammetry studies of the surface redox characteristics of an oxide electrode are useful for diagnosing changes in the surface structure prior to the onset of catalytic activity, whereas in situ and ex situ spectroscopic studies can provide insight into the nature of the catalytically active surface. Modern studies of the OER often rely on such a multidisciplinary approach and several prominent examples are discussed in Sect. 2.5.

Nevertheless, the kinetic approach described here is of fundamental value in the study of the OER on transition metal oxides. Although it is not possible to isolate a single reaction pathway with the exclusion of all others, identification of the rate-determining step provides an important mechanistic foundation. Optimising electrocatalytic activity is a matter of achieving the highest possible current at the lowest possible overpotential. Low Tafel slopes are therefore an essential requirement for OER electrocatalysts. From this simple observation one can begin to formulate the principles for catalyst design. With reference to the mechanisms in Table 2.1 low Tafel slopes are associated with steps other than the primary discharge step being rate-determining. From this perspective, facilitating the discharge step offers an opportunity for catalyst enhancement; if the stability of the adsorbed intermediate formed in the primary discharge step can be increased then the rate control will shift to a later step. Thus, low Tafel slopes can be achieved through the interaction of strongly adsorbed surface intermediates (Hoare 1967, 1968). An alternative view point is highlighted in the mechanism of O'Grady et al. (1974) in Table 2.1, who emphasised the importance of oxide redox transitions in the course of the OER. In this concept, the OER is facilitated by the cyclic formation and decomposition of an unstable intermediate containing a metal centre in a higher valence state ($z+1$) than its initial state ($z$) (O'Grady et al. 1974; Burke et al. 1982a). The significance of this approach lies in its ability to account for low experimental Tafel slopes without the need to consider strong surface adsorption; instead the activity of the surface oxide is linked to its stability. Despite their humble origins, these electrocatalytic concepts remain at the forefront of modern mechanistic interpretations of OER activity and will be explored in greater detail throughout the remainder of this chapter. In Sect. 2.3, the electrochemical activity of transition metal oxides for the OER is examined from a theoretical perspective.

## 2.3 Thermochemical Perspectives

In recent years, the oxygen evolution reaction on metal oxide surfaces has been widely studied using theoretical methods. As a complex, multistep, inner sphere process, the OER depends critically on the formation and stabilisation of high-energy surface-confined intermediates. Detailed knowledge of the interaction between the catalyst surface and the reaction intermediates is therefore essential if a fundamental understanding of the OER is to be developed. These issues are often difficult to address experimentally due to short lifetimes of reaction intermediates, harsh reaction conditions and extensive gas evolution at the surface of interest. Theoretical approaches, on the other hand, are very constructive as they allow one to access individual reaction steps and reaction intermediates. Although the complexity of the oxide surface makes accurate computation a challenging task, a simple thermochemical framework has emerged providing insights into the origin of the OER overpotential and the viable mechanistic paths.

### 2.3.1 *The Potential-Determining Step*

Over the past 10 years, extensive work by the groups of Norskov and Rossmeisl has led to the development of an effective method for modelling the thermochemistry of electrochemical reactions (Rossmeisl et al. 2005; Rossmeisl et al. 2007a; Nørskov et al. 2009; Valdés et al. 2012). This model utilises quantum chemical calculations based on Density Functional Theory (DFT) to determine the thermodynamics of the separate reaction steps in a catalytic mechanism. Applying this method to the OER, they considered the following model reaction mechanism, termed the "associative" mechanism (Rossmeisl et al. 2007a):

$$M + H_2O \overset{\Delta G_1}{\rightarrow} MOH + H^+ + e^- \tag{2.33}$$

$$MOH \overset{\Delta G_2}{\rightarrow} MO + H^+ + e^- \tag{2.34}$$

$$MO + H_2O \overset{\Delta G_3}{\rightarrow} MOOH + H^+ + e^- \tag{2.35}$$

$$MOOH \overset{\Delta G_4}{\rightarrow} M + O_2 + H^+ + e^- \tag{2.36}$$

where $\Delta G_i$ is the Gibbs free reaction energy of reaction step $i$. Note each reaction step involves the transfer of one electron implying the free energy change for each step will be potential dependent. Also, while this mechanism describes the OER under acidic conditions, the thermodynamic conclusions are independent of pH as the free energies deduced from Eqs. (2.33) to (2.36) vary in the same way with pH (Valdés et al. 2012). In their theoretical framework, Rossmeisl et al. (2007a) calculated the Gibbs free adsorption energies of the surface intermediates $\Delta G_{MOH}$, $\Delta G_{MO}$, and $\Delta G_{MOOH}$ as a function of the electrode potential. At a given applied potential, the reaction energy of each elementary mechanistic step was given by the difference between the adsorption energies of two intermediates. For example, $\Delta G_3$ depends on the relative free energies of the MO and MOOH intermediates according to: $\Delta G_3 = \Delta G_{MOOH} - \Delta G_{MO} - qV$. In this way, reaction free energy diagrams were generated allowing the identification of the thermodynamically least favourable step in the reaction path (Rossmeisl et al. 2005, 2007a).

Sample free energy diagrams for two hypothetical catalysts are plotted at different potentials in Fig. 2.5. The horizontal lines represent the free adsorption energies of the individual intermediates and the reaction coordinate moves from free water on the left to free $O_2$ on the right. For a typical electrocatalyst, the reaction energies for each step will differ due to irregular variations in the adsorption energies of the intermediate species (Dau et al. 2010). This situation is depicted by the "Real" catalyst trace in Fig. 2.5 where $\Delta G_3 > \Delta G_1 = \Delta G_2 > \Delta G_4$. As one might expect, all steps in the reaction are thermodynamically unfavourable at potentials below the reversible potential, $V_1 - V^0 = \eta < 0$, with $\Delta G_i > 0$. As the potential is increased, the free energies of the intermediates shift negatively so that the reaction steps eventually become thermodynamically favourable with $\Delta G_i \leq 0$. However, due to differences in the adsorption energies of the intermediates, each individual $\Delta G_i$ value will turn negative at a different potential. In Fig. 2.5 it can be

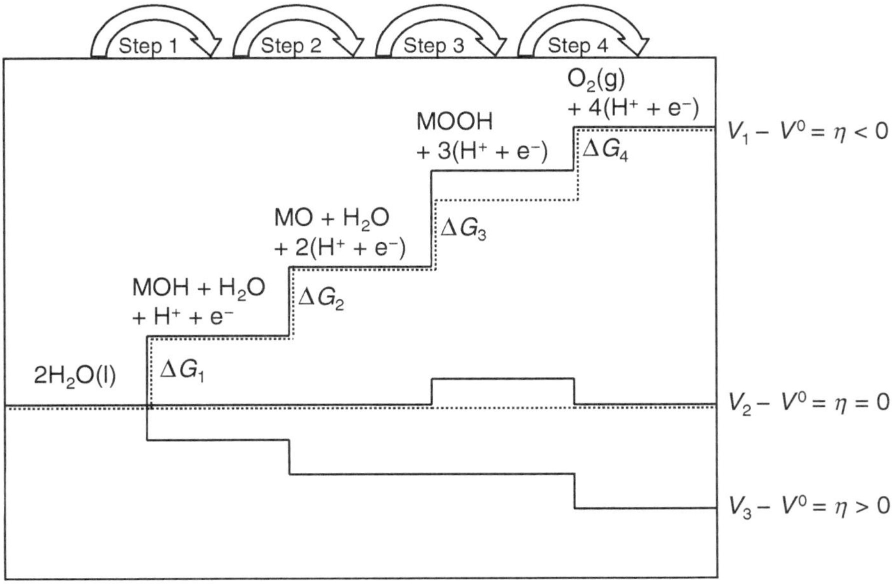


**Fig. 2.5** Schematic plot of the Gibbs free energy of the reactive species and intermediates in the oxygen evolution reaction versus the reaction coordinate. *Solid lines* indicate the energetics of a 'Real' catalyst at three different potentials. *Dotted lines* represent the energetics of an 'Ideal' catalyst. Adapted from Dau et al. (2010)

seen that $\Delta G_3$ remains positive at the reversible potential, $V_2 - V^0 = \eta = 0$. This step only becomes thermodynamically favourable when the potential $V_3 - V^0$ is reached. This potential is termed the thermodynamic overpotential as it is the potential at which all thermodynamic barriers have been removed and the overall rate of the reaction will be independent of potential (Rossmeisl 2013).

Clearly, then, the overall thermodynamics of the OER are determined by the last elementary reaction step to become thermodynamically favourable. This step is referred to as the potential-determining step (PDS) (Koper and Heering 2010), and its magnitude is simply the largest $\Delta G_i$ value such that (Rossmeisl 2013):

$$\Delta G^{\mathrm{OER}} = \max\{\Delta G_1, \Delta G_2, \Delta G_3, \Delta G_4\} \tag{2.37}$$

In this respect, the thermodynamic overpotential is just the additional potential, past the reversible potential, required by the PDS and is given by (Rossmeisl 2013):

$$\eta^{\mathrm{OER}} = \left(\Delta G^{\mathrm{OER}}/q\right) - 1.23\ V \tag{2.38}$$

Importantly, this analysis allows the properties of an "ideal" catalyst to be specified. From Eq. (2.38), the optimum $\Delta G^{\mathrm{OER}}$ value is 1.23 eV giving $\eta^{\mathrm{OER}} = 0$. Now, since the total work to be provided by the four electrons in the OER is 4.92 eV (Rossmeisl 2013), this condition implies that $\Delta G_1 = \Delta G_2 = \Delta G_3 = \Delta G_4 = 1.23$ eV. Thus, the ideal catalyst will exhibit equally spaced adsorption energies for the OER intermediates (Rossmeisl 2013). This situation is illustrated by the "Ideal" catalyst trace in Fig. 2.5. In contrast to the "Real" catalyst, where Eq. (2.35) is the PDS, all the reaction steps become thermodynamically favourable at the same potential—the reversible potential. Furthermore, a comparison of the PDS for the "Real" catalyst with the same step in the "Ideal" case suggests that the MOOH intermediate is less stable or bound "too weakly" on the "Real" catalyst (Dau et al. 2010). Along these lines, the OER properties of the "Real" catalyst could be enhanced through stronger binding or stabilisation of the MOOH intermediate, thereby allowing the OER to proceed at a potential closer to the reversible potential. However, as will be discussed in Sect. 2.4.2, scaling relations between the free energies of related intermediates limit the effectiveness of this approach (Rossmeisl 2013).

In view of the implications for catalyst design, it is important here to distinguish between the latter concept of a potential-determining step and the kinetic concept of a rate-determining step. The approach outlined above only accounts for the thermochemistry of the elementary reaction steps, assuming that each elementary step must be thermodynamically favourable for the OER to proceed at an appreciable rate. Yet various activation barriers will also be important for the overall rate of the reaction. Indeed, the PDS and RDS need not necessarily be the same. Koper (2013) provides an instructive illustration of the significance of this distinction. Consider the two free energy plots in Fig. 2.6 for a two-step reaction involving a single intermediate. In both cases, step 1 is thermodynamically limiting with $\Delta G_1 > 0$ and so, is potential-limiting. Now, if the RDS is identified as the step that passes over

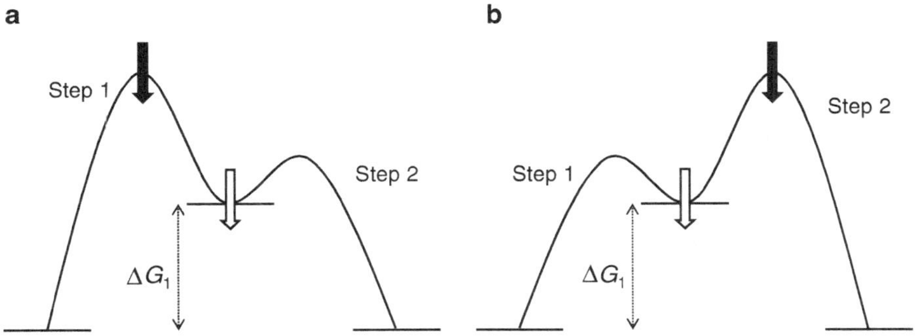


**Fig. 2.6** Free energy diagrams for a two-step reaction involving a single intermediate. (**a**) Step 1 is potential determining and rate-determining. (**b**) Step 1 is potential-determining and step 2 is rate determining. Adapted from Koper (2013)

the highest energy barrier then, step 1 is rate-determining in Fig. 2.6a and step 2 is rate-determining in Fig. 2.6b. From a purely kinetic viewpoint, the rate of the reaction could be increased by facilitating the RDS. This could be achieved by lowering the energy of the transition states for step 1 and step 2, respectively, as indicated by the filled arrows in Fig. 2.6. But, if a linear relation between the activation energy and the reaction energy exists, as described by the Bronsted–Evans–Polanyi relation (Bligaard et al. 2004; Nørskov et al. 2008), then the easiest way to lower the transition states is to lower the energy of the intermediate state (Koper 2013). This approach is indicated by the unfilled arrows in Fig. 2.6. Koper (2013) notes that the real problem is not necessarily the high energies of the transition states but the high energy of the intermediate state. In this way, identifying the thermodynamically limiting step—the PDS—offers a more direct approach for catalyst design.

In this respect, it should be noted that a direct comparison between predicted thermodynamic limits and the results from electrokinetic studies is not possible, as DFT calculations do not provide overpotential values that are directly comparable with experiment (Montoya et al. 2015). Nevertheless, these interpretations are linked. Relating the PDS to the experimental observables, Rossmeisl (2013) and Rossmeisl et al. (2008) provide an expression for the kinetic current density at the anode:

$$j_a = j_0 \exp\left[\alpha\left(qV^0 - qV\right)/kT\right] = j_{\lim} \exp\left[\alpha\left(qV^{OER} - qV\right)/kT\right] \tag{2.39}$$

where,

$$j_{\lim} = j_0 \exp\left[\alpha\left(qV^0 - qV^{OER}\right)/kT\right] \tag{2.40}$$

Here, $V^{OER}$ is the potential at which the kinetic current is maximised and $j_{\lim}$ represents the current density achieved if all surface reactions are thermodynamically downhill, with the highest possible turn-over-frequency per site and minimal diffusion limitations. The term $j_{\lim}$ is clearly dependent on a range of factors, including the electrode structure and its interface with the liquid electrolyte, the concentration of active sites, and potential independent surface reactions. On the other hand, $V^{OER}$ is a parameter that depends only on the catalyst material (Rossmeisl 2013). In particular, $\Delta G^{OER}$, and therefore the thermodynamic overpotential $\eta^{OER}$, represents a lower-limit estimate of this limiting potential. Likewise, it has been noted that the experimentally observed "onset" potential for a catalytic reaction is expected to occur ca. 0.15 V prior to the predicted overpotential (Rossmeisl et al. 2008; Su et al. 2012). Taking this into consideration, Rossmeisl (2013) reaches a similar conclusion to Koper (2013) stating that differences in electrocatalytic performance between different materials are mostly described by differences in overpotentials.

### 2.3.2 Reaction Pathways and Surface Effects

In a strictly thermochemical description of the OER, the overall rate depends on the free energy of reaction steps involving catalyst bound intermediates. Since the energies of the reaction steps depend on the nature of the catalyst, it is important to develop a fundamental understanding of these reactions on different materials (Valdés et al. 2012). Various surface structures, types of adsorbate and solvent effects can have a strong influence on the relative stability of the reaction intermediates. Thus, the potential-determining steps and viable reaction paths can vary from one surface to another.

In their early work, Rossmeisl et al. (2007a) compared the OER activity of rutile-type oxides of $RuO_2$, $IrO_2$ and $TiO_2$. In agreement with the kinetic studies in Sect. 2.2.3, the surface coverage of intermediates was shown to be a significant factor in determining the activity of these surfaces. The authors compared the activity and stability of two different surface conditions: one surface with all bridge and coordinately unsaturated (CUS) sites occupied by oxygen (O) and one surface with these sites occupied by hydroxyl groups (OH). It was found that the O-covered surface showed the highest activity. In the case of $RuO_2$, an OER overpotential of 0.37 V was predicted for the O-covered surface, whereas the HO-covered surface needed an overpotential of 0.50 V for the OER to become thermodynamically downhill. Interestingly, these limiting potentials did not correspond to the same PDS. On the O-covered surface the formation of the MOOH intermediate Eq. (2.35) was associated with the largest free energy change. But on the HO-covered surface the MO intermediate was the least stable, emphasising the influence of the local environment on the overall energy profile. Accordingly, it was suggested that hydrogen bonds between the hydroxyl groups on the HO-covered surface stabilised the MOH and MOOH intermediates relative to the MO intermediate. In any case, the O-covered surface was found to be the most stable surface at the potentials required for the OER to proceed and a comparison of the three materials under these conditions showed an activity trend of $RuO_2 > IrO_2 \gg TiO_2$.

This type of analysis has since been extended to describe the thermodynamics of the OER at a range of well-defined planar oxide surfaces, including rutile, perovskite, spinel, rock salt and bixbyite oxides (Valdés et al. 2008, 2012; Man et al. 2011). Taking into account the self-consistent surface for each oxide, that is the surface that is thermodynamically stable at the necessary reaction overpotential (Su et al. 2012), the nature of the PDS is shown to be largely dependent on the oxygen binding energy at each surface (Valdés et al. 2012). On surfaces that bind oxygen strongly, the PDS is the formation of the MOOH intermediate. This is easily understood from a chemical bonding point of view; this is the $O-O$ bond formation step and therefore, involves a reduction in the oxygen bond order with the surface. Hence, strong binding of oxygen to the surface will significantly increase the difficulty of this step. Correspondingly, for weak oxygen binding, the formation of the MO intermediate, which involves an increase in the oxygen bond order to the surface, becomes potential-limiting. It should be noted that with this transition in the PDS other reaction mechanisms become possible. This point is illustrated by the case of rutile $MnO_2$ (Su et al. 2012). It was found that $\beta$-$MnO_2$ exhibited intermediate oxygen binding energies, close to the transition point, with the result that the formation of the MO intermediate was slightly more potential-limiting than the formation of MOOH. In this case, a direct $O_2$ recombination mechanism was proposed:

$$M + H_2O \rightarrow MOH + H^+ + e^- \tag{2.41}$$

$$MOH \rightarrow MO + H^+ + e^- \tag{2.42}$$

$$MO \rightarrow M + \frac{1}{2}O_2 \tag{2.43}$$

This mechanism was found to be slightly lower in free energy by 0.08 eV than the associative mechanism in Eqs. (2.33)–(2.36). However, Rossmeisl (2013) notes that the use of alternative reaction mechanisms does not change the analysis of the PDS. It can be seen that both pathways begin with the same two steps: water splitting on the active site followed by MOH oxidation to MO. But for the direct recombination mechanism to be relevant, the free energy of oxygen binding on the surface has to be close to the free energy of oxygen in the gas phase (Valdés et al. 2012). This can only be the case for weak oxygen binding, essentially when Eq. (2.43) is thermodynamically downhill and will play no major role in the reaction kinetics. Under these conditions both mechanisms predict the same PDS, (Eq. (2.34)) or (Eq. (2.42)). Thus, for determining the overpotential to a good approximation, Rossmeisl (2013) argues that only the associative mechanism needs to be considered even though the actual reaction path might not involve MOOH formation at all.

Interestingly, the self-consistent surfaces for the range of oxides considered in the latter studies were all quite similar. At the high potentials required to drive the OER, the thermodynamically most stable surfaces were those covered by oxygen

(Valdés et al 2012). Based on this result, Rossmeisl et al. (2007a) note that the interactions between water and the adsorbed reaction intermediates can be neglected, as there are no active sites available for water adsorption. However, the influence of the hydrogen bonding network surrounding such intermediates can be significant (Tuckerman et al. 1995; Marx et al. 1999; Vassilev et al. 2002). Molecular dynamic (MD) simulations of surface adsorbed hydroxyl groups indicate that the hydrogen bonding surrounding these surface bound OH species is perturbed from that of a surface water molecule (Vassilev et al. 2002). Specifically, it is shown that the oxygen-oxygen distances in the MOH hydrogen bonded complex are shorter than for water molecules in the extended network. As a result, the rate-limiting step for OH migration on a surface is the rearrangement of the local oxygen environment and not the proton transfer event itself (Tuckerman et al. 1995; Marx et al. 1999; Vassilev et al. 2002). In this way, the initial discharge of water/hydroxide at the surface could be viewed analogously as an electron transfer facilitated migration of $OH^-$ from solution onto the surface, with a fast initial discharge step and a slow hydrogen bonding restructuring process as follows:

$$M + H_2O \rightarrow MOH^* + H^+ + e^- \tag{2.44}$$

$$MOH^* \rightarrow MOH \tag{2.45}$$

where MOH* represents a higher energy surface bound hydroxyl group and Eq. (2.45) depicts the restructuring process to accommodate this energetically inequivalent species. This type of process has recently been invoked to account for the kinetic parameters observed for an $IrO_2$ catalyst in acidic media (Doyle et al. 2013). When the purely chemical step in Eq. (2.44) is rate limiting, a Tafel slope of 60 mV $dec^{-1}$ and a reaction order $m_{H^+} = -1$ are predicted, in excellent agreement with experimental results (Lyons and Floquet 2011).

With this in mind, Muckerman's group have combined DFT calculations with MD simulations to explicitly account for solvation interactions at GaN, ZnO and GaN/ZnO alloy surfaces (Shen et al. 2010; Akimov et al. 2013; Kharche et al. 2014). The aqueous interface chemistry was found to profoundly affect the structure of water molecules in contact with the surface and hence, the mechanism of the OER (Shen et al. 2010; Akimov et al. 2013). Water adsorption is primarily dissociative on these surfaces such that surface anions, located at N or O sites, act as bases accepting protons from dissociated water molecules while the corresponding hydroxide ions bond with surface cations, Ga or Zn. This acid–base chemistry results in a substantially hydroxylated surface (Kharche et al. 2014) and it is at these sites that the molecular transformations involved in the OER are thought to take place (Shen et al. 2010; Akimov et al. 2013). Accordingly, the following reaction path was proposed for the OER at a GaN surface (Shen et al. 2010):

$$MOH^- \rightarrow MO^{\bullet -} + H^+ + e^- \tag{2.46}$$

$$MO^{\bullet -} + H_2O \rightarrow MOOH^- + H^+ + e^- \quad (2.47)$$

$$MOOH^- \rightarrow MO_2^{\bullet -} + H^+ + e^- \quad (2.48)$$

$$MO_2^{\bullet -} + H_2O \rightarrow MOH^- + O_2 + H^+ + e^- \quad (2.49)$$

While certainly similar to the associative mechanism outlined in Eqs. (2.33)–(2.36), the distinguishing feature of this mechanism is the absence of any vacant sites. Since MD studies showed that the surface active site was unstable when vacant, the reaction is initiated at a surface attached hydroxide ion and no vacant sites are generated during the OER. This is in contrast to the associative mechanism where the OER begins and ends at a vacant surface site M. Indeed, as noted by Shen et al. (2010), it remains to be seen whether such sites would be available under more realistic experimental conditions. That said, the hydroxide oxidation step (Eq. (2.46)) is again shown to be the PDS, although in this case it represents the first step in the overall reaction.

In summary, theoretical studies of the OER provide valuable insight into the microscopic properties and thermodynamic barriers associated with efficient catalysis. Though an attempt has been made here to integrate important theoretical results with experimental observations, the true power of computational studies, particularly those based on DFT, lies in their ability to describe trends in catalytic activity. In Sect. 2.4, a combined theoretical and empirical trends based analysis of the OER is discussed.

## 2.4 Trends in Activity

Over the past 50 years considerable research efforts have been devoted to the kinetic and mechanistic study of the OER. While significant progress has been made in terms of a theoretical understanding of the OER (see Sect. 2.3), little mechanistic consensus has been achieved. Thermodynamic and kinetic analyses of the OER are complicated by the fact that it is the metal oxide and not the metal that catalyses oxygen evolution (Conway 1995; Rossmeisl et al. 2005). These electrocatalytic materials are complex experimental systems whose catalytic activity is highly sensitive to a multitude of parameters including the surface and bulk structure of the oxide film, the manner in which the oxides are prepared and even the experimental history of the electrode (Surendranath and Nocera 2012; Doyle et al. 2013). Moreover, the way in which these properties affect the activity of electrocatalysts for the OER is not yet fully understood.

In light of this, many studies have attempted to rationalise the relative electrocatalytic activities of different electrode materials in terms of a given physicochemical property or “descriptor”. The concept of a “descriptor” involves the hypothesis that a single microscopic parameter may be a key controlling factor in a catalytic process (Dau et al. 2010; Marinia et al. 2012). This hypothesis arises

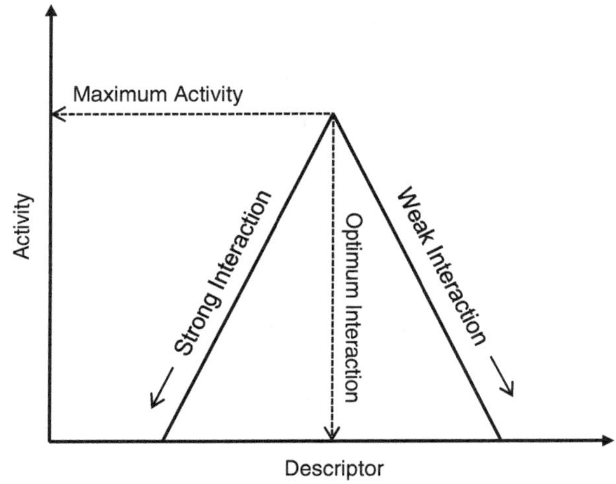


**Fig. 2.7** Schematic representation of a volcano plot which shows the variation in catalytic activity as a function of a reactivity descriptor. Maximum activity is achieved at an optimum descriptor value where the interaction of a key intermediate with the surface is neither too strong nor too weak

from the observation that the rate of a heterogeneous reaction shows a maximum when considered as a function of the reactivity of the catalytic surface (Sabatier 1911). This type of plot, known as a volcano plot, has been widely applied in the study of heterogeneous chemical catalysis (Dumesic et al. 2008) and a schematic representation is presented in Fig. 2.7. In principle, a descriptor could be any of a number of properties of the catalyst or catalyst–reactant interaction (Koper 2013). However, the most successful descriptors generally describe the interaction between a key reaction intermediate and the catalytic surface. In this sense, the observation of a maximum in Fig. 2.7 suggests that an optimal interaction exists. This is the Sabatier principle which states that the best catalyst binds the intermediates "just right", neither too weakly nor too strongly (Sabatier 1911; Dumesic et al. 2008). If the interaction is too weak, the catalyst fails to activate the reactant whereas if the interaction is too strong, the surface can become blocked by intermediates or product which fails to dissociate. Thus, the identification of suitable descriptors provides knowledge of the intrinsic factors governing the experimental behaviour. This approach is particularly attractive as it lays the basis for possible predictions of catalytic activity and enables meaningful catalyst comparisons across a range of materials. In this section, the details and implications of some of the most important examples of this approach are discussed.

### 2.4.1 Bulk Thermochemistry

In terms of the OER, the challenge is in quantifying what is meant by the "reactivity" of an oxide surface. Early studies examined the relationship between activity and various thermodynamic data (Ruetschi and Delahay 1955; Parsons 1958; Trasatti 1980, 1984; Bockris and Otagawa 1983, 1984). The first volcano relation

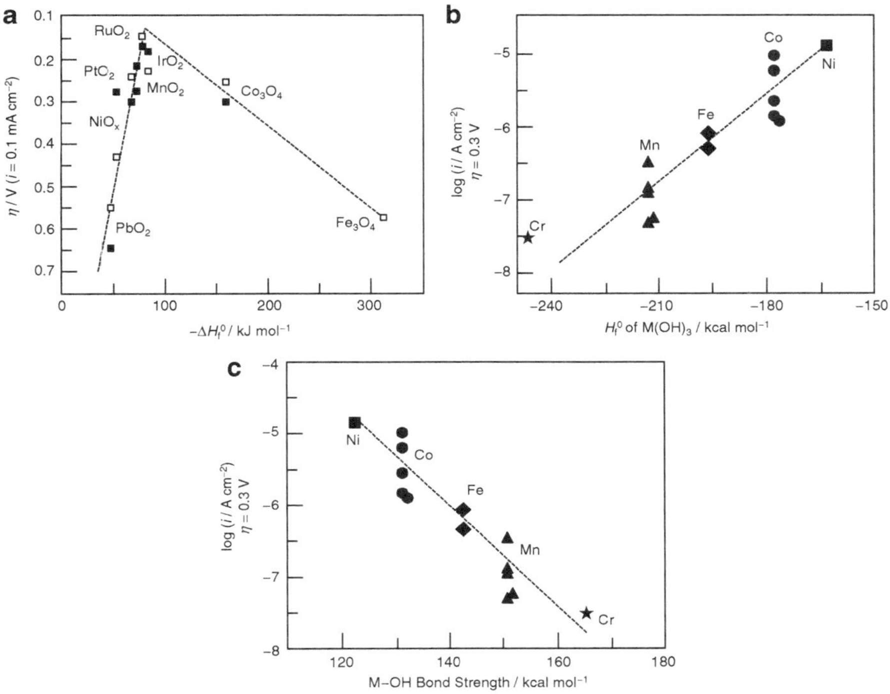


**Fig. 2.8** (**a**) Volcano plot of OER activity for various oxide catalysts in acidic (*unfilled*) and basic (*filled*) media versus the enthalpy of a lower to higher valent oxide transition. Adapted from Trasatti (1980). (**b**) OER current density (based on real surface area) at an overpotential of 0.3 V for a series of perovskites versus the enthalpy of formation of the corresponding $M(OH)_3$ and (**c**) the same quantity versus the M–OH bond strength. Adapted from Bockris and Otagawa (1984) and Surendranath and Nocera (2012)

for the OER was reported by Trasatti (1980) who showed a correlation between the overpotential and the standard enthalpy change for a lower to higher valent oxide transformation, $MO_x \rightarrow MO_{x+1}$. The success of this study lies in the fact that it correctly predicted the high activity of $RuO_2$ and $IrO_2$ catalysts. As illustrated in Fig. 2.8a, the volcano plot generated for a range of binary oxides showed a peak near these state-of-the-art catalysts. This correlation can be rationalised if one considers that the OER is expected to involve one or more surface bound oxygen moieties such as MOH, MO or MOOH. The interaction of these intermediates with the surface of a given oxide $MO_x$ could result in an increase in the coordination sphere of M, and therefore their formation would be expected to parallel the heat of formation of the oxide in the next valency state (Trasatti 1980, 1984). Along these lines, the OER can be viewed as the formation and subsequent decomposition of high-valent metal surface oxides (Dau et al. 2010).

Bockris and Ottagawa (1983, 1984) found that the activity of a series of first-row transition metal perovskites ($AMO_3$) correlated linearly with both the standard enthalpy of formation of the corresponding metal hydroxide $M(OH)_3$ and the metal–OH bond energy. In Fig. 2.8b, the current density at a fixed overpotential is shown to increase as the enthalpy of hydroxide formation decreases. Although a volcano peak was not observed in this case, the inverse relationship between the OER activity of perovskites and the enthalpy of $M(OH)_3$ formation suggested that the stronger the MOH bond, the weaker the catalysis. This was confirmed using calculated MOH bond energies which were also found to correlate inversely with activity, as shown in Fig. 2.8c. Interestingly, a similar trend has been observed by Subbaraman et al. (2012) who found that the oxophilicity or OH bond strength of a series of well-defined first-row transition metal hydroxide nanoclusters was a suitable descriptor for their activity. These authors showed that the OER overpotential at a fixed current density of 5 mA $cm^{-2}$ increased in the order $Ni > Co > Fe > Mn$ indicating that the activity was inversely proportional to the OH bond strength on these catalyst surfaces. Therefore, in both cases the authors proposed that the rate of oxygen evolution was determined by the degree of difficulty in removing OH intermediates from the surface. That is, the rate determining step may be the desorption of OH or other oxygenated species from the surface, or their recombination to form the $O - O$ bond.

### 2.4.2 Binding Energies and Scaling Relations

Advances in DFT calculations have enabled an explicit description of OER activity in terms of calculated adsorption energies. In Sect. 2.3.1 it was shown that OER activity is to a large extent determined by the binding strength of the reaction intermediates to the electrode surface. Depending on the number of different important surface intermediates several descriptors may be identified. For the OER at metal oxide surfaces the suggested intermediates are MOH, MO and MOOH. Indeed, subject to whichever reaction step is potential determining the binding energy of any one of these species could prove a suitable descriptor. That said, it has been shown that the binding energies of these three intermediates are strongly correlated (Rossmeisl et al. 2005, 2007a, b; Hansen et al. 2010). In general, the binding energies of intermediates which bind to a surface through the same kind of atom are found to scale linearly with each other as the electrode material is varied (Rossmeisl 2013). This point is illustrated in Fig. 2.9a, where schematic scaling relations between the binding energies of the OER intermediates and $\Delta G_{MOH}$ are presented. The slopes of these scaling relations are related to the number of bonds to the surface each intermediate partakes in. For example, the MOH and MOOH intermediates both involve a single oxygen bond to the surface and so, the slope of their scaling relation is one, whereas the slope of the scaling relation between the MOH and MO binding energies will be two, as the MO intermediate involves two bonds with the surface. Significantly, it has been found that $\Delta G_{MOH}$ and $\Delta G_{MOOH}$ are further related to each other by a constant of approximately 3.2 eV (Koper 2011):

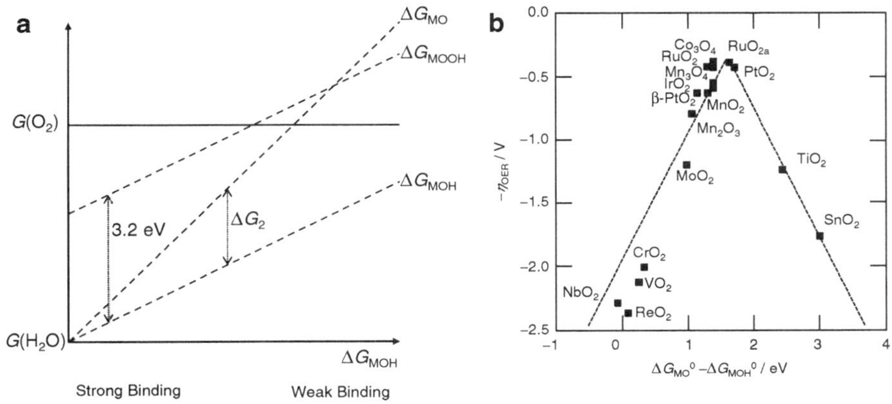


**Fig. 2.9** (**a**) Schematic representation of the scaling relations for $\Delta G_{\mathrm{MOH}}$, $\Delta G_{\mathrm{MO}}$ and $\Delta G_{\mathrm{MOOH}}$ as functions of $\Delta G_{\mathrm{MOH}}$. The constant difference between $\Delta G_{\mathrm{MOH}}$ and $\Delta G_{\mathrm{MOOH}}$ of 3.2 eV and the universal activity descriptor $\Delta G_2$ are also shown. Adapted from Rossmeisl (2013). (**b**) OER activity trends for rutile, anatase, $Co_3O_4$ and $Mn_xO_y$ oxides. The negative values of the theoretical overpotential are plotted against the standard free energy difference of $\Delta G_{\mathrm{MO}} - \Delta G_{\mathrm{MOH}}$. Adapted from Man et al. (2011)

$$\Delta G_{\mathrm{MOOH}} = \Delta G_{\mathrm{MOH}} + 3.2\ \mathrm{eV} \tag{2.50}$$

This relationship has been confirmed both for metals and for a wide range of oxide surfaces (Man et al. 2011; Koper 2011), implying that there is a universal scaling relation between the MOH and MOOH intermediates regardless of the binding site.

Taking this into consideration, the energy of the second reaction step in the associative path, $\Delta G_2$, has been proposed as a universal descriptor of oxygen evolving catalytic activity (Man et al. 2011). It was shown in Sect. 2.3.2 that the PDS for the OER is either the hydroxyl oxidation step (Eq. (2.34)) or the O – O bond formation step (Eq. (2.35)). Noting that the MO intermediate is involved in both steps, and taking into account the universal scaling relation, the expression:

$$\Delta G_2 = \Delta G_{\mathrm{MO}} - \Delta G_{\mathrm{MOH}} - qV \tag{2.51}$$

clearly contains information on the binding energies for all three important OER intermediates. Indeed this parameter has been shown be a good general descriptor of the overpotential trends for a wide variety of oxides. In particular, it allows for a comparison between different families of oxides using a single parameter (Man et al. 2011). In Fig. 2.9b, the overpotential trends for a series of rutile, anatase, cobalt and manganese oxides are presented in a single volcano plot. Again $RuO_2$ and $IrO_2$ are positioned near the peak. Moreover, excellent agreement, in terms of trends, was found between the calculated overpotentials and the experimental results reported in the literature (Man et al. 2011).

The implications of these findings for catalyst optimisation are significant. It was shown in Sect. 2.3.1 that the ideal catalyst is defined by a free energy reaction diagram in which the four charge transfer steps have identical reaction free energies of 1.23 eV. This can only be achieved at a specific binding of all intermediates. However, due to the scaling relations it is not possible to independently tune the binding energy of each intermediate on a surface to achieve this optimal situation. Altering the binding energy of one intermediate will also change the binding energies of all other intermediates. Moreover, the universal scaling relation between MOH and MOOH defines a lower limit estimate for the thermodynamic overpotential. Regardless of the binding energy of the MO intermediate, there is a constant difference between the binding energies of these intermediates of 3.2 eV. This is considerably more than the optimal separation of 2.46 eV, which would be expected for the transfer of two electrons and two protons, 1.23 eV $\times$ 2. This discrepancy predicts a minimum overpotential of $0.37 \pm 0.20$ V ($=[3.20 - 2.46$ eV]/2e) (Man et al. 2011; Koper 2011). Interestingly, as can be seen from Fig. 2.9b, the best known catalysts are already very close to this minimum value which raises the question: can OER electrocatalysis be improved? Even if the MO level can be situated optimally, half-way between the MOH and MOOH levels, within this framework there will always be an appreciable OER overpotential. This question is addressed in Sect. 2.5.1.

### 2.4.3 *Electronic Structure and Activity*

While the identification of descriptors has been greatly facilitated by the accurate calculation of adsorption energies, it is important to develop an understanding of the relationship between this adsorption behaviour and the fundamental properties of the catalytic material. Since the activation energies for elementary surface reactions are strongly correlated with adsorption energies (Nørskov et al. 2009), knowledge of the surface's ability to form bonds is essential. In principle, the catalytic properties of a material are completely determined by its electronic structure. In the case of pure transition metals, the "d-band" model provides a useful account of the ability of surface atoms to form bonds to an adsorbate (Nørskov et al. 2008); the higher in energy the d-states are relative to the metal Fermi level, the stronger the interaction with the adsorbate. Arising from this, the d-band centre is widely used as a descriptor for the activity of transition metals and their alloys (Mavrikakis et al. 1998; Bligaard and Norskov 2007; Chen et al. 2008; Inoglu and Kitchin 2011). In the case of transition metal oxides, however, it is unclear whether such an interpretation can be realistically applied. The complexities of the oxide surface such as the configuration of the metal atoms and their ligands, the oxidation state of the metal and the nature of the interaction between the active site and the adsorbates, can all influence the adsorption energies (Calle-Vallejo et al. 2013). Instead, motivated by the success of d-band theory, several groups have sought a molecular level understanding of OER activity using the concepts of orbital occupancy and electron counting.

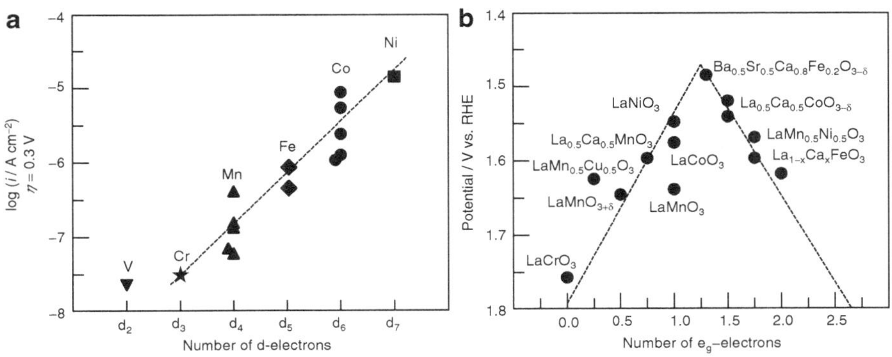


**Fig. 2.10** (**a**) OER current density (based on real surface area) at an overpotential of 0.3 V for a series of perovskites versus the number of d-electrons of the transition metal cations. Adapted from Bockris and Otagawa (1984). (**b**) The relation between the OER activity, defined as the potential at 50 mA $cm^{-2}$ of OER current, and the occupancy of the $e_g$-orbital of the transition metal. $x = 0$, 0.25 and 0.5 for Fe. Adapted from Suntivich et al. (2011)

In their study of the OER on perovskite oxides Bockris and Otagawa (1984) correlated the experimental trends in activity with the number of 3d-electrons of the transition metal M(III) cations. A linear increase in OER activity was observed with increasing 3d-occupancy, as shown in Fig. 2.10a. This trend is the opposite of that observed for the metal–OH bond strength in the same study, that is to say the bond strength is inversely proportional to the number of d-electrons. To account for this, these authors constructed molecular orbital diagrams for the bonding of an OH entity to an $M(III)O_5$ complex in an octahedral geometry. From this model of the MOH bond, they showed that the electrons from the d-orbitals of the surface transition metal in the perovskite occupy antibonding orbitals when involved in OH bonding. In this way, the decreasing bond strength was accounted for by the increasing number of d-electrons in anti-bonding orbitals. Building on this work, Suntivich et al. (2011) propose that the occupancy of a single set of molecular orbitals is a universal activity descriptor for perovskite type oxides. Noting that the octahedral splitting of the d-orbitals of atoms in the transition metal M sites results in a low energy triplet with $t_{2g}$ symmetry and a high energy doublet with $e_g$ symmetry, they proposed that the filling of the surface transition metal states of $e_g$-orbital parentage is a more appropriate descriptor for catalysis than the total number of d-electrons. This correlation is based on the idea that $e_g$-orbitals participate in σ-bonding with surface-anion adsorbates (Ballhausen and Gray 1962; Betley et al. 2008), and can therefore form stronger overlaps with the oxygenated adsorbates than the π-bonding $t_{2g}$-orbitals. Thus, the level of occupancy of the $e_g$-orbital at the surface transition metal site can greatly influence the binding of the OER intermediates, and consequently the OER activity. In Fig. 2.10b, the $e_g$-occupancy is shown to be a good descriptor of OER activity for a range of perovskite oxides. A remarkable achievement of this analysis was the prediction of the high activity catalyst $Ba_{0.5}Sr_{0.5}Co_{0.8}Fe_{0.2}O_{3-\delta}$ (BSCF). The authors found that a near-unity occupancy of the $e_g$-orbital would enhance the intrinsic OER activity of perovskite transition metal oxides. Noting that BSCF had a near optimal $e_g$-occupancy, this catalyst was subsequently shown to have the highest experimental activity of the oxides studied (Suntivich et al. 2011).

Regardless of the approach taken, these studies show a clear correlation between OER activity and the electronic structure of the oxide. A notable study from Calle-Vallejo et al. (2013) helps to shed light on the nature of this relationship. These authors examined the relationship between the number of transition metal "outer" electrons, defined as the number of valence electrons remaining on the metal in a particular oxidation state, and the calculated adsorption energies for the OER intermediates, $\Delta G_{MOH}$, $\Delta G_{MO}$ and $\Delta G_{MOOH}$. Two distinct cases were considered. In the first case, the transition metal in a given type of oxide was varied in order to determine the effect of changing the number of outer electrons. In Fig. 2.11a, the trends observed for a series of monoxides are presented. The adsorption energies increased linearly with the number of outer electrons indicating a systematic weakening of the interaction. Analogous to the explanation provided by Bockris and Ottagawa (1984), this was attributed to increasing electron density in anti-bonding orbitals. The only exception was that of CaO which is easily justified as both components—$Ca^{2+}$ and $O^{2-}$—have complete valence shells, making CaO inert with regard to adsorption or surface bonding (Calle-Vallejo et al. 2013). Importantly, though, it was found that the slope for $\Delta G_{MO}$ in Fig. 2.11a was twice that of both $\Delta G_{MOH}$ and $\Delta G_{MOOH}$. This feature explains the occurrence of the adsorption scaling relations outlined in Sect. 2.4.2 and was observed for all systems examined including the pure metals as well as a range of perovskite oxides. In the second case, the oxidation state of a given transition metal was varied in order to determine the effect of changing the valence shell filling. In Fig. 2.11b, the trends observed for a range of Ni based systems are shown. Again the same linear scaling relations are present except that this time an increase in oxidation state corresponds to a weakening of the interaction. To understand this, consider a surface Ni atom coordinated to five oxygen atoms. Noting that the oxygen adsorbate completes the octahedral arrangement, the authors propose that when a Ni cation at the surface coordinates with an electrophilic ligand it must donate electrons to it in order to form the bond (Calle-Vallejo et al. 2013). In this respect, the more oxidised the metal, the fewer electrons available for bonding implying that the strength with which highly oxidised cations create bonds that require further oxidation is weak, compared with atoms of the same species in a lower oxidation state.

Taking this into consideration, the number of outer electrons is seen to be the primary electronic-structure factor influencing the surface energetics of the OER intermediates (Calle-Vallejo et al. 2013). Moreover, the fact that linear relations were observed in all cases is significant as it indicates that the number of outer electrons can be used as a descriptor to understand differences in adsorption energies not only within the same type of compound but also between different structural families. Interestingly, the number of outer electrons has also been shown to smoothly describe trends in the formation energy of various bulk oxides. Calle-Vallejo et al. (2015) showed that the formation energy for a particular oxide

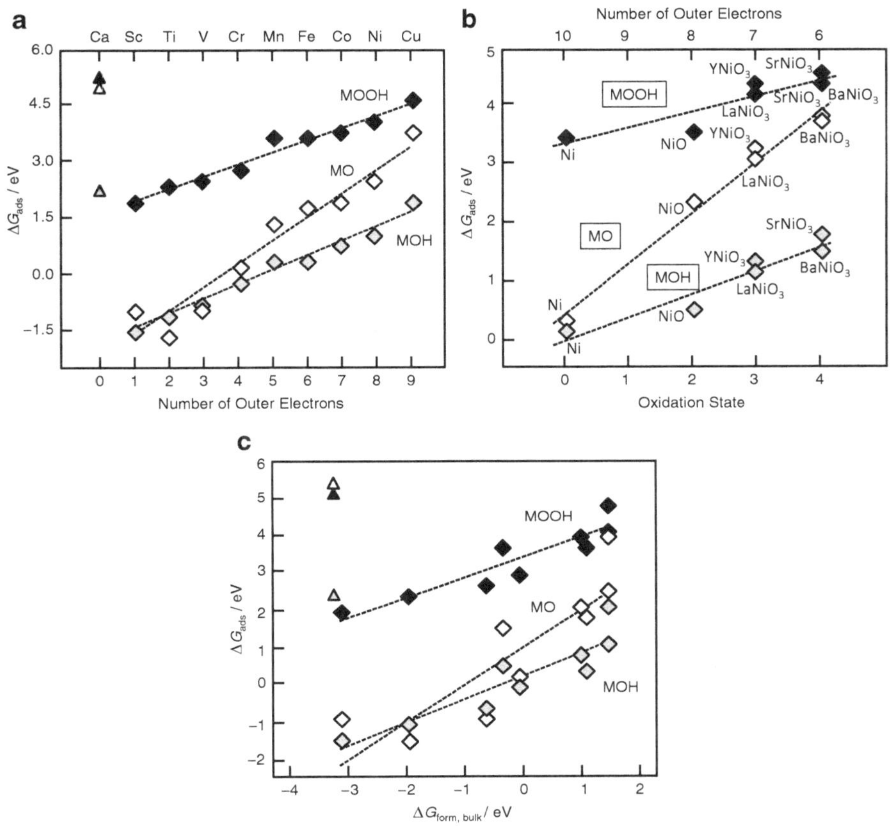


**Fig. 2.11** Trends in $\Delta G_{MOH}$, $\Delta G_{MO}$ and $\Delta G_{MOOH}$ for (**a**) a series of transition metal monoxides versus the number of outer electrons (**b**) a series of Ni compounds versus the metal oxidation state and (**c**) the same series of monoxides versus the bulk oxide formation energy. Adapted from Calle-Vallejo et al. (2013, 2014)

increases linearly with the number of outer electrons. By extension, the same linear scaling relations were observed between the formation energy and the adsorption energies of the OER intermediates. The corresponding trends observed for a series of first row transition metal monoxides are shown in Fig. 2.11c This direct correlation between oxide formation energy and intermediate adsorption energy is fortuitous as it not only explains the early success of the bulk thermochemical descriptors, it validates the use of these quantities in describing OER activity trends. In fact, given the large availability of bulk thermochemical data this approach offers a simple yet chemically sound predictive model of OER catalysis.

In essence, these studies suggest that the number of valence electrons is the hidden parameter behind all successful OER descriptor models. Broadly speaking, the main driving force for the reactivity of transition metals and their oxides can be viewed in terms of their search for maximum stability on a molecular scale (Calle-Vallejo et al. 2013). That is, the fulfilment of simple electron-counting rules by each component in the system. In this way, the varying OER activity of metal oxides can begin to be understood on a molecular level. However, while this is most certainly an important step towards a fully predictive model of catalyst design, it must be stated that any approach utilising a single parameter description of OER activity will be subject to the scaling relations outlined in Sect. 2.4.2 (Halck et al. 2014). Indeed, trends based on the number of outer electrons literally give rise to the universal scaling relations. Thus, a major challenge in OER catalysis is to find a way to modify oxide surfaces such that these design limitations can be broken, enabling the independent optimisation of the reaction parameters. In Sect. 2.5, significant developments in OER catalyst design are discussed with a view towards enhancing catalytic activity and gaining atomic level insight into the nature of the active site.

## 2.5 Tailoring Activity

The goal in OER catalyst development is to have enough knowledge of the factors determining catalytic activity to be able to tailor catalysts atom-by-atom (Nørskov et al. 2009). In this respect, the single descriptor models detailed in Sect. 2.4 represent an important step forward, bringing simplicity and generality to the search for improved catalysts. Yet it can be argued that this simplicity imposes artificial restrictions on catalyst design. Breaking the adsorption scaling relations on well-defined planar oxides requires a three dimensional description of activity (Halck et al. 2014)—a situation that is likely already present in real high surface area catalyst systems. Nevertheless, OER catalyst design is a complex endeavour. Prospective catalysts must not only possess high activities, but should also be electronically conductive and exhibit long term stability under the harsh operating conditions of an electrolyser unit (Chang et al. 2014). In this section, several multi-parameter frameworks to develop new catalysts with enhanced OER properties are presented. These studies provide fundamental insight into the reactivity of the oxide surface at the atomic scale—yielding useful design principles. Moreover, a significant challenge in electrocatalyst development is the identification of the catalytically active species on the catalytic surface. This section then concludes with a molecular level consideration of the nature of the active site.

### *2.5.1 Catalyst Synergy*

Multi-component catalyst systems are one of the most promising avenues of research for OER catalyst enhancement. Various mixed oxides, dopants and catalyst substrates have been examined with a view towards garnering synergistic effects. In many cases significant improvements in catalytic activity have been observed. The beneficial effect of Fe impurities on the OER activity of Ni hydroxides was reported over 25 years ago (Desilvestro et al. 1986; Corrigan 1987; Corrigan et al. 1987) and since then Ni–Fe based oxide catalysts have been shown to be some of the most active catalysts in alkaline media (Landon et al. 2012; Trotochaud et al. 2012; McCrory et al. 2013; Gong et al. 2013; Smith et al. 2013a, b; Louie and Bell 2013; Kim and Choi 2014). Notably, $Ni_{0.9}Fe_{0.1}OOH$ was found to have an OER turnover frequency greater than tenfold higher than that of $IrO_x$ catalysts (Trotochaud et al. 2012). Similarly, the activity of manganese (El-Deab et al. 2007; Mohammad et al. 2008; Gorlin et al. 2014), cobalt (Yeo and Bell 2011; Lu et al. 2014) and nickel (Yeo and Bell 2012) based oxides towards oxygen evolution can be drastically increased in the presence of gold. Impurity doping in semiconductor electrodes has also been shown to lead to activity enhancements. For example, improvements in the activity of hematite (Liao et al 2012) and rutile $TiO_2$ (García-Mota et al. 2011) can be observed with Ni/Co and Mn/Mo doping, respectively. Even in the case of state-of-the-art ruthenium oxide based catalysts significant increases in OER activity have been obtained by mixing with Ni or Co (Petrykin et al. 2010).

In an effort to better understand this catalytic enhancement, several models have emerged describing the reactivity of these catalysts at the atomic scale. Halck et al. (2014) proposed a hydrogen acceptor model to account for the enhanced OER activity of Ni and Co modified $RuO_2$. This model assumes an associative reaction pathway (Eqs. (2.33)–(2.36)) and suggests a stabilisation of the MOOH intermediate through localised hydrogen bonding or full proton transfer. A schematic representation of this model is presented in Fig. 2.12. For rutile type oxide catalysts, the OER is thought to occur on the surface CUS sites, as these sites provide reactive "atop" positions essential for the formation of strongly adsorbed intermediates (Halck et al. 2014). In the hydrogen acceptor model, it is envisaged that the local CUS site architecture is modified by the incorporation of the heterovalent Ni/Co cations. The oxygen bridge positions, which are generally considered to be inactive, get activated by the presence of the nearby Ni or Co such that the bridging O accepts a strong hydrogen bond or even adsorbs the hydrogen from the MOH and MOOH intermediates, as shown in Fig. 2.12. This results in a significant deviation in the overall energetics of the OER from the conventional analysis outlined in Sect. 2.3. In Fig. 2.12 it can be seen that the free energy changes for step 1 and step 3 on the Ni and Co modified $RuO_2$ are lowered compared with those on pure $RuO_2$, but not by equal amounts. This is an important observation as it indicates that the universal scaling relation between the binding energies of the MOH and MOOH intermediates no longer applies. That is, the thermodynamic limitation of the overpotential to ca. 0.37 V has been removed and the OER can proceed at potentials closer to the reversible potential. Accordingly it was found that the largest reaction energies for the Ni and Co modified $RuO_2$ were 1.49 eV (step 2 potential-determining) and 1.33 eV (step 1 potential-determining) respectively, which gave rise to corresponding thermodynamic overpotentials of ca. 0.26 and 0.1 V. This trend was confirmed experimentally where a series of $Ru_xNi_{1-x}O_2$ and $Ru_xCo_{1-x}O_2$ catalysts were all shown to outperform the pure $RuO_2$ catalyst (Halck et al. 2014).

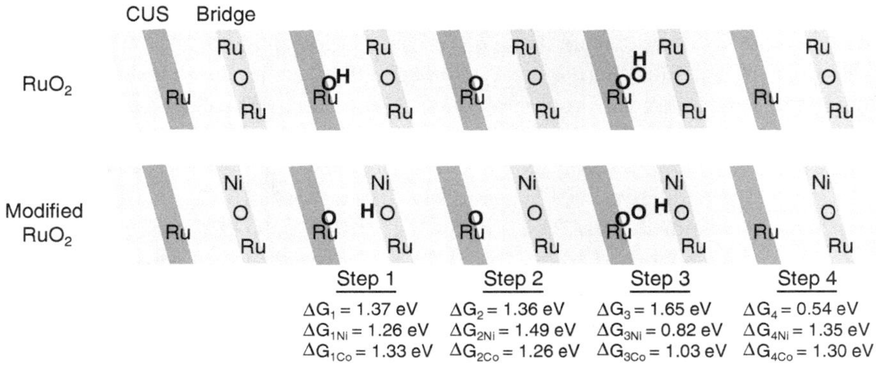


**Fig. 2.12** Schematic representation of the reaction mechanism for the oxygen evolution reaction on conventional rutile $RuO_2$ and Ni/Co modified $RuO_2$ illustrating the role of the two binding sites. The *dark grey row* represents the CUS sites and the *lighter grey* the bridge sites. Adsorbed intermediates are represented in *bold font*. The Gibbs free energies obtained from DFT calculations for each of the reaction steps are included. Adapted from Halck et al. (2014)

In this way, decoupling of the MOOH and MOH binding energies makes it possible to tune the catalytic properties of a material by varying the hydrogen acceptor. Indeed, Halck et al. (2014) have successfully extended this analysis to model the beneficial interaction of both Mn and Co with gold. Various literature reports have highlighted significant improvements in OER activity when Au is added to Mn or Co based catalysts. For example, El-Deab et al. (2007) and Mohammad et al. (2008) compared the OER electrocatalytic activity of γ-MnOOH nanorods deposited on Au, Pt and glassy carbon substrates. They found that the lowest overpotential, by more than 200 mV, was obtained for a manganese oxide modified Au substrate. Similarly, it has been shown that the OER turnover frequency, measured at an overpotential of 400 mV, for $MnO_x$ nanoparticles is increased 20-fold when they are co-deposited with Au nanoparticles (Gorlin et al. 2014). Likewise, gold nanoparticles have also been found to enhance the OER activity of mesoporous $Co_3O_4$ (Lu et al. 2014). Applying the hydrogen acceptor model, Frydendal et al. (2015) suggest that Au=O functionalities on the gold nanoparticles/substrates act to lower the energy of the MOOH intermediate in the same way as the rutile oxygen bridge sites. The inclusion of this gold hydrogen acceptor in the theoretical analysis predicted a lowering of the thermodynamic overpotential for several Mn and Co based catalysts by ca. 0.1–0.3 V, in line with the experimental observations (El-Deab et al. 2007; Mohammad et al. 2008; Gorlin et al. 2014; Lu et al. 2014). Thus, the addition of a proton donor-acceptor functionality represents a simple multi-dimensional approach for the optimisation of the electrocatalytic surface.

Busch et al. (2011a, b, 2013a, b) provide an alternative interpretation of the origin of such catalytic enhancement, arguing that mixing transition metal oxides with opposite reactivity can lead to improved catalytic performance. In contrast to the hydrogen acceptor model, these authors assumed a binuclear reaction path similar to the Bockris Electrochemical Oxide (Table 2.1)/direct recombination mechanism (Eqs. (2.41)–(2.43)). The potential catalytic performance of a series of homo-(bi)nuclear transition metal catalysts was evaluated with respect to the two primary steps in the mechanism: the oxidation of two MOH hydroxyl groups to form two MO oxo groups and the subsequent μ-peroxo bond formation between the MO groups. The choice of transition metal was shown to have a significant effect on the stabilities of the MO intermediates. In particular, two distinct classes of compound were identified depending on the degree of single bond character in the $M = O$ bond of the MO intermediate (Busch et al. 2011a, 2013a). The energetics of these two classes are typified by the Fe–Fe and Co–Co reaction energy diagrams in Fig. 2.13. For the Co–Co (Mn–Mn, Ni–Ni) catalyst the MO moieties contain considerable radical character and so, their formation is endothermic requiring an overpotential of approximately 0.5 eV (1.06 eV/2) relative to a tyrosine reference potential.[4] As a result, the subsequent O–O bond formation becomes strongly exothermic of the order of −0.75 eV. On the other hand, the $M = O$ bonds are stable for the Fe-Fe (Cr-Cr, V-V) catalyst with the result that the two MOH oxidation steps are exothermic. This stabilisation leads to an O–O bond formation step that is strongly endothermic of the order of 0.75 eV. In both cases, it is clear that significant overpotential or thermal activation is necessary for the reaction to proceed. Based on these findings, Busch et al. (2011b) proposed a hetero-nuclear Fe-Co catalyst as an attractive compromise which could overcome the shortcomings of the individual homonuclear catalysts. While this catalyst can follow two pathways—forming Fe=O or Co=O first—it can be seen from Fig. 2.13 that the inertness of Fe=O was counteracted by the reactivity of the Co=O to the extent that the di-hydroxo species displays a negligible endothermicity or overpotential of 0.11 eV and the μ-peroxo formation step is slightly exothermic by −0.14 eV. This finding is consistent with experimental observations that the OER performance of $Fe_3O_4$ based catalysts can be improved by combining with Mn and Co oxides (Singh et al. 1996; Rios et al. 1999; Merrill and Dougherty 2008; McCrory et al. 2013). Therefore, by mixing oxides from opposing reactivity groups it may be possible to selectively tune the relative energies of important intermediates, leading to minimal OER reaction barriers.

On the other hand, several authors have attributed the catalytic enhancement to changes in the local electronic properties of the active sites. Yeo and Bell (2011, 2012) compared the OER activity of Ni and Co oxides prepared on a range of substrates. They observed that the OER activity of a submonolayer of cobalt oxide decreased with decreasing electronegativity of the substrate in the order $CoO_x/Au > CoO_x/Pt > CoO_x/Pd > CoO_x/Cu > CoO_x/Co$ (Yeo and Bell 2011). Similarly, it was found that a monolayer of nickel oxide deposited on Au exhibited greater electrocatalytic activity when compared with an equivalent layer deposited

[4] The significance of this reference potential is beyond the scope of this discussion but interested readers are directed to Busch et al. (2011a, 2011b) for more details.

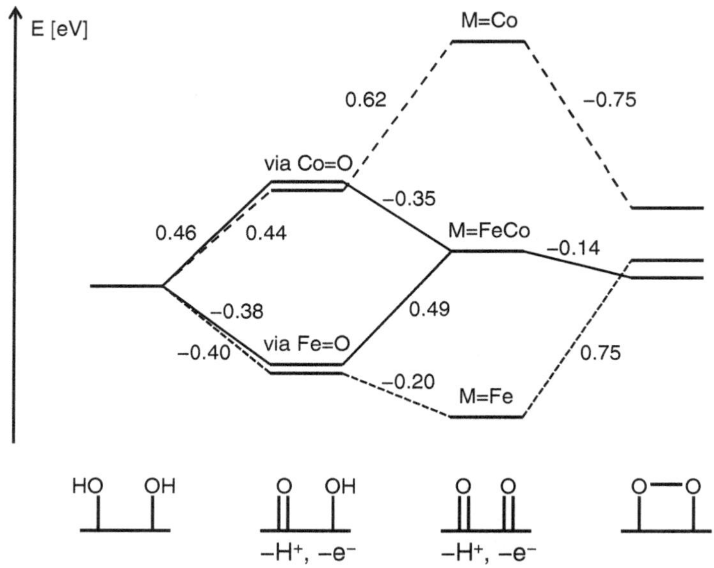


**Fig. 2.13** The reaction energetics for hydroxide oxidation and μ-peroxo bridge formation at the homo-binuclear Fe (*short dash*) and Co (*long dash*) sites and hetero-binuclear FeCo (*solid lines*) sites showing the step-by-step dehydrogenation processes. All energy values are given in eV. Adapted from Busch et al. (2011a, b)

on a Pd substrate (Yeo and Bell 2012). In all cases the increase in activity follows the trend in electronegativities and may be the result of partial electron transfer from the oxide to the more electronegative metal substrate. This concept has been elaborated on by Trotochaud et al. (2014) to account for the extraordinary synergy in Ni–Fe based catalysts. It has been shown that the Fe in these catalysts is predominantly found in the $Fe^{3+}$ oxidation state (Corrigan et al. 1987; Kim et al. 1994; Balasubramanian et al. 2000). However, at potentials associated with active oxygen evolution, *V*–pH (Pourbaix) diagrams predict that both $Fe^{3+}$ and $Fe^{4+}$ could be stable (Pourbaix 1974; Schweitzer and Pesterfield 2010). In principle, partial-charge transfer can therefore occur between the Fe and Ni centres. For example, Corrigan et al. (1987) found evidence from in situ Mossbauer spectroscopy of partial electron transfer away from $Fe^{3+}$ centres when $Ni^{3+}$ centres were further oxidised to $Ni^{4+}$. In addition, the potential of the $Ni^{2+}/Ni^{3+}$ redox couple has been shown to depend on the Fe content of the film (Louie and Bell 2013; Trotochaud et al. 2014). The presence of Fe effectively makes it more difficult to oxidise $Ni^{2+}$. Based on these observations, Trotochaud et al. (2014) suggest that partial electron transfer between the Ni and Fe centres leads to $Ni^{3+}$ or $Ni^{4+}$ species with more oxidising power and thus, enhanced OER activity. In this way, altering the electrophilicity of the active centres, through the relative electronegativities of the component metals, represents an experimentally facile approach to catalyst optimisation.

### 2.5.2 Activity–Stability Relations

The requirement of long-term stability for an OER catalyst is clear and widely acknowledged (Merrill and Dougherty 2008; McCrory et al. 2013; Fabbri et al. 2014). Even if it could be born economically, the frequent replacement of electrode materials makes little sense from both an operational and energy supply standpoint. However, a key question that is rarely addressed is: What is the relationship between the stability of an oxide material and its activity for the OER? While this relationship is addressed implicitly in the volcano relations detailed in Sect. 2.4—simply put, the most active catalyst is neither too reactive nor too stable—an explicit understanding of the atomic scale stability of oxide materials during the OER is lacking.

The potential impact of such knowledge on the development of OER catalysts is highlighted by the work of Markovic's group (Chang et al. 2014, 2015; Danilovic et al. 2014b). These authors studied the fundamental links between the activity and stability of well-defined monometallic and bimetallic oxides, arguing that it is the stability, or lack thereof, of surface atoms that controls the OER activity of oxide materials. In these studies, the activity of a particular oxide was given by the overpotential measured at 5 mA $cm^{-2}$ of OER current density and the stability of each oxide was defined by its degree of dissolution, as quantified by the concentration of corresponding metal ions in solution following each OER polarisation experiment. The trends observed for a series of monometallic oxides in acidic media are presented in Fig. 2.14. In order to discern any structural effects, two types of oxide with very different morphologies were examined: for each metal a crystalline thermally prepared oxide (TC) and a highly defective amorphous electrochemically prepared oxide (EC) were studied. It can be seen from Fig. 2.14 that the OER activity of both types of oxide increased with the oxophilicity of the metal in the order Os $\gg$ Ru > Ir > Pt > Au and that the activity was inversely proportional to the stability of the oxide. Notably, Os was found to be considerably more active than Ru but was also extremely unstable. Furthermore, the amorphous oxides all showed higher activity but less stability than the crystalline oxides. Given the higher density of defects in the EC-oxides relative to the TC-oxides, it was noted that the density of surface defects may play a significant role in the relationships between activity and stability (Danilovic et al. 2014b). Similar trends were also observed for a range of polycrystalline and single crystalline oxides (Chang et al. 2014; Danilovic et al. 2014b). In all cases the most defective surfaces were the most active and the least stable, suggesting that the inverse relationship between activity and stability is a general phenomenon in OER electrocatalysis.

Taken together, these studies show that both the nature of the oxide and the density of surface defects play an important role in controlling the stability and reactivity of surface atoms (Chang et al. 2015). This relationship has been explained in terms of a potential induced transformation from a stable low-valence oxide into an unstable higher-valence oxide, reminiscent of the O'Grady pathway in Table 2.1. It was observed that the onset of Ru and Ir dissolution for polycrystalline surfaces

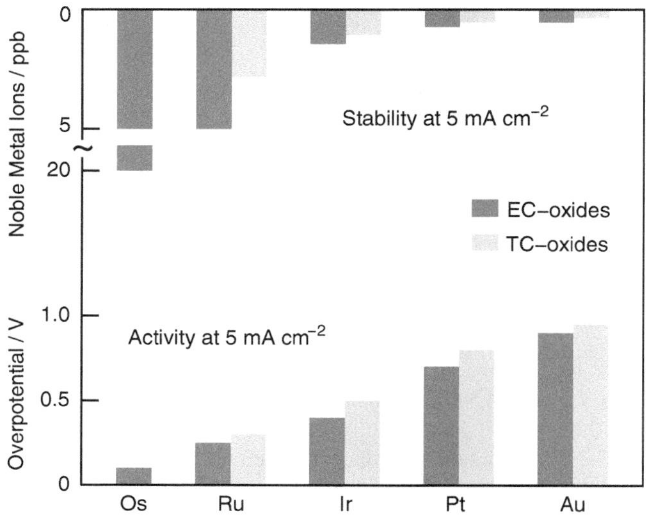


**Fig. 2.14** Relationships between activity and stability for the oxygen evolution reaction on monometallic oxides. Activity: overpotential at 5 mA cm$^{-2}$. Stability: the concentration of dissolved metal ions in the electrolyte solution following OER polarisation up to 5 mA cm$^{-2}$. Adapted from Danilovic et al. (2014b)

was always accompanied by the onset of OER current density (Danilovic et al. 2014b). Using in-situ X-ray near edge structure (XANES) analysis these processes were found to coincide with a change in the average metal oxidation state from $n = 4$ to a less well-defined $n > 4$ state. Significantly, the extent to which this transition occurred mirrored the activity and stability trends. Polycrystalline Ru showed a greater rate of oxidation than polycrystalline Ir, consistent with its higher activity and lower stability. On this basis the active sites for these catalysts were identified as locations on the electrode surface where the transition from the stable $n = 4$ state to the unstable $n > 4$ state is facilitated, such as defect sites (see Sect. 2.5.4), highlighting how potential dependent variations in the oxidation state of surface atoms govern the stability of, and induce activity in, these oxides (Danilovic et al. 2014b; Chang et al. 2015).

This link between the intrinsic activity and stability of oxide materials provides an interesting interpretation of OER catalyst optimisation. In this framework, the best materials for the OER should balance activity and stability in such a way that the dissolution rate is neither too fast nor too slow (Danilovic et al. 2014a; Chang et al 2015). This is seen clearly in the nanoscaling effects reported for Ru catalysts. Reier et al. (2012) compared the activity of bulk and nanoparticulate Ru catalysts in acidic media. While these catalysts exhibited similar activities, the Ru nanoparticles suffered from strong corrosion at the applied OER potentials and were unable to sustain their activity. Even on Au “the noblest of metals” the OER has been shown to follow a decomposition mechanism. Diaz-Morales et al. (2013) performed online mass spectrometry analysis of the oxygen evolved from polycrystalline Au surfaces. Using $^{18}O$ labelling of the surface oxide, they showed that the very first oxygen evolved was $^{18}O_2$, which could only have arisen from the surface oxide layer, indicating that the initial oxygen evolution process was accompanied by decomposition of the surface oxide. In an effort to overcome these stability limitations the Markovic group have developed a new synthesis strategy that is based on the concept of surface segregation or surface enrichment (Ruban et al. 1999; Novell-Leruth et al. 2013; Danilovic et al. 2014a). Thermal annealing of $Ru_{0.5}Ir_{0.5}$ alloys and nanoparticles was shown to enrich the catalyst surface with the more stable Ir atoms (Danilovic et al. 2014a). This resulted in the formation of a nano-segregated surface domain that enhanced the stability of the catalyst four-fold without compromising its OER activity. Thus, tuning the near-surface composition of mixed oxide catalysts affords a viable strategy for balancing catalyst activity and stability.

### 2.5.3 Conductivity Effects

The conductivity properties of the catalyst film can strongly impact its performance as an oxygen evolution electrocatalyst. Efficient charge transport to the active site is crucial if the reaction is to proceed at a practical rate. This can be a major limitation for many metal oxides as they lack the conductivity of the parent metal (Bockris and Otagawa 1984; Matsumoto and Sato 1986). As a result, the extent to which the applied potential is translated to the oxide–electrolyte interface can vary over a large range. In cases where the catalyst film is not sufficiently conductive, a portion of the potential will drop across the catalyst film in order to drive the current transport, leading to a lower apparent activity relative to a more-conductive film (Trotochaud et al. 2014). Therefore, from the perspective of catalyst optimisation, it is important to be able to decouple these conductivity limitations from the intrinsic catalytic activity.

A number of authors have proposed a dual energy barrier model to account for conductivity limitations in thin surface oxide layers (Meyer 1960; McDonald and Conway 1962; Lyons and Brandon 2010; Doyle and Lyons 2014b). Electrokinetic studies of the OER rely on the assumption that the measured current is a direct result of the interfacial charge transfer process. However, in situations where the current is also dependent on the charge-transport efficiency through the oxide film, interpretation of the kinetic parameters can be more complicated. In this respect, the dual energy barrier model assumes that only a fraction $\Delta V_{FS}$ of the total potential difference between the metal support and the electrolyte $\Delta V_{MS}$ determines the electrochemical charge transfer kinetics in the double layer (McDonald and Conway 1962). The remainder $\Delta V_{MF}$ appears across an electronically semi-conducting "barrier" oxide, through which the charge passed must migrate under the influence of an electric field. In this way, the process can be considered equivalent to a complex electrode reaction with two rate limiting steps following one another: charge migration through the oxide film and charge transfer at the electrolyte

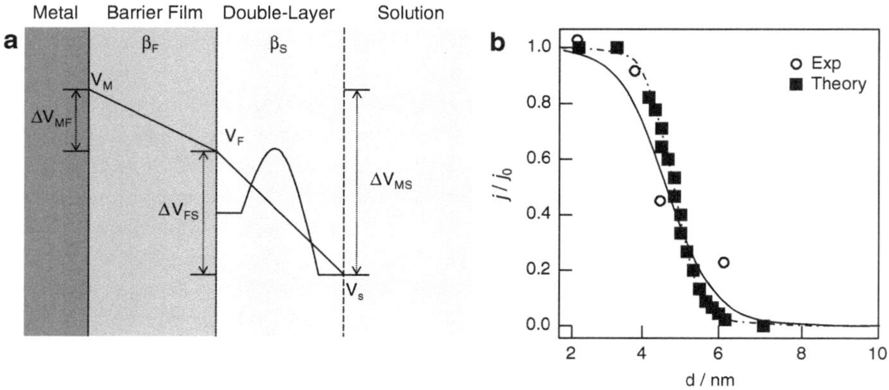


**Fig. 2.15** (**a**) Schematic representation of the dual energy barrier model showing associated potential gradients through the barrier film and double-layer interface. Adapted from McDonald and Conway (1962). (**b**) Comparison of theoretical (*squares*) and experimental (*circles*) data for the oxygen evolution current density at 2.0 V vs. RHE. The current density is normalised relative to a 1.2 nm $TiO_2$ sample and a roughness of about 0.5 nm is included in the analysis. Adapted from Viswanathan et al (2014)

interface, as shown schematically in Fig. 2.15a. The result of this is a composite symmetry factor for the overall process across the two barriers as follows:

$$\beta_{\Sigma} = \frac{\beta_{\mathrm{F}}\beta_{\mathrm{S}}}{\beta_{\mathrm{F}} + \beta_{\mathrm{S}}} \tag{2.52}$$

where $\beta_{\mathrm{F}}$ is the symmetry factor for field assisted charge transport through the oxide and $\beta_{\mathrm{S}}$ represents the symmetry factor for the electrocatalytically rate-determining electron transfer reaction. A similar dual barrier expression can also be derived for the effective reaction order with respect to the activity of a particular reactant $a_{\mathrm{S}}$:

$$\frac{\mathrm{dlog}i}{\mathrm{dlog}a_{\mathrm{S}}} = \frac{m_{\mathrm{S}}\beta_{\mathrm{F}}}{\beta_{\mathrm{F}} + \beta_{\mathrm{S}}} \tag{2.53}$$

where $m_{\mathrm{S}}$ is the expected value of the reaction order in the absence of dual barrier conditions. In this way the true mechanistically significant kinetic parameters can be extracted from the overall current response. Although this model was originally suggested for monolayer Au oxides, McDonald and Conway (1962) suggest that it can be logically extended to layers thicker than a monolayer and so, should be generally applicable for thin compact oxides through which facile charge transport cannot be assumed.

In relation to this, Viswanathan et al. (2014) propose that there is a critical thickness for such compact oxide films beyond which charge-transport should always be considered. Utilising $TiO_2$ as a model system, these authors explored the role of coherent tunnelling as a mode of charge transport in electrocatalytic materials with low conductivity. To account for the possibility of electron tunnelling, the equation for the kinetic current density (Eq. (2.39)) can be modified as follows (Viswanathan et al. 2014):

$$j = j\,\min\left(\exp\left\{-\frac{q\left(V - V_{\text{bias}} - V^{\text{OER}}\right)}{kT}\right\}, 1\right)\Big|\left[j_{\text{t}}\left(qV_{\text{bias}}\right) = j\right] \qquad (2.54)$$

Here, the tunnelling current density $j_t$ is dependent on the potential difference across the film and must be sufficient to sustain the electrochemical current density $j$. This condition gives rise to a minimum potential bias $V_{\text{bias}}$ which can have a significant influence on the observed electrocatalytic activity—appearing as an apparent increase in overpotential or a decrease in the current density at a fixed potential. For example, calculated and experimentally measured current densities at a fixed OER potential are plotted as a function of the $TiO_2$ layer thickness in Fig. 2.15b. For sufficiently thin films it was found that tunnelling can contribute in a dominant way to charge transport with the result that below a thickness of ca. 4 nm, $V_{\text{bias}}$ was negligible and the OER kinetics were unchanged. However, for thicker layers, charge-transport begins to limit the OER current density. The tunnelling current density at a fixed $V_{\text{bias}}$ was shown to decay rapidly with increasing layer thickness. Accordingly, a greater proportion of the applied potential is necessary to drive the tunnelling process resulting in a decrease in the observed OER current density as the layer thickness increases. Based on these observations the authors suggest that nanostructuring offers a potential solution; by lowering the distance required for the charge to travel and allowing for additional charge transport mechanisms such as tunnelling to become active, the intrinsic activity of the catalyst film can be studied irrespective of its bulk conductivity (Viswanathan et al. 2014).

While these studies have focussed on the electronic conductivity of compact anhydrous oxides, alternative charge transport mechanisms will become important for more porous structures. In electrochemically prepared hydrous oxides the electric field decays rapidly within a few nanometers of the electrode support (Surendranath and Nocera 2012). Redox centres immediately adjacent to the support electrode are directly affected by the electrode potential, whereas charge is further propagated along the oxy-metal strands through an electron hopping mechanism (Fachinotti et al. 2007; Lyons et al. 2011; Doyle et al. 2013). Importantly, this electron hopping process is accompanied by the transport of ions along the same path and as such, the ionic conductivity of these films can also be a key determinant of their electrocatalytic performance.[5] However, even for these highly porous films, simple attempts at facilitating electron transport within the film, such as the incorporation of conductive carbon architectures, have yielded significant

[5] IS and CV studies can provide a useful qualitative and quantitative characterisation of the charge-transport processes in mixed ionic/electronic conductors. The IS analysis of such processes is discussed in detail by Bisquert et al. (2000a, 2000b) and Terezo et al. (2001), whereas the CV characteristics have been comprehensively reviewed by Doyle et al. (2013).

catalytic enhancements. An electrochemically prepared hydrous Fe oxide/single walled carbon nanotube composite catalyst showed a 50 mV reduction in OER overpotential relative to the pure hydrous oxide, with a three-fold increase in turnover frequency (Doyle and Lyons 2014a). This enhancement was attributed to the superior conductivity of the composite film—IS analysis of the film during active oxygen evolution showed an eight-fold decrease in charge transfer resistance compared with the pure hydrous oxide film. Similar enhancements in catalytic activity have been reported for a range of chemically synthesised spinel oxide/graphene composite catalysts including $Co_3O_4$ (Liang et al. 2011), $MnCo_2O_4$ (Wang et al. 2011; Liang et al. 2012), $NiCo_2O_4$ (Lee et al. 2013) and $CoFe_2O_4$ (Bian et al. 2014).

Additionally, nanostructuring has also been shown to enhance the activity of amorphous electrodeposited oxides, although in this case it appears that increasing rather than decreasing the charge transport network is necessary. This is illustrated clearly for the cobalt catalyst films developed by Nocera's group (Farrow et al. 2013). Catalyst films formed in borate (CoBi) and phosphate (CoPi) electrolytes showed marked differences in activity with increasing film thickness. In both cases the performance of the layer increased with thickness reflecting an increase in effective surface area, as would be expected for a porous layer. However, whereas the CoBi films displayed a monotonic increase in activity with film thickness, the activity of CoPi reached a limiting activity. This observation correlated with structural differences in the catalyst films: CoBi films exhibited coherent domains consisting of 3–4 nm cobaltate clusters (see Sect. 2.5.4) with up to three layers, whereas CoPi films comprised significantly smaller clusters that were not coherently stacked. It was suggested that the larger domain size and lower disorder of CoBi relative to CoPi facilitated intralayer and interlayer hopping of electrons, providing for greater charge transport mobility in the CoBi films. In this way the effective concentration of catalytic sites could be more easily increased for the CoBi films while still maintaining an efficient charge transport network. Thus, an effective OER catalyst design protocol should also include the careful engineering of the material structure, so that conductivity can be maximised while preserving the surface structures necessary for efficient catalysis. In Sect. 2.5.4 the specific nature of these catalytically active surface structures is discussed.

### 2.5.4 The Active Site

It is generally accepted that surface reactions occur predominantly at active sites (Schwab 1981; Burke et al. 2000, 2003). This concept, first introduced almost a century ago by Taylor (1925), is based on the idea that heterogeneous reactions do not take place over the entire available catalytic surface but are instead localised at specific catalytically active surface sites. Although pinpointing these active sites experimentally has proven difficult, some consensus has been achieved regarding their general properties. In the case of metal and metal oxide surfaces, the active sites are thought to be located at surface defects in the lattice structure. Somorjai (1996) pointed out that, for the same surface, catalytic processes occur much more rapidly at defects, such as kinks and ledges, than on terraces, stressing that more defective surfaces are most active—a point supported by the work of Markovic's group discussed in Sect. 2.5.2. Similarly, Sun et al. (2012) found, using a combined Hückel theory and DFT approach, that the most thermodynamically stable oxygen adsorption sites on Au are those with the potential for multi-fold coordination. Moreover, Nowicka et al. (2010) have shown that the active sites on Au can be selectively "knocked out" by treatment with hydroxyl radicals, suggesting that surface defects are the loci of partially filled d-orbitals that can stabilise free-radical intermediates. From these observations it seems clear that sites with low lattice coordination at the atomic level control the surface reactivity. However, the specific nature of these sites is still not well understood, hence the somewhat ambiguous representation of the OER active sites by M in this work. A fundamental challenge in OER catalyst design is understanding the processes occurring at these sites at the microscopic level.

An intuitively chemical description of OER catalysis has been provided by the groups of Burke and Lyons (Burke and O'Sullivan 1981; Lyons and Burke 1987; O'Sullivan and Burke 1990; Rebouillat et al. 2011; Lyons et al. 2011; Doyle et al. 2013). These authors contend that the specific nature of the active site can be envisaged if attention is paid to the underlying surface redox chemistry of the oxide. In aqueous solution, metal oxide surfaces have a significant hydrophilic character with the result that surface oxy groups can become extensively hydrated or hydroxylated (Burke et al. 1982b; Egdell et al. 1983). For compact crystalline oxides this hydration is likely restricted to coordinately unsaturated surface defect sites, whereas for more porous oxides the hydration can permeate deeper into the layer. Indeed, for anodically generated hydrous oxide films it appears that the oxy-metal cations are hydrated at the individual level, virtually intermingling with the solution phase (Burke and Lyons 1986). In this way, it is suggested that metal oxides share a common surface–electrolyte interface, only differing in the extent of their hydration/hydroxylation, as illustrated in Fig. 2.16. Importantly, insight into the structure of these hydrated surface species is afforded by an examination of their distinctive redox behaviour. Negative *super-Nernstian V*-pH shifts have been observed for the redox transitions of a number of transition metal oxide surfaces including Ni (Burke and Twomey 1984; Lyons et al. 2012a, b) Fe (Burke and O'Sullivan 1978; Burke and Lyons 1986; Doyle and Lyons 2013a, b) Au (Burke et al. 1982b) Rh (Burke et al. 1981) Ir (Burke and Whelan 1984) and Ru (Lyons and Burke 1987; Lyons and Floquet 2011) implying that the oxidised state of these oxides acquires a net negative charge relative to the reduced state. This anionic oxide formation arises from well-known hydrolysis reactions which accompany the oxidation of the metal cations (Burke and Lyons 1986; Sato 1998), and can be equivalently regarded in terms of the adsorption of excess $OH^-$ ions, proton loss from coordinated water molecules or the formation of hydroxyl surface complexes. Based on these considerations, a general structure for the hydrated surface species

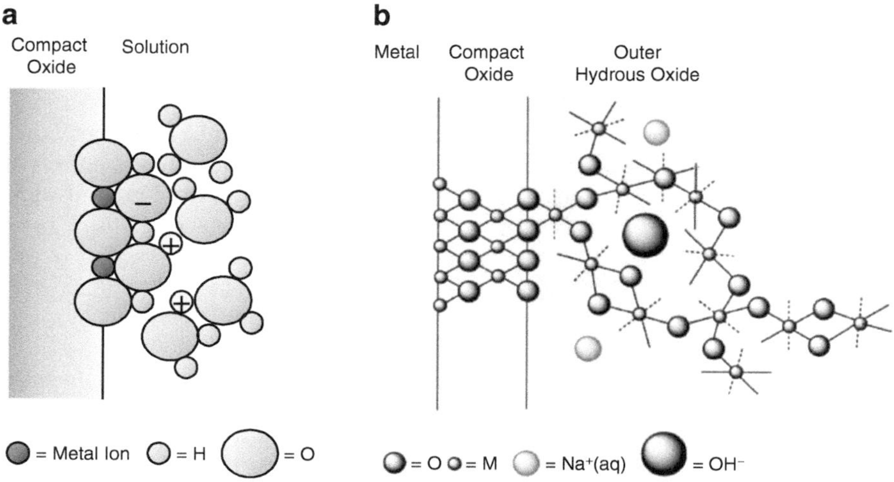


**Fig. 2.16** Schematic representation of the oxide–solution interface for (**a**) compact thermally prepared oxides such as $RuO_2$ and (**b**) anodically prepared hydrous oxides. In both cases the electrocatalytically active region is hydrated containing octahedrally co-ordinated surfaquo groups either individually dispersed or interlinked in a more extended network. Adapted from Lyons et al. (2014)

has been proposed: $\left[M(z)O_m(OH)_n(OH_2)_y\right]^{p-}$ where $p = 2m + n - z$ and $z$ is the oxidation state of the metal M, which should apply to any oxide phase known to acquire a net negative charge in aqueous solution. Accordingly, the electrocatalytic activity of the various oxide films can be ascribed to the presence of complex anionic surface clusters, consisting of octahedrally co-ordinated metal complexes—the *surfaquo* group.

It is important to emphasise here that the formulae for these anionic species are devices presented to rationalise the observed $V$-pH behaviour; the exact surface stoichiometry is often unknown and will depend on the nature of the metal cation and its energetically stable coordination states. However, it is believed that this chemically intuitive approach is more realistic than thinking in terms of discrete stoichiometric oxide species. In their original work on thermally prepared Ru based oxides, Lyons and Burke (1987) proposed that the OER took place at Ru surfaquo groups according to the following reaction sequence:

$$(-O-)_2RuO^-(OH)_3 + OH^- \rightleftharpoons (-O-)_2RuO.O^-(OH)_2 + H_2O + e^- \quad (2.55)$$

$$(-O-)_2RuO.O^-(OH)_2 \rightarrow (-O-)_2RuO_2(OH)_2 + e^- \quad (2.56)$$

$$(-O-)_2RuO_2(OH)_2 \rightleftharpoons (-O-)_2RuO_2 + 2H^+ + O_2 + 2e^- \quad (2.57)$$

$$(-O-)_2RuO_2 + OH^- + H_2O \rightleftharpoons (-O-)_2RuO^-(OH)_3 \tag{2.58}$$

Here the surfaquo groups are linked to the oxide surface by two bridging oxygen species (−O−). With the second electron transfer step being rate-determining, this mechanism was shown to be in good agreement with the observed electrokinetic data: $b = 60$ mV dec$^{-1}$ and $m_{OH^-} = 1$. Building on this concept, surfaquo groups have been specifically implicated as the OER catalytic centres on several thermal oxide (Lyons and Burke 1987; Lyons and Floquet 2011; Godwin et al. 2014), passive oxide (Lyons and Brandon 2010; Doyle and Lyons 2014b) and hydrous oxide surfaces (O'Sullivan and Burke 1990; Lyons and Brandon 2009; Lyons et al. 2012a, b; Doyle and Lyons 2013a, b). In particular, the pH dependence of the OER on hydrous Fe oxides was shown to mirror that of the surface redox processes, as shown in the inset of Fig. 2.17 (Doyle and Lyons 2013a, b). Rather than the expected *V*-pH shift of −59 mV/pH unit for a process involving equivalent proton/hydroxide and electron transfer, the OER onset potential exhibited a −93 mV/pH super-Nernstian shift of similar magnitude to the primary hydrous oxide redox transition A3. On this basis it was suggested that the anionic surfaquo groups located throughout the hydrous layer actively partake in oxygen evolution, and the mechanism outlined in Fig. 2.17 was shown to be in excellent agreement with the electrokinetic data obtained for a range of hydrous Fe oxide catalysts (Doyle and Lyons 2013a, b, 2014a).

In the context of the current discussion, it is worth pointing out the molecular nature of these mechanisms. Essentially a modification of the Kobussen Pathway (Table 2.1) to take into account the nature of the surfaquo groups on the Fe surface, the mechanism in Fig. 2.17 also draws heavily on molecular concepts, including Fe (V) oxo catalysis (McDonald and Que 2011), and is analogous to those depicted for various homogeneous catalyst systems (Wang et al. 2010; Busch et al. 2011a, b; Duan et al. 2012). This is not unexpected given the very dispersed and somewhat tenuous nature of the catalytically active hydrous oxide layer. A common feature of these systems is a lack of any significant contribution to double-layer charging with increasing hydrous layer thickness indicating that a distinct oxide–solution interface is not present (Burke and Lyons 1986). In contrast, the electrocatalytic currents associated with these films often increase with layer thickness (Doyle and Lyons 2013a, b; O'Sullivan and Burke 1990), reflecting an increased concentration of active sites. As noted by O'Sullivan and Burke (1990), the behaviour of the open polymeric network is probably not greatly different from that of its oxymetal components. Thus, the hydrous oxide film may be regarded as surface bound polynuclear species consisting of linked surfaquo groups arranged in a dispersed three dimensional structure with each group being electrocatalytically active. Even in the case of the largely anhydrous thermal oxides, surfaquo group coverage can be considerable. Lyons and Burke (1987) have estimated that the fractional surfaquo group coverage on $RuO_2$ can reach as much as 30 %. Indeed, it has been noted that the origin of the electrocatalytic properties of thermally prepared $RuO_2$ films is not to be sought in the intrinsic rutile structure of the oxide, rather the catalytic

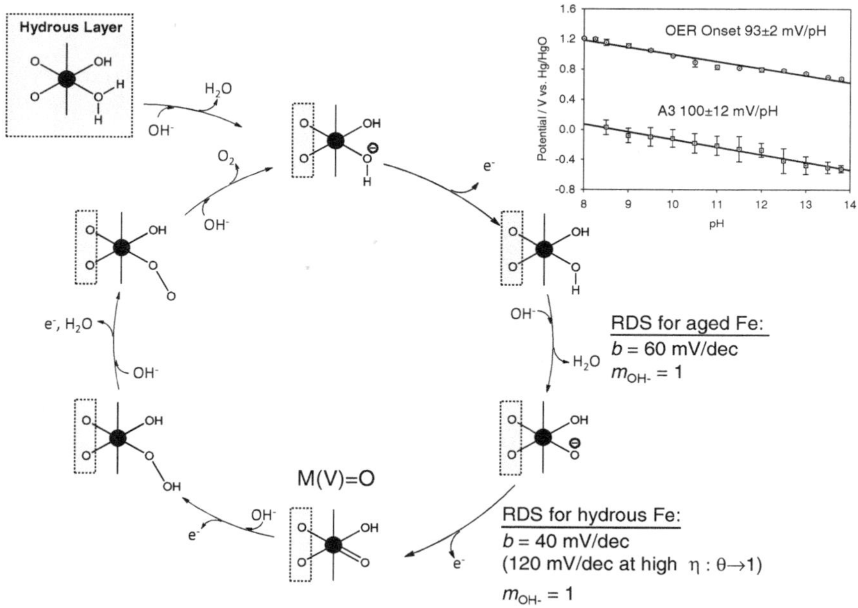


**Fig. 2.17** Schematic representation of the mechanism for oxygen evolution at hydrous Fe oxides in alkaline media. Note that octahedrally co-ordinated oxymetal surfaquo groups are identified as the catalytically active species and are located throughout the hydrous layer. Deprotonation of co-ordinated water is assumed to be facile at high pH and so the reaction is initiated at co-ordinated hydroxide moieties. The *inset* shows the pH dependence of the OER onset potential and the hydrous oxide redox potential A3 associated with an Fe(II)/Fe(III) transition. Adapted from Doyle et al. (2013)

properties are more likely to originate from the state of surface atoms (Lodi et al. 1978). In this way, the surfaquo group is a potentially unifying concept, bringing together current understanding of the OER at various metal oxide catalysts while providing a bridge with the allied field of water oxidation in homogeneous catalytic systems via transition metal complexes.

While the direct spectroscopic determination of the structure and composition of these anionic entities has not been achieved, a number of properties underpinning the surfaquo group concept have been independently observed by other researchers. Hydrated surface structures have also been implicated as the catalytic active sites for the OER by the group of Nocera. These authors combined electrokinetic studies with detailed spectroscopic analysis to provide a mechanistic picture of the OER at several Co (Kanan and Nocera 2008; Surendranath et al. 2010; Kanan et al. 2010; McAlpin et al 2010) Mn (Huynh et al. 2014) and Ni (Bediako et al. 2012, 2013) based oxygen evolution catalysts. X-ray absorption spectroscopy (XAS) studies of electrodeposited Co oxides showed that the catalyst films were composed of molecular cobaltate clusters (MCC) consisting of edge-sharing $CoO_6$ octahedra (Kanan et al. 2010). Specifically, oxygen isotope studies suggested that solvent

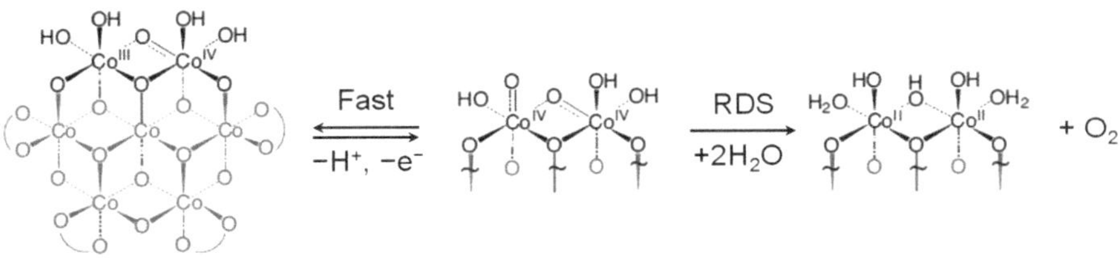


**Fig. 2.18** Reaction pathway for the OER at CoPi films. A proton coupled electron transfer equilibrium is followed by a rate-determining O–O bond forming step. The structure of the MCC is shown on the *left*. *Curved lines* denote phosphate, $OH_x$ or terminal bridging ligands. Adapted from Surendranath and Nocera (2012)

exposed Co centres at the edges of these MCCs were the active sites for oxygen evolution catalysis. Furthermore, X-ray absorption near-edge spectroscopy (XANES) (Kanan et al. 2010) and EPR (McAlpin et al. 2010) studies indicated that the resting state of the catalyst at potentials associated with active oxygen evolution—the high surface coverage species in the terminology of Sect. 2.2—was a mixed valent Co(III)/Co(IV) surface unit, as shown in Fig. 2.18. Similar octahedral mixed valent catalytic resting states have been proposed for electrodeposited Mn (Huynh et al. 2014) and Ni (Bediako et al. 2012) oxides as well as Ni–Fe mixed oxides (Trotochaud et al. 2014). On this basis, the key steps in the OER at the Co based catalysts involved the oxidation of hydroxylated surface Co(III)–OH units to Co(IV)–O moieties followed by slow $O_2$ evolution as part of a chemical rate-determining step (Surendranath et al. 2010), in agreement with the experimentally observed 60 mV $dec^{-1}$ Tafel slope and reaction order $m_{H^+} = -1$.

The amorphisation of metal oxide surfaces has been observed by a number of research groups. Implicit in the surfaquo group description of the oxide surface is the presence of an amorphous oxide–solution interface irrespective of the crystallinity of the bulk oxide. In line with this, Lee et al. (2012) reported significant changes in the surface structure of lithium ion battery materials during oxygen evolution. Transmission electron microscopy (TEM) showed that the crystalline surface structure of $LiCoPO_4$ materials became progressively more amorphous with potential cycling under oxygen evolution conditions. At neutral pH, these structural changes correlated with a significant increase in OER current density. Using energy dispersive X-ray (EDX) spectroscopy coupled with scanning TEM it was found that the change in surface structure with potential cycling was accompanied by phosphorous leaching from the layer. In particular, the amorphous surface structure of the activated $LiCoPO_4$ was compositionally similar to that of the CoPi film, suggesting the increased activity could be attributed to the formation of hydrated molecular scale domains at the oxide–solution interface. Similar structural changes have also been observed for perovskite oxide catalysts. May et al. (2012) showed that a series of high activity perovskites such as BSCF and $SrCo_{0.8}Fe_{0.2}O_{3-\delta}$ can quickly undergo amorphisation of their surface at OER potentials. These structural changes were also accompanied by reduced surface concentrations of $Ba^{2+}$ and $Sr^{2+}$ ions as well as increased OER current densities. Interestingly, perovskites with lower activities did not undergo similar structural changes, indicating that these amorphous surface structures play an important role in the catalytic activity of these oxides.

Certainly, knowledge of the chemical and structural properties of the surface oxide phase is important for a complete description of OER catalysis. As a whole, these studies highlight how restructuring of the oxide surface during the OER can lead to active chemistry which is significantly different from the bulk material. Understanding how these changes impact the OER activity is critical for the design of new and improved catalysts. In this regard, the surfaquo group concept offers a molecular like understanding of the activity of oxide catalysts; the amphoteric character of the anodic oxides implies that it is more realistic to view the OER active sites in terms of anionic surface complexes with molecular properties rather than the traditional viewpoint of stoichiometric units of the bulk oxide material. Within this framework catalyst design principles such as the hydrogen acceptor and partial-charge transfer models may be productively understood along the lines of well-established molecular concepts, such as intramolecular hydrogen bonding and electron withdrawing groups. Insights such as these, bridging the fields of homogeneous and heterogeneous OER catalysis, hold the key to further advancement in the mechanistic understanding of the OER and a more intuitive design of enhanced oxygen evolution catalysts.

## 2.6 Conclusion

The OER is a critical obstacle to the widespread development of a solar to fuel energy conversion infrastructure. Optimising the kinetics of the OER is crucial if the commercial potential of promising water-splitting technologies for the generation of hydrogen as a renewable fuel are to be maximised. Accordingly, the development of efficient, abundant, and inexpensive oxygen evolution catalysts is essential and the formulation of a function led design strategy would have a broad impact on renewable energy research.

The central challenge for oxygen evolution research is understanding the mechanistic details and structural motifs required for efficient catalysis. Catalyst design requires knowledge not only of the intrinsic activity of the catalytic material but also the structure and stability of the oxide surface. Significant breakthroughs will come only from an open-minded multidisciplinary approach. The OER remains a fundamental challenge for electrochemistry and a thorough electrokinetic analysis should underpin any mechanistic study of the OER. From this platform, a combined theoretical and spectroscopic study can provide vital insight into the reactive structures. Theoretical calculations have contributed significantly to our understanding of reactivity at well-defined oxides and will continue to be an important tool. But the active oxides are not well-defined, undergoing dynamic structural changes, and direct spectroscopic analysis of the reactive surface is currently lacking. In this respect it is envisaged that ongoing improvements in in-situ spectroscopic techniques will be a major source of advancement in the understanding of structure–function relationships.

Ultimately, uncovering the structural and chemical features of the active site is the key to revealing oxygen evolution catalyst design principles. Throughout this chapter the discussion has signposted a move towards a microscopic consideration of catalyst activity in terms of the atomic or molecular scale properties of the layer. Certainly there is an emerging opinion that future developments in the study of the OER will come from the integration of the allied fields of heterogeneous surface and homogeneous molecular catalysis. Indeed, it seems that the molecular chemistry of the oxide–solution interface—the surfaquo group?—determines the chemistry of the oxygen evolution catalyst.

**Acknowledgements** The research described here has emanated in part from projects conducted with the financial support of Science Foundation Ireland (SFI) under grant number SFI/10/IN.1/I2969. RLD also wishes to acknowledge the Irish Research Council (IRC) for a Government of Ireland Postdoctoral Fellowship GOIPD/2014/120.

## References

Akimov AV, Muckerman JT, Prezhdo OV (2013) Nonadiabatic dynamics of positive charge during photocatalytic water splitting on GaN(10-10) surface: charge localization governs splitting efficiency. J Am Chem Soc 135:8682–8691

Albery WJ (1975) Electrode kinetics. Clarendon, Oxford, p 41

Balasubramanian M, Melendres CA, Mini S (2000) X-ray absorption spectroscopy studies of the local atomic and electronic structure of iron incorporated into electrodeposited hydrous nickel oxide films. J Phys Chem B 104:4300–4306

Ballhausen CJ, Gray HB (1962) The electronic structure of the vanadyl ion. Inorg Chem 1:111–122

Bard AJ, Faulkner LR (2000) Electrochemical methods: fundamentals and applications, 2nd edn. Wiley, New York, pp 87–136

Bediako DK, Lassalle-Kaiser B, Surendranath Y, Yano J, Yachandra VK, Nocera DG (2012) Structure-activity correlations in a nickel-borate oxygen evolution catalyst. J Am Chem Soc 134:6801–6809

Bediako DK, Surendranath Y, Nocera DG (2013) Mechanistic studies of the oxygen evolution reaction mediated by a nickel–borate thin film electrocatalyst. J Am Chem Soc 135:3662–3674

Betley TA, Wu Q, Van Voorhis T, Nocera DG (2008) Electronic design criteria for O-O bond formation via metal-oxo complexes. Inorg Chem 47:1849–1861

Bian W, Yang Z, Strasser P, Yang R (2014) A $CoFe_2O_4$/graphene nanohybrid as an efficient bi-functional electrocatalyst for oxygen reduction and oxygen evolution. J Power Sources 250:196–203

Bisquert J, Garcia-Belmonte G, Fabregat-Santiago F, Ferriols NS, Bogdanoff P, Pereira EC (2000a) Doubling exponent models for the analysis of porous film electrodes by impedance. Relaxation of $TiO_2$ nanoporous in aqueous solution. J Phys Chem B 104:2287–2298

Bisquert J, Garcia Belmonte G, Fabregat Santiago F, Ferriols NS, Yamashita M, Pereira EC (2000b) Application of a distributed impedance model in the analysis of conducting polymer films. Electrochem Commun 2:601–605

Bligaard T, Norskov JK (2007) Ligand effects in heterogeneous catalysis and electrochemistry. Electrochim Acta 52:5512–5516

Bligaard T, Nørskov JK, Dahl S, Matthiesen J, Christensen CH, Sehested J (2004) The Brønsted-Evans-Polanyi relation and the volcano curve in heterogeneous catalysis. J Catal 224:206–217
Bockris JO'M, Khan SUM (1993) Surface electrochemistry. Plenum Press, New York, pp 218–223
Bockris JO'M, Otagawa T (1983) Mechanism of oxygen evolution on perovskites. J Phys Chem 87:2960–2971
Bockris JOM, Otagawa T (1984) The electrocatalysis of oxygen evolution on perovskites. J Electrochem Soc 131:290–302
Bockris JO'M, Reddy AKN, Gamboa-Aldeco M (2002) Modern electrochemistry 2A, 2nd edn. Kluwer Academic Publishers, New York, p 1412
Bourgault PL, Conway BE (1960) The electrochemical behavior of the nickel oxide electrode. Part 2. Quasi-equilibrium behavior. Can J Chem 38:1557–1575
Brimblecombe R, Dismukes GC, Swiegers GF, Spiccia L (2009) Molecular water oxidation catalysts for photoelectrochemical cells. Dalton Trans 43:9374–9384
Burke LD, Lyons MEG (1986) The formation and stability of hydrous oxide films on iron under potential cycling conditions in aqueous solution at high pH. J Electroanal Chem 198:347–368
Burke LD (1986) M.E.G. Lyons. In: Bockris JOM, White RE, Conway BE (eds) Modem aspects of electrochemistry, vol 18. Plenum Publ. Corp, New York, pp 169–248
Burke LD, O'Sullivan EJM (1978) Enhanced oxide growth at a rhodium surface in base under potential cycling conditions. J Electroanal Chem 93:11–18
Burke LD, O'Sullivan EJM (1981) Oxygen gas evolution on hydrous oxides – an example of three-dimensional electrocatalysis? J Electroanal Chem 117:155–160
Burke LD, Twomey TAM (1984) Influence of the acid/base character of the surface on the electrocatalytic behavior of both nickel and nickel oxide anodes, with particular reference to oxygen gas evolution. J Electroanal Chem 167:285–290
Burke LD, Whelan DP (1984) A voltammetric investigation of the charge storage reactions of hydrous iridium oxide layers. J Electroanal Chem 162:121–141
Burke LD, Lyons MEG, O'Sullivan EJM, Whelan DP (1981) Influence of hydrolysis on the redox behavior of hydrous oxide films. J Electroanal Chem 122:403–407
Burke LD, Lyons MEG, McCarthy M (1982a) Oxygen evolution and corrosion at ruthenium dioxide-based anodes. Adv Hydrogen Energy 3:267
Burke LD, Lyons MEG, Whelan DP (1982b) Influence of pH on the reduction of thick anodic oxide films on gold. J Electroanal Chem 139:131–142
Burke LD, Collins JA, Horgan MA, Hurley LM, O'Mullane AP (2000) The importance of the active states of surface atoms with regard to the electrocatalytic behaviour of metal electrodes in aqueous media. Electrochim Acta 45:4127–4134
Burke LD, O'Connell AM, O'Mullane AP (2003) The role of defects, or active states, in surface electrochemistry with particular reference to gold in neutral solution. J Appl Electrochem 33:1125–1135
Busch M, Ahlberg E, Panas I (2011a) Hydroxide oxidation and peroxide formation at embedded binuclear transition metal sites; TM = Cr, Mn, Fe, Co. Phys Chem Chem Phys 13:15062–15068
Busch M, Ahlberg E, Panas I (2011b) Electrocatalytic oxygen evolution from water on a Mn (III–V) dimer model catalyst—A DFT perspective. Phys Chem Chem Phys 13:15069–15076
Busch M, Ahlberg E, Panas I (2013a) Validation of binuclear descriptor for mixed transition metal oxide supported electrocatalytic water oxidation. Catal Today 202:114–119
Busch M, Ahlberg E, Panas I (2013b) Water oxidation on MnOx and IrOx: why similar performance? J Phys Chem C 117:288–292
Calle-Vallejo F, Inoglu NG, Su H-Y, Martinez JI, Man IC, Koper MTM, Kitchin JR, Rossmeisl J (2013) Number of outer electrons as descriptor for adsorption processes on transition metals and their oxides. Chem Sci 4:1245–1249
Calle-Vallejo F, Diaz-Morales O, Kolb M, Koper MTM (2015) Why Is bulk thermochemistry a good descriptor for the electrocatalytic activity of transition metal oxides? ACS Catal 5:869–873

Carugati A, Lodi G, Trasatti S (1981) Fractional reaction orders in oxygen evolution from acidic solutions at ruthenium oxide anodes. Mater Chem 6:255–266
Chang SH, Danilovic N, Chang K-C, Subbaraman R, Paulikas AP, Fong DD, Highland MJ, Baldo PM, Stamenkovic VR, Freeland JW, Eastman JA, Markovic NM (2014) Functional links between stability and reactivity of strontium ruthenate single crystals during oxygen evolution. Nat Commun 5:4191–4199
Chang SH, Connell JG, Danilovic N, Subbaraman R, Chang K-C, Stamenkovic VR, Markovic NM (2015) Activity–stability relationship in the surface electrochemistry of the oxygen evolution reaction. Faraday Discuss 176:125–133
Chen JG, Menning CA, Zellner MB (2008) Monolayer bimetallic surfaces: experimental and theoretical studies of trends in electronic and chemical properties. Surf Sci Rep 63:201–254
Conway BE (1995) Electrochemical oxide film formation at noble metals as a surface-chemical process. Prog Surf Sci 49:331–452
Conway BE, Bourgault PL (1962) Significance of E.M.F. decay measurements. Applications to the nickel oxide electrode. Trans Faraday Soc 58:593–607
Conway BE, Bourgault PL (1959) The electrochemical behavior of the nickel – nickel oxide electrode part 1. Kinetics of self-discharge. Can J Chem 37:292–307
Conway BE, Salomon M (1964) Electrochemical reaction orders: applications to the hydrogen- and oxygen-evolution reactions. Electrochim Acta 9:1599–1615
Conway BE, Gileadi E (1962) Kinetic theory of pseudo-capacitance and electrode reactions at appreciable surface coverage. Trans Faraday Soc 58:2493–2509
Conway BE, Bai L, Sattar MA (1987) Role of the transfer coefficient in electrocatalysis: applications to the H2 and O2 evolution reactions and the characterization of participating adsorbed intermediates. Int J Hydrogen Energy 12:607–621
Corrigan DA (1987) The catalysis of the oxygen evolution reaction by iron impurities in thin film nickel oxide electrodes. J Electrochem Soc 134:377–384
Corrigan DA, Conell RS, Fierro CA, Scherson DA (1987) In situ mossbauer study of redox processes in a composite hydroxide of iron and nickel. J Phys Chem 91:5009–5011
Crabtree GW, Dresselhaus MS, Buchanan MV (2004) The hydrogen economy. Phys Today 57:39–44
Damjanovic A, Dey A, Bockris JO'M (1966) Kinetics of oxygen evolution and dissolution on platinum electrodes. Electrochim Acta 11:791–814
Damjanovic A, Genshaw MA, Bockris JO'M (1967a) The role of hydrogen peroxide in oxygen reduction at platinum in $H_2SO_4$ solution. J Electrochem Soc 114:466–472
Damjanovic A, Genshaw MA, Bockris JO'M (1967b) The mechanism of oxygen reduction at platinum in alkaline solutions with special reference to H2O2. J Electrochem Soc 114:1107–1112
Danilovic N, Subbaraman R, Chang K-C, Chang SH, Kang Y, Snyder J, Paulikas AP, Strmcnik D, Kim YT, Myers D, Stamenkovic VR, Markovic NM (2014a) Using surface segregation to design stable Ru-Ir oxides for the oxygen evolution reaction in acidic environments. Angew Chem Int Ed 53:14016–14021
Danilovic N, Subbaraman R, Chang K-C, Chang SH, Kang YJ, Snyder J, Paulikas AP, Strmcnik D, Kim Y-T, Myers D, Stamenkovic VR, Markovic NM (2014b) Activity–stability trends for the oxygen evolution reaction on monometallic oxides in acidic environments. J Phys Chem Lett 5:2474–2478
Dau H, Limberg C, Reier T, Risch M, Roggan S, Strasser P (2010) The mechanism of water oxidation: from electrolysis via homogeneous to biological catalysis. ChemCatChem 2:724–761
Desilvestro J, Corrigan DA, Weaver MJ (1986) Spectroelectrochemistry of thin nickel hydroxide films on gold using surface-enhanced raman spectroscopy. J Phys Chem 90:6408–6411
Diaz-Morales O, Calle-Vallejo F, de Munck C, Koper MTM (2013) Electrochemical water splitting by gold: evidence for an oxide decomposition mechanism. Chem Sci 4:2334–2343
Dinga GP (1985) Hydrogen: the ultimate fuel and energy carrier. J Chem Ed 65:688–691

Doyle RL, Lyons MEG (2013a) Kinetics and mechanistic aspects of the oxygen evolution reaction at hydrous iron oxide films in base. J Electrochem Soc 160:H142–H154
Doyle RL, Lyons MEG (2013b) An electrochemical impedance study of the oxygen evolution reaction at hydrous iron oxide in base. Phys Chem Chem Phys 15:5224–5237
Doyle RL, Lyons MEG (2014a) Redox and oxygen evolution properties of nafion and single walled carbon nanotube/hydrous iron oxide composite films. Electrocatalysis 5:114–124
Doyle RL, Lyons MEG (2014b) The mechanism of oxygen evolution at superactivated gold electrodes in aqueous alkaline solution. J Solid State Electrochem 18:3271–3286
Doyle RL, Godwin IJ, Brandon MP, Lyons MEG (2013) Redox and electrochemical water splitting catalytic properties of hydrated metal oxide electrodes. Phys Chem Chem Phys 15:13737–13783
Duan L, Bozoglian F, Mandal S, Stewart B, Privalov T, Llobet A, Sun L (2012) A molecular ruthenium catalyst with water-oxidation activity comparable to that of photosystem II. Nature 4:418–423
Dumesic JA, Huber GW, Boudart M (2008) In: Ertl G, Knözinger H, Schüth F, Weitkamp J (eds) Handbook of heterogeneous catalysis, 2nd edn. Wiley, Wienheim, p 1445
Egdell RG, Goodenough JB, Hamnett A, Naish CJ (1983) Electrochemistry of ruthenates part 1. Oxygen reduction on pyrochlore ruthenates. J Chem Soc Faraday Trans 79:893–912
El-Deab MS, Awad MI, Mohammad AM, Ohsaka T (2007) Enhanced water electrolysis: electrocatalytic generation of oxygen gas at manganese oxide nanorods modified electrodes. Electrochem Commun 9:2082–2087
Fabbri E, Habereder A, Waltar K, Kötz R, Schmidt TJ (2014) Developments and perspectives of oxide-based catalysts for the oxygen evolution reaction. Catal Sci Technol 4:3800–3821
Fachinotti E, Guerrini E, Tavares AC, Trasatti S (2007) Electrocatalysis of H2 evolution by thermally prepared ruthenium oxide. Effect of precursors: Nitrate vs. chloride. J Electroanal Chem 600:103–112
Farrow CL, Bediako DK, Surendranath Y, Nocera DG, Billinge SJL (2013) Intermediate-range structure of self-assembled cobalt-based oxygen-evolving catalyst. J Am Chem Soc 135:6403–6406
Frydendal R, Busch M, Halck NB, Paoli EA, Krtil P, Chorkendorff I, Rossmeisl J (2015) Enhancing activity for the oxygen evolution reaction: the beneficial interaction of gold with manganese and cobalt oxides. ChemCatChem 7:149–154
Galán-Mascarós JR (2015) Water oxidation at electrodes modified with earth-abundant transition-metal catalysts. ChemElectroChem 2:37–50
García-Mota M, Vojvodic A, Metiu H, Man IC, Su H-Y, Rossmeisl J, Nørskov JK (2011) Tailoring the activity for oxygen evolution electrocatalysis on Rutile TiO2(110) by transition-metal substitution. ChemCatChem 3:1607–1611
Gileadi E (1993) Electrode Kinetics, VCH, New York, pp 140–144
Godwin IJ, Doyle RL, Lyons MEG (2014) The mechanism of oxygen reactions at porous oxide electrodes III. Water oxidation catalysis at RuO2/NiO mixed oxide electrodes. J Electrochem Soc 161:F906–F917
Gong M, Li Y, Wang H, Liang Y, Wu JZ, Zhou J, Wang J, Regier T, Wei F, Dai H (2013) An advanced Ni-Fe layered double hydroxide electrocatalyst for water oxidation. J Am Chem Soc 135:8452–8455
Gorlin Y, Chung C-J, Benck JD, Nordlund D, Seitz L, Weng T-C, Sokaras D, Clemens BM, Jaramillo TF (2014) Understanding interactions between manganese oxide and gold that lead to enhanced activity for electrocatalytic water oxidation. J Am Chem Soc 136:4920–4926
Grätzel M (2005) Mesoscopic solar cells for electricity and hydrogen production from sunlight. Chem Lett 34:8–13
Greeley J, Markovic NM (2012) The road from animal electricity to green energy: combining experiment and theory in electrocatalysis. Energy Environ Sci 5:9246–9256
Guerrini E, Chen H, Trasatti S (2007) Oxygen evolution on aged $IrO_x$/Ti electrodes in alkaline solutions. J Solid State Electrochem 11:939–945

Guidelli R, Compton RG, Feliu JM, Gileadi E, Lipkowski J, Schmickler W, Trasatti S (2014a) Defining the transfer coefficient in electrochemistry: an assessment (IUPAC Technical Report). Pure Appl Chem 86:245–258

Guidelli R, Compton RG, Feliu JM, Gileadi E, Lipkowski J, Schmickler W, Trasatti S (2014b) Definition of the transfer coefficient in electrochemistry (IUPAC Recommendations 2014). Pure Appl Chem 86:259–262

Halck NB, Petrykin V, Krtil P, Rossmeisl J (2014) Beyond the volcano limitations in electrocatalysis - oxygen evolution reaction. Phys Chem Chem Phys 16:13682–13688

Hall DE (1983) Ni(OH)2-impregnated anodes for alkaline water electrolysis. J Electrochem Soc 130:317–321

Hansen HA, Man IC, Studt F, Abild-Pedersen F, Bligaard T, Rossmeisl J (2010) Electrochemical chlorine evolution at Rutile oxide 110 surfaces. Phys Chem Chem Phys 12:283–290

Harrington DA, van den Driessche P (2011) Mechanism and equivalent circuits in electrochemical impedance spectroscopy. Electrochim Acta 56:8005–8013

Häussinger P, Lohmüller R, Watson AM (2000) Ullmann's encyclopedia of industrial chemistry. Wiley, Weinheim

Hoare JP (1967) Oxygen electrode on noble metals. Adv Electrochem Electrochem Eng 6:201–288

Hoare JP (1968) The electrochemistry of oxygen. Interscience, New York, pp 82–91

Hrussanova A, Guerrini E, Trasatti S (2004) Thermally prepared Ti/RhOx electrodes IV: O2 evolution in acid solution. J Electranal Chem 564:151–157

Huynh M, Bediako DK, Nocera DG (2014) A functionally stable manganese oxide oxygen evolution catalyst in acid. J Am Chem Soc 136:6002–6010

Inoglu N, Kitchin JR (2011) Identification of sulfur-tolerant bimetallic surfaces using DFT parametrized models and atomistic thermodynamics. ACS Catal 1:399–407

Joya KS, Joya YF, Ocakoglu K, van de Krol R (2013) Water-splitting catalysis and solar fuel devices: artificial leaves on the move. Angew Chem Int Ed 52:10426–10437

Kanan MW, Nocera DG (2008) In situ formation of an oxygen-evolving catalyst in neutral water containing phosphate and $Co2^{+}$. Science 321:1072–1075

Kanan MW, Yano J, Surendranath Y, Dinca M, Yachandra VK, Nocera DG (2010) Structure and valency of a cobalt–phosphate water oxidation catalyst determined by in situ X-ray spectroscopy. J Am Chem Soc 132:13692–13701

Katsounaros I, Cherevko S, Zeradjanin AR, Mayrhofer KJ (2014) Oxygen electrochemistry as a cornerstone for sustainable energy conversion. J Angew Chem Int Ed 53:102–121

Kharche N, Hybersten MS, Muckerman JT (2014) Computational investigation of structural and electronic properties of aqueous interfaces of GaN, ZnO, and a GaN/ZnO alloy. Phys Chem Chem Phys 16:12057–12066

Kim S, Tryk DA, Antonio MR, Carr R, Scherson DJ (1994) In situ X-ray absorption fine structure studies of foreign metal ions in nickel hydrous oxide electrodes in alkaline electrolytes. Phys Chem 98:10269–10276

Kim TW, Choi K-S (2014) Nanoporous $BiVO_4$ photoanodes with dual-layer oxygen evolution catalysts for solar water splitting. Science 343:990–994

Kinoshita K (1992) Electrochemical oxygen technology. Wiley, New York

Klahr B, Gimenez S, Fabregat-Santiago F, Hamann T, Bisquert J (2012a) Water oxidation at hematite photoelectrodes: the role of surface states. J Am Chem Soc 134:4294–4302

Klahr B, Gimenez S, Fabregat-Santiago F, Bisquert J, Hamann TW (2012b) Photoelectrochemical and impedance spectroscopic investigation of water oxidation with "Co-Pi"-coated hematite electrodes. J Am Chem Soc 134:16693–16700

Kobussen AGC, Broers GHJ (1981) The oxygen evolution on $La_{0.5}Ba_{0.5}CoO_3$: theoretical impedance behaviour for a multi-step mechanism involving two adsorbates. J Electroanal Chem 126:221–240

Koper MTM (2011) Thermodynamic theory of multi-electron transfer reactions: implications for electrocatalysis. J Electroanal Chem 660:254–260

Koper MTM (2013) Analysis of electrocatalytic reaction schemes: distinction between rate-determining and potential-determining steps. J Solid State Electrochem 17:339–344
Koper MTM, Heering HA (2010) In: Wiekowski JKNA (ed) Fuel cell science: theory, fundamentals and bio-catalysis. Wiley, New York, pp 71–110
Krasil'shchikov AI (1963) Intermediate stages in the anodic evolution of oxygen. Zh Fiz Khim 37:531
Landon J, Demeter E, Inoglu N, Keturakis C, Wachs IE, Vasic R, Frenkel AI, Kitchin JR (2012) Spectroscopic characterization of mixed Fe-Ni oxide electrocatalysts for the oxygen evolution reaction in alkaline electrolytes. ACS Catal 2:1793–1801
Lee DU, Kim BJ, Chen ZW (2013) One-pot synthesis of a mesoporous $NiCo_2O_4$ nanoplatelet and graphene hybrid and its oxygen reduction and evolution activities as an efficient bi-functional electrocatalyst. J Mater Chem A 1:4754–4762
Lee SW, Carlton C, Risch M, Surendranath Y, Chen S, Furutsuki S, Yamada A, Nocera DG, Shao-Horn Y (2012) The nature of lithium battery materials under oxygen evolution reaction conditions. J Am Chem Soc 134:16959–16962
Lewis NS, Nocera DG (2006) Powering the planet: chemical challenges in solar energy utilization. Proc Natl Acad Sci U S A 103:15729–15735
Liang YY, Li YG, Wang HL, Zhou JG, Wang J, Regier T, Dai HJ (2011) $Co_3O_4$ nanocrystals on graphene as a synergistic catalyst for oxygen reduction reaction. Nat Mater 10:780–786
Liang YY, Wang HL, Zhou JG, Li YG, Wang J, Regier T, Dai HJ (2012) Covalent hybrid of spinel manganese-cobalt oxide and graphene as advanced oxygen reduction electrocatalysts. J Am Chem Soc 134:3517–3523
Liao P, Keith JA, Carter EA (2012) Water oxidation on pure and doped hematite (0001) surfaces: prediction of Co and Ni as effective dopants for electrocatalysis. J Am Chem Soc 134:13296–13309
Lodi G, Sivieri E, De Battisti A, Trasatti S (1978) Ruthenium dioxide-based film electrodes. III. Effect of chemical composition and surface morphology on oxygen evolution in acid solutions. J Appl Electrochem 8:135–143
Louie MW, Bell AT (2013) An investigation of thin-film Ni–Fe oxide catalysts for the electrochemical evolution of oxygen. J Am Chem Soc 135:12329–12337
Lu X, Ng YH, Zhao C (2014) Gold nanoparticles embedded within mesoporous cobalt oxide enhance electrochemical oxygen evolution. ChemSusChem 7:82–86
Lyons MEG, Brandon MP (2009) Redox switching and oxygen evolution electrocatalysis in polymeric iron oxyhydroxide films. Phys Chem Chem Phys 11:2203–2217
Lyons MEG, Brandon MP (2010) A comparative study of the oxygen evolution reaction on oxidised nickel, cobalt and iron electrodes in base. J Electroanal Chem 641:119–130
Lyons MEG, Burke LD (1987) Mechanism of oxygen reactions at porous oxide electrodes. Part 1. Oxygen evolution at ruthenium dioxide and ruthenium tin oxide ($Ru_xSn_{1-x}O_2$) electrodes in alkaline solution under vigorous electrolysis conditions. J Chem Soc Faraday Trans 83:299–321
Lyons MEG, Floquet S (2011) Mechanism of oxygen reactions at porous oxide electrodes. Part 2—Oxygen evolution at $RuO_2$, $IrO_2$ and $IrxRu_{1-x}O_2$ electrodes in aqueous acid and alkaline solution. Phys Chem Chem Phys 13:5314–5335
Lyons MEG, Doyle RL, Brandon MP (2011) Redox switching and oxygen evolution at oxidized metal and metal oxide electrodes: iron in base. Phys Chem Chem Phys 13:21530–21551
Lyons MEG, Russell L, O'Brien M, Doyle RL, Godwin I, Brandon MP (2012a) Redox switching and oxygen evolution at hydrous oxy-hydroxide modified nickel electrodes in aqueous alkaline solution: effect of hydrous oxide thickness and base concentration. Int J Electrochem Sci 7:2710–2763
Lyons MEG, Cakara A, O'Brien P, Godwin I, Doyle RL (2012b) Redox, pH sensing and electrolytic water splitting properties of electrochemically generated nickel hydroxide thin films in aqueous alkaline solution. Int J Electrochem Sci 7:11768–11795

Lyons MEG, Doyle RL, Fernandez D, Godwin IJ, Browne MP, Rovetta A (2014) The mechanism and kinetics of electrochemical water oxidation at oxidised metal and metal oxide electrodes. Part 2. The surfaquo group mechanism. A mini review. Electrochem Commun 45:56–59

Man IC, Su H-Y, Calle-Vallejo F, Hansen HA, Martinez JI, Inoglu NG, Kitchin J, Jaramillo TF, Nørskov JK, Rossmeisl J (2011) Universality in oxygen evolution electrocatalysis on oxide surfaces. ChemCatChem 3:1159–1165

Marinia S, Salvi P, Nelli P, Pesentia R, Villa M, Berrettoni M, Zangaric G, Kiros Y (2012) Advanced alkaline water electrolysis. Electrochim Acta 82(384–391)

Markovic NM, Ross PN (2002) Surface science studies of model fuel cell electrocatalysts. Surf Sci Rep 45:121–229

Marx D, Tuckerman ME, Hutter J, Parrinello M (1999) The nature of the hydrated excess proton in water. Nature 397:601–604

Matsumoto Y, Sato E (1986) Electrocatalytic properties of transition metal oxides for oxygen evolution reaction. Mater Chem Phys 14:397–426

Mavrikakis M, Hammer B, Nørskov JK (1998) Effect of strain on the reactivity of metal surfaces. Phys Rev Lett 81:2819–2822

May KJ, Carlton CE, Stoerzinger KA, Risch M, Suntivich J, Lee Y-L, Grimaud A, Shao-Horn Y (2012) The influence of oxygen evolution upon water oxidation on the surface of perovskite oxide catalysts. J Phys Chem Lett 3:3264–3270

McAlpin JG, Surendranath Y, Dinca M, Stich TA, Stoian SA, Casey WH, Nocera DG, Britt RD (2010) EPR evidence for Co(IV) species produced during water oxidation at neutral pH. J Am Chem Soc 132:6882–6883

McCrory CCL, Jung SH, Peters JC, Jaramillo TF (2013) Benchmarking heterogeneous electrocatalysts for the oxygen evolution reaction. J Am Chem Soc 135:16977–16987

McDonald AR, Que L (2011) Iron–oxo complexes: elusive iron(V) species identified. Nature 3:761–762

McDonald JJ, Conway BE (1962) The role of surface films in the kinetics of oxygen evolution at Pd + Au alloy electrodes. Proc Roy Soc Lond A 269:419–440

Merrill MD, Dougherty RC (2008) Metal oxide catalysts for the evolution of $O_2$ from $H_2O$. J Phys Chem C 112:3655–3666

Meyer RE (1960) Cathodic processes on passive zirconium. J Electrochem Soc 107:847–853

Michas A, Andolfatto F, Lyons MEG, Durand R (1992) Gas evolution reactions at conductive metallic oxide electrodes for solid polymer electrolyte water electrolysis. Key Eng Mater 72–74:535–549

Mohammad AM, Awad MI, El-Deab MS, Okajima T, Ohsaka T (2008) Electrocatalysis by nanoparticles: optimisation of the loading level and operating pH for the oxygen evolution at crystallographically orientated manganese oxide nanorods modified electrodes. Electrochim Acta 53:4351–4358

Montoya JH, Garcia-Mota M, Nørskov JK, Vojvodic A (2015) Theoretical evaluation of the surface electrochemistry of perovskites with promising photon absorption properties for solar water splitting. Phys Chem Chem Phys 17:2634–2640

Nocera DG (2012) The artificial leaf. Acc Chem Res 45:767–776

Nørskov JK, Bligaard T, Hvolbaek B, Abild-Petersen F, Chorkendorff I, Christensen CH (2008) The nature of the active site in heterogeneous metal catalysis. Chem Soc Rev 37:2163–2171

Nørskov JK, Bligaard T, Rossmeisl J, Christensen CH (2009) Towards the computational design of solid catalysts. Nat Chem 1:37–46

Novell-Leruth G, Carchini G, Lópeza N (2013) On the properties of binary rutile $MO_2$ compounds, M = Ir, Ru, Sn, and Ti: A DFT study. J Chem Phys 138:194706–194715

Nowicka AM, Hasse U, Sievers G, Donten M, Stojek Z, Fletcher S, Scholz F (2010) Selective knockout of gold active sites. Angew Chem Int Ed 49:3006–3009

O'Grady W, Iwakura C, Huang J, Yeager E (1974) In: Breiter MW (ed) Proceedings of the symposium on electrocatalysis. The Electrochemical Society Inc., Pennington, NJ, p 286

Ohi J (2005) Hydrogen energy cycle: an overview. J Mater Res 20:3180–3187

O'Sullivan EJM, Burke LD (1990) Kinetics of oxygen gas evolution on hydrous rhodium oxide films. J Electrochem Soc 137:466–471
Parsons R (1951) General equations for the kinetics of electrode processes. Trans Faraday Soc 47:1332–1344
Parsons R (1958) Rate of electrolytic hydrogen evolution and the heat of adsorption of hydrogen. Trans Faraday Soc 54:1053–1063
Parsons R (1961) In: Delahay P (ed) Advances in electrochemistry and electrochemical engineering, vol 1. Interscience, New York, p 1
Petrykin V, Macounova K, Shlyakhtin OA, Krtil P (2010) Tailoring the selectivity for electrocatalytic oxygen evolution on ruthenium oxides by zinc substitution. Angew Chem Int Ed 49:4813–4815
Pourbaix M (1974) Atlas of electrochemical equilibria in aqueous solutions, 2nd edn. National Association of Corrosion Engineers, Houston, TX
Rebouillat S, Lyons MEG, Brandon MP, Doyle RL (2011) Paving the way to the integration of smart nanostructures Part 2 - Nanostructured metal oxides for electrocatalysis and energy conversion. Int J Electrochem Sci 6:5830–5917
Reier T, Oezaslan M, Strasser P (2012) Electrocatalytic Oxygen Evolution Reaction (OER) on Ru, Ir, and Pt catalysts: a comparative study of nanoparticles and bulk materials. ACS Catal 2:1765–1772
Rios E, Chartier P, Gautier JL (1999) Oxygen evolution electrocatalysis in alkaline medium at thin $Mn_xCo_{3-x}O_4$ ($0 \leq x \leq 1$) spinel films on glass/$SnO_2$:F prepared by spray pyrolysis. Solid State Sci 1:267–277
Rossmeisl J (2013) In: Schlogl R (ed) Chemical energy storage. De Gruyter Graduate, Berlin, pp 151–162
Rossmeisl J, Logadottir A, Nørskov JK (2005) Electrolysis of water on (oxidized) metal surfaces. Chem Phys 319:178–184
Rossmeisl J, Qu Z-W, Zhu H, Kroes G-J, Nørskov JK (2007a) Electrolysis of water on oxide surfaces. J Electroanal Chem 607:83–89
Rossmeisl J, Dimitrievski K, Siegbahn P, Nørskov JK (2007b) Comparing electrochemical and biological water splitting. J Phys Chem C 111:18821–18823
Rossmeisl J, Karlberg GS, Jaramillo T, Norskov JK (2008) Steady state oxygen reduction and cyclic voltammetry. Faraday Discuss 140:337–346
Ruban AV, Skriver HL, Nørskov JK (1999) Surface segregation energies in transition-metal alloys. Phys Rev B 59:15990–16000
Ruetschi P, Delahay P (1955) Potential at zero charge for reversible and ideal polarized electrodes. J Chem Phys 23:697–699
Sabatier P (1911) Hydrogenation and dehydrogenation by catalysis. Ber Dtsch Chem Ges 44:1984–2001
Sato N (1998) Electrochemistry at metal and semiconductor electrodes. Elsevier, Amsterdam, pp 181–184
Schlogl R (2010) The role of chemistry in the energy challenge. ChemSusChem 3:209–222
Schwab GM (1981) In: Anderson JR, Boudart M (eds) Catalysis – science and technology, vol 2. Springer, Berlin, p 4
Schweitzer GK, Pesterfield LL (2010) The aqueous chemistry of the elements. Oxford University Press, Oxford
Shen X, Small YA, Wang J, Allen PB, Fernandez-Serra MV, Hybertsen MS, Muckerman JT (2010) Photocatalytic water oxidation at the GaN (1010) – water interface. J Phys Chem C 114:13695–13704
Sheng WC, Gasteiger HA, Shao-Horn Y (2010) Hydrogen oxidation and evolution reaction kinetics on platinum: acid vs alkaline electrolytes. J Electrochem Soc 157:B1529–B1536
Singh NK, Tiwari SK, Anitha KL, Singh RN (1996) Electrocatalytic properties of spinel-type $Mn_xFe_{3-x}O_4$ synthesized below 100°C for oxygen evolution in KOH solutions. J Chem Soc Faraday Trans 92:2397–2400

Smith RDL, Prévot MS, Fagan RD, Zhang Z, Sedach PA, Siu MKJ, Trudel S, Berlinguette CP (2013a) Photochemical route for accessing amorphous metal oxide materials for water oxidation catalysis. Science 340:60–63

Smith RDL, Prévot MS, Fagan RD, Trudel S, Berlinguette CP (2013b) Water oxidation catalysis: electrocatalytic response to metal stoichiometry in amorphous metal oxide films containing iron, cobalt, and nickel. J Am Chem Soc 135:11580–11586

Somorjai GA (1996) Modern surface science and surface technologies: an introduction. Chem Rev 96:1223–1235

Su H-Y, Gorlin Y, Man IC, Calle-Vallejo F, Nørskov JK, Jaramillo TF, Rossmeisl J (2012) Identifying active surface phases for metal oxide electrocatalysts: a study of manganese oxide bi-functional catalysts for oxygen reduction and water oxidation catalysis. Phys Chem Chem Phys 14:14010–14022

Subbaraman R, Tripkovic D, Chang K-C, Strmcnik D, Paulikas AP, Hirunsit P, Chan M, Greeley J, Stamenkovic V, Markovic NM (2012) Trends in activity for the water electrolyser reactions on 3d M(Ni,Co,Fe,Mn) hydr(oxy)oxide catalysts. Nat Mater 11:550–557

Sun K, Kohyama M, Tanaka S, Takeda S (2012) Theoretical study of atomic oxygen on gold surface by Hückel theory and DFT calculations. J Phys Chem A 116:9568–9573

Suntivich J, May KJ, Gasteiger H, Goodenough JB, Shao-Horn Y (2011) A perovskite oxide optimized for oxygen evolution catalysis from molecular orbital principles. Science 334:1383–1385

Surendranath Y, Nocera DG (2012) Progress. In: Karlin KD (ed) Inorganic chemistry, vol 57. Wiley, New York, pp 505–560

Surendranath Y, Kanan MW, Nocera DG (2010) Mechanistic studies of the oxygen evolution reaction by a cobalt-phosphate catalyst at neutral pH. J Am Chem Soc 132:16501–16509

Tafel J (1904) Polarization in cathodic hydrogen evolution. Z Phys Chem 50:641–712

Taylor HS (1925) A theory of the catalytic surface. Proc Roy Soc Lond A 108:105–111

Terezo AJ, Bisquert J, Pereira EC, Garcia-Belmonte G (2001) Separation of transport, charge storage and reaction processes of porous electrocatalytic $IrO_2$ and $IrO_2/Nb_2O_5$ electrodes. J Electroanal Chem 508:59–69

Thomas JGN (1961) Kinetics of electrolytic hydrogen evolution and the adsorption of hydrogen by metals. Trans Faraday Soc 57:1603–1611

Trasatti S (1980) Electrocatalysis by oxides - attempt at a unifying approach. J Electroanal Chem 111:125–131

Trasatti S (1984) Electrocatalysis in the anodic evolution of oxygen and chlorine. Electrochim Acta 29:1503–1512

Trasatti S (1994) In: Lipkowski J, Ross PN (eds) Electrochemistry of novel materials. VCH, New York, pp 207–295

Tributsch H (2008) Photovoltaic hydrogen generation. Int J Hydrogen Energy 33:5911–5930

Trotochaud L, Ranney JK, Williams KN, Boettcher SW (2012) Solution-cast metal oxide thin film electrocatalysts for oxygen evolution. J Am Chem Soc 134:17253–17261

Trotochaud L, Young SL, Ranney JK, Boettcher SW (2014) Nickel-iron oxyhydroxide oxygen-evolution electrocatalysts: the role of intentional and incidental iron incorporation. J Am Chem Soc 136:6744–6753

Tuckerman M, Laasonen K, Sprik M, Parrinello M (1995) Ab initio molecular dynamics simulation of the solvation and transport of hydronium and hydroxyl ions in water. J Chem Phys 103:150–161

Turner JA (2004) Sustainable hydrogen production. Science 305:972–974

Valdés Á, Qu Z-W, Kroes G-J, Rossmeisl J, Nørskov JK (2008) Oxidation and photo-oxidation of water on $TiO_2$ surface. J Phys Chem C 112:9872–9879

Valdés Á, Brillet J, Grätzel M, Gudmundsdóttir H, Hansen HA, Jónsson H, Klüpfel P, Kroes G-J, Le Formal F, Man IC, Martins RS, Nørskov JK, Rossmeisl J, Sivula K, Vojvodicf A, Zäch M (2012) Solar hydrogen production with semiconductor metal oxides: new directions in experiment and theory. Phys Chem Chem Phys 14:49–70

Vassilev P, Koper MTM, van Santen RA (2002) Ab initio molecular dynamics of hydroxyl-water coadsorption on Rh(111). Chem Phys Lett 359:337–342

Viswanathan V, Pickrahn KL, Luntz AC, Bent SF, Nørskov JK (2014) Nanoscale limitations in metal oxide electrocatalysts for oxygen evolution. Nano Lett 14:5853–5857

Wang L, Zhao X, Lu Y, Xu M, Zhang D, Ruoff RS, Stevenson KJ, Goodenough JB (2011) $CoMn_2O_4$ spinel nanoparticles grown on graphene as bifunctional catalyst for lithium-air batteries. J Electrochem Soc 158:A1379–A1382

Wang L-P, Wu Q, Van Voorhis T (2010) Acid-base mechanism for ruthenium water oxidation catalysts. Inorg Chem 49:4543–4553

Willems H, Kobussen AGC, De Wit JHW, Broers GHJ (1984) The oxygen evolution reaction on cobalt. Part 1. Reaction order experiments and impedance measurements. J Electroanal Chem 170:227–242

Yeo BS, Bell AT (2011) Enhanced activity of gold-supported cobalt oxide for the electrochemical evolution of oxygen. J Am Chem Soc 133:5587–5593

Yeo BS, Bell AT (2012) In situ raman study of nickel oxide and gold-supported nickel oxide catalysts for the electrochemical evolution of oxygen. J Phys Chem C 116:8394–8400

Zeng K, Zhang D (2010) Recent progress in alkaline water electrolysis for hydrogen production and applications. Prog Energy Combust Sci 36:307–326

Zhou L (2005) Progress and problems in hydrogen storage methods. Renew Sustain Energy Rev 9:395–408

# Chapter 3
# Hydrogen and $CO_2$ Reduction Reactions: Mechanisms and Catalysts

**Pitchaimuthu Sudhagar, Nitish Roy, Raman Vedarajan, Anitha Devadoss, Chiaki Terashima, Kazuya Nakata, and Akira Fujishima**

## 3.1 Introduction

*Winning the future with clean energy* has been the global juggernaut to meet the galloping energy consumption. The most abundantly available, solar energy offers a desirable alternative to the existing fossil fuel based technologies. Solar energy is also considered as the panacea for all the energy trouble due to its inherently decentralized and inexhaustible nature (Turner 2004; Lewis and Nocera 2006). Solar energy can be transformed into chemical energy by means of natural photosynthesis. For fuel generation, solar energy can be utilized to drive the photovoltaic cells, resulting in electricity generation that can be used to power the vehicles (Heller 1981; Maugh 1983; Turner 2014). However, increasing the production volume, storage, and transportation still remains challenging. Alternatively, the electricity produced from photovoltaic cells can be converted into fuels through electrocatalysis. Thus, the development of superficial solar energy conversion systems that could convert solar-to-chemical energy will offer solution to the conventional photovoltaic electricity by offering more storage options. Recent

P. Sudhagar (✉) • N. Roy • A. Devadoss
Photocatalysis International Research Center (PIRC), Tokyo University of Science, 2641 Yamazaki, Noda, Chiba 278-8510, Japan
e-mail: vedichi@gmail.com

R. Vedarajan
School of Materials Science, Japan Advanced Institute of Science and Technology (JAIST), 1-1 Aasahidai, Nomi, Ishikawa 923-1211, Japan

C. Terashima • K. Nakata (✉) • A. Fujishima
Photocatalysis International Research Center (PIRC), Tokyo University of Science, 2641 Yamazaki, Noda, Chiba 278-8510, Japan

Research Institute for Science & Technology, Tokyo University of Science, 2641 Yamazaki, Noda, Chiba 278-8510, Japan
e-mail: nakata@rs.tus.ac.jp



S. Giménez, J. Bisquert (eds.), *Photoelectrochemical Solar Fuel Production*, DOI 10.1007/978-3-319-29641-8_3

advances in coupling solar energy with electrochemical process into photoelectrocatalytic devices have produced fuel either from water or biomass. The advantages of storing chemical energy in the form of molecules (hydrogen, oxygen, carbon monoxide, and methane) compared to electricity storage opens a new paradigm in renewable fuels not only for transport and electricity generation but also for the chemical feedstock used in industrial sectors ranging from petrochemicals and fertilizers to plastics and pharmaceuticals.

Since the pioneering discovery of the photoelectric effect by Becquerel in 1839, photoelectrochemistry has drawn substantial research scrutiny in various fields. However, the possibility of producing hydrogen and oxygen from water using photoelectrochemical systems was first demonstrated by Honda and Fujishima in the year 1972 by using $TiO_2$ as the photoanode (Fujishima and Honda 1972). Following the seminal report by Fujishima et al. on PEC water oxidation, also termed as "artificial photosynthesis," many see hydrogen as a viable and clean source of power due to its extraordinarily high energy density and the possibility to be stored in pressurized bottles or as metal hydrides. As a comparison, 3 kg of $H_2$ corresponds to about 100 kWh of chemical energy, whereas 450 kg of lithium ion battery is needed to produce same quantity of electrical energy. Nonetheless, there still exist significant challenges to design sustainable and economic means of producing and storing hydrogen. As a conception, if all the fossil fuel based electricity generation can be replaced with a renewable source, and thus carbon emission into atmosphere would be avoided; perhaps, the global warming threat can be greatly minimized. This has opened a new paradigm in research on the development of low-cost processes for generating fuels from sunlight, water and $CO_2$ for a truly sustainable energy supply for mankind by chemical means.

In order to achieve the potential benefits of integrating hydrogen into the energy infrastructure, several challenges must be overcome. Developing low cost catalytic components with substantial stability is main stream in artificial photosynthesis. In this chapter we focus on *advanced dark catalysts* as well as *light driven photocatalytic materials* for applying in renewable solar fuel cells using water and $CO_2$ as feedstock, respectively. On this ground, this chapter is designed in two parts. The first part discusses the recent strategies in developing hydrogen evolution reaction materials which can act as nonprecious catalysts for solar fuel cells. In the second part, we describe photocatalytic materials and their underlying mechanism for carbon dioxide reduction.

## 3.2 Theory: Storage of Fuel from PEC Water Splitting

The photoelectrocatalytic water splitting was discovered by Fujishima and Honda (Fujishima and Honda 1972). Figure 3.1 depicts the schematic illustration of Fujishima–Honda photoelectrochemical (PEC) water splitting cell. A conventional PEC cell consists of two indispensable parts (a) light driven photocatalytic materials assembled on current collectors (mostly used transparent conducting oxide (TCO) substrates) as working electrodes (anode compartment), and (b) metal

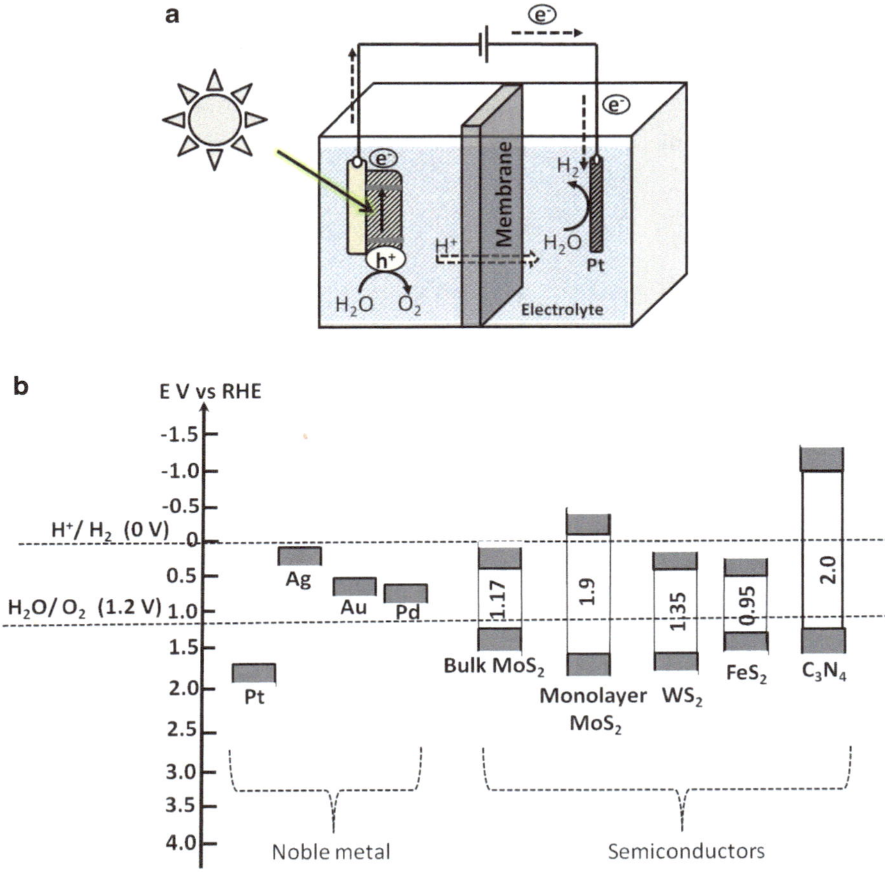


**Fig. 3.1** (**a**) Schematic illustration of solar light driven water splitting cells [photocatalytic semiconductor is placed in the anode compartment and metal based hydrogen evolution catalyst (platinum) connected in the cathode compartment]; (**b**) energetic structure of metal and semiconductor based HER catalyst in producing hydrogen evolution at 0 V vs. RHE at pH = 7

electrocatalyst (platinum) as counter electrode (cathode compartment). In between the anode and cathode compartments aqueous electrolyte is filled and separate by a proton conducting membrane. Upon photo irradiation, charge carriers are photogenerated at the semiconductor material, and subsequently the photoexcited electrons are injected from the conduction band of semiconductor to charge collector terminal.

Concurrently, the photoholes at valence band (VB) of semiconductor are scavenged by electrolyte. In accord to the photoelectrocatalytic water oxidation process, the high energy of photoholes can be utilized to effectively oxidize water molecules. The conventional photoelectrocatalytic water oxidation process is as follows;

$$2H_2O + 4h^+ \Leftrightarrow 4H^+ + O_2 \left(E^0{}_{ox} = -1.229\ V\ \text{vs. NHE}\right) \quad (3.1)$$

From the above relation (3.1), four photoholes dissociate the water molecules into four protons and liberate oxygen gas. On the other hand, the extracted photoelectrons from semiconductor are further transported to the counter electrode, where they reduce the protons into hydrogen.

$$4H^+ + 4e^- \Leftrightarrow 2H_2 \left(E^0{}_{red} = +0.000V\ \text{vs. NHE}\right) \quad (3.2)$$

The applied potential to the circuit 1.23 V is thermodynamically equivalent to Gibbs free energy 237 kJ/mol that is sufficient to oxidize the water molecules.

Though platinum based catalytic counter electrode shows effective hydrogen evolution performance over a long period, the redox energy level position of platinum (1.8 V vs. RHE) is lower compared to that of water reduction potential (0 V vs. RHE) (Fig. 3.1b). Therefore, more kinetic energy ($V_{bias}$) could require operating the proton reduction into hydrogen compare to other semiconductors.

## 3.3 Hydrogen Evolution Reaction (HER)

As discussed above, direct solar-to-hydrogen technology using photochemical and photoelectrochemical water splitting is a promising strategy for sustainable hydrogen generation. But, the extensive use of Pt as the hydrogen evolution reaction (HER) catalyst constraints their practical applications due to its high cost and rarity. Subject to these necessities, the preferred choice for providing cost-competitive hydrogen relies on the development of non-noble metal catalysts and earth-abundant elements.

Exciting progress has been made in both the design and development of non-noble metal catalysts and earth abundant elements. Figure 3.2a shows the general view of the elements used as HER electrocatalysts. These elements can be categorized into three groups based on their chemical and physical properties: (a) commonly used, Pt-containing noble metal HER catalyst, (b) transition metals based HER catalyst such as Fe, Co, Ni, Cu, Mo, and W, and (c) non-metals based HER catalyst such as B, C, N, P, S, and Se. Extensive research has been devoted to produce noble-metal free HER catalyst since early '80s to match the performance of the state-of-the-art electrocatalyst, Pt. For instance, a wide range of earth abundant elements have been used for the synthesis of HER electrocatalyst (Lu et al. 2015; McCrory et al. 2015; Vesborg et al. 2015; Zou and Zhang 2015). Figure 3.2b depicts the crustal abundance of the HER electrocatalysts in weight percentage (Vesborg et al. 2015). Despite extensive effort, the electrocatalytic performance and HER efficiency were still low in their bulk form. Specifically, the past decade has seen a huge leap in reduction of the onset potential of hydrogen evolution reaction using nonprecious metals based on the metal sulfides, carbides, and

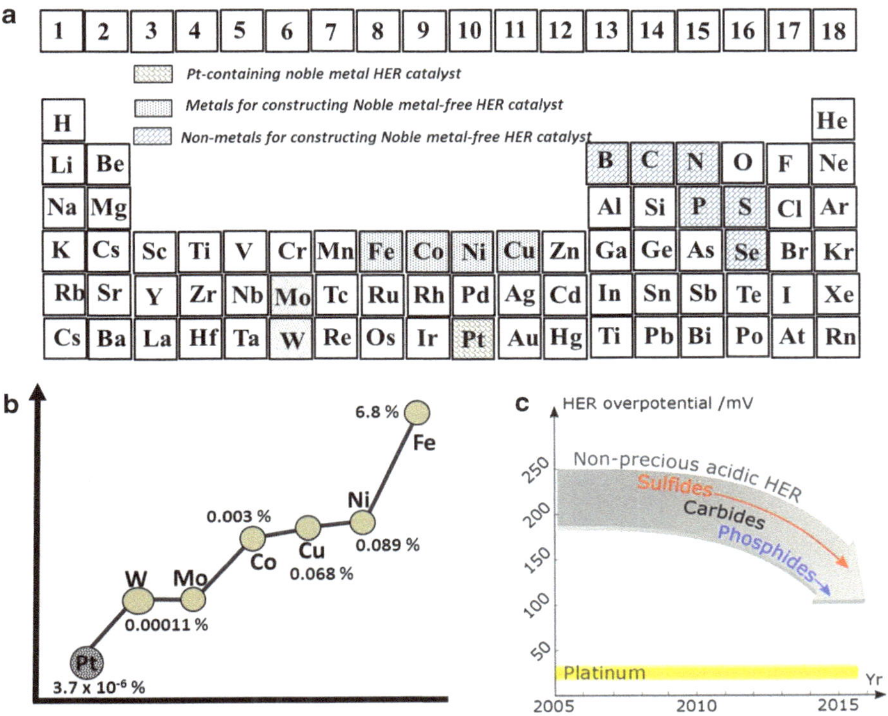


**Fig. 3.2** (**a**) Periodic table highlighted with futuristic HER metals; (**b**) strategy of crustal abundance of metals for constructing HER electrocatalysts ((**a**) and (**b**) redrawn from (Zou and Zhang 2015); (**c**) road map of nonprecious metal catalyst in reducing overpotential towards hydrogen evolution reaction (adopted from (Vesborg et al. 2015) with permission from American Chemical Society)

nitrides (Fig. 3.2c), and these results encourage replacing the platinum catalyst in solar water splitting cells.

### *3.3.1 Historical Background and Theory of HER*

Electrolysis of water to produce oxygen and hydrogen was introduced to the world in the nineteenth century. During this period, this technology gradually gained significance and became a convenient method to produce high purity hydrogen and oxygen. The splitting of water is carried out employing a two-electrode system where, hydrogen gas is produced in the cathode chamber, while oxygen gas in the anode chamber. Therefore, electrolysis of water to produce hydrogen and oxygen is the combination of two half reaction as follows:

(a) Cathodic reaction.

$$2H_3O^+ \,(l) + 2e \rightarrow H_2\,(g) + H_2O\,(l) \tag{3.3}$$

or

$$2H_2O\,(l) + 2e \rightarrow H_2\,(g) + 2OH^-\,(l) \tag{3.4}$$

(b) Anodic reaction.

$$2H_2O\,(l) \rightarrow O_2\,(g) + 4H^+ + 4e \tag{3.5}$$

or

$$4OH^-\,(l) \rightarrow O_2\,(g) + 2H_2O\,(l) + 4e \tag{3.6}$$

The overall reaction is therefore:

$$2H_2O\,(l) \rightarrow H_2\,(g) + O_2\,(g) \tag{3.7}$$

The standard potential of reaction (3.7) is 1.229 V at 25 °C which corresponds to 237.2 kJ/mol. But additional 48.65 kJ/mol thermal energy is required for water electrolysis. Therefore, a total of 285.83 kJ/mol or 1.48 V must be supplied for electrolysis of water under room temperature and pressure. In this section, focus will be only on the cathodic hydrogen evolution reaction.

Electrolysis of water takes place generally on noble metal catalysts. For example, Pt is a state-of-the-art cathode material and $IrO_2$ acts as a superior anode material. Therefore, Pt and $IrO_2$ are extensively used in electrochemical fuel cells as cathode or anode materials or as both of them. The overall Faradic efficiency depends on both the catalytic activities of cathode and anode catalysts. In theory, the total amount of hydrogen produced should be double than the amount of oxygen produced, though in general, these values may vary and hence the Faradic efficiency becomes less than unity (Conway and Tilak 2002).

Hydrogen evolution reaction takes place in the following three different steps. The first one is Volmer reaction which is nothing but the electrochemical adsorption of hydrogen on the catalyst surface and a mandatory step while other two correspond to desorption of adsorbed hydrogen. This atomic adsorbed hydrogen either desorbs via electrochemical desorption, known as Heyrovsky reaction or chemical desorption known as Tafel reaction. A step-by-step hydrogen evolution reaction is shown here:

$$M + H_3O^+ + e \rightarrow M\text{-}H_{ads} + H_2O \text{ (Volmer step in acid media)} \tag{3.8}$$

$$M + H_2O + e \rightarrow M\text{-}H_{ads} + OH^- \text{ (Volmer step in basic media)} \tag{3.9}$$

$$H_3O^+ + e + M\text{-}H_{ads} \rightarrow M + H_2 + H_2O \text{ (Heyrovsky step in acid media)} \quad (3.10)$$

$$H_2O + e + M\text{-}H_{ads} \rightarrow M + H_2 + H_2O \text{ (Heyrovsky step in basic media)} \quad (3.11)$$

$$M\text{-}H_{ads} + M\text{-}H_{ads} \rightarrow 2M + H_2 \text{ (Tafel recombination)} \quad (3.12)$$

Hydrogen evolution reaction on Pt surface by means of electrolysis of water has been extensively studied during the last century. J. Tafel corroborated the reactions pertaining to the adsorption of atomic hydrogen onto Pt surface and their subsequent desorption and recombination to produce hydrogen in acidic media or basic media; reactions (3.8) or (3.9) and (3.12). In fact, the activity of a catalyst under certain conditions is measured by Tafel slope, $b$ and exchange current density, $j_0$. The recombination of atomic hydrogen was explained by the rate-determining step on Pt surface. However, the Volmer step is necessary for HER; Tafel recombination may not follow in all cases. Tafel recombination primarily deals with few noble metals cathode catalysts with rate-limiting step with Tafel slope 29 mV/dec which is minimum potential in HER. As mentioned, Tafel recombination only follows over few noble metal catalysts as the rate-determining step, therefore, hydrogen evolution on other catalyst surfaces follows via either through Volmer step (Tafel slope ~118 mV/decade) or Heyrovsky step (Tafel slope ~39 mV/decade) as the rate-determining step, respectively.

As mentioned above, the Volmer–Tafel mechanism follows on Pt surface at low overpotentials where Tafel recombination is the rate-determining step followed by Volmer step, but most of the cathode catalysts deviate from the Tafel recombination as the rate-determining step. It is believed that in all HER mechanism, the Volmer step where adsorption of protons on the cathode surface takes place is mandatory. The Tafel slope for Volmer step is quite high (116 mV/dec) while it is 29 mV/dec and 44 mV/dec for Tafel recombination and Heyrovsky step, respectively. Therefore, the values of the potential in the HER activity are very important for measuring the rate-determining step which explains the nature of the materials and HER dependent pathways (Li et al. 2011).

A suitable HER catalyst should have the capability to easily desorb the adsorbed atomic hydrogen or proton or hydride anion making a chemical desorption Tafel process as the rate-determining step (reaction (3.12)). A catalyst may have higher affinity towards adsorption with the hydrogen species (reaction (3.8) or (3.9)), but may not be suitable for desorption of the adsorbed species (reaction (3.12)) and suffers from superior HER catalysis. Therefore, binding with hydrogen (reaction (3.8) or (3.9) and easy release of the surface adsorbed hydrogen species (reaction (3.12)) only provides a good HER catalyst. In addition, a hydrogen evolution catalyst should not undergo corrosion in acidic as well as in basic media, the catalysts should also be stable under a wide window of applied potentials for reliable applications. The noble metal catalysts respond very well towards sustainable hydrogen production, while the transition metal oxides, hydroxide, sulfides, carbides, phosphides are found to follow mostly the Volmer step as the slowest step (reaction (3.8) or reaction (3.9)) or in a few cases, the Heyrovsky step (reaction (3.10) or (3.11)) as the rate-limiting step. This is due to the lower affinity of the non-noble materials towards adsorption of hydrogen species in acidic or basic media (reaction (3.8) or (3.9)) or very poor dissociation of surface adsorbed hydrogen (reaction (3.10) or (3.11)) or lack of the stability under applied potential. Based on the materials and their affinity for bonding with hydrogen which is known as hydrogen bonding energy (HBE), Volcano plot is suitable to predict a good or bad HER catalyst. Figure 3.3a shows a Volcano plot where, y-axis reveals the exchange current density based on logarithm (log $j_0$) and $x$-axis shows the free energy change of adsorption of hydrogen with catalyst surface, in eV. In spite of three different steps of HER on different cathode catalyst surfaces, Parsons et al. have shown that the extent of exchange current density is the measure of HER activity (Parsons 1958). The higher the exchange current density, $j_0$, the higher the HER activity is. The volcano plot in Fig 3.3b shows that the HER catalyst with Gibbs free change of hydrogen adsorption close to zero exhibits maximum exchange current density. Pt, Ir, Pd, $MoS_2$ fall near zero free energy change, hence these catalysts are suitable for HER. Materials like Mo, W, Nb, Ni, or Co have high negative free energy of hydrogen adsorption, and hence, the dissociation of adsorbed hydrogen is the rate-limiting step with high overpotential and poorly act as HER catalysts. Similarly, Ag, Au, Cu, or transition metal oxides have

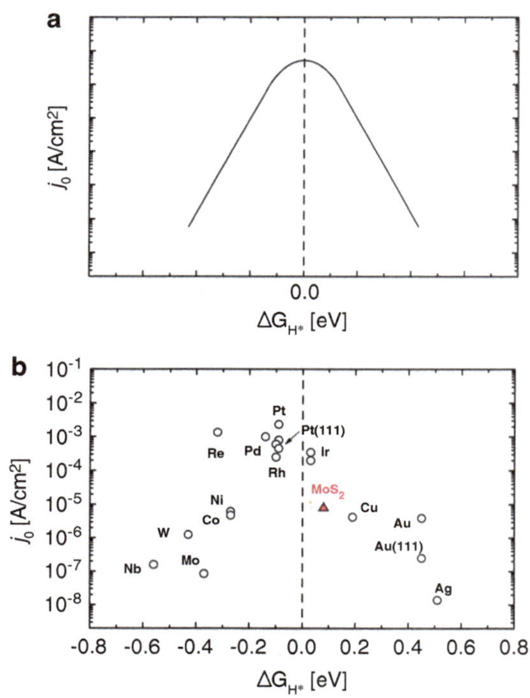


**Fig. 3.3** (**a**) The theoretical HER volcano plot; (**b**) exchange current density of different HER catalysts as a function of hydrogen adsorption free energy (Volcano plots) ((**a**) and (**b**) were adapted from Benck et al. 2014, with permission from American Chemical Society)

positive free energy change of adsorption of hydrogen, which reveals that the adsorption of hydrogen or Volmer step is the rate-deterring step followed by high Tafel slope and hence these catalysts act as deficient HER materials (Wendt et al. 2005; Morales-Guio et al. 2014; Quaino et al. 2014; Roduner 2014).

In the following section, the recent progress on metal sulfide, metal nitride, metal carbide, and metal phosphide based HER catalysts are briefly discussed. Among the abovementioned electrocatalytic materials more focus will be paid on $MoS_2$ catalysts.

### 3.3.2 Metal Sulfides Based HER Catalysts

*Molybdenum sulfide ($MoS_2$)*: $MoS_2$ is known to be robust in many strong acids, which is one of the prerequisite for a practical acid electrolyte based HER catalyst. The investigation on electrochemical and photoelectrochemical activity of $MoS_2$ has progressed since 1970s and was realized as non-reactive toward HER (Tributsch and Bennett 1977). Therefore, $MoS_2$ has not been considered as a potential HER electrocatalyst for a long time.

Recently, Heinemann et al. (2005) estimated the free energy of atomic hydrogen boding on $MoS_2$ edge as ~0.08 eV, which is very close to the optimum value of ideal HER catalyst $\Delta G_H{}^{\circ} = 0$ eV like Pt (Fig. 3.4a). Particularly, (1010) plane Mo edge structure showed 50 % of H coverage, analogue to the nitrogenase catalyst (Fig. 3.4b). Figure 3.4b illustrates that in both $MoS_2$ and nitrogenase structures, the sulfur atom, which binds the hydrogen, is twofold coordinated to metal atoms, either to molybdenum or to iron. The edges of $MoS_2$ are interesting in this context, as the basal plane of $MoS_2$ is catalytically inactive. This phenomenon was further verified with experimental results using $MoS_2$ nanoparticles supported on graphite membrane and showed promising performance at 10 mA/cm$^2$ in electrochemical hydrogen evolution at a moderate overpotential of 0.175 V. These experiments suggested that better HER catalytic performance could be achieved with nanoscale $MoS_2$ crystallites. This criterion triggered the search for nonprecious metal HER catalysts, and revived the $MoS_2$ nanostructure utilization in HER applications.

The bulk $MoS_2$ mostly exposes closed planes of threefold coordinated sulfur atoms to the electrolyte, and therefore, it is expected that only few of the active twofold coordinated moly-edge sulfur atoms can participate in the reaction. It was experimentally confirmed that the $MoS_2$ edges are active catalyst for the HER. The monolayer of $MoS_2$ nanoparticles were deposited on Au (111) synthesized by physical vapor deposition of Mo under $H_2S$ environment. The nanoparticle area and edge length of resultant $MoS_2$ were quantified with scanning tunneling microscope. This helped to identify the catalytic edge of $MoS_2$ in HER. Further, electrocatalytic activity measurements reveal that the catalytic reaction rate of $MoS_2$ relay on their edge state length (Jaramillo et al. 2007). Another important point is the turnover frequency (TOF) value (per-site activity) $MoS_2$ edges were extremely high, but still 50–1000 times less than Pt (Marković et al. 1997).

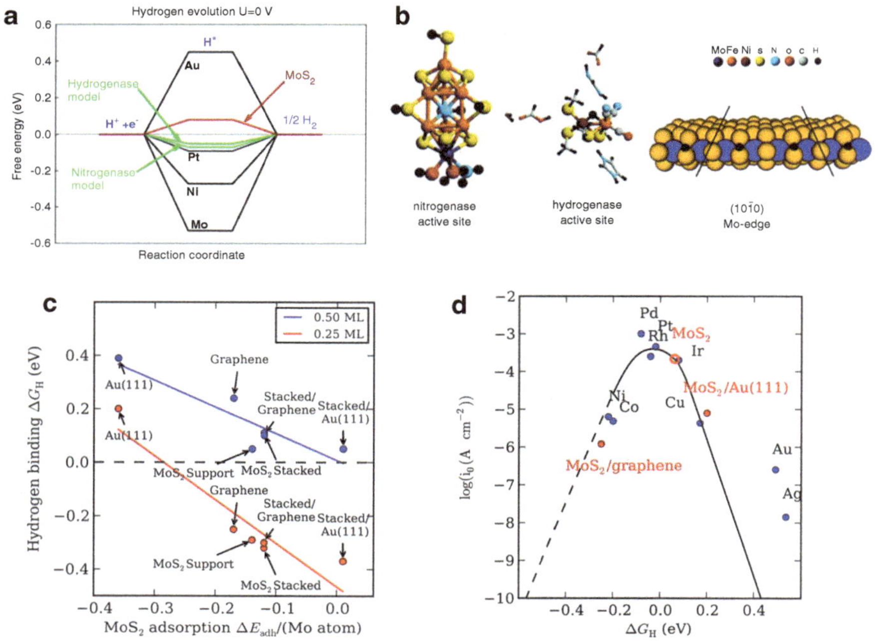


**Fig. 3.4** (**a**) Calculated free energy diagram for hydrogen evolution relative to the standard hydrogen electrode at pH 0 and (**b**) $MoS_2$ slab with Mo (1010) edge (*right*) compared with nitrogenase and hydrogenase active site (*left*) ((**a**) and (**b**) were adapted from Hinnemann et al. 2005, with permission from American Chemical Society); (**c**) estimated hydrogen adsorption free energy on the Mo-edge at each coverage with $MoS_2$ adsorption onto the support and (**d**) Activity map for the hydrogen evolution reaction (HER) showing the exchange current density as a function of the calculated free energy of hydrogen adsorption ((**c**) and (**d**) were adopted from (Tsai et al. 2014) with permission by American Chemical Society)

The HER activity of $MoS_2$ edges can be improved by choosing appropriate substrate. Recently, Tsai et al. found that the supporting materials provided for $MoS_2$ growth will influence the van der Waals force and lead to significant changes in the hydrogen binding energy, resulting in several orders of magnitude difference in HER activity. Figure 3.4c, d explains the variation of hydrogen adsorption free energy on the Mo-edge and their influence on exchange current density (Tsai et al. 2014).

*State of the art $MoS_2$ catalysts*: The exploration of active site enriched $MoS_2$ catalysts has triggered huge body of research in HER applications. Recent years have seen many efforts to establish nanoscale $MoS_2$ in several forms, including crystalline, amorphous and molecular clusters. The improvement strategies can be divided into two categories (a) promoting active sites and (b) improving electronic conductivity. The active sites can be improved by modifying the physical and chemical properties of the $MoS_2$ through (1) increasing the number of exposed active sites, (2) enhancing the reactivity of active sites and (3) improving the electrical contact to active sites (Laursen et al. 2012). The electronic conductivity of $MoS_2$ can be improved by either doping appropriate heteroatoms into the $MoS_2$ lattice or providing highly conducting scaffold such as carbon nanotubes, graphene etc. In the following section the recent efforts on $MoS_2$ catalysis will be summarized based on the aforementioned categories.

Most of the recent work focuses on crystalline $MoS_2$ nanostructures such as nanosheet, nanowire, and nanoparticle. It is necessary to understand the crystallite phase structure of $MoS_2$ for improving the intrinsic density of $MoS_2$ active sites. $MoS_2$ has a two-dimensional layered structure similar to graphene with the individual layers stacked upon each other to form the bulk (see Fig. 3.5a) (Deng et al. 2015a) and each 2D crystal layer is about 0.65 nm thick. These layers are composed together by weak van der Waals forces. The STEM image of trigonal prismatic (2H) and octahedral geometry phase (1T) $MoS_2$ is illustrated in Fig. 3.5b. Owing to this layered structure, it is possible to fabricate single to few layer crystal flakes through chemical (solvent based exfoliation), and physical exfoliation methods (scotch tape mechanical exfoliation). Each 2D crystal layer of $MoS_2$ consists of a plane of hexagonally arranged molybdenum atoms sandwiched between two planes of hexagonally arranged sulfur atoms. These sulfur atoms are covalently bonded with Mo atoms (S–Mo–S) in a trigonal prismatic arrangement forming a hexagonal crystal structure (Deng et al. 2015b). The most commonly found crystal symmetry configurations for $MoS_2$ are hexagonal and octahedral structures. The hexagonal is semiconducting, and the octahedral is metallic. The hexagonal (2H) structure was discovered much earlier than the octahedral structure (1T). In theory, the 2H structure is more stable than the 1T $MoS_2$ (Wypych and Schollhorn 1992). Dimensionally, the Mo–S bond length is 2.4 Å, the crystal lattice constant is 3.2 Å, and the distance between the upper and lower sulfur atoms is 3.1 Å (Ataca et al. 2011, 2012).

*$MoS_2$ nanostructures*: As discussed above, the $MoS_2$ unique layered crystal structure has a strong tendency to form sheet-like nanocrystals. Chemical exfoliation of bulk $MoS_2$ by lithium intercalation method is widely applied to synthesize single or few layered $MoS_2$ nanocrystals. This method generally involves an intercalation process of lithium components between $MoS_2$ sheets. However, these chemical exfoliated $MoS_2$ nanosheets result in unexpected phase transformation from thermodynamically favored 2H phase to metastable 1T polymorph. Jin and coworkers (2013) for the first time explored the influence of $MoS_2$ phase transformation on hydrogen evolution performance. They demonstrated that the 1T-$MoS_2$ chemically exfoliated via lithium intercalation from semiconducting 2H-$MoS_2$ nanostructures were grown directly on graphite. The 2H $MoS_2$ nanosheets decorated graphite electrode showed inferior HER performance compared to lithium (butyllithium-treated) intercalated 1T $MoS_2$ nanosheets. The 1T $MoS_2$ nanosheet coated graphite electrode show less overpotential of −187 mV vs. RHE and a Tafel slope of 43 mV/decade (inset of Fig. 3.4c). From Fig. 3.4c, the large current density and low Tafel slope infers that the catalytic activity is substantially enhanced in 1T phase $MoS_2$. Similar kind of low Tafel slope has been reported from amorphous $MoS_x$ deposited

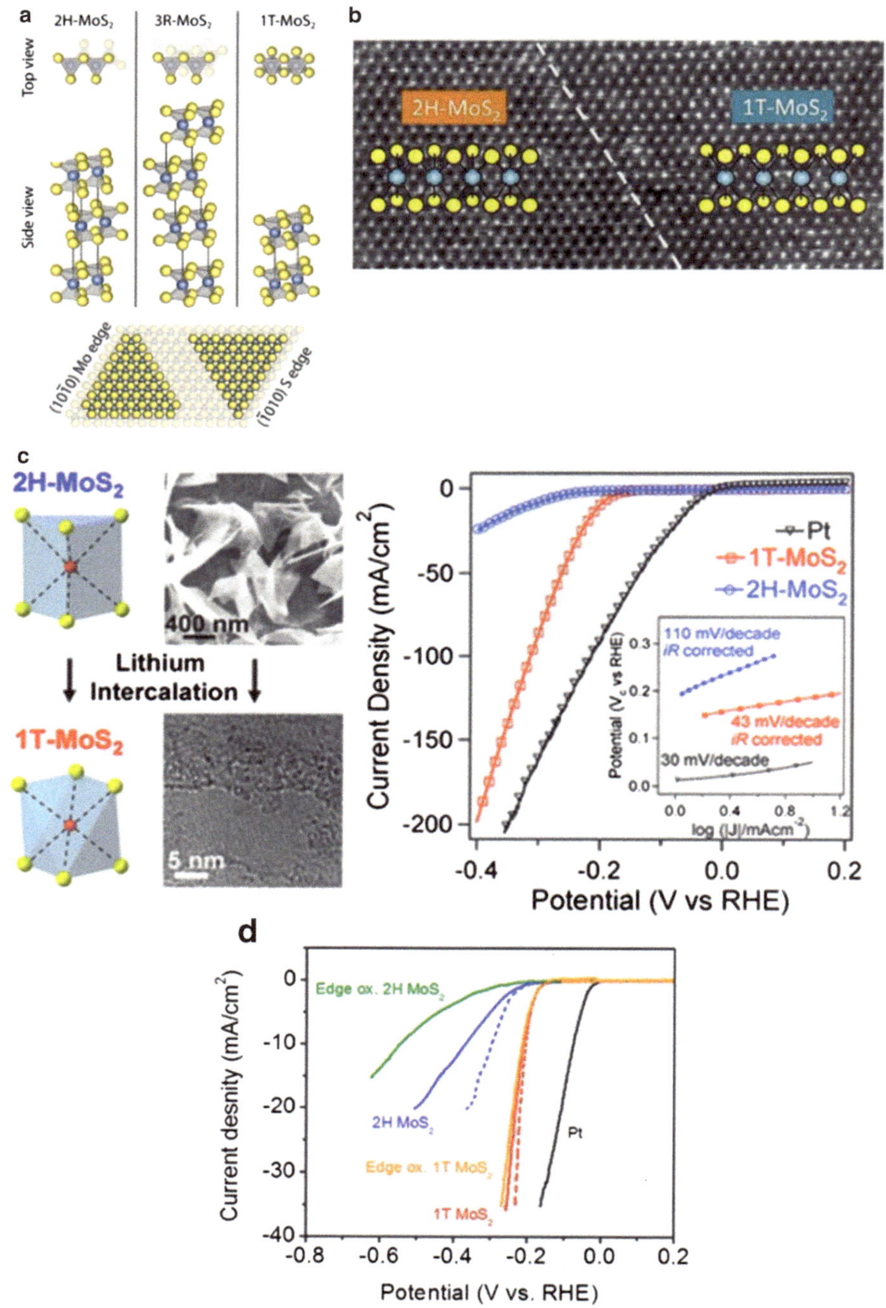


**Fig. 3.5** (**a**) Schematic illustration of the 2H and 1T-type structure of $MoS_2$ (adapted from Benck et al. 2014, with permission from American Chemical Society); (**b**) STEM image of single-layer $MoS_2$ where a boundary between 2H and 1T phases is distinguished with solid line (adapted from Eda et al. 2012, with permission from American Chemical Society; (**c**) Schematic illustration of

by graphene-protected nickel foam on nickel (Chang et al. 2013). They concluded that high electrical conductivity of 1T compared to 2H phase is the reason for higher HER activity rather the improving the edge sites. Chhowalla and coworkers (Voiry et al. 2013a) strongly endorse that the charge transfer kinetics in metallic $MoS_2$ is a key parameter for high performance of hydrogen evolution (Fig. 3.5d). They demonstrated that the HER activity of electrochemically oxidized edge sites of 2H $MoS_2$ was significantly reduced, but 1T-$MoS_2$ remained unaffected after oxidation. It clearly explains that edges of the nanosheets are not only the main active sites. Further, they realized that the catalytic activity of the 2H phase was significantly improved by increasing its conductivity through doping with SWNTs, however, the reaction kinetics is limited by the inefficient adsorption of $H^+$ on the basal plane. Therefore, the high performance of 1T-$MoS_2$ compared to 2H-$MoS_2$ is originated from effective charge transfer kinetics. However, practical implementation of metallic 1T-$MoS_2$ is limited at the high temperature conditions due to its quick conversion back to the 2H phase, since 1T-$MoS_2$ is metastable only below 95 °C.

In addition to lithium intercalation methods, ball-milling and sonication techniques have been also used to obtain highly active $MoS_2$ nanosheets. In ball-milling method $MoO_3$ and S microparticles were used as the starting material. The advantage of introducing lattice defects and dislocations by mechanical milling enhances the catalytic activity of $MoS_2$ for hydrogen evolution (Wang et al. 2013b; Wu et al. 2013a, b). Recently, Gopalakrishnan et al. (Gopalakrishnan et al. 2014) demonstrated the heterodimensional nanostructured $MoS_2$ by combining $MoS_2$ quantum dots dispersed few-layered sheets of $MoS_2$, using a liquid exfoliation technique. This unique hybrid morphology results from the optimized experimental conditions involving bath sonication followed by ultrasound probe sonication. The post-treated $MoS_2$ hybrid structures with less polar volatile solvents such as chloroform showed excellent electrocatalytic activity with low overpotential ~0.190 mV. The large exchange current density about 3.2 A $cm^{-2}$ was achieved due to improved electron transfer rate between $MoS_2$ sheets, and enhanced edge to plane ratio. In this line, hydrothermal synthesis is a promising approach for synthesizing defect introduced $MoS_2$ nanosheets. Particularly sulfur source of starting material plays a key role on the HER performance (Chung et al. 2014). Xie et al. explored that thiourea functionalized synthesis results in defect rich $MoS_2$ nanosheets with a thickness ~5.9 nm (Xie et al. 2013a, b). The unsaturated sulfur sites can bridge the $S_2{}^{2-}$ or apical $S_2{}^{-}$ in the $MoS_2$ and promote the HER activity (Yan et al. 2013). Surface wettability modified $MoS_2$ films pave a new way for

**Fig. 3.5** (continued) 2H-$MoS_2$ phase transformation into 1T-$MoS_2$ by lithium intercalation (*left side*), and electrocatalytic performance of chemically exfoliated and as-grown $MoS_2$ nanosheets at higher potentials (corresponding Tafel plots were presented as inset) (adopted from Lukowski et al. 2013, with permission from American Chemical Society); (**d**) Polarization curves of 1T and 2H-$MoS_2$ nanosheet electrodes before and after edge oxidation (adopted from (Voiry et al. 2013a, b) with permission from American Chemical Society)

designing more efficient gas evolution electrodes. Lu et al. (2014) fabricated vertically aligned $MoS_2$ nanosheets to the substrates by hydrothermal technique. Interestingly, this resulted in "superaerophobic" based $MoS_2$ surface. This could benefit the quick removal of small gas bubbles from the electrode surface, and thus achieving a faster HER current increasing and more steadily working state under high overpotentials.

It is noteworthy to mention that from the above discussed top-down approaches an adverse stacking effect on $MoS_2$ nanosheets may exists through unstable reaction-intermediates and tend to form quasi-0 D nanoparticles or 3D bulk materials during the preparation process (Mdleleni et al. 1998; Song et al. 2015). This might affect catalytic effect of $MoS_2$ in hydrogen evolution performance. In this case, assembling $MoS_2$ layers onto wide porous structured electrodes without altering their edge sites will be appropriate choice for producing effective HER catalyst.

Jaramillo and coworkers (Chen et al. 2011) demonstrated the high aspect ratio and high surface area based $MoS_2/MoO_3$ core–shell nanowire arrays (Fig. 3.6a). The post-sulfurizing process onto vertically grown $MoO_3$ nanowire leads to a thin layer of $MoS_2$ shell layer. This conformal $MoS_2$ shell layer serves as both HER catalyst and as protective layer. In the $MoS_2$ shell layer, the parts of basal planes tend to be parallel to the nanowire axis, exposing few $MoS_2$ edge sites at the surface of the nanowires. As a result this $MoS_2$ edge sites can drive approximately 8 mA/cm$^2$ of current density at −0.3 V (−0.24 V iR-corrected) (Fig. 3.6b). This quantity roughly corresponds to a 10 % efficient solar-to-hydrogen conversion efficiency in a solar water splitting device. Also this core shell NW electrodes exhibit a reliable catalytic performance in acid based electrolyte at 10,000 cycles. Moreover this electrode shows comparative exchange current with precious metal based HER catalysts (Fig. 3.6c). Yet increasing the proportion of edge sites at the shell surface could lead to improvements in performance. Based on this ground, the same group (Kibsgaard et al. 2012) exposed a double-gyroid (DG) based $MoS_2$ bicontinuous network in hydrogen evolution reaction. This unique DG structure encompasses with highly ordered pore channels (~3 nm) prepared by electrodepositing Mo onto a silica template. The removal of silica template after $MoS_2$ thin layer coating promotes the exposure of $MoS_2$ edge sites (Fig. 3.6d). In close proximity of TEM images, it reveals that the S–Mo–S layer-to-layer spacing is 6.6 Å. This value is higher compared to that of the bulk structure of $MoS_2$ (6.15 Å), which results in curvature of the $MoS_2$. As a result DG $MoS_2$ structure exposes a large fraction of edge sites and yields high catalytic performance in HER compared to $MoO_3$ nanowire templates (Fig. 3.6e).

Though porous structured electrode provides edge site controllability, the number of $MoS_2$ layers involved in the catalytic reaction also influences the overall HER performance. Yu et al. (2014) quantitatively demonstrated the layer-dependent electrocatalysis of $MoS_2$ and elucidated that the layer dependence is originated in the interlayer hopping of electrons. They found that catalytic activity of controlled growth of atomically thin $MoS_2$ films in HER decreases by a factor of

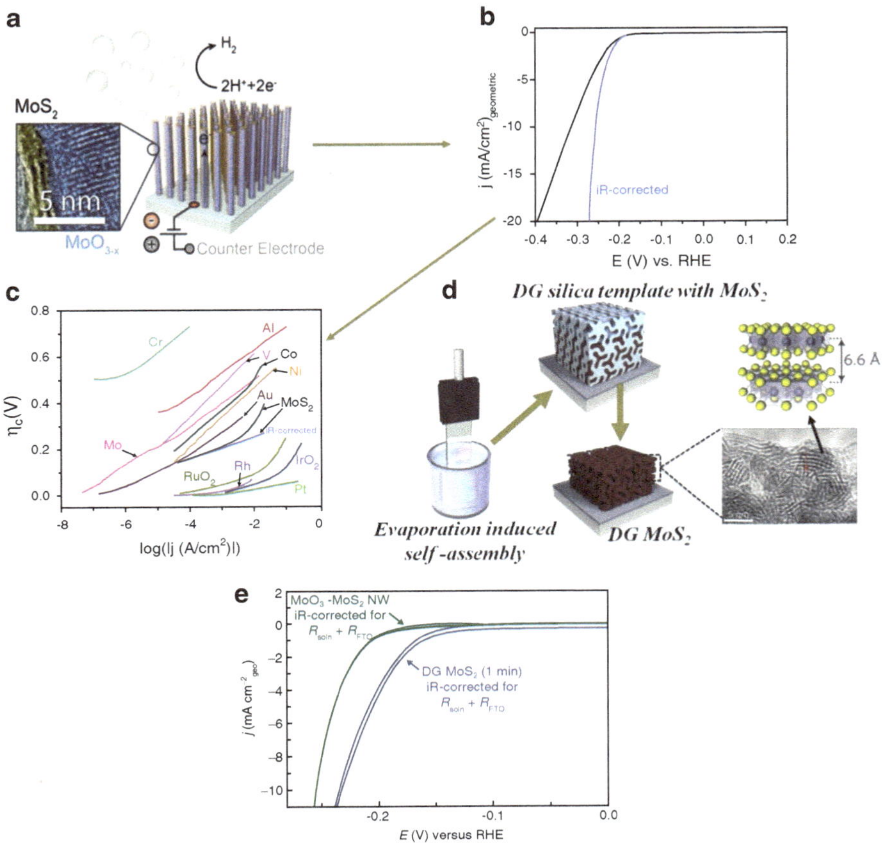


**Fig. 3.6** (**a**) TEM image of $MoS_2/MoO_{3-x}$ and (**b**) corresponding linear sweep voltagram plot in hydrogen evolution reaction; (**c**) Tafel plot of sulfurized $MoS_2$ electrodes and compare their performance with other HER catalyst (Note that $MoS_2$ electrodes were tested in 1.0M of $H_2SO_4$ electrolyte) ((**a–c**) adopted from (Chen et al. 2011) with permission from American Chemical Society); (**d**) schematic illustration of fabrication stages of mesoporous $MoS_2$ with a double-gyroid (DG) morphology and (**e**) cyclic voltammogram of DG$MoS_2$ and core–shell $MoO_3$–$MoS_2$ nanowires at 5 mV s−1 ((**d**) and (**e**) were adopted from (Kibsgaard et al. 2012) with permission by Nature Publishers)

$\sim$4.47 with the addition of each layer (Fig. 3.7a). The conventional theory of $MoS_2$ describes that edge site is catalytically active and may provide an easier path for the hopping of electrons compared to the atoms in the basal plane. But their results show that atoms in the basal plane can be active sites as well. Further this layer dependence phenomenon was verified with edge-riched $MoS_2$ pyramid platelets. These results suggest that increasing the hopping efficiency of electrons is key for the rational design of $MoS_2$ materials with optimal catalytic activities. The efficiency of hopping can be further promoted by coupling of electron orbitals at interlayers. Therefore, strategies that can increase the interlayer coupling, such as

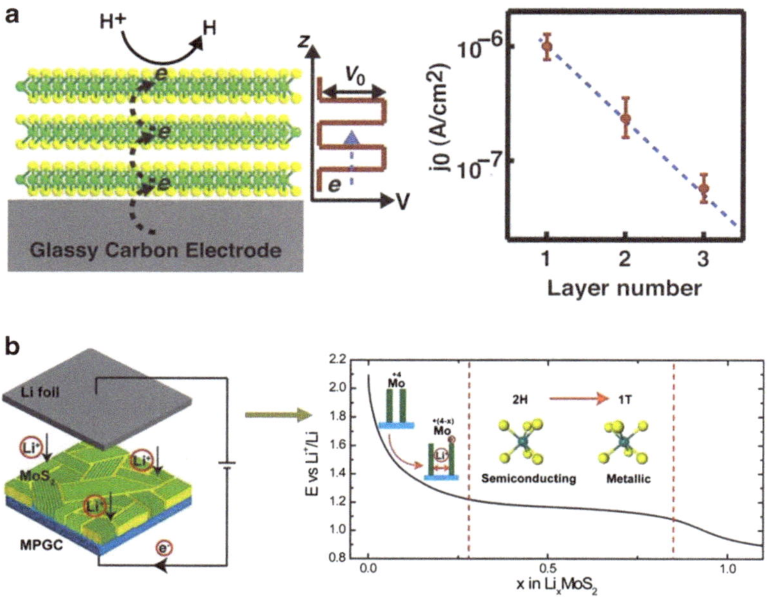


**Fig. 3.7** (**a**) Schematic illustration of electrons hopping route in vertically grown $MoS_2$ sheets onto glassy carbon electrode (*left*) and inverse proportion relationship between number of $MoS_2$ layer and exchange current density (*right*) (adopted from (Yu et al. 2014) with permission by American Chemical Society); (**b**) Schematic illustration of vertically grown lithium intercalated $MoS_2$ nanosheet by electrochemical deposition (left) and the potential drop indicates Li intercalation into the van der Waals gaps of $MoS_2$ and undergoes a 2H to 1T $MoS_2$ transformation (adopted from (Wang et al. 2013a, b, c) with permission by PNAS)

intercalation of metal ions or atoms, are expected to enhance the electrocatalytic performance of $MoS_2$ materials.

*Doping $MoS_2$ with foreign atoms*: As discussed above doping foreign atoms in Mo–S lattice either promote the edge sites or modify the physical properties (optical and electrical behavior) which affords the opportunity to improve the HER performance. The band gap energy of $MoS_2$ varies between 1.2 eV (bulk) and 1.8 eV (monolayer), depending on layer thickness (Lee et al. 2012; Yin et al. 2012). This implies that $MoS_2$ sheet-like structure could allow the heteroatoms doping based on its ionic radii. So far, metal elements including Co, Ni, V and Li have been successfully doped into the $MoS_2$ crystal structure. In $MoS_2$ structure, the Mo-edge has been considered to be active for HER whereas the S-edge is inert. Therefore, most of the doping strategies will make an attempt on modifying the S-edge. The computational descriptor based approach (Hou et al. 2014; Wang et al. 2015a, b, c, d) facilitates the search for suitable dopants to modify the S-edge for the improved HER activity. The recent reports on metals and non-metal doping and their consequences towards $MoS_2$ HER activity are summarized in Table 3.1. Altering the inert S-edge sites of $MoS_2$ surface by metal doping rely on $\Delta G_sH$ value (refer volcano plot in Fig. 3.3a) of dopant atoms. It is found that the positive $\Delta G_sH$ values of metals such as V, Ti, Fe, Mn, and Cr tend to bind with four S atoms, while the negative $\Delta G_sH$ value based metals Pt, Ag, Pd, Co, Ni, etc. prefer to bind with six S atoms. In the latter case, the remaining two S atoms unsaturated and the resultant structural difference could cause distinctly different adsorption behaviors of H atoms and thereby different HER activity. In this view, Pt, Co, and Ni, are found to effectively promote the HER performance of $MoS_2$. Apart from the metal doping, the 2D $MoS_2$ sheets serve as synthetic templates for the solution-phase preferred epitaxial growth of metal nanoparticles (Pt, Pd and Ag), for enhanced HER catalytic activity compared to commercial metal nanoparticles (Huang et al. 2013). The influence of heteroatom doping at $MoS_2$ is classified in Table 3.1.

The recent reports on $MoS_2$ nanosheet synthesis from their bulk material using different exfoliation techniques such as aqueous surfactant solutions under shear (Laursen et al. 2012), electrochemical method (Liu et al. 2014a, b), mechanical method (Wu et al. 2013a, b), and liquid $N_2$ quenching (Van Thanh et al. 2014) are interesting in view of industrial development.

In recent years, $MoS_2$ has been identified as promising co-catalyst assembled on either photocatalyst particles or photoelectrocatalytic electrodes for effective hydrogen evolution. The deposition of $MoS_2$ nanostructures onto semiconductors offers rapid photoelectron separation from conduction band of semiconductor to electrolyte (Fig. 3.8a). Subsequently, these energetic photoelectrons will conduct the proton reduction into hydrogen. Sudhagar and coworkers (Chang et al. 2014; Raja et al. 2015) demonstrated the PEC hydrogen evolution using 1T-$MoS_2$ coated

**Table 3.1** Influence of foreign atom doping at $MoS_2$ lattice

| Doping atoms | Effect on $MoS_2$ | References |
|---|---|---|
| *Metal* | | |
| Co | Modifies S-edge site and promotes HER | (Bonde et al. 2009; Deng et al. 2015a; Zhang et al. 2015) |
| Ni | Modifies S-edge site and promotes HER | (Bonde et al. 2009; Deng et al. 2015a) |
| V | Enhance the in-plane conductivity | (Sun et al. 2014) |
| Li | Intercalate in between Mo and S, and produce multiple effects such as (a) varying Mo oxidation state; (b) 2H to 1T phase transformation and (c) enhance the HER activity | (Voiry et al. 2013a; Wang et al. 2013b) |
| Au | (a) Plasmonic effect, (b) localized phase transition of $MoS_2$ enhances HER activity | (Kang et al. 2015) |
| *Non-metal* | | |
| Oxygen | Band gap narrowing from 1.7 eV to 1.3 eV and enhances intrinsic conductivity | (Xie et al. 2013a) |
| Nitrogen | Improves the electronic conductivity and stability | (Qin et al. 2014; Zhou et al. 2014) |

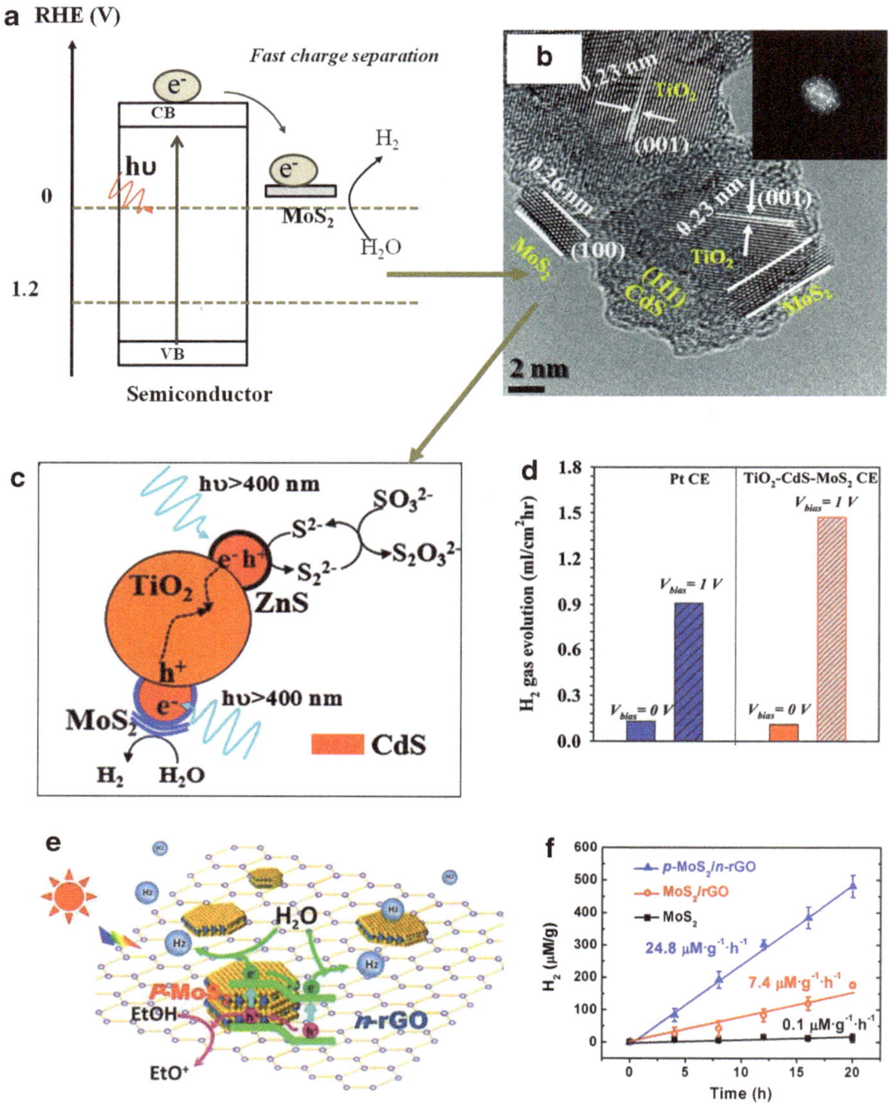


**Fig. 3.8** (**a**) Schematic illustration of photoelectron transfer from illuminated semiconductor to $MoS_2$ HER catalyst; (**b**) TEM image of 1T-$MoS_2$ coated $TiO_2$-CdS mesoporous electrode and (**c**) schematic illustration of charge separation at $TiO_2$–CdS–$MoS_2$ interfaces (**d**) corresponding quantity of hydrogen evolution and compare with Pt electrode ((**b**), (**c**) and (**d**) adopted from (Raja et al. 2015) with permission by RSC publishers); (**e**) $MoS_2$ HER catalyst coated reduced graphene oxide sheets and (**f**) the hydrogen evolution quantity from $MoS_2$ coated graphene sheets ((**e**) and (**f**) adopted from (Meng et al. 2013) with permission by American chemical society)

CdS QDs which was assembled onto mesoporous $TiO_2$ electrodes in the PEC cells (Fig. 3.8c, d). The overall hydrogen evolution from this PEC cells at 1 V vs. RHE using 1T-$MoS_2$ is markedly higher compared to similar system based on Pt counterelectrodes (Fig. 3.8d). This implies that 1T-$MoS_2$ can compete with Pt counter electrode under applied bias conditions (Chang et al. 2014; Raja et al. 2015). Recently graphene material is realized as appropriate supporting scaffold for $MoS_2$ HER catalyst to enhance the photocatalytic hydrogen generation quantity as graphene layers controlling nanoscopic few layer $MoS_2$ aggregation in solution (Fig. 3.8e). As a result $MoS_2$/Graphene-based hybrid material possesses abundance of exposed edges and shows superior performance in HER (Fig. 3.8f) (Min and Lu 2012; Meng et al. 2013; Deng et al. 2015a, b). Similar reports on $MoS_2$ coated photocatalytic solids such as $TiO_2$ (Bai et al. 2015), Si nanowires (Zhang et al. 2015), $C_3N$ (Tian et al. 2014) also highlights the versatility of $MoS_2$ as a potential co-catalyst in hydrogen evolution.

*Tungsten sulfide ($WS_2$)*: Among monolayer transition-metal dichalcogenides, tungsten disulfide ($WS_2$) is of particular interest as its electronic band gap tunes from indirect (1.4 eV) to direct (2 eV) transition, when its size shrinks from bulk to monolayer (Wang et al. 2012). $WS_2$ has similar layer structure and electronic features as $MoS_2$. Therefore, significant attention has been paid on this system in recent years. However, controlled structure and tunable properties are required to achieve high catalytic performance in hydrogen evolution. In this line, different $WS_2$ nanostructures have been developed with appropriate modification. One significant approach is strained chemically exfoliated $WS_2$ nanosheets (Voiry et al. 2013a). The lithium intercalated $Li_xWS_2$ was prepared by forced hydration. The as-exfoliated $WS_2$ nanosheets contain a high concentration of strained 1T phase metallic regions. The STEM results reveal that 1T $WS_2$ phase is not uniformly strained and it contains both tensile and compressive regions (zigzag pattern). The zigzag chain super lattice regions having two distinct W–W distances about 2.7 Å and 3.3 Å that are substantially different from pristine 2H $WS_2$ (3.15 Å). The catalytic activity of $WS_2$ nanosheets originates from tension region, whereas those in compression are inert. In addition, electrical properties of the nanosheets are also influenced by the variation in the 1T phase and the strain, which plays an important role on the catalytic performance of the nanosheets. Interestingly, zero free energy can be observed in the strained sheets at 2.7 % by assuming a linear relation between free energy and strain. As a result, the as-exfoliated $WS_2$ with 1T phase shows higher catalytic activity that the 2H phase (annealing at 300 °C) and is comparable to Pt nanoparticle (Voiry et al. 2013a). Following this report, a template based hydrothermal or solution phase method were successfully demonstrated in synthesizing sheet-like $WS_2$. The 2D morphology based templates such as graphene oxide (Yang et al. 2013; Cheng et al. 2014a, b), $WS_2$ nanotubes (Choi et al. 2013) were utilized in these routes. For instance, the resultant $WS_2$ flakes prepared onto graphene sheets showed higher edge sites and catalytic activity compared to bulk $WS_2$ (Fig. 3.9a). Ball milling and electrochemical techniques were also effective in producing $WS_2$ nanostructures (Wu et al. 2012). The

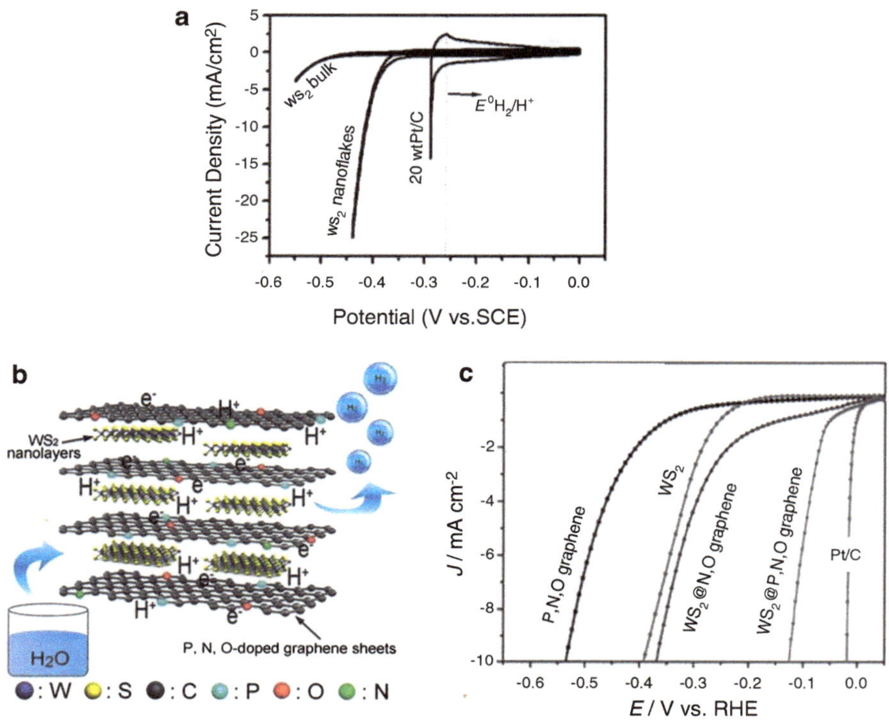


**Fig. 3.9** (**a**) Polarization curves of bulk and as-exfoliated $WS_2$ along with those corresponding to Pt nanoparticles and bulk $WS_2$ powder (Figure adopted from (Cheng et al. 2014a, b) with permission by Wiley-VCH Publishers); (**b**) Schematic structure of non-metals doped 1T-$WS_2$ sheets decorated onto dimensional graphene in hydrogen evolution reaction and (**c**) corresponding polarization plots compare with Pt/C electrode ((**b**) and (**c**) were adapted from Duan et al. 2015, with permission from Wiley-VCH Publishers)

heteroatom doped $WS_2$ by electrochemical technique produce $MWS_x$ sheets (M=Ni, Co). In this system, M-S-W clusters were occupied in the MS centers and enhanced the catalytic activity of $WS_2$ nanosheets (Tran et al. 2013) in hydrogen evolution reactions.

Song Jin and coworkers (Lukowski et al. 2014) show that exfoliation and polymorph control is a vital tool for enhancing the catalytic activity of $MX_2$ materials. Unprecedentedly, microwave assisted intercalated 1T-phase $WS_2$ nanosheets showed less overpotential −142 mV vs. RHE at $j = 10$ mAcm$^{-2}$ compared to other previous HER materials like $MoS_2$, 2H-$WS_2$. The non-metals such as P, N, O doped $WS_2$ sheets decorated onto three dimensional graphene scaffold (Fig. 3.9b) without binders exhibited extraordinary HER performance. Comparing individual non-metals doped 1T-$WS_2$, the synergistic doping of three species P, N, and O at 1T-$WS_2$@3D-graphene leads to significantly lower overpotential of 125 mV vs. RHE at $j = 10$ mAcm$^{-2}$, which is very close to that of commercial Pt/C (Fig. 3.9c) (Duan et al. 2015). This 3D self-supported graphene electrode based HER architecture present many advantages compared to flat type films (a) exposed active centers on 2D $WS_2$ nanolayers (b) highly expanded surface, and continuous conductive network, and (c) strong synergistic effects between the components.

*Metal pyrites ($NiS_2$, $CoS_2$, $FeS_2$)*: Owing to the composition of first-row transition metals and rock-forming chalcogenides, the transition metal pyrites are very abundant and inexpensive, making them particularly interesting materials for energy conversion applications. The cubic pyrite-phase transition metal dichalcogenides ($MX_2$, M = Fe, Co, or Ni and X = S) have recently emerged as efficient HER electrocatalysts (Ivanovskaya et al. 2013; Kong et al. 2013; Faber et al. 2014a, b). Though it has progressed two decades ago (Onuchukwu 1982; Jaegermann and Tributsch 1988; Drobner et al. 1990), recent developments on metal pyrites nanostructures revisiting their usage in hydrogen evolution reactions. Song Jin and coworkers (Lukowski et al. 2014) established a detailed analysis on metal pyrites and their performance in HER catalyst.

The electrochemical hydrogen generation was explored from neutral water using FeS (pyrrhotite) nanoparticles as a bio-inspired catalyst (Di Giovanni et al. 2014). This system shows weak catalytic response with substantial stability for six days. However, $FeS_2$ exhibits relatively better catalytic activity in hydrogen evolution compared to $NiS_2$ in acidic electrolyte. Further their HER activity was gradually increased by modifying their electronic structure with Co doping (Kong et al. 2013; Wang et al. 2015a, b, c, d). In the case of $NiS_2$, when coated on carbon cloth, their catalytic activity was improved (Tang et al. 2015). For overall comparison, both Fe and Ni based sulfide catalyst still exhibit significantly lower catalytic activity compared to that of other nonprecious HER catalysts, such as $MoS_2$. More attention is required to promote the catalytic activity of Fe and Ni based sulfides since it shows excellent material stability in acetic electrolyte.

Cobalt sulfides ($CoS_2$) are now promising HER catalysts, though Co has less abundance than Fe or Ni. Recent research work on $CoS_2$ based catalyst outperforms $FeS_2$ and $NiS_2$ for HER in both acidic and neutral media (Kong et al. 2013) (Wang et al. 2015a, b, c, d). Sun et al. (2013) tested electrocatalytic acidity of Co–S film directly coated onto p-Si electrode by simple potentiodynamic deposition. The Co–S films show low onset overpotential, 100 % Faradaic efficiency, at photoelectrochemical reactions and exhibit high current densities that can persist for at least 40 h. Faber et al. (2014a, b) explored the influence of $CoS_2$ morphology on HER activity. They examined three different morphologies such as film, microwire, and nanowire structures (Fig. 3.10a). They found that $CoS_2$ nanowire electrodes achieved geometric current densities of $-10$ mAcm$^{-2}$ at overpotentials as low as $-145$ mV vs. the RHE. Furthermore this micro-nanowire architecture synergistically increased its operational stability, cyclability, and maximum achievable rate of hydrogen generation by promoting the release of evolved gas bubbles from the electrode surface. Currently, the mixed phase of $Fe_{1-x}Co_xS$ electrodes shows competitive HER activity compared to pristine $CoS_2$ (Wang et al. 2015a, b, c, d). In particular $Fe_{0.4}8Co_{0.52}S$ shows less overpotential of $-196$ V RHE at 10 mAcm$^{-2}$

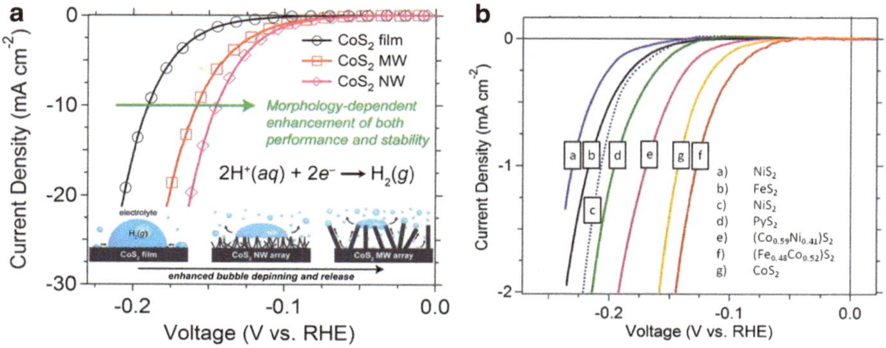


**Fig. 3.10** (**a**) Polarization plots of $CoS_2$ toward the HER; (**b**) Polarization plots of Fe, Co, S alloy based electrodes and compared to pure $FeS_2$, $NiS_2$, and $CoS_2$ ((**a**) and (**b**) adapted from (Faber et al. 2014a, b) with permission from American Chemical Society)

and Tafel slope 47.5 mV decade$^{-1}$ (Fig. 3.10b). The high HER performance may originate from structural disorder introduced through Fe and Co metal alloying. This could lead proliferation of active sites for catalysis, and the electronic interaction of multiple metal species may synergistically enhance intrinsic activity (Faber et al. 2014a, b). This suggests that alloying may also be an effective method for enhancing the electrocatalytic activity.

### *3.3.3 Metal Phosphide Based HER Catalysts*

Recently, several first and second row transition metal phosphides are utilized as potential electrochemical HER catalysts. The transition metal phosphides include nickel phosphide ($Ni_2P$), cobalt phosphide (CoP), iron phosphide ($Fe_2P$), molybdenum phosphide (MoP), tungsten phosphide (WP), copper phosphide ($Cu_3P$), etc. These phosphides are stable in ambient conditions and have high conductivity as well as mechanical stability. The origin of HER by the transition metal based phosphides are ascribed to the intrinsic hydrogen adsorption on the catalyst surface which upon applied negative potential, gets desorbed from the catalyst surface forming molecular hydrogen. The phosphide anions are partially induced with negative charge while the metal sites are with positive charge, which act as the proton and hydride acceptor sites, respectively thereby extending the HER mechanism with flexibility. Therefore, the phosphide based electrocatalysts are found to be active for stable HER in all pH conditions (Callejas et al. 2014; Zou and Zhang 2015). The size of the phosphide anion is comparatively high (0.109 nm) which is favorable for more isotropic crystal structure (Fig. 3.11a) unlike the sulfides which prefer to form layered structures. Such isotropic crystal structures generally have higher degree of surface unsaturated sites thereby making the phosphide based catalysts more promising for electrocatalysts (Liu and Rodriguez 2005; Oyama

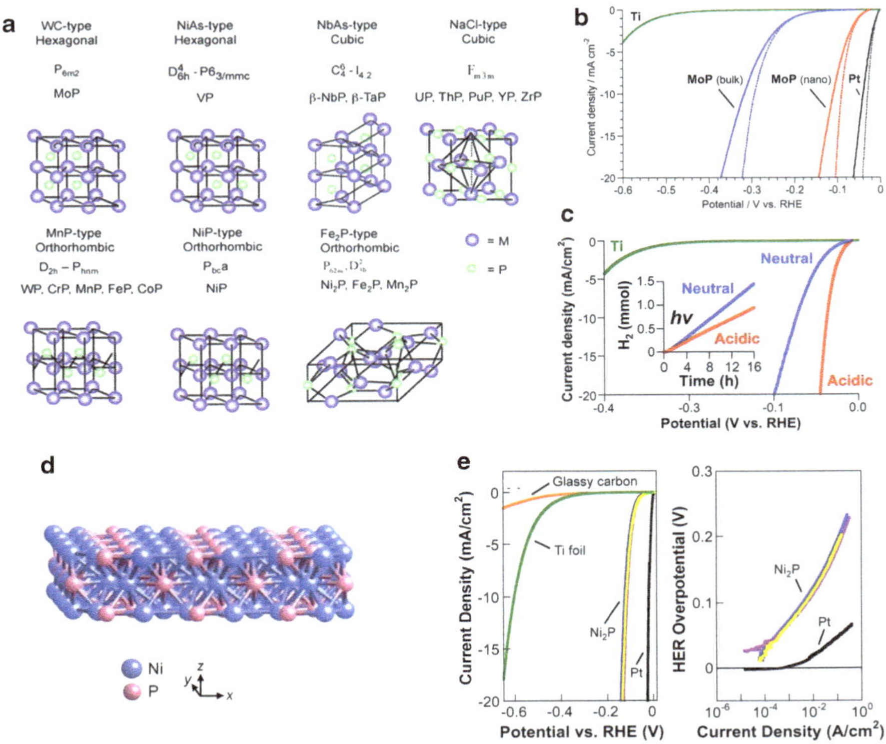


**Fig. 3.11** (**a**) Unit cells of few HER active metal phosphide electrocatalysts (adapted from Oyama et al. 2009, with permission from Elsevier Publishers); (**b**) Polarization curves of MoP/Ti electrode at a mass loading of ~1 mg $cm^{-2}$ compare with crystalline bulk MoP, Pt, and Ti curves measured at 0.50 M $H_2SO_4$ electrolyte (adopted from (McEnaney et al. 2014) with permission by American Chemical Society); (**c**) polarization data for three distinct FeP/Ti electrodes, along with Pt and Ti, in both acidic (0.50 M $H_2SO_4$) and neutral conditions (1.0 M PBS) (*left*), (adapted from Callejas et al. 2014, with permission from American Chemical Society); (**d**) a two-dimensional piece of $Ni_2P$, showing the (001) surface on top; and (**e**) Polarization plots of $Ni_2P$ electrodes in 0.5 M $H_2SO_4$, along with glassy carbon, Ti foil, and Pt (*left side*) and corresponding Tafel plots for the $Ni_2P$ and Pt electrodes (*right side*) ((**d**) and (**e**) were adapted from Popczun et al. 2013, with permission from American Chemical Society)

et al. 2009). Few recent developments of phosphide material for HER catalysts are discussed in this section.

Experimental evidence of HER by $Ni_2P$ was first reported by the Popczun et al. (Popczun et al. 2013). $Ni_2P$ exhibits a crystal structure of hexagonal $Fe_2P$ like structure where both the unsaturated Ni and P atom are exposed to the surface (Fig. 3.11b). As the electronegativity of P is higher than Ni, partial negative and positive charges are found to locate over the respective unsaturated surface elements providing a proton and hydride acceptor sites which facilitates HER. Such synergy effect of $Ni_2P$ surface for HER is known as ensemble effect (Liu and

Rodriguez 2005). HER activity of the $Ni_2P$ nanoparticles are found close to Pt nanoparticles supported on Vulcan carbon (Pt/C) and thereby making phosphide based catalysts more promising and low cost robust catalyst with high degree stability and durability. The overpotential is found to ~180 mV at current density of 20 mA/cm$^2$ and the Tafel slope is found 46 mV/decade in the overpotential range 25–125 mV which resembles similar to $MoS_2$ nanostructures. Similarly, FeP and CoP nanorods are found to exhibit a promising HER catalytic activity with Tafel slope 60 mV/decade in the range of overpotential ~20–100 mV (Du et al. 2014; Liu et al. 2014a, b). Callejas et al. report the HER activity of FeP nanoparticles in acidic and neutral pH (Callejas et al. 2014). The overpotentials of 50 and 61 mV are found at current densities of 10 and 20 mA/cm$^2$ respectively in acidic solution. The same is obtained at the overpotential 102 and 136 mV in alkaline media (Fig. 3.11d, e). The MoP nanoparticles are also found to active as HER catalyst with Tafel slope ~60 mV/decade and overpotential of 110 mV at current density of 10 mA/cm$^2$ with high stability and durability (McEnaney et al. 2014).

### *3.3.4 Metal Carbide Based HER Catalysts*

Primarily (II)–(VI) transition metal carbides have been studied as HER catalysts due to their high stability, corrosion resistance, mechanical strength and conductivity. Carbon inclusion in the lattice structure of the transition metals increases the lattice volume. Such increase in crystal volume results in deforming transition metal d-orbitals in metal carbides. This is due to the interaction of the d-orbitals with s and p orbitals of carbon. Such deformation or broadening of d-orbitals resembles like d-orbitals of Pt. Levy and Boudart in 1973 revealed that WC has Pt-like properties (Levy and Boudart 1973). Many other transition metal carbides are found to exhibit surface and catalytic properties similar to Pt as well (Chen 1996; Oyama 1996; Hwu and Chen 2005). Therefore, Pt-like behavior of the transition metal carbides has been utilized for HER activity (Chen et al. 2013a, b, c). WC has high electrical conductivity, non-reactivity towards carbon monoxide or hydrogen sulfide which is suitable for fuel cell applications. Thereby WC attracts huge attention and has been studied for HER catalyst compared to other transition metal carbides. Moreover, WC is much more stable than metal rich $W_2C$. WC undergoes oxidation and forms a thin insulating layer of $WO_3$ which prevents further oxidation in aqueous solution, though in strong acid solution it goes slow dissolution process. Such corrosion in strong acid solution can be accelerated or slowed down by addition of suitable ions. For example, the presence of phosphate ions greatly increases the corrosion while Co(II) ion prevents it. Metal rich $W_2C$ is less stable and gets oxidized at an onset potential of 0.4 V while WC is found to be more stable up to 0.8 V vs. NHE. The pH dependent HER activity of different carbide based transition metal and their oxidation potentials is shown in Fig. 3.12a. Amongst the different carbides, WC is found to more stable and superior HER catalyst compared to other carbide based materials. Esposito

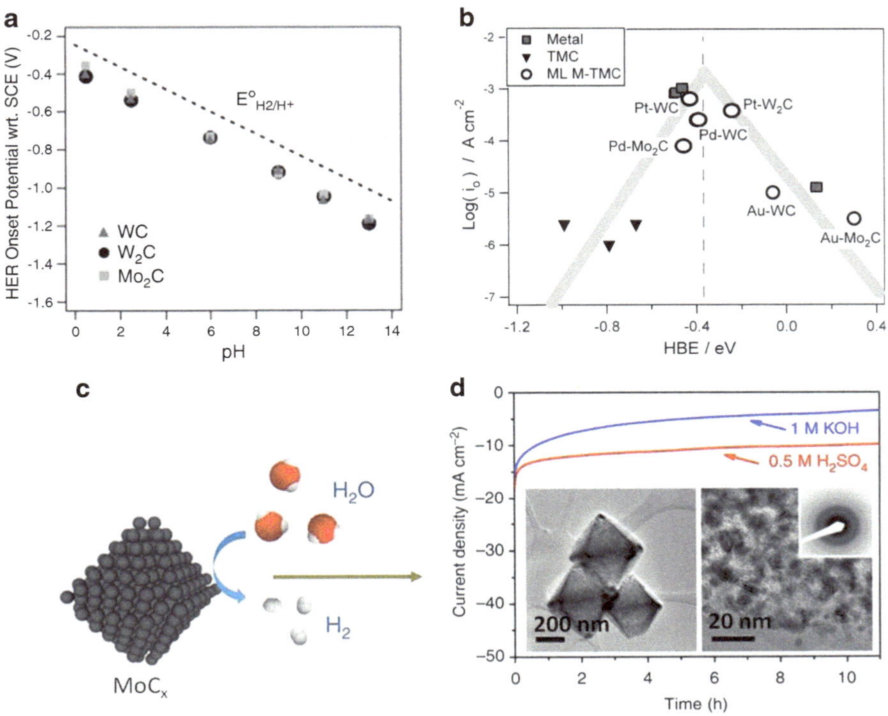


**Fig. 3.12** (**a**) pH dependent HER onset and oxidation onset potentials for WC, $W_2C$, and $Mo_2C$ (adapted from Weidman et al. 2012, with permission from Elsevier Publishers); (**b**) Volcano relationship between exchange current density and the HBE for metal modified carbide surfaces (adapted from Esposito et al. 2012, with permission from American Chemical Society); (**c**) schematic structure of porous $MoC_x$ nano-octahedrons and (**d**) time-dependent current density curves of porous $MoC_x$ nano-octahedrons measured under overpotentials $\eta = 170$ mV in 0.5 M $H_2SO_4$ and $\eta = 180$ mV in 1 M KOH (insets: TEM images and SAED pattern after experiment in 0.5 M $H_2SO_4$) (adapted from Wu et al. 2015, with permission from Nature Publishers)

et al., report the relationship between Pt and monolayer covered Pt-WC and Pt-$W_2C$ and found that these monolayer covered Pt-modified WC and $W_2C$ are excellent materials for cathode HER catalyst which are comparable with that of bulk Pt (Esposito et al. 2012). They have also studied the effect of covering monolayer of other noble metals like Pd, Au or they have modified the core material with MoC and the results are shown in Fig. 3.12b.

Harnish et al. studied the stability of WC at near neutral pH and found that dissolution occurs slowly. The rate of dissolution is found to increase with increase in pH and also with increase in phosphate ion concentration (Harnisch et al. 2009a, b). The HER activity of WC is found to be less by a factor than Pt while it exhibits temperature dependence increased HER activity which is much superior than Pt (Nikolov et al. 1983; Nikiforov et al. 2012). Tafel slope and exchange current density of the HER mechanism by the $W_2C$ catalyst were reported to be 118 mV/decade and −0.35 mA/cm$^2$, thereby revealing that Volmer–Heyrovsky HER mechanism takes place on the $W_2C$ nanoparticles with activation energy ~35 kJ/mol (Garcia-Esparza et al. 2013). Further charge transfer from Tungsten Carbide–Nitride was improved with graphene nanoplatelets (Chen et al. 2014).

Wu et al. (2015) report the metal organic framework (MOF) assisted synthesis of mesoporous MoC octahedron nanoparticles with high specific surface area and HER activity in both acidic and basic media. The fine nanocrystallites are obtained in such MOF assisted carburization method which is due to the interaction and inclusion of metal guest precursor into the host organic ligand and framework in molecular level. Therefore, during carburization method, the metal sites are separated from each other and hence results in producing fine nanoparticles (Fig. 3.12c, d).

Cheng et al. (2013a, b, c) report the decoration of fine β-MoC nanoparticles on CNT with size ~15 nm and linked with carbon black. These β-MoC nanoparticles linked with carbon black and CNT exhibit very small overpotential (63 mV) at 1 mA/cm$^2$ current density of HER. The conjunction of CNT and carbon black in their synthesis prevents the formation of bigger MoC particles and also the composite matrix provides a higher degree of conductivity and hence the facile charge transport properties for HER. In addition to the charge transport properties, the d-band center of W is reported to shift downwards which reflects on the hydrogen binding energy (HBE) on the composite catalyst surface. The HBE decreases during such composite formation and hence the easy release of surface adsorbed atomic hydrogen from the catalyst surface in the course of HER takes place. Chromium carbide has also been studied for electrochemical HER. The activity of chromium based carbides is composition dependent like other HER active transition metal carbides. However, the exact role of C/Cr is ambiguous. At higher ratio value, the number of metal sites exposed to surfaces decreases and hence the HER activity. On the other hand, at lower ratio values, the stability and catalytic site might decrease which reflects on the HER. For example, $Cr_7C_3$ exhibits twofold higher activity than $Cr_3C_2$ at overpotential of 180 mV (Chen et al. 2013a, b, c).

### *3.3.5 Nitride Based HER Catalysts*

Nonmetal nitride based compounds primarily graphitic carbon nitride (g-$C_3N_4$) and boron nitride (BN) have recently drawn much attention by the scientific community towards electrochemical energy conversion processes which can be an alternative to state-of-art Pt catalyst. g-$C_3N_4$ is prepared by polymerization of amine and carbon containing compounds in both wet chemical as well as physical methods. Depending on the reaction conditions and molecular precursors used in the synthesis process, the degree of condensation can be tuned and hence the physical and chemical properties. BN is isoelectronic to carbon and thereby has similar allotropic properties of carbon. Hexagonal BN is similar to graphite while cubic BN is similar to diamond. One dimensional BN nanotubes resembles like carbon nanotube while flat two dimensional BN resembles to graphene, though their chemical as well as physical properties are distinct from the corresponding carbon allotropes. Hexagonal BN has high thermal and mechanical strength and thereby generally used in high temperature equipment.

Zheng et al. (2014) recently report the g-$C_3N_4$@nitrogen doped graphene (g-$C_3N_4$@NG) nanocomposite as superior HER electrocatalyst. The elemental interaction and chemical bonding between the g-$C_3N_4$ and NG during the synthesis are described to increase the active site as well as the conductivity. The metal free g-$C_3N_4$@NG nanocomposite is found to be stable in acidic and basic conditions thereby extending its long term use. The origin of HER activity at g-$C_3N_4$@NG nanocomposite was ascribed to the impregnation of NG into the framework of g-$C_3N_4$ which not only increases the conductivity, but also the catalytic sites. Theoretical studies show that the Gibb's free energy of protonation on g-$C_3N_4$ is much negative compared to Pt or NG indicating the poor atomic hydrogen release upon applied negative potential. Though, Gibb's free energy for protonation of g-$C_3N_4$@NG nanocomposite is found to be slightly more negative compared to Pt thereby making it one of the promising metal free robust HER catalyst (Fig. 3.13a, b). Thus, a Pt-like electrocatalytic HER behavior emerges in the HER of $C_3N_4$. They have also measured the HER activity with different atomic hydrogen coverage. The reaction coordinate is found to be favorable at low coverage of atomic hydrogen (Fig. 3.13c) (Zheng et al. 2014). In addition to the recent development of $C_3N_4$ based catalysts in energy conversion processes, hexagonal BN also features interest due to its high stability and longevity. Few works have been forwarded in catalytic activity of BN based materials. Hexagonal BN can be tuned to nanotube like CNT, or graphene like structures, and therefore, suitable manipulation of the chemical and structural composition of BN would make a giant step in the near future for energy conversion and electronic devices as the BN based compounds are semiconducting nature with high mechanical and thermal tolerance (Huang et al. 2015; Roy et al. 2015; Shen et al. 2015).

The above section has shed light on the influence of various HER catalysts, their activity, stability and mechanism of hydrogen evolution. An in-depth investigation into different catalysts have however, pointed out the mechanism pertaining to hydrogen evolution basically in three steps (1) Volmer, (2) Heyrovsky, and (3) Tafel. The activity of the catalyst briefly lies on the rate-determining step associated with one of the three steps listed above. A systematic study of the relationship between catalyst properties (e.g., chemical composition, phase structure, morphology, size distribution, and surface area) and photocatalytic performance should be mapped into an atlas, which will aid identifying the catalyst of need. Also, the electrochemical capabilities of the catalysts for HER should be tabulated. Considering the vast amount of catalysts, co-catalysts and dopants available in this arena, only some of them have been discussed in this section. However, considering the lacunae and demerits of the available catalysts, concerted advancement in development of new catalysts is required.

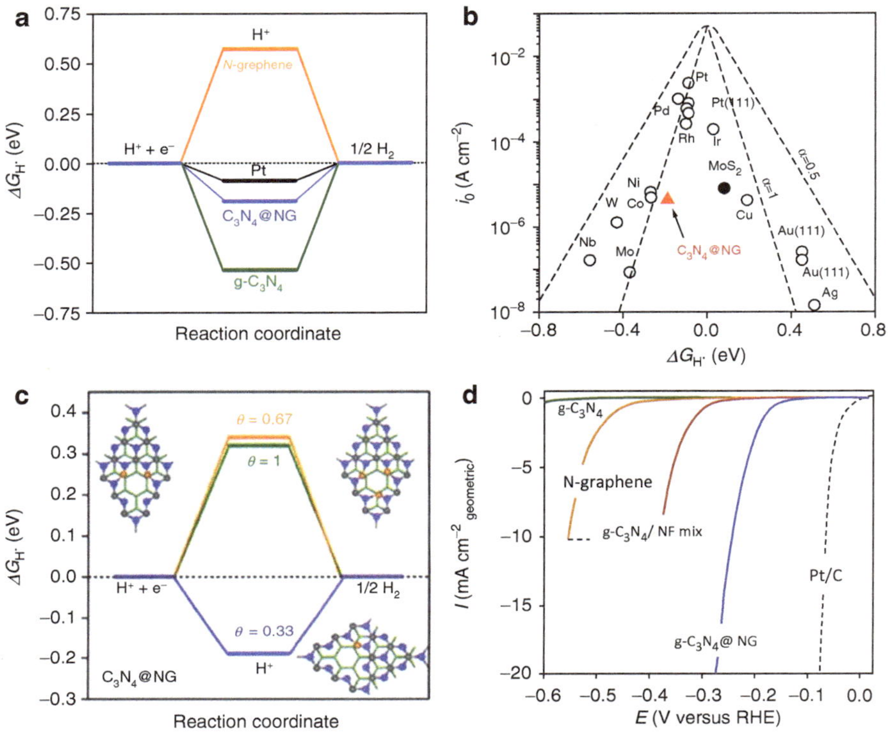


**Fig. 3.13** (**a**) The calculated free-energy diagram of HER at the equilibrium potential for three metal-free catalysts and compare with Pt; (**b**) Volcano plots of $i_0$ as a function of the HBE for $C_3N_4$@NG (indicated with triangle) compare with common metal catalysts and nanostructured $MoS_2$ catalysts; (**c**) Free-energy diagram of HER on the surface of $C_3N_4$@NG under different H* coverage conditions; (**d**) The HER polarization curves for four metal-free electrocatalysts and 20% Pt/C (electrolyte: 0.5M $H_2SO_4$, scan rate: 5 $mVs^{-1}$. (**a–d**) have been adapted from Zheng et al. 2014, with permission from Nature Publishers

## 3.4 $CO_2$ Reduction

A perennial quest for managing $CO_2$ effectively through capture, sequestration, utilization, and conversion to other hydrocarbon has become inevitable due to the fast-growing $CO_2$ discharge leading to the twentieth century cliché "global warming." Hence, the research on processes involving recycling of $CO_2$ into various usable forms has gathered significant attention. This field of science has been in the limelight since the '90s. Processes like biomass (Berndes et al. 2003), (Schmid et al. 2001; Panwar et al. 2011) biocatalysis (Schmid et al. 2001), thermochemical(Lovegrove et al. 1999), electrochemical (Hammouche et al. 1991; Augugliaro et al. 2008; Ahmed et al. 2012; Costentin et al. 2013; Akhter et al. 2015), and photocatalysis (Augugliaro et al. 2008; Ahmed et al. 2012; Akhter

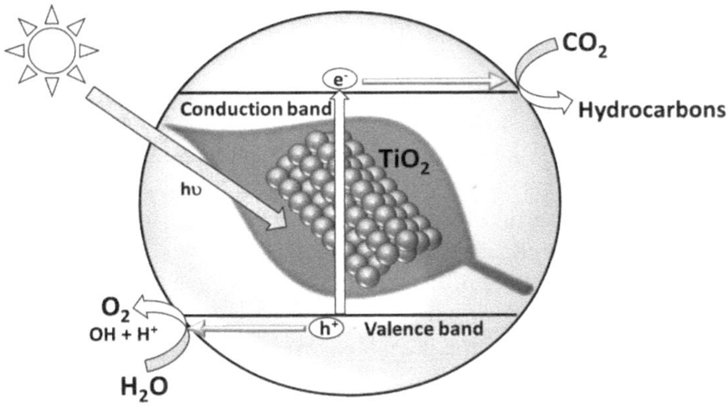


**Fig. 3.14** Schematic depiction of a $TiO_2$ photocatalyst (artificial leaf), its electron flow from conduction band to subsequent $CO_2$ reduction

et al. 2015) are well known and have been understood well by the research community. Although $CO_2$ reduction by photocatalysts has sprinted far ahead of the other processes. Electrochemical reduction technique, though it can convert $CO_2$, can attain only low efficiencies resulting from deactivation of electrodes. Although $CO_2$ reduction to other forms of fuel is an endothermic step, tapping solar energy would be a versatile method as this form of energy is readily available in abundance. Fuels derived out of $CO_2$ reduction in combination with water should be feasible by employing photocatalytic semiconductors which can utilize solar energy for this endothermic $CO_2$ conversion.

Among the semiconductor materials $TiO_2$ has attained the epitome since its first reports by Fujishima and Honda in 1972 (Fujishima and Honda 1972) and Inoue et al. in 1979. However, a large variety of semiconductor materials have been developed since then and each one them having a distinct feature capable of a particular function (Fan et al. 2011; An et al. 2012; Gallo et al. 2012; Chen et al. 2013a, b, c; Agu et al. 2014; Aresta et al. 2014; Cao et al. 2014; Cheng et al. 2014a, b; Gao et al. 2014; Gui et al. 2014; Akhter et al. 2015; Akimov et al. 2015). By all means, if the photocatalytic steps involved in conversion of $CO_2$ are to be described in three steps: (a) generation of charge carriers ($e^-$ and $h^+$) through excitation, (b) charge separation and transportation, and (c) reduction of $CO_2$ by photoelectrons. This process resembles photosynthesis to considerable extent and hence conveniently termed as artificial photosynthesis. Figure 3.14 schematically represents the electron flow chart when a semiconductor is irradiated with photons with energy capable of exciting the valence electron into the conduction band. The separated charge in the form of electrons and holes migrate to the surface and reduce or oxidize the appropriate moiety (Carp et al. 2004).

### 3.4.1 Selecting Ideal Photocatalyst for $CO_2$ Reduction

In order to create the rendezvous between the catalyst and $CO_2$, theoretically a less important but practically a vital step, simply addressed as "adsorption" forms the undercurrent for improving the photoactivity of photocatalysts. The better the absorption of the molecules on the surface of the photocatalyst the easier and faster will be the reach of electrons to reduce the moiety. This also facilitates lower electron–hole recombination. The number of adsorbed molecules under steady state equilibrium is determined by equilibrium uptake. However, $CO_2$ uptake does not reach equilibrium due to the constant conversion of $CO_2$. The determining factors or the driving force controlling the actual adsorption of $CO_2$ molecules will be the equilibrium uptake and the abundance of the molecules in the vicinity of the photocatalyst. In general, equilibrium uptake is expressed in terms of the partial pressure of an adsorbate, temperature, surface area, and surface energetics. In all, the main parameters pertaining to the materials aspect of $CO_2$ reduction will be surface area and surface energetics. Notably, for any catalyst, surface area to larger extents determines the efficiency. In the case of $CO_2$ reduction as well, the higher the number of molecules permitted to be adsorbed, the quicker the reduction and the higher the product yield. Also, another attribute of larger surface area is that the charge carrier density per unit area lowers leading to reduced possibility of recombination (Lee et al. 2014; Sang et al. 2014; Wang et al. 2014). Considering the aspect of surface energetics of a photocatalyst, which is the second most important factor itself is governed by other secondary factors such as structure, composition and chemistry. Tailoring these parameters substantial changes in $CO_2$ adsorption and subsequent photoreduction can be achieved (Liu et al. 2013; Xie et al. 2013a, b; Manzanares et al. 2014). Carbon dioxide being an acidic molecule shows strong affinity to basic surfaces which can lead to potentially stronger interaction and adsorption. A second strategy often considered is tuning the planes of the surface with high-energy facets. Although the natural tendency for any catalyst would be to possess predominantly the low-energy facets on its surface in order to minimize the surface free energy, specialized routes do yield the desired plane. A widely employed strategy in achieving high-energy facets on surface is by protecting the required facet with specific capping agents to mask the activity of the facet and grow the photocatalyst along with the low-energy facet. The removal of the capping agents results in the exposure of the high-energy facets. Apart from surface area and surface energetics, mass transfer phenomena appear to control the $CO_2$ reduction after initialization of the reduction reaction itself. The adsorption process leads to a pressure gradient between the surface proximity and the bulk gas phase, which drives new molecules to move towards the surface. When the diffusion and adsorption kinetics of $CO_2$ are poor, the product yield is also low. Hence, a common strategy is to use nanostructured photocatalysts possessing mesopores (2–50 nm), providing both high surface area and good mass transport properties. In contrast, very small pore sized nanomaterials generally lead to lower diffusion of $CO_2$ and lower product yield. Another vital advantageous aspect of mesopored photocatalysts is that the formed products naturally diffuse out of the active site allowing the next set of $CO_2$ to undergo reduction. Hence, desorption or mass transport of products will also impact on the product yield as they block the active sites and secondly the products can be susceptible to side reactions with new reactants.

Hence, an ideal catalyst should satisfy all the above mentioned factors of surface area, surface energetics and mass transfer aspect.

### *3.4.2 Theory of Photocatalytic $CO_2$ Reduction*

Now, dwelling deeper into the materials aspect of $CO_2$ reduction, the major share of importance is taken away by the photocatalysts. Tailoring the catalysts to readily facilitate reduction reaction is the key to success. In order to generate a wonderful material for $CO_2$ reduction, in-depth understanding of band tuning is required. The positions of the valence and conduction bands should both match the oxidation and reduction potentials of water splitting and $CO_2$ reduction. Only a countable number of semiconductors satisfy this condition. Figure 3.15 shows the band edge position of a few selected semiconductor photocatalysts relative to the energy of various redox couples at pH 7 in aqueous solution (Hoffmann et al. 2011; Cowan and Durrant 2013; Li et al. 2014).

$CO_2$ being chemically stable due to the covalently bonded carbon and oxygen (C=O) with a bond enthalpy of +805 kJmol$^{-1}$, a substantial amount of energy is required for the bond breaking process. Hence, a photocatalyst with sufficient reduction potential for $CO_2$ will enable the formation of a spectrum of products

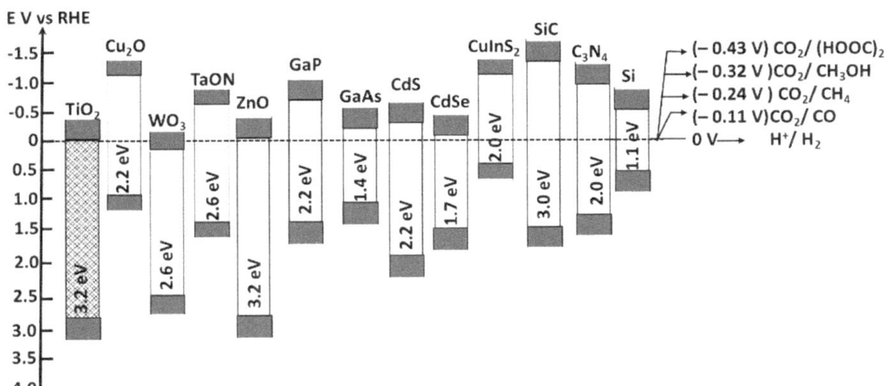


**Fig. 3.15** Band edge positions of various semiconductors relative to the redox potentials at pH 7 of compounds involved in $CO_2$ reduction

depending on the photocatalytic reduction potential as given in the following equations; (Indrakanti et al. 2009; Yang et al. 2010; Mao et al. 2013; Tahir and Amin 2013)

$$CO_2 + e^- \rightarrow CO_2^{\bullet} \quad E^O = -1.90 \tag{3.13}$$

$$CO_2 + H^+ + 2e^- \rightarrow HCO_2^{\bullet} \quad E^O = -0.49 \tag{3.14}$$

$$CO_2 + 2H^+ + 2e^- \rightarrow CO + H_2O \quad E^O = -0.53 \tag{3.15}$$

$$CO_2 + 4H^+ + 4e^- \rightarrow HCHO + H_2O \quad E^O = -0.48 \tag{3.16}$$

$$CO_2 + 6H^+ + 6e^- \rightarrow CH_3OH + H_2O \tag{3.17}$$

$$CO_2 + 8H^+ + 8e^- \rightarrow CH_4 2H_2O + H_2O \quad E^O = -0.24 \tag{3.18}$$

$$2CO_2 + 8H^+ + 12e^- \rightarrow C_2H_4 + 2H_2O \tag{3.19}$$

$$2CO_2 + 9H^+ + 12e^- \rightarrow C_2H_4OH + 3H_2O \quad E^O = -0.33 \tag{3.20}$$

Three plausible mechanisms have been widely accepted for the conversion of one mole of $CO_2$ to methane over a photocatalyst. One the formaldehyde route second, the carbine route and finally the glyoxal route. The three mechanisms are depicted in Fig. 3.16 (Koci et al. 2008; Habisreutinger et al. 2013)

Although a few commonly available semiconductors qualify as photocatalysts owing to their bandgap energies, they suffer other limitations. Photocorrosion is one major issue which is accepted as significant drawback among metal chalcogenides semiconductors, e.g., CdS, PbS, CdSe. These materials are highly susceptible to photocorrosion and show very low stability in aqueous media. Environmental toxicity is another limiting factor amongst these semiconductors (Mills and LeHunte 1997; Koci et al. 2008).

$TiO_2$ is a very suitable photocatalyst for $CO_2$ photoreduction due to sufficient positive and negative redox potentials in the valence band and the conduction band, respectively. It will be worthwhile to consider other notable research on $CO_2$ reduction where a few efficient semiconductors have been employed viz., pure $ZrO_2$, catalytic dyads such as Cu-Fe/$TiO_2$-$SiO_2$(Kohno et al. 2000; Nguyen et al. 2008) and Cu-ZnO/Pt-$K_2Ti_6O_{13}$ (Guan et al. 2003), metal and compound oxides such as CuO, $LaCoO_3$, $Ga_2O_3$, and $NaTaO_3$. The main criterion looked upon in these semiconductor materials however remains the conventional parameters looked for in any catalyst, i.e., large surface area, cost effectiveness, accessibility, resistance to corrosion or production of toxic by-products. In addition, the bandgap tuning is a vital criterion which is intensely looked upon (Malato et al. 2009; Bellardita et al. 2013; Marszewski et al. 2015; Yuan and Xu 2015).

As mentioned at the beginning of the section, $TiO_2$ continues to dominate the research arena due to its abundance, chemical inertness, low cost, high catalytic activity, and meager photocorrosion. The only major drawback which limits its anonymous claim as the panacea of photocatalyst is that $TiO_2$ requires ultraviolet light for excitation which is <5 % of sunlight. In order to tap the visible part of the

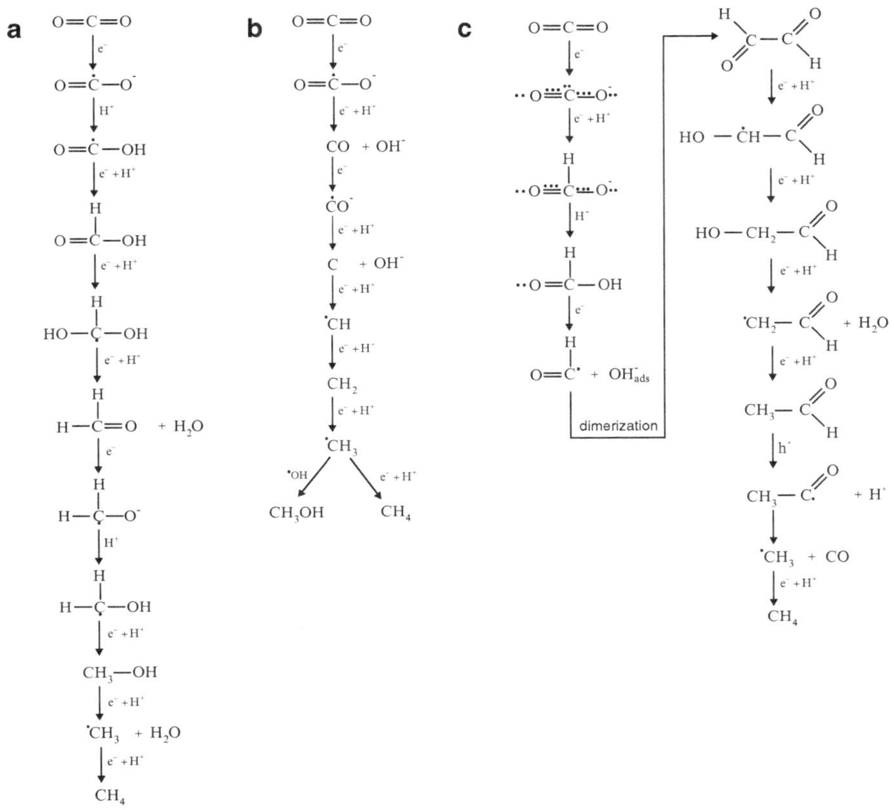


**Fig. 3.16** $CO_2$ reduction mechanism following (**a**) formaldehyde route; (**b**) carbene route and (**c**) gloyoxal route ((**a**)–(**c**) adapted from Koci et al. 2008; Habisreutinger et al. 2013)

solar spectrum, modification of these catalysts is required. In order to improve the visible light absorption, the photon harvesting points working as antenna for trapping sunlight and transferring the electrons to the photocatalyst are embedded on to the parent semiconductor material. Various possible harvesters or antenna have been reported. Organic molecules working as sensitizers or dyes are highly efficient. However, one main disadvantage among them is that they decompose easily though they have very high activity. Organic transition metal complexes exhibit good stability compared to the previous organic moieties however, considering the inadequate long term stability also arise concern. Hence, noble metal nanoparticles (NPs) having a plasmonic resonance in the visible region have exhibited high photocatalytic stability. In the following section a brief description on each of the above mentioned modifications will be provided.

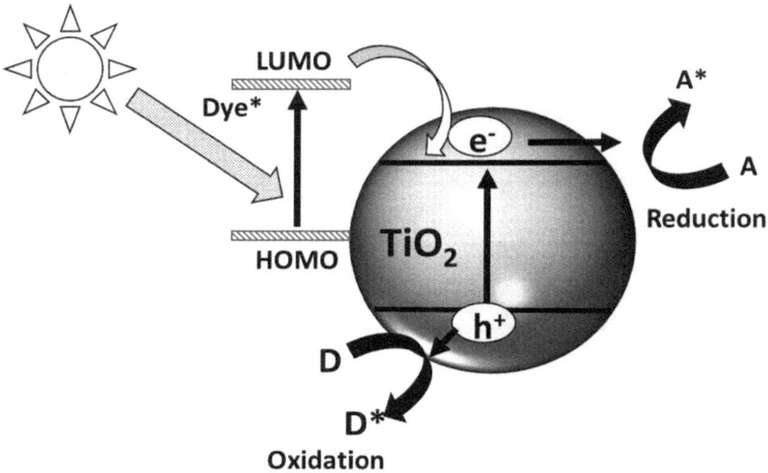


**Fig. 3.17** Schematic depiction of electron flow with introduction of a sensitizer on metal oxide photocatalyst ("A" represents acceptor and "D" represents donor)

### *3.4.3 Dye Sensitized $TiO_2$ for $CO_2$ Reduction*

An extension of the dye sensitized solar cells (DSSC) is the currently discussed dye sensitized photoelectrohemical $CO_2$ reduction cell (please refer to Chap. 13 for more details). The dye moiety effectively absorbs visible light and transfers the photoelectron to the conduction band of the host semiconductor while the conduction band energy level of the dye molecule should be more negative than the host semiconductor. (Carp et al. 2004) (Ni et al. 2007; Malato et al. 2009).

Figure 3.17 illustrates the electron flow chart involved in this process. Thampi et al, (Thampi et al. 1987) employed Ru/RuOx as sensitizer for $TiO_2$ and reported 900 $\mu Lh^{-1}$ of methane under 0.08 $Wcm^{-2}$ intensity simulated solar irradiation. The use of commercially available N3 dye ($Ru_{II}^{*}$2.2(-bipyridyl-4,4(-dicarboxylate)$_2$-$(NCS)_2$) have also been investigated by Nguyen et al. (2008) and have found that theses dyes when coated with Cu-Fe/$TiO_2$ were capable of visible light absorption, producing 0.617 $\mu molg(c_{atal}h)^{-1}$of $CH_4$ after 5.5 h. Instability, photo and heat degradation of dye molecules are factors which are potential area of high impact research.

### *3.4.4 Photocatalytic Dyads*

The position of conduction and valence bands dictates the flow of charge in a dyad type semiconductor containing a heterojunction (Ma et al. 2014). In general, two

a

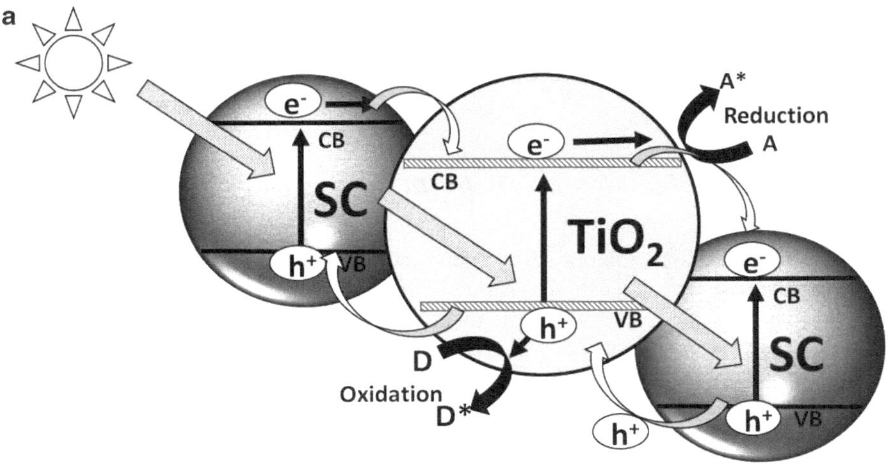


b

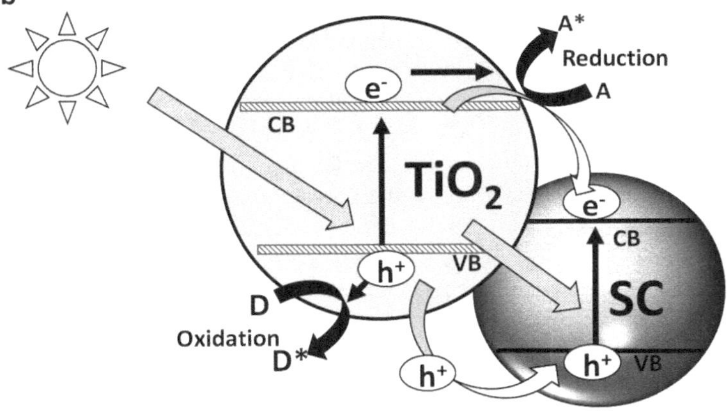


**Fig. 3.18** Schematic depiction of electron flow with introduction of second photocatalyst to enhance absorption (**a**) direct Z-scheme method and (**b**) indirect Z-scheme method

techniques are adapted in designing the semiconductor dyads, viz., the direct Z-scheme or the indirect Z-scheme.

In both methods the semiconductor with lower Fermi level possesses higher density of negative charges and the semiconductor with higher Fermi level possess more positive charge due to charge transfer (Fujii et al. 1998; Malato et al. 2009; Liu et al. 2010; Wang et al. 2010, 2013b). However, the only difference between the methods lies on the spatial charge separation occurring in direct Z-scheme and no charge separation taking place in indirect Z-scheme as shown in Fig. 3.18.

As it can be observed from Fig. 3.18b, dyads enjoy the merits of longer band widths in the visible spectrum. Wang et al. reported yields of 48 $ppmg^{-1}\ h^{-1}$ (methane) and 3.3 $ppmg^{-1}\ h^{-1}$ (methanol) in gas phase with CdSe quantum dots loaded with Pt impregnated $TiO_2$ films. They reported the charge injection into $TiO_2$ due to the shift of the conduction band of CdSe into higher energy which initiated $CO_2$ reduction. Process efficiency was also increased via charge separation due to electron transfer from CdSe to $TiO_2$. No activity was recorded when both semiconductors were used independently and using the same wavelength of light.

### 3.4.5 $TiO_2$-Noble Metal Nanoparticle Doping

Doping of $TiO_2$ like semiconductor material with noble metal nanoparticles have been developed more recently and is much more preferred than the former methods due to the broad visible range surface plasmonic activity showing high photocatalytic stability. These materials have also shown higher efficiency and selectivity. For instance, gold nanoparticles on $TiO_2$ has gained importance as a thermal catalyst for various reactions involving low temperature CO oxidation, highly selective oxidation of alcohols, nitro group reduction and epoxide rearrangement (Daniel and Astruc 2004; Mayer and Hafner 2011). This material has gained the benchmark catalyst status and widely used for comparative investigation (Cunningham et al. 1993; Haruta and Date 2001; Kudo and Miseki 2009).

The excellent photocatalytic activity can be attributed to the visible light absorption of gold nanoparticles. On absorption of one photon, hot electrons are injected into the conduction band of $TiO_2$ (Fig. 3.19a). Upon doping, the band gap of the semiconductor is altered by inserting new levels in the band gap. These levels can correspond to empty orbitals below the conduction band energy of the semiconductor or by introducing additional occupied orbitals above the valence band of the semiconductor. One of the major issues related to doping is the reproducibility of the photocatalytic activity. The maximum catalytic activity is achieved only at an optimal doping concentration. A higher concentration of dopants shows detrimental effect as these metal centers also take part in charge recombination. Most of the doping procedures undergo heat treatment procedures after which the dopant metal can either be embedded deep in to the semiconductor framework or completely get removed (Bamwenda et al. 1995; Chen and Mao 2007). Determining in advance the amount of metal dopant embedded or removed from the semiconductor is not possible. Further, the long term photocatalytic stability should be reliable as the activity of a doped semiconductor material decays with operation.

### 3.4.6 $TiO_2$-Noble Metal-Metal Oxide Doping

Yet another variation of composite material is incorporation of metal oxide in a catalyst already doped with noble metal nanoparticles. These metal oxides generally possess smaller band gaps than $TiO_2$. Nanostructured metal oxides provide better structural parameters—surface area and porosity. In bulk form, the composites will exhibit detrimental effect. Depending on the framework and metal oxide may take advantage of one or more these benefits. The most commonly used metal

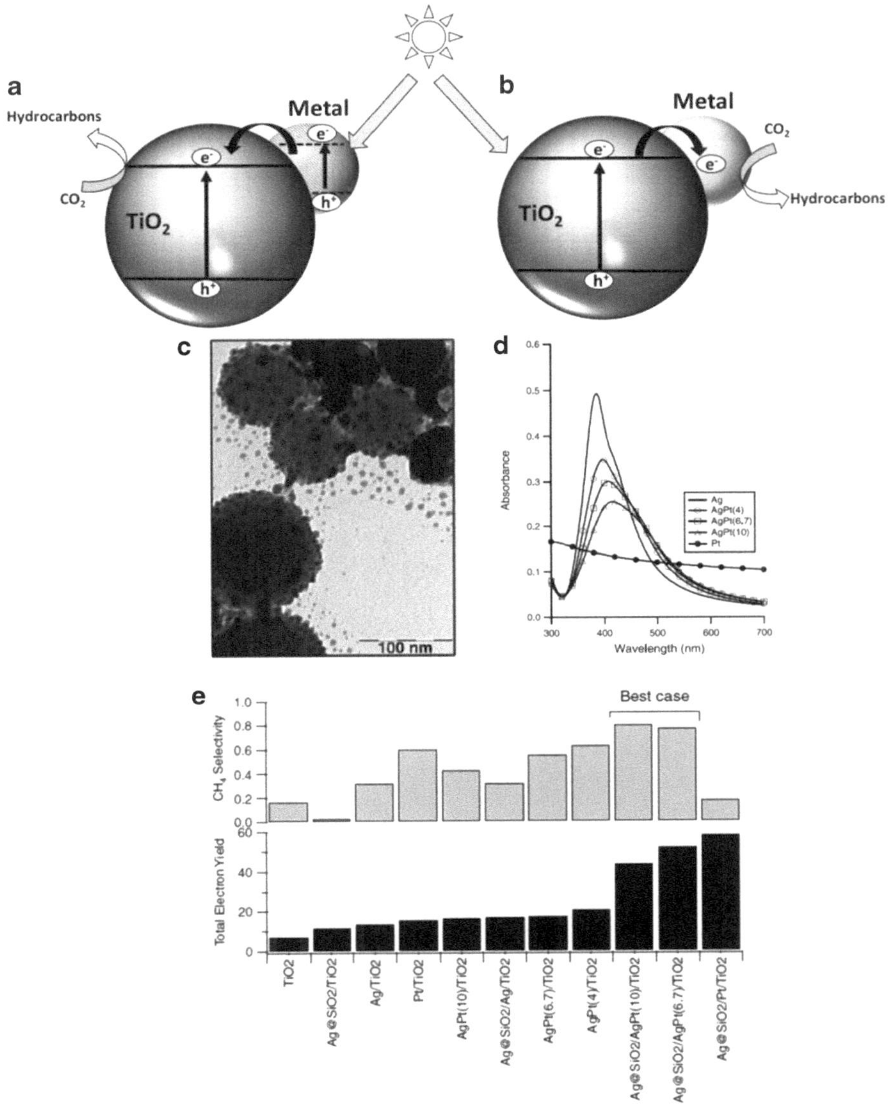


**Fig. 3.19** Schematic illustrations of (**a**) plasmonic metal nanoparticle decorated $TiO_2$ and (**b**) metal nanoparticle catalyst decorated $TiO_2$ in $CO_2$ reduction process; (**c**) TEM image of Ag–Pt nanoparticles on $TiO_2$; (**d**) optical absorption properties of Ag, Pt, and bimetallic Ag–Pt nanoparticles in aqueous solutions (**e**) photoreduction performance of $CO_2$ using Ag-Pt decorated $TiO_2$ ((**c**–**e**) adopted from (Mankidy et al. 2013)with permission by IOP publishers)

oxides are MgO (Liu et al. 2013; Xie et al. 2013a, b) and $Cu_xO$ (In et al. 2012; Zhai et al. 2013; Xu et al. 2014). A study by Xie et al. (2013a, b) showed $CO_2$ reduction over metal oxides deposited on $TiO_2$/Pt. The metal oxides dealt in this report consisted of MgO SrO, CaO, BaO, $La_2O_3$, and $Lu_2O_3$ they found that the basicity had a strong impact on the $CH_4$ yield. Increased $CH_4$ yields were obtained with MgO–$TiO_2$ composite. The authors attributed this result to chemisorption of $CO_2$ on MgO patches that concentrated and destabilized $CO_2$ molecules to be more reactive.

### 3.4.7 Heterogeneous Semiconductors as Photocatalysts

Pure semiconductors and matrix-dispersed photocatalysts are the two types of semiconductor materials widely used in $CO_2$ reduction (Dhakshinamoorthy et al. 2012; Izumi 2013; Navalon et al. 2013). Having briefly seen pure semiconductor materials in the previous sections, we will glance what matrix-dispersed photocatalysts are. A photoactive material dispersed in an inert material or matrix forms this class of photocatalyst. The inert matrix provides enhanced absorption characteristics and higher surface area. In 1997, Anpo et al. (1997) reported the photocatalytic reduction of $CO_2$ with $H_2O$ on titanium oxides anchored within micropores of zeolites. Also, they discussed the effect of the structure of the active sites and addition of Pt to this material. The highly dispersed titanium oxides included within zeolite cavities (Ti-oxide/zeolite) were prepared using an ion-exchange method. Figure 3.20a depicts the schematic structure of $TiO_2$ entrapment inside the micropores of Y-zeolites (Neatu et al. 2014). UV-irradiation of the photocatalyst in the presence of a mixture of $CO_2$ and $H_2O$ leads to the evolution of $CH_4$ and $CH_3OH$ at 328 K, as well as trace amounts of CO, $C_2H_4$, and $C_2H_6$. Figure 3.20b displays the distribution of the photocatalytic reduction of $CO_2$ from $H_2O$ using anatase $TiO_2$ powder coated with zeolite templates. It clearly shows that Pt co-catalyst coated $TiO_2$ onto zeolite template perform high photocatalytic $CO_2$ reduction compare to pristine $TiO_2$.

### 3.4.8 Graphene-Based Photocatalysts

Although the amount of work heading towards efficient $CO_2$ reduction is appreciable and is at a brisk phase, due to the comparatively complex mechanism involved, the achieved efficiency of solar to carbon-based fuel is only 1 % for a single semiconductor material. As described in the previous section, it would be now clear that $CO_2$ reduction is a multielectron transfer mechanism requiring at least 4–8 electrons and high negative reduction potential. Effective charge separation leads to efficient catalysis, this is the criterion which decides how well the photocatalyst will perform. The enhancement of photocatalytic activity is ascribed

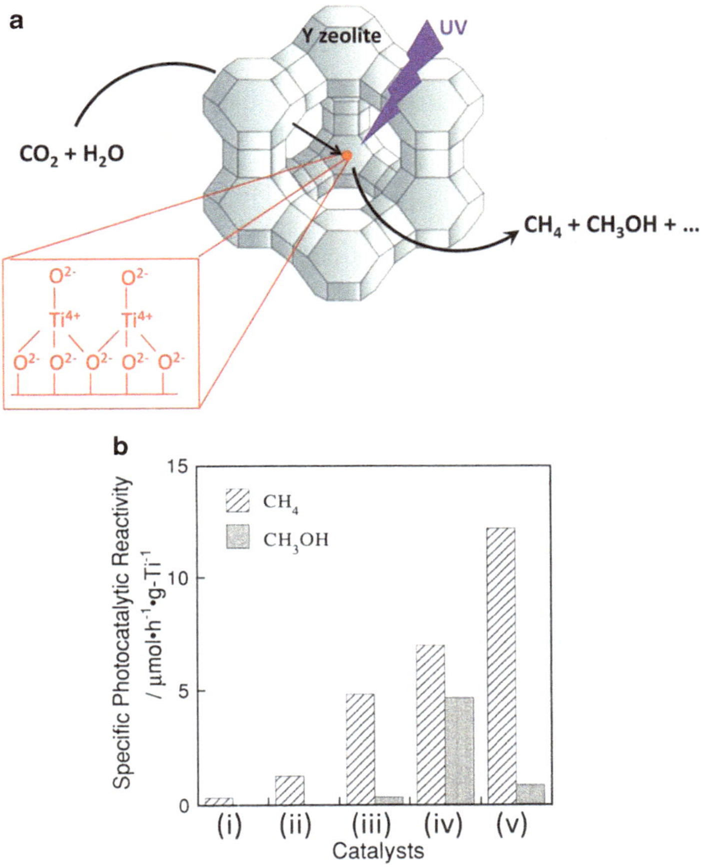


**Fig. 3.20** (**a**) Illustration of $TiO_2$ entrapped in Zeolite Y matrix acting as photocatalyst (adapted from Neatu et al. 2014, with permission from MDPI publishers); (**b**) product distribution of the photocatalytic reduction of $CO_2$ from $H_2O$ using anatase $TiO_2$ powder coated with zeolite templates for different cases: (*i*) the imp-Ti-oxide/Y-zeolite (10.0 wt % as $TiO_2$), (*ii*) the imp-Ti-oxide/Y-zeolite (1.0 wt % as $TiO_2$), (*iii*) the ex-Ti-oxide/Y-zeolite, (*iv*) and the Pt-loaded ex-Ti-oxide/Y-zeolite, (*v*) catalysts (adopted from (Anpo et al. 1997) with permission by American Chemical Society)

to suppressing the recombination of the photogenerated electrons and migration of charge. One key strategy which is catching up the race with other photocatalyst is the use of graphene. Many investigations in this line have proved the high performance of the graphene-based photocatalysts (Williams et al. 2008; Lightcap 2010; Rawalekar and Mokari 2013; Julkapli and Bagheri 2015). It is recognized that the high workfunction and higher conductivity of graphene enables acceptor/transporter to induce electron transfer from the conduction band of semiconductors to graphene, leading to an efficient charge separation (Hsu et al. 2013).

Graphene also acts as co-catalyst to catalyze $H_2$ production and $CO_2$ reduction. This can be attributed to the reduction potential of graphene being more negative than the reduction potential of water to hydrogen and of $CO_2$ to hydrocarbon fuels. In principle, the intrinsic structural and electronic properties of graphene can be tuned by heteroatom doping or by an electrostatic field. Graphene, graphene oxide and other derivatives have the potential to act as small bandgap semiconductors after properly adjusting their degree of oxidation. In general, composites of a photocatalyst and graphene have been demonstrated high catalytic efficiency.

Liang et al., very recently demonstrated exfoliated graphene/P25 $TiO_2$ nanocomposite thin film as a highly efficient visible light photocatalyst for $CO_2$ reduction. In general, the interface between photocatalysts and graphene in the graphene-based composite photocatalysts strongly affects the photocatalytic activity. A large contact area and strong interaction between the components can promote the transfer of electrons in the photocatalytic reaction and improve the performance (Liang et al. 2012).

Tu et al., recently developed a hollow sphere composite consisting of $TiO_2$ nanosheets and graphene oxide. A large number of $TiO_2$ nanoparticles are uniformly dispersed and tightly packed on the surface of graphene sheets to form a 2D sandwich-like structure. This close neighborhood of $TiO_2$ and graphene in the hybrid can favor the transfer of photogenerated electrons from $TiO_2$ to graphene sheets, and thus facilitate the charge separation efficiency (Fig. 3.21a, b). Furthermore, the presence of abundant $Ti^{3+}$ sites on the surface of $TiO_2$ creates a low-coordinated oxygen species, which leads to an increase of the number of basic sites in $TiO_2$. The graphene content has a great effect on the photocatalytic activity. With increasing graphene content, the production rate of $C_2H_6$ noticeable increases. This change is attributed to two factors: (1) The lower conversion rate of $CH_4$ for graphene/$TiO_2$ (G-$TiO_2$) nanocomposites is because methyl radicals ($^{\cdot}CH_3$) are intermediates and can be absorbed on the surface of graphene by π-conjugation. (2) The enhanced $C_2H_6$ production rate is due to the chemical bond interactions in the G-$TiO_2$ nanocomposites which favor the transfer of excited-state electrons to the graphene surface where they can react with $CO_2$. However, there is an optimized graphene content which gives the greatest overall production rate of $CH_4$ and $C_2H_6$ in the photocatalytic $CO_2$ conversion (Tu et al. 2013).

### *3.4.9 Graphene: Based Photoelectrocatalysts*

In contrast to photocatalytic $CO_2$ reduction, photoelectrocatalysts requires an applied external bias which is an added energy input and is undesirable for technological applications. However, the role of an electrical bias can increase the space charge layer and the magnitude of the electric fields, which can increase the lifetime of photogenerated charges, leading to higher photocatalytic efficiency. For instance, Cheng et al., reported a photoelectrocatalytic reduction of $CO_2$ to chemical fuels using Pt-modified reduced graphene oxide and Pt-modified $TiO_2$

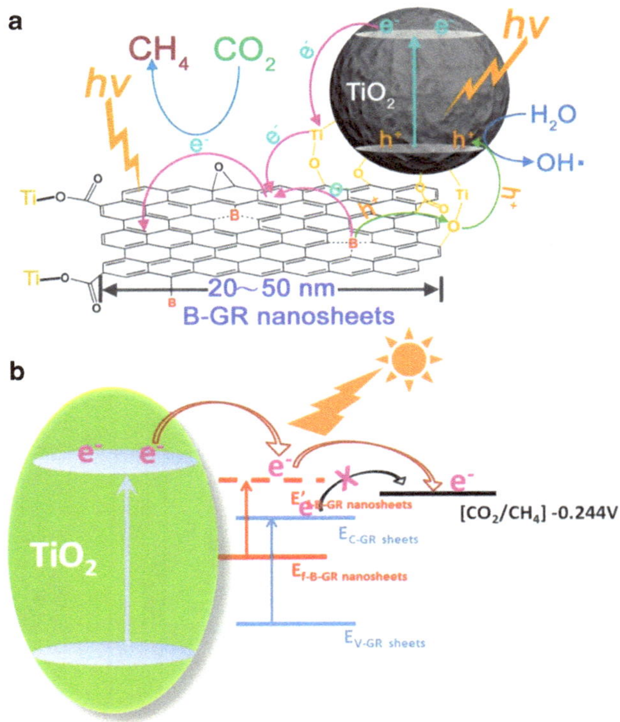


**Fig. 3.21** Schematic diagrams of (**a**) photo-excited electrons and holes transfer among $TiO_2$ nanoparticles and boron doped graphene nanosheets and (**b**) photo-generated electrons transfer between $TiO_2$ and graphene materials ((**a**) and (**b**) adopted from (Xing et al. 2014) with permission by Nature Publishers)

nanotubes as cathode and photoanode catalysts respectively. The studies on graphene-based photocatalysts for reduction of $CO_2$ are at its infancy. Considering the potentiality of graphene and the need for an efficient photocatalyst any step in this direction will be a significant leap (Tripkovic et al. 2013; Cheng et al. 2014a, b).

*Alternative non-$TiO_2$ Photocatalysts*: Although $TiO_2$ gobbles the limelight without any doubt, with advancement in technology and sophistication of instruments, other photocatalytic materials are being explored. The number of publication related to non-$TiO_2$ (Wang et al. 2015c) photocatalysts is gradually increasing. The current efforts in the development of inorganic photocatalysts are focused on gallium, germanium, tungsten-based materials. Niobium, vanadium, copper, and cerium oxides are also widely considered in this game. In addition, silver–silver halide compounds are reported as plasmonic systems, photocatalytic metal-organic frameworks, organic–inorganic composite photocatalysts are also widely studied. The recent reports on $TiO_2$ and non-$TiO_2$ based photocatalyst and their $CO_2$ reduction is summarized in Table 3.2.

**Table 3.2** Summary of $CO_2$ reductive photocatalyst materials and corresponding output products

| Photocatalyst | Synthesis method | Co-catalyst | Output products | Yield (μmol/g.h) (*ppm/g.h) (**mmol/h) | References |
|---|---|---|---|---|---|
| *Pure titania materials* | | | | | |
| Mesoporous $TiO_2$ NFs | Electrospinning and solvothermal treatment | – | $CH_4$ | 3.08–19.55 | (Fu et al. 2014) |
| Anatase $TiO_2$ with coexposed {001} and {101} facets | Solvothermal method | – | $CH_4$ | – | (Yu et al. 2014) |
| Hollow anatase $TiO_2$ single crystals and mesocrystals with dominant {101} facets | Hydrothermal | – | $CH_4$ | 18 | (Jiao et al. 2012) |
| *Titania-metal oxide composites* | | | | | |
| $TiO_2$/Pt-MgO | Photo-deposition | Pt-MgO | $CH_4$, CO | 0.7 ($CH_4$); 0.28 (CO) | (Xie et al. 2013b) |
| $Cu_2O$@Pt core-shell loaded onto $TiO_2$-P25 | Photo-deposition | Pt and Cu | $H_2$, CO, $CH_4$ | 25($H_2$); 8.3(CO); 33($CH_4$) | (Zhai et al. 2013) |
| $Cu_2O/TiO_2$ | Microwave-solvothermal | $Cu_2O$ | $CH_4$ | 25* | (Xu et al. 2014) |
| $CeO_2$-$TiO_2$ | Nano-casting | $CeO_2$ | $CH_4$, CO | 12000-13100 (CO);442($CH_4$) | (Wang et al. 2013c) |
| $Ni^{2+}$-$TiO_2$ | Sol-gel | $Ni^{2+}$ | $CH_4$, $CH_3OH$ | 7.7 ($CH_4$); 19.5 ($CH_3OH$) | (Ola and Mercedes Maroto-Valer 2014) |
| TiO2, ZnO and WO3 | – | – | CH2O | 16-17.14 | (Inoue et al. 1979) |
| *Titania-metal composites* | | | | | |
| Cu/$TiO_2$-P25 | Precipitation and calcination | Cu | $CH_4$, CO | 12-24 (CO); 0.46-0.65 ($CH_4$) | (Liu et al. 2013) |
| Ag doped $TiO_2$ | Sol-gel | Ag | $CH_4$, $CH_3OH$ | 0.44 ($CH_4$ + $CH_3OH$) | (Krejčíková et al. 2012) |
| Cu and Pt loaded $TiO_2$ NTs | Adonization | Cu and Pt | $CH_4$, $C_xH_y$ | 2.6 mL/g.h ($CH_4$); 3.5 mL/g.h ($C_xH_y$) | (Zhang et al. 2012) |
| Au/Pt NPs-$TiO_2$ | Electrospinning | Au/Pt | $CH_4$, CO | 114 ($CH_4$), 17 (CO) | (Wu et al. 2013) |

| | | | | | |
|---|---|---|---|---|---|
| *Titania-carbon composite* | | | | | |
| $TiO_2$-graphene 2D sandwiched like hybrid sheets | In situ simultaneous reduction-hydrolysis technique | Graphene | $CH_4$, $C_2H_6$ | 4.6-10.1 ($CH_4$), 7.2-16.8 ($C_2H_6$) | (Tu et al. 2013) |
| Hollow spheres consisting of alternate $TiO_2$ and Graphene | Layer-by-layer technique | Graphene | $CH_4$, CO | 1.1 ($CH_4$), 8.9 (CO) | (Tu et al. 2012) |
| MWCNT/$TiO_2$ core-shell | – | MWCNT | $CH_4$ | 0.17 ($CH_4$) | (Gui et al. 2014) |
| Solvent exfoliated graphene loaded Ti nanosheets | – | Graphene | $CH_4$ | 9.3 ($CH_4$) | (Liang et al. 2012) |
| *Non-titania photocatalysts* | | | | | |
| Mesoporous 4.5 ($ZnGa_2O_4$):($Zn_2GeO_4$) | Hydrothermal ion exchange reaction | – | $CH_4$, $O_2$ | 3.8($CH_4$), 22 ($O_2$) | (Yan et al. 2013) |
| $ZnGa_2O_4$ nanocubes with exposed {100} facets | Hydrothermal ion exchange reaction | – | $CH_4$, $O_2$ | 1.2($CH_4$), 12 ($O_2$) | (Yan et al. 2013) |
| Zn-doped, Ag-modified $Ga_2O_3$ | – | – | CO, $H_2$, $O_2$ | 117 (CO), 16.9 ($H_2$), 70.1 ($O_2$) | (Teramura et al. 2014) |
| $Cd_2Ge_2O_6$ nanowires ($RuO_2$ loaded) | Solution-chemical route | – | $CH_4$ | 0.72 ($CH_4$) | (Liu et al. 2014) |
| $NaNbO_3$ photocatalyst | Furfural alcohol derived polymerization oxidation (FAPO) process | – | $CH_4$ | 0.45–5.7 ($CH_4$) | (Li et al. 2012) |
| $In_2O_3$/g-$C_3N_4$ hybrid photocatalysts | Hydrothermal | – | $H_2$ | 1990* (with Pt) | (Lu et al. 2015) |
| Layered double hydroxides | Co-precipitation and hydrothermal | – | CO, $O_2$ | 0.5–3.7 (CO), 0.4–6.4 (O2) | (Teramura et al. 2012) |
| Plasmonic shaped AgX:Ag | glycerol-mediated solution route | – | $CH_3OH$ | 8 (AgCl:Ag, $CH_3OH$), 22 (AgBr:Ag, $CH_3OH$) | (An et al. 2012) |
| $W_{18}O_{49}$ | One-pot solution phase | | $CH_4$ | 666* | (Xi et al. 2012) |
| $CuGaO_2$ | Stoichiometrix mixing | – | CO | 9* | (Lekse et al. 2012) |
| $LiNbO_3$ | Pulvarized with a mortar and pestle to produce particles | – | $CH_2O_2$ | 7.7** | (Stock and Dunn 2011) |

(continued)

**Table 3.2** (continued)

| Photocatalyst | Synthesis method | Co-catalyst | Output products | Yield (μmol/g.h) (*ppm/g.h) (**mmol/h) | References |
|---|---|---|---|---|---|
| $Bi_2WO_6$ | Solid state reaction and hydrothermal method | – | CH4 | 1.1 | (Zhou et al. 2011) |
| $BiVO_4$ | Microwave assisted hydrothermal method | – | $C_2H_5OH$ | 110 | (Liu et al. 2009) |
| $Ga_2O_3$ | cleaning at 13.3 kPa of oxygen atmosphere at 1073 K for 1 h | – | $H_2$, CO | 4.13 (H2); 1.8 (CO) | (Yuliati et al. 2008) |
| $ZrO_2$ | $ZrO_2$ is immobilized by coated glass pellets | – | CO | 0.51 | (Lo et al. 2007) |
| MgO and $ZrO_2$ | Hydrated, filtered and calcinated in air at 873 K for 2 h | – | CO | 2.4 | (Teramura et al. 2004) |
| NiO | – | – | $CH_3OH$ | 393 | (Yahaya et al. 2004) |

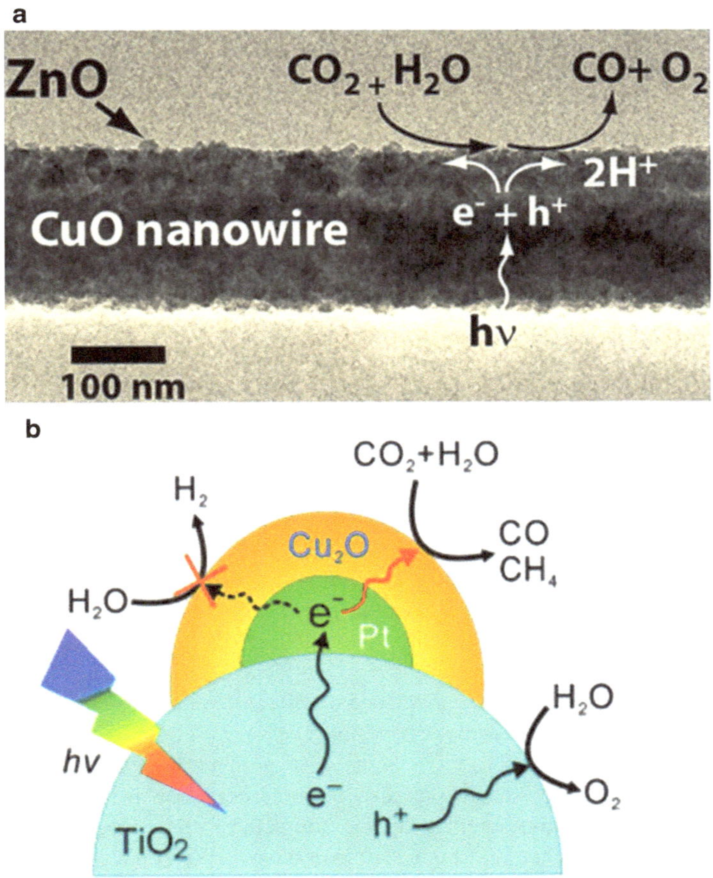


**Fig. 3.22** (**a**) schematic structure of CuO nanowire surface engineered by atomic layer deposition with ZnO (adapted from Wang et al. 2015c, with permission from American Chemical Society); (**b**) Core–shell structure of $Cu_2O$ shell coated on TiO2-Pt nanostructures (adapted from Rajeshwar et al. 2013, with permission from Wiley Publishers)

Recently Wang et al. reported the design and synthesis of CuO nanowires surface engineered with dense ZnO islands using a few pulsed cycles of atomic layer deposition (ALD) (Wang et al. 2015a, b, c, d). They utilized this material for $CO_2$ photoreduction and observed the formation of CO (Fig. 3.22a). They reported a peak conversion of 1.98 mmol$(g_{cat}h)^{-1}$ and a quantum efficiency of 0.0035%. The bilayered structure of CuO/$Cu_2O$ enhance the photoreduction of $CO_2$ to methanol by forming favorable band-edge alignment of both oxide phases (Rajeshwar et al. 2013). A thin layer of $Cu_2O$ can act as shell layer on metal oxide and could provide the sites for the preferential activation and conversion of $CO_2$. Zhai et al. (Rajeshwar et al. 2013) showed that the binary co-catalysts of Pt and $Cu_2O$ coated on $TiO_2$ forming a core–shell structure markedly promote the photocatalytic reduction of $CO_2$ with $H_2O$ to $CH_4$ and CO (Fig. 3.22b). Interestingly, deposition of $Cu_2O$ shell on Pt nanoparticles strongly suppresses the reduction of $H_2O$ to $H_2$ from 51 to 25 μmol $g^{-1}$ $h^{-1}$ and thus enhance the selectivity of $CO_2$ reduction and produce the high quantity methane yield 33 μmol $g^{-1}$ $h^{-1}$.

## 3.5 Conclusions

In this chapter, highly active and robust HER catalysts processed with inexpensive and earth-abundant elements are summarized and discussed. To date, a wide range of nonprecious HER catalysts have been successfully developed for photocatalytic and or photoelectrocatalytic $H_2$ production, which can be categorized into transition metals, transition metal compounds (e.g., oxides, hydroxides, sulfides, and carbides), nanocarbons, nanocarbon-based composites, natural hydrogenases, and artificial organometallic molecules. Yet the charge transfer mechanism at HER catalyst/electrolyte is not much understood. In particular, investigation on atomically well-bonded nanojunctions at catalyst–electrolyte interface has to be explored. The main challenges of (a) inhibiting recombination (b) lowering activation energy and (c) suppressing corrosion play comprehensive roles on achieving high-efficiency, stable, and cost-effective HER catalysts. New compositions of bimetallic alloys also a good choice for active HER catalysts. Apart from the rare earth material composition, new electrode architectures have to be encouraged for designing effective electrode–electrode interfaces in hydrogen evolution reaction.

In order to realize a fossil fuel free society it would not be ideal to look for direct energy from renewable energy but also to metamorphose the high quantity of pollutants scavenging the serenity of our atmosphere. Scientists have identified precise points in the chemical cycles of our biosphere, tapping which will be a vulnerable support not only to the energy system but to the whole ecosystem. A huge number of publications on $CO_2$ reduction over $TiO_2$ has been presented recently. A number of improvements over $TiO_2$ have been reported. However, the main disadvantage of this material is the wide band-gap and thereby poor visible light absorption. The growing number of types of modification with $TiO_2$ and without $TiO_2$ in the field of $CO_2$ photoreduction is outstanding and the deeper insights thrown up on these materials by scientists will lead to a breakthrough in the near future.

## References

Agu UA, Oliva MI, Marchetti SG, Heredia AC, Casuscelli SG, Crivello ME (2014) Synthesis and characterization of a mixture of CoFe2O4 and MgFe2O4 from layered double hydroxides: Band gap energy and magnetic responses. J Magn Magn Mater 369:249–259

Ahmed N, Morikawa M, Izumi Y (2012) Photocatalytic conversion of carbon dioxide into methanol using optimized layered double hydroxide catalysts. Catal Today 185:263–269

Akhter P, Hussain M, Saracco G, Russo N (2015) Novel nanostructured-TiO2 materials for the photocatalytic reduction of CO2 greenhouse gas to hydrocarbons and syngas. Fuel 149:55–65

Akimov AV, Jinnouchi R, Shirai S, Asahi R, Prezhdo OV (2015) Theoretical Insights into the Impact of Ru Catalyst Anchors on the Efficiency of Photocatalytic CO2 Reduction on Ta2O5. J Phys Chem B 119:7186–7197

An C, Wang J, Jiang W, Zhang M, Ming X, Wang S, Zhang Q (2012) Strongly visible-light responsive plasmonic shaped AgX:Ag (X = Cl, Br) nanoparticles for reduction of $CO_2$ to methanol. Nanoscale 4:5646–5650

An CH, Wang JZ, Qin C, Jiang W, Wang ST, Li Y, Zhang QH (2012) Synthesis of Ag@AgBr/AgCl heterostructured nanocashews with enhanced photocatalytic performance via anion exchange. J Mater Chem 22:13153–13158

Anpo M, Yamashita H, Ichihashi Y, Fujii Y, Honda M (1997) Photocatalytic reduction of CO2 with H2O on titanium oxides anchored within micropores of zeolites: Effects of the structure of the active sites and the addition of Pt. J Phys Chem B 101:2632–2636

Aresta M, Dibenedetto A, Baran T, Angelini A, Labuz P, Macyk W (2014) An integrated photocatalytic/enzymatic system for the reduction of CO2 to methanol in bioglycerol-water. Beilstein J Org Chem 10:2556–2565

Ataca C, Topsakal M, Aktürk E, Ciraci S (2011) A comparative study of lattice dynamics of three- and Two-dimensional MoS2. J Phys Chem C 115:16354–16361

Ataca C, Şahin H, Ciraci S (2012) Stable, single-layer MX2 transition-metal oxides and dichalcogenides in a honeycomb-like structure. J Phys Chem C 116:8983–8999

Augugliaro V, Kisch H, Loddo V, Lopez-Munoz MJ, Marquez-Alvarez C, Palmisano G, Palmisano L, Parrino F, Yurdakal S (2008) Photocatalytic oxidation of aromatic alcohols to aldehydes in aqueous suspension of home-prepared titanium dioxide 1. Selectivity enhancement by aliphatic alcohols. Appl Catal A Gen 349:182–188

Bai S, Wang L, Chen X, Du J, Xiong Y (2015) Chemically exfoliated metallic MoS2 nanosheets: a promising supporting co-catalyst for enhancing the photocatalytic performance of TiO2 nanocrystals. Nano Res 8:175–183

Bamwenda GR, Tsubota S, Nakamura T, Haruta M (1995) Photoassisted hydrogen-production from a water-ethanol solution—a comparison of activities of Au-Tio2 and Pt-Tio2. J Photochem Photobiol A Chem 89:177–189

Bellardita M, Di Paola A, Garcia-Lopez E, Loddo V, Marci G, Palmisano L (2013) Photocatalytic CO2 reduction in Gas-solid regime in the presence of bare, SiO2 supported or Cu-loaded TiO2 samples. Curr Org Chem 17:2440–2448

Benck JD, Hellstern TR, Kibsgaard J, Chakthranont P, Jaramillo TF (2014) Catalyzing the hydrogen evolution reaction (HER) with molybdenum sulfide nanomaterials. ACS Catal 4:3957–3971

Bonde J, Moses PG, Jaramillo TF, Norskov JK, Chorkendorff I (2009) Hydrogen evolution on nano-particulate transition metal sulfides. Faraday Discuss 140:219–231

Berndes G, Hoogwijk M, van den Broek R (2003) The contribution of biomass in the future global energy supply: a review of 17 studies. Biomass Bioenergy 25:1–28

Callejas JF, McEnaney JM, Read CG, Crompton JC, Biacchi AJ, Popczun EJ, Gordon TR, Lewis NS, Schaak RE (2014) Electrocatalytic and photocatalytic hydrogen production from acidic and neutral-pH aqueous solutions using iron phosphide nanoparticles. ACS Nano 8:11101–11107

Cao SW, Liu XF, Yuan YP, Zhang ZY, Liao YS, Fang J, Loo SCJ, Sum TC, Xue C (2014) Solar-to-fuels conversion over In2O3/g-C3N4 hybrid photocatalysts. Appl Catal B-Environ 147:940–946

Carp O, Huisman CL, Reller A (2004) Photoinduced reactivity of titanium dioxide. Prog Solid State Chem 32:33–177

Chang Y-H, Lin C-T, Chen T-Y, Hsu C-L, Lee Y-H, Zhang W, Wei K-H, Li L-J (2013) Highly efficient electrocatalytic hydrogen production by MoSx grown on graphene-protected 3D Ni foams. Adv Mater 25:756–760

Chang K, Mei Z, Wang T, Kang Q, Ouyang S, Ye J (2014) MoS2/graphene cocatalyst for efficient photocatalytic H2 evolution under visible light irradiation. ACS Nano 8:7078–7087
Chen JG (1996) Carbide and nitride overlayers on early transition metal surfaces: preparation, characterization, and reactivities. Chem Rev 96:1477–1498
Chen X, Mao SS (2007) Titanium dioxide nanomaterials: synthesis, properties, modifications, and applications. Chem Rev 107:2891–2959
Chen Z, Cummins D, Reinecke BN, Clark E, Sunkara MK, Jaramillo TF (2011) Core–shell MoO3–MoS2 nanowires for hydrogen evolution: a functional design for electrocatalytic materials. Nano Lett 11:4168–4175
Chen JS, Xin F, Qin SY, Yin XH (2013a) Photocatalytically reducing CO2 to methyl formate in methanol over ZnS and Ni-doped ZnS photocatalysts. Chem Eng J 230:506–512
Chen W-F, Muckerman JT, Fujita E (2013b) Recent developments in transition metal carbides and nitrides as hydrogen evolution electrocatalysts. Chem Commun 49:8896–8909
Chen WF, Wang CH, Sasaki K, Marinkovic N, Xu W, Muckerman JT, Zhu Y, Adzic RR (2013c) Highly active and durable nanostructured molybdenum carbide electrocatalysts for hydrogen production. Energy Environ Sci 6:943–951
Chen W-F, Schneider JM, Sasaki K, Wang C-H, Schneider J, Iyer S, Iyer S, Zhu Y, Muckerman JT, Fujita E (2014) Tungsten carbide–nitride on graphene nanoplatelets as a durable hydrogen evolution electrocatalyst. ChemSusChem 7:2414–2418
Cheng J, Zhang M, Wu G, Wang X, Zhou JH, Cen KF (2014a) Photoelectrocatalytic reduction of CO2 into chemicals using Pt-modified reduced graphene oxide combined with Pt-modified TiO2 nanotubes. Environ Sci Technol 48:7076–7084
Cheng L, Huang W, Gong Q, Liu C, Liu Z, Li Y, Dai H (2014b) Ultrathin WS2 nanoflakes as a high-performance electrocatalyst for the hydrogen evolution reaction. Angew Chem Int Ed 53:7860–7863
Choi C, Feng J, Li Y, Wu J, Zak A, Tenne R, Dai H (2013) WS2 nanoflakes from nanotubes for electrocatalysis. Nano Res 6:921–928
Chung DY, Park S-K, Chung Y-H, Yu S-H, Lim D-H, Jung N, Ham HC, Park H-Y, Piao Y, Yoo SJ, Sung Y-E (2014) Edge-exposed MoS2 nano-assembled structures as efficient electrocatalysts for hydrogen evolution reaction. Nanoscale 6:2131–2136
Conway BE, Tilak BV (2002) Interfacial processes involving electrocatalytic evolution and oxidation of H2, and the role of chemisorbed H. Electrochim Acta 47:3571–3594
Costentin C, Robert M, Saveant JM (2013) Catalysis of the electrochemical reduction of carbon dioxide. Chem Soc Rev 42:2423–2436
Cowan AJ, Durrant JR (2013) Long-lived charge separated states in nanostructured semiconductor photoelectrodes for the production of solar fuels. Chem Soc Rev 42:2281–2293
Cunningham D, Tsubota S, Kamijo N, Haruta M (1993) Preparation and catalytic behavior of subnanometer gold deposited on Tio2 by vacuum calcination. Res Chem Intermed 19:1–13
Daniel MC, Astruc D (2004) Gold nanoparticles: assembly, supramolecular chemistry, quantum-size-related properties, and applications toward biology, catalysis, and nanotechnology. Chem Rev 104:293–346
Deng J, Li H, Xiao J, Tu Y, Deng D, Yang H, Tian H, Li J, Ren P, Bao X (2015a) Triggering the electrocatalytic hydrogen evolution activity of the inert two-dimensional MoS2 surface via single-atom metal doping. Energy Environ Sci 8:1594–1601
Deng ZH, Li L, Ding W, Xiong K, Wei ZD (2015b) Synthesized ultrathin MoS2 nanosheets perpendicular to graphene for catalysis of hydrogen evolution reaction. Chem Commun 51:1893–1896
Dhakshinamoorthy A, Navalon S, Corma A, Garcia H (2012) Photocatalytic CO2 reduction by TiO2 and related titanium containing solids. Energy Environ Sci 5:9217–9233
Di Giovanni C, Wang W-A, Nowak S, Grenèche J-M, Lecoq H, Mouton L, Giraud M, Tard C (2014) Bioinspired iron sulfide nanoparticles for cheap and long-lived electrocatalytic molecular hydrogen evolution in neutral water. ACS Catal 4:681–687

Drobner E, Huber H, Wachtershauser G, Rose D, Stetter KO (1990) Pyrite formation linked with hydrogen evolution under anaerobic conditions. Nature 346:742–744

Du H, Liu Q, Cheng N, Asiri AM, Sun X, Li CM (2014) Template-assisted synthesis of CoP nanotubes to efficiently catalyze hydrogen-evolving reaction. J Mater Chem A 2:14812–14816

Duan J, Chen S, Chambers BA, Andersson GG, Qiao SZ (2015) 3D WS2 nanolayers@heteroatom-doped graphene films as hydrogen evolution catalyst electrodes. Adv Mater 27:4234–4241

Eda G, Fujita T, Yamaguchi H, Voiry D, Chen M, Chhowalla M (2012) Coherent atomic and electronic heterostructures of single-layer MoS2. ACS Nano 6:7311–7317

Esposito DV, Hunt ST, Kimmel YC, Chen JG (2012) A New class of electrocatalysts for hydrogen production from water electrolysis: metal monolayers supported on Low-cost transition metal carbides. J Am Chem Soc 134:3025–3033

Faber MS, Dziedzic R, Lukowski MA, Kaiser NS, Ding Q, Jin S (2014a) High-performance electrocatalysis using metallic cobalt pyrite (CoS2) micro- and nanostructures. J Am Chem Soc 136:10053–10061

Faber MS, Lukowski MA, Ding Q, Kaiser NS, Jin S (2014b) Earth-abundant metal pyrites (FeS2, CoS2, NiS2, and their alloys) for highly efficient hydrogen evolution and polysulfide reduction electrocatalysis. J Phys Chem C 118:21347–21356

Fan J, Liu EZ, Tian L, Hu XY, He Q, Sun T (2011) Synergistic effect of N and Ni2+ on nanotitania in photocatalytic reduction of CO2. J Environ Eng-Asce 137:171–176

Fu J, Cao S, Yu J, Low J, Lei Y (2014) Enhanced photocatalytic $CO_2$-reduction activity of electrospun mesoporous $TiO_2$ nanofibers by solvothermal treatment. Dalton Trans 43:9158–9165

Fujii H, Ohtaki M, Eguchi K, Arai H (1998) Preparation and photocatalytic activities of a semiconductor composite of CdS embedded in a TiO2 gel as a stable oxide semiconducting matrix. J Mol Catal A Chem 129:61–68

Fujishima A, Honda K (1972) Electrochemical photolysis of water at a semiconductor electrode. Nature 238:37–38

Gallo A, Montini T, Marelli M, Minguzzi A, Gombac V, Psaro R, Fornasiero P, Dal Santo V (2012) H2 production by renewables photoreforming on Pt-Au/TiO2 catalysts activated by reduction. Chemsuschem 5:1800–1811

Gao MY, Jiang D, Sun DK, Hou B, Li DB (2014) Synthesis of Ag/N-TiO2/SBA-15 photocatalysts and photocatalytic reduction of CO2 under visible light irradiation. Acta Chim Sin 72:1092–1098

Garcia-Esparza AT, Cha D, Ou Y, Kubota J, Domen K, Takanabe K (2013) Tungsten carbide nanoparticles as efficient cocatalysts for photocatalytic overall water splitting. ChemSusChem 6:168–181

Gopalakrishnan D, Damien D, Shaijumon MM (2014) MoS2 quantum Dot-interspersed exfoliated MoS2 nanosheets. ACS Nano 8:5297–5303

Guan GQ, Kida T, Yoshida A (2003) Reduction of carbon dioxide with water under concentrated sunlight using photocatalyst combined with Fe-based catalyst. Appl Catal B-Environ 41:387–396

Gui MM, Chai SP, Mohamed AR (2014) Modification of MWCNT@TiO2 core-shell nanocomposites with transition metal oxide dopants for photoreduction of carbon dioxide into methane. Appl Surf Sci 319:37–43

Habisreutinger SN, Schmidt-Mende L, Stolarczyk JK (2013) Photocatalytic reduction of CO2 on TiO2 and other semiconductors. Angew Chem-Int Ed 52:7372–7408

Hammouche M, Lexa D, Momenteau M, Saveant JM (1991) Chemical catalysis of electrochemical reactions—homogeneous catalysis of the electrochemical reduction of carbon-dioxide by iron(0) porphyrins—role of the addition of magnesium cations. J Am Chem Soc 113:8455–8466

Harnisch F, Schröder U, Quaas M, Scholz F (2009a) Electrocatalytic and corrosion behaviour of tungsten carbide in near-neutral pH electrolytes. Appl Catal B Environ 87:63–69

Harnisch F, Sievers G, Schröder U (2009b) Tungsten carbide as electrocatalyst for the hydrogen evolution reaction in pH neutral electrolyte solutions. Appl Catal B Environ 89:455–458

Haruta M, Date M (2001) Advances in the catalysis of Au nanoparticles. Appl Catal A Gen 222:427–437
Heller A (1981) Conversion of sunlight into electrical power and photoassisted electrolysis of water in photoelectrochemical cells. Acc Chem Res 14:154–162
Hinnemann B, Moses PG, Bonde J, Jørgensen KP, Nielsen JH, Horch S, Chorkendorff I, Nørskov JK (2005) Biomimetic hydrogen evolution: MoS2 nanoparticles as catalyst for hydrogen evolution. J Am Chem Soc 127:5308–5309
Hoffmann MR, Moss JA, Baum MM (2011) Artificial photosynthesis: semiconductor photocatalytic fixation of CO2 to afford higher organic compounds. Dalton Trans 40:5151–5158
Hou Y, Zhang B, Wen Z, Cui S, Guo X, He Z, Chen J (2014) A 3D hybrid of layered MoS2/ nitrogen-doped graphene nanosheet aerogels: an effective catalyst for hydrogen evolution in microbial electrolysis cells. J Mater Chem A 2:13795–13800
Hsu H-C, Shown I, Wei H-Y, Chang Y-C, Du H-Y, Lin Y-G, Tseng C-A, Wang C-H, Chen L-C, Lin Y-C, Chen K-H (2013) Graphene oxide as a promising photocatalyst for CO2 to methanol conversion. Nanoscale 5:262–268
Huang X, Zeng Z, Bao S, Wang M, Qi X, Fan Z, Zhang H (2013) Solution-phase epitaxial growth of noble metal nanostructures on dispersible single-layer molybdenum disulfide nanosheets. Nat Commun 4:1444
Huang C, Chen C, Zhang M, Lin L, Ye X, Lin S, Antonietti M, Wang X (2015) Carbon-doped BN nanosheets for metal-free photoredox catalysis. Nat Commun 6
Hwu HH, Chen JG (2005) Surface chemistry of transition metal carbides. Chem Rev 105:185–212
In SI, Vaughn DD, Schaak RE (2012) Hybrid CuO-TiO2-xNx hollow nanocubes for photocatalytic conversion of CO2 into methane under solar irradiation. Angew Chem-Int Ed 51:3915–3918
Indrakanti VP, Kubicki JD, Schobert HH (2009) Photoinduced activation of CO2 on Ti-based heterogeneous catalysts: Current state, chemical physics-based insights and outlook. Energy Environ Sci 2:745–758
Inoue T, Fujishima A, Konishi S, Honda K (1979) Photoelectrocatalytic reduction of carbon dioxide in aqueous suspensions of semiconductor powders. Nature 277:637–638
Ivanovskaya A, Singh N, Liu R-F, Kreutzer H, Baltrusaitis J, Van Nguyen T, Metiu H, McFarland E (2013) Transition metal sulfide hydrogen evolution catalysts for hydrobromic acid electrolysis. Langmuir 29:480–492
Izumi Y (2013) Recent advances in the photocatalytic conversion of carbon dioxide to fuels with water and/or hydrogen using solar energy and beyond. Coord Chem Rev 257:171–186
Jaegermann W, Tributsch H (1988) Interfacial properties of semiconducting transition metal chalcogenides. Prog Surf Sci 29:1–167
Jaramillo TF, Jørgensen KP, Bonde J, Nielsen JH, Horch S, Chorkendorff I (2007) Identification of active edge sites for electrochemical H2 evolution from MoS2 nanocatalysts. Science 317:100–102
Jiao W, Wang L, Liu G, Lu GQ, Cheng H-M (2012) Hollow anatase $TiO_2$ single crystals and mesocrystals with dominant {101} facets for improved photocatalysis activity and tuned reaction preference. ACS Catalysis 2:1854–1859
Julkapli NM, Bagheri S (2015) Graphene supported heterogeneous catalysts: An overview. Int J Hydrog Energy 40:948–979
Kibsgaard J, Chen Z, Reinecke BN, Jaramillo TF (2012) Engineering the surface structure of MoS2 to preferentially expose active edge sites for electrocatalysis. Nat Mater 11:963–969
Koci K, Obalova L, Lacny Z (2008) Photocatalytic reduction of CO2 over TiO2 based catalysts. Chem Pap 62:1–9
Kohno Y, Tanaka T, Funabiki T, Yoshida S (2000) Reaction mechanism in the photoreduction of CO2 with CH4 over ZrO2. Phys Chem Chem Phys 2:5302–5307
Kong D, Cha JJ, Wang H, Lee HR, Cui Y (2013) First-row transition metal dichalcogenide catalysts for hydrogen evolution reaction. Energy Environ Sci 6:3553–3558

Krejčíková S, Matějová L, Kočí K, Obalová L, Matěj Z, Čapek L, Šolcová O (2012) Preparation and characterization of Ag-doped crystalline titania for photocatalysis applications. Appl Catal Environ 111–112:119–125
Kudo A, Miseki Y (2009) Heterogeneous photocatalyst materials for water splitting. Chem Soc Rev 38:253–278
Laursen AB, Kegnaes S, Dahl S, Chorkendorff I (2012) Molybdenum sulfides-efficient and viable materials for electro—and photoelectrocatalytic hydrogen evolution. Energy Environ Sci 5:5577–5591
Lee HS, Min S-W, Chang Y-G, Park MK, Nam T, Kim H, Kim JH, Ryu S, Im S (2012) MoS2 nanosheet phototransistors with thickness-modulated optical energy gap. Nano Lett 12:3695–3700
Lee K, Mazare A, Schmuki P (2014) One-dimensional titanium dioxide nanomaterials: nanotubes. Chem Rev 114:9385–9454
Lekse JW, Underwood MK, Lewis JP, Matranga C (2012) Synthesis, characterization, electronic structure, and photocatalytic behavior of $CuGaO_2$ and $CuGa_1$–$xFexO_2$ (x = 0.05, 0.10, 0.15, 0.20) Delafossites. J Phy Chem C 116:1865–1872
Levy RB, Boudart M (1973) Platinum-like behavior of tungsten carbide in surface catalysis. Science 181:547–549
Lewis NS, Nocera DG (2006) Powering the planet: chemical challenges in solar energy utilization. Proc Natl Acad Sci 103:15729–15735
Li Y, Wang H, Xie L, Liang Y, Hong G, Dai H (2011) MoS2 nanoparticles grown on graphene: an advanced catalyst for the hydrogen evolution reaction. J Am Chem Soc 133:7296–7299
Li KF, An XQ, Park KH, Khraisheh M, Tang JW (2014) A critical review of CO2 photoconversion: catalysts and reactors. Catal Today 224:3–12
Liang YT, Vijayan BK, Lyandres O, Gray KA, Hersam MC (2012) Effect of dimensionality on the photocatalytic behavior of carbon-titania nanosheet composites: charge transfer at nanomaterial interfaces. J Phys Chem Lett 3:1760–1765
Lightcap IV, Kosel TH, Kamat PV (2010) Anchoring semiconductor and metal nanoparticles on a Two-dimensional catalyst Mat. Storing and shuttling electrons with reduced graphene oxide. Nano Lett 10:577–583
Liu P, Rodriguez JA (2005) Catalysts for hydrogen evolution from the [NiFe] hydrogenase to the Ni2P(001) surface: the importance of ensemble effect. J Am Chem Soc 127:14871–14878
Liu Y, Huang B, Dai Y, Zhang X, Qin X, Jiang M, Whangbo M-H (2009) Selective ethanol formation from photocatalytic reduction of carbon dioxide in water with BiVO4 photocatalyst. Catalysis Commun 11:210–213
Liu G, Wang LZ, Yang HG, Cheng HM, Lu GQ (2010) Titania-based photocatalysts-crystal growth, doping and heterostructuring. J Mater Chem 20:831–843
Liu LJ, Zhao CY, Zhao HL, Pitts D, Li Y (2013) Porous microspheres of MgO-patched TiO2 for CO2 photoreduction with H2O vapor: temperature-dependent activity and stability. Chem Commun 49:3664–3666
Liu N, Kim P, Kim JH, Ye JH, Kim S, Lee CJ (2014a) Large-area atomically thin MoS2 nanosheets prepared using electrochemical exfoliation. ACS Nano 8:6902–6910
Liu R, Gu S, Du H, Li CM (2014b) Controlled synthesis of FeP nanorod arrays as highly efficient hydrogen evolution cathode. J Mater Chem A 2:17263–17267
Lo C-C, Hung C-H, Yuan C-S, Wu J-F (1765–1774) Photoreduction of carbon dioxide with $H_2$ and $H_2O$ over $TiO_2$ and $ZrO_2$ in a circulated photocatalytic reactor. Sol Energ Mat Sol Cells 2007:91
Lovegrove K, Luzzi A, Kreetz H (1999) A solar-driven ammonia-based thermochemical energy storage system. Sol Energy 67:309–316
Lu Y, Chu D, Zhu M, Du Y, Yang P (2015) Exfoliated carbon nitride nanosheets decorated with NiS as an efficient noble-metal-free visible-light-driven photocatalyst for hydrogen evolution. Phys Chem Chem Phys 17:17355–17361

Lu Z, Zhu W, Yu X, Zhang H, Li Y, Sun X, Wang X, Wang H, Wang J, Luo J, Lei X, Jiang L (2014) Ultrahigh hydrogen evolution performance of under-water "superaerophobic" MoS2 nanostructured electrodes. Adv Mater 26:2683–2687

Lu Q, Hutchings GS, Yu W, Zhou Y, Forest RV, Tao R, Rosen J, Yonemoto BT, Cao Z, Zheng H, Xiao JQ, Jiao F, Chen JG (2015) Highly porous non-precious bimetallic electrocatalysts for efficient hydrogen evolution. Nat Commun 6

Lukowski MA, Daniel AS, Meng F, Forticaux A, Li L, Jin S (2013) Enhanced hydrogen evolution catalysis from chemically exfoliated metallic MoS2 nanosheets. J Am Chem Soc 135:10274–10277

Lukowski MA, Daniel AS, English CR, Meng F, Forticaux A, Hamers RJ, Jin S (2014) Highly active hydrogen evolution catalysis from metallic WS2 nanosheets. Energy Environ Sci 7:2608–2613

Ma Y, Wang XL, Jia YS, Chen XB, Han HX, Li C (2014) Titanium dioxide-based nanomaterials for photocatalytic fuel generations. Chem Rev 114:9987–10043

Malato S, Fernandez-Ibanez P, Maldonado MI, Blanco J, Gernjak W (2009) Decontamination and disinfection of water by solar photocatalysis: recent overview and trends. Catal Today 147:1–59

Mankidy BD, Joseph B, Gupta VK (2013) Photo-conversion of CO 2 using titanium dioxide: enhancements by plasmonic and co-catalytic nanoparticles. Nanotechnology 24:405402

Manzanares M, Fabrega C, Osso JO, Vega LF, Andreu T, Morante JR (2014) Engineering the TiO2 outermost layers using magnesium for carbon dioxide photoreduction. Appl Catal B-Environ 150:57–62

Mao J, Li K, Peng TY (2013) Recent advances in the photocatalytic CO2 reduction over semiconductors. Catal Sci Technol 3:2481–2498

Marković NM, Grgur BN, Ross PN (1997) Temperature-dependent hydrogen electrochemistry on platinum low-index single-crystal surfaces in acid solutions. J Phys Chem B 101:5405–5413

Marszewski M, Cao SW, Yu JG, Jaroniec M (2015) Semiconductor-based photocatalytic CO2 conversion. Mater Horizons 2:261–278

Maugh TH 2nd (1983) Catalysis in Solar Energy: Photoelectrochemical cells approach the efficiency of photovoltaics, but durability and cost are still major hurdles. Science 221:1358–1361

Mayer KM, Hafner JH (2011) Localized Surface Plasmon Resonance Sensors. Chem Rev 111:3828–3857

McCrory CCL, Jung S, Ferrer IM, Chatman SM, Peters JC, Jaramillo TF (2015) Benchmarking Hydrogen Evolving Reaction and Oxygen Evolving Reaction Electrocatalysts for Solar Water Splitting Devices. J Am Chem Soc 137:4347–4357

McEnaney JM, Crompton JC, Callejas JF, Popczun EJ, Biacchi AJ, Lewis NS, Schaak RE (2014) Amorphous Molybdenum Phosphide Nanoparticles for Electrocatalytic Hydrogen Evolution. Chem Mater 26:4826–4831

Mdleleni MM, Hyeon T, Suslick KS (1998) Sonochemical Synthesis of Nanostructured Molybdenum Sulfide. J Am Chem Soc 120:6189–6190

Meng F, Li J, Cushing SK, Zhi M, Wu N (2013) Solar Hydrogen Generation by Nanoscale p–n Junction of p-type Molybdenum Disulfide/n-type Nitrogen-Doped Reduced Graphene Oxide. J Am Chem Soc 135:10286–10289

Mills A, LeHunte S (1997) An overview of semiconductor photocatalysis. J Photochem Photobiol A Chem 108:1–35

Min S, Lu G (2012) Sites for High Efficient Photocatalytic Hydrogen Evolution on a Limited-Layered MoS2 Cocatalyst Confined on Graphene Sheets—The Role of Graphene. J Phys Chem C 116:25415–25424

Morales-Guio CG, Stern L-A, Hu X (2014) Nanostructured hydrotreating catalysts for electrochemical hydrogen evolution. Chem Soc Rev 43:6555–6569

Navalon S, Dhakshinamoorthy A, Alvaro M, Garcia H (2013) Photocatalytic CO2 Reduction using Non-Titanium Metal Oxides and Sulfides. Chemsuschem 6:562–577

Neatu S, Macia-Agullo JA, Garcia H (2014) Solar Light Photocatalytic CO2 Reduction: General Considerations and Selected Bench-Mark Photocatalysts. Int J Mol Sci 15:5246–5262
Nguyen T-V, Wu JCS, Chiou C-H (2008) Photoreduction of CO2 over Ruthenium dye-sensitized TiO2-based catalysts under concentrated natural sunlight. Catal Commun 9:2073–2076
Ni M, Leung MKH, Leung DYC, Sumathy K (2007) A review and recent developments in photocatalytic water-splitting using TiO2 for hydrogen production. Renew Sustain Energy Rev 11:401–425
Nikiforov AV, Petrushina IM, Christensen E, Alexeev NV, Samokhin AV, Bjerrum NJ (2012) WC as a non-platinum hydrogen evolution electrocatalyst for high temperature PEM water electrolysers. Int J Hydrog Energy 37:18591–18597
Nikolov I, Petrov K, Vitanov T, Guschev A (1983) Tungsten carbide cathodes for electrolysis of sulphuric acid solutions. Int J Hydrog Energy 8:437–440
Ola O, Mercedes Maroto-Valer M (2014) Role of catalyst carriers in $CO_2$ photoreduction over nanocrystalline nickel loaded $TiO_2$-based photocatalysts. J Catal 309:300–308
Onuchukwu AI (1982) Hydrogen evolution on iron pyrite (FeSy) in sodium chloride solution. Electrochim Acta 27:529–533
Oyama ST (1996) The Chemistry of Transition Metal Carbides and Nitrides. Springer, New York
Oyama ST, Gott T, Zhao H, Lee Y-K (2009) Transition metal phosphide hydroprocessing catalysts: A review. Catal Today 143:94–107
Panwar NL, Kaushik SC, Kothari S (2011) Role of renewable energy sources in environmental protection: A review. Renew Sustain Energy Rev 15:1513–1524
Parsons R (1958) The rate of electrolytic hydrogen evolution and the heat of adsorption of hydrogen. Trans Faraday Soc 54:1053–1063
Popczun EJ, McKone JR, Read CG, Biacchi AJ, Wiltrout AM, Lewis NS, Schaak RE (2013) Nanostructured Nickel Phosphide as an Electrocatalyst for the Hydrogen Evolution Reaction. J Am Chem Soc 135:9267–9270
Qin S, Lei W, Liu D, Chen Y (2014) In-situ and tunable nitrogen-doping of $MoS_2$ nanosheets. Sci Rep 4. doi:10.1038/srep07582
Quaino P, Juarez F, Santos E, Schmickler W (2014) Volcano plots in hydrogen electrocatalysis—uses and abuses. Beilstein J Nanotechnol 5:846–854
Raja R, Sudhagar P, Devadoss A, Terashima C, Shrestha LK, Nakata K, Jayavel R, Ariga K, Fujishima A (2015) Pt-free solar driven photoelectrochemical hydrogen fuel generation using 1T MoS2 co-catalyst assembled CdS QDs/TiO2 photoelectrode. Chem Commun 51:522–525
Rajeshwar K, de Tacconi NR, Ghadimkhani G, Chanmanee W, Janáky C (2013) Tailoring copper oxide semiconductor nanorod arrays for photoelectrochemical reduction of carbon dioxide to methanol. ChemPhysChem 14:2251–2259
Rawalekar S, Mokari T (2013) Rational Design of Hybrid Nanostructures for Advanced Photocatalysis. Adv Energy Mater 3:12–27
Roduner E (2014) Understanding catalysis. Chem Soc Rev 43:8226–8239
Roy AK, Park SY, In I (2015) Mussel-inspired synthesis of boron nitride nanosheet-supported gold nanoparticles and their application for catalytic reduction of 4-nitrophenol. Nanotechnology 26:105601
Sang LX, Zhao YX, Burda C (2014) TiO2 Nanoparticles as Functional Building Blocks. Chem Rev 114:9283–9318
Schmid A, Dordick JS, Hauer B, Kiener A, Wubbolts M, Witholt B (2001) Industrial biocatalysis today and tomorrow. Nature 409:258–268
Shen H, Duan C, Guo J, Zhao N, Xu J (2015) Facile in situ synthesis of silver nanoparticles on boron nitride nanosheets with enhanced catalytic performance. J Mater Chem A 3:16663–16669
Song I, Park C, Choi HC (2015) Synthesis and properties of molybdenum disulphide: from bulk to atomic layers. RSC Adv 5:7495–7514
Stock M, Dunn S (2011) $LiNbO_3$—a polar material for solid-gas artificial photosynthesis. Ferroelectrics 419:9–13

Sun Y, Liu C, Grauer DC, Yano J, Long JR, Yang P, Chang CJ (2013) Electrodeposited Cobalt-Sulfide Catalyst for Electrochemical and Photoelectrochemical Hydrogen Generation from Water. J Am Chem Soc 135:17699–17702

Tahir M, Amin NS (2013) Advances in visible light responsive titanium oxide-based photocatalysts for CO2 conversion to hydrocarbon fuels. Energy Convers Manag 76:194–214

Tang C, Pu Z, Liu Q, Asiri AM, Sun X (2015) NiS2 nanosheets array grown on carbon cloth as an efficient 3D hydrogen evolution cathode. Electrochim Acta 153:508–514

Teramura K, Tanaka T, Ishikawa H, Kohno Y, Funabiki T (2004) Photocatalytic reduction of $CO_2$ to CO in the presence of $H_2$ or $CH_4$ as a reductant over MgO. J Phys Chem B 108:346–354

Teramura K, Iguchi S, Mizuno Y, Shishido T, Tanaka T (2012) Photocatalytic conversion of $CO_2$ in water over layered double hydroxides. Angew Chem Int Ed 51:8008–8011

Teramura K, Wang Z, Hosokawa S, Sakata Y, Tanaka T (2014) A doping technique that suppresses undesirable $H_2$ evolution derived from overall water splitting in the highly selective photocatalytic conversion of $CO_2$ in and by water. Chem Eur J 20:9906–9909

Thampi KR, Kiwi J, Gratzel M (1987) Methanation and photo-methanation of carbon dioxide at room temperature and atmospheric pressure. Nature 327:506–508

Tian Y, Ge L, Wang K, Chai Y (2014) Synthesis of novel MoS2/g-C3N4 heterojunction photocatalysts with enhanced hydrogen evolution activity. Mater Charact 87:70–73

Tran PD, Chiam SY, Boix PP, Ren Y, Pramana SS, Fize J, Artero V, Barber J (2013) Novel cobalt/nickel-tungsten-sulfide catalysts for electrocatalytic hydrogen generation from water. Energy Environ Sci 6:2452–2459

Tributsch H, Bennett JC (1977) Electrochemistry and photochemistry of MoS2 layer crystals. I. J Electroanal Chem Interfacial Electrochem 81:97–111

Tripkovic V, Vanin M, Karamad M, Bjorketun ME, Jacobsen KW, Thygesen KS, Rossmeisl J (2013) Electrochemical CO2 and CO Reduction on Metal-Functionalized Porphyrin-like Graphene. J Phys Chem C 117:9187–9195

Tsai C, Abild-Pedersen F, Nørskov JK (2014) Tuning the MoS2 Edge-Site Activity for Hydrogen Evolution via Support Interactions. Nano Lett 14:1381–1387

Tu WG, Zhou Y, Zou ZG (2013) Versatile Graphene-Promoting Photocatalytic Performance of Semiconductors: Basic Principles, Synthesis, Solar Energy Conversion, and Environmental Applications. Adv Funct Mater 23:4996–5008

Tu W, Zhou Y, Liu Q, Tian Z, Gao J, Chen X, Zhang H, Liu J, Zou Z (2012) Robust hollow spheres consisting of alternating titania nanosheets and graphene nanosheets with high photocatalytic activity for $CO_2$ conversion into renewable fuels. Adv Funct Mater 22:1215–1221

Turner JA (2004) Sustainable Hydrogen Production. Science 305:972–974

Turner JA (2014) Shining a light on solar water splitting--response. Science 344:469–470

Van Thanh D, Pan C-C, Chu C-W, Wei K-H (2014) Production of few-layer MoS2 nanosheets through exfoliation of liquid N2-quenched bulk MoS2. RSC Adv 4:15586–15589

Vesborg PCK, Seger B, Chorkendorff I (2015) Recent Development in Hydrogen Evolution Reaction Catalysts and Their Practical Implementation. J Phys Chem Lett 6:951–957

Voiry D, Salehi M, Silva R, Fujita T, Chen M, Asefa T, Shenoy VB, Eda G, Chhowalla M (2013a) Conducting MoS2 Nanosheets as Catalysts for Hydrogen Evolution Reaction. Nano Lett 13:6222–6227

Voiry D, Yamaguchi H, Li J, Silva R, Alves DCB, Fujita T, Chen M, Asefa T, Shenoy VB, Eda G, Chhowalla M (2013b) Enhanced catalytic activity in strained chemically exfoliated WS2 nanosheets for hydrogen evolution. Nat Mater 12:850–855

Wang CJ, Thompson RL, Baltrus J, Matranga C (2010) Visible Light Photoreduction of CO2 Using CdSe/Pt/TiO2 Heterostructured Catalysts. J Phys Chem Lett 1:48–53

Wang QH, Kalantar-Zadeh K, Kis A, Coleman JN, Strano MS (2012) Electronics and optoelectronics of two-dimensional transition metal dichalcogenides. Nat Nano 7:699–712

Wang D, Wang Z, Wang C, Zhou P, Wu Z, Liu Z (2013a) Distorted MoS2 nanostructures: An efficient catalyst for the electrochemical hydrogen evolution reaction. Electrochem Commun 34:219–222

Wang H, Lu Z, Xu S, Kong D, Cha JJ, Zheng G, Hsu P-C, Yan K, Bradshaw D, Prinz FB, Cui Y (2013b) Electrochemical tuning of vertically aligned MoS2 nanofilms and its application in improving hydrogen evolution reaction. Proc Natl Acad Sci 110:19701–19706
Wang YG, Li B, Zhang CL, Cui LF, Kang SF, Li X, Zhou LH (2013c) Ordered mesoporous CeO2-TiO2 composites: Highly efficient photocatalysts for the reduction of CO2 with H2O under simulated solar irradiation. Appl Catal B-Environ 130:277–284
Wang XD, Li ZD, Shi J, Yu YH (2014) One-Dimensional Titanium Dioxide Nanomaterials: Nanowires, Nanorods, and Nanobelts. Chem Rev 114:9346–9384
Wang D-Y, Gong M, Chou H-L, Pan C-J, Chen H-A, Wu Y, Lin M-C, Guan M, Yang J, Chen C-W, Wang Y-L, Hwang B-J, Chen C-C, Dai H (2015a) Highly Active and Stable Hybrid Catalyst of Cobalt-Doped FeS2 Nanosheets–Carbon Nanotubes for Hydrogen Evolution Reaction. J Am Chem Soc 137:1587–1592
Wang H, Tsai C, Kong D, Chan K, Abild-Pedersen F, Nørskov J, Cui Y (2015b) Transition-metal doped edge sites in vertically aligned MoS2 catalysts for enhanced hydrogen evolution. Nano Res 8:566–575
Wang W-N, Wu F, Myung Y, Niedzwiedzki DM, Im HS, Park J, Banerjee P, Biswas P (2015c) Surface Engineered CuO Nanowires with ZnO Islands for CO2 Photoreduction. ACS Appl Mater Interfaces 7:5685–5692
Wang Y, Tang J, Kong B, Jia D, Wang Y, An T, Zhang L, Zheng G (2015d) Freestanding 3D graphene/cobalt sulfide composites for supercapacitors and hydrogen evolution reaction. RSC Adv 5:6886–6891
Weidman MC, Esposito DV, Hsu Y-C, Chen JG (2012) Comparison of electrochemical stability of transition metal carbides (WC, W2C, Mo2C) over a wide pH range. J Power Sources 202:11–17
Wendt H, Spinacé EV, Oliveira Neto A, Linardi M (2005) Electrocatalysis and electrocatalysts for low temperature fuel cells: fundamentals, state of the art, research and development. Quim Nova 28:1066–1075
Williams G, Seger B, Kamat P (2008) V. "TiO2-graphene nanocomposites. UV-assisted photocatalytic reduction of graphene oxide". ACS Nano 2:1487–1491
Wu Z, Fang B, Bonakdarpour A, Sun A, Wilkinson DP, Wang D (2012) WS2 nanosheets as a highly efficient electrocatalyst for hydrogen evolution reaction. Appl Catal B Environ 125:59–66
Wu J, Li H, Yin Z, Li H, Liu J, Cao X, Zhang Q, Zhang H (2013a) Layer Thinning and Etching of Mechanically Exfoliated MoS2 Nanosheets by Thermal Annealing in Air. Small 9:3314–3319
Wu Z, Fang B, Wang Z, Wang C, Liu Z, Liu F, Wang W, Alfantazi A, Wang D, Wilkinson DP (2013b) MoS2 Nanosheets: A Designed Structure with High Active Site Density for the Hydrogen Evolution Reaction. ACS Catal 3:2101–2107
Wu HB, Xia BY, Yu L, Yu X-Y, Lou XW (2015) Porous molybdenum carbide nano-octahedrons synthesized via confined carburization in metal-organic frameworks for efficient hydrogen production. Nat Commun 6
Wypych F, Schollhorn R. 1T-MoS2, a new metallic modification of molybdenum disulfide. J Chem Soc Chem Commun 1992, 1386–1388.
Xi G, Ouyang S, Li P, Ye J, Ma Q, Su N, Bai H, Wang C (2012) Ultrathin W18O49 nanowires with diameters below 1 nm: synthesis, near-infrared absorption, photoluminescence, and photochemical reduction of carbon dioxide. Angew Chem Int Ed 51:2395–2399
Xie J, Zhang H, Li S, Wang R, Sun X, Zhou M, Zhou J, Lou XW, Xie Y (2013a) Defect-Rich MoS2 Ultrathin Nanosheets with Additional Active Edge Sites for Enhanced Electrocatalytic Hydrogen Evolution. Adv Mater 25:5807–5813
Xie SJ, Wang Y, Zhang QH, Fan WQ, Deng WP, Wang Y (2013b) Photocatalytic reduction of CO2 with H2O: significant enhancement of the activity of Pt-TiO2 in CH4 formation by addition of MgO. Chem Commun 49:2451–2453
Xing M, Shen F, Qiu B, Zhang J (2014) Highly-dispersed Boron-doped Graphene Nanosheets Loaded with TiO2 Nanoparticles for Enhancing CO2 Photoreduction. Sci Rep 4

Xu H, Ouyang SX, Liu LQ, Wang DF, Kako T, Ye JH (2014) Porous-structured Cu2O/TiO2 nanojunction material toward efficient CO2 photoreduction. Nanotechnology 25
Yahaya AH, Gondal MA, Hameed A (2004) Selective laser enhanced photocatalytic conversion of $CO_2$ into methanol. Chem Phys Lett 400:206–212
Yan Y, Xia B, Ge X, Liu Z, Wang J-Y, Wang X (2013) Ultrathin MoS2 Nanoplates with Rich Active Sites as Highly Efficient Catalyst for Hydrogen Evolution. ACS Appl Mater Interfaces 5:12794–12798
Yang CC, Yu YH, van der Linden B, Wu JCS, Mul G (2010) Artificial Photosynthesis over Crystalline TiO2-Based Catalysts: Fact or Fiction? J Am Chem Soc 132:8398–8406
Yang J, Voiry D, Ahn SJ, Kang D, Kim AY, Chhowalla M, Shin HS (2013) Two-Dimensional Hybrid Nanosheets of Tungsten Disulfide and Reduced Graphene Oxide as Catalysts for Enhanced Hydrogen Evolution. Angew Chem Int Ed 52:13751–13754
Yin Z, Li H, Li H, Jiang L, Shi Y, Sun Y, Lu G, Zhang Q, Chen X, Zhang H (2012) Single-Layer MoS2 Phototransistors. ACS Nano 6:74–80
Yu Y, Huang S-Y, Li Y, Steinmann SN, Yang W, Cao L (2014) Layer-Dependent Electrocatalysis of MoS2 for Hydrogen Evolution. Nano Lett 14:553–558
Yuan L, Xu YJ (2015) Photocatalytic conversion of CO2 into value-added and renewable fuels. Appl Surf Sci 342:154–167
Yuliati L, Itoh H, Yoshida H (2008) Photocatalytic conversion of methane and carbon dioxide over gallium oxide. Chem Phy Lett 452:178–182
Zhai QG, Xie SJ, Fan WQ, Zhang QH, Wang Y, Deng WP, Wang Y (2013) Photocatalytic conversion of carbon dioxide with water into methane: platinum and copper(I) oxide co-catalysts with a core-shell structure. Angew Chem-Int Ed 52:5776–5779
Zhang L, Liu C, Wong A, Resasco J, Yang P (2015) MoS2-wrapped silicon nanowires for photoelectrochemical water reduction. Nano Res 8:281–287
Zhang X, Han F, Shi B, Farsinezhad S, Dechaine GP, Shankar K (2012) Photocatalytic conversion of diluted $CO_2$ into light hydrocarbons using periodically modulated multiwalled nanotube arrays. Angew Chem Int Ed 51:12732–12735
Zheng Y, Jiao Y, Zhu Y, Li LH, Han Y, Chen Y, Du A, Jaroniec M, Qiao SZ (2014) Hydrogen evolution by a metal-free electrocatalyst. Nat Commun 5
Zhou W, Hou D, Sang Y, Yao S, Zhou J, Li G, Li L, Liu H, Chen S (2014) $MoO_2$ nanobelts@-nitrogen self-doped $MoS_2$ nanosheets as effective electrocatalysts for hydrogen evolution reaction. J Mater Chem A 2:11358–11364
Zhou Y, Tian Z, Zhao Z, Liu Q, Kou J, Chen X, Gao J, Yan S, Zou Z (2011) High-yield synthesis of ultrathin and uniform Bi2WO6 square nanoplates benefitting from photocatalytic reduction of $CO_2$ into renewable hydrocarbon fuel under visible light. ACS Appl Mater Interfaces 3:3594–3601
Zou X, Zhang Y (2015) Noble metal-free hydrogen evolution catalysts for water splitting. Chem Soc Rev 44:5148–5180

# Part II
# Methods

# Chapter 4
# Photoelectrochemical Cell Design, Efficiency, Definitions, Standards, and Protocols

**Wilson A. Smith**

## 4.1 Introduction

The storage of solar energy into chemical energy through photoelectrochemical water splitting offers a long-term, sustainable, and effective solution to the global energy and environmental problems (Lewis and Nocera 2006). It has been over 40 years since the discovery of electrochemical photolysis of water (Fujishima and Honda 1972), and yet today no commercial or industrial device exists that is effectively producing solar hydrogen. The major limitations of the technological advancement of this field are due to the complicated physical, chemical, and engineering feats that are required to convert photons into electrons that can directly drive electrochemical reactions with well-defined, separated, and transportable products. While a commercial solar fuel device is not presently contributing to the global energy supply, many attempts have been made to understand the mechanisms and limitations of the photo-physical and chemical problems, and many different arrangements of lab-scale devices have been explored.

The two biggest hurdles to accomplish the development of a practical PEC water-splitting device are both scientific and technical. From a scientific standpoint, a PEC device needs to be able to manage solar irradiation, transport electric and ionic charges, and perform oxidative and reductive catalytic reactions simultaneously. From a technological point of view, these challenges must be all addressed using materials and fabrication processes which are cheap and scalable, putting severe limitations on the methods and compounds that can be used. However, since the scientific challenges to achieve stable, efficient, and cost-effective PEC water splitting have still not been overcome to a sufficient level, it may be pre-emptive to begin the design of a practical system larger than a lab-scale device. Therefore,

W.A. Smith (✉)
Materials for Energy Conversion and Storage (MECS), Department of Chemical Engineering, Faculty of Applied Sciences, Delft University of Technology, 2628 BL Delft, The Netherlands
e-mail: w.smith@tudelft.nl



S. Giménez, J. Bisquert (eds.), *Photoelectrochemical Solar Fuel Production*,
DOI 10.1007/978-3-319-29641-8_4

there is still a need to focus on the basic understanding, system and materials diagnostics, and fundamental mechanisms involved in PEC water splitting.

For the purposes of this chapter, only the PEC water-splitting reaction will be discussed and will not be compared to a solar fuel device that also carries out $CO_2$ reduction (where the corresponding oxidation reaction is still water oxidation). The overall water-splitting reaction can be summarized in the following equation:

$$2H_2O\,(l) \rightarrow 2H_2\,(g) + O_2\,(g) \quad E^o = 1.23\text{ V } (\Delta G = +237.22\text{ kJ/mol}) \tag{4.1}$$

which shows that, theoretically, it takes a minimum of 1.23 V to split water into molecular hydrogen and oxygen at the standard temperature ($T^0 = 298$ K) and pressure ($P^0 = 1$ bar). In practice, it takes several hundreds of mV overpotential to drive the water-splitting reaction, mainly due to overpotentials associated with water oxidation (Rossmeisl et al. 2007; Koper 2011), but can also depend on the electrode material(s) used, the electrolyte, the distance between the electrodes, and the device geometry.

The overall reaction takes place simultaneously at two different sites, which mediate the oxidation reaction at an anode and the reduction reaction at a cathode. In an acidic environment (pH = 0), the two relevant half-reactions can be written:

$$2H_2O\,(l) \rightarrow 4H^+\,(aq) + 4e^- + O_2\,(g) \quad E^o = 1.23\text{ V vs. NHE} \tag{4.2}$$

$$4H^+\,(aq) + 4\,e^- \rightarrow 2H_2\,(g) \quad E^o = 0.00\text{ V vs. NHE} \tag{4.3}$$

In an alkaline environment (pH = 14), these red-ox equations then become:

$$4OH^-\,(aq) \rightarrow 2H_2O\,(l) + 4e^- + O_2\,(g) \quad E^o = -0.401\text{ vs. NHE} \tag{4.4}$$

$$2H_2O\,(l) + 2e^- \rightarrow 2OH\text{-}\,(aq) + H_2\,(g) \quad E^o = -0.828\text{ V vs. NHE} \tag{4.5}$$

In the simplest form, these two oxidation and reduction reactions can occur over metal electrodes (oxygen is produced at the anode, and hydrogen is produced at the cathode), with current and voltage supplied by an external power supply. The challenge of PEC water splitting is to create this external voltage and current directly from converted solar energy in a monolithic device. The actual means to do such a conversion can be accomplished in many different ways, which are described in detail in the following sections.

In this chapter, several PEC device designs will be considered with respect to the architecture of the components used, the management of different photo-absorbing and catalyst materials, and the general operating principle that governs the synergy of these materials. Furthermore, the focus of materials used will be inorganic, meaning the light-absorbing compounds discussed will be semiconductors, which have shown numerous applications in PEC devices that are able to achieve overall solar water splitting, as opposed to molecular absorbers, which have shown poor stability and conversion efficiency (to date), which has limited their applications in practical devices.

## 4.2 The Photoelectrochemical Cell

### *4.2.1 Cell Design*

A photoelectrochemical reaction takes place in a photoelectrochemical cell. At the very least, the PEC cell hosts the (photo)anode, the (photo)cathode, and electrolyte solution. The major difference between a PEC cell and a (dark) electrochemical cell is that one or more of the electrodes is photoactive, and thus needs to absorb light to drive one or both of the chemical reactions associated with water splitting. Therefore, a PEC cell must have at least one transparent window in order to allow light to penetrate the cell and be absorbed by one or both electrodes. Furthermore, the anode and cathode must be in electrical contact, and so a conductive wire is needed to connect the two electrodes, or the electrodes must be monolithically integrated on opposing sides of an electrically conductive substrate, as shown in the design of the 'artificial leaf' (Nocera 2012). For the sake of clarity, the electrode (either photoanode or photocathode) that absorbs light can be called the working electrode, while the electrode that drives the opposing half reaction that is not light-activated is called the counter electrode. In the case that both electrodes are photoelectrodes, they can both be referred to as the working electrodes and are further clarified by being referenced as the working photoanode and the working photocathode.

In an ideal case where bias-free water splitting can be achieved, the aforementioned materials are the only components of a PEC device that are needed to convert solar energy and water into hydrogen and oxygen. However, for systems that require an extra bias to drive the water splitting reaction, or for detailed diagnostics of the mechanisms for achieving water splitting, more components are needed. For example, to accurately examine the potential of the working electrode, a reference electrode is needed. In addition, to maintain the electrolyte concentration, minimize pH gradients, and aid in reactant/product mobility, magnetic stirrers and gas circulation are necessary. An illustration of a PEC cell with the aforementioned components is shown in Fig. 4.1a. If the two electrodes (working and counter) are too close to each other, it is possible that the gas evolved at one surface may contaminate or back-react with the reactants/intermediates of the other electrode. Therefore, a membrane may be used to prevent crossover of gaseous products, which directly allows hydrogen and oxygen to be evolved into their own separate containers. Such a system would then require two separate circulation systems, and an illustration of this configuration is shown in Fig. 4.1b.

These PEC cell designs are useful for an electrochemical system where the working and counter electrode are spatially separated; however, many potential PEC device designs favor the approach of a monolithic device that has the working and counter electrode on one substrate. An illustration of a PEC cell design for a monolithic system is shown in Fig. 4.2.

As can be seen from the previous figures, a working PEC cell used for overall solar water splitting requires many components, each of which has a large variability in terms of materials that can be used, stability, and standardization.

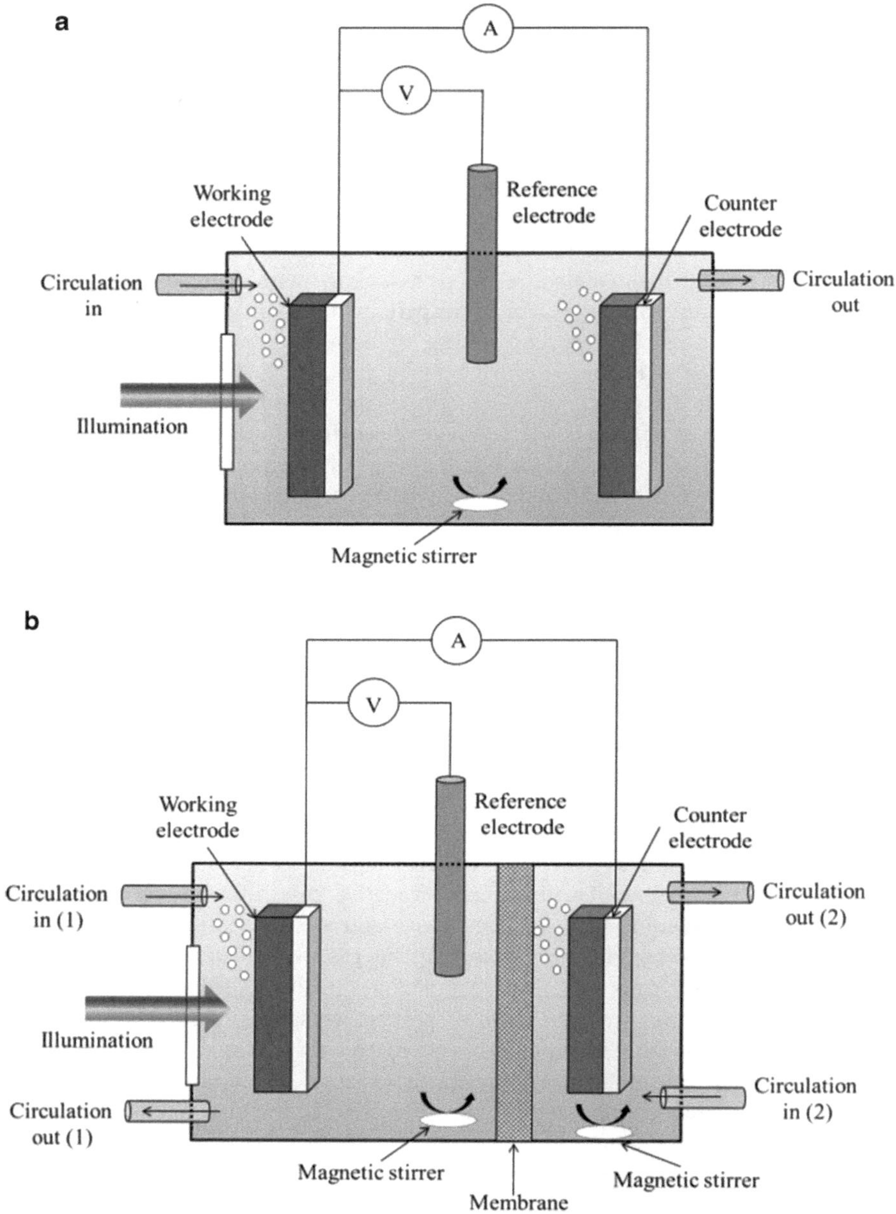


**Fig. 4.1** Illustration of the basic components of a photoelectrochemical cell where there is (**a**) a single compartment for the working and counter electrode and (**b**) two compartments that separate the working and counter electrodes by a membrane

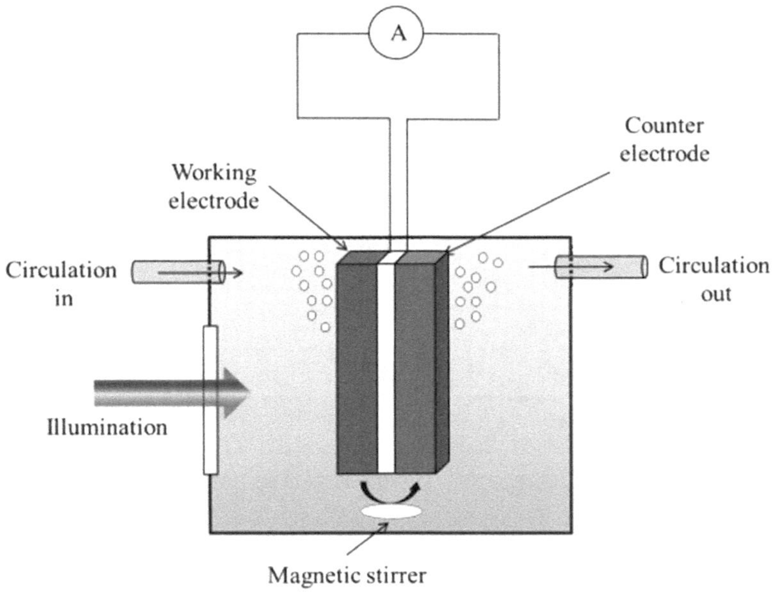


**Fig. 4.2** Illustration of the basic components of a photoelectrochemical cell with a monolithic device combining the working and counter electrode onto one substrate

The following sections will briefly describe the different elements of an overall PEC cell, and how they can be used to determine the performance and efficiency of a solar water-splitting device.

Before mentioning the individual components, it is useful to illustrate an actual working PEC cell in slightly more detail than in Figs. 4.1 and 4.2. A conceptual design of a working PEC cell designed by van de Krol is shown in Fig. 4.3 (van de Krol 2012).

This cell is made from PTFE and is custom-designed to fit a working electrode, counter electrode, electrolyte, reference electrode, and transparent window to allow solar irradiation to penetrate the cell and hit the photoactive electrode. In this design, the sample (deposited on a flat and conductive substrate) is mounted on the left side of the cell and makes an airtight seal with the body of the PEC cell. There is a small chamber in the bottom of the cell to allow a magnetic stirrer to be used, which can help distribute reactants and disperse products during electrochemical reactions. More details about this PEC cell, its design, and functionality can be found in ref (van de Krol 2012).

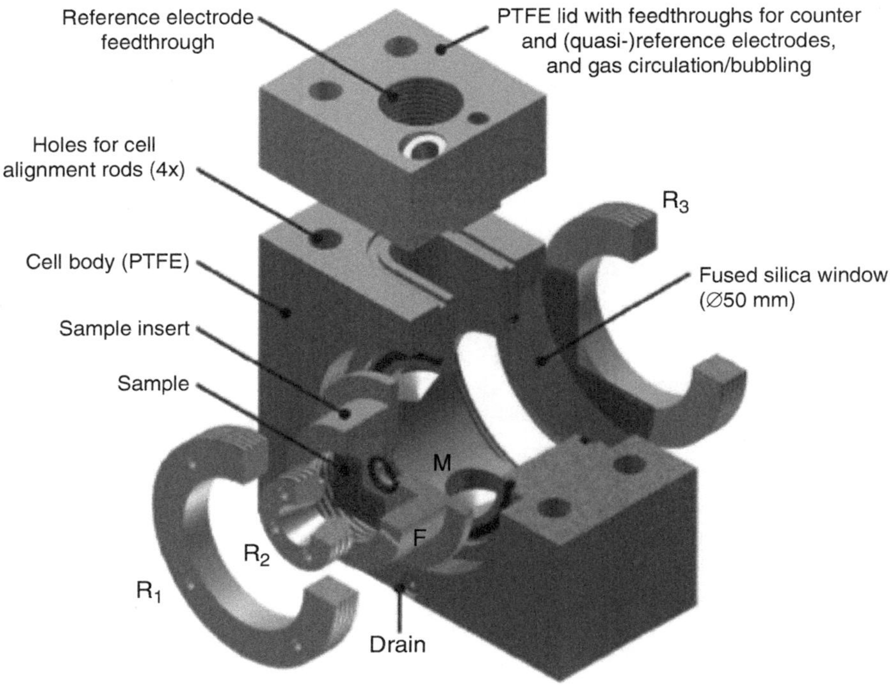


**Fig. 4.3** A practical PEC cell designed by van de Krol, used with permission from (van de Krol 2012)

### *4.2.2 Electrodes*

A practical PEC device needs to be able to manage optical, electronic, and catalytic functionalities all at the same time. For a true photoelectrochemical device, the light-absorbing material should be immersed in the liquid, significantly increasing the difficulty and complexity of cell design considerations. For example, depending on the configuration of the device, the incident light may need to travel through the water/electrolyte first, which can reduce the photon flux that is received at the semiconductor surface, thus decreasing the possible maximum photocurrent that can be obtained. This in turn decreases the potential solar to hydrogen conversion efficiency (STH) of a practical device. Conversely, if the light does not need to go through the electrolyte first, and instead goes through the back of the substrate, different opto-electronic requirements for the substrate are required to maximize the efficiency of the device. This section describes the different configurations that the materials of a PEC device can have, and how the semiconductors and catalysts can be arranged in different ways that can affect the stability, efficiency, and practicality of a solar fuel system. The optimization required to achieve high STH efficiencies with tandem device configurations, i.e. band gap matching, spectral utilization, etc., will not be discussed in this chapter, as it is elaborated on in explicit detail in Chap. 12. Furthermore, the separation of anode and cathode into separate spatial components (i.e. a wired device) versus a monolithic-integrated device (i.e. a wireless device) will not be discussed in this chapter, as it is also discussed in Chap. 12.

#### 4.2.2.1 General Considerations

The following discussion about PEC device design and considerations is focused on semiconductor thin films and will not elaborate on particle-based systems that lack an electrically wired configuration. Furthermore, the discussion will focus on lab-scale devices and architectures used mainly for testing efficiencies of materials and device configurations and not emphasize the up-scaling towards reactor and industrial level designs, which have been excellently elaborated in several key publications (James et al. 2009; Pinaud et al. 2013; Sathre et al. 2014). Finally, the working principle for the different architectures will be briefly discussed in order to speculate on the possible performance limitations of each configuration, which in turn may be used to choose a different cell design or orientation in order to obtain the maximum possible overall solar water-splitting efficiency of a device.

For both single-component photoanode and photocathode films (i.e. without buried junctions which will be discussed later in this chapter), the light-absorbing material needs to be deposited on a highly conductive substrate (the current collector) that allows charges to be extracted or injected between the working electrode and counter electrode. For an n-type photoanode, photogenerated holes migrate towards the surface to perform water oxidation, and thus electrons should flow through the bulk of the semiconductor, to the back contact, through a conductive wire, arrive at the counter electrode, and there reduce water/protons. For a p-type photocathode, photogenerated electrons migrate towards the semiconductor surface where they reduce protons/water, and holes must migrate through the bulk of the material to the back contact, through a conductive wire, arrive at the counter electrode, where they must oxidize water. In both cases, an ohmic contact is required at the semiconductor/back-contact interface, and thus highly conductive layers are typically used to form the top layer of the substrate before depositing a photoelectrode. However, if the device requires light to be incident from the back-side of the sample (i.e. light hits the substrate before the photoelectrode), the ohmic contact must also be transparent to light. It is again important to note that these considerations are only valid for a single-absorber system and do not hold if a tandem absorber electrode is constructed as the light path would need to travel through more than one light-absorbing material, and thus have different optimization criteria.

With the aforementioned requirements for an ohmic back-contact, which may or may not be transparent, it is possible to find materials which fit such a specific criteria. The most widely used materials for this application when it is necessary to have back-side illumination (or for a tandem device) are transparent conducting oxides (TCO's) such as F-doped $SnO_2$ (FTO), In-doped $SnO_2$ (ITO), and Al-doped ZnO (AZO). These TCO layers have a relatively high conductivity (with respect to typical semiconductor photoelectrodes) and allow a large amount of spectral transmission so that the incident solar irradiation is maximized when it hits the light-absorbing photoelectrodes. Typical conductivities (in S/cm) for FTO, ITO, and AZO are $1 \times 10^3$, $1 \times 10^4$, and $7 \times 10^3$, respectively.

In addition to the requirements that are necessary for a transparent ohmic contact, there are serious implications for the practical efficiency of a PEC device by using front-side or back-side illumination. An example of these implications is shown in Fig. 4.4, which illustrates the photogenerated charge carriers created for an n-type photoanode. For the case of front-illumination (Fig. 4.4a), most of the absorption in the photoanode will occur near the surface of the electrode, and as the light is absorbed through the thickness of the material, less light reaches the back of the electrode. The result of this is a greater density of photogenerated charge carriers near the surface of the semiconductor than at the back of the electrode. Near the surface, photogenerated holes are very close to the semiconductor-liquid junction (SLJ), and thus the hole diffusion length does not need to be very long. On the other hand, the photogenerated electrons created near the surface need to diffuse through the bulk of the electrode to the back-contact where they are extracted and transported to the counter electrode for hydrogen evolution. Therefore, for a photoanode being subjected to front illumination, it is important that the electron diffusion length is greater than or equal to the thickness of the films. Conversely, for a photoanode illuminated from the back-side, as shown in Fig. 4.4b, there is a higher density of photogenerated charge carriers closer to the back-contact than the

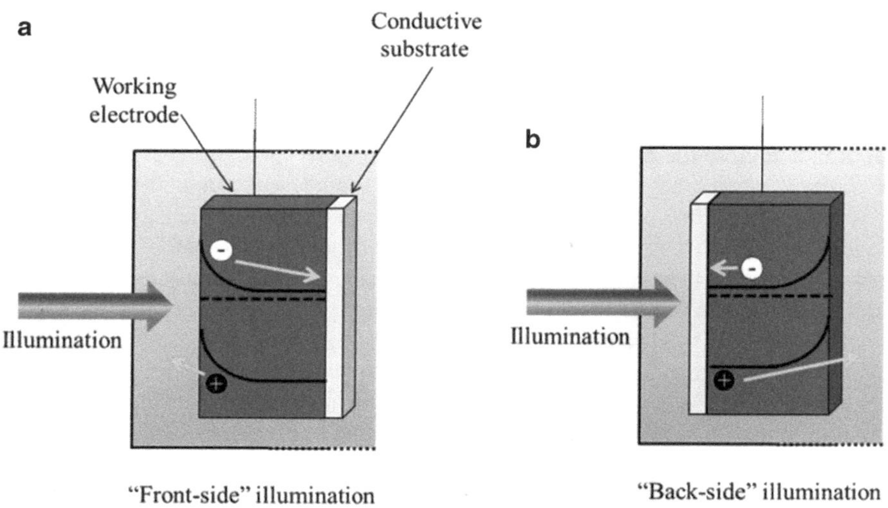


**Fig. 4.4** Schematic illustration of a system where (**a**) light is incident on the semiconductor surface first, i.e. "front-side" illumination, and (**b**) where the light is incident on the substrate side first, i.e. "back-side" illumination

**Table 4.1** Typical metal oxide photoelectrodes and their associated carrier mobility, carrier lifetime, and diffusion length

| Photoelectrode material | Carrier mobility ($cm^2\ V^{-2}\ s^{-1}$) | Carrier lifetime (s) | Diffusion length (nm) |
|---|---|---|---|
| $Fe_2O_3$ | 0.5 | $3 \times 10^{-12}$ | 2~4 |
| $WO_3$ | 10 | $1\sim9 \times 10^{-9}$ | 150~500 |
| $Cu_2O$ | 6 | $40 \times 10^{-12}$ | 25 |
| $BiVO_4$ | 0.044 | $40 \times 10^{-9}$ | 70 |
| TaON | 0.01 | $1 \times 10^{-3}$ | ~31 |
| $Ta_3N_5$ | 0.07 | $1 \times 10^{-3}$ | ~84 |

SLJ. Therefore, for this case, the electrons only need to diffuse a very short length to reach the back-contact, and the holes must be able to diffuse through the bulk of the material without recombining in order to reach the SLJ and oxidize water. These two cases can be flipped for a configuration with a photocathode, where the diffusion of charge carriers is opposite, i.e. the photogenerated electrons need to reach the SLJ and the photogenerated holes need to diffuse to the back-contact.

The importance of this formalization is that the electron and hole diffusion length and mobility is different for many materials. An example of the mobility of electrons and holes for commonly used photoelectrode materials is shown below in Table 4.1.

These values are important because they can help to determine how thick a film should be to balance the maximum absorption of light with the transport of photogenerated charge carriers, and thus which type of illumination, i.e. front-side or back-side, should be used based on the diffusion length of the minority and majority charge carriers.

In addition to the selection of back-contact material, the actual photoelectrode materials must be fabricated and deposited on the substrate. The choice of materials and device configuration is not straight forward, and a number of device geometries and compositions have been reported in the literature. In general, it is possible to de-couple or distribute the task of light harvesting and catalysis into different materials that together make up a working photoelectrode. The litany of configurations of photoactive and catalytic materials can be divided into three main categories: (1) photovoltaic cells with electrocatlayst layers deposited on top of them, (2) photovoltaic cells with photoelectrode layers deposited on top of them, and (3) a fully photoelectrochemical device with either/both a photoanode and/or photocathode, i.e. one or both water oxidation and/or reduction are photo-driven reactions. The working principles and outlook for each design configuration is given in the following sections.

#### 4.2.2.2 Photovoltaic Cell + Electrocatalyst (PV + EC)

The most straightforward approach to achieve efficient solar-driven water splitting is to completely separate the light absorption and catalysis functions. This can be done most effectively by having a PV panel to convert solar energy into electricity, which can then be connected in series to an electrolyzer, which can perform the water-splitting reaction. Such a configuration may be accomplished on a labscale with the PV and electrolysis components very close to each other and with small dimensions, (Luo et al. 2014; Cox et al. 2014). However, such an approach on a large scale would require inverters to convert the DC current generated by the solar cells to usable current for the electrochemical cell. This system integration may be the simplest from a practical standpoint and can use already developed off-the-shelf PV components with industrial scale electrochemical cells. However, as of now this approach is not cost-effective, as the cost of the total system integration would produce hydrogen that is not competitive with the current price of fossil fuels (James et al. 2009; Pinaud et al. 2013; Sathre et al. 2014).

This approach actually makes the production of oxygen and hydrogen from water completely separate from any light absorption and dependent only on a current/voltage supply. This allows the separate optimization of solar to electricity conversion, as well as the dark catalytic reactions for the OER and HER, for which recent benchmarking of OER and HER catalysts has been assembled for acidic and alkaline environments (McCrory et al. 2013, 2015). The advantage of such a system is that any source of electricity (preferably renewable) can be used to power the electrolysis reactions. One logical extension of this device configuration is to use large-scale utility PV panels connected to grid-powered electrolyzers. Using such a large scale system, it is possible to tackle the terawatt scale demand the world has and is thus the most likely and technologically advanced way to store solar energy through water splitting. However, industrial electrolyzers are run at large current densities (hundreds of $mA/cm^2$) and require precious metal catalysts that are only stable under these conditions and easily corrode when their current/voltage source is removed. This is a strong set-back if this system is to be used with solar energy as the input, as sunlight is intermittent, and thus the electrolyzers would experience significant downtime and thus corrode. Furthermore, since the light-absorbing PV units are not immersed in the electrolyte, this is technically not a photoelectrochemical device and is only an electrochemical cell powered by renewable electricity. For these reasons, this device architecture will not be discussed in more detail in this chapter.

While a large-scale PV cell coupled to an electrolyzer has several potential disadvantages as listed above, the direct integration of the two components into a monolithic device offers a potential solution to the combination of materials. An overall schematic for such a device using a 3-jn a-Si PV cell is shown in Fig. 4.5a, with the associated theoretical electronic band diagram given in Fig. 4.5b.

In this configuration, a multi-junction PV cell is coated with an ohmic contact on each side in order to prevent corrosion and offer excellent electronic charge transfer. A multi-junction PV cell is shown here, as it is more likely to provide the necessary photovoltage to drive the water-splitting reaction, taking into account the thermodynamic potential for water splitting plus necessary overpotentials, when compared to a single junction solar cell. On top of the ohmic contacts, a dedicated oxygen evolution catalyst (OEC) and hydrogen evolution catalyst (HEC) are

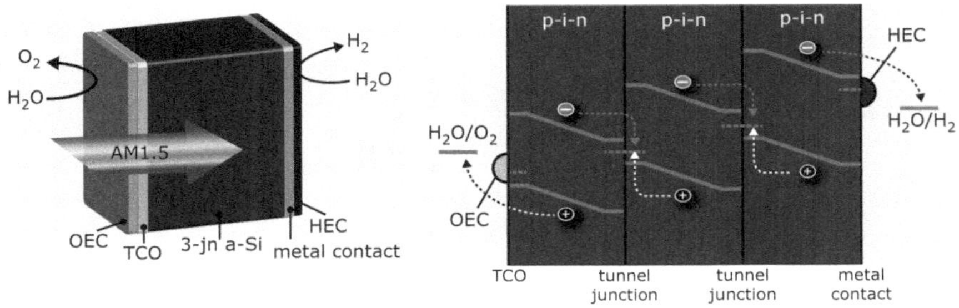


**Fig. 4.5** A cartoon representation of a wireless monolithic PV-electrolysis cell (a), and the associated electronic band diagram for a potential triple junction PV device that powers the water-splitting reaction, where OEC is the oxygen evolution catalyst and HEC is the hydrogen evolution catalyst

deposited, which carry out the respective redox reactions efficiently. Such a device architecture has been successfully used extensively in the literature (Appleby et al. 1985; Sakai et al. 1988; Lin et al. 1989; Rocheleau et al. 1998; Khaselev et al. 2001; Licht et al. 2001; Reece et al. 2011). One distinct advantage of such a system design is that it is possible to combine state-of-the-art materials for each component, i.e. you can potentially use a high performance PV cell with excellent optoelectronic properties (high $V_{oc}$, high $J_{sat}$), and couple this to high performance hydrogen and oxygen evolution catalysts (low overpotential, high turnover frequency). However, one practical limitation of such a device is that due to the architecture, the light-absorbing PV cell is buried beneath both hydrogen and oxygen evolution catalysts, and thus the incident solar irradiation must first pass through one of the catalyst materials. This may introduce more optical losses in the overall system, since light can either be reflected or absorbed by these extra layers. This is particularly troubling since many of the state-of-the-art HEC's are made from metallic precious metals such as Pt and Ir, and the state of the art OEC's become dark when a potential (in this case a photovotlage) is applied (Bendert and Corrigan 1989; Corrigan and Knight 1989; Conell et al. 1992; Trotochaud et al. 2013). Therefore, the combination of materials to make a monolithic PV-electrocatalyst device is not as straight forward as simply connecting efficient PV cells and electrocatalysts, but their optical properties must be taken into account. A technical summary of the effect of the different configurations of the light path going through either the OEC side or the HEC side on the photoelectrochemical performance was recently developed by Seger et al. (Seger et al. 2014). The true implications for the described PEC device configurations are more closely tied to the optimization of a tandem PEC device, with multiple absorbing materials, and are discussed in further detail in Chap. 12.

A slight modification of this approach can be achieved by changing the arrangement of the PV cell, while still maintaining the same architecture, i.e. having a horizontal protected multi-junction PV cell, and having a separate dedicated HEC's

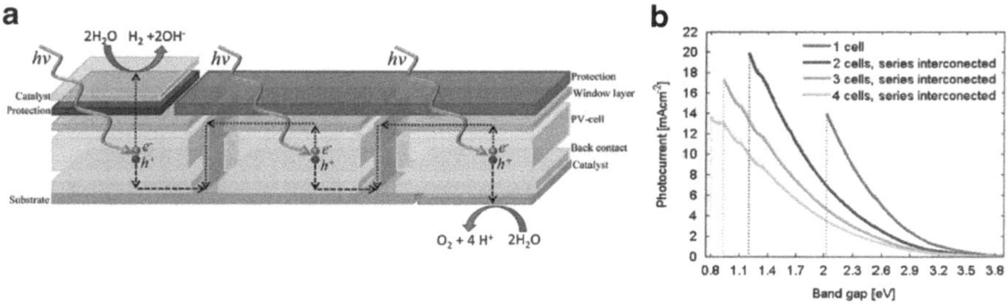


**Fig. 4.6** (**a**) A schematic of a side-by-side PV-EC system, and (**b**) theoretical maximum STH efficiencies as a function of the PV band gap energy, adapted reference (Jacobsson et al. 2015)

and OEC's spatially separated. In this device configuration, PV cells can be placed side-by-side and connected in series, as shown in Fig. 4.6.

This side-by-side PV system has two practical advantages compared to the monolithic device previously described. Since this architecture places the PV cells side by side, the optical absorption by each PV junction does not interfere with the others' ability to absorb light, i.e. there is no parasitic light absorption for the light going through consecutive layers as is necessary in the device shown in Fig. 4.5. This opens the potential for using series-combined PV cells that are each optimized for the entire solar spectrum, removing the requirement to have buried PV junctions that are optimized for the transmitted spectra of light that pass through top layers, i.e. having to match top-cell and bottom-cell band gap energies, as described in Chap. 12. The drawback of such a device architecture is that it extends the solar irradiation surface area, thus making the current density (i.e. photocurrent density) that travels to the catalyst surface area much more dilute. The limitations of such a device configuration were discussed by Jacobsson et al. (2015), who found the theoretical potential of such devices can be up to ~20 % STH, slightly lower than a traditional monolithic device (see Chap. 12). An important conclusion of this study showed that a high theoretical STH conversion efficiency is obtainable in this architecture, implying that this device configuration is still valuable to be explored.

For the consideration of the forthcoming devices, it is important to note that a PV-electrolysis configuration has no semiconductor liquid junctions (SLJ), because the light-absorbing semiconductor is not in direct contact with the electrolyte. This is an important feature to note, since the interfacial band edge energetics at the semiconductor-liquid junction differ significantly than those determined by ohmic contacts and Schottky barriers that are associated with PV-electrolysis devices. It is also important to note here that the band edge positions of a PV cell are irrespective of the applied potential/redox potentials in the solution (determined by electrocatalysts in contact with the electrolyte).

In order to facilitate understanding of how this and other device configurations work, it is useful to look at the operating mechanisms of this system. In particular, by combining the current vs. potential plots for the two separated systems (PV and electrocatalyst), it is possible to extract an operating point for the combined system,

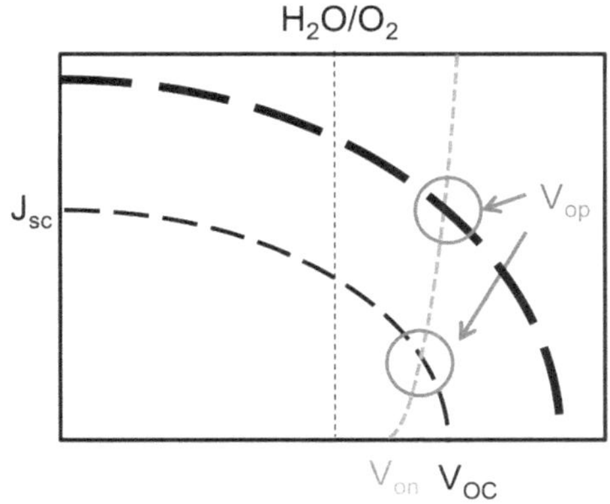


**Fig. 4.7** A hypothetical current versus potential plot of two solar cells and one oxygen evolution catalyst, with a reference water oxidation potential shown to illustrate that an oxygen evolution catalyst must have more potential applied than the thermodynamic potential of this reaction

and from this estimate the potential and limitations of a 'best case' scenario, forming a future prospective for further technological and scientific development.

The *J–V* curves of two hypothetical PV cells are shown in Fig. 4.7 (black dashed lines), where the short circuit density ($J_{sc}$) and open circuit potential ($V_{oc}$) are shown. Typically, *J–V* curves of PV cells are measured in a two-electrode configuration, since they are solid state devices that are connected by Ohmic contacts, and thus the potential is directly measured across the donor and acceptor regions of the photovoltaic device. In the same figure, a *J–V* curve of a hypothetical oxygen evolution catalyst is shown (gray dashed curve) with its corresponding onset potential ($V_{on}$) for catalysis. Where the two curves intersect is the operational point ($V_{op}$) of the PV-powered electrolysis device. For reference, the water oxidation potential is shown (while its real value is 1.23 V vs. RHE, here the figure is illustrative and thus does not represent the actual potential value, only to demonstrate it is to the left, i.e. at a lower potential than the *J–V* curve of the electrocatalyst.). It is again useful to note that for an electrochemically derived *J–V* curve (either for an electrocatalyst or a photoelectrocatalyst), the potential is typically measured in a three-electrode configuration, where the potential of the working electrode is measured with respect to a third (reference) electrode with a known redox potential. More details about two-electrode and three-electrode measurements will be discussed in Sect. 4.4. From this figure, several important limitations can be extracted. First, from the electrocatalyst side, a dark (i.e. not photocatalytically active) electrocatalyst can never have an onset potential less than 1.23 V vs. RHE ($V_{H2O/O2}$), because this is the thermodynamic equilibrium potential of water oxidation. This illustrates that the dashed gray line corresponding to the *J–V* performance of the OEC will never become more cathodic (i.e. be to the left of) to the water oxidation potential of 1.23 V. The implications of this are that the PV cell used to provide current and potential to the OEC must then satisfy the requirement of having a $V_{oc}$ higher than (i.e. to the right of) the water oxidation potential. In practice, the actual state-of-the-art OEC materials still require a minimum of 200~300 mV overpotential to drive this reaction (McCrory et al. 2013, 2015), meaning the $V_{oc}$ of a practical PV must be at least 1.4~1.5 V. in order to operate in a bias-free device. This puts a strict limit on the performance characteristics of the PV cell, which can influence the potential applicability of an efficient, scalable, and cost-effective device. First, there is an inherent tradeoff for PV cells between the $J_{sc}$ and $V_{oc}$, i.e. the higher the $J_{sc}$ the lower the $V_{oc}$ and vice versa. This means for a PV cell to have a $V_{oc}$ higher than 1.23 V, the $J_{sc}$ will be reduced and thus have a lower performance. In addition to the current density limitations that a high $V_{oc}$ places on a PV cell, the ability to achieve such a performance in a stable and cheap material may be limited. For example, the leading PV materials that can achieve such high current densities and large $V_{oc}$'s are either made from very expensive materials (GaAs) or are unstable (perovskites), and thus may not be practical for a cost-effective system to produce solar hydrogen. At the moment, perovskite PV cells have gained significant attention due to their rapid growth in cell efficiency (over 20 % as of 2015); however, the materials are inherently unstable and thus the use of such a material class, at the moment, seems unlikely. The most practical material that can be used to balance cost and efficiency is silicon, with crystalline silicon, c-Si, having a band gap energy of 1.1 eV and amorphous silicon, a-Si, having a band gap energy of 1.8 eV. However, the overall efficiency of single and multi-junction Si solar cells may be limited due to the low $V_{oc}$ obtainable for these materials, thus limiting the potential $V_{op}$. While this seemingly puts many restrictions on a PV + electrolysis cell, there are other device configurations that can have a more beneficial *J*–*V* performance operating point, which will be discussed in the next sections.

#### 4.2.2.3 Photovoltaic Cell + Photoelectrode (PV + PEC)

While PV-electrocatalyst devices offer the ability to directly combine state-of-the-art PV materials with state-of-the-art electrocatalyst materials in a straight forward manner, there are many possible limitations of using such a device in a practical application. A variation of this architecture is to couple a PV cell with a photocatalyst material that can directly photo-drive either the water oxidation or reduction reaction. A sketch of a wireless monolithic PV/PEC system with a photoanode (i.e. a photocatalyst driving the water oxidation reaction) is shown in Fig. 4.8a, with the associated band diagrams shown in Fig. 4.8b.

This PV/PEC system has several advantages and disadvantages when compared to a direct PV-electrocatalyst system. From a fabrication and cost perspective, this architecture is simpler, and thus possibly can be more promising for upscaling to a large area device. This is because the single photoanode layer (in this particular configuration) replaces one *p-i-n* PV junction and tunnel layer. From a manufacturing point of view, this means depositing one layer instead of four, which obviously can reduce overall device costs, provided the photoanode material and fabrication process are cheaper than the 4-layer *p-i-n* and tunnel junction layer depositions. Furthermore, since the photoanode (in this case, though the same is true for the opposite case with a photocathode) is in contact with the electrolyte, a semiconductor liquid junction is formed. This means that the interfacial band edge

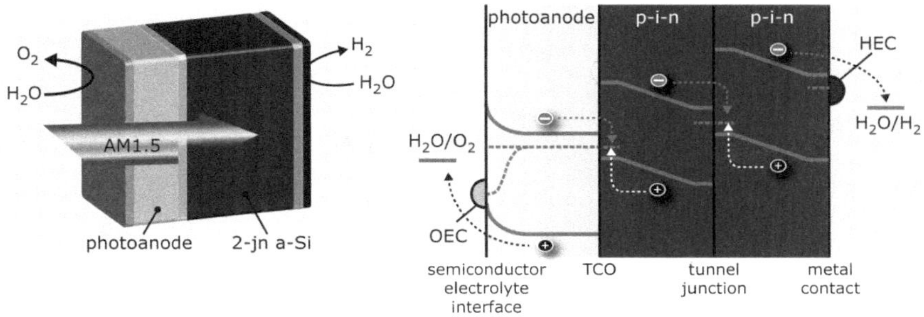


**Fig. 4.8** A cartoon representation of a wireless monolithic PV-PEC cell with a double-junction PV cell attached to a photoanode to drive the water oxidation reaction (a), and the associated electronic band diagram for a potential double-junction PV/PEC device that powers the water-splitting reaction

alignment of the Fermi-level in the photocatalyst should, in principle, align with the relevant redox reaction potentials. In an ideal case, this alignment can happen directly with no losses or added overpotentials, but in practice, overpotentials exist due to kinetic-driving forces required to carry out the 4-step water/$OH^-$ oxidation and 2-step water/proton reduction reactions, as well as the presence of electronic surface states that may pin the electronic band energies at potentials less than the highest achievable photovoltage.

While the band edge positions of a PV-cell are not dependent on the redox potentials in the solution, for a photoelectrode that absorbs light and drives a chemical reaction, the valence band ($E_v$) and conduction band ($E_c$) positions must be favorable relative to the water oxidation and reduction potentials. In particular, the valence band should be lower (more positive) than the oxygen evolution potential, and the conduction band should be higher (more negative) than the hydrogen evolution potential. Ideally, a single material could be used to drive the overall reaction, with conduction and valence bands that straddle the hydrogen and oxygen evolution potentials; however, such a material has not yet been found or developed to an efficient device. More on the practical utilization of single absorber materials is described in Chap. 12.

Similar to the previous section, it is useful to compare the *J*–*V* characteristics of the different components of this system to see its operational principle, and its inherent advantages and disadvantages. A current vs. potential plot is shown below in Fig. 4.9, where the photocurrent is shown for the photoelectrode (in this case a photoanode) in the dashed gray, and the *J*–*V* characteristics of the buried junction PV are shown in the dashed black line.

Similar to the PV-electrolysis case, the intersection of the two *J*–*V* curves is the operational point of the device. The most striking difference to the previous case is that here the intersection point can come at a lower potential than the water redox potential, which gives more flexibility in terms of system optimization. In particular, using this PEC/PV approach offers significant flexibility in lowering the $V_{op}$ of a practical device, and thus may offer a more realistic pathway towards a high

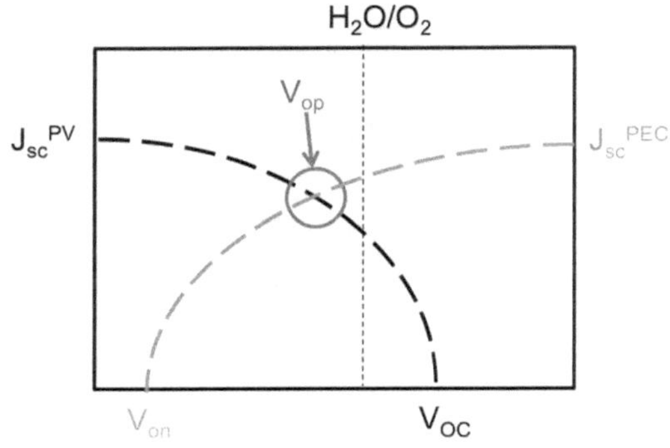


**Fig. 4.9** A hypothetical *J–V* plot of a photoanode and a series connected PV cell

efficiency, low-cost PEC device, as many benchmark device performances have been shown using this device configuration (Morisaki et al. 1976; Khaselev and Turner 1998; Miller et al. 2005; Park and Bard 2006; Arakawa et al. 2007; Gaillard et al. 2010; Brillet et al. 2012; Abdi et al. 2013; Han et al. 2014).

#### 4.2.2.4 Dual Photoelectrodes (Photoanode + Photocathode, i.e. PEC)

While a PV-PEC device offers some advantages and disadvantages compared to a PV-electrocatalyst system, further consideration can be applied to a fully PEC system composed of a photoanode and/or photocathode. In such a device configuration, the water/$OH^-$ oxidation and water/$H^+$ reduction reactions can both be photodriven, giving the system 1 or 2 semiconductor liquid junctions. In a 2 SLJ system, compared to a PV-PEC system, this may provide more losses at the SLJ interfaces due to catalytic overpotentials for each reaction. In addition, this configuration (2 photoelectrodes) requires light to pass through the electrolyte in all cases, giving rise to further optical losses that may limit the potential efficiency achievable.

From an operational point of view, the current matching for a fully PEC photoanode/photocathode system is identical to the PV-PEC system, as shown in Fig. 4.10. The main difference is that the fully PEC system has 2 SLJ's, and thus the obtainable operating current is limited by the photovoltages that can be obtained by both the photoanode and photocathode, which in practice will be smaller than those of a PEC and PV material. In short, the PV material will obtain its maximum $V_{oc}$ by having ohmic contacts on either side, while the PEC is limited by a SLJ, which requires equilibration of the Fermi level and dominant redox potential associated with either desired chemical reaction. From a practical perspective, this may limit the overall potential of such a system, which is shown by the poor demonstrated efficiencies in the literature (Nozik 1976; Kainthla et al. 1987; Mor et al. 2008; Sato et al. 2011; Arai et al. 2013; Bornoz et al. 2014).

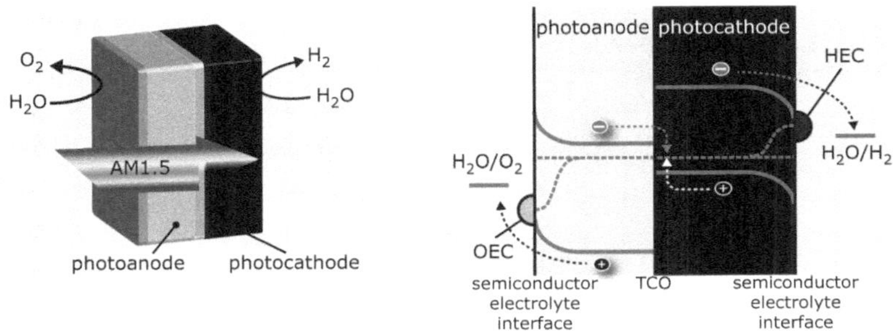


**Fig. 4.10** A cartoon representation of a wireless monolithic PEC cell with a photoanode to drive the water oxidation reaction and a photocathode to drive the water reduction reaction (a), and the associated electronic band diagram for a potential single-junction PEC/PEC device that powers the water-splitting reaction

### *4.2.3 The Electrolyte*

The composition of the electrolyte used in a PEC cell is essential to the performance and stability of the overall device. For the considerations of this chapter where we only discuss PEC water splitting, the electrolyte solution will be composed of liquid water with different solvated ions. Water itself is a poor conductor, so it is necessary to dissolve charged ions to aid in the charge transfer process between the working and counter electrode. Many considerations need to be accounted for in using a particular ionic species in an electrolyte including the materials stability, the ionic conductivity, and the diffusion of each ion through a potential membrane. The role of the electrolyte, which is an ionic liquid solution, is to transfer charge between the surfaces of the working and counter electrodes. Positive charge is passed through protons ($H^+$), while negative charge is passed through hydroxide ions ($OH^-$). In an aqueous solution with the standard used ionic species, $H^+$ and $OH^-$ have the highest limiting ionic conductivities of 349.8 and 197 ($10^{-4}\ \Omega^{-1}\ mol^{-1}\ m^2$), respectively. Other commonly used cations such as $K^+$ and $Na^+$ have lower limiting ionic conductivities of 73.5 and 50.1 ($10^{-4}\ \Omega^{-1}\ mol^{-1}\ m^2$), respectively, while commonly used anions such as $Cl^-$ and $SO_4^{2-}$ also have lower limiting ionic conductivities of 76.4 and 162 ($10^{-4}\ \Omega^{-1}\ mol^{-1}\ m^2$), respectively. A table of commonly used acid, base, and neutral solutions of various concentrations are shown in Table 4.2, where the conductivity of the electrolyte, $\kappa$, the electrolyte resistance, $R_E$, and potential loss at 5 mA/cm$^2$ are given for each electrolyte composition.

While the measured ionic conductivities for the different ionic species may seem high, relative to the conductivity of electrons through a conductive wire (for copper, conductivity, $\sigma$, ~ $6 \times 10^7$ S/m), they are several orders of magnitude smaller. The relatively low ionic conductivities can lead to large ohmic losses, which increase necessary overpotentials to drive the water splitting half reactions, thus decreasing the overall device efficiency. The comparison of the conductivity of a metal wire and an ionic solution is important when considering the design of a monolithic PEC device. For example, if the working and counter electrode are spatially separated, as

**Table 4.2** Typical electrolyte compositions and acidity/pH with the associated conductivity, resistances, and potential losses at 5 mA/cm$^2$, adapted from (van de Krol 2012)

| pH | Electrolyte composition | $\kappa$ ($\Omega^{-1}$m$^{-1}$) | $R_E$ ($\Omega$) | $V_{loss}$ @ 5 mA/cm$^2$ (mV) | $T$°C |
|---|---|---|---|---|---|
| Neutral | Distilled water | $10^{-3}$~$10^{-4}$ | $10^5$~$10^6$ | ∞ | 20 |
| Neutral | Purified water | ~5.5 × $10^{-6}$ | ~18 × $10^6$ | ∞ | 25 |
| Acid | 0.5 M $K_2SO_4$ | 6.2 | 16 | 81 | 20 |
| | 1.0 M $H_2SO_4$ | 36.6 | 2.7 | 14 | 18 |
| | 3.5 M $H_2SO_4$ | 73.9 | 1.4 | 7 | 18 |
| Neutral | 0.1 M NaCl | 1.07 | 93 | 467 | 18 |
| | 0.5 M NaCl | 3.8 | 26 | 132 | 18 |
| | 1.0 M NaCl | 7.44 | 13 | 67 | 18 |
| Base | 0.1 M KOH | 2.26 | 44 | 221 | 18 |
| | 0.5 M KOH | 10.7 | 9.3 | 47 | 18 |
| | 1.0 M KOH | 20.1 | 5.0 | 25 | 20 |

shown in Fig. 4.1a, b, the distance for ionic diffusion, and thus the ionic conductivity is decreased compared to the integrated monolithic device shown in Fig. 4.5. The detailed comparison between a wired versus wireless PEC device is discussed explicitly in Chap. 12.

In addition to the bulk composition of an electrolyte designed to transfer positive and negative charges, the addition of a buffer to an electrolyte can have a very positive effect. For example, when $OH^-$ or $H^+$ are consumed at an electrode/electrolyte interface, the local concentration (i.e. pH) of charged ions in the solution becomes slightly more acidic or basic, respectively. This small change can alter the kinetics and possibly thermodynamics of the desired chemical reaction. Therefore, buffer salts are used to react with the increased/decreased pH layers in order to maintain a steady pH balance in the entire solution, but most importantly near the electrode/electrolyte interface. Common buffers are phosphate ($KH_2PO_4$/$K_2HPO_4$) and borate ($H_2BO_3$/$HBO_3$), which maintain a solution pH at ~7 and ~9, respectively.

The composition of the electrolyte is important in determining the ionic transport in the PEC cell, but is also critical in determining the stability of the working and counter electrodes used during operation. In particular, the materials used in a PEC cell should not corrode during prolonged exposure to light and the electrolyte. A detailed chart of typical PEC photoelectrode materials and their associated self-reduction (black horizontal lines) and self-oxidation (dark grey horizontal lines) potentials have been accumulated by Chen and Wang (Chen and Wang 2012), shown in Fig. 4.11.

In view of this reference, it is clear that the electrolyte for a PEC cell must not only be chosen for its favourable charge/ionic transport properties between the working and counter electrodes, but must also be favourable for the long-term stability of the electrode materials. Therefore, the (photo)electrodes and electrolyte

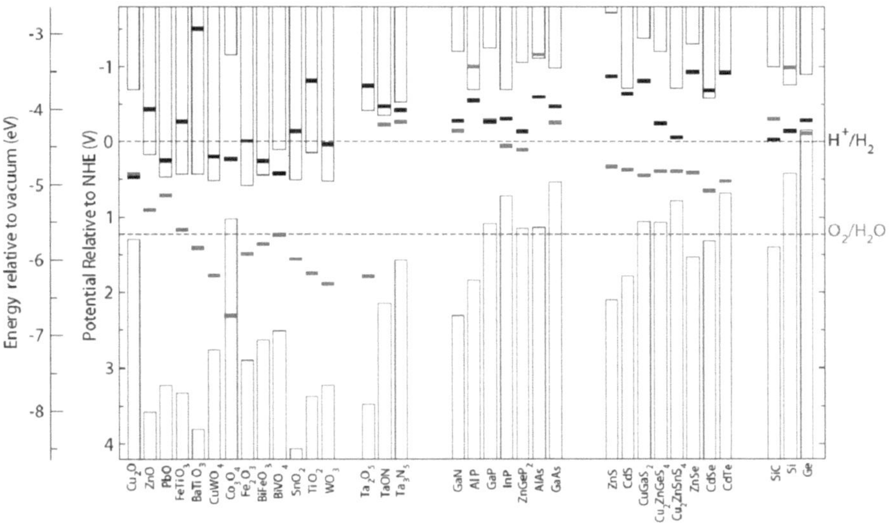


**Fig. 4.11** The electronic band diagram and associated self-reduction and oxidation potentials for selected semiconductor materials at pH = 0, figure taken with permission from reference (Chen and Wang 2012)

solutions must be chosen and carefully selected in tandem and never considered irrespective of each other.

## *4.2.4 Ion Exchange Membranes*

To avoid mixing of the produced fuel (hydrogen or carbon-based products) with oxygen gas, ion exchange membranes are often applied to separate the anodic and cathodic electrolyte in proposed solar fuel designs. Three types of ion exchange membranes are discussed: proton exchange membranes, anion exchange membranes and bipolar membranes. In addition, also membrane-less designs that avoid mixing produced hydrogen and oxygen gasses have been suggested in literature and will be discussed briefly.

### 4.2.4.1 Proton Exchange Membranes

Based on the traditional fuel cells and electrolysers, a proton exchange membrane (PEM) can be used in solar fuel devices to allow the transport of $H^+$ from the anode to the cathode (Haussener et al. 2012; Roy et al. 2010). The most commonly used proton exchange membrane is Nafion®, which is known for its high conductivity, high chemical stability, and optical transparency. Other proton exchange membranes have been proposed and investigated (Hickner et al. 2004; Peckham and Holdcroft 2010), but are still less widely used.

Two issues related to proton exchange membranes can be identified for the use in solar fuel devices. First, the price of these membranes is high, which is in particular an issue because the low current density requires a large area. Second, in contrast to what the name suggests, proton exchange membranes also allow the transport of other cations than protons (Chae et al. 2007), albeit with lower conductivities. Consequently, when other cations (e.g., $K^+$ or $Na^+$) dominate the concentration of protons, which is usually the case at pH > 1, these cations partly account for the charge transport through the membrane, while protons are consumed at the cathode and produced at the anode. In the long term, the pH at the cathode will increase while the pH at the anode will decrease, which increases the required voltage for water splitting (polarisation) (McKone et al. 2014). Modestino et al. (Hashemi et al. 2015) have shown that partly mixing the anodic and cathodic electrolyte can limit this effect to a single pH unit, with a minor compromise in gas purity.

#### 4.2.4.2 Anion Exchange Membranes

As an alternative for the proton exchange membranes, anion exchange membranes are proposed for solar fuel devices as well, which only allow the transport of anions such as $OH^-$ (McKone et al. 2014). In order to facilitate $OH^-$ transport rather than transport of other anion species, the use of anion exchange membranes is in particular attractive for alkaline electrolytes, which matches the activity of earth abundant oxygen evolution catalysts (McCrory et al. 2015). For near-neutral solutions, anion species other than $OH^-$ are transported as well, which creates additional voltage loss after multiple hours of operation (Hernandez-Pagan et al. 2012), similar to cation or proton exchange membranes. In addition to that, anion exchange membranes suffer from limited chemical stability in strongly alkaline environments, lower conductivity (although that may not be an issue at current densities $\leq 10\ mA/cm^2$), and limited selectivity leading to cation crossover instead of $OH^-$ transport (Varcoe et al. 2014; Hickner et al. 2013). Hence, the development of stable, selective, and possibly transparent anion exchange membranes is an on-going challenge for the solar fuel development.

#### 4.2.4.3 Bipolar Membranes

Electrolyte restrictions for the stability and activity of photoelectrodes and (co-) catalysts limit the options of an integrated practical solar fuel device. To enlarge the compatibility of (photo-)anodes and cathodes, a bipolar membrane (BPM) can be used to separate the anodic and cathodic electrolyte. A bipolar membrane dissociates water into $H^+$ and $OH^-$ due to the two-layered ion membrane structure, which allows maintaining a different pH at either side of the membrane (Simons 1993). Compared to the other ion exchange membranes, the use of such membrane provides the additional advantage of using different (stable) pH at either electrode. This idea has been explored for dark electrolysis (McDonald et al. 2014; Vargas-Barbosa et al. 2014), in which water splitting with a cathode at pH 0 and an anode at pH 14 did show no increased voltage with respect to the traditional case at a single pH without a bipolar membrane. Recently, a BPM has been used in a photodriven system consisting of a $BiVO_4$ photoanode in pH 7 or 14 and a Pt cathode in pH 0 (Vermaas et al. 2015). The milder conditions, i.e. pH 7 versus pH 0, yield insignificant potential losses after 80 h of operation at current densities estimated to be close to those needed to be produced in large-scale solar fuel devices.

#### 4.2.4.4 Membrane-Less Systems

To avoid costs for membranes and to avoid polarisation over the membrane at near-neutral pH, membrane-less solar fuel systems have been proposed as well. Examples of membrane-less system with proven separation of hydrogen and oxygen gasses are based on mesh electrodes with divergent convective flow (Gillespie et al. 2015) or devices with fast tangential water flow along plate electrodes (Hashemi et al. 2015). Although the latter system offers promising low hydrogen and oxygen crossover (only a few percent), only microscale systems have been tested as of yet. Similar for all membrane and membrane-less designs, the type of system strongly depends on the electrode and catalyst requirements. Hence, as no consensus is achieved for an integrated design for solar fuels, the options for one of the mentioned membranes or membrane-less designs are all open for development.

## 4.3 Measurement Protocols

While the previous section describes the components and configurations of PEC cells, it is also important to have well-defined protocols for measuring the performance and efficiency of PEC materials and systems. This is most important to aid in the comparison of materials and devices made in different laboratories in different countries around the world. Therefore, several performance benchmark metrics are described in the following section, along with standard measurement protocols and equipment so that the performance and efficiency of PEC materials and devices can be normalized across the field.

### *4.3.1 Simulated Solar Irradiation Measurements*

The most obvious measurement to consider for standard protocols is how to observe the performance of a photoelectrode under solar irradiation. While the solar spectrum is constant from its source 93,000,000 miles away, there is a variance in the

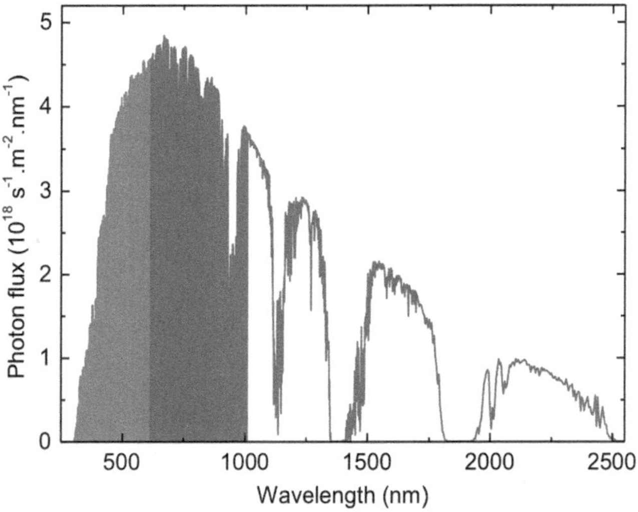


**Fig. 4.12** The AM 1.5 global solar spectrum with the indicated areas that correspond to the light energy of 1.23 eV (*dark grey*) and 2.0 eV (*light grey*), which indicate the water splitting potential and theoretical potential needed to drive actual water-splitting including losses, respectively

location where you measure its power, which also depends on the time of day and season you are measuring. Therefore, a normalized standard solar spectrum and power density has to be introduced in order to have a metric by which to standardize materials performance. Such a standard has been used extensively for decades in the photovoltaic field, and the same conditions are applied to the PEC field. The agreed upon standard metric for simulated solar irradiation is global air mass 1.5 (AM 1.5), as shown in Fig. 4.12.

This illumination source must be calibrated in each lab by means of a photodiode to ensure that the spectral distribution and power density is closely related to the specifications. An extensive comparison between light sources and their specifications has been organized by R. van de Krol (2012), which the readers are guided for reference.

For practical purposes, solar irradiation measurements are generally used while performing linear sweep or cyclic voltammetry measurements, where the photocurrent density is measured as a function of applied potential. The information gained from such a measurement is enormous as it can dictate the flatband potential, saturated photocurrent density, and fill factor of a photoelectrode. An example of a typical linear sweep voltammogram for a photoanode ($BiVO_4$) and a photocathode (a-SiC) is shown in Fig 4.13a, b, respectively. These materials and figures are used to show the general trends for each class of material, i.e. to show that photoanodes produce a positive (photo)current density when a positive potential is applied, and that photocathodes produce a negative (photo)current density when a negative potential is applied. For the following sections, the $BiVO_4$ photoanodes were deposited by a spray pyrolysis technique (as detailed in Abdi et al. 2013), and the

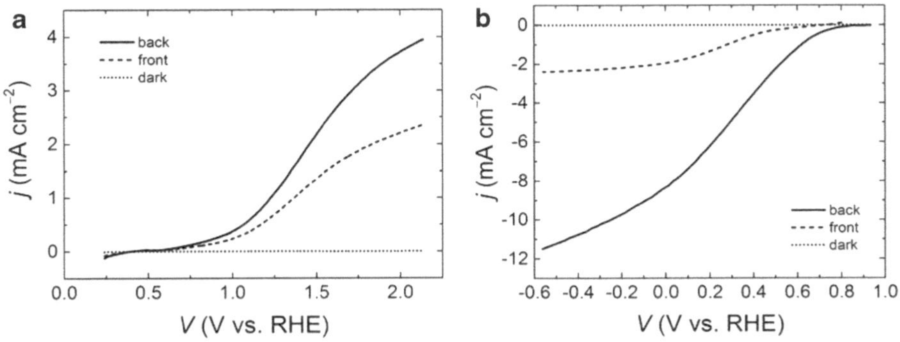


**Fig. 4.13** Typical photocurrent vs. voltage plot for (**a**) an n-type $BiVO_4$ photoanode and (**b**) a p-type a-SiC photocathode

a-SiC photocathodes were grown by plasma-enhanced chemical vapour deposition (PECVD) (as detailed in Digdaya et al. 2015).

In these plots, several interesting features can be observed. For each figure, the dotted line represents the dark current, which is the current measured at different potentials when no light is incident on the photoelectrodes. If a photoelectrode shows any dark current at a given potential, it is usually a sign of corrosion, and thus instability. Therefore, each of the dotted line plots in Fig. 4.13 indicate that the materials are stable (i.e. do not corrode) within the potential range they are swept. It is also interesting to observe that when each of the samples is illuminated by back- and front-side illumination, the photocurrent generated shows different trends. For example, with the $BiVO_4$ photoanode, there is a higher photocurrent generated when the sample is illuminated from the front-side, while for the a-SiC photocathode, there is a higher photocurrent generated when the sample is illuminated from the backside. These measurements can be an indication of the performance limiting photo-generated carrier diffusion length (see Sect. 4.2.2.1).

To clearly understand how much of the current density is due to the absorption and conversion of sunlight (i.e. photocurrent), it is necessary to make measurements under solar irradiation and in the dark. If the current density in the dark is subtracted from the current density under illumination, this is called the photocurrent density, where the density term applies to the areal coverage of the photoelectrode (usually in $cm^2$ for laboratory measurements). One approach to make such a plot is make two (or more) separate measurements and plot them on the same axis as shown in the dashed line of Fig. 4.13. An alternative is to make a chopped illumination measurement, where a timed light-chopper is placed in between the light source and the photoelectrode in timed intervals to give a single measurement profile that alternates between light and dark measurements during a potentiodynamic sweep, as shown in Fig. 4.14a for a $BiVO_4$ photoanode and (b) for an a-SiC photocathode.

This measurement clearly shows that when light is able to reach the photoelectrode, there is a sharp increase in the current density, and when the light

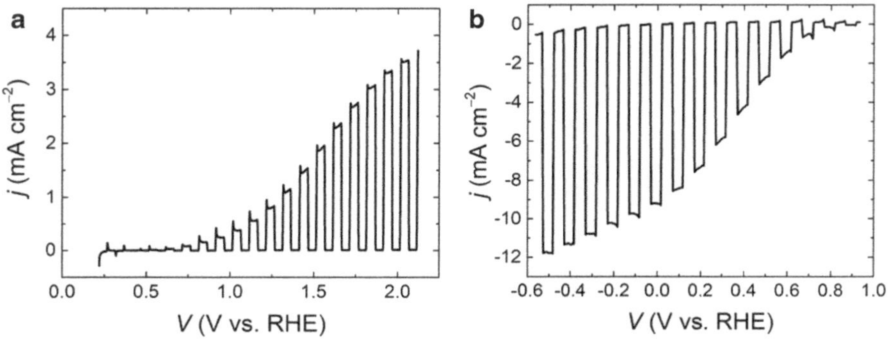


**Fig. 4.14** Typical chopped illumination photocurrent vs. voltage plot for (**a**) an n-type $BiVO_4$ photoanode and (**b**) a p-type a-SiC photocathode

path is blocked, there is a sharp decrease in the measured current density, which relates directly to the dark current measurements as shown previously in Fig. 4.13.

Several important pieces of information can be extracted from both the *J–V* curves (shown in Fig. 4.13) and the chopped illumination curves (shown in Fig. 4.14). When sweeping anodically/cathodically for photoanodes/photocathodes, the potential where the photocurrent generation begins is called the onset potential, $V_{on}$. According to Fig. 4.13, the $V_{on}$ for $BiVO_4$ is ~0.6 V vs. RHE, while the $V_{on}$ for a-SiC is ~ 0.8 V vs. RHE. While the $V_{on}$ are similar for the two materials, it is important to again note that the trends are different for photoanodes and photocathodes. In particular, the $V_{on}$ for $BiVO_4$ implies that photocurrent will begin to increase at potentials more positive than $V_{on}$, while for the a-SiC photocathode the photocurrent generation will increase at potentials more negative than $V_{on}$. In addition, at potentials much larger than $V_{on}$ (more positive for photoanodes, and more negative for photocathodes), the photocurrent density eventually saturates at a maximum value, called the saturated photocurrent density, $J_{sc}$. Similar to the PV-field, the slope of the *J–V* curve as it moves from $V_{on}$ to the $J_{sc}$ can give an indication of the electronic properties and strength of the semiconductor used. However, unlike in the PV-field where this fill factor is determined solely by the intrinsic bulk properties of the semiconductor and not limited by the ohmic contacts where charge carriers are extracted, for PEC materials, the "fill-factor" is determined by the SLJ, where electrons/holes are less easily exchanged due to poor kinetics and the associated overpotentials. This is observed in the relative large amount of potential that is required to reach $J_{sc}$ after the $V_{on}$ (for the aforementioned $BiVO_4$ this potential is $>$ 1.5 V vs. RHE, while for the a-SiC photocathode this potential is $>$ 1.2 V vs. RHE).

### 4.3.2 Determining the Flatband Potential

Similar to the $V_{on}$ mentioned in the previous section, an essential component of a PEC material is the so-called flatband potential. The flatband potential, as indicated by the name, is the potential at which no band bending occurs at the SLJ, and thus the conduction and valence bands are flat, as shown in Fig. 4.15a for an n-type photoanode. The actual potential is measured in a three-electrode configuration and is defined as the potential between the Fermi level of the semiconductor and the reference electrode.

These flatband conditions do not hold when a potential is applied that is greater than the $V_{fb}$, and/or when the photoelectrode is illuminated, as shown in Fig. 4.15b. In this case, the Fermi level of the photoanode is brought down below the previously determined flatband potential, either by the addition of an external bias, or by the relative change in the electron-hole concentration due to illumination which drives photoelectrocatalysis at the SLJ.

In order to actually measure the flatband potential, the most powerful technique is impedance spectroscopy (IS), or more specifically, Mott-Schottky analysis (Klahr et al. 2012). Using this technique, the capacitance of the space charge layer, $C_{SC}$, is measured, and $1/C_{SC}^2$ is plotted against the applied potential, as shown in Fig. 4.16 for a thin film of $TiO_2$ grown by ALD (Digdaya et al. 2015). From this plot, a linear slope can be made through the measured inverse capacitance squared, and where the linear regression crosses the $x$-axis is the flatband potential. In particular, the plot and subsequent linear region can be extrapolated from the Mott-Schottky equation, given as:

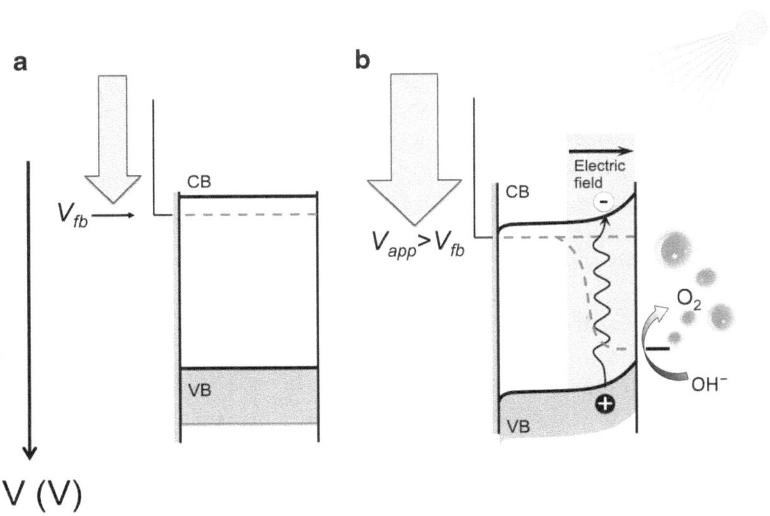


**Fig. 4.15** Semiconductor electronic band positions for (**a**) the flatband condition, and (**b**) with an applied potential greater than the flatband potential, and illuminated

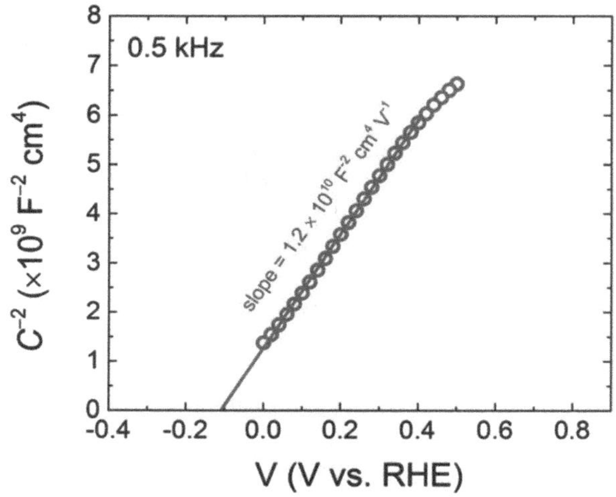


**Fig. 4.16** A representative Mott-Schottky plot for a $TiO_2$ film grown by atomic layer deposition (ALD) on an FTO substrate at 150 °C, used with permission from (Digdaya et al. 2015)

$$\frac{1}{C_{SC}^2} = \frac{2}{\varepsilon_0 \varepsilon_r e N_D A^2}\left(V_{app} - V_{fb} - \frac{kT}{e}\right) \tag{4.6}$$

The change of the flatband as a function of the pH of the electrolyte has been found to be especially pronounced in metal oxide photoelectrodes, though it may also hold for nonoxide semiconductors as well.

### *4.3.3 Evolved Gas Quantification*

To ensure faradaic efficiency in photoelectrodes and to measure this in absolute terms for dispersed photocatalyst materials, it is absolutely necessary to measure and quantify the evolved products for the water-splitting reaction, i.e. to quantify the amount of oxygen and hydrogen evolved. Similar to previous sections, here we will not discuss the quantification of other solar fuel products, such as those made from $CO_2$ reduction, which may vary significantly, as it has been recently shown that up to 16 gaseous and liquid products may be formed from a single $CO_2$ electroreduction reaction (Kuhl et al. 2012). Therefore, the discussion in this section only deals with the quantification of hydrogen and oxygen gases from solar water splitting devices.

In order to determine the faradaic efficiency of either the hydrogen evolution or oxygen evolution reactions, it is necessary to know three primary characteristics of a film; (1) the active surface area of the catalyst, (2) the amount of current density passing through the electrode, and (3) the number of moles of hydrogen/oxygen produced as a function of time. If you can obtain all three of these criteria, then it is possible to accurately describe how much of the current density measured goes to the desired reduction/oxidation reaction, and how much goes to other processes (i.e. side-reactions, back-reactions, corrosion, etc.). However, obtaining an accurate estimation of the active sites in a chemical reaction, or even the amount of active surface area, especially for a nanostructured (photo)electrode (Osterloh 2013), may be very difficult to obtain. Therefore, in general, the actual surface area used in most reports for semiconductor photoelectrodes is the projected surface area, or the amount of area of the electrode exposed to the electrolyte, and does not include nano-, micro-, or other sized features in the determination of the active surface area. Thus, it may even be harder to compare current densities of different semiconductor photoelectrodes, especially comparing planar electrodes to nanostructured electrodes.

Furthermore, a large difference may be seen from making either static or dynamic measurements of current density/gas production, and thus it is suggested to make static voltage/current density measurements for more accurate measurements to allow for a more controlled production of oxygen/hydrogen. Using a fixed potential and measuring the (photo)current density over time can also be a good way to show stability/instability, as the current density will decrease if the sample is unstable and generally remains constant if the system is stable (though the current could also remain constant if there is a constant corrosion process).

## 4.4 Efficiency Definitions

In order to quantify the performance and efficiency of PEC materials and devices, it is necessary to have well-defined benchmark metrics of assessment. Many reports list the photocurrent density for photoanodes at 1.23 V vs. RHE and for photocathodes at 0 V vs. RHE as benchmark performance metrics. However, these metrics by themselves are irrelevant for a practical device, since the operational potential, as outlined in the previous sections, will never be at 0 V or 1.23 V vs. RHE and only show the half-cell potential of a given working electrode and neglecting the (over)-potentials used to drive the counter electrode and ionic conductivity losses in the solution. Therefore, normalized metrics are required to establish a benchmarking for the performance of different materials in order to make fair comparisons between materials and systems that are made and tested in different labs across the globe.

### *4.4.1 Solar-to-Hydrogen (STH) Conversion Efficiency*

Perhaps the most significant metric for measuring the performance and efficiency of a solar fuel device is the solar-to-hydrogen conversion efficiency (STH). This efficiency directly relates the input energy (solar irradiation) to output energy (electric/chemical energy via hydrogen evolution minus the input-applied potential) via the following equation:

$$\eta_{STH} = \frac{P_{out}}{P_{in}} = \frac{\Delta P_{electrical}}{P_{light}} = \left[\frac{\Phi_{H_2}(mol/s\,m^2) \times G^{o}_{f,H_2}(kJ/mol)}{P_{light}(W/m^2)}\right]_{AM\ 1.5G} \quad (4.7)$$

where the numerator contains the output in terms of the rate of gas evolved, $\Phi_{H2}$ (mol $H_2$/s $m^2$) times the Gibbs free energy of formation for hydrogen ($G^{o}_{f,H2} = 237$ kJ/mol), divided by the total solar irradiation input in terms of the power density of the incident illumination ($P_{light}$ in W/$m^2$, or more commonly for PEC devices, mW/$cm^2$). This expression only holds true when the illumination source is the direct (or simulated) solar irradiation-matched spectra equal to air mass global (AM) 1.5. Furthermore, it is only possible to use this equation to measure the STH of a solar-driven water-splitting reaction when it is possible to directly measure $H_2$ accurately as a function of time, most importantly for particle-based photocatalysts. When this is not available, for example, it is possible to convert this equation to a different form that can use a modified version:

$$\eta_{STH} = \left[\frac{(j_{sc}(mA/cm^2) \times (V_{redox}) \times \eta_f}{P_{light}(mW/cm^2)}\right]_{AM\ 1.5G} \quad (4.8)$$

where the numerator now has the power output in terms of the measured current density $j_{sc}$ in mA/$cm^2$ times the effective potential required to run the desired reaction (the redox potential of interest, $V_{redox}$, which here is the potential converted from the previously used $\Delta G = 237$ kJ/mol $= 1.23$ V), times the faradaic efficiency of the hydrogen evolution reaction, $\eta_f$. The denominator does not need to have a term to include the illuminated area of the electrode, since the numerator has the current density in terms of current per unit area already included.

It is important to note that the STH is measured in a 2-electrode configuration, and all the potentials applied must be taken between the working and counter electrode, i.e. it is not possible to use a 3-electrode system and use the potential applied to a working electrode against a reference electrode.

While the focus of this chapter and the discussion is on the solar to hydrogen conversion efficiency of the solar water-splitting reaction, a similar metric can be applied to general solar fuel systems, where hydrogen is not the reduction product via water splitting, but where, for example, the reduction of $CO_2$ to different chemical fuels is achieved. In such a case, it is straightforward to calculate the solar to fuel conversion efficiency, $\eta_{SFE}$, by the following equation:

$$\eta_{SFE} = \left[\frac{j_{OP}(mA/cm^2) \times V_{redox} \times \eta_f}{P_{total}(mW/cm^2)}\right]_{AM\ 1.5G} \quad (4.9)$$

where $J_{op}$ is the operational current density that is directed towards a specific product. The potential is correlated to the thermodynamic potential for a different fuel-forming reaction, $V_{redox}$. This metric is much more difficult to extract from the current densities, as it is likely that many products are formed during electroreduction of $CO_2$, and therefore the faradaic efficiency and partial current density towards a particular chemical reaction are needed, which is very complicated from a practical perspective and is thus not discussed further in this chapter.

### 4.4.2 Applied Bias Photon to Current Conversion Efficiency (ABPE)

An additional tool to determine how a photoelectrode is able to convert photons into usable electrons via a chemical reaction is to observe how the photon to current conversion efficiency changes with an applied bias using the so-called applied bias photon to current conversion efficiency (ABPE). This technique is an obvious extension to the STH efficiency, with the notable difference that this technique uses an applied bias between the working electrode and counter electrode, while the STH is measured without the application of any external bias potential. Therefore, the ABPE can be written as follows;

$$\mathrm{ABPE} = \left[\frac{j_{\mathrm{sc}}(\mathrm{mA/cm^2}) \times \left(V_{\mathrm{redox}} - V_{\mathrm{app}}\right) \times \eta_{\mathrm{f}}}{P_{\mathrm{light}}(\mathrm{mW/cm^2})}\right]_{\mathrm{AM\ 1.5G}} \tag{4.10}$$

where $V_{app}$ is the applied potential between the working and counter electrode. The utility of using the ABPE measurement is that it uses extra potential to drive the water-splitting reaction for a given photoelectrode, which may be useful for estimating how a particular photoanode or photocathode may operate in a tandem device where an extra potential can be supplied by a second photoelectrode or a photovoltaic cell connected in series. This allows the measurement of a single component of a tandem device to be used to estimate the overall photocurrent density and efficiency that could be drawn if it is used in a tandem absorbing device. The practical aspects of a tandem absorbing device are briefly discussed in Sect. 4.2.2 of this chapter, and in more detail in Chap. 12.

### 4.4.3 Spectral Response Measurements

To measure overall performance and conversion efficiencies of a photoelectrode, it is necessary to use the entire solar spectrum to excite photogenerated charge carriers in a photoelectrode. However, it is also useful to understand where these photogenerated charge carriers come from during photoexcitation, i.e. to be able to tell which photons produce a certain amount of electrons. Therefore, making photocurrent measurements as a function of individual wavelengths of light is necessary. Such a measurement can be accomplished with a light source, a monochromator, and a potentiostat.

#### 4.4.3.1 Incident Photon to Current Conversion Efficiency (IPCE)

While STH remains the single most important figure of merit to measure the performance of a PEC material/device, other techniques can be used to provide essential information of how the material/device works. These metrics are essential to assess the origin of how a material performs, so that its practical limits can be defined, and hopefully then overcome with optimized engineering. One such diagnostic technique is the incident photon to current conversion efficiency (IPCE), which may also be referred to as the external quantum efficiency (EQE). The IPCE/EQE measures the efficiency of converting an individual photon to an extractable electron via the following formula:

$$\begin{aligned} \text{IPCE} = \text{IPCE}(\lambda) = \text{EQE}(\lambda) &= \frac{\text{electron flux (mol/s)}}{\text{photon flux (mol/s)}} \\ &= \frac{|j_{\text{ph}}(\text{mA/cm}^2)| \times hc(\text{Vm})}{P_\lambda(\text{mW/cm}^2) \times \lambda(\text{nm})} \end{aligned} \quad (4.11)$$

where $j_{ph}$ is the photocurrent density, $h$ is Plank's constant, $c$ is the speed of light, (therefore $hc$ can be simplified to 1239.8 Vm), $P_\lambda$ is the power of light at a particular wavelength, and $\lambda$ is the wavelength of irradiation. To make accurate IPCE measurements, a light source, monochromator, and potentiostat are required in order to have a spectral distribution that is selective by wavelength, while at the same time the current density generated at each wavelength needs to be measured. In addition, it is required that such a measurement takes place in a 3-electrode configuration, so that the potential of the working electrode can be varied and measured against a reference electrode. This is in sharp contrast to the measurement configuration needed for obtaining the STH, which is most important for defining the overall efficiency of a material, while measuring IPCE is more of a diagnostic tool to tell more detailed information about an electrode and to help determine the performance limiting factors.

The technique of obtaining IPCE is very useful and relevant for PEC materials characterization, but has its limitations for what it can tell about the total efficiency of a system. For example, it is assumed that for the output of the IPCE measurements, i.e. the electron flux, 100 % is used for the evolution of hydrogen and oxygen and not for a side or back-reaction. Therefore, it is necessary to couple IPCE measurements with $H_2$ and $O_2$ quantification to ensure that the water oxidation/reduction reactions being driven by the individual photons show faradaic efficiency, and thus all the converted photons are only consumed in the water-splitting reaction. A typical IPCE plot for a $BiVO_4$ photoanode and an a-SiC photocathode illuminated from the front-side and back-side are shown in Fig. 4.17a, b, respectively.

Interestingly, the IPCE can be used to estimate the maximum obtainable photocurrent under AM 1.5 irradiation by the following relationship:

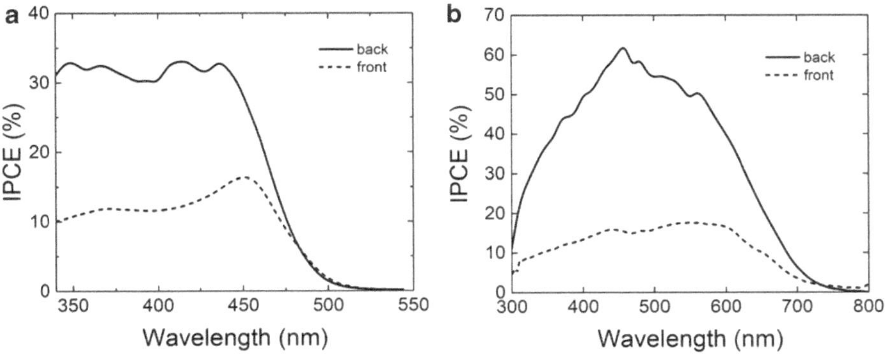


**Fig. 4.17** IPCE data for (**a**) a $BiVO_4$ photoanode illuminated from the front and backside, and held at a potential of 1.23 V vs. RHE, and (**b**) a a-SiC photocathode illuminated from the front and backside, and held at a potential of 0 V vs. RHE

$$J_{\mathrm{AM\ 1.5}} = \int (\mathrm{IPCE}_\lambda \times \Phi_\lambda \times e) \mathrm{d}\lambda \tag{4.12}$$

where $J_{\mathrm{AM\ 1.5}}$ is the total photocurrent density under solar irradiation (mA/cm$^2$), $\Phi_\lambda$ is the photon flux of the solar irradiation (photons/(m$^2$s)), and $e$ is the elementary charge ($C$). While this is not a direct or 100 % accurate way to estimate the photocurrent density of a material under AM 1.5 solar irradiation, it can give a close estimate if a solar simulator is not available in a particular laboratory, and only IPCE testing equipment is available. A correlation between IPCE (integrated photocurrent) and information provided by *J*–*V* measurements is essential for ensuring consistency of measurements.

#### 4.4.3.2 Absorbed Photon to Current Conversion Efficiency (APCE)

The IPCE measures the total amount of electrons converted from all of the incident photons (broken down into individual wavelengths), and thus is useful to estimate the maximum possible current that can be extracted by a photon source. However, this technique inherently takes into account all of the photons that are incident on a photoelectrode (i.e. light that is either reflected or transmitted through the sample) and converted to usable (i.e. able to drive the water redox reactions) electrons. This is certainly not the case for a practical semiconductor material, and therefore it is also useful to normalize the IPCE by the absorbed spectrum of a sample, which results in the absorbed photon to current conversion efficiency (APCE), or internal quantum efficiency (IQE).

$$\mathrm{APCE} = \mathrm{APCE}(\lambda) = \mathrm{IQE}(\lambda) = \frac{\mathrm{IPCE}}{A_\lambda} = \frac{\left|j_{\mathrm{ph}}(\mathrm{mA/cm^2})\right| \times hc(\mathrm{Vm})}{P_\lambda(\mathrm{mW/cm^2}) \times \lambda(\mathrm{nm}) \times A_\lambda} \tag{4.13}$$

where $A_\lambda$ is the absorptance as a function of wavelength. The APCE is primarily used as a tool to determine the optimal thickness of a material to maximize the light-absorbing path length through a semiconductor.

## 4.5 Summary and Conclusions

This chapter serves to introduce the reader to the important aspects of measuring the performance and efficiency of photoelectrochemical water-splitting materials. In particular, the considerations for designing a PEC cell are discussed in the context of the materials used (electrodes, electrolyte, membranes) and the different configurations that photo- and electrocatalysts can be combined to make an overall water-splitting device. In addition, standard measuring equipment and techniques are summarized to aid the reader in the basic materials used in PEC testing. Finally, several important efficiency and performance metrics are established to determine the actual usefulness of the measured data, and how this should be compared to other samples made in different labs across the world. It is hoped that this chapter serves as a general introduction to the testing and efficiency definitions for PEC water splitting so that the following chapters are more accessible and understandable on a fundamental level.

**Acknowledgments** The author gratefully acknowledges Bartek J. Trzesniewski, Ibadillah A. Digdaya and Fatwa F. Abdi for assistance with several of the figures, Dr. David Vermaas for contributions to the membrane section, and the MECS group at TU Delft for helpful discussions. The author is also very grateful for generous funding from Towards BioSolarCells (grant FOM 03), the NWO VENI scheme, and the $CO_2$-neutral Fuel program of NWO/FOM/Shell (project APPEL).

## References

Abdi FF, Han L, Smets AHM, Zeman M, Dam B, van de Krol R (2013) Efficient solar water splitting by enhanced charge separation in a bismuth vandate-silicon tandem photoelectrode. Nat Commun 4:2195

Appleby J, Delahoy AE, Gau SC, Murphy OJ, Bockris JOM (1985) An amorphous silicon-based one-unit photovoltaic electrolyzer. Energy 10:871

Arai T, Sato S, Kajino T, Morikawa T (2013) Solar $CO_2$ reduction using $H_2O$ by a semiconductor/metal-complex hybrid photocatalyst: enhanced efficiency and demonstration of a wireless system using $SrTiO_3$ photoanodes. Energ Environ Sci 6:1274

Arakawa H, Shiraishi C, Tatemoto M, Kishida H, Usui D, Suma A, Takamisawa A, Yamaguchi T (2007) Solar hydrogen production by tandem cell system composed of metal oxide semiconductor film photoelectrode and dye-sensitized solar cell. Proc SPIE 6650:665003

Bendert RM, Corrigan DA (1989) Effect of coprecipitated metal ions on the electrochemistry of nickel hydroxide thin films: cyclic voltammetry in 1M KOH. J Electrochem Soc 136:1369

Bornoz P, Abdi FF, Tilley SD, Dam B, van de Krol R, Graetzel M, Sivula K (2014) A bismuth vanadate-cuprous oxide tandem cell for overall solar water splitting. J Phys Chem C 118:16959

Brillet J, Yum J-H, Cornuz M, Hisatomi T, Solarska R, Augustynski J, Graetzel M, Sivula K (2012) Highly efficient water splitting by a dual-absorber tandem cell. Nat Photonics 6:824

Chae KJ, Choi M, Ajayi FF, Park W, Chang IS, Kim IS (2007) Mass transport through a proton exchange membrane (nafion) in microbial fuel cells. Energy Fuel 22:169

Chen S, Wang L-W (2012) Thermodynamic oxidation and reduction potentials of photocatalytic semiconductors in aqueous solution. Chem Mater 24:3659

Conell RS, Corrigan DA, Powell BR (1992) The electrochromic properties of sputtered nickel oxide films. Sol Energy Mater Sol Cells 25:301

Corrigan DA, Knight SL (1989) Electrochemical and spectroscopic evidence on the participation of quadrivalent nickel in the nickel in the nickel hydroxide redox reaction. J Electrochem Soc 136:613

Cox CR, Lee JZ, Nocera DG, Buonassisi T (2014) Ten percent solar-to-fuel conversion with nonprecious materials. Proc Natl Acad Sci U S A 111:14057

Digdaya IA, Han L, Buijs TWF, Zeman M, Dam B, Smets AHM, Smith WA (2015) Extracting large photovoltages from a-SiC photocathodes with an amorphous $TiO_2$ front surface field layer for solar hydrogen evolution. Energ Environ Sci 8:1585

Fujishima A, Honda K (1972) Electrochemical photolysis of water at a semiconductor electrode. Nature 238:37

Gaillard N, Chang Y, Kaneshiro J, Deangelis A, Miller EL (2010) Status of research on tungsten oxide-based photoelectrochemical devices at the University of Hawai'I. Proc SPIE 7770:77700V

Gillespie MI, van der Merwe F, Kriek RJ (2015) Performance evaluation of a membraneless divergent electrode-flow-through (DEFT) alkaline electrolyser based on optimisation of electrolytic flow and electrode gap. J Power Sources 293:228

Han L, Abdi FF, van de Krol R, Liu R, Huang Z, Lewrenz H-J, Dam B, Zeman M, Smets AHM (2014) Efficient water-splitting device based on a bismuth vanadate photoanode and thin-film silicon solar cells. ChemSusChem 7:2832

Hashemi SMH, Modestino MA, Psaltis D (2015) A membrane-less electrolyzer for hydrogen production across the pH scale. Energ Environ Sci 8:2003

Haussener S, Xiang C, Spurgeon JM, Ardo S, Lewis NS, Weber AZ (2012) Modeling, simulation, and design criteria for photoelectrochemical water-splitting systems. Energ Environ Sci 5:9922

Hernandez-Pagan EA, Vargas-Barbosa NM, Wang T, Zhao Y, Smotkin ES, Mallouk TE (2012) Resistance and polarization losses in aqueous buffer-membrane electrolytes for water-splitting photoelectrochemical cells. Energ Environ Sci 5:7582

Hickner MA, Ghassemi H, Kim YS, Einsla BR, McGrath JE (2004) Alternative polymer systems for proton exchange membranes (PEMs). Chem Rev 104:4587

Hickner MA, Herring AM, Coughlin EB (2013) Anion exchange membranes: current status and moving forward. J Poly Sci B: Polymer Phys 51:1727

Jacobsson TJ, Fjallstrom V, Edoff M, Edvinsson T (2015) A theoretical analysis of optical absorption limits and performance of tandem devices and series interconnected architectures for solar hydrogen production. Sol Energy Mater So Cells 138:86

James BD, Baum GN, Perez J, Baum KN (2009) Technoeconomic analysis of photoelectrochemical (PEC) hydrogen production. Directed Technologies Inc., Arlington, VA, US DOE Contract no GS-10F-009J

Kainthla RC, Zelenay B, Bockris JO'M (1987) Significant efficiency increase in self-driven photoelectrochemical cell for water photoelectrolysis. J Electrochem Soc 134:841

Khaselev O, Turner J (1998) A monolithic photovoltaic-photoelectrochemical device for hydrogen production via water splitting. Science 280:425

Khaselev O, Bansal A, Turner J (2001) High-efficiency integrated multijunction photovoltaic/electrolysis systems for hydrogen production. Int J Hydrogen Energ 26:127

Klahr B, Gimenez S, Fabregat-Santiago F, Hamann TW, Bisquert J (2012) Water oxidation at hematite photoelectrodes: The role of surface states. J Am Chem Soc 134:4294

Koper MTM (2011) Thermodynamic theory of multi-electron transfer reactions: Implications for electrocatalysis. J Electroanal Chem 660:254

Kuhl KP, Cave ER, Abram DN, Jaramillo TF (2012) New insights into the electrochemical reduction of carbon dioxide on metallic copper surfaces. Energ Environ Sci 5:7050

Lewis NS, Nocera DG (2006) Powering the planet: Chemical challenges in solar energy utilization. Proc Natl Acad Sci U S A 103:15792

Licht S, Wang B, Mukerji S, Soga T, Umeno M, Tributsch H (2001) Over 18% solar energy conversion to generation of hydrogen fuel; theory and experiment for efficient solar water splitting. Int J Hydrogen Energ 26:653

Lin GH, Kapur M, Kainthla RC, Bockris JO'M (1989) One step method to produce hydrogen by a triple stack amorphous silicon solar cell. Appl Phys Lett 55:386

Luo J, Im J-H, Mayer MT, Schreier M, Nazeeruddin MK, Park N-G, Tilley SD, Fan HJ, Gratzel M (2014) Water photolysis at 12.3% efficiency via perovskite photovoltaics and earth-abundant catalysts. Science 345:1593

McCrory CCL, Jung S, Peters JC, Jaramillo TF (2013) Benchmarking heterogeneous electrocatalysts for the oxygen evolution reaction. J Am Chem Soc 135:16977

McCrory CCL, Jung S, Ferrer IM, Chatman SM, Peters JC, Jaramillo TF (2015) Benchmarking hydrogen evolving reaction and oxygen evolving reaction electrocatalysts for solar water splitting devices. J Am Chem Soc 137:4347

McDonald MB, Ardo S, Lewis NS, Freund MS (2014) Use of bipolar membranes for maintaining steady-state pH gradients in membrane-supported, solar-driven water splitting. ChemSusChem 7:3021

McKone JR, Lewis NS, Gray HB (2014) Will solar-driven water-splitting devices see the light of day? Chem Mater 26:407

Miller EL, Paluselli D, Marsen B, Rocheleau RE (2005) Development of reactively sputtered metal oxide films for hydrogen-producing hybrid multijunction photoelectrodes. Sol Energy Mater Sol Cells 88:131

Mor GK, Varghese OK, Wilke RHT, Sharma S, Shankar K, LaTempa TJ, Choi K-S, Grimes CA (2008) p-tpye Cu-Ti-O nanotube arrays and their use in self-biased heterojunction photoelectrochemical diodes for hydrogen generation. Nano Lett 8:1906

Morisaki H, Watanabe T, Iwase M, Yazawa K (1976) Photoelectrolysis of water with TiO2-covered solar-cell electrodes. Appl Phys Lett 29:338

Nocera DG (2012) The artificial leaf. Acc Chem Res 45:767

Nozik AJ (1976) p-n photoelectrolysis cells. Appl Phys Lett 29:150

Osterloh FE (2013) Inorganic nanostructures for photoelectrochemical and photocatalytic water splitting. Chem Soc Rev 42:2294

Park JH, Bard AJ (2006) Photoelectrochemical tandem cell with bipolar dye-sensitized electrodes for vectorial electron transfer for water splitting. Electrochem Solid-State Lett 9, E5

Peckham TJ, Holdcroft S (2010) Structure-morphology-property relationships of non-perfluorinated proton-conducting membranes. Adv Mater 22:4660

Pinaud BA, Benck JD, Seitz LC, Forman AJ, Chen Z, Deutsch TG, James BD, Baum KV, Baum GN, Ardo S, Wang H, Miller E, Jaramillo TF (2013) Technical and economic feasibility of centralized facilities for solar hydrogen production via photocatalysis and photoelectrochemistry. Energ Environ Sci 6:1983

Reece SY, Hamel JA, Sung K, Jarvi TD, Esswein AJ, Pijpers JJH, Nocera DG (2011) Wireless solar water splitting using solicon-based semiconductors and earth-abundant catalysts. Science 334:645

Rocheleau RE, Miller EL, Misra A (1998) High-efficiency photoelectrochemical hydrogen production using multijuntion amorphous silicon photoelectrodes. Energy Fuel 12:3

Rossmeisl J, Qu Z-W, Zhu H, Kroes G-J, Norskov JK (2007) Electrolysis of water on oxide surfaces. J Electroanal Chem 607:83

Roy SC, Varghese OK, Paulose M, Grimes CA (2010) Towards solar fuels: photocatalytic conversion of carbon dioxide to hydrocarbons. ACS Nano 4:1259

Sakai Y, Sugahara S, Matsumura M, Nakato Y, Tsubomura H (1988) Photoelectrochemical water splitting by tandem type and heterojunction amorphous silicon electrodes. Can J Chem 66:1853

Sathre R, Scown CD, Morrow WR III, Stevens JC, Sharp ID, Ager JW III, Walczak K, Houle FA, Greenblatt JB (2014) Life-cycle net energy assessment of large-scale hydrogen production via photoelectrochemical water splitting. Energ Environ Sci 7:3264

Sato S, Arai T, Morikawa T, Uemura K, Suzuki TM, Tanaka H, Kajino T (2011) Selective $CO_2$ conversion to formate conjugated with H2O oxidation utilizing semiconductor/complex hybrid photocatalysts. J Am Chem Soc 133:15240

Seger B, Castelli IE, Vesborg PCK, Jacobsen KW, Hansen O, Chorlendorff I (2014) 2-Photon tandem davice for water splitting: comparing photocathode first versus photoanode first designs. Energ Environ Sci 7:2397

Simons R (1993) Preparation of a high performance bipolar membrane. J Membrane Sci 78:13

Trotochaud L, Mills TJ, Boettcher SW (2013) An optocatalytic model for semiconductor-catalyst water-splitting photoelectrodes based on in situ optical measurements on operational catalysts. J Phys Chem Lett 4:931

van de Krol R (2012) in Photoelectrochemical Hydrogen Production ed. by R. van de Krol and M. Grätzel and. Springer, Berlin, p 73

Varcoe JR, Atanassov P, Dekel DR, Herring AM, Hickner MA, Kohl PA, Kucernak AR, Mustain WE, Nijmeijer K, Scott K (2014) Anion-exchange membranes in electrochemical energy systems. Energ Environ Sci 7:3135

Vargas-Barbosa NM, Geise GM, Hickner MA, Mallouk TE (2014) Assessing the utility of bipolar membranes for use in photoelectrochemical water-splitting cells. ChemSusChem 7:3017

Vermaas DA, Sassenberg M, Smith WA (2015) Photo-assisted water splitting with bipolar membrane induced pH gradients for practical solar fuel devices. J Mater Chem A 3:19556–19562

# Chapter 5
# Interface Engineering of Semiconductor Electrodes for Photoelectrochemical Water Splitting: Application of Surface Characterization with Photoelectron Spectroscopy

**Wolfram Jaegermann, Bernhard Kaiser, Jürgen Ziegler, and Joachim Klett**

## 5.1 Introduction

The "holy grail" of renewable energy research is the investigation and possibly realization of Artificial Photosynthesis, in which an "artificial leaf" would transform solar energy in the form of light directly into a storable fuel. The given discontinuities in the energy supply provided by renewable sources as solar or wind energy must be overcome with efficient storage devices. Chemical fuels produced from solar radiation will play a major role for the future energy mix as it can be easily stored and transported. The possibility if this fuel will be $H_2$ ($H_2$-technology) or if $H_2$ will be further transformed to hydrocarbons in subsequent dark reactions can be kept open at the given moment (Romm 2005). The expected successful production of storable fuels such as $H_2$ with high energy content would provide a sustainable and secure renewable primary energy carrier, which provides similar advantages in application as carbon based fossil fuels produced by plant photosynthesis. Therefore, it is very clear that an efficient route to a "solar fuel" like $H_2$ by splitting of $H_2O$ must be developed in the coming decades.

The scientific community has notified this scientific and engineering challenge again recently, and a large number of research actions have been restarted in the last few years following an activity wave in the respective research area in the 1970s of the last century after the first oil crisis. But the artificial systems identified so far do not reach sufficient conversion efficiencies or are far too expensive in their

W. Jaegermann (✉) • B. Kaiser • J. Ziegler • J. Klett
Institute of Materials Science and Excellency Graduate School for Energy Science and Engineering, Jovanka-Bontschits-Straße 2, Technical University Darmstadt, 64287 Darmstadt, Germany
e-mail: jaegermann@surface.tu-darmstadt.de; kaiser@surface.tu-darmstadt.de; jziegler@surface.tu-darmstadt.de; jklett@surface.tu-darmstadt.de



S. Giménez, J. Bisquert (eds.), *Photoelectrochemical Solar Fuel Production*,
DOI 10.1007/978-3-319-29641-8_5

technological realization attempts, which makes further intensive research efforts mandatory covering the whole range from fundamental studies to technological developments in its interrelation on every step. So far it seems that artificial photosynthesis devices based on solid state inorganic semiconductors may provide advantages from a fundamental point of view and from accessible device engineering concepts. However, there still exist different device setups, e.g., semiconducting absorbers as photoelectrochemical devices based on semiconductor electrodes, photocatalytic devices using particulate systems, and photochemical arrangements based on organic dyes, which of course need different research approaches even if the involved mechanisms for optimizing the elementary processes are not that different (see for example the following review articles and books on solar water splitting (Krol and Grätzel 2012; Grimes et al. 2008; Lewerenz and Peter 2013; Braun et al. 2010; Maeda and Domen 2010; Peter and Upul Wijayantha 2014; Prévot and Sivula 2013; Valdes et al. 2012; Styring 2012; Zhang 2011; Osterloh and Parkinson 2011; Walter et al. 2010; Mallouk 2010; Rajeshwar et al. 2008)). For this reason a solid physical understanding of the different scientific concepts is needed to realize scientific breakthroughs, which may lead to advanced engineering strategies. We restrict our interest to artificial fuel producing systems, for which energy conversion efficiencies above the 10 % range can possibly be realized, which is a benchmark already reached today by combining solar cells with electrolyzers as separate devices.

Efficient artificial photosynthesis can only be realized if a number of succeeding elementary processes are coupled to each other with minimized losses: broad band light absorption, optimized charge carrier generation and separation, as well as the efficient electrocatalytic production of $H_2$ and $O_2$ from $H_2O$ in separated compartments (hydrogen evolution reaction HER, oxygen evolution reaction OER). For a technological realization adapted device structures must be manufactured using preferentially low cost, abundant, and nonhazardous materials. From the given diversity of possible research strategies we concentrate in our discussion on the combination of complementary components, i.e., combining efficient and stable photovoltaic converter arrangements with efficient catalysts, which provides, to our understanding, the most realistic artificial photosynthetic system in the foreseeable future. For light absorption semiconductors are preferentially used as they provide best results for broadband charge carrier generation. To generate the photovoltage needed for water splitting wide band gap compound semiconductors or low band gap tandem (triple) structures must be considered. To catalyze the subsequent $H_2O$ redox reactions (nano-sized) noble metals, earth abundant solid-state compounds, or molecular (biomimetic) coordination compounds are generally discussed as electrocatalysts. One of the key constraints for a promising system is a loss minimized charge transfer from the photovoltaic converter to the catalysts. Promising device structures can only be expected based on knowledge based interface engineering strategies as will be discussed in detail as main subject of this contribution. We expect that the needed requirements can only be realized with a detailed analysis and deep understanding of the involved elementary processes and the materials and device related conditions that must be fulfilled for their successful implementation. Therefore, our understanding of the decisive mechanism will be presented and discussed based on experiments on well-defined model systems using surface sensitive experimental characterization procedures namely photoelectron spectroscopy of the core level and valence band region. In addition, an analysis of the surface and interface potentials will be provided, which can be used to deduce the electronic structure of the active interfaces in relation to contact formation and surface/interface density of states as needed to understand electron transfer processes. These experiments are very helpful to achieve an improved understanding of the interface conditions needed for efficient light induced water splitting. The generated knowledge should provide design strategies for promising materials and device structures, which have already been tested in first experiments.

If we compare semiconductor based photoelectrochemical systems based on inorganic solid state device structures to each other there are on a first sight three fundamentally different approaches namely photoelectrochemical device semiconductor–electrolyte junctions, photoelectrochemical junctions using buried solid state photovoltaic contacts immersed into the electrolyte, and photovoltaic cells connected to an electrolyzer (see Fig. 5.1 and Chap. 4). Even if the abovementioned semiconductor devices are usually discussed in different contexts it seems clear to us, that with respect to the involved elementary processes and the inherent need of performance optimization, which is equivalent to a minimization of the losses by appropriate materials and device structures, they are not very unlike and physically mostly identical. Most strategic concepts can and must be transferred from one device to the other and they must be designed in an equivalent way. Because of our background in thin film photovoltaic research, we discuss in our contribution mostly thin film solar cells compared to thin film based photoelectrochemical cells as they may provide the most probable route to an economically competitive "artificial leaf" for solar fuel production (compare Fig. 5.1 and Chap. 4) However, the concepts derived for these structures are also valid for single

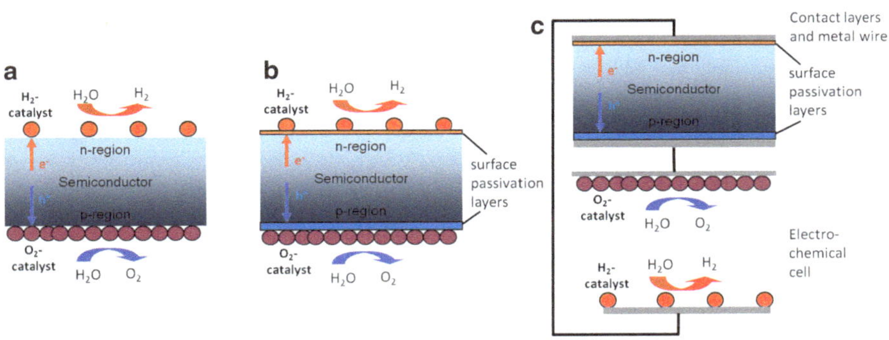


**Fig. 5.1** Schematic draft of different semiconductor based "artificial leaf" structures: (**a**) semiconductor–electrolyte contacts including cocatalysts for the HER and OER reactions, (**b**) buried photoelectrochemical solar cell with cocatalysts, (**c**) solar cell connected to an electrochemical cell with the HER and OER half-cell reactions. The absorbing semiconductor is approximated by a thin film based p-i-n structure

crystalline or nanostructured devices. The challenges and research needs are similar as will hopefully become clear in our contribution: The photovoltage of the water splitting device structure must be in the range of 1.5–1.9 eV (at the maximum power point) depending on the involved overvoltage losses, and the photocurrent quantum efficiency must approach 1 for all photons absorbed above the band gap.

## 5.2 Theoretical Considerations on Photovoltaic and Photoelectrochemical Cells

As is shown in Fig. 5.2 the optimization needs of a photovoltaic converter are basically comparable to those of a photoelectrochemical solar cell as will be elaborated in more detail in this chapter from a theoretical point of view.

Starting from a black box approach on the absorber properties we neglect at first all details related to the semiconductor bulk properties as important for light absorption, charge carrier generation and recombination as well as charge carrier separation, for more details on bulk PV effects see below and refs. (Lewerenz and Jungblut 1995; Würfel 2009; Luque and Hegedus 2011). The absorber must provide the maximum number of photogenerated electron–hole pairs on its surface to the electric contacts, which leads to the attained photocurrent $i_{ph}$ (for a good semiconducting absorber quantum efficiencies close to 1 can be reached for all photons with energies above the band gap). In addition, the difference of the electrochemical potentials of electrons and holes under illumination, given by their quasi-Fermi levels ${}_{n}E_F^*$ and ${}_{p}E_F^*$ at the contacts must be maximized in their energy difference, which is equivalent to the maximum open circuit photovoltage $U_{ph}$ the device can

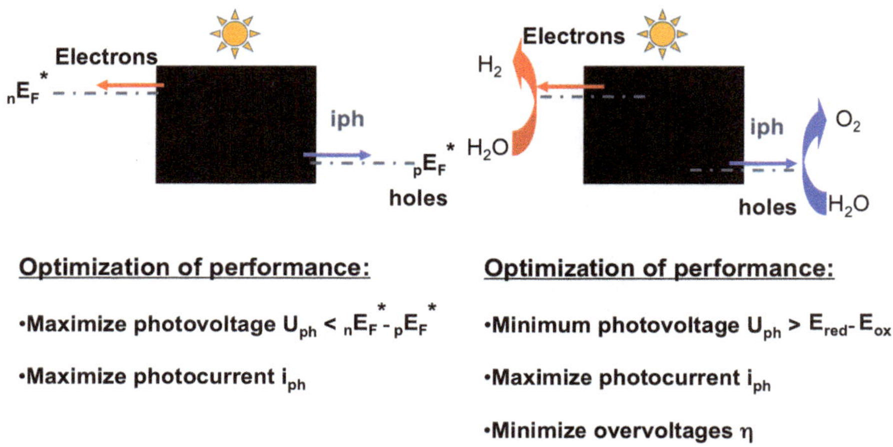


**Fig. 5.2** Schematic black box approach comparing a photovoltaic (PV) converter for electric power generation with an electrochemical converter (EC) for fuel generation and the needed optimization of PV performance values

**Table 5.1** Characteristic values of various single absorber solar cells (Green et al. 2015)

| Classification | Efficiency [%] | $V_{OC}$ [V] | $J_{SC}$ [mA/cm$^2$] | Fill factor [%] |
|---|---|---|---|---|
| c-Si | 25.6 | 0.740 | 41.8 | 82.7 |
| GaAs | 28.8 | 1.122 | 29.7 | 86.5 |
| CIGS | 20.5 | 0.752 | 35.3 | 77.2 |
| CdTe | 21.0 | 0.876 | 30.3 | 79.4 |
| a-Si | 10.2 | 0.896 | 16.36 | 69.8 |
| $RPbI_3$ | 20.1 | 1.059 | 24.65 | 77.0 |

deliver. With that the energy efficiency of the PV device in the conversion of the light power $P_{h\nu}$ coming from the sun to electric power is given by:

$$\eta_{el} = U_{ph} \cdot i_{ph} \cdot {}^{FF}\big/_{P_{h\nu}}, \tag{5.1}$$

FF is the fill factor of the PV converter. The current voltage curve must be rather rectangular for obtaining FF values above 70 % as is essential for high-energy conversion efficiencies. The FF is strongly dependent on the parallel or shunt resistance $R_p$ (describing charge carrier recombination pathways within the device, $R_p$ must be close to infinity) and the series resistance $R_s$ (describing limitations in charge carrier separation transport, $R_s$ must be close to 0). Typical performance data for good solar cells of established semiconductors are given in Table 5.1. As it is clear from these values thin films (CIGS, CdTe, $RPbI_3$) may become competitive with high performance Si or GaAs solar cells at a reduced prize. Therefore, one may expect that thin film devices will also play a role for competitive water splitting devices.

The photovoltaic converter either as solar cell or as part of the photoelectrochemical cell must provide the photovoltage (electrochemical potential of electrons and holes under illumination) as driving force for the HER and OER reaction. If we again use the terminus of solid state physics this means that the difference of the quasi Fermi levels of electrons and holes $\Delta\mu$, which is the maximum photovoltage produced by the PV converter, must exceed the difference of the half-cell potentials of the HER and OER:

$$\Delta\mu = {}_nE_F^* - {}_pE_F^* > U_{ph} > E_{red} - E_{ox} = -(E(H_2/H_2O) - E(O_2/H_2O)) \tag{5.2}$$

(for sign convention we take a positive energy scale for electrons, which means the vacuum level $E_{vac}$ is at the top with a binding energy = 0 eV and more strongly bound electrons are found below with a more negative binding energy). Thus the value of $U_{ph}$ must exceed a value of about 1.5–1.9 eV for efficient water splitting with adjusted values of the reaction overpotentials ensuring reasonable exchange currents and avoiding further resistive losses in the circuit $i_{ph} \times R$ (see below or Sect. 5.3). In contrast there is no lower limit of $U_{ph}$ for photovoltaic converters producing electric power. As is immediately clear, most of the prominent PV cells with high solar efficiencies do not provide the open circuit photovoltage $U_{ph}$ needed for $H_2O$ splitting (see Fig. 5.3).

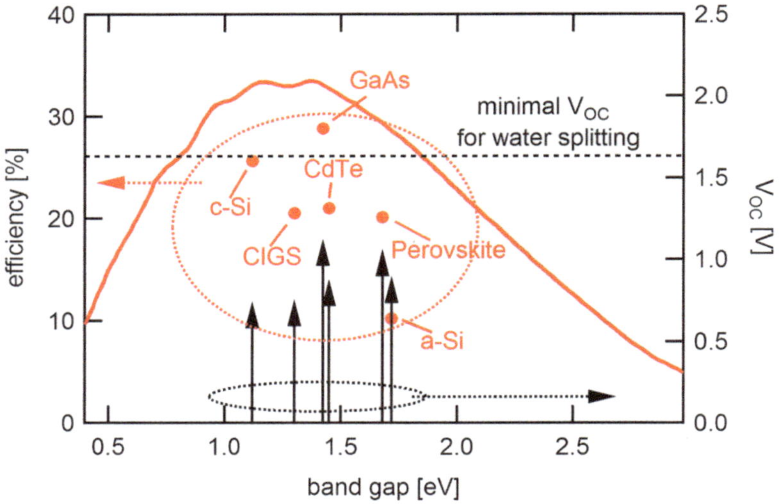


**Fig. 5.3** Maximum efficiencies (*dots*) and photovoltages (*arrows*) reached for different solar cell absorbers. The Shockley–Queisser limit is depicted as a *continuous line* (Peter 2011; Shockley and Queisser 1961). The *horizontal dotted line* shows the minimum open circuit photovoltage needed for the water splitting reaction assuming an overall overvoltage of 0.4 eV

For photoelectrochemical cells the conditions for PV performance optimization are basically the same as for electricity producing solar cells as is schematically indicated in Fig. 5.2. Also in this case the photovoltaic converter part must be optimized with respect to $U_{ph}$, $i_{ph}$ and fill factor (FF) in the same way. However, there are additional conditions, which must be fulfilled when the electrochemical redox reactions of $H_2O$ splitting defined by HER and OER are directly coupled to the device. For efficient devices cocatalysts must be added, which are highly active for $H_2$ and $O_2$ evolution, respectively, to reduce the overvoltages to a minimum value. Any additional resistance in the circuit either due to low ionic conductance or charge transfer or transport resistances in the device will lead to additional overvoltage losses given by $i_{ph} \times R$. In addition one has to take into account, that for reaching high operational photocurrent, the PV device/component must be operated close to the maximum power point, which will add an additional voltage difference $U_{ph} - U_{MPP}$ of at least 0.3–0.5 V depending on the quality of the photovoltaic converter (MPP = maximum power point, equivalent to the operational voltage). For bad photovoltaic converters with low fill factor the photocurrent will only slowly rise, which translates into an additional strong loss of the operational photovoltage compared to the open circuit photovoltage $U_{ph}$. Thus one may define an operational photovoltage $U_{ph}^{op}$ of the PEC system, which is needed for bias free splitting of $H_2O$ and which must be close to the maximum power point photovoltage $U_{MPP}$ of the used PV converter. From these considerations it is clear, that the performance of a water splitting device is determined by the photocurrent flowing at this operational photovoltage. This voltage must be large enough to drive the $H_2O$ splitting reaction including all kinetic losses (overpotentials) without the application of an additional electric (using potentiostats) or chemical bias (using concentration differences). Thus from a detailed analysis of the performance data of a PV vs. a $H_2O$ splitting device, it follows that the photovoltage of the PV converter at the MPP in relation to the thermodynamic $H_2O$ splitting potential multiplied with the achieved photocurrent provides a realistic upper value for an estimate of the solar to $H_2$ (STH) conversion efficiency:

$$\eta_{\mathrm{STH}} = \Delta E_{\mathrm{H_2O}} \cdot {}^{i_{\mathrm{ph}}}/_{P_{h\nu}} < U_{\mathrm{MPP}} \cdot {}^{i_{\mathrm{ph}}}/_{P_{h\nu}} \quad \text{with } U_{\mathrm{MPP}} > U_{\mathrm{H_2O}} = 1.23\ \mathrm{V} \tag{5.3}$$

$$\Delta E^{\mathrm{op}} = \Delta E_{\mathrm{H_2O}} + \eta_{\mathrm{c}} + \eta_{\mathrm{a}} + i_{\mathrm{ph}} \cdot R + \left(U_{\mathrm{ph}}^{\mathrm{oc}} - U_{\mathrm{MPP}}\right) \tag{5.4}$$

These results indicate that for highly efficient photoelectrochemical $H_2O$ splitting devices the PV converter's $U_{MPP}$ must be larger, but as close as possible to $\Delta E^{op}$ (with $\eta_c$, $\eta_a$ as cathodic and anodic overvoltages and $R$ the electrolyte resistance). The last term ($U_{ph}{}^{oc} - U_{MPP}$) giving the difference of the open circuit photovoltage to the photovoltage of the maximum power point is often neglected, but it is needed to ensure reasonable photocurrents.

On the other hand, if the photovoltage $U_{MPP} = U_{ph}{}^{oc}$ produced by the PV converter is much too large compared to the difference $\Delta E_{H2O}$ of the reversible half-cell reaction potentials of $E°(H_2O/O_2)$ and $E°(H_2/H_2O)$, any kinetic limitations of the water splitting reactions may be avoided. But this extra potential is translated directly into an energy transformation loss as the thermodynamic potential difference $E°(H_2O/O_2) - E°(H_2/H_2O)$ of 1.23 V, which is equivalent to the energy gain of $H_2$ formation (fuel production) will not be changed. This behavior is different to an electric power producing PV converter, where both $i_{ph}$ as well as $U_{ph}$ can be used to increase the performance value (see Eq. (5.1)).

Additionally, if we compare the PV converter to a photoelectrochemical converter (PEC), contact formation will be different and must be discussed in its specific properties (see also Sect. 5.4). For a PV converter metallic conductors must be used as contacts to withdraw the minority carriers at their respective electrochemical potentials without any additional potential loss, which will only be possible with ideal Ohmic contacts. However, for a PEC cell the electrolyte may provide the contacts, which must then lead to similar loss-free charge exchange properties. But most semiconductors are nonefficient catalysts for HER and OER in itself. In addition, most semiconductors will strongly interact with electrolyte species forming new surface "phases" or in extreme cases even chemically corrode by unwanted side reactions with the electrolyte. Thus it may be required to use surface protecting buffer layers with negligible losses in photopotentials and photocurrents. To overcome any drawbacks related to kinetic overpotentials on the bare semiconductor surface it is of utmost importance to increase the reactivity for forming the desired products and avoiding corrosive side reactions by the implementation of appropriate cocatalysts (electrocatalysts) on the semiconducting substrates. If these cocatalysts may be deposited directly onto the semiconductor surface or if additional protecting buffer layers may be needed, this will depend on the electronic and chemical properties of the semiconductor surface. These cocatalysts are on one hand needed for providing efficient reaction kinetics for HER and OER, but in addition the contact properties of these cocatalyst layers must be adjusted with respect to the electrolyte phase. Therefore, the selection of good cocatalysts depends not only on their catalytic properties but also on their contact properties.

Thus novel materials and material combinations must be identified in future research. At first, they shall provide the potential for high performance in the electric power conversion yield as defined by the PV part of the overall system and as must be proven by the realization of competitive PV systems. This has been shown to be possible with costly single crystalline solar cells but should also be manageable with cost effective solutions as with novel thin film materials. In addition, for PECs the chemical reaction conditions for HER and OER must be implemented with minimizing any additional performance loss. Therefore, specifically adjusted cocatalysts on specifically engineered passivation layers must be included in the device arrangement. So far best results have been obtained with noble metal clusters for electrocatalysis. To reduce costs new cocatalysts are under investigation as transition metal oxides or as chemically defined mononuclear or multinuclear organometallic molecular coordination compounds which are also used as precursors to form appropriate electrocatalysts (monomeric species, clusters, or nanoparticles) on the surface of semiconductors. The bonding of the electrocatalysts to the semiconductor substrates should allow for an efficient charge transfer between the bulk and the catalysts. The electronic conditions which must be fulfilled for efficient coupling of the semiconductor to the electrocatalyst are central aspects of interface engineering of photoelectrochemical cells and are discussed in further detail below.

We expect that photovoltaic converters coupled to an electrolyzer or photoelectrochemical solar cells using semiconductor–electrolyte contacts will provide the best performance for water splitting compared to any other approach as the underlying mechanism are equivalent with regard to the PV- and catalytic functionality either decoupled or coupled directly to each other. The best PEC results so far have been obtained with III–V multi-junction epitaxial cells (without bias voltage), proving that direct fuel production with solar light is possible with overall conversion efficiencies above 10 % STH (Khaselev and Turner 1998; May et al. 2015; Licht et al. 2000). For PV coupled to electrolyzers even higher performance data of 18 % have been quoted which is considered as benchmark for any other approach (Peharz et al. 2007).

## 5.3 Energetic Conditions of PEC Based Water Splitting

Both half-cell reactions involved in water splitting, the HER and the OER, are multi-electron reactions in which two or four electrons, respectively, are involved and which are not fully understood so far with respect to the involved intermediate reaction steps. Overall four electrons must be transferred to form from two molecules of $H_2O$ two molecules of $H_2$ and one molecule of $O_2$:

$$4H^+ + 4e^- \rightarrow 2H_2 \quad \varepsilon^\circ(H_2/H_2O) = 0\ \text{eV} \tag{5.5}$$

$$2H_2O \rightarrow 4H^+ + 4e^- + O_2 \quad \varepsilon^\circ(O_2/H_2O) = -1.23\ \text{eV} \tag{5.6}$$

$$2H_2O \xrightarrow{h\nu} 2H_2 + O_2 \quad E^\circ = \varepsilon^\circ(O_2/H_2O) - \varepsilon^\circ(H_2/H_2O) = -1.23\ \text{eV} \tag{5.7}$$

The HER and OER shift negatively in their redox potential from their values at pH = 0 (1 N [$H^+$]) with pH by 59 mV/pH. This shift must be considered when the position of the band edges is aligned to the OER or HER half-cell redox potentials. The band edge positions of the semiconductor may also shift by 59 mV/pH for an oxidic surface; a reduced or no shift of band edges is expected for less ionic or more covalent semiconductors. In addition, the band edges tend to shift due to the trapping of charge carriers during illumination (Jaegermann and Tributsch 1988; Meissner and Memming 1988; Sinn et al. 1990; Kühne and Tributsch 1986; Schefold and Kühne 1991). The PV converter under illumination must provide the overall voltage needed to drive this reaction. The photovoltage must at least exceed the thermodynamic potential related to water splitting $\Delta E_{H2O}$ given by 1.23 V. Additionally potential losses for driving noticeable photocurrents have to be considered, which are the gradients of the electrochemical potentials of electrons and holes grad ${}_{n}E_F{}^*$ and grad ${}_{p}E_F{}^*$ to drive the charge carrier separation, the overvoltages $\eta_c$ and $\eta_a$ for the HER and OER respectively, and finally the Ohmic losses in the circuit $i_{ph}R$, e.g., the resistance of the electrolyte solution between cathode and anode to carry the photocurrent $i_{ph}$ as it is provided by the photovoltaic converter (see for example Eq. (5.4)). The energetic conditions are schematically shown in Fig. 5.4 for a one band gap absorber system, which in principle is equivalent for tandem or multi-junction structures.

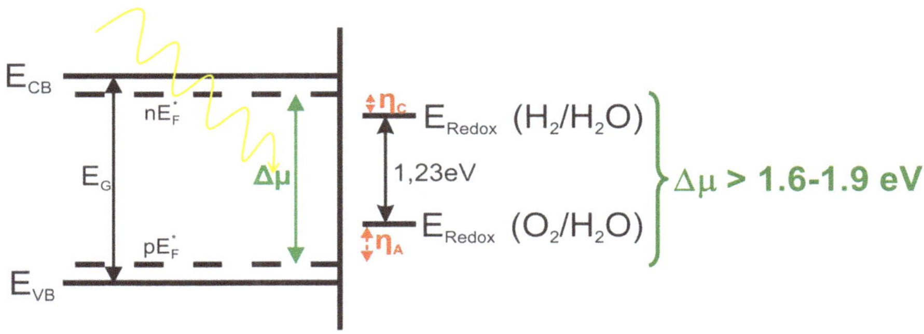


**Fig. 5.4** Schematic energy diagram for a PV absorber coupled to electrochemical reactions under illumination. In most real cases the electron and hole contact would be situated at opposite parts of the absorber. A minimum $\Delta\mu$ value of 1.6 eV may be calculated for open circuit conditions; for achieving reasonable photocurrents an additional 0.3 eV kinetic driving force for efficient charge separation is needed for MPP conditions

Within this context a remark considering the relative position of the valence and conduction band edge with respect to the half-cell potentials of the OER and the HER shall be made. Very often the positions of the conduction band edge and the valence band edge are plotted as a criterion for materials selection in the electrochemical scale vs. the normal hydrogen electrode (NHE) to deduce which semiconductor will be able to drive the water splitting reaction. It is of course true that for working conditions with the semiconductor immersed in the electrolyte solution and under illumination the relative positions of the energy bands must be aligned to the HER and OER redox potentials as schematically shown in Fig. 5.4. However, the ionization potentials and electron affinity of semiconductors as measured with respect to the vacuum level are no constant values but depend on a number of parameters as for example surface orientation and adsorbate coverage. The possible shifts of the band edge position vs the vacuum level may easily reach values exceeding 1 eV due to variations of surface dipole potentials. The situation is even more complicated by the fact that in electrolyte solutions adsorbates and electrochemical double layers may provide additional electrostatic potential drops due to different charging of the electrochemical double layers, with strongest effects expected from the inner and outer Helmholtz layer. Very often the band edge positions are measured from Mott-Schottky plots in the dark not taking into account that charging of surface states or other light induced effects as low charge transfer kinetics may considerably shift the band edge positions of the semiconductor under real working conditions of PEC water splitting. This has been proven by a number of experiments, e.g., Mott–Schottky plots under illumination or time resolved voltammetric experiments (Jaegermann and Tributsch 1988; Meissner and Memming 1988; Sinn et al. 1990; Kühne and Tributsch 1986; Schefold and Kühne 1991). It is very clear from these data that for most semiconductors under operation and with illumination a dynamic adjustment of energy levels (valence and conduction band positions) must be taken into account especially when catalytic metal layers, passivation layers, and/or charge trapping surface states are involved in the photochemical reactions.

The maximum photovoltage which can be gained by the semiconducting absorber neglecting at first all loss processes related to the actual design of contacts is given by the splitting of the quasi Fermi levels under steady state illumination in the bulk (compare Eq. (5.8)). $\Delta\mu$ is given by (Würfel 2009):

$$\Delta\mu = {}_nE_F^* - {}_pE_F^* = kT\ln\left(n^* \cdot p^*/n_i^2\right) > eU_{ph}^{oc} \tag{5.8}$$

With $n^*$ and $p^*$ as maximum level of electron and hole concentration within the PV device under illumination, $n_i$ as intrinsic charge carrier concentration in the dark ($n_i^2 = n_0\, p_0$; mass action law of electrons and holes). The open circuit photovoltage $U_{ph}{}^{oc}$ of the device is usually smaller than $\Delta\mu$ due to extra energetic losses related to surface contact properties including surface recombination. For idealized systems with perfect contacts these additional losses are small and the photovoltage delivered may approach the $\Delta\mu$-value provided by the absorber materials. For bad contacts using non-properly adjusted contact phases, e.g., with large work function difference or being dominated by active surface/interface states, the photovoltage may be considerably smaller than the value of $\Delta\mu$ the absorber material may deliver. As most semiconductor surfaces will provide a large concentration of surface states a high value of the photovoltage can only be expected for electronically passivated surfaces, which must be specifically designed either by adsorbates or electronic passivation layers. For most absorber materials the design of the contacts and surface passivation layers is the crucial point of cell development (Jaegermann et al. 2009). Generally, surfaces and interfaces are detrimental for high conversion efficiencies. The maximum photovoltage reached by the PV-absorber structure must exceed the potential differences needed for driving both half-cell reactions. Here, one may assume that for good electrocatalysts the HER needs an overpotential of less than 0.1 V and the OER of about 0.3 V for photocurrents in the range of 20 mA/cm$^2$. The loss due to the Ohmic resistance of the electrolyte should possibly be below 0.1 V. This would translate into an operational photovoltage of $U_{ph}^{op} > 1.6$ eV, which must be delivered by the PV converter. If in addition, for reaching reasonable photocurrents the gradient of the Quasi-Fermi level positions between bulk and surface is also considered, which is the potential difference between open circuit and maximum power point photovoltage, an additional voltage of >0.3 eV may be needed, which leads to an overall open circuit photovoltage of at least 1.9 eV, which must be provided by the PV converter, as also shown in Fig. 5.4. The above given rather optimistic values, however, can only be expected for rather ideal assumptions. In most cases higher values of $\eta_c$, $\eta_a$, $iR$, and $U_{ph}^{oc} - U_{MPP}$ are obtained especially for non-noble electrocatalysts and thin film absorbers, which set harsh conditions for the optoelectronic quality of the photovoltaic converter. These estimates on the needed $U_{ph}^{op} \approx U_{MPP}$ vs. $U_{ph}^{oc}$ sets the conditions for choosing promising absorber materials.

Considering an empirical rule, which is based on the thermodynamics of PV converters and which has been proven by many long term optimized PV solar cells (Green 1986; Luque and Hegedus 2011; Lewerenz and Jungblut 1995; Würfel 2009) (see Fig. 5.3), $U_{ph}^{oc}$ may at best reach a value given by $eU_{ph}^{oc} = E_g - 0.4$ eV which translates to the condition that the band gap $E_g$ of a single high quality absorber material must reach at least 2.3 eV to be applicable for water splitting. For most thin film materials there is the relation $U_{ph}^{oc} \leq 0.6\ E_g$ often found, which would lead to required band gaps clearly above 3 eV. As a consequence it can hardly be expected that any efficient PEC system can be developed, which is solely based on a single absorber layer.

In any case by using either absorber layers of wide band gap semiconductors or of tandem or multi-junction cells the needed photovoltage at the operational point must exceed a value of above 1.6 eV (including overvoltages) for the water splitting reaction, which may translate to an open circuit photovoltage of more than 1.9 eV, which strongly limits the possible PV absorber materials which may be used for water splitting. The overall theoretical PV performance is strongly dependent on the energy gap (e.g., Shockley–Queisser limit with a maximum value of 31 % at around $E_g = 1.2$ eV for AM1.5 conditions, see Fig. 5.5). In related curves, for example for

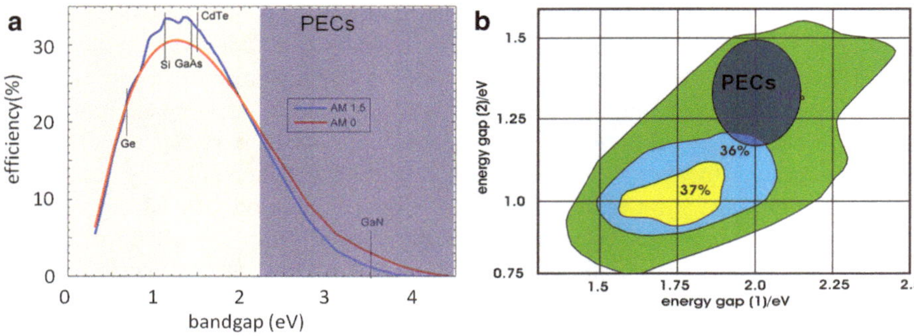


**Fig. 5.5** Estimates of the electric power conversion efficiencies of water splitting PEC devices. (**a**) Single absorber with a minimum band gap of 2.3 eV. (**b**) Tandem cells with a combination of two absorber layers around 1.3 and 2 eV. The efficiency for $H_2$ formation is about 30 % below the tabulated power conversion efficiencies. Figures adapted from Lewerenz and Jungblut (1995), Shockley and Queisser (1961)

$H_2O$ splitting the maximum values which have been calculated depend on additional assumptions on possible loss mechanisms (Bolton et al. 1985; Hu et al. 2013). As already stated above the conversion efficiency to a solar fuel, i.e., $H_2$ (Eq. 5.3) will not increase with larger photovoltages in contrast to electric power conversion (Eq. (5.1)) after a sufficient $U_{ph}{}^{op}$ has been reached ($U_{ph}{}^{op} \approx 1.6$–1.9 V). Even higher values for $U_{ph}{}^{op}$ would lead to a reduced solar to fuel conversion efficiency as the Gibbs free energy or reversible redox potential differences of the $H_2O$ splitting reaction are not changed but the photocurrent is reduced. From the presented curves the electric power conversion efficiency is expected to reach a maximum theoretical value below 18 % for band gaps above 2.3 eV. As an estimate of the solar to $H_2$ efficiency $\eta_{STH}$ from the electric power efficiency $\eta_{el}$ one may just take the ratio of $U_{ph}{}^{op}/1.23$ V assuming identical photocurrents, which leads to a nearly 30 % reduction (1.6 or 1.9 V / 1.23 V) of conversion efficiency. However, as elaborated before the specific losses to be considered for a water splitting PEC cell and realistic assumptions on the possibly obtainable maximum open circuit photovoltages and overvoltages will probably need band gaps clearly exceeding these values (probably approaching 3 eV) and thus lead to clearly reduced efficiencies (below 3 % STH). For this reason, any single absorber material does not seem to be appropriate for any competitive technological device.

Thus, it seems clear to us, that only tandem or multi-junction arrangements of absorber materials may provide reasonable STH conversion efficiencies by providing high enough $U_{MPP}$ values and reasonable photocurrents. This may already be deduced from natural photosynthesis where two chlorophyll absorber molecules are coupled to each other in a Z-scheme arrangement adding up the electrochemical potentials of the two absorbing molecules (Birch 2009). However, the overall performance is limited due to the reduced quantum efficiencies of at best of 50 % in producing photocurrents. As both dyes absorb in the same light energy range, two photons are needed to produce one electron–hole pair. This limitation is also principally valid for using two identical semiconductors in a tandem arrangement. The highest STH performance may thus be expected for specifically adjusted tandem arrangements of two different absorber materials harvesting photons with different wavelength. For tandem cells a maximum electric conversion yield of about 40 % can be reached using a bottom cell with a band gap of 1.1 eV, e.g., crystalline Si (μc-Si) and a top cell with a band gap of 1.7 eV, e.g., amorphous Si (a-Si) (Keppner et al. 1999). However, a tandem cell like this may just fall short to deliver the photovoltage needed for bias-free water splitting (Calvet et al. 2014; Kaiser et al. 2015); therefore, semiconductor tandems with larger single band gaps may be needed, which reduces the theoretical electric conversion limits into the range of about 35 % (see Fig. 5.5). The maximum photovoltages which can realistically be delivered may be taken from solar cells of the same materials, e.g., 0.74 eV for c-Si and 0.9 eV for a-Si which adds up to an open circuit voltage $U_{ph}{}^{oc}$ of 1.64 V, which falls short of the needed value of $U_{MPP} = U_{ph}{}^{op}$ of 1.6 eV or $U_{ph}{}^{oc}$ of 1.9 eV. A possible tandem configuration would be the combination of CIGS ($E_g = 1.2$–$1.3$ eV, $U_{ph}{}^{oc} = 0.75$ eV) and alloyed Cd(Zn)Te(Se) cells ($E_g = 1.9$–$2.1$ eV, $U_{ph}{}^{oc} = 1.1$ eV), further promising combinations could be Si, InP, or GaAs and GaInPN alloys, which however may be very expensive. In summary, the above given considerations would suggest that only tandem or multi-junction arrangements, where two or more photons are used to produce one electron–hole pair, would allow to obtain reasonable STH efficiencies. The tandem partners preferentially should absorb light of a different wavelength to maximize $j_{ph}$.

However, most of the presented calculations on theoretically expected conversion efficiencies seem to be too optimistic in estimating the operational voltage needed for driving the $H_2O$ splitting reaction with high photocurrents (see above). As the value of $j_{ph}$ under operational conditions defines the amount of $H_2$ produced, it is important to deduce which value of the operational photovoltage in analogy to the maximum power point of a solar cell is needed to drive the reaction in a bias free two electrode arrangement for obtaining reasonable conversion efficiencies exceeding 10 %. For systems, which may approach a practical maximum value of about 25 % STH all overvoltage losses must be made as small as possible and new absorber combinations are needed to fulfill the PV criteria. Triple junction cells may not be an appropriate solution as any surplus photovoltage produced by the PV converter exceeding the $U_{ph}{}^{op}$ limit needed for water splitting may be lost. Therefore, highly efficient triple junction solar cells based on III–V semiconductors which produce a photovoltage of about 3.6 eV (Green et al. 2015) will not provide favorable solutions as the additional photovoltage will be completely lost or may only be used for compensating non-optimized overvoltage losses of the electrochemical reaction.

## 5.4 Promising Integrated PV/PEC Device Structures

We have recently suggested a n-i-p device structure as promising PV device structure for most thin film solar cells (Jaegermann et al. 2009) following the theoretical considerations of Würfel (2009). A structure like that (see Fig. 5.6) would allow to use intrinsic absorber layers without defined doping which has been proven to be a problem for many compound semiconductor absorber materials to build-up the diffusion voltage in a homojunction or heterojunction arrangement. The arrangement is schematically shown in Fig. 5.7. Such device structures have already proven their applicability in Si thin film solar cells (Luque and Hegedus 2011), and also the HIT™ cell based on single crystalline Si profits from this principle (Mishima et al. 2011). The relation to photoelectrochemical devices is evident when the contact materials are substituted by electrochemical contacts providing the conditions for charge transfer reaction to the electrolyte. We see later on that this will not be that easy possible for real devices, since chemical and electronic passivation layers are needed for efficient PV devices as well as for PEC conversion with electrolyte contacts. As a consequence contact layers with adjusted properties will be needed for PV as well as for PEC systems, which may be composed by electronic and chemical passivation layers and the cocatalyst. If you may call these devices surface engineered solar cells for $H_2O$ splitting or surface modified photoelectrochemical cells seems to us only a semantic difference (compare Fig. 5.1).

In a n-i-p cell as suggested above including a naturally doped intrinsic absorber layer between wide band gap front ($n^+$-doped) and back ($p^+$-doped) contacts high theoretical efficiencies can be reached. For absorber layers with high absorption coefficients beyond $\alpha > 10^4\ \mathrm{cm}^{-1}$ their thickness d (equivalent to the absorption length $L_{ph}$) may be reduced to typical values of 1–2 µm ($L_{ph} > 3/\alpha$). The charge carriers can be expected to reach the contacts without severe recombination losses

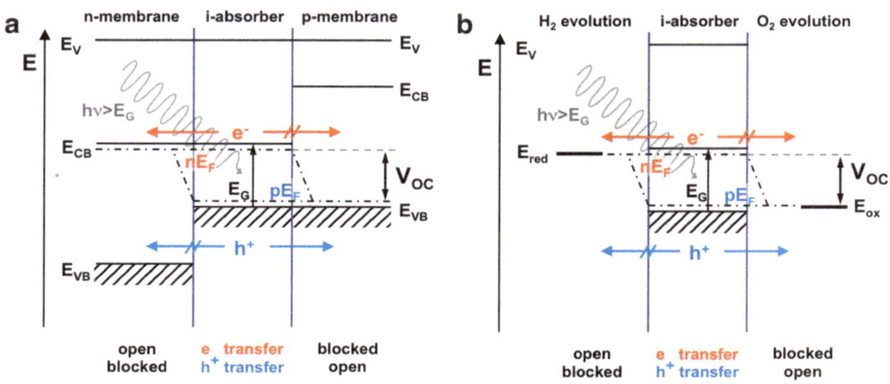


**Fig. 5.6** (**a**) Schematic n-i-p PV or n-i-p PEC structure formed from a semiconductor absorber layer with either n-doped and p-doped wide band gap window contact layers (n/p-membranes) or (**b**) with selective electrolyte contacts for electron transfer (HER) and hole transfer (OER)

when the minority carrier diffusion lengths given by $L_d = (D\tau)^{1/2} = ((kT/e)\mu\tau)^{1/2}$ (with $D$, $\mu$, $\tau$ as minority carrier diffusion constant, mobility, and lifetime) exceed the thickness d of the absorber layer. It should be noted at this point that the band energy diagram of Fig. 5.6 is shown under optimal open circuit conditions under illumination. For any efficient system the operation photovoltage must be close to the open circuit photovoltage of the device. It is misleading to show and discuss working devices in the dark and arguing with the space charge layer and band bending supporting charge carrier separation. Additionally, using the Gärtner equation (Gärtner 1959; Reichmann 1980) for charge carrier transport and surface concentrations, when the potential maximum efficiencies are discussed, is inappropriate under those circumstances. Such settings will not represent the working conditions of an efficient device close to the maximum power point. Important, however, are proper energy alignments of the contact materials at the interface as shown in Fig. 5.6a. The conduction band offset must be close to 0 at the front contact to the n-contact and maximized at the back contact; and vice versa for the valence band back contact to minimize voltage losses. In addition, such an alignment will lead to reduced dark currents due to the large additional barriers for reverse current $j_0 \propto \exp(-\Phi_B/kT)$ (with $\Phi_B$ adding up the value for the energy distance of the Fermi level to the respective band edge position plus the additional value for the band discontinuity, which leads to strongly reduced reverse saturation currents $j_0$). Similar conditions must be reached for the PEC device. As a consequence the electrolyte density of states must allow an efficient charge transfer by tunneling close from the conduction/valence band edge for electron/hole transfer, respectively; the conditions of which will be discussed in more details later on. The interface must be kept free of surface (interface) states to avoid recombination losses for the PV as well as for the PEC device, which are the additional requirements for potentially high conversion yields (which favors wide band gap contacts as oxides and disregards directly deposited metals). As a consequence, assuming ideal conditions concerning recombination losses in the bulk of the absorber layer (parallel or shunt resistance $R_p$ = infinity) and concerning charge transfer and transport resistances (series resistance $R_s = 0$), a maximum photovoltage close to the semiconductor band gap and approaching the Shockley–Queisser limit can be expected:

$$U_{ph} = (kT/e)\ln\left(j_{ph}/j_0\right) \tag{5.9}$$

For proper electrolyte contacts similar conditions may be realized, if the redox potentials of the half-cell reactions provide comparable contact properties as the idealized TCO (transparent conductive oxide) membranes assumed for the PV converter. This is in principle possible as has been proven by Lewis et al. at well defined and specifically designed PEC cells with reversible redox couples as contact phases (Rosenbluth and Lewis 1986; Abrahams et al. 1987). Again it becomes clear from this comparison that PEC devices follow the same design rules as for example thin film solar cells. However, in both arrangements either

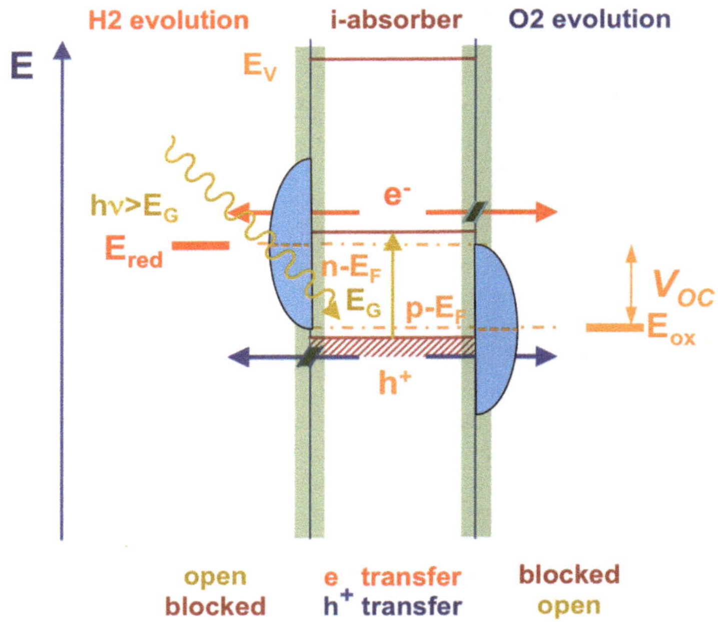


**Fig. 5.7** Schematic representation of a surface modified PV or PEC device consisting of a highly absorbing semiconductor layer, a surface passivation layer and metallic cocatalysts for HER and OER

using a n-i-p PV cell or an $E$(red(I)/ox(I))-i-$E$(red(II)/ox(II)) PEC device structure with electrochemical contacts, in cases where instead of reversible redox couples multi-electron transfer reactions as for $H_2O$ splitting are considered the need of cocatalysts for achieving an efficient charge transfer, makes the application of additional solid state layers for surface/interface engineering necessary. These must be aligned in their electronic coupling to the above presented base materials and their cell arrangements as it is shown schematically in Fig. 5.7.

Therefore, the device will probably consist of electronic and chemical passivation layers (in rare cases one layer may provide both functionalities) and a cocatalyst layer (for metals, its DOS distribution in energy will be even broader than shown in Fig. 5.7). As the catalyst must transfer the electrons and holes close to the band edges its contact properties defined by its Fermi level position after contact must allow a charge transfer without potential loss. How this can be reached for metals of non-adjusted work functions will be discussed in more detail below in Sect. 5.5.

Following the given argumentation we suggest two different device structures for the photoelectrochemical water splitting application. One is based on only a single semiconductor absorber material and may be designed according to the structure shown in Fig. 5.8. In this figure we use the standard presentation of a photoelectrochemical cell with the thickness of the semiconductor being much

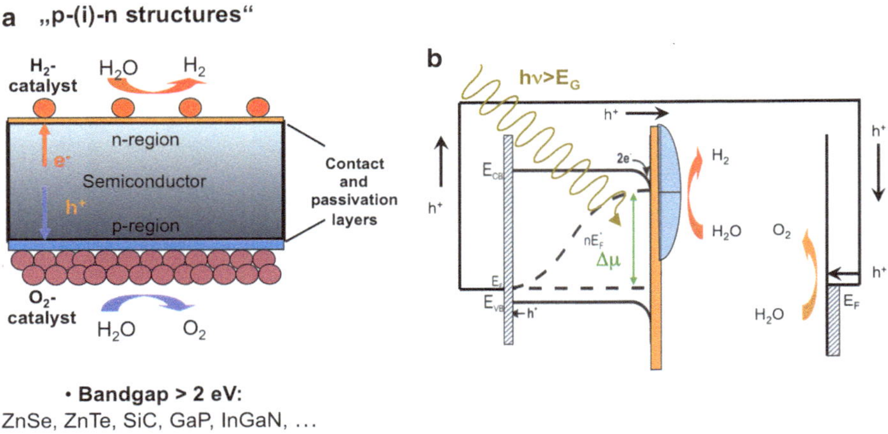


**Fig. 5.8** Schematic structure of a photoelectrochemical solar cell for water splitting (**a**) composed of a p-i-n absorber layer with n/p doped surface layers, chemical and/or electronic passivation layers on front and back sides as well as cocatalysts for HER and OER. (**b**) Schematic energy band diagram under illumination (approximating flatband conditions) and the needed photovoltage for $H_2O$ splitting given by the splitting of the electron–hole chemical potential (difference of electron and hole quasi-Fermi level) under illumination

larger that the absorption length; for this reason the splitting of the quasi Fermi levels is restricted to the thickness of the absorption plus diffusion layer (illumination from the electrolyte side). We suggest using the semiconductor junction as a photocathode with the transfer of electrons as minority carriers across the electrolyte junction. Compared to photoanodes this arrangement may have a strong advantage, because holes as minority carriers tend to have a much stronger corrosion tendency. In the schematic picture we have added extra passivation layers between the semiconductor surface and the cocatalysts and the electrolyte. From the given examples of applicable semiconductor materials with the right band gap larger than 2.3 eV and sufficient charge carrier mobilities it is clear that these passivation layers are needed to stabilize the materials from corrosive side reactions. It should be noted however, that none of the suggested absorber layers as represented in the extensive literature have reached photovoltages above 1.6–1.9 eV until now, which would be required for efficient water splitting.

As an alternative device structure the combination of two different absorber materials with clearly different band gaps may be considered, which will absorb different parts of the solar spectrum (see also Chaps. 4 and 12). Such tandem structures (see Fig. 5.9) may be arranged in the form of classical semiconductor tandem solar cells as shown in our example (b) but may also be composed of two different materials electrically connected to each other and being exposed to the same electrolyte, which has often been presented as a favorable solution for photoelectrochemical water splitting arrangements (a).

Systems like this would have the advantage, that the final photovoltage will be the sum of the two coupled PV converters without loosing too much of photocurrent,

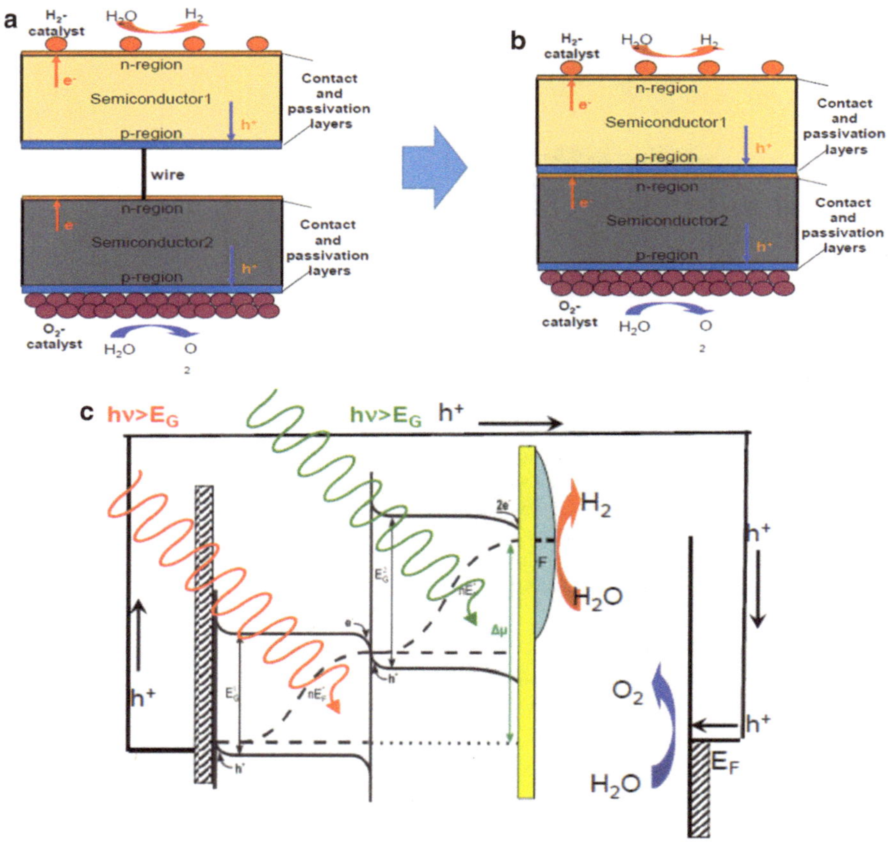


**Fig. 5.9** Schematic structure of a buried junction solar cell used as PEC device either as two independent PV converters (**a**) or as integrated tandem cells deposited on each other using a tunnel junction for charge carrier recombination (**b**), applying p-i-n junctions as shown in Fig. 5.6. (**c**) Schematic energy band diagram under illumination (approximating flat band conditions) and the needed photovoltage for $H_2O$ splitting as added up from the two solar cell components and given by the splitting of the electron–hole chemical potential (difference of electron and hole quasi-Fermi level) under illumination

since photons with different energies will be utilized. This is schematically shown in the energy diagram of Fig. 5.9. The photovoltages of both systems add up to the operating photovoltage for $H_2O$ splitting. Also shown are the possibly needed passivation layers as well as the cocatalysts with the favorable energy alignments. Like PV tandem solar cells such devices are able to provide higher conversion efficiencies with the further advantage that also small band gap semiconductors can be used to deliver part of the needed potential difference of electrons and holes for the water splitting reaction under illumination. To achieve high efficiencies the two semiconductors must be matched to each other with respect to their band alignment to guarantee that the electrons and holes recombine along a tunnel junction without loosing part of their chemical potential. In addition the currents from the two separate PV junctions must be matched to each other, since the cell with the lower photocurrent determines the overall current through the device. These side conditions are also true for all other arrangements in which two photoelectrochemical cells are combined as photoanode and photocathode to the final water splitting device.

Otherwise, all the arguments given above for the one absorber-based cell are also true for the tandem cell. The PV converter part must be optimized in a similar way independent of a possible application as electricity producing solar cell or fuel producing photoelectrochemical cell. Furthermore, the specific need to optimize the multi electron hydrogen and oxygen evolution reactions with the help of cocatalysts, the need to stabilize and passivate the semiconductor chemically and electronically with specifically designed intermediate wide band gap layers, and finally the requirement of a loss free coupling of the cocatalysts to the valence and conduction band states are equivalent challenges for all possible arrangements.

## 5.5 A Survey of Semiconductor Junctions Incorporated in PEC Cells

To gain atomic-scale insights into the processes occurring during the photocatalytic conversion of water by light the elementary processes must be studied and understood in detail by the development of optimization procedures and recipes in the preparation of the PV converter part and the electrocatalytic part of the photoelectrochemical device. It is clear from the given knowledge in literature that this duty has not been fulfilled yet in a satisfying way during the last decades in spite of a large number of related studies on fundamental and applied aspects of light induced water splitting. Especially, the relation of the structural entity of surface and bulk defects as well as of surface molecular bonding arrangements of involved surface reactions are hardly understood in relation to their electron energy states. Within our focus for this contribution we concentrate on surface and interface related aspects of semiconductor–electrolyte interaction to deduce strategies for further interface engineering. At first, one needs to understand semiconductor–electrolyte contacts of bare semiconductor surfaces. For more complex interface engineered junctions one must consider in addition semiconductor passivation layers, and semiconductor metal junctions with and without passivation layers. For such buried interfaces possible modifications induced by the electrolyte must be considered as well. Finally, well-defined model system for such studies should consist of the semiconducting PV converter to generate and separate the electron–hole pairs and the cocatalyst for catalyzing the multi-electron transfer reactions as required for the HER and OER. The goal should be to understand the processes in the energetic coupling of the electronic states across the junctions and to identify the limiting factors in interface charge transfer and the involved chemical reactions. There are, in principle, very many surface sensitive techniques, which may be applied to study interface related phenomena with specific advantages and disadvantages for the considered application in non-UHV environment. Because of our background we concentrate on our experimental approach, specifically on the application of photoelectron spectroscopy (PES). It must be emphasized that we mostly describe a given actually not yet satisfying status of knowledge. We mostly raise questions and suggest possibilities for more detailed future investigations. The photoemission from electrochemical junctions is still not investigated in in situ experiments, despite recent advantages of applying high pressure PES to wet chemical interfaces (Casalongue et al. 2013; Lichterman et al. 2015; Arrigo et al. 2013). But the recently renewed awareness of the challenges in understanding PEC devices for finding an efficiently working artificial photosynthetic system will foster further efforts in this area.

## 5.5.1 Formation of Semiconductor Junctions: Metal Contact

The contact formation at semiconductor junctions has been a long lasting topic of semiconductor physics and in the photovoltaic literature, see for example Green (1986), Lewerenz and Jungblut (1995), Luque and Hegedus (2011), Würfel (2009). Specifically the dominating factors for the formation of space charge layers, which are considered as a precondition for photo-induced charge carrier separation, have been of main interest. Photoelectrochemical semiconductor–electrolyte junctions are usually considered to be similar to semiconductor–metal junctions. But in this case the specific aspects, which govern the obtained potential distributions of "real contacts," are neglected. These are mainly due to additional electronic interface states and to the influence of electrochemical double layers. They lead to considerable deviations from the idealized devices described in the Schottky limit (see below).

For an idealized contact within the Schottky limit a possible energetic difference of the Fermi level position $E_F$ of the metal to the semiconductor leads to a charge transfer across the interface until the Fermi levels (electrochemical potentials of electrons) are equilibrated (Fig. 5.10a) (Sze 1981; Rhoderick and Williams 1988; Milnes and Feucht 1972; Brillson 1982). For the given case of an n-doped semiconductor in contact to a high work function metal, electrons will flow from the SC to the metal to ensure electronic equilibrium. The contact potential is given by the band bending $eV_{bb}$ across the space charge layer in the semiconductor being equivalent to the original difference in the work functions. The barrier height for electron flow is given by the difference between the metal work function $\phi_m$ and the semiconductor electron affinity $E_A^{SC}$:

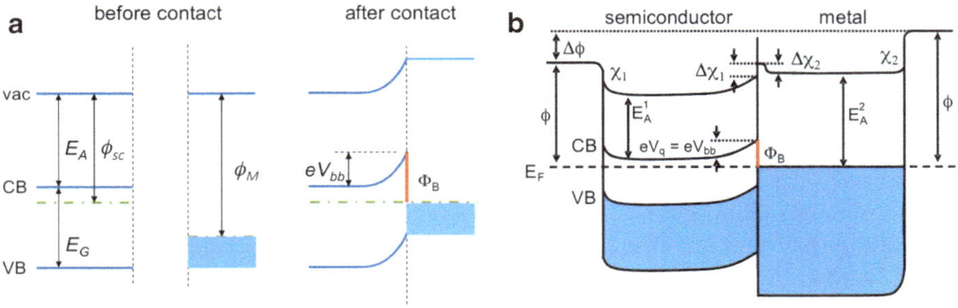


**Fig. 5.10** Semiconductor–metal junctions in the Schottky limit. (**a**) Standard description for which the band bending is simply defined by the difference of metal to SC work function. (**b**) More complex description considering the additional effect induced by dipolar potential changes on both metal and semiconductor surfaces

*Barrier height*:

$$\Phi_B = \varphi_M - E_A^{SC} = \Delta E_A \tag{5.10}$$

*Band bending*:

$$eV_{bb} = \phi_M - \phi_{SC} = \Delta\phi \tag{5.11}$$

The space charge layer follows from the integration of the Poisson equation across the junction boundaries, which is ranging from inside the semiconductor beyond the space charge layer (typically 1 μm in depth, depending on doping) across the interface regime up to the bulk of the metal beyond the range of the surface electron density variation (typically below 0.1 nm).

$$\mathrm{d}\varphi^2 / \mathrm{d}x^2 = -\frac{1}{\varepsilon_0 \varepsilon_r} \cdot \rho(x) \tag{5.12}$$

The solution of the Poisson equation for the idealized Schottky barrier assuming only one donor level leads to the exchanged surface charge $Q_{SC}$ and the width $W$ of the space charge layer:

$$Q_{SC} = \sqrt{2\varepsilon_0\varepsilon_r e N_D \left(\phi_{SC} - \frac{kT}{e}\right)}. \tag{5.13}$$

$$W = \sqrt{\frac{2\varepsilon_0\varepsilon_r}{eN_D}\left(\phi_{SC} - \frac{kT}{e}\right)}. \tag{5.14}$$

The width of the space charge layer is inversely dependent on the doping level $N_D$, i.e., leading to a smaller space charge width for higher dopant concentrations.

So far it has not been considered that in addition to the formation of the space charge layer also modifications of the surface dipolar layers of the semiconductor and the metal during contact formation may lead to shifts of the energy bands to each other. These surface dipole potentials $\Delta\chi$ exist on all solids also on solid/vacuum boundary layers, see Lang and Kohn (1970), Egelhoff (1987), Jaegermann (1996) and references therein: for metals they result from a penetration (tunneling) of the electron cloud beyond the atomic core potentials into vacuum, for semiconductors the electron cloud of dangling bonds and/or structural dipoles due to polar bonds may contribute to this dipole potential drop. The work function $\phi$ equivalent to the electrochemical potential of the electron $-\eta_e$ summarizes an inner bonding potential of electrons within the solid, the chemical potential $-\mu_e$, and a surface dipole potential contribution $e\Delta\chi$, which is related to the surface structure and thus to the electrostatic potential drop induced by an inhomogeneous distribution of charges at the solid/vacuum phase boundary (electron tunneling from solid to vacuum, polar surfaces, strongly bound adsorbates).

$$\phi = -\eta_e = -\mu_e + e\Delta\chi \tag{5.15}$$

As a consequence the position of energy bands and related surface potentials as ionization potential, electron affinity, and work function are different for different surface orientations as is well known for metals. For semiconductors of different surface orientation the influence of changed surface dipolar double potential drops $e\Delta\chi$ must be discriminated from any possible band bending effect $eV_{bb}$ induced by surface states. For semiconductors it holds in general:

$$\Delta I_P = \Delta E_a = e\Delta\chi \tag{5.16}$$

$$\Delta\phi = eV_{bb} + e\Delta\chi \tag{5.17}$$

The first term $\Delta I_P$ shifts the energetic position of the band edges due to the influence of modified dipolar potential drops vs. the vacuum level, e.g., for the semiconductor surface before contact formation. The second term $\Delta\phi$ describes the change of the Fermi level position, which is due to a shift of the valence and the conduction band position vs. the vacuum level as given by the changed ionization potential $\Delta I_p$ and electron affinity $\Delta E_a$, respectively, and the change of the surface position of the Fermi level within the band gap of the semiconductor due to band bending possibly induced by surface states. Since these surface dipole layers will change from the situation having contact to vacuum to having contact to another material, the band bending and the barrier height have to be modified by the related changes of the surface dipolar contribution:

$$eV_{bb} = \phi_M - \phi_{SC} - (\Delta\chi_M - \Delta\chi_{SC}) \tag{5.18}$$

Therefore, one may describe the formation of an idealized semiconductor metal interface (as to be discussed for the interaction of the semiconductor substrate to the metal cocatalysts) as shown in Fig. 5.10b taking into account these inherent changes of the dipolar layers.

Only the part of band bending $eV_{bb}$ of Fermi level alignment forming the space charge layer is related to the extended electron transfer. As a consequence the Schottky limit cannot be expected to be exactly given by Eqs. (5.10) and (5.11) and is better described by a linear change of band bending $eV_{bb}$ with varied metal work functions or metal electronegativity assuming a similar additional dipole potential shift across all junctions (Rhoderick and Williams 1988; Cowley and Sze 1965; Kurtin et al. 1969). Systematic studies of metal junctions of the past have shown that strongly ionic semiconductors as wide band gap oxides as ZnO or $TiO_2$ tend to follow the Schottky limit. However, also in this case strong deviations from the Schottky limit may be found in dependence of surface processing depending on the structural arrangement of the surface and defect states in the band gap leading to a partial Fermi level pinning (Klein 2013). This is also true for transition metal oxides with filled d electron states as for example $Fe_2O_3$ but has not yet been explored to our knowledge.

In most cases with traditional compound semiconductors, one may encounter Fermi level pinning (see also Chaps. 1 and 6), which is due to the influence of extra electron states situated in the band gap of the semiconductor. These may either already exist on the surface of the SC in contact to vacuum (dangling bond states of covalent SCs, defect states of ionic SC as for example oxides) or may be formed by contact formation with the metal (tunnel states of the contact layer or metal induced gap states). The contact formation in the Bardeen limit is schematically shown in Fig. 5.11. Due to surface states the clean semiconductor surface already shows the formation of a space charge layer with a defined value of band bending $eV_{bb}$, which

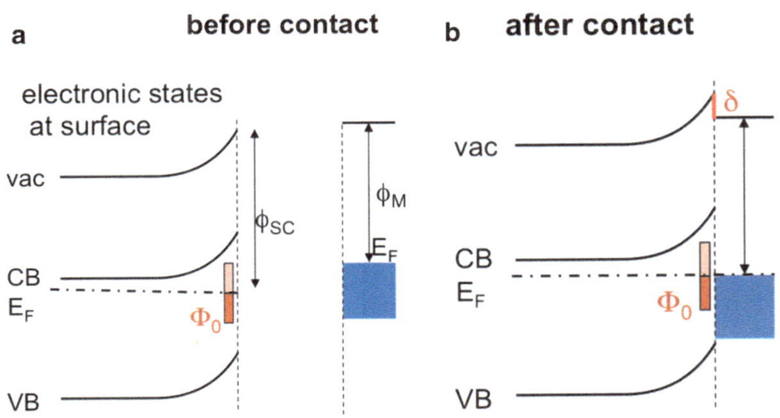


**Fig. 5.11** Semiconductor–metal junctions in the Bardeen limit (Fermi level pinning). (**a**) Due to a high concentration of surface states the space charge layer is defined by their intrinsic charge neutrality level $\Phi_0$. (**b**) The difference to the metal work function leads to a dipolar double layer potential drop $\delta$ between the semiconductor and the metal surface (any additional dipolar effects $\Delta\chi$ as described in Fig. 5.10b cannot be discriminated and are included in $\delta$)

is determined by the charge neutrality level CNL $\Phi_0$ of the given distribution of surface states (the CNL $\Phi_0$ corresponds to the level of half occupation of the surface states equivalent to a surface Fermi level position). For a high concentration of surface states contact formation to a metal of deviating $E_F$-level position leads again to charge transfer for adjusting $E_F$, but in this case the charge is mostly exchanged between the surface states and the metal.

$$eV_{bb} = \phi_0 - \phi_{SC}(fb) \quad \text{or} \quad \Phi_B = \phi_0 - E_A^{SC} \tag{5.19}$$

With $\phi_{SC}$ (fb) as semiconductor work function in a hypothetical flatband condition; $\phi_0$ is the work function of the surface states. In the original description by Sze (Cowley and Sze 1965) $\Phi_0$ was defined as energetic distance of the charge neutrality level to the valence band maximum $\Phi_0 = (E_{CNL} - E_{VB}) = IP - \phi_0$.

As a consequence after equilibrium formation to the contact phase the space charge layer in the semiconductor is nearly unchanged and still defined by $\Phi_0$ for a high concentration of surface states; the contact is established by a double layer potential drop δ formed between the surface states and the metal (see Fig. 5.11). This double layer potential drop δ due to charge transfer and given by the original difference of work functions has the same spatial extension as possible modifications of the dipole layer potentials which may occur in addition, as discussed above, but which cannot be discriminated from each other and are therefore not shown in this figure.

In solid state contacts the electronic density of states of the contact phases are well defined and can be measured by photoelectron spectroscopy for occupied states and inverse photoemission for unoccupied states. In many cases these surface states can also be measured in their energy distribution and can be discriminated from bulk electron states due to a missing dispersion along their k-vector normal to the surface. However, this may be a spectroscopic challenge in those cases when intense bulk valence band emissions dominate the spectra.

A specific problem may arise when metal cluster contacts are formed leading to an inhomogeneous interface layer composed of covered and non-covered areas. For the application in photoelectrochemical devices such nanoparticle deposition of cocatalysts is one of the design strategies to optimize the performance of the device by minimizing the amount of noble metal needed. In this case the electric potential distribution (distribution of the space charge layer) will also be inhomogeneous on the surface as is schematically shown in Fig. 5.12 (Schlaf et al. 1993; Fritsche et al. 2004) as has been measured for junctions between Au or In clusters and layered semiconductors, which are free of surface states. One may expect areas which are not affected by the metal contacts in parallel to areas which show a full band bending depending on the size and the distribution of metal nanoparticles, and depending on the doping of the semiconductors. This is schematically shown along the line (a) where no band bending occurs and thus also no photoeffects can be expected in comparison to line (b) where a Schottky barrier is formed below the deposited In cluster and surface photovoltage effects are expected (modifications of barrier height with illumination) (Schlaf et al. 1993). Such cluster size effects and

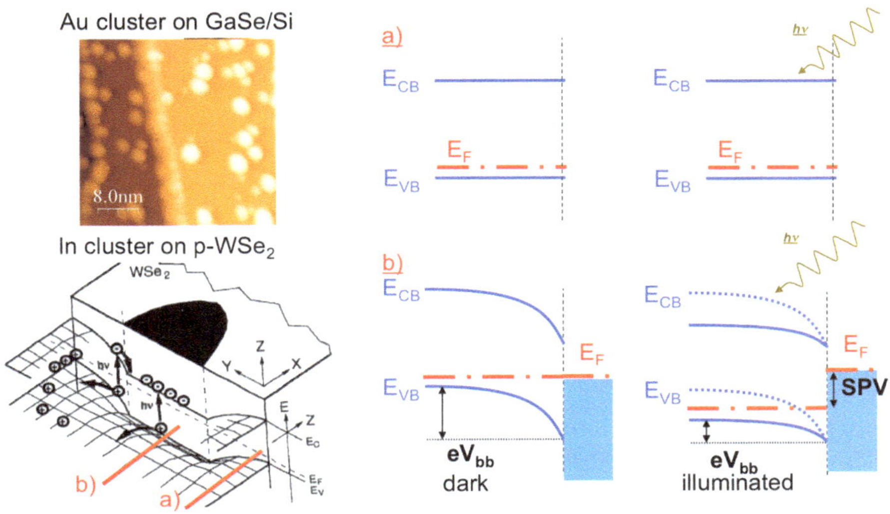


**Fig. 5.12** Schematic representation of a 3D space charge layer formed below metal clusters on idealized semiconductor surfaces (no pinning states) as p-$WSe_2$(0001) (Schlaf et al. 1993). 1D space charge layers along lines (*a*) and (*b*) in the dark and with illumination showing surface photovoltage effects (compare Fig. 5.25)

most probably also an influence of the electrolyte environment on the actual given distribution of space charge layers around such metal clusters are expected when an electrolyte contact is applied on nano-sized metal cluster modified semiconductor junctions (pinch-off effects (Nakato et al. 1988; Nakato and Tsubomura 1992; Rossi et al. 2000)). How such metal cluster interfaces in contact to electrolyte contacts will behave is still not clear from previous experimental studies of their electrochemical behavior and to our knowledge well defined surface science studies have hardly been performed, yet.

### 5.5.2 *Formation of Semiconductor Junctions: Electrolyte Contact*

The semiconductor–electrolyte contact is often described in analogy to the semiconductor metal contact (see Fig. 5.13). In this contribution we do not give an extended survey of the thermodynamics and related kinetics of semiconductor–electrolyte contacts which has been extensively discussed in the literature (see for example Memming 2001; Gerischer 1970; Morrison 1980; Myamlin and Pleskov 1967; Jaegermann 1996; Schmickler 2010), but we just want to emphasize important but often not realized misconceptions and still given limits of understanding on the electronic properties of photoelectrochemical contacts, which are important for electron (hole) transfer reactions in water splitting devices.

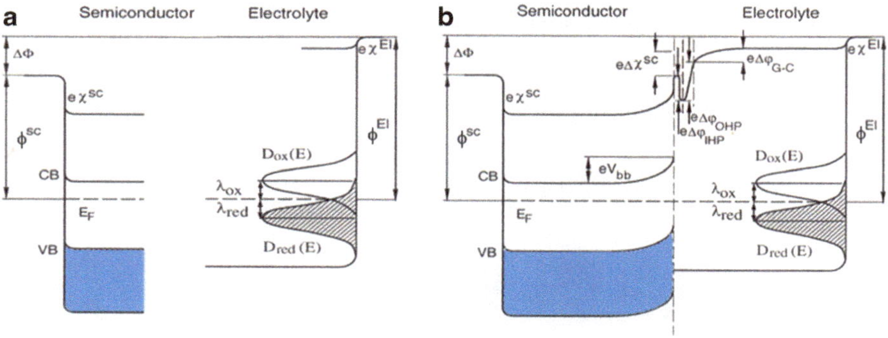


**Fig. 5.13** Contact formation of semiconductor–electrolyte interfaces containing a well defined one-electron transfer reversible outer sphere redox couple with its redox potential $E$(red/ox) given by its work function $\phi^{\mathrm{El}}$. (**a**) Before contact, both work functions contain a surface dipole potential drop of e$\chi$, which will be modified after contact formation (**b**) as schematically shown as contributions of different dipolar or electrochemical double layers (semiconductor dipole, inner Helmholtz, outer Helmholtz, Gouy–Chapman). As a consequence the original given difference in work function is divided into an extended space charge layer $eV_{\mathrm{bb}}$ and double layer potential drops

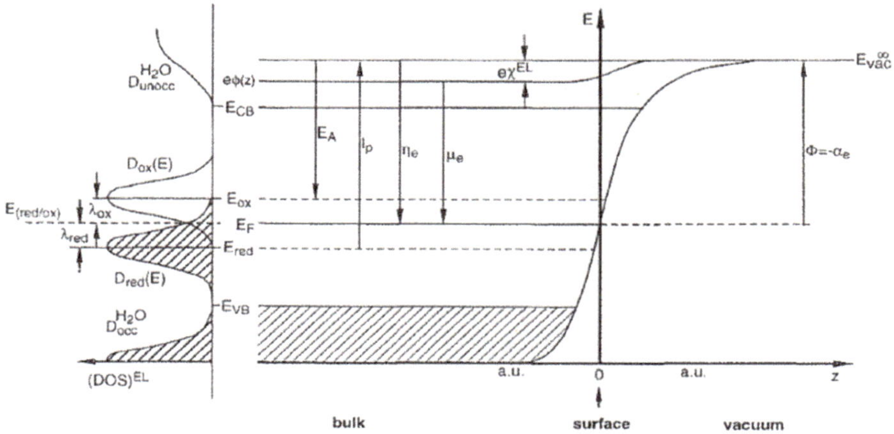


**Fig. 5.14** Schematic representation of the surface electronic density of states and electronic structure of an electrolyte (Jaegermann 1996) in analogy to semiconductor surfaces. The DOS is due to intrinsic states of $H_2O$ and the DOS derived from a one-electron reversible redox couple

In Fig. 5.14 the electron density of states is sketched in the electrolyte as it is often found in the recent literature on semiconductor–electrolyte junctions. This picture is also of use for discussing the charge transfer reactions from and to the electrolyte. This conceptual picture is based on the Marcus–Gerischer theory in which Gerischer has adopted the electron transfer kinetics of Marcus to the semiconductor–electrolyte interface (Gerischer 1969, 1970). As the charge transfer kinetics of solid–electrolyte junctions is probably the most difficult topic in semiconductor photoelectrochemistry, we are not able to cover this topic within this chapter to any extent. Interesting readers may refer to more competent summaries written by experts in the field (Memming 2001; Gerischer 1969, 1970). However, as any electron transfer reaction at phase boundaries will depend on a tunneling process from and to the semiconductor the distribution of electron states across the phase boundary should be known. For solid state contacts the surface and bulk density of electron states on either side of the junction will determine the contact properties but also the charge transport across the interface. The latter depends on the relative position of the Fermi levels modified, e.g., by applied potentials and by their occupation as given in the dark or under illumination. This will also be the case for the electron transfer reactions at electrochemical contacts. However, whereas the electronic states at solids are well defined on both sides of the interface and can be measured, e.g., by electron spectroscopies, this is so far not the case for electrolytes, where the electron density of states is mostly unknown from direct measurements and is only deduced from kinetic experiments. As a consequence any quantitative kinetic evaluation for the optimization of photoelectrochemical cells is hindered by the fact that any rate constant will be determined by the probability of electron transfer and the relative density of states and their occupation on both sides of the junction, which cannot easily be discriminated from each other. Thus experimental values and/or good approximations of the electronic structure of the active junction in photoelectrochemical cells especially for the $H_2O$ splitting reaction are needed for a better understanding of the involved processes. This is even more the case for complicated multilayered buried junctions as discussed above as promising systems, which include the semiconductor, possibly surface or interface states and different passivation layers and the cocatalysts. In principle, surface science approaches, particularly electron spectroscopy may provide experimental data but must be further developed for measurements of wet contacts (see below).

The usually given distribution of electrolyte electron states for a redox couple as shown in Fig. 5.13 is based on the Marcus–Gerischer theory. In this approach the electrolyte is composed of a one-electron redox couple (e.g., $Fe^{2+/3+}$ or $Fe(CN)_6{}^{3-/4-}$) dissolved in molar quantities in an inert solvent. The solvent, e.g., $H_2O$, may be considered as a wide band gap insulator, which will be doped by the redox active component (see Fig. 5.14). The valence ($D_{occ}{}^{H2O}$) and conduction band states ($D_{unocc}{}^{H2O}$) of $H_2O$ result from the occupied 1b1 HOMO and the unoccupied 4a1 level of molecular $H_2O$ and the valence band and conduction band edges are situated at about 9 and 1.5 eV below the vacuum level, respectively (Jaegermann 1996). The additional DOS of the reversible redox couple may then be considered as an electronic doping of the insulating solvent, in this case $H_2O$, by its additionally introduced electronic states. These additionally introduced electronic density of states (DOS) depend on the concentration of the redox couple and the number and energetic positions of its occupied and unoccupied electron states. For one-electron reversible redox couples there is only one electronic state which will be considered for the electron transfer reactions: this electronic state will be occupied during electron transfer to the redox couple and will be unoccupied during electron transfer from the redox couple. Due to the different electrostatic stabilization of the occupied and the unoccupied state by the surrounding solvent molecules the bound state is shifted downward in energy (the ionization potential $I_p$ is increased by the reorganization energy $\lambda_{red}$ due to electrostatic stabilization) and the unoccupied state is increased in energy (the electron affinity $E_a$ is reduced by the reorganization energy $\lambda_{ox}$). As a consequence of the thermal excitation of solvent vibrations around the redox couple the related fluctuation of the electrostatic interaction will lead to a broadening of energy states which is also shown schematically in Fig. 5.14.

The resulting Gaussian distributions of electron states depend on the reorganization energy, which, in general, will be different for the reduced and oxidized state (or any other state depending on the electrostatic stabilization energy), but often are also approximated to be equal. Their analytical forms are:

$$W_{ox}(E) = \frac{1}{\sqrt{4\pi\lambda kT}} \exp\left(\frac{-\left(E - E^0_{redox} + \lambda\right)^2}{4kT\lambda}\right) \tag{5.20}$$

$$W_{red}(E) = \frac{1}{\sqrt{4\pi\lambda kT}} \exp\left(\frac{-\left(E - E^0_{redox} - \lambda\right)^2}{4kT\lambda}\right). \tag{5.21}$$

It should be noted that these DOS of the electrolyte electron states are not equivalent to the electronic DOS of a solid. The state distribution at any singular redox species (either occupied or unoccupied) is only valid in the time domain for electron transfer giving its temporal probability distribution in energy (in its occupied or unoccupied state). However, if we consider the large number of redox species in a ca. 1 M concentration this distribution function can also be considered as a spatial distribution of energy states available in the electrolyte. But as any redox species at a specific time and a specific site will only be in one defined energy level, an energetic transition from occupied states to unoccupied states will not be possible by optical transitions but only by electron transfer reactions (self-exchange and bimolecular exchange). But in any case the original electron state must be in the energetic range achievable for electron transfer, which may not be the case for many multi-electron redox-couples. For example involving $H_2O$ as redox-couple the original electron states are out of reach for single electron tunneling transfer (see below).

For contact formation at electronic equilibrium the electrochemical potential of the electrons in the solid semiconductor equivalent to the Fermi level ${E_F}^{SC} = {\eta_e}^{SC}$ is aligned with the electrochemical potential of the electrons in the electrolyte equivalent to the redox potential of the electrolyte $E(\text{red/ox}) = {\eta_e}^{el}$ (see Fig. 5.13). As in the case of semiconductor–metal contacts (see above) the original given difference in the electrochemical potential is often related to the vacuum level and in this case given by the work function difference $\Delta\phi$ of both phases.

In principle, also the work function of the electrolyte is conceptually well defined but hard to measure in comparison to solids (see above) as most measurement techniques (e.g., UV photoelectron spectroscopy, Kelvin probe techniques) which can be applied will only lead to reliable results for solid samples. For electrolytes changes in the arrangements of surface species for the possibly chosen measurement conditions will lead to modified values of the surface dipole potential $e\Delta\chi^{el}$ of the electrolyte. This is the reason why there does not exist a defined and broadly accepted value for the absolute electrode potential related to the vacuum level of the normal hydrogen electrode's (NHE) electrochemical reference potential $eE_0$(red/ox), which is used for electrochemical standard redox potentials. The relation is given as:

$$\phi^{el} = -\eta_e^{el} = -\mu_e^{el} + e\Delta\chi^{el} = -eE_{abs}^0 - eE_0(\text{red/ox}) = -eE_{abs}\,(\text{red/ox}) \quad (5.22)$$

The signs result from the different sign conventions of the "absolute" and the "relative" electrochemical scale, which is positive for the work function and negative for the electrochemical potential of electrons as well as for the absolute electrode potential. A more strongly bound electron has a larger negative binding energy on the vacuum scale but a more positive work function, which translates to a more positive redox potential. The most often used value for ${E_{abs}}^0$ is 4.5 V (Memming 2001; Jaegermann 1996) but also values of 4.31, 4.44, and 4.73 eV have been published based on Bosch–Haber calculations. More recent values based on experimental work function measurements suggest values in the range of 4.7–4.85 V.

The changes of interfacial potential distribution induced by the contact formation semiconductor/electrolyte and equilibration of the electrochemical potential of electrons must be described in a similar way as for a metal–semiconductor junction (Fig. 5.13). The contact potential after equilibrium formation leads for Schottky contacts to an electron transfer induced potential drop (the Volta potential) equivalent to band bending in the semiconductor $eV_{bb}$, which is related to an extended space charge layer in the semiconductor for originally flat band conditions as in the shown case. Alternatively, an existing band bending induced by intrinsic surface states (see Fig. 5.11) might be modified. In addition, a double layer potential drop $e\Delta\chi$ (including double layer potential drops δ in case of Fermi level pinning) is formed across the surface of the semiconductor to the electrolyte, which results from the modified intrinsic double layer and the electrochemical double layer potential drop from the electrolyte mostly across the inner (induced by specifically adsorbed electrolyte components) or outer Helmholtz plane (induced by solvated ions; any contribution of diffuse double layer potential drops as for example Gouy–Chapman layers will not be considered). It should be noted that such interfacial dipole layers $e\Delta\chi$ (even without a contribution of δ with no Fermi level pinning) are inherently connected to the formation of any solid/electrolyte contact. Changes in the potential distribution of the electrons in the solid from the vacuum to the electrolyte contact (intrinsic double layer potential drops), but also in the dipolar and ionic contribution of the interfacial arrangement of electrolyte species or adsorbates (extrinsic double layer potential drops), will always occur during contact formation (see above).

This rearrangement of the double layer potential drop without any extended charge transfer (Volta potential change) will shift the valence and conduction band edges with respect to the electron states and the redox potential of any electrolyte species. It is clear, of course, that only part of the double layer potential drop will be operative for those adsorbate species or electrolyte components, which are situated in its spatial positions between the bulk potential of the electrolyte and the bulk solid. This may cause problems for porous solids or inhomogeneous surfaces, where the double layer is inhomogeneously or gradually changed; in these cases the specific position of any species across the interface must be known for the assignment of its energy state.

Due to the spatial distribution of electron states along the surface, the level of half occupancy of redox electron states can be related to the mean electrochemical potential of electrons $\eta_e^{el}$ or set equivalent to the potential $E$(red/ox) for this redox couple. As a consequence these values can be taken for the formation of electronic equilibrium at solid/electrolyte contacts aligning these to the electrochemical potential of the electrons in the solid $\eta_e^{sol}$ equivalent to the Fermi level. However, it must be kept in mind that the involved electronic states may not be given by the DOS of reversible redox-couples as shown in Figs. 5.13 and 5.14 for more complicated multi-electron transfer redox couples. In this case the redox potential $E$(red/ox) is still defined by the equilibrium of forward and backward reaction for equilibrium conditions. But the density of states can hardly be deduced in a similar simple way and must be related to the original given electron energy position of the active species in the electrolyte including intermediate states and their change in electron occupation with charge transfer. For the case of the $H_2O$ oxidation and reduction reactions the sequence of electron states involved and their energetic positions are not yet known for the intermediate one-electron redox reaction steps and any pictorial representation as shown in Figs. 5.13 and 5.14 are not valid. Therefore, it is also clear that for a discussion of the charge transfer reactions the intermediates and their involved electronic states must be known for any mechanistic and quantitative conclusions. Some of the first results on the possibly involved electron states $X_2/X^-$ ($X =$ halogene) redox couples may be found in Mayer et al. (1992b, 1992b, 1996) Jaegermann 1986). It can be stated that in general the electrolyte DOS of the multi-electron transfer redox-couples are not yet known.

As already mentioned above the DOS of $H_2O$ is probably outside the energy range which can be reached for most semiconductors by outer sphere electron tunneling across the junction. This is already evident from the redox potentials for the one electron transfer $OH^{\bullet}$ radical formation from $H_2O$ which is found at $E_0 = 2.73$ V vs. NHE in acidic solution. Additional evidence is found for the energetic positions of the occupied electron states of adsorbed $H_2O$ layers as measured in UHV (Jaegermann and Mayer 2005; Thiel and Madey 1987). As a typical example valence band spectra of $H_2O$ reversibly adsorbed onto $WSe_2$ (0001) surfaces are shown in Fig. 5.15 (Mayer et al. 1992a; Jaegermann 1996). The formed ice layer has its valence band maximum at a binding energy of about 6 eV below the Fermi level, which is a typical value deduced for the lone pair $1b_1$

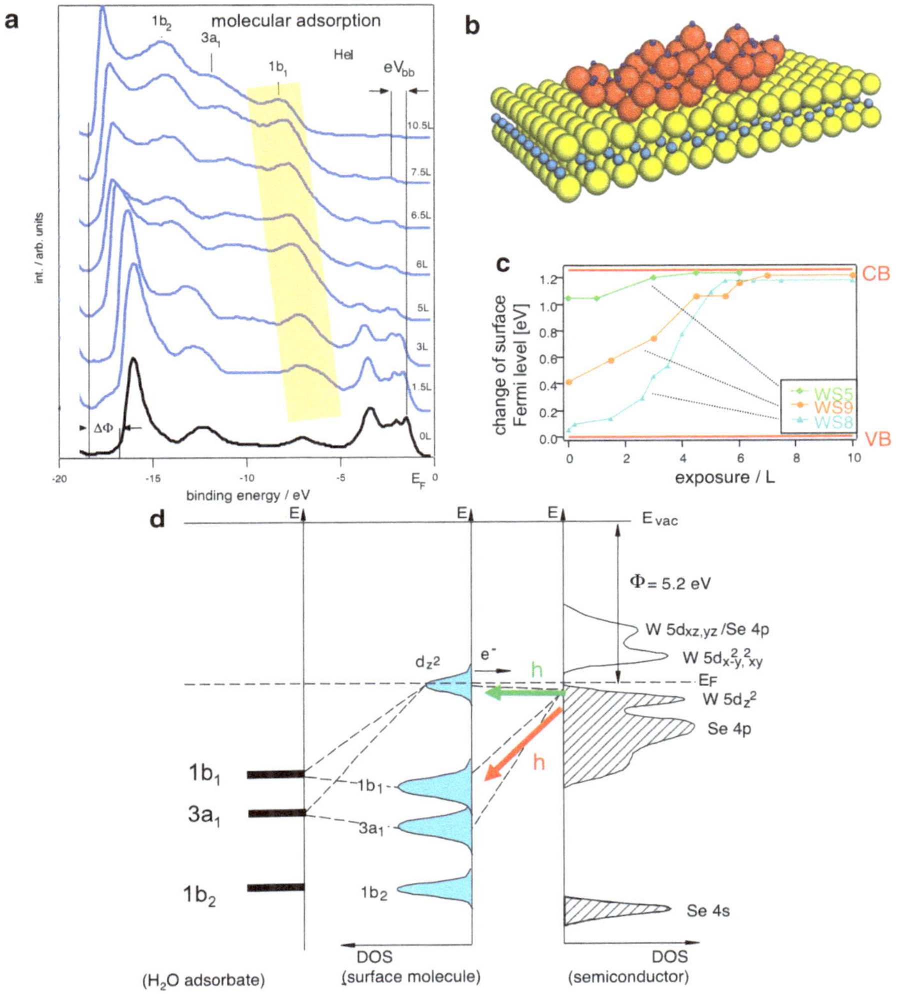


**Fig. 5.15** (**a**) Valence band spectra showing the adsorption of a $H_2O$ ice layer onto $WSe_2$ (0001) surfaces. The energetic position of the $H_2O$ valence band maximum $1b_1$ state is indicated. (**b**): Schematic arrangement of $H_2O$ molecules on the $WSe_2$ surface. The energy diagram (**c**) shows the shift of the Fermi level position in differently doped samples to the conduction band edge indicating the donor character of $H_2O$. (**d**) Schematic LCAO scheme to explain the formation of occupied surface molecular states from the interaction of the lone pair $1b_1$ $H_2O$ orbitals with the W $5d_z2$ valence band maximum. As a consequence of this interaction hole transfer from the valence band is possible by tunneling into this surface molecular state

HOMO level of adsorbed $H_2O$. Despite this fact adsorbed $H_2O$ behaves as a donor with electron injection into the semiconductor, which is seen from to a downward bending of the band edges shifting $E_F$ towards the conduction band edge. This discrepancy can be dissolved by assuming that $H_2O$ interacts with the occupied valence band d-states of the semiconductor surface and the resulting occupied surface molecular states, which are situated above the Fermi level, will be able to inject electrons into the solid. This picture can also be turned around. For outer sphere hole transfer into $H_2O$ the occupied valence band states are out of reach. However, any surface molecular state formed by the electronic interaction of the solid with $H_2O$ lone pair states will be at the right energy range for hole transfer forming a strongly bound OH-species on the surface. The above presented picture is in good agreement with the fact that $H_2O$ oxidation reactions approaching the redox potential of $E_0 = 1.23$ V in acidic solution need strongly interacting catalysts. As a consequence of the schematic model hole transfer reactions will depend on newly formed electron states from the interaction of the lone pair O $1b_2$ states of $H_2O$ with the solid substrate.

### 5.5.3 Fermi Level Pinning and Surface (Interface) States

In very many cases the semiconductor–electrolyte contact as well as the semiconductor–metal contact must be modified by the use of additional electronic and/or chemical passivation layers in order to avoid Fermi level pinning by surface states (see below) or to avoid chemical corrosion reactions. These passivation layers may be ultrathin (1–2 nm) but dense layers, preferentially of oxides or other wide band gap compound semiconductor or insulators. The special effect of intermediate layers is well known in semiconductor electronics (metal oxide (insulator) semiconductor junctions MOS or MIS) and its junction behavior has been described by the famous Cowley–Sze model (Cowley and Sze 1965) which can easily be adapted to semiconductor electrolyte interface with intermediate surface phases (Jaegermann 1996). In dependence of the density of defect states in the intermediate phase (surface reaction phase or passivation phase) covering the energy range of the semiconductor band gap these additional states may or may not be involved in equilibrium contact formation or charge transfer reactions.

For the MOS model it holds (Cowley and Sze 1965)

$$\Phi_B = S \cdot \left(\varphi_M - E_A^{SC}\right) + (1 - S)\ \left(E_G - \Phi_0\right) \tag{5.23}$$

with $S = \left[1 + \left(e^2/\varepsilon_i\varepsilon_0\right)\ D_{is} \cdot d_{dl}\right]^{-1}$ two limiting cases can be considered (n-type semiconductor):

1. $D_{is} \to 0$; ($D_{is}$ = interface state density) for this case, $S$ goes to 1 and the barrier height $\Phi_B$ is given by the Schottky limit, Eq. (5.10)
2. $D_{is} \to \infty$; for this case, $S$ goes to 0 and $\Phi_B$ is given by $E_G - \Phi_0$, see also Eq. (5.19).

For an intermediate density of interface states, the potential drop is distributed between the space-charge layer, $V_{bb}$, and the surface double layer, $\Delta\Psi_{dl}$

(or as denoted before δ). The density of interface states $D_{is}$ for real semiconductor–metal contacts can be determined from the slope $S$ of a plot of $\Phi_B$ versus $\Phi_M$ in accordance with Eq. (5.23). The reduced value of band bending $eV_{bb}$ or barrier hight $\Phi_B$ is dependent on the total charge $Q_{is}^{'} = Q_{is} + Q_{SC}$ (due to the formation of the space-charge layer and the metal contact) residing on the interface states (Rhoderick and Williams 1988):

$$eV_{bb} = \phi_M - \phi_{SC} - \left(\frac{e}{\varepsilon_0 \varepsilon_i}\right) d_{dl} \cdot Q_{is}^{'} \tag{5.24}$$

$$\Phi_B = \phi_M - E_A^{SC} - \left(\frac{e}{\varepsilon_0 \varepsilon_i}\right) d_{dl} \cdot Q_{is}^{'} \tag{5.25}$$

The last term accounts for the potential drop $\Delta\Psi_{dl}$ across the surface double layer of width $d_{dl}$ and permittivity $\varepsilon_i$. The interface charge $Q'_{is}$ is given by the integral over $e{\cdot}D_{is}(E)$, as given by Eq. (5.24) or (5.25), from $\Phi_0$ up to the equilibrium position of $E_F \equiv \Phi_B$, approximately determined by $\Phi_M$:

$$Q_{is}^{'} = e \cdot \int_{\Phi_0}^{\Phi_M} D_{is}(E) \cdot dE \tag{5.26}$$

The degree of Fermi level pinning is thus dependent on the density and energy distribution of the interface states. A deviation from the Schottky limit ($S$ considerably smaller than 1) can only be expected when the surface concentration of interface states is larger than $10^{13}$ eV$^{-1}$ cm$^{-2}$, which corresponds approximately to about 1 % of a monolayer (for a discussion of Fermi level pinning see also Sect. 5.5.1).

It should be mentioned that a similar expression can be derived when donor and acceptor surface state distributions $D_{is}^{D}$ and $D_{is}^{A}$ are assumed, which may be separated by a gap as in the case of polar semiconductors (Rhoderick and Williams 1988). In this case the plot of barrier height $\Phi_B$ versus metal work function $\phi_M$ would lead to areas with $S$ approaching 0, when the Fermi level intersects with $D_{is}^{D}$ or $D_{is}^{A}$. In other cases $\Phi_B$ follows the Schottky limit ($S \rightarrow 1$).

As already mentioned above, the magnitude of the electronic barrier $\Phi_B$ formed at the interface varies between the extremes of the Schottky limit ($S = 1$) and the Bardeen limit ($S = 0$). This dependence on the chemical nature of the semiconductor clearly indicates that the type of interface interaction and the related nature of interface states play an important role. A more ideal behavior ($S = 1$) is observed for ionic semiconductors, such as ZnSe and ZnO (closer relation to the Schottky limit), whereas a higher tendency for Fermi level pinning ($S = 0$) is observed for covalent semiconductors, such as Si, GaAs, and InP (closer relation to the Bardeen limit) (Sze 1981; Rhoderick and Williams 1988; Milnes and Feucht 1972; Brillson 1982).

Semiconductor–electrolyte interface may also be treated according to the Cowley–Sze formalism (Cowley and Sze 1965): In this case the formed barrier heights or band bending may be rewritten as (Jaegermann 1996):

$$eV_{\text{bb}} = S \cdot e\left[E_{\text{fb}}^{0} - E(\text{red/ox})\right] + (1 - S)\ (E_{\text{G}} + \xi - \Phi_0) \tag{5.27}$$

$$\Phi_{\text{B}} = S \cdot e\left[E_{\text{fb}}^{0} - \xi - E(\text{red/ox})\right] + (1 - S)\ (E_{\text{G}} - \Phi_0) \tag{5.28}$$

Where $S = \left[1 + (e^2/\varepsilon_{\text{i}}\varepsilon_0)\ D_{\text{SS}} \cdot d_{\text{H}}\right]^{-1}$ and $E_{\text{fb}}^{0}$ is the flatband potential of the semiconductor with no excess electronic charge in the surface state ($Q_{\text{SS}} \rightarrow 0,\ \Psi_{\text{dl}} \rightarrow 0$), $\xi$ is the bulk difference $E_{\text{CB}}$ to $E_{\text{F}}$. However, shifts of $E_{\text{fb}}^{0}$ due to charging of the surface by adsorbed ions and due to dipolar effects must still be considered. Care must be taken in applying Eqs. (5.27) and (5.28) due to the different reference levels used for the first term in brackets (electrochemical scale, in volts versus the reference electrode), and the second term (absolute scale, in electron volts versus the valence band maximum), considered as an intrinsic property of the semiconductor independent of the contact with the electrolyte.

The shift of the band edges due to surface-state charging $\Delta\Psi_{\text{H}}$, is given by:

$$\Delta\Psi_{\text{H}} = \left(\frac{1}{\varepsilon_0\varepsilon_{\text{i}}}\right) \cdot Q_{\text{SS}}^{'} \cdot d_{\text{H}} \tag{5.29}$$

with

$$Q_{\text{SS}}^{'} = \int_{\Phi_0}^{\Phi_{\text{B}}} D_{\text{SS}}(E)\ \text{d}E \tag{5.30}$$

Thus, the double layer potential shift is given by the additional charge residing in the surface states as given by the original position of $E_{\text{F}}$ given by $\Phi_0$ and the final position as given by $\Phi_{\text{B}}$. Expressing $\Phi_{\text{B}}$ and $\Phi_0$ of the Cowley–Sze model in electrochemical terms and assuming uniform distribution of $D_{\text{SS}}$ leads to

$$Q_{\text{SS}}^{'} = eD_{\text{SS}} \cdot \left[E_{\text{F}}^{\text{SS}} - E(\text{red/ox})\right] \tag{5.31}$$

with $E_{\text{F}}^{\text{SS}}$ as position of the Fermi level in the surface states before equilibrium formation to the redox couple as given in the electrochemical scale (see Fig. 5.16 assuming a high density of surface states). The term Fermi level pinning in the original sense is related to an equilibrium situation. Therefore, its definition in terms of deviations of the linearity of photovoltage versus redox potential is, at least, misleading and may also lead to incorrect results, as the influence of slow charge-transfer rates in nonequilibrium or light induced formation and reoccupation of surface states may lead to additional shifts of band edges due to additional charge carriers trapped in the surface and changing $Q_{\text{SS}}$ and $\Delta\Psi_{\text{M}}$ (kinetic charging of the

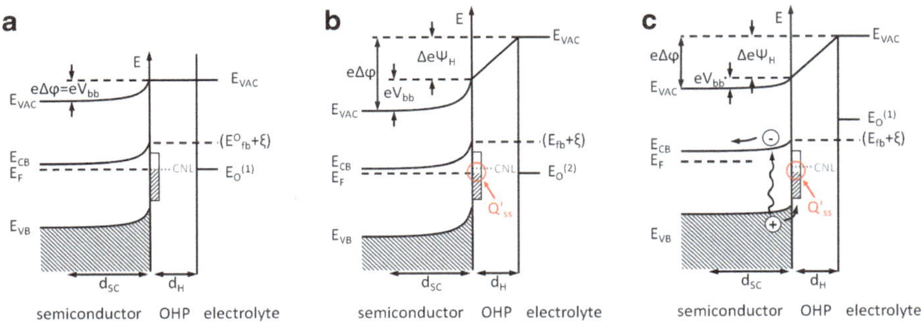


**Fig. 5.16** Influence of surface states on the electric potential drop at semiconductor–electrolyte interfaces (Fermi-level pinning). (**a**) For energetic alignment of the charge neutrality level with the redox potential of the electrolyte no electrostatic double-layer potential drop (see $\Delta e\Psi_H$) develops. (**b**) For a deviation the space charge layer mainly remains unaffected and the additional exchange charge (see $Q'_{SS}$) leads to a potential drop $\Delta e\Psi_H$ across the double layer. (**c**) A similar effect can also happen when the surface states are charged during illumination for slow charge transfer to the electrolyte (dynamic Fermi-level pinning). *OHP* outer Helmholtz plane

semiconductor surface or dynamic Fermi level pinning). As for semiconductor–metal contacts, the barrier properties should also be investigated with *n*- and *p*-doped samples of the same material, since the effect of Fermi level pinning is considered to be independent of doping. The barrier heights for n- ($\Phi_B^n$) and p-doped samples ($\Phi_B^p$) should add up to the band gap $E_G$: $\Phi_B^n + \Phi_B^p = E_G$. A deviation may be related to different values of the surface dipoles ${}^{SC}\Delta^{EL}\chi$.

As it is evident from these considerations for Schottky barriers and semiconductor–electrolyte interfaces intrinsic or extrinsic interface states may strongly change such interface energy diagrams as shown in Fig. 5.16. As already stated above intrinsic states may be formed on the surface of semiconductors due to the loss of bulk translational symmetry. For most covalently bound semiconductors such surface states exist at different energy positions in the band gap dependent on surface orientation and influenced by surface relaxation and reconstruction (see for example a summary in Chiarotti (1993)). However, due to contact formation with the electrolyte these intrinsic surface states are modified by chemical bond formation to ambient or electrolyte components forming extrinsic surface states or interface states. As a consequence the electronic structure of most semiconductor surfaces will deviate from the bulk and as measured for UHV processed surfaces and needs to be investigated in very detail before contact potential distributions and electron transfer reactions can be discussed and understood. Surface science techniques have been used to elucidate the microscopic details of semiconductor–metal interface formation and to determine the decisive factors governing the barrier height. The results indicate that the "interphase" may be of considerable complexity in both structure and composition and is strongly influenced by the interface chemistry of the semiconductor–metal combination (Rhoderick and Williams 1988; Sharma 1984; Sze 1981; Brillson 1982; Lüth 1993; Mönch 1993). For this reason, a number of different theoretical models are still being discussed in the literature to account for the experimental facts regarding nonideal Schottky barrier heights. The original model of Bardeen attributes the pinning effect to intrinsic surface states of clean semiconductor surfaces (Bardeen 1947). In more recent models, extrinsic surface states resulting from the interaction with the contact metal are favored. In the defect model the interfacial pinning states are due to near-surface lattice defects induced by the deposited contact metal, which only reflect the intrinsic bulk properties of the semiconductor (Spicer et al. 1986, 1988). In the metal-induced gap states (MIGS) model, the surface states result from the penetration of metal electron states into the semiconductor band gap (Heine 1965; Louie et al. 1977; Tersoff 1984a, b).

As for metal contacts, many semiconductor–electrolyte contacts show deviations in their experimentally observed behavior from the idealized picture of contact formation due to Fermi level pinning. In both cases the deviation is related to the electronic influence of extrinsic surface states. The term Fermi level pinning, introduced into the field by Bard, Wrighton, and coworkers, (Bard et al. 1980; Bocarsly et al. 1980; Aruchamy and Wrighton 1980) derives from the nonlinear response of photovoltage to changes in redox potentials. As the conclusion drawn from this work are based on nonequilibrium properties involving charge-transfer rates, they may be misleading (Rosenbluth and Lewis 1989). A large number of subsequent investigations (Jaegermann and Tributsch 1988; Memming 2001; Morrison 1980; Harrison 1989; Jaegermann 1996; Krol and Grätzel 2012; Lewerenz and Peter 2013; Peter and Upul Wijayantha 2014; Walter et al. 2010) have shown that, for most semiconductor–electrolyte interfaces, surface states and their response must be considered and may lead to deviations from the idealized Schottky limit. As discussed above, surface states are expected to have an influence only in the energy range in which $E_F$ intersects the energetic position of the surface states $E_{SS}$ due to contact formation with the electrolyte (see Fig. 5.16 and compare to Figs. 5.10 and 5.11). In this case the charge-neutrality condition for electrochemical contacts is modified to take into account the charge stored in the surface states: $Q_{SC} + Q_{SS} + Q_{El} = 0$.

A schematic energy diagram for semiconductor–electrolyte junctions with two different electrochemical potentials for the case of Fermi level pinning is shown in Fig. 5.16. For simplicity, it was assumed that no surface dipolar effects must be considered in addition. In principle, the charging of surface states and the related double-layer potential drop must be added to the changes in surface dipoles. It cannot be expected that these changes can be treated independently of each other, which makes the interface charge and electric potential distribution extremely complex. When the available surface state density far exceeds the typical charge carrier numbers of $10^{11}$–$10^{13}$ $cm^{-2}$ exchanged for the formation of space-charge layers, the band bending remains fixed (the semiconductor Fermi level pinned versus the surface) when electronic equilibrium is established. The contact potential drops across the electrostatic double layer (Helmholtz layer) between the surface and the electrolyte ($\approx\Delta\Psi_H$) by changing the occupancy of surface states. In many real cases the situation may be more complex as different types and lower concentrations of surface states may be involved and the contact potential difference is distributed inside the semiconductor and across the double layer.

There are only a limited number of detailed spectroscopic investigations on the physical and chemical nature of surface states at semiconductor–electrolyte contacts. As in the investigation of semiconductor–metal interfaces, surface science techniques as photoelectron spectroscopy may provide valuable additional information on the structural and electronic properties of these extrinsic surfaces states, as will be discussed in detail in the next section.

However, one important difference with respect to semiconductor–metal contacts should be emphasized again. When the semiconductor band bending of a metal–semiconductor Schottky barrier is changed, the occupancy of surface states is affected in accordance with Fermi statistics. For electrolyte contacts, however, also the type and concentration of surface states at the surface may change due to electrochemical reactions occurring at certain threshold electrochemical potentials. This problem is even more severe when currents flow due to illumination. Photocorrosive decomposition reactions of semiconductors are in most cases multistep electron-transfer processes. The intermediates may form different distributions of transient surface states with electron occupations not in equilibrium with bulk states, which in turn will change the electric potential distribution at the interface. This problem of dynamic Fermi level pinning involving transport of charges and reactants from and to the semiconductor surfaces may lead also to additional double layer potential drops $\Delta\Psi_H$ across the electrochemical double layer as is shown in Fig. 5.16.

We can not give here an extensive summary of the intrinsic or extrinsic surface electronic structure of semiconductors, for example see Brillson (1982), Jaegermann (1996), Many et al. (1965), Frankl (1967), Morrison (1977), Hansson and Uhrberg (1988), Moss and Balkanski (1994), but some general conclusions shall be given. In addition, for very many novel semiconducting materials as used for photoelectrochemical cells no widespread and elaborate surface studies have been performed yet to conclude on their surface electronic properties. There are only few general remarks to be made here:

1. Generally all semiconductors form intrinsic surface states due to the loss of translational bulk symmetry depending on surface orientation and influenced by surface relaxation and reconstruction.
2. These intrinsic surface states will be modified due to chemical bond formation/ surface interaction with contact phases forming extrinsic surface states.
3. Due to intrinsic surface states and their modification to extrinsic surface states band bending can be induced or removed depending on the bulk doping and the surface/interface DOS and their electron occupation.
4. The concentration of surface/interface states is usually very high in the range of $10^{14}$–$10^{15}$ cm$^{-2}$; if they are active they will lead to Fermi level pinning and in dependence on energy positions to enhanced surface recombination.

5. Strong parts of the research efforts in semiconductor (opto)electronics is related to the electronic passivation or controlled design of surface/interface states; they must be situated in an energy regime favorable for contact formation and the demanded electron transfer reactions.

It can even be expected that the $H_2O$ redox reactions, if these shall happen directly on the semiconductor surface, will only be possible if strongly interacting intermediates are formed. On the other hand intermediate states which are energetically or kinetically unfavorable will lead to undesirably formation, occupation and side reactions of surface/interface states, which will lead to bad performing photoelectrochemical cells. Due to these minority carriers being trapped on the surface there is very often, in addition, a dynamic shift of the energy bands with an increased potential drop across the double layer. As a consequence it is not important to discuss the straddling of the HER and OER potentials by the semiconductor energy bands based on values of the valence and conduction band positions as measured in the dark. It is only important that the energy bands are in the right position under illumination without forming a high concentration of detrimental (midgap) interface states, which would be efficient recombination centers.

### 5.5.4 Interface Engineering: Formation of Semiconductor–Passivation Layer–Cocatalyst–Electrolyte Interfaces

Proper interface engineering of the active semiconductor junction in contact to the electrolyte is the main challenge of the water splitting device. The duty is to form semiconductor–passivation layer–cocatalyst–electrolyte interfaces, which are able to perform the desired reactions without potential and current loss. Of course also dry contacts as for example the metal back contact may strongly influence the performance of the overall device, but here the classical knowledge on solid state semiconductors applies. We do not speculate on the band energy diagrams of the very many photoelectrochemical junctions, which have been investigated for water splitting, showing unfavorable performance. However, it is clear that these junctions are dominated by a high concentration of interface states either as extrinsic states on the semiconductor or as defect states in the chemically formed or deliberately deposited passivation layers. Whenever the surface photovoltage is strongly reduced from the expectation of good semiconductors or when recombination and/or charge trapping kinetics is evident rather high concentrations of interface states residing close to midgap position will be active. Therefore, an electronic passivation layer is needed, which shifts by chemical bond formation nearly all of the detrimental states out of the band gap region. Otherwise, the occupation of defects across the junction, which depends on the electron transfer rate constants, will lead to strongly modified junction arrangements. These may lead to strongly deviating junction properties compared to traditional descriptions especially during

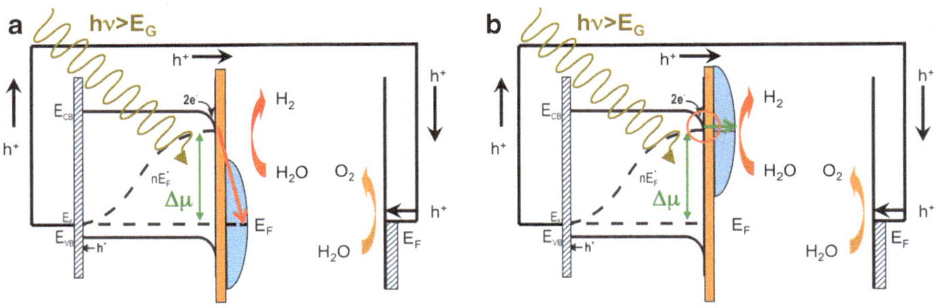


**Fig. 5.17** Schematic energy diagram of a p-semiconductor–passivation layer–metal cocatalyst interface. In (**a**) the passivation layer has no effect on contact formation and a high work function catalyst as may lead to the loss of photovoltage. In (**b**) the passivation layer defines the contact: In this case the also high work function metals are aligned to the Fermi level of the contact layer, if their doping level is high enough, and the photovoltage is operative

charge transfer reactions as dynamic pinning effects would shift the position of the energetic states of each layer in dependence of dynamic charging (Lin and Boettcher 2014). Therefore, a proper adjusting of energy levels across the junction during the photoelectrochemical HER or OER reaction would be essential for obtaining high conversion efficiencies.

This shall be exemplified for a p-semiconductor–passivation layer–cocatalyst–electrolyte junction assuming perfect passivation properties, which means that no detrimental interface states/defect states are involved. In Fig. 5.17 we show the band energy diagram, which would be expected for a p-semiconductor junction for HER. If we use bulk Pt layers as the $H_2$ evolving catalyst and a passivation layer not involved in the contact formation (wide band gap oxide in a MOS structure), the high work function would lead to a position of the metal Fermi level close to the valence band edge of the semiconductor. As a consequence the photovoltage gained in the semiconductor would be completely lost as the electron transfer will take place first to the Fermi level of the Pt catalyst close in energy to the valence band edge. As a consequence no effective junction will be obtained and the onset of photocurrent will hardly be shifted by any photovoltage. This can clearly be avoided, if the passivation layer also works as a contact layer, e.g., as to be expected for a low work function, highly conductive TCO layer as $SnO_2$, $TiO_2$ etc. In this case an energy-converting junction with large band bending is built up at the p-SC/TCO interface and the high work function catalyst will align to the Fermi level of the TCO layer with no potential loss. As a consequence the electron transfer will happen at the potential of the conduction band edge, which means that the inner layer forms a buried junction, which defines the energetic conditions of the junction and thus the operative photovoltage, whereas the cocatalyst will not disturb the junction but will just allow for low overpotentials in the electrochemical reaction. Such buried junctions to the passivation layers evidently characterize the envisaged interface design of any efficient electrolyte junction with the subsequent cocatalyst coupled to the Fermi level of the highly conductive passivation layer. This may be a TCO or degenerately doped SC-layer of appropriate work function. In this case the cocatalyst layer and its energy states involved in electron transfer will not lead to voltage losses. Examples of such interface designs are $pn^+$-Si or p-Si/n-$TiO_2$ junctions as recently published (Seger et al. 2013b; Warren et al. 2014).

As a consequence the buried junction to be used for photoelectrochemical $H_2O$ splitting systems may be rather complex. It consists at first of an electronic passivation and/or a contact layer to the semiconductor. The next layer may be an appropriate chemical passivation layer before the cocatalyst for HER and OER is added. This sequence of layers must allow for electron (hole) transfer close to the conduction (valence) band edge at operation voltage under illumination for the HER (OER). It may of course also be possible that some of the layers may provide different properties needed in such engineered interface at once, but this must be checked for every case.

## 5.6 Fundamentals and Applications of Photoelectron Spectroscopy

The external photoelectric effect which is the basis of photoelectron spectroscopy has been discovered by Hertz in 1887 (Hertz 1887) and explained in its quantum mechanical implication by Einstein (Nobel prize 1905 (Einstein 1905a)). The effect has been developed to a widely applicable spectroscopic technique to study the electronic properties of solid materials in UHV environment after the development of applicable electron spectrometers by Siegbahn (Nobel prize 1981 (Siegbahn et al. 1967)). Since that time a huge number of studies has been published on the fundamental physics and the analytical application of photoemission or photoelectron spectroscopy (PES), which depending on the excitation energy is abbreviated in different ways as XPS and UPS or SXPS (photoelectron spectroscopy using different excitation sources as X-rays, UV-light, or synchrotron light, respectively). The availability of commercial spectrometers has made PES one of the most often applied surface analytical technique, which gives information about the chemical composition (qualitatively and quantitatively) of a surface and also on the chemical bonding due to small binding energy shifts related to a partial charging of the analyte atom due to its chemical bonding interactions (ESCA: electron spectroscopy for chemical analysis). In addition, physical properties as the density of occupied states in the valence band region as well as surface potentials and contact formation at interfaces can be obtained. A number of books have been published in the meantime on the different aspects of PES as on the fundamental physics of photoelectron excitation, on the experimental spectrometer set-up and the basic processes of photoelectron measurement, as well as on the physical and chemical information to be obtained from photoelectron spectroscopy (Briggs and Seah 1983; Cardona and Ley 1978, 1979; Eberhardt 1995; Egelhoff 1987; Ertl and Küppers 1985; Hüfner 1995; Moulder et al. 1995; Reinert and Hüfner 2005). Therefore, there is no need to describe these basic principles in any more detail and the interested reader may refer to the given literature. However, if the applicability of PES for the analysis of semiconductor junctions is considered, there is already much more lack of information on how to proceed. This is even truer, if electrochemical interfaces are under consideration as for the devices discussed here. Here exists still a lack of awareness about the procedures and the type of information, which may be drawn from the application of PES in its different forms (UPS, XPS, SXPS). We therefore concentrate our discussion on the application of PES for characterizing semiconductor junctions as they are of interest in photoelectrochemical or photocatalytic devices. Valuable information on the composition and the electronic structure of surfaces, the effects of doping, chemical reactions at solid-gas-, solid-liquid- and solid-solid interfaces, growth modes of thin films, the chemical state of adsorbates, barrier heights at contacts and even charge transport mechanisms and diffusion processes at interfaces can be obtained. Some more information about our approach may be found in some previously published review articles or book contributions cited in the text given below, which has been modified and adjusted to the analysis of $H_2O$ splitting devices from Klein et al. (2008b).

## 5.6.1 *Experimental Procedure*

### 5.6.1.1 Basic Setup and Operation Principle

Photoelectron spectroscopy, or photoemission, is based on the external photoelectric effect, where illumination of a sample with photons of defined energy larger than the ionization energy causes electrons to be emitted from the sample (Einstein 1905b). Various photon sources are available for lab-based photoemission setups. Among them are X-ray tubes, which typically utilize Mg or Al anodes, providing photons of 1253.6 and 1486.6 eV energy with full widths at half-maximum (FWHM) of 0.7 and 0.85 eV, respectively. When X-rays are used as the excitation source, the technique is called **X**-ray **P**hotoelectron **S**pectroscopy: **XPS**. With single crystal X-ray monochromators coupled to Al anodes, the energy resolution of the X-ray source can be drastically improved to be around 300 meV. A monochromatized X-ray source has the additional advantage of avoiding the typical satellite X-ray emissions (Moulder et al. 1995). This becomes particularly useful for recording valence band spectra, which is important for the analysis of the electronic structure of surfaces and interfaces.

Photoemission spectra of valence bands, which are formed by the filled electronic (chemically bonding) states of materials with binding energies of ~15 eV below the Fermi energy, can also be excited with vacuum ultraviolet radiation. Photoemission using UV radiation is called **U**ltraviolet **P**hotoelectron **S**pectroscopy: **UPS**. The UV radiation is provided from a differentially pumped gas discharge, duoplasmatron or electron cyclotron resonance (ECR) type sources, typically operated with high purity He gas. The gas discharge gives rise to a number of emission lines with different intensities. The He-I line with energy of 21.2 eV is most often used. To employ the He-II line with energy of 40.6 eV, the He pressure has to be reduced to the lowest level at which the discharge is maintained.

The difference between XPS and UPS is related to the different excitation energies. With UPS only valence electron states can be accessed. Direct information on the chemical bonding to the substrate can be obtained from UPS of molecular adsorbates. Band structure measurements of single crystalline solids further provide a detailed description of the electronic structure of materials (Ertl and Küppers 1985; Cardona and Ley 1978, 1979; Hüfner 1995; Reinert and Hüfner 2005). In addition to the valence band region, XPS can access deeply bound core-level states, which are characteristic for the respective chemical element and do not contribute to the chemical bonding in a compound. The binding energy of the core levels, however, depends sensitively on the charge at the atomic site and therefore varies with the chemical bonds of the atom (Egelhoff 1987). This sensitivity is the mostly exploited feature of XPS and the origin of the frequently used expression **E**lectron **S**pectroscopy for **C**hemical **A**nalysis: **ESCA**.

Synchrotron radiation is extremely useful for photoemission experiments. It extends the capabilities of photoemission by providing intense monochromatic light with variable photon energies from the infrared up to several keV. This enables for example studies with variable information depths, band structure and resonant photoemission measurements. Describing the benefits of synchrotron radiation for photoemission experiments, although extensively used by the authors, is beyond the scope of this chapter and the reader is referred to literature (Briggs and Seah 1983; Cardona and Ley 1978, 1979; Eberhardt 1995; Ertl and Küppers 1985; Hüfner 1995; Moulder et al. 1995; Reinert and Hüfner 2005; Mutoro et al. 2007).

In photoelectron spectroscopy the information is contained in the energetic and spatial distribution of the emitted electrons. In vacuum the mean free path of the emitted primary electron is large enough to travel through an energy analyzer to a detector without scattering. In recent years, the restriction for high vacuum environments has also been partially overcome. The development of high-pressure photoemission currently allows photoemission of samples stored in a pressure of up to 1000 Pa. The experimental setups used in such experiments have recently been reviewed (Bluhm et al. 2007; Lichterman et al. 2015). It is for example possible to place the sample a few millimeters from the entrance aperture of a differentially pumped electrostatic lens system. Systems like this are nowadays used to develop PES approaches, which can be used for the analysis of wet chemical interfaces.

Photoelectron spectroscopy is a highly surface sensitive technique with an information depth of only a few nanometers. The surface sensitivity is related to the inelastic mean free path of the primary electrons in solids, which obeys a rather general behavior (Rhodin and Gadzuk 1979; Tanuma et al. 1991). Mostly the primary electrons contribute to the important chemical and electronic information. Inelastically scattered electrons contribute to the background intensity or also to satellite emissions. The most surface sensitive information is reached at the minimum path length of 0.5 nm for kinetic energies around 50 eV of the primary photoelectrons. This minimum is mainly caused by excitation of plasmons. At higher kinetic energies, the mean free path $\lambda$ follows roughly a square root dependence on energy. There are basically two methods to vary the surface sensitivity: either by changing the electron emission angle or by changing the kinetic energy when variable energy photon sources (synchrotrons) are available, or when several core-levels of one element (e.g., Ga 2p with binding energy ~1120 eV and Ga 3d with binding energy ~20 eV) can be excited.

The lifetime of the excited state, the energy width of the photon source, and the resolution of the electron analyzer limit the energy resolution in standard photoemission experiments. In XPS, the light source and lifetime are most important. In UPS, or with synchrotron radiation, the energy resolution of the analyzer may also limit the resolution due to the narrow energy width of the photon line, which is only ~2 meV for He-I, and of similar magnitude for synchrotron radiation (~2 meV) or even better for UV laser sources (<0.5 meV).

The kinetic energy of the photoelectrons $E_{\text{kin}}$ is given by

$$E_{\text{kin}} = h\nu - E_{\text{B}} - \phi_{\text{SP}}, \tag{5.32}$$

where $h\nu$ is the photon energy, $E_{\text{B}}$ the binding energy with respect to the Fermi energy of the sample holder, and $\phi_{\text{SP}}$ the work function of the analyzer. $\phi_{\text{SP}}$ is an instrumental parameter, which is adjusted by setting the Fermi edge emission of a clean metallic sample to zero binding energy. For conducting samples the Fermi energies of the samples and the spectrometer are aligned when electrical contact is established, thereby keeping the energy calibration. The measurement of binding energies with respect to the Fermi energy is a highly useful feature of photoelectron spectroscopy when variations of the chemical potentials of electrons (shifts of $E_{\text{F}}$) are involved, e.g., due to contact formation.

As an example a typical XPS spectrum recorded from a $SrTiO_3$ single crystal surface is displayed in Fig. 5.18. The survey spectrum shows emissions from all core levels at characteristic binding energies for the respective elements, which can be found in reference books (Moulder et al. 1995) or online databases (Wagner et al.). Due to spin-orbit interactions, the electrons emitted from shells with an orbital momentum (p, d, f) show a typical doublet with a $L+1/2$ and $L-1/2$ component. The intensity ratio of the two components follows from the degeneracy of the levels: $(L+1)/L$, which is 2:1, 3:2, and 4:3 for p-, d-, and f-orbitals, respectively.

Photoelectron lines originate from a transition of a bound electron to vacuum. Their kinetic energy is given by Eq. (5.32). In addition to photoelectron lines, also Auger transitions occur in a photoelectron spectrum. Two of them, the Ti LMM and O KLL transitions are indicated in the survey spectrum in Fig. 5.18b. Auger transitions are induced by the recombination of an electron from a lower bound energy level with the core hole created by the original photoemission process. The energy from the recombination is then transferred to another electron, which can also be emitted if it has sufficient energy. Auger transitions are indicated as

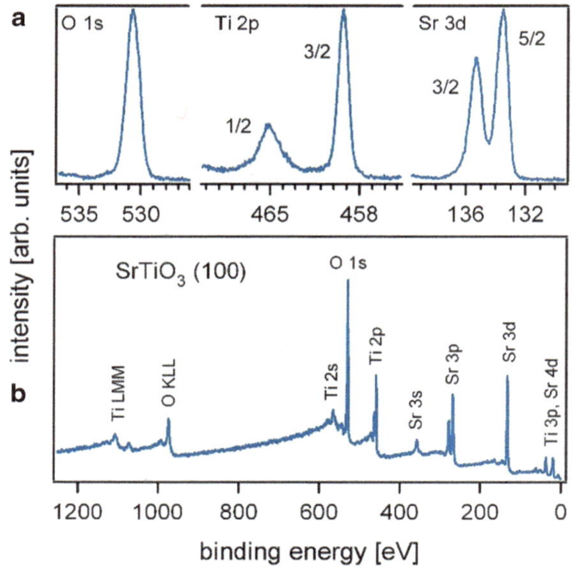


**Fig. 5.18** XPS survey (**b**) and detail (**a**) spectra of a $SrTiO_3(100)$ single crystal recorded with monochromatized Al Kα radiation. The Ti 2p and Sr 3d levels show splitting into spin-orbit doublets. The sample has been cleaned by heating at 600 °C in 1 Pa oxygen atmosphere to remove the typical hydrocarbon contaminations from ambient conditions

"XYZ"-transitions where the electron shell of the initial core hole is indicated as "X," the shell of the electron, which recombines with the core hole as "Y," and the shell of the electron which adopts the recombination energy as "Z." In our example, the O KLL emission in Fig. 5.18b is caused by a photo-excitation of the K-shell (1s), a recombination from the L-shell ($L_1$: 2s, $L_{2,3}$: 2p) and transfer of the energy difference to a second electron within the L-shell. According to Eq. (5.32) the kinetic energy of a photoelectron depends on the excitation energy. In contrast, the kinetic energy of an Auger electron is fixed by the energy of the recombination process. Auger lines can thus be easily distinguished from photoelectron lines if the photon energy can be varied. The Auger process is an intra-atomic process and leaves the samples with two empty states on a single atom. The doubly charged final state of the Auger process, compared to the singly charged final state of the photoemission process, leads to considerable differences in the energy shifts of photoelectron and Auger lines, which are sometimes indispensable for the analysis of the chemical state (Briggs and Seah 1983; Egelhoff 1987; Moulder et al. 1995; Klein et al. 2008a), in cases when the chemical shift in XPS is small.

#### 5.6.1.2 Sample Preparation: Vacuum Conditions

Photoemission experiments are typically performed in ultrahigh vacuum, i.e., at pressures below $10^{-7}$ Pa. The vacuum conditions are necessary to keep the inelastic mean free path of the electrons large enough in order to reach the detector without energy loss. In addition, vacuum is required to maintain clean or unchanged surface conditions during the experiment. Once a surface is completely clean, it will take up a monolayer of adsorbates in 1 s in a vacuum of $10^{-6}$ Torr, if the sticking coefficient of the adsorbates is one (Ertl and Küppers 1985). The gas dose is quantified in Langmuir units, where $1\text{L} = 10^{-6}\text{Torr s} = 1.3 \cdot 10^{-4}\text{Pa s}$. Hence, a gas dose of 1 L corresponds roughly to one monolayer coverage for a unity sticking coefficient. In a vacuum of $10^{-7}$ Pa, a monolayer may adsorb in 1000 s, which is less than a typical experiment time. Fortunately, the room temperature sticking coefficient of adsorbates from the residual gas in the vacuum chamber ($N_2$, $H_2$, $H_2O$, CO, $CO_2$) is usually much smaller than one.

The sticking coefficient depends crucially on the gas species, the substrate material and temperature. In particular for transition metal surfaces, it is more difficult to maintain clean surfaces, while chalcogenides are much less reactive. With oxide surfaces it can also be difficult to avoid hydroxide formation, as water is typically one dominant species in the residual gas. Hydroxide formation can be very difficult to identify on oxide surfaces, since the substrate lattice oxygen dominates the oxygen and valence band emissions (Weiss and Ranke 2002; Kunat et al. 2002). In addition, peroxide species, which are also likely on oxide surfaces, can lead to very similar photoemission features (Klein and Säuberlich 2008). In general, even a short exposure to ambient air is detrimental for keeping the surface conditions in the state of interest. This may not be immediately clear from XPS, but UPS and SXPS usually show severe changes due to contamination and reaction layers. Especially, the value measured for the work function will strongly be changed due to a contamination layer of hydrocarbons.

#### 5.6.1.3 Preparation of Solid Surfaces for Photoemission Experiments

In photoemission, one can measure samples prepared either ex situ or in situ, i.e., outside or inside the vacuum system of the spectrometer unit. In situ experiments are possible either directly inside the spectrometer chamber or in vacuum chambers, which are connected by an ultra-high vacuum transfer system to the spectrometer. The layout of such a system, which is used in the surface science group at TU Darmstadt, is shown in Fig. 5.19.

In situ UHV sample deposition or processing is a straightforward way to prepare clean surfaces of various materials. It is almost indispensable for the study of interface properties, which can be performed using stepwise deposition of a contact material onto a substrate. A large number of deposition methods can be used for such studies. With the system shown in Fig. 5.19, deposition of films using magnetron sputtering, metal-organic chemical vapor deposition, and molecular beam epitaxy is possible. However, the preparation of thin films with any of these techniques will usually involve extensive developments, so that in situ deposition is not always the first choice. Also single or multiple adsorbates, which represent electrolyte components, can be deposited in a controlled manner, in order to simulate the interaction with wet electrolytes (see below).

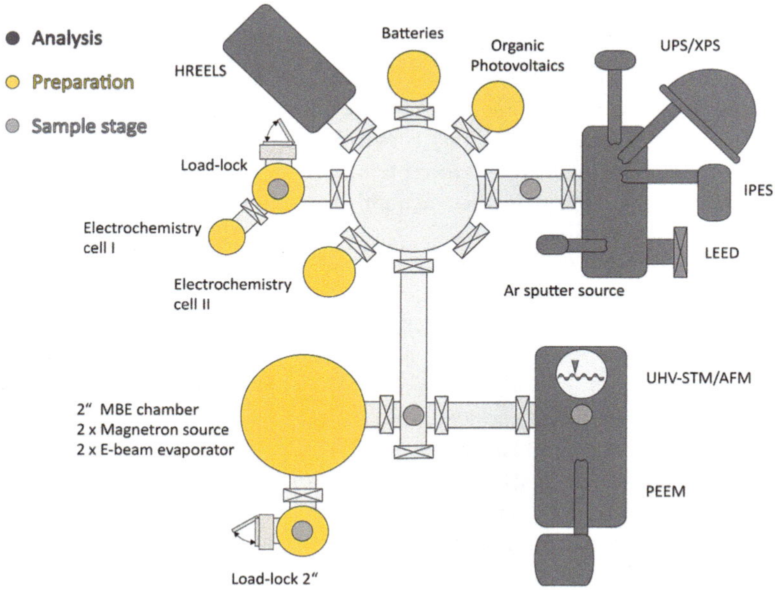


**Fig. 5.19** Layout of an integrated ultrahigh vacuum system (DAISY-FUN) combining a multi-technique surface analysis system including a photoelectron spectrometer for XPS and UPS and several chambers for surface preparation and thin film deposition like molecular beam epitaxy (MBE), metal-organic chemical vapor deposition (MOCVD), and glass chambers for electro-chemical experiments

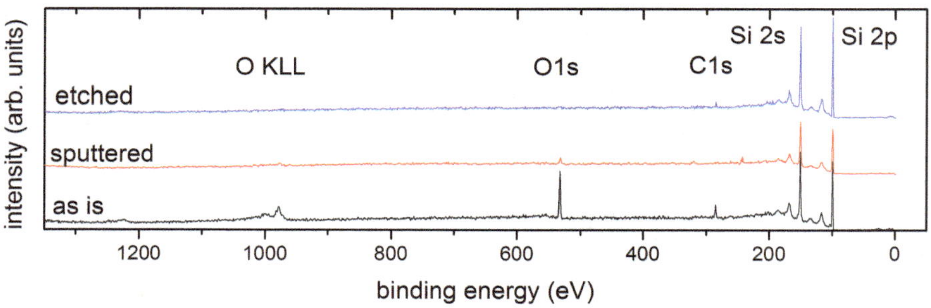


**Fig. 5.20** XPS-survey spectra of air exposed (as is), Ar ion sputtered, and $NH_4F$-etched samples. The air exposed sample shows additional emission lines of oxygen and carbon

Samples prepared outside the vacuum are always contaminated by adsorbed hydrocarbons and water leading to typical oxygen and carbon emissions at binding energies near 533 and 285 eV (see Fig. 5.20). For oxide surfaces, these can be removed by heating to moderate temperatures of 400–600 °C. The presence of oxygen during heating can considerably promote the removal of hydrocarbons (Gassenbauer et al. 2006). A widely used alternative technique for surface cleaning is Argon ion bombardment (sputtering, IB), or Argon ion bombardment combined with sample annealing (IBA). Ion sputtering can effectively remove surface contaminations but disrupts chemical bonds and therefore destroys the crystallinity of the sample surface and leads to the formation of defects. Due to preferential sputtering cross sections, also the composition of a compound surface can be modified severely. Sputtering of $TiO_2$ surfaces for example leads to a reduction of the surface by preferential removal of oxygen (Göpel et al. 1983). While subsequent annealing can often restore the crystalline state, this is usually not possible for the altered composition. Care must therefore be taken in the interpretation of results of sputtered samples.

Argon ion sputtering can also be used to perform depth profiling with XPS. In such cases low ion energies (<1 kV) should be used to avoid the problems mentioned above as preferential sputtering or defect formation or even amorphization. For the cases where these effects can be avoided, depth profiling offers additional opportunities as it could be used to study the effect of post-deposition treatments on interface properties (Fritsche et al. 2005), which is not possible in an evaporation experiment.

Alternatives for preparing clean surfaces of ex situ prepared materials are cleaving, fracturing or scratching in vacuum. Cleaving can be particularly useful for preparing selected single crystal surfaces, as for example the (110) surfaces of III–V semiconductors like GaAs and InP. An elegant way for surface preparation is offered by the decapping procedure, where a cap layer is deposited in situ onto a freshly prepared sample. When the sample is exposed to air, the cap layer oxidizes and becomes contaminated with adsorbates. When the sample is then heated in a vacuum system, the cap layer evaporates and a clean substrate surface is exposed again. This technique has been successfully used with As as a cap layer on epitaxial GaAs (Vitomirov et al. 1992) and with Se cap layers on polycrystalline Cu(In,Ga)$Se_2$ films (Schulmeyer et al. 2004).

#### 5.6.1.4 Analysis of Solid–Solid Interfaces Using Photoemission

Interfaces between solids are particularly important for a large number of devices. Due to the high surface sensitivity of photoemission, the analysis of the chemical and electronic properties of interfaces in a completed structure is usually not possible as they are buried under subsequently deposited layers. To overcome this restriction, fundamental properties of interfaces can be studied during their formation in an experiment, which repeats photoemission and film deposition in a number of steps as sketched in Fig. 5.21. An experiment starts with a clean substrate surface. Then, the contact material is deposited in several steps until a complete attenuation of the substrate emission is obtained. For a layer-by-layer growth of the deposited material, this corresponds to a film thickness a few times the inelastic mean free path, i.e., typically 5–10 nm. Several deposition steps starting from sub-monolayer coverage are typically required in order to derive a complete set of interface properties. With such experiments it is possible to derive growth modes of the deposited films, chemical interactions at the interface, and electronic

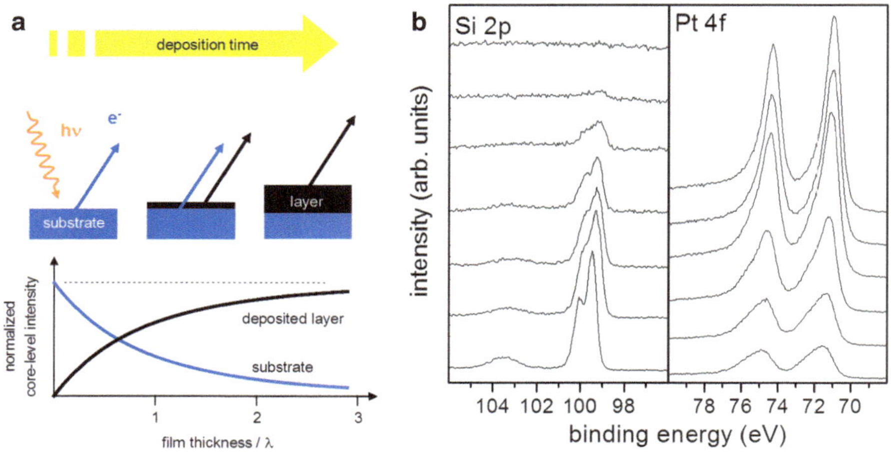


**Fig. 5.21** Schematic procedure of a stepwise deposition experiment for the analysis of chemical and electronic properties of interfaces between solids (**a**). XP-Spectra recorded during the stepwise deposition of Pt onto Si surfaces (**b**) ($h\nu = 1484$ eV). Changes of substrate binding energies are due to band bending, whereas changes of the Pt-lines are due to the modification of its electronic structure (final state size effects) (Klett et al. 2014)

properties as Schottky barrier heights, band discontinuities, changes of Fermi level positions and change of the work function. Examples will be given below. With suitable experimental setups, different deposition techniques and thus a wide range of materials have become accessible for such experiments.

#### 5.6.1.5 Study of Solid-Liquid Interfaces Using Photoemission

Wet chemical processing is widely applied in chemistry, materials science, and device fabrication. As discussed above elementary processes occurring at the solid-electrolyte interface are detrimental for PEC water splitting devices. A fundamental understanding of the solid-liquid interface on an atomic level is therefore strongly needed. In order to apply photoelectron spectroscopy to study electrochemical interfaces, we have developed the experimental station SoLiAS especially designed as a **So**lid/**Li**quid **A**nalysis **S**ystem, which is run at the synchrotron light source BESSY (Jaegermann and Mayer 2004; Mayer et al. 2005) (for a schematic representation of its set-up, see Fig. 5.22).

The SoLiAS allows for the SXPS analysis of electrochemical interfaces via three complementary experimental routes:

1. Emersion of substrates (working electrodes) from liquid electrolytes and transfer to UHV under controlled, clean, and inert atmosphere (Beerbom et al. 2000c).
2. Freezing of thin layers of the electrolyte by cooling emersed samples with liquid nitrogen during pumping to UHV to avoid loss of volatile species (Mayer et al. 2006).

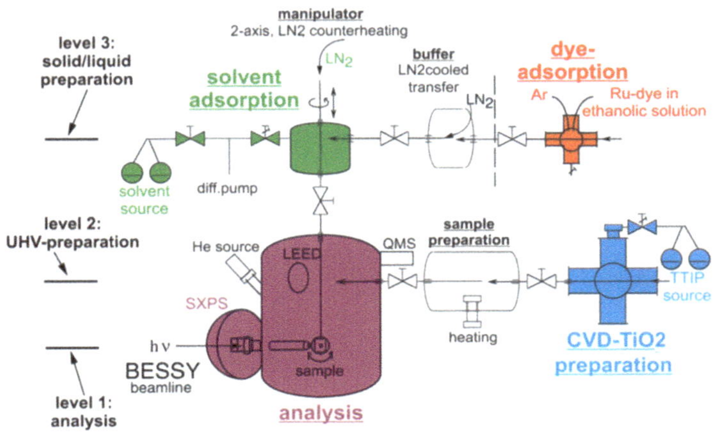


**Fig. 5.22** SoLiAS experimental station at BESSY II. For a detailed description see text

3. Synthesis of frozen models of electrolytes in vacuo by co-adsorption of solvent and redox active species onto liquid nitrogen cooled electrode surfaces (Jaegermann and Mayer 1995).

Detailed insights into the final surface composition as a function of applied potential, intermediate reaction steps and reaction products of electrochemical processes as etching (Beerbom et al. 2000c; Lebedev et al. 2004), chemical bath deposition (Hunger et al. 2005), and pulse plating (Ensling et al. 2003) have been accomplished for a number of semiconductor–electrolyte systems. Also the interaction of solvent molecules with different crystallographic surfaces of semiconductor electrodes (Henrion et al. 1996) and with adsorbed organic sensitizing dye molecules (Schwanitz et al. 2007a) has been studied. In addition, results on fundamental but yet unsolved questions of electrochemistry have been achieved (Jaegermann 1996; Mayer and Jaegermann 2000), e.g., concerning the redox potential of an electrolyte, the density of electronic states distribution in an redox electrolyte, and the solvation of redox active species. The analysis of the electronic structure of the Helmholtz layer and of the bulk electrolyte are still ongoing challenges; also investigations on simple one-electron redox pairs, which have been studied in UHV, are rare (Henrion and Jaegermann 1997).

There are different layers, which may contribute to the overall contact properties and which need to be designed in a proper way for efficient devices. This is schematically shown in Fig. 5.23 assuming a III–V semiconductor in contact to an electrolyte containing $X_2$ and $X^-$ (X = Cl, Br, I) as the two electron redox couple. First experiments on the interaction of such electrolytes with III–V semiconductors have been published, see Henrion et al. (1996, 2000), Beerbom et al. (2000a, b), Mayer et al. (2005), Lebedev et al. (2008). Exemplary, two different surfaces are compared: A surface state free (110) surface on the left and a (100) or (111) surface with intrinsic surface states on the right. The latter is in

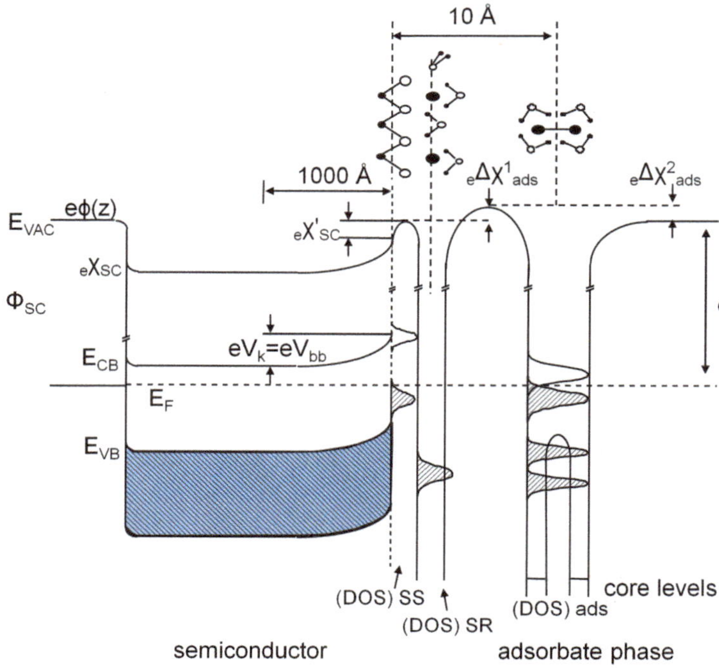


**Fig. 5.23** Schematic energy diagram of a hypothetical III–V semiconductor with (110) surface (left) and (111) or (100) surface (right). The right side is in contact with a halogen containing electrolyte, showing surface states, their modification due to adsorption and the electrolytes DOS. Also shown are possibly formed surface dipoles and double layers

contact with the electrolyte showing different interface states and double layer potential drops as well as an assumed electrolyte DOS. This schematic picture may be compared to model experiments of halogen redox couples (Jaegermann and Mayer 1995, 2004; Mayer et al. 1992b, 1996). We argue that a detailed analysis of such interface energy diagrams is needed to understand the electrochemical reactions of semiconductors on an atomistic level. A detailed analysis of the surface/interface properties of semiconductor devices is needed for the design of appropriate contact layers detrimental for any efficient water splitting devices as presented above, especially when new material combinations shall be considered. The possible information and experimental approach for such studies based on the surface science methodology will be presented in the next section.

We expect that the further development of high pressure electron analyzers (Bluhm et al. 2007) and dedicated setups for analyzing liquid surfaces will open up new routes to overcome the restriction to use frozen models of the electrolyte in the near future. The liquid-vacuum interface has also been analyzed using nonvolatile electrolytes (Siegbahn et al. 1981; Morgner 1994). Also, the surface composition of volatile electrolytes has been analyzed recently using a liquid micro jet within an UHV chamber (Weber et al. 2004). Recently a number of in situ and quasi in situ experiments have been published (Lichterman et al. 2015; Casalongue et al. 2013; Lu et al. 2012; Axnanda et al. 2015; Arrigo et al. 2013).

## 5.6.2 *Information Available from XPS*

In X-ray photoelectron spectroscopy, the absolute values of intensities and binding energies contain the most important information. The extraction of this information is, at least partially, independent and will thus be outlined separately in the following subsections. But, for consistent experiments both sets of data have to be considered together. It often happens e.g., that the observed binding energies do not allow for a unique assignment of the emissions to a particular compound at the surface. In this case, also the relative intensities are helpful to support or exclude possible explanations. Therefore, any detailed analysis typically requires taking into account emissions from all of the elements.

### 5.6.2.1 Analysis of Peak Lines and Intensities

Elemental Analysis

The first step in evaluating XPS data is the analysis of the elements present at the surface of a sample. This can already provide very useful information, since unexpected species are quite often observed, in particular for ex situ prepared samples. Elemental analysis is accomplished by identification of all peaks in a survey spectrum, which spans the maximum possible energy range defined by the excitation source. An example is provided in Fig. 5.20. Reference spectra for the different elements, as displayed for example in the *Handbook of X-ray Photoelectron Spectroscopy* (Moulder et al. 1995), are indispensable tools for elemental analysis. Many elements exhibit more than one emission line with a characteristic intensity pattern. Noticeable deviation from the intensity pattern typically indicates the presence of additional elements. Excitation of levels with other photon energies will lead to ghost lines, which may originate from a contaminated X-ray anode (to be inferred from calibration measurements using clean samples).

Compositional Analysis

XPS is the most important technique for the analysis of the chemical composition of surfaces. Although a quantitative analysis is possible with XPS, a large number of effects can contribute to the peak intensities and an absolute determination with an accuracy of 10 % or better can be reached only in a few cases. The relative composition, however, is highly reproducible and the technique can thus provide valuable information on compositional changes in dependence on sample treatment.

The first step in the analysis of a surface composition is the extraction of peak position and intensities, which starts with a background subtraction. While fitting of the inelastic background by linear or polynomial functions is only a mathematical operation, the Shirley (Shirley et al. 1972) and the Tougaard (1986) methods for background subtraction are based on physical approaches.

Curve fitting can be required if intensities of different chemical species of an element are to be evaluated. For nonmetallic species, the lines are generally symmetric and reproduced by a convolution of Lorentzian and Gaussian curves. Emissions from metallic species exhibit an asymmetry towards higher binding energy due to interactions of the conduction electrons with the core hole (Doniach and Sunjic 1970). The intensity of a photoelectron line of a particular atom depends on the density of atoms $n$ in the detected area, their photoionization cross-section $\sigma$, the photon flux, the angle between the photon source and the detector, the detection efficiency, the mean free path of the photoelectrons and for single crystalline samples also on the orientation of the sample with respect to the detector and the polarization plane of the photons (Hüfner 1995; Cardona and Ley 1978, 1979). A simplified analysis of sample composition is based on tabulated sensitivity factors $S$ for individual photoelectron lines (Moulder et al. 1995). The amount of a species in the sample $n$ is given by the intensity of the line $I$ divided by the sensitivity factor $S$:

$$n_x = \frac{I_x}{S_x} \tag{5.33}$$

and the atomic fraction $c_x$ of a multi-element sample by

$$c_x = \frac{I_x/S_x}{\Sigma_n I_n/S_n} \tag{5.34}$$

Although this works quite well in a number of examples, the absolute values can differ significantly from the given composition in other cases. It is further explicitly pointed out that Eq. (5.34) is only appropriate for homogeneous samples due to the intensity dependence of emissions on the inelastic mean free path.

The surface sensitivity of XPS can affect the determination of composition in different ways. If a sample is covered by adsorbates, e.g., the attenuation of the substrate signal depends on the mean free path $\lambda$ of the photoelectrons in the adsorbate layer and thus on the kinetic energy. A composition analysis involving levels with strongly different energies is therefore only appropriate for clean surfaces. The compositional analysis of very thin films is similarly affected (see Fig. 5.21).

Compositional analysis of single crystal samples can be strongly affected by photoelectron diffraction. As (photo-)electrons are propagating as waves through a material, they can be scattered and diffracted at the electron shells of the crystal lattice. X-ray photoelectron diffraction (XPD) can lead to considerable variation of peak intensities of single crystal surfaces with excitation energy and emission angle. XPD can be very useful in the determination of the structure of surfaces (Chambers 1992; Fadley 1993), but compositional analysis can become impossible due to this effect.

#### 5.6.2.2 Analysis of Energy Shifts

The detailed analysis of core level binding energies in photoemission can provide detailed information on the chemical and electronic structure of the surface. However, since a large number of different effects can contribute to changes of binding energies, the identification of the origin of the observed shifts can be quite difficult. In the following we give an overview and a few examples on the most important contributions to measured values of binding energies in photoelectron spectroscopy, which are relevant in the context of materials science and physical chemistry of energy converting devices.

##### Chemical Shifts

Photoelectron spectroscopy is famous mostly because it can detect differences in chemical bonding of atoms and the literature is full of examples. The binding energy of a core electron in a solid roughly depends on the oxidation state of the atom. In most cases, a more positive oxidation state leads to higher binding energy. However, as there are also exceptions from this behavior, one has to restrict to reference databases for identification of chemical species (Moulder et al. 1995; Wagner et al.). It is sometimes useful to distinguish between initial and final state contributions to the binding energy of core electrons (Egelhoff 1987; Hüfner 1995; Reinert and Hüfner 2005). The initial state is described by the occupied electronic level, whose energy depends on the local valence and core charge distribution at the atomic site. The energy of this level, however, cannot be measured as it changes with the removal of the electron. This contribution to the binding energy results from the screening of the positive charge of the core hole produced by photoemission by a polarization of the atoms' surrounding charges, leading to a reduction of the binding energy of the core level. As the screening occurs simultaneously with the excitation of the electron, the photoelectron contains in its kinetic energy the energy gain of the screening.

##### Surface Potential Changes

The binding energies of solids in XPS and UPS are measured with respect to the Fermi energy of the spectrometer. If the sample is in electronic equilibrium and in electrical contact with the sample holder the Fermi energy at the surface of the sample is aligned with the Fermi level of the sample holder and represents hence the

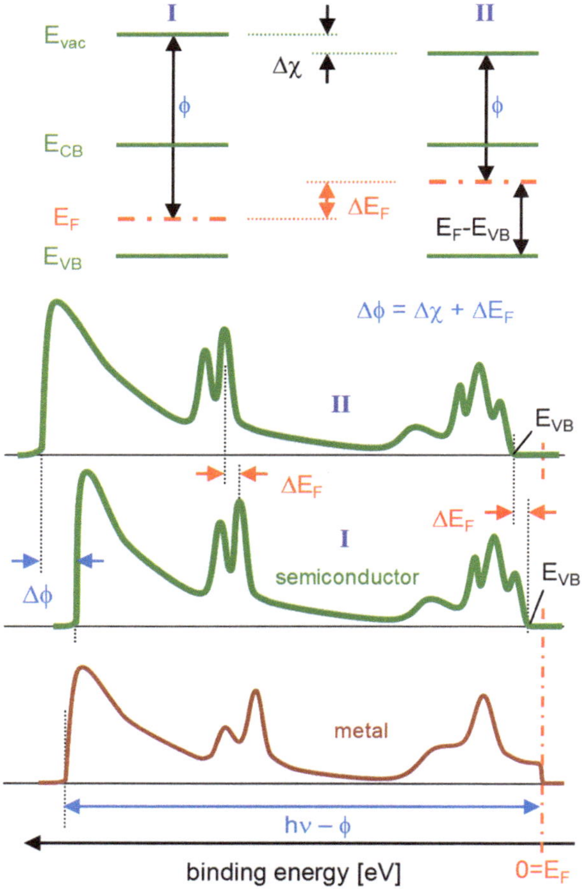


**Fig. 5.24** Changes of surface potential for semiconducting samples. The binding energy reference is given by the Fermi energy of the spectrometer (binding energy = 0 eV), which can be directly measured from a metallic sample. Shifts of $E_F$ within the material will change all binding energy values and the secondary onset (providing information about the work function $\phi$). Changes of surface dipole $e\Delta\chi$ will only change the work function

reference energy (binding energy = 0 eV) (Jaegermann 1996). The reference energy can be easily measured by recording the topmost (lowest binding energy) emission of a clean metallic sample or by measuring the well-known core level binding energies of metals such as Cu, Ag, or Au (Moulder et al. 1995). For nonmetallic samples as for example semiconductors, the Fermi energy can vary inside the energy gap (Sze 1981). Any variation of the Fermi level position inside the gap, caused for example by variation of the electrostatic potential at the surface, leads to a change of the Fermi level ($\Delta E_F$) with respect to the occupied states, as illustrated in Fig. 5.24. As the Fermi level of the sample is aligned with the spectrometer reference energy, the position of the Fermi energy on the binding energy scale of the experiment is not affected by changes of its position inside the gap. A change of the Fermi energy with respect to the band edges of the sample is instead observed by a change of the binding energies of all levels, i.e., all valence, core, and Auger levels of the sample.

Hence, if all levels exhibit parallel shifts upon a sample treatment, the shifts can be explained by changes in $E_F$.

A change of the Fermi level $\Delta E_F$ also changes the work function of the sample, which is given by the difference between the Fermi level and the vacuum level at the surface. The work function of a sample is furthermore affected by changes of the surface structure (Zangwill 1988). This can modify the surface dipole $\Delta\chi$ and hence the electron affinity $E_A$ or ionization potential $I_p$, which are given by the difference between the vacuum level and the conduction band minimum/valence band maximum. The change of the work function is hence given by sum of the change of the Fermi level and the change of the electron affinity, compare to Eqs. (5.16) and (5.17):

$$\Delta\phi = \Delta E_F + e\Delta\chi = eV_{bb} + e\Delta\chi \tag{5.35}$$

The energetic position of the valence band maximum ($E_{VB}$) is usually determined by a linear extrapolation of the leading edge of the valence band emission. More refined models, utilizing calculated density of states in the valence band have also been described (Kraut et al. 1983). Values with an accuracy of 100 meV can only be obtained if high-resolution photon sources such as UV lamps, X-ray monochromators or synchrotron beam lines are used. A good indication for a reliable determination of $E_{VB}$ is provided by the differences between $E_{VB}$ and the different core level binding energies, which should be constant for a given material. This is very useful, as it allows following changes of the Fermi energy by monitoring the binding energy shifts of the core levels. To separate from shifts caused by chemical changes, it is important to compare the changes for all core levels of a compound material.

The variation of the Fermi energy at a semiconductor surface can be related to band bending induced by adsorbates or interface formation, or to different doping of a material. Photoelectron spectroscopy offers two possibilities to distinguish between the different situations. First, depth profiling is partially possible by changing the photon energy. Even though this has become more useful with high energy XPS, the different information depth leads only to an asymmetric broadening of the peaks and not directly to a shift of the maximum. A direct identification of surface band bending is possible by exploiting surface photovoltage induced shifts (see below and Fig. 5.25) (Jacobi et al. 1990; Schlaf et al. 1993; Schellenberger et al. 1992). In the presence of a surface band bending, creation of electron hole pairs leads to reduction of band bending due to charge separation in the space charge field (Sze 1981) and hence to a rigid shift of binding energies of all levels. Photovoltages are particularly expected at low temperatures due to the thermally activated nature of the recombination current.

The determination of Schottky barrier heights (see Sect. 5.5.1) at metal–semiconductor interfaces is directly possible by following the Fermi level at the substrate surface in the course of metal deposition (Schlaf et al. 1993; Klein et al. 1998; Schafranek et al. 2008). A proper determination of the barrier height may require taking surface photovoltages into account, in particular if the experiments are carried out at low

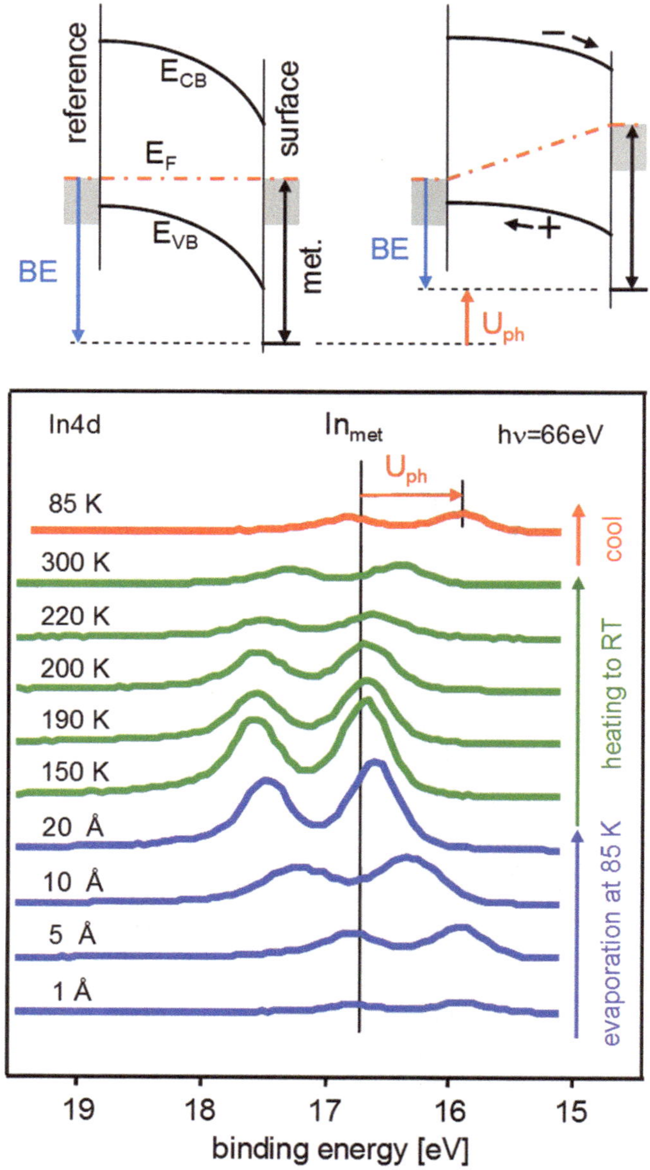


**Fig. 5.25** In 4d core level recorded during deposition and temperature cycling of a p-$WSe_2$–In interface experiment (Schlaf et al. 1993). The deviation of the binding energy of the In 4d level from the indicated reference for metallic indium is caused by a surface photovoltage, which is induced by the photoemission light source (synchrotron radiation)

substrate temperatures (Schlaf et al. 1993; Klein et al. 1998) or if the barrier heights are large (Schafranek et al. 2008). The photovoltage can be induced by the photoemission photon source itself as inelastic scattering of primary photoelectrons creates electron–hole pairs (Alonso et al. 1990; Hecht 1990). It is indicated by a deviation of the core level or Fermi edge emission from reference binding energies (zero in the case of the Fermi energy) and its direction depends on the band bending in the substrate. In principle, photovoltages can occur at all semiconductor surfaces and interfaces. If no metallic surface layers are present, they are however difficult to identify. As already mentioned above, a temperature or intensity dependent measurement can resolve this issue. Furthermore, cluster size effects can lead to comparable shifts of metallic binding energies (Wertheim 1989) and are often difficult to distinguish from source induced photovoltages for n-type semiconductors (compare Figs. 5.25 and 5.12).

The photovoltage can be calculated using the general expression

$$U_{\mathrm{ph}} = \frac{k_{\mathrm{B}}T}{q} \cdot \left( \ln\frac{j_{\mathrm{ph}}}{j_0} + 1 \right), \tag{5.36}$$

$$j_0 = A^{*}\mathrm{T}^2 \exp(-\Phi_{\mathrm{B}}/k_{\mathrm{B}}T), \tag{5.37}$$

where $j_{\mathrm{ph}}$ and $j_0$ are the photocurrent density and the reverse saturation current density of the diode, respectively. Equation (5.37) is valid for semiconductor–metal interfaces when the current transport is dominated by thermionic emission over the barrier (Sze 1985). Here $A^*$ is the effective Richardson constant and $\Phi_{\mathrm{B}}$ the barrier height of the interface.

An example for a source-induced photovoltage at the p-$WSe_2$/In interface is presented in Fig. 5.25 (see also Fig. 5.12) (Schlaf et al. 1993; Klein et al. 1998). Photovoltages are indicated by the deviation of the binding energy of the In 4d core level from its reference value of 16.7 eV. The experiment starts with a vacuum cleaved single crystal $WSe_2$ surface, which exhibits flat band conditions, since the surface is free of surface states. In a first step, Indium is deposited onto the sample at liquid nitrogen temperature, which suppresses island formation of the deposited In. The photovoltage for low In coverage is very large and amounts to ~0.8 eV. With increasing metal film thickness, indicated by the increase of the In peak intensity, the photovoltage gradually disappears. This is attributed to a short circuit between the sample surface and the sample holder formed by the growing In film (Horn et al. 1992; Klein et al. 1998).

Afterwards, the sample is subsequently heated to room temperature, which leads to the clustering of the In film as indicated by the decrease of the In intensity. Consequently, the metallic film is no more continuous. This eliminates the short circuit leading to a reappearance of the photovoltage. As predicted by Eq. (5.36), the photovoltage is lower at higher temperature and the subsequent cooling of the sample increases the photovoltage again. Using Eq. (5.36), the temperature dependence of the photovoltage allows for an independent determination of the barrier height (Klein et al. 1998; Schlaf et al. 1993). On the other hand, a quantitative evaluation of the photovoltage allows verifying that the transport across the interface is indeed accomplished by thermionic emission. In other cases it was possible to follow changes of the transport properties, caused for example by the intercalation of metal atoms (Schellenberger et al. 1992). It should be noted that, although there are exceptions when the barrier height is very large (Schafranek et al. 2008), photovoltages induced by laboratory sources at room temperature can be neglected in most cases.

In the presence of photovoltages, the Fermi level at the surface is no longer aligned with the sample holder. As a consequence all emission lines on the surface are shifted by the same amount as given by the surface photovoltage. This might also occur if nonconducting samples are investigated. Due to the emitted electrons, the sample surface will charge up to a certain amount, which can lead to energy shifts of several hundred volts. In principle, charge neutralizers (low energy electron guns) are usually employed to overcome charging effects. However, an accurate determination of binding energies becomes very difficult and requires the use of spectator lines as the typically adsorbed hydrocarbons or intentionally added silicone (Moulder et al. 1995). Analysis of Fermi level positions is no longer possible.

It is finally noted that changes of core level line shapes can also be caused by inhomogeneous surface potentials, whenever the surface potential varies along the surface. Band bending might vary across a surface due to inhomogeneous composition, structure, or adsorption. Charging and photovoltages (Schlaf et al. 1993) can also be inhomogeneous. Comparing the line shapes of the different core levels of a compound can often identify the presence of inhomogeneous surface potentials.

### *5.6.3 Interface Analysis*

Electronic properties of interfaces including surface potentials are particularly interesting for materials with band gaps between the highest occupied and the lowest unoccupied electronic states like semiconductors and insulators. Interfaces are crucial for all electronic devices as they determine the electric contact properties (see Sect. 5.5). In many cases, interfaces between two different or differently doped materials provide the essential functional element. In solar cells e.g., a rectifying junction between an n- and a p-doped semiconductor is responsible for the photovoltaic effect. The n- vs. p-type contact can be built up between the same material as in silicon or GaAs or between two different semiconductors as in Cu(In,Ga)$Se_2$ and CdTe heterojunction solar cells (Luque and Hegedus 2003). In the dye sensitized (Grätzel) solar cell, the heterojunction is formed between a nanocrystalline $TiO_2$ film on a fluorine-doped $SnO_2$, an adsorbed monolayer of a dye molecule and an electrolyte solution (O'Reagan and Grätzel 1991). In photoelectrochemical cells the junction may be defined between the semiconductor and the electrolyte or may include passivation layers and/or metal cocatalysts. These examples illustrate that many different types of interfaces can be present in a single device: (1) an interface

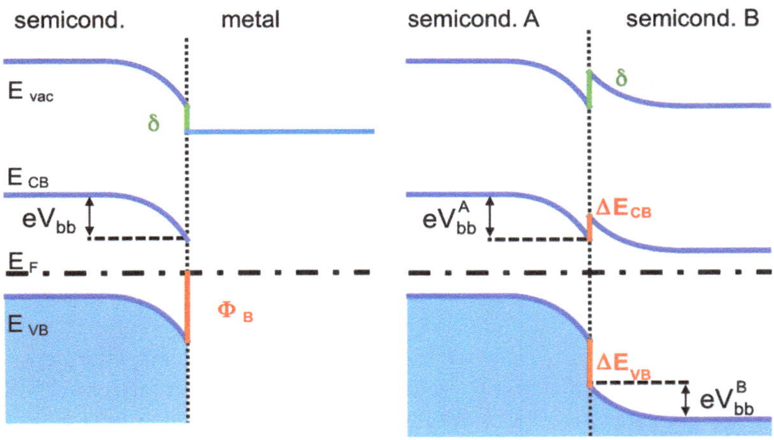


**Fig. 5.26** Energy band diagrams for semiconductor–metal and semiconductor hetero-interfaces indicating formation of the experimentally to be determined space charge layer $eV_{bb}$ and the interface dipole $\delta$

between two semiconductors, (2) an interface between an oxide semiconductor and an adsorbed molecule (the adsorption is accomplished within the solvent), (3) the interface between the adsorbed molecules and the electrolyte, and (4) the interface between the electrolyte and a metallic conductor.

Energy band diagrams for two rectifying semiconductor contacts, a semiconductor–metal and a semiconductor p/n-heterocontact, are shown in Fig. 5.26 (compare to Sect. 5.5). The important parameter of a semiconductor–metal interface is the Schottky barrier height $\Phi_B$, which is given as the distance between the Fermi energy and the valence band maximum $E_{VB}$ or the conduction band minimum $E_{CB}$ for p-type or n-type semiconductors, respectively. Corresponding barrier heights for hetero-interfaces are the valence and conduction band discontinuities $\Delta E_{VB}$ and $\Delta E_{CB}$. Together with the doping profile, the barrier heights determine the potential distribution and thus the bending of energy bands due to the formation of space charge layers across an interface. In general, a localized charge transfer dipole δ exists across the phase boundary due to the polarity of the interface bonds in addition to the extended space charge layer $eV_{bb}$. This leads to an interface dipole, which shifts the energy bands of both materials relative to each other by a double layer potential drop $\delta$. This is drawn as a step in the vacuum level at the phase boundary, which is used to indicate the spatial variation of the electric potential perpendicular to the interface.

Extensive research has been devoted in the past to understand the mechanisms governing barrier formation at interfaces in order to enable prediction and possible modification of barrier heights (Mönch 2003; Franciosi and Van de Walle 1996). Photoelectron spectroscopy has been extensively used in this context, because it allows for a direct determination of barrier heights and interface dipole potentials from the analysis of work functions. More recently, photoemission has also been used to monitor changes of barrier heights at semiconductor–metal interfaces upon annealing in different oxygen partial pressures, which induces changes in the interfacial defect concentration (Mosbacker et al. 2005; Schafranek et al. 2008).

The determination of Schottky barrier heights is straightforward from a stepwise deposition experiment. Once the valence band maximum binding energy of the clean substrate surface is known, one has to follow the changes in binding energy of the substrate core levels as a function of the deposition time. As example, a detailed analysis of the interface formation between (Ba,Sr)$TiO_3$ and Pt is presented in Schafranek et al. (2008). Of course, chemical reactions might also occur during interface formation (McGilp 1984), which complicates the determination of surface potential changes.

The determination of barrier heights at semiconductor hetero-interfaces is also well described in literature (see for example Klein and Säuberlich (2008), Yu et al. (1992), Kraut et al. (1980)). In principle, the valence band offset $\Delta E_{\mathrm{VB}}(\mathrm{A, B})$ between two semiconductors A and B is determined according to

$$\Delta E_{\mathrm{VB}}(\mathrm{A,B}) = \mathrm{BE}_{\mathrm{CL}}^{\mathrm{VB}}(\mathrm{A}) - \mathrm{BE}_{\mathrm{CL}}^{\mathrm{VB}}(\mathrm{B}) - \Delta\mathrm{BE}_{\mathrm{CL}}, \tag{5.38}$$

where $\mathrm{BE}_{\mathrm{CL}}^{\mathrm{VB}}(\mathrm{A\ or\ B})$ are the core level to valence band maximum binding energy differences of semiconductors A and B, and $\Delta\mathrm{BE}_{\mathrm{CL}}$ the binding energy difference of the core levels of the substrate and the overlayer determined during a stepwise deposition experiment. To extract the requested information the evolution of the core level binding energies of the substrates and the overlayer as a function of deposition time have to be displayed. It is convenient to subtract the $\mathrm{BE}_{\mathrm{CL}}^{\mathrm{VB}}(\mathrm{A\ or\ B})$ values as done in Fig. 5.27. From such a graph, the evolution of the valence band maxima of the substrate and overlayer can be directly obtained. The valence

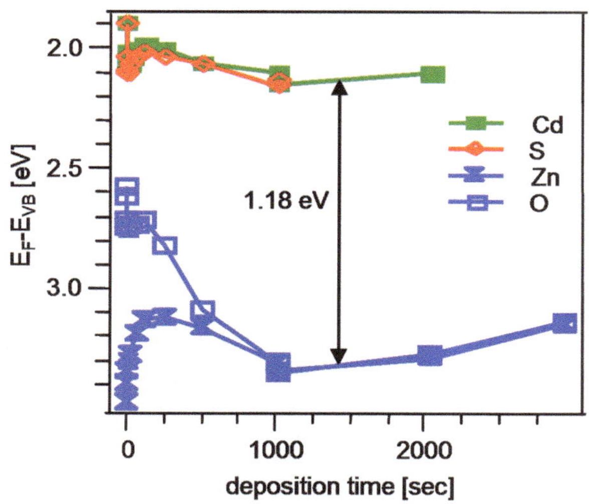


**Fig. 5.27** Evolution of valence band maxima recorded during deposition of ZnO onto CdS by magnetron sputtering (Klein and Säuberlich 2008). A deposition time of 1000 s corresponds to a ZnO film thickness of ~2 nm

band discontinuity is then derived from the energy difference of the two valence band maxima.

In principle, the VBM of the substrate and the overlayer should change in parallel in the course of deposition. In addition, if compound materials are studied, the VBM positions can be derived from different core levels, which all should exhibit parallel shifts. This behavior is exemplified by the Cd 3d and S 2p lines of the CdS substrate in Fig. 5.27 (Klein and Säuberlich 2008). In contrast, the Zn 2p and O 1s lines show a strongly different evolution at low coverage, indicating a different chemical bonding between Zn and O in the first two nanometers of the film. This illustrates that it is often essential to follow the evolution of core levels in dependence on film thickness. It is also mandatory to consider all core levels of a material.

Similar experiments can also be performed on semiconductor–adsorbate interfaces, which may be taken as model interfaces of electrochemical junctions (see Sect. 5.6.1). Such experiments provide information about band bending (Fermi-level shifts), work function changes (interface dipoles) as well as on the newly formed surface electronic structure. Including surface photovoltage measurements such studies give a lot of valuable information on semiconductor–electrolyte junctions. Extending such experiments to real electrolyte interfaces would provide a large step to a better understanding of photoelectrochemical water splitting.

### *5.6.4 Information from Valence Band Spectra*

Ultraviolet photoelectron spectroscopy can also be used to determine surface potentials and partially also chemical interactions at interfaces. Although it is also possible with XPS (Schlaf et al. 2001), the determination of work functions is usually performed by UPS, which is straightforward following the scheme in Fig. 5.24. UPS is also of particular importance when the valence band density of states is in the center of interest. In addition, energy dependent and angle-resolved UPS (ARUPS) can be used to determine the energy band structure $E(\mathrm{k})$ of solids as energy and momentum are conserved in the photoemission process (Himpsel 1983; Reinert and Hüfner 2005; Cardona and Ley 1978, 1979). Valence band spectra are of utmost importance for all cases where an electron transfer process is under consideration. As discussed above in Sect. 5.5.2 the contact formation but also the electron transfer at solid-electrolyte interfaces in PEC devices are determined by the density of states distribution in the solid and in the electrolyte. For this reason the energy diagram of the contact must be known in very detail in the dark for contact formation but also under illumination or applied potential for current flow. Thus the electrostatic potential distribution as well as the DOS should be given for all possible layers across the contact ranging from the bulk of the solid (semiconductor), to the surface, involving possible passivation layers and cocatalysts, and finally to the electrolyte double layers. At the end this knowledge is

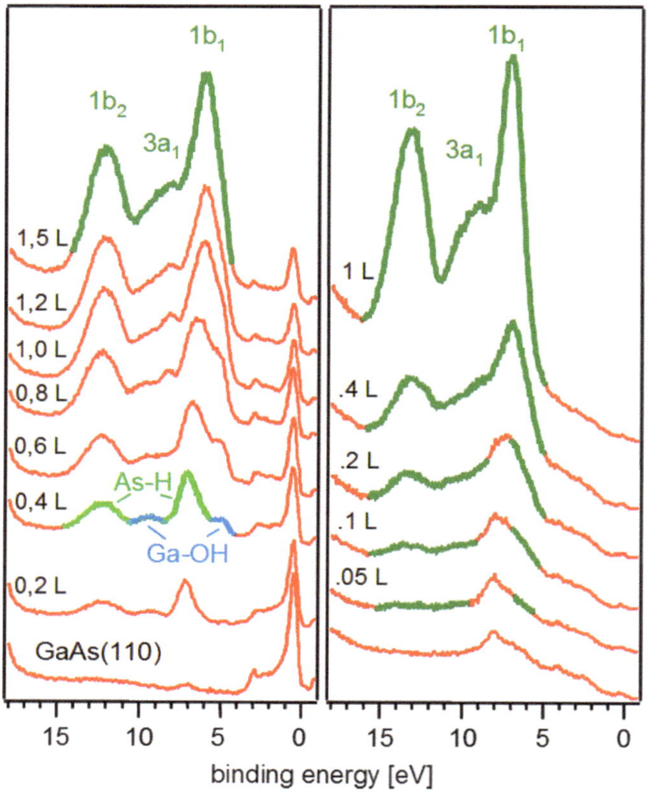


**Fig. 5.28** Valence band spectra recorded during low temperature adsorption of $H_2O$ on vacuum cleaved GaAs(110) (*left*) (Henrion et al. 1996) and decapped GaAs(100) (*right*). Water dosage is given in Langmuir ($1\ L = 10^{-6}$ Torr s)

needed to conclude on the electron and hole transfer reaction and possible involved loss processes.

Valence band spectra recorded during the adsorption of water from the gas phase onto GaAs surfaces are displayed in Fig. 5.28 as an example for UPS spectra applied for semiconductor studies (Henrion et al. 1996). Water adsorption was performed by dosing purified water using a leak valve onto samples cooled to ~100 K. After a water dosage of more than one Langmuir the typical emission of adsorbed water molecules ($1b_1$, $3a_1$, and $1b_2$) in an ice layer are observed. On the GaAs(110) surface, which was prepared by cleaving an oriented single crystal in vacuum prior to the adsorption experiment, different emission features are observed for lower water dosage. These can be attributed to dissociated water species, leading to the formation of surface Ga-OH and As-H species (Henrion et al. 1996). No dissociative adsorption is observed on GaAs(100), which was prepared by decapping of an As-capped GaAs(100) wafer. The different adsorption behavior can be explained by the different surface atomic structures (Moll et al. 1996). On GaAs(110), Ga and As dangling bonds exist on neighboring surface atoms. This makes water dissociation feasible as the –OH and –H fractions of the $H_2O$ molecule can be bound on the surface in close vicinity. This is not possible on GaAs(100), as a single substrate surface species dominates for this surface orientation.

Additional experiments on water adsorption have already been described above (see Fig. 5.15), but we expect that further systematic experiments in relation to water adsorption, oxidation (OER) and reduction (HER) reactions are needed to obtain detailed insights in possibly involved reaction mechanisms during water splitting.

## 5.7 Case Study: Si Based Photoelectrochemical Cells

In the following section we exemplify our approach in developing buried junction photoelectrochemical solar cells using Si as absorber material. We do not intend to give a complete literature survey of the investigations performed in the past, but mostly present our recent studies by combining photoelectrochemical investigations of Si single crystal and Si thin film tandem cells in combination with surface science experiments. Our cooperation partners from the Forschungszentrum Jülich have provided the thin film tandems, and the obtained performance values for water splitting efficiencies are mostly due to their expertise in preparing well defined and highly efficient thin film solar cells.

### *5.7.1 Water Splitting with Si Based Photoelectrodes*

The most simplistic approach for a photo electrochemical device is the immersion of the semiconductor directly into the electrolyte. In this case the band bending and thus the photovoltage is defined by the semiconductor–electrolyte contact. Due to the occurrence of chemical reactions at the surface of Si, the properties of this contact are hard to predict and depend on surface orientation, electrolyte composition, reaction sequence etc. For example, in silicon the corrosion of the surface in aqueous media leads to defects that pin the Fermi level in the middle of the bandgap. Therefore, the achievable photovoltage is only small. Furthermore charge transfer through a possibly formed oxide layer leads to an additional voltage loss as the layer grows. In Fig. 5.29 cyclovoltammetric measurements of bare Si(111) and Si(111) with additional platinum catalysts are shown. The platinum was deposited electrochemically from hexachloroplatinate solution. Due to the photovoltage the onset of the HER is expected to shift to positive potentials, which is hardly the case. The current saturates at 2.2 $mA/cm^2$ due to the low intensity of the illumination. After electrochemical deposition of platinum particles, the whole curve is shifted

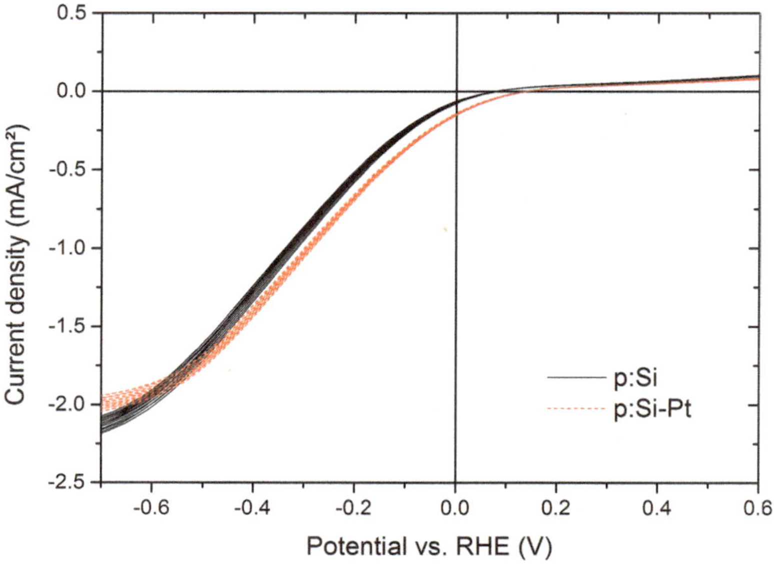


**Fig. 5.29** CV-characteristic of p-Si in 0.1 M $H_2SO_4$ solution and illumination with a white light LED (10 mW/cm$^2$). The *black curve* is the blank wafer and in *red* the Si wafer after platinum deposition

only slightly further to the right. This can be expected when considering the reduced overvoltage, due to the deposition of a catalyst.

Clearly the lack of photovoltage accompanied by the corrosion process makes such a device unfeasible as a photoelectric converter. In order to extract the maximum achievable photovoltage Fermi level pinning must be overcome by appropriate surface passivation layers. One easy way to achieve such a surface is to degenerately dope the surface region of the wafer, thus building a buried junction (Seger et al. 2013b; Warren et al. 2012).

Applying a homocontact into the silicon electrode provides several advantages: The charge separation is detached from the highly defective electrolyte interface, which promotes a high recombination rate of the light induced charge carriers. Additionally, the Si homo junction is known for its highly efficient photovoltaic power conversion proven in many PV power plants. A silicon PV device supplies an open circuit photovoltage of >0.6 V (Green et al. 2012), which was almost reached in a photoelectrochemical arrangement by Seger et al. (2013c) and Warren et al. (2012). These photoelectrodes exhibit a considerably higher fill factor when compared to other photoelectrodes made from oxide semiconductors. This is mainly due to the steeper current onset in homo junction silicon electrodes that leads to higher fill factors and improved efficiencies.

Thus the buried homo junction incorporated into the semiconductor is able to separate the catalytic processes at the surface from photovoltaic action with the defect sensitive charge separation. But with respect to the water splitting reaction itself a self-sustaining device is not possible without additional absorbers, because one silicon cell alone does not deliver the necessary photovoltage.

### 5.7.2 *Tandem Solar Cell as a Buried Junction Photoelectrode*

As a single buried junction electrode with a crystalline silicon homo contact is not suitable for light induced water electrolysis as shown above due to its low photovoltage, a series connection of three Si cells would be required to produce the necessary potential of >1.8 V. Another strategy is to employ a multi-junction solar cell made from thin film silicon, which is capable of a higher photovoltage. (For additional information on tandem devices see Chap. 12). We employ this type of solar cells as photocathodes. The standard μc-Si/a-Si tandem solar cells are in a p-i-n/p-i-n superstrate geometry and their silver back contact is in contact to the electrolyte.

The a-Si:H/μc-Si:H tandem solar cell provides a remarkable photovoltage of 1.4 V vs. the reversible Hydrogen electrode (RHE) as a photoelectrode at open circuit potential under AM1.5 illumination Fig. 5.30a. This potential would be in principle high enough to split water, but to achieve a high STH efficiency the $U_{MPP}$ or $U^{op}$ of the photoelectrode has to be situated at a potential, where a suitable counter electrode can deliver an equal current for the OER. We have employed commercial $RuO_2$ as a dimensionally stable electrode. The electrochemical behavior of this counter electrode is also depicted in Fig. 5.30a mirrored at the potential axis. The crossing point between the curve of the photo electrode and the counter electrode will define the working point of the water splitting PEC. It is obvious, that the CV of the a-Si:H/μc-Si:H photoelectrode has no intersection with the counter electrode at a reasonable current density. Therefore, the efficiency of the PEC would be almost zero. To improve the performance of the photoelectrode a cocatalyst has to be applied, because the catalytic efficiency of Ag for the HER is rather low. It has been shown, that a potential shift of 200 mV in anodic direction is possible when a Pt catalyst is applied (Calvet et al. 2014; Kaiser et al. 2015). But this improvement would be still too low to split water efficiently. A photovoltaic converter is required, which supplies a higher photovoltage than the a-Si:H/μc-Si:H tandem solar cell. One possible candidate with a higher photovoltage is a a-Si:H/a-Si:H tandem solar cell (Urbain et al. 2014b). In Fig. 5.30b an electrochemical measurement of an a-Si:H/a-Si:H photoelectrode with a Pt catalyst deposited on the silver back contact is shown in addition. The intersection with the counter electrode is at $-4.0\ mA/cm^2$. Assuming a Faradaic efficiency of 100 %, the PEC will generate hydrogen with a STH efficiency of 4.9 %. The crossing point of the *i-V* curves of the photocathode and the $RuO_2$ OER electrode is still not situated at the potential where the photocurrent saturates. Applying a more efficient catalyst system may lead to an

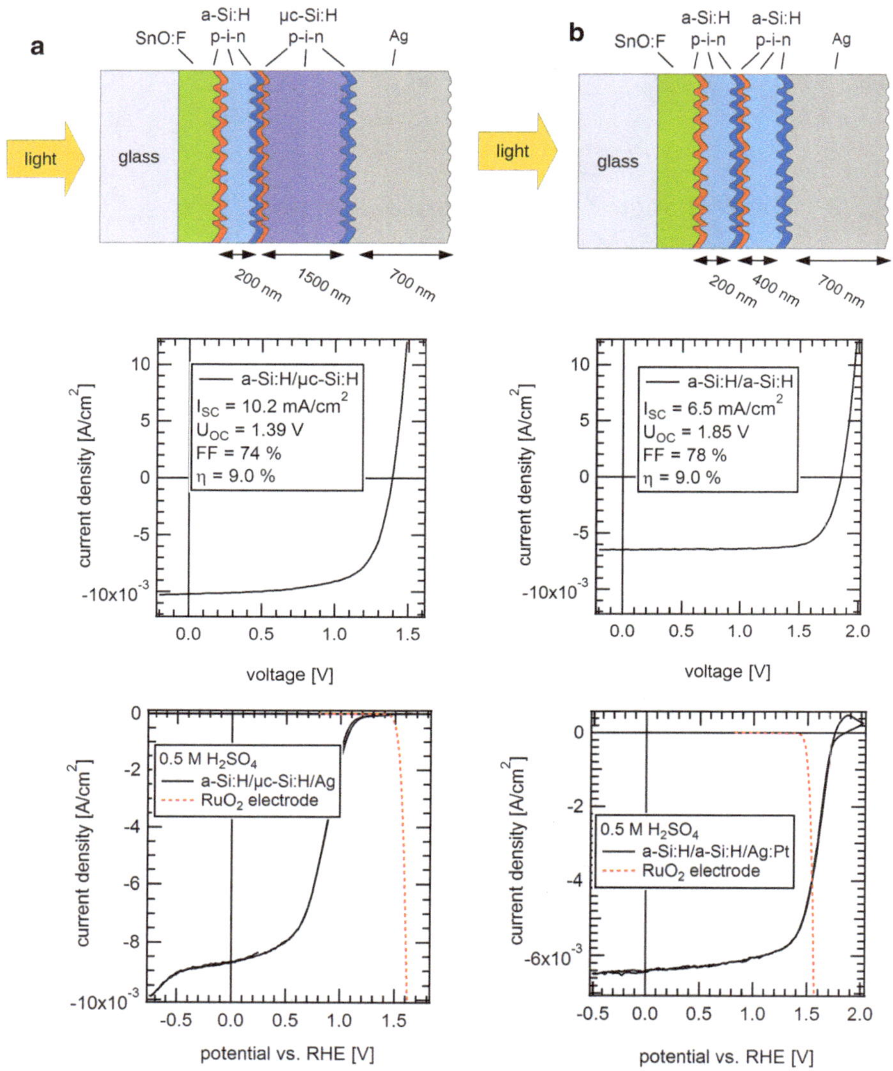


**Fig. 5.30** *Top*: Arrangement of the buried μc-Si/a-Si and a-Si/a-Si tandem cells. *Middle*: PV current voltage curves. *Bottom*: Cyclic voltammogram of the a-Si:H/μc-Si:H/Ag and $RuO_2$ electrode in 0.5 M $H_2SO_4$ (**a**). Cyclic voltammogram of a-Si:H/a-Si:H/Ag with Pt catalyst in 0.5 M $H_2SO_4$ (**b**)

improved efficiency. However, the expected gain in efficiency would be rather low, because Pt and $RuO_2$ are already considered as the best catalysts for water splitting. On the other hand, applying earth abundant catalysts with higher overvoltages would even shift the operation voltage to more positive potentials, thus drastically reducing the STH conversion efficiency. As an alternative an additional slight

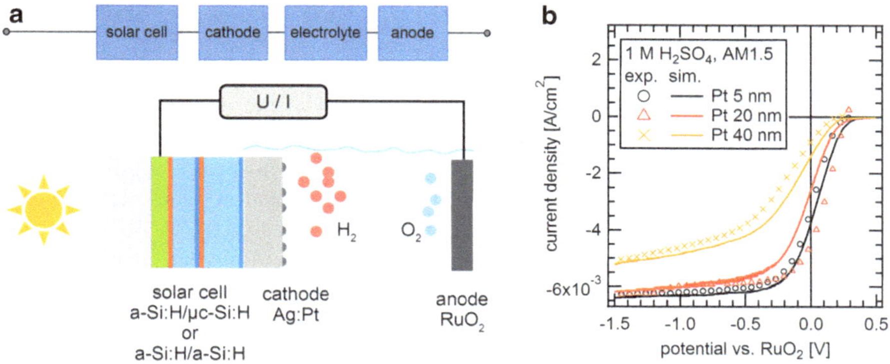


**Fig. 5.31** (**a**) Schematic representation of a photoelectrochemical cell as a series connection of its components. (**b**) Cyclic voltammograms of a-Si:H/a-Si:H/Ag:Pt with various Pt loading against $RuO_2$ in 1 M $H_2SO_4$ under AM1.5 illumination (*symbols*); and calculated curve from CV of Ag:Pt, $RuO_2$ in 3 electrode configuration and photovoltaic characteristic of a-Si:H/a-Si:H/Ag tandem solar cell (*lines*)

increase in the photovoltage would lead to a strong increase in efficiency. Therefore, a further improvement of the photovoltaic converter systems is crucial for optimizing the overall efficiency of these systems. This has to be done always in relation to the applied catalytic materials. For example Urbain et al. discuss this topic for the application of tandem structures prepared by silicon thin film technology (Urbain et al. 2014a, b).

A tandem solar cell made from thin film silicon is in principle a buried junction electrode if it is used in a photoelectrochemical cell. The photovoltage is generated inside both p-i-n diodes, separated from the electrolyte interface as discussed in detail in Sect. 5.4. Therefore, the charge separation is not influenced by the semiconductor–electrolyte interface directly. It is possible to separate the total photoelectrochemical cell in its functional components, like it is depicted in Fig. 5.31a. The photovoltaic curve of the solar cell can be measured directly by taking advantage of the metallic back contact layer. Thereby the complete charge separation characteristic of the photovoltaic converter is accessible, which is not always possible for photoelectrodes using, for example, the semiconductor–electrolyte junction as the photoactive converter. The electrochemical behavior of the catalysts on the photoelectrode can be obtained by preparing model electrodes with the same top layer. In this case we use silver foil as a model system for the silver back contact of the solar cell. With the electrochemical behavior of the $RuO_2$ counter-electrode the PEC characteristics are available in its separated functional parts. To proof this consideration we combine the potential current characteristics of these parts in a series circuit: This should lead again to the measured curve of the complete PEC. This is shown in Fig. 5.31b, where the photoelectrochemical characteristics of tandem solar cells with different amounts of Pt catalyst are plotted against a $RuO_2$ counter electrode. Additionally, the simulated curves created by the series connection of the separated parts are depicted. They fit quite well to the measured ones and proof that a separation of the buried PEC into the components (solar converter and electrolyzer) is indeed possible. For a more detailed description please refer to Ziegler et al. (2014). This approach can help to achieve a better understanding of the photoelectrochemical characteristics of photoelectrodes, which is usually necessary to optimize them in a proper way. Another elegant way to distinguish between charge separation and catalytic properties is the acceleration of the chemical reaction by applying charge scavengers to the electrolyte (Abdi et al. 2013; Dotan et al. 2011). The latter can also be applied for photoelectrodes without a buried junction where the photovoltaic performance is unknown, preferentially to check for the photovoltaic performance of the junction. If the PV conversion parameters do not reach the required values, the PEC cell cannot work efficiently for light induced water splitting either. Without such step-by-step strategies to optimize the PV component, the interfaces, and the cocatalysts independently a targeted research seems difficult or even impossible for achieving efficient photoelectrochemical water splitting devices.

It should be mentioned that the highest STH conversion recently reached uses a-Si/a-Si/μc-Si p-i-n triple cells (Urbain et al. 2015). With Pt/Ag layer stack as HER electrode on the PEC a-Si contact and using $RuO_2$ as the OER electrode an overall STH efficiency of 9.5 % has been reached in 1 M KOH solution. This high value is a result of the targeted design of the PV component and the electrocatalytic component of this buried junction PEC cell. In turning this promising result around, we expect that any efficient system even when started as original semiconductor–electrolyte junction needs to be modified by additional processing steps to form a buried PV junction in series connected to optimized electrolyzer surface components.

### *5.7.3 Si Photoelectrode Interface Analysis*

In order to understand the contact properties in the aforementioned cells and their impact on the performance of the device, model experiments of electrochemical interfaces are a valuable tool. Starting point of these experiments is an n-type silicon wafer (representing the n-layer of the p-i-n junction) that is not pinned, thus giving insight into the energetics of an ideal contact. Since a bare silicon surface always shows dangling bond defect states, the wafer was passivated with a thin, thermally grown $SiO_2$ layer. In order to understand the band lineup, thin films of the different contact materials were deposited via sputtering onto the silicon wafer while monitoring the change in binding energy with XPS.

The first simple example is the direct deposition of the platinum catalyst onto the silicon surface. Given the high work function of platinum of 5.8 eV (von Oertzen et al. 1991) the formation of a depletion layer in our silicon substrate is to be expected. We can visualize contact formation by monitoring the binding energy of the Si 2p line (see Figs. 5.32 and 5.21). As we can exclude chemical reactions, the change in binding energy we see in the Si 2p line, is equivalent to the change in the

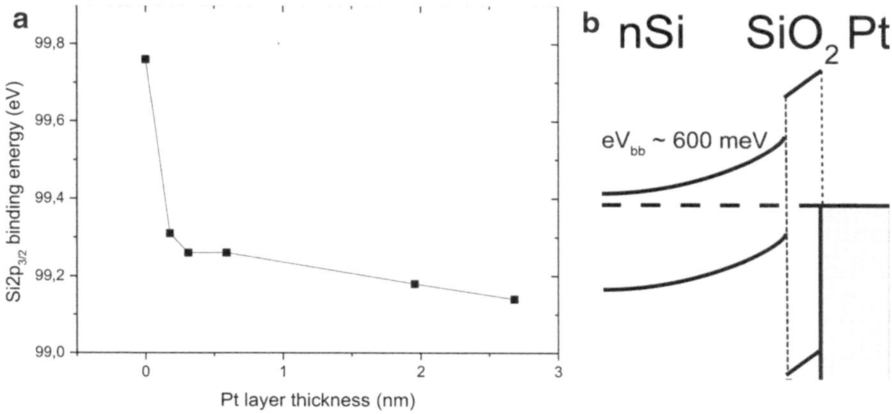


**Fig. 5.32** In (**a**) the change in binding energy of the Si 2p core line upon platinum deposition is shown. In (**b**) the derived band diagram is schematically depicted

valance band position. In Fig. 5.32a the change in binding energy of the Si 2p line is depicted against the thickness of the platinum overlayer. The shift of 600 meV to a smaller binding energy is consistent with the theory of the barrier height for a MOS contact. Knowing this shift and the valence band offset of the pristine surface we can derive the band diagram of this interface, see Fig. 5.32b.

Equivalently one can deduce the energy band diagram for a direct Pt/p-Si or Pt/$SiO_2$/p-Si interface, which would be part of a photocathode for HER. In this case the high work function of Pt leads to a position of $E_F$ close to the valence band edge and nearly no photovoltage can be induced (compare also with Fig. 5.29). As a consequence p-Si/Pt junctions in contact to electrolytes show only very bad HER performance in photoelectrochemical cells. One solution to overcome the problem would be to take advantage of a buried p-Si/$n^+$-Si contact layer. But limitations are also expected for p–$n^+$/Pt junctions. Considering that the HER is occurring at the platinum catalyst, it is necessary to transfer electrons to the platinum. Clearly, a possible formation of the depletion layer as seen in Fig. 5.32b is not beneficial for this process. This can only be avoided if the $n^+$-layer is degenerately doped. Alternatively, an appropriate intermediate layer is required to match the work function difference between the two materials.

One material of possible use is titanium dioxide. $TiO_2$ has a sufficiently high conductivity and is also chemically stable in most electrolytes. A number of experiments have already been performed to prove the beneficial effects of an $n^+$-doped $TiO_2$ interface layer (Seger et al. 2013a; Hu et al. 2014). In order to investigate the energetic band lineup, analogous to Si–Pt, an interface experiment has been performed. Figure 5.33 shows the band diagram derived for a Si–$TiO_2$–Pt contact. In this case only a minor depletion zone is formed within the silicon. Titania is strongly n-type, and therefore, no electron injection barrier is formed between the silicon and the conduction band of titania. Additionally platinum was deposited on the titania layer, thus also probing the contact to the catalyst.

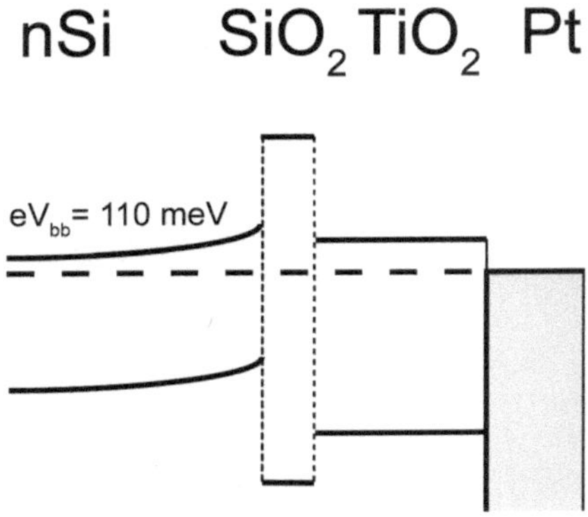


**Fig. 5.33** Band diagram of a n-Si substrate to a $TiO_2$ protective layer, as well as to the platinum catalyst as determined by photoemission experiments

Since there is no change in binding energy in the Ti2p line upon platinum deposition we can deduce that no additional band bending occurs. This is a result of the high charge carrier density in $TiO_2$ (Forro et al. 1994). These results indicate that the electronic structure and surface properties depend strongly on the details of thin film preparation and surface treatment. In past experiments the surface density of states of $TiO_2$ has been analyzed by resonant enhanced photoemission (Reckers et al. 2015; Schwanitz et al. 2007b). Recent studies quote that a high defect density in the bulk lead to specific contact properties with hole conduction valuable for PEC cells (Hu et al. 2014).

### 5.7.4 *Interface Studies on p-i-n Tandem Cell Photoelectrodes*

In addition to the a-Si:H/a-Si:H photoelectrodes with silver we have employed cells without a metallic back contact. In this arrangement the last $n^+$-doped layer is in contact to the electrolyte with additional Pt clusters as cocatalyst. In this arrangement there is no direct influence of the semiconductor–electrolyte interface on the charge separation of the buried junction electrode, but the interface and possible side reaction may still play an important role for the overall performance. Silicon tends to oxidize in contact to aqueous solutions and forms a highly defective oxide layer. On the one hand this leads to Fermi level pinning that in return gives rise to a barrier for electrons traveling from the PV component of the photoelectrode to the electrolyte. On the other hand, the natural oxide layer usually grows to thicknesses above 1 nm, which prevents a tunneling of the charge carriers through the oxide interface. This leads to a high transfer resistance, which reduces the generated photovoltage visible in the photoelectrochemical measurement, see Fig. 5.34a. From the PES measurements one can deduce the chemical composition of the

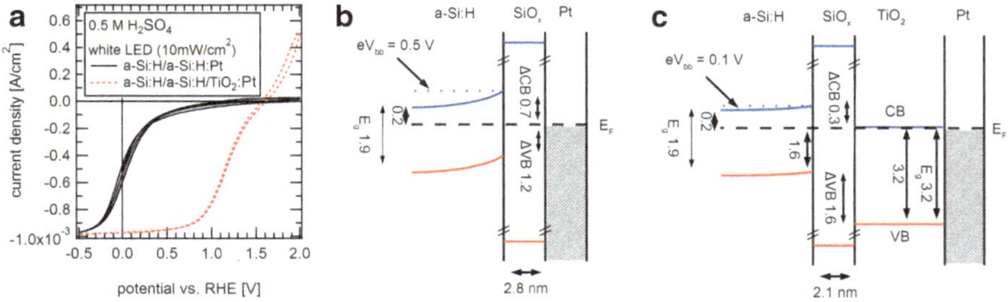


**Fig. 5.34** Cyclic voltammograms of a-Si:H/a-Si:H photoelectrodes with and without a $TiO_2$ interface layer between the solar cell and the Pt catalyst (**a**). Band alignment of a-Si:H to Pt (**b**) and of a-Si:H/$TiO_2$:Pt (**c**) obtained from PES measurements

electrode surface and the band alignment of the n-doped a-Si top layer to the Pt catalyst as schematically summarized in Fig. 5.34b. The measurements confirm the existence of a thick silicon oxide layer and the unfavorable band bending in the n-doped a-Si layer.

To prevent the oxide formation on the a-Si a suitable contact/buffer layer is required. We employ $TiO_2$ because of its chemically stability over a broad pH range. Additionally, it is an intrinsically n-doped semiconducting material and fulfills the requirements for electron transport into the hydrogen redox couple, which has been confirmed by the model experiments on crystalline silicon (see above). In an experiment with a stepwise deposition of $TiO_2$ by magnetron sputtering and coupled PES analysis the contact formation can be monitored. Upon $TiO_2$ deposition the band bending in the a-Si is consecutively reduced to a band bending of 100 mV (Fig. 5.34b). The discontinuity in the conduction band from the a-Si to the $TiO_2$ is negative, hence there is no barrier for electrons moving in this direction. Energetically the band alignment of the buffer layer to the buried junction electrode is favorable. However, in the PES spectra a growing silicon oxide component is visible, which increases during sputter deposition. This can be assigned to a $SiO_x$ interface layer between the a-Si and $TiO_2$ with a layer thickness of 2 nm, which will reduce the charge transfer rate across the interface and will introduce an additional series resistance. The electrochemical deposition of the catalyst forms metallic Pt on the buffer layer. The Fermi level defined by the Pt is situated 3.2 eV above the valence band of the $TiO_2$. Consequently there is no barrier for the electrons at the interface to the catalyst. The corresponding electrochemical measurement in Fig. 5.34a shows the improvement due to the interface engineering by applying the $TiO_2$ buffer layer. Still, a reduction of the $SiO_x$ thickness must be realized in order to improve the photoelectrochemical performance of these electrodes even further. Previously published data on the deposition of $TiO_2$ by atomic layer deposition (ALD) show that with this technique the formation of an interfacial $SiO_2$ layer can mostly be avoided (Hu et al. 2014; Malizia et al. 2014; Seger et al. 2013a, b).

## 5.8 Summary and Conclusion

In this chapter we have tried to sum up a number of criteria to understand how efficient photoelectrochemical water splitting devices may be designed. Most of our argumentation results from our past research orientation on thin film solar cells and related interface and thin film engineering approaches. Despite this fact, we are convinced that some of our arguments must be considered for all device structures and materials choices, if efficient and stable devices are in the focus of research, which may lead to feasible technologic solutions with the next few decades.

At first, all photoelectrochemical $H_2O$ splitting devices combine the functionality of a photovoltaic device in the generation of photocurrent and photovoltage with an electrocatalytic system for the hydrogen and oxygen evolving reaction (Sects. 5.1 and 5.2). Whether they are coupled as an integrated system or in a series connection does not change the physical and chemical boundary conditions. High solar to $H_2$ conversion efficiencies can only be reached, if the PV component also allows high electric power conversion efficiencies. This is evident, if a solar cell is coupled to an electrolyzer; it is also true for buried junction photoelectrochemical cells, as well as for device structures where a semiconductor–electrolyte junction not only defines the photovoltaic efficiency of the cell but is also responsible for catalyzing the kinetically demanding multi electron transfer reaction of the HER and OER. If there will be any bare semiconductor–electrolyte system available once, which in its physicochemical properties are optimized for both functionalities, cannot be finally concluded today. However, if an absorber material and its junctions, formed either in situ or ex situ, do not provide high PV efficiencies (to solid state contacts or for a semiconductor–electrolyte junction to reversible fast redox couples) then also no high STH efficiencies can be expected. Therefore, before any complex optimization research is performed for the water splitting reaction, the achievable power conversion efficiency should be checked first by applying solid state or power converting electrochemical junctions. Or as already stated above, if an absorber material does not deliver high electric power conversion, it is not worth to investigate its water splitting performance.

Secondly, the selection of the absorber materials and the design of the PV component of the overall system is a crucial point for high performance. The maximum photovoltage an absorber material is able to deliver is given by the maximum splitting of the electron–hole pair difference in chemical potential (difference in quasi-Fermi levels) inside the PV converter. Depending on the assumptions of the needed photovoltage at the operational point of the converter considering losses induced by the charge separation and transfer to the contact sites and kinetic losses for the electrochemical reactions, an operational photovoltage larger than approximately 1.6 V is needed, which translates to an open circuit photovoltage of about 1.9 eV (Sect. 5.3). As the open circuit photovoltage is related to the band gap of the absorber by the relation $U_{ph}{}^{oc} = E_G - 0.4$ for the best single crystalline semiconductors and the relation $U_{ph}{}^{oc} \leq (0.5–0.6)\,E_G$ for most thin film absorbers it is evident that it will be hard to find a single material of band gap between 2.2 and 2.5 eV, which will provide the needed photovoltage and is still able to absorb most of the visible light for competitive STH efficiencies. Therefore, it seems safe to conclude that most probably only multi-junction cells, e.g., tandem cells, combining a band gap $E_{G1}$ of about 1.2 and $E_{G2}$ of about 2 eV are needed. In this case high theoretical efficiencies are possible but the current matching conditions must be followed, which means it is not reasonable to combine a highly efficient absorber with a lousy absorber.

Thirdly, proper interface engineering is needed to couple the PV converter functionality to the catalytic functionality (Sects. 5.4 and 5.5). Here, different additional, possibly ultra-thin layers may be needed, as an electronic passivation layer, a contact formation layer, a chemical passivation layer, and as cocatalyst being in contact to the electrolyte. It may be possible that in some cases one layer serves more than one duty. The electronic passivation layer is needed to passivate the detrimental surface states of the semiconductor. In very many cases the contact layer is needed to define the junction properties of the PV device. A chemical passivation layer may be needed to prevent corrosion reactions. The catalyst shall provide efficient conditions for HER and OER. For efficient devices the electrons (holes) must be transferred at the conduction (valence) band edge without loss of their electrochemical potentials nor by any photocurrent loss due to recombination. If these challenging interface-engineering requirements can be fulfilled by inherently stable oxide semiconductors, as for example $Fe_2O_3$ or $CuO_2$ used as absorbers, may be questioned, since it is evident from the given investigations in the literature that detrimental surface states reduce severely the obtainable surface photovoltage.

So far there is not sufficient knowledge about those surface and interface properties, which lead to severe losses in performance, and how these interfaces may be designed to overcome the evident problems. For this reason we have discussed in very detail our surface science approach, mostly based on photoelectron spectroscopy to get detailed insights on surface/interface related properties of semiconductor junctions (Sect. 5.6). However, as the experiments are still not based on real wet chemical junctions, further improvements in the applicability of the techniques are strongly needed to possibly study electrochemical interfaces in future experiments in situ.

As last part of our contribution we have presented in Sect. 5.7 as a case study of our work related to investigations of PEC-systems based on Si. It is evident from these studies that buried multi-junction solar cells specifically designed for the purpose of $H_2O$ splitting and used as photoelectrode in contact to the electrolyte may provide high STH conversion efficiencies. For this reason we are optimistic that highly efficient and stable photoelectrochemical devices for light induced $H_2O$ splitting and solar fuel ($H_2$) formation are within reach. There are, however, still severe research efforts needed, which must be channeled to develop tandem or multi-junction converters (solar cells) preferentially based on thin film technology and further engineered in a proper way to be used as buried photoelectrochemical cells. Besides the efforts related to wide band gap III–V semiconductors, we expect that wide band gap perovskites (Snaith 2013; Gao et al. 2014; Bi et al. 2015; Sun et al. 2015; Hanusch et al. 2014), but also newly developed absorber materials as for example (Zn Cd)(Te Se) (Carmody et al. 2010; Garland et al. 2011; Grigoryev et al. 2014) must be considered for future work.

**Acknowledgements** The authors would like to acknowledge funding by the BMBF and the DFG within the priority program SPP 1613 "Solar Fuels," which was essential for our research activities in this topic. We also would like to thank our cooperation partners from TU Darmstadt, FZ Jülich, Evonik, and HZ Berlin, who were involved in part of the work, which was cited in the references. We also like to acknowledge the contribution of our coworkers A. Klein, T. Mayer, and A. Thissen, who were part of the team who has developed our experimental infrastructure and measurement expertise.

## References

Abdi FF, Han L, Smets AHM, Zeman M, Dam B, van de Krol R (2013) Efficient solar water splitting by enhanced charge separation in a bismuth vanadate-silicon tandem photoelectrode. Nat Commun 4:2195. doi:10.1038/ncomms3195

Abrahams JL, Casagrande LG, Rosenblum MD, Rosenbluth ML, Santangelo PG, Tufts BJ, Lewis NS (1987) Opportunities in semiconductor photoelectrochemistry. New J Chem 11:157–165

Alonso M, Cimino R, Horn K (1990) Surface photovoltage effects in photoemission from metal-GaP(110) interfaces: importance for band bending evaluation. Phys Rev Lett 64:1947

Arrigo R, Hävecker M, Schuster ME, Ranjan C, Stotz E, Knop-Gericke A, Schlögl R (2013) In situ study of the gas-phase electrolysis of water on platinum by NAP-XPS. Angew Chem Int Ed 52 (44):11660–11664. doi:10.1002/anie.201304765

Aruchamy A, Wrighton MS (1980) A comparison of the interface energetics for n-type cadmium sulfide/- and cadmium telluride/nonaqueous electrolyte junctions. J Phys Chem 84 (22):2848–2854. doi:10.1021/j100459a004

Axnanda S, Crumlin EJ, Mao B, Rani, S, Chang R, Karlsson PG, Edwards MOM, Lundqvist M, Moberg R, Ross P, Hussain Z, Liu Z (2015) Using "Tender" X-ray ambient pressure X-Ray photoelectron spectroscopy as a direct probe of solid-liquid interface. Sci Rep 5. doi:10.1038/srep09788

Bard AJ, Bocarsly AB, Fan FRF, Walton EG, Wrighton MS (1980) The concept of Fermi level pinning at semiconductor/liquid junctions. Consequences for energy conversion efficiency and selection of useful solution redox couples in solar devices. J Am Chem Soc 102 (11):3671–3677. doi:10.1021/ja00531a001

Bardeen J (1947) Surface states and rectification at a metal semiconductor contact. Phys Rev 71 (10):717–727

Beerbom M, Henrion O, Klein A, Mayer T, Jaegermann W (2000a) XPS analysis of wet chemical etching of GaAs(110) by Br2–H2O: comparison of emersion and model experiments. Electrochim Acta 45(28):4663–4672, doi:http://dx.doi.org/10.1016/S0013-4686(00)00618-6

Beerbom M, Mayer T, Jaegermann W (2000b) Synchrotron-induced photoemission of emersed GaAs electrodes after electrochemical etching in $Br_2/H_2O$ solutions. J Phys Chem B 104 (35):8503–8506. doi:10.1021/jp0011342

Beerbom M, Mayer T, Jaegermann W (2000c) Synchrotron-induced photoemission of emersed GaAs electrodes after electrochemical etching in Br-2/$H_2O$ solutions. J Phys Chem B 104 (35):8503–8506

Bi C, Yuan Y, Fang Y, Huang J (2015) Low-temperature fabrication of efficient wide-bandgap organolead trihalide perovskite solar cells. Adv Energy Mater 5(6). doi:10.1002/aenm.201401616

Birch H (2009) The artificial leaf. Chem World(May):42–45

Bluhm H, Hävecker M, Knop-Gericke A, Kiskinova M, Schlögl R, Salmeron M (2007) In-situ X-ray photoelectron spectroscopy studies of gas-solid interfaces at near-ambient conditions. Mater Res Soc Bull 34(12):1022–1030

Bocarsly AB, Bookbinder DC, Dominey RN, Lewis NS, Wrighton MS (1980) Photoreduction at illuminated p-type semiconducting silicon photoelectrodes. Evidence for Fermi level pinning. J Am Chem Soc 102(11):3683–3688. doi:10.1021/ja00531a003

Bolton JR, Strickler SJ, Connolly JS (1985) Limiting and realizable efficiencies of solar photolysis of water. Nature 316(6028):495–500

Braun A, Augustynski J, Chandler EA, Mao SS, Miller EL, Turner JA, Ye J (2010) Photocatalysis for energy and environmental sustainability. J Mater Res 25(1–2)

Briggs D, Seah MP (1983) Practical surface analysis by Auger and X-Ray photoelectron spectroscopy. Wiley, New York

Brillson LJ (1982) The structure and properties of metal-semiconductor interfaces. Surf Sci Rep 2 (2):123–326, doi:http://dx.doi.org/10.1016/0167-5729(82)90001-2

Calvet W, Murugasen E, Klett J, Kaiser B, Jaegermann W, Finger F, Hoch S, Blug M, Busse J (2014) Silicon based tandem cells: novel photocathodes for hydrogen production. Phys Chem Chem Phys 16:12043–12050. doi:10.1039/c3cp55198a

Cardona M, Ley L (eds) (1978) Photoemission in solids, vol I. Springer, Berlin

Cardona M, Ley L (eds) (1979) Photoemission in solids, vol II. Springer, Berlin

Carmody M, Mallick S, Margetis J, Kodama R, Biegala T, Xu D, Bechmann P, Garland JW, Sivananthan S (2010) Single-crystal II-VI on Si single-junction and tandem solar cells. Appl Phys Lett 96(15):153502, doi:http://dx.doi.org/10.1063/1.3386529

Casalongue HS, Kaya S, Viswanathan V, Miller DJ, Friebel D, Hansen HA, Nørskov JK, Nilsson A, Ogasawara H (2013) Direct observation of the oxygenated species during oxygen reduction on a platinum fuel cell cathode. Nat Commun 4:2817. doi:10.1038/ncomms3817

Chambers SA (1992) Elastic scattering and interference of backscattered primary. Auger and X-ray photoelectrons at high kinetic energy: principles and applications. Surf Sci Rep 16:261–331

Chiarotti G (ed) (1993) Physics of solid surfaces. Springer, Heidelberg

Cowley AM, Sze SM (1965) Surface states and barrier height of metal-semiconductor systems. J Appl Phys 36:3212

Doniach S, Sunjic M (1970) Many-electron singularity in X-Ray photoemission and X-Ray line spectra from metals. J Phys C3:285–291

Dotan H, Sivula K, Gratzel M, Rothschild A, Warren SC (2011) Probing the photoelectrochemical properties of hematite (alpha-Fe2O3) electrodes using hydrogen peroxide as a hole scavenger. Energy Environ Sci 4(3):958–964. doi:10.1039/c0ee00570c

Eberhardt W (ed) (1995) Applications of synchrotron radiation. Springer, Berlin

Egelhoff WF Jr (1987) Core-level binding-energy shifts at surfaces and in solids. Surf Sci Rep 6:253–415

Einstein A (1905a) Über einen die Erzeugung und Verwandlung des Lichtes betreffenden heuristischen Gesichtspunkt. Ann Phys 322(6):132–148. doi:10.1002/andp.19053220607

Einstein A (1905b) Über einen die Erzeugung und Verwandlung des Lichtes betreffenden heuristischen Gesichtspunkt. Ann Phys 17:132

Ensling D, Hunger R, Kraft D, Mayer T, Jaegermann W, Rodriguez-Girones M, Ichizli V, Hartnagel HL (2003) Pulse plating of Pt on n-GaAs (100) wafer surfaces: synchrotron induced photoelectron spectroscopy and XPS of wet fabrication processes. Nucl Inst Methods Phys Res B 200:432–438

Ertl G, Küppers J (1985) Low energy electrons and surface chemistry. Verlag Chemie, Weinheim

Fadley CS (1993) Diffraction and holography with photoelectrons and Auger electrons: some new directions. Surf Sci Rep 19:231–264

Forro L, Chauvet O, Emin D, Zuppiroli L, Berger H, Lévy F (1994) High mobility n-type charge carriers in large single crystals of anatase (TiO2). J Appl Phys 75(1):633–635, doi:http://dx.doi.org/10.1063/1.355801

Franciosi A, Van de Walle CG (1996) Heterojunction band offset engineering. Surf Sci Rep 25:1–140

Frankl DR (1967) Electrical properties of semiconductor surfaces. Elsevier, Oxford

Fritsche R, Jaeckel B, Schulmeyer T, Klein A, Jaegermann W (2004) Modification of the Si(111)/ Au interface by a GaSe half-sheet termination layer. Appl Surf Sci 234(1–4):321–327

Fritsche J, Klein A, Jaegermann W (2005) Thin film solar cells: materials science at interfaces. Adv Eng Mater 7(10):914–920

Gao P, Gratzel M, Nazeeruddin MK (2014) Organohalide lead perovskites for photovoltaic applications. Energy Environ Sci 7(8):2448–2463. doi:10.1039/C4EE00942H

Garland JW, Biegala T, Carmody M, Gilmore C, Sivananthan S (2011) Next-generation multijunction solar cells: the promise of II-VI materials. J Appl Phys 109(10):102423, doi: http://dx.doi.org/10.1063/1.3582902

Gärtner WW (1959) Depletion-layer photoeffects in semiconductors. Phys Rev 116(1):84–87

Gassenbauer Y, Schafranek R, Klein A, Zafeiratos S, Hävecker M, Knop-Gericke A, Schlögl R (2006) Surface states, surface potentials and segregation at surfaces of tin-doped $In_2O_3$. Phys Rev B 73:245312

Gerischer H (1969) Charge transfer processes at semiconductor-electrolyte interfaces in connection with problems of catalysis. Surf Sci 18(1):97–122, doi:http://dx.doi.org/10.1016/0039-6028(69)90269-6

Gerischer H (1970) Semiconductor electrochemistry. In: Physical chemistry: an advanced treatise, vol 9. Academic, New York, pp 463–542

Göpel W, Rocker G, Feierabend R (1983) Intrinsic defects of TiO2(110): interaction with chemisorbed $O_2$, $H_2$, CO, and $CO_2$. Phys Rev B 28(6):3427–3438

Green MA (1986) Solar cells: operating principles, technology and system applications. University of New South Wales, Sydney

Green MA, Emery K, Hishikawa Y, Warta W, Dunlop ED (2012) Solar cell efficiency tables (version 39). Prog Photovolt Res Appl 20(1):12–20. doi:10.1002/pip.2163

Green MA, Emery K, Hishikawa Y, Warta W, Dunlop ED (2015) Solar cell efficiency tables (Version 45). Prog Photovolt Res Appl 23(1):1–9. doi:10.1002/pip.2573

Grigoryev DV, Lozovoy KA, Pishchagin AA (2014) Analysis of efficiency of solar energy conversion by tandem Cd x Zn 1-x Te/Si solar cell. J Phys Conf Ser 541(1):012048

Grimes CA, Varghese OK, Ranjan S (2008) Light, water, hydrogen. Springer, New York

Hansson GV, Uhrberg RIG (1988) Photoelectron spectroscopy of surface states on semiconductor surfaces. Surf Sci Rep 9(5–6):197–292, doi:http://dx.doi.org/10.1016/0167-5729(88)90003-9

Hanusch FC, Wiesenmayer E, Mankel E, Binek A, Angloher P, Fraunhofer C, Giesbrecht N, Feckl JM, Jaegermann W, Johrendt D, Bein T, Docampo P (2014) Efficient planar heterojunction perovskite solar cells based on formamidinium lead bromide. J Phys Chem Lett 5 (16):2791–2795. doi:10.1021/jz501237m

Harrison WA (1989) Electronic structure and the properties of solids. Dover Publications, New York

Hecht MH (1990) Role of photocurrent in low-temperature photoemission studies of Schottky barrier formation. Phys Rev B 41(11):7918–7921

Heine V (1965) Theory of surface states. Phys Rev 138(6):A1689–A1696

Henrion O, Jaegermann W (1997) Surface redox reactions of cobaltocene adsorbed onto pyrolytic graphite (HOPG). Surf Sci 387(1–3):L1073–L1078, doi:http://dx.doi.org/10.1016/S0039-6028 (97)00514-1

Henrion O, Löher T, Klein A, Pettenkofer C, Jaegermann W (1996) Low temperature adsorption of water on cleaved GaAs(110) surfaces. Surf Sci Lett 366:L685–L688, doi:http://dx.doi.org/10. 1016/0039-6028(96)00886-2

Henrion O, Klein A, Jaegermann W (2000) Water adsorption on UHV cleaved InP(110) surfaces. Surf Sci 457(1–2):L337–L341, doi:http://dx.doi.org/10.1016/S0039-6028(00)00417-9

Hertz H (1887) Ueber einen Einfluss des ultravioletten Lichtes auf die electrische Entladung. Ann Phys 267(8):983–1000

Himpsel FJ (1983) Angle-resolved measurements of the photoemission of electrons in the study of solids. Adv Phys 32:1–51

Horn K, Alonso M, Cimino R (1992) Non-equilibrium effects in photoemission from metal-semiconductor interfaces. Appl Surf Sci 56–58:271–289

Hu S, Xiang C, Haussener S, Berger AD, Lewis NS (2013) An analysis of the optimal band gaps of light absorbers in integrated tandem photoelectrochemical water-splitting systems. Energy Environ Sci 6(10):2984–2993. doi:10.1039/C3EE40453F

Hu S, Shaner MR, Beardslee JA, Lichterman M, Brunschwig BS, Lewis NS (2014) Amorphous $TiO_2$ coatings stabilize Si, GaAs, and GaP photoanodes for efficient water oxidation. Science 344(6187):1005–1009. doi:10.1126/science.1251428

Hüfner S (1995) Photoelectron spectroscopy, Springer series in solid-state sciences. Springer, Berlin

Hunger R, Schulmeyer T, Klein A, Jaegermann W, Lebedev M, Sakurai K, Niki S (2005) SXPS investigation of the Cd partial electrolyte treatment of CuInSe2 absorbers. Thin Solid Films 480–481:218–223

Jacobi K, Myler U, Althainz P (1990) Determination of band bending at the Si(113) surface from photovoltage-induced core-level shifts. Phys Rev B 41(15):10721–10726

Jaegermann W (1986) Adsorption of $Br_2$ on n-$MoSe_2$: modelling photoelectrochemistry in UHV. Chem Phys Lett 126(3/4):301–305

Jaegermann W (1996) The semiconductor/electrolyte interface: a surface science approach. In: White RE (ed) Modern aspects of electrochemistry, vol 30. Plenum Press, New York, pp 1–185

Jaegermann W, Mayer T (1995) What do we learn from model experiments of semiconductor/electrolyte interfaces in UHV - coadsorption of Br-2 with Na and $H_2O$ on $WSe_2$(0001). Surf Sci 335(1-3):343–352

Jaegermann W, Mayer T (2004) Surface science studies of elementary processes in photoelectrochemistry: adsorption of electrolyte components on layered transition metal dichalogenides. Sol Energy Mater Sol Cells 83(4):371–394

Jaegermann W, Mayer T (2005) Water adsorption on semiconductor surfaces. In: Bonzel H (ed) Landolt Börnstein, vol III/42. Springer, Berlin, pp 226–298

Jaegermann W, Tributsch H (1988) Interfacial properties of semiconducting transition metal chalcogenides. Prog Surf Sci 29(1/2):1–167

Jaegermann W, Klein A, Mayer T (2009) Interface engineering of inorganic thin-film solar cells – materials-science challenges for advanced physical concepts. Adv Mater 21(42):4196–4206. doi:10.1002/adma.200802457

Kaiser B, Calvet W, Murugasen E, Ziegler J, Jaegermann W, Pust SE, Finger F, Hoch S, Blug M, Busse J (2015) Light induced hydrogen generation with silicon-based thin film tandem solar cells used as photocathode. Int J Hydrog Energy 40(2):899–904, doi:http://dx.doi.org/10.1016/j.ijhydene.2014.11.012

Keppner H, Meier J, Torres P, Fischer D, Shah A (1999) Microcrystalline silicon and micromorph tandem solar cells. Appl Phys A 69(2):169–177. doi:10.1007/s003390050987

Khaselev O, Turner JA (1998) A monolithic photovoltaic-photoelectrochemical device for hydrogen production via water splitting. Science 280(5362):425–427. doi:10.1126/science.280.5362.425

Klein A (2013) Transparent conducting oxides: electronic structure – property relationship from photoelectron spectroscopy with in-situ sample preparation. J Am Ceram Soc 96:331

Klein A, Säuberlich F (2008) Surfaces and interfaces of sputter-deposited ZnO films. In: Ellmer K, Klein A, Rech B (eds) Transparent conductive zinc oxide: basics and applications in thin film solar cells. Springer, Berlin, pp 125–185

Klein A, Tomm Y, Schlaf R, Pettenkofer C, Jaegermann W, Lux-Steiner MC, Bucher E (1998) Photovoltaic properties of $WSe_2$ single crystals studied by photoelectron spectroscopy. Sol Energy Mater Sol Cells 51:181–191

Klein A, Dieker H, Späth B, Fons P, Kolobov A, Steimer C, Wuttig M (2008a) Changes in electronic structure and chemical bonding upon crystallization of the phase change material GeSb2Te4. Phys Rev Lett 100:016402

Klein A, Mayer T, Thissen A, Jaegermann W (2008b) Photoelectron spectroscopy in materials science and physical chemistry: analysis of composition, chemical bonding and electronic structure of surfaces and interfaces. Bunsenmagazin 10(4):124–139

Klett J, Krähling S, Elger B, Schäfer R, Kaiser B, Jaegermann W (2014) The electronic interaction of Pt-clusters with ITO and HOPG surfaces upon water adsorption. Z Phys Chem 228 (4–5):503–520. doi:10.1515/zpch-2013-0499

Kowalczyk SP, McFeely FR, Ley L, Gritsyna VT, Shirley DA (1977) The electronic structure of $SrTiO_3$ and some simple related oxides (MgO, $Al_2O_3$, SrO, $TiO_2$). Solid State Commun 23:161

Kraut EA, Grant RW, Waldrop JR, Kowalczyk SP (1980) Precise determination of the valence-band edge in X-Ray photoemission spectra: application to measurement of semiconductor interface potentials. Phys Rev Lett 44:1620

Kraut EA, Grant RW, Waldrop JR, Kowalczyk SP (1983) Semiconductor core-level to valence-band maximum binding-energy differences: precise determination by x-ray photoelectron spectroscopy. Phys Rev B 28(4):1965–1977

Kühne HM, Tributsch H (1986) Energetics and dynamics of the interface of $RuS_2$ and implications for photoelectrolysis of water. J Electroanal Chem 201(2):263–282, doi:http://dx.doi.org/10.1016/0022-0728(86)80054-7

Kunat M, Girol SG, Becker T, Burghaus U, Wöll C (2002) Stability of the polar surfaces of ZnO: a reinvestigation using He-atom scattering. Phys Rev B 66:081402

Kurtin S, McGill TC, Mead CA (1969) Fundamental transition in the electronic nature of solids. Phys Rev Lett 22:1433–1436

Lang ND, Kohn W (1970) Theory of metal surfaces: charge density and surface energy. Phys Rev B 1:4555

Lebedev MV, Ensling D, Hunger R, Mayer T, Jaegermann W (2004) Synchrotron photoemission spectroscopy study of ammonium hydroxide etching to prepare well-ordered GaAs(100) surfaces. Appl Surf Sci 229(1-4):226–232. doi:10.1016/j.apsusc.2004.01.067

Lebedev MV, Mankel E, Mayer T, Jaegermann W (2008) Wet etching of GaAs(100) in acidic and basic solutions: a synchrotron−photoemission spectroscopy study. J Phys Chem C 112 (47):18510–18515. doi:10.1021/jp805568t

Lewerenz H-J, Jungblut H (1995) Photovaltaik. Springer, Berlin

Lewerenz H-J, Peter L (eds) (2013) Photoelectrochemical water splitting. Royal Soc. Chem, Cambridge

Licht S, Wang B, Mukerji S, Soga T, Umeno M, Tributsch H (2000) Efficient solar water splitting, exemplified by $RuO_2$-catalyzed AlGaAs/Si photoelectrolysis. J Phys Chem B 104 (38):8920–8924. doi:10.1021/jp002083b

Lichterman MF, Hu S, Richter MH, Crumlin EJ, Axnanda S, Favaro M, Drisdell W, Hussain Z, Mayer T, Brunschwig BS, Lewis NS, Liu Z, Lewerenz H-J (2015) Direct observation of the energetics at a semiconductor/liquid junction by operando X-ray photoelectron spectroscopy. Energy Environ Sci. doi:10.1039/C5EE01014D

Lin F, Boettcher SW (2014) Adaptive semiconductor/electrocatalyst junctions in water-splitting photoanodes. Nat Mater 13(1):81–86. doi:10.1038/nmat3811

Louie SG, Chelikowsky JR, Cohen ML (1977) Ionicity and the theory of Schottky barriers. Phys Rev B 15(4):2154–2162

Lu Y-C, Crumlin EJ, Veith GM, Harding JR, Mutoro E, Baggetto L, Dudney NJ, Liu Z, Shao-Horn Y (2012) In situ ambient pressure X-ray photoelectron spectroscopy studies of lithium-oxygen redox reactions. Sci Rep 2

Luque A, Hegedus S (eds) (2003) Handbook of photovoltaic science and engineering. Wiley, Chichester

Luque A, Hegedus S (eds) (2011) Handbook of photovoltaic science and engineering, 2nd edn. Wiley, Chichester

Lüth H (1993) Surfaces and interfaces of solids. Springer, Berlin

Maeda K, Domen K (2010) Photocatalytic water splitting: recent progress and future challenges. J Phys Chem Lett 1(18):2655–2661. doi:10.1021/jz1007966

Malizia M, Seger B, Chorkendorff I, Vesborg PCK (2014) Formation of a p-n heterojunction on GaP photocathodes for H2 production providing an open-circuit voltage of 710 mV. J Mater Chem A 2(19):6847–6853. doi:10.1039/C4TA00752B

Mallouk TE (2010) The emerging technology of solar fuels. J Phys Chem Lett 1(18):2738–2739. doi:10.1021/jz101161s

Many A, Goldstein Y, Grover NB (1965) Semiconductor surfaces. North-Holland Pub Co, New York

May MM, Lewerenz HJ, Lackner D, Dimroth F, Hannappel T (2015) Efficient direct solar hydogen conbersion by in situ interface transformation of a tandem structure. Nat Commun 6:8286

Mayer T, Jaegermann W (2000) A photoemission study of solute-solvent interaction: coadsorption of Na and $H_2O$ on $WSe_2$ (0001). J Phys Chem B 104(25):5945–5952

Mayer T, Klein A, Lang O, Pettenkofer C, Jaegermann W (1992a) $H_2O$ adsorption on the layered chalcogenide semiconductors $WSe_2$, InSe and GaSe. Surf Sci 269(270):909–914

Mayer T, Lehmann J, Pettenkofer C, Jaegermann W (1992b) Coadsorption of Na and $Br_2$ on $WSe_2$ (0001). Creating a surface redox couple? Chem Phys Lett 198(6):621–627, doi:http://dx.doi.org/10.1016/0009-2614(92)85039-D

Mayer T, Pettenkofer C, Jaegermann W (1996) Synchrotron-induced photoelectron spectroscopy of semiconductor/electrolyte model interfaces: coadsorption of $Br_2$ and $H_2O$ on $WSe_2$(0001). J Phys Chem 100(42):16966–16977. doi:10.1021/jp961116d

Mayer T, Lebedev M, Hunger R, Jaegermann W (2005) Elementary processes at semiconductor/electrolyte interfaces: perspectives and limits of electron spectroscopy. Appl Surf Sci 252 (1):31–42, doi:http://dx.doi.org/10.1016/j.apsusc.2005.01.110

Mayer T, Lebedev MV, Hunger R, Jaegermann W (2006) Synchrotron photoemission analysis of semiconductor/electrolyte interfaces by the frozen-electrolyte approach: interaction of HCl in 2-propanol with GaAs(100). J Phys Chem B 110:2293–2301

McGilp JF (1984) On predicting the chemical reactivity of metal-semiconductor interfaces. J Phys C 17:2249–2254

Meissner D, Memming R (1988) Unpinning of energy bands in PEC cells: a consequence of surface chemistry and surface charge. In: Grassi G, Hall DO (eds) Photocatalytic production of energy-rich compounds. Elsevier, London, pp 138–147

Memming R (2001) Semiconductor electrochemistry. Wiley, Weinheim

Milnes AG, Feucht DL (1972) Heterojunctions and metal-semiconductor junctions. Academic, New York

Mishima T, Taguchi M, Sakata H, Maruyama E (2011) Development status of high-efficiency HIT solar cells. Sol Energy Mater Sol Cells 95(1):18–21, doi:http://dx.doi.org/10.1016/j.solmat.2010.04.030

Moll M, Kley A, Pehlke E, Scheffler M (1996) GaAs equilibrium crystal shape from first principles. Phys Rev B 54(12):8844–8855

Mönch W (1993) Semiconductor surfaces and interfaces. Springer, Heidelberg

Mönch W (2003) Electronic properties of semiconductor interfaces. Springer, Heidelberg

Morgner H (1994) The investigation of liquid surfaces by electron-spectroscopy. J Electron Spectrosc Relat Phenom 68:771–777

Morrison SR (1977) The chemical physics of surface. Plenum Press, New York

Morrison SR (1980) Electrochemistry at semiconductor and oxidized metal electrodes. Plenum Press, New York

Mosbacker HL, Strzhemechny YM, White BD, Smith PE, Look DC, Reynolds DC, Litton CW, Brillson LJ (2005) Role of near-surface states in ohmic-Schottky conversion of Au contacts to ZnO. Appl Phys Lett 87:012102

Moss TS, Balkanski M (eds) (1994) Handbook on semiconductors. Elsevier, Amsterdam

Moulder JF, Stickle WF, Sobol PE, Bomben KD (1995) Handbook of X-ray photoelectron spectroscopy. Physical Electronics, Inc., Eden Prairie

Mutoro E, Luerßen B, Günther S, Janek J (2007) Photoelectron microscopy. Bunsenmagazin 9:18–33

Myamlin VA, Pleskov YV (1967) Electrochemistry of semiconductors. Plenum, New York

Nakato Y, Tsubomura H (1992) Silicon photoelectrodes modified with ultrafine metal islands. Electrochim Acta 37(5):897–907, doi:http://dx.doi.org/10.1016/0013-4686(92)85041-I

Nakato Y, Ueda K, Yano H, Tsubomura H (1988) Effect of microscopic discontinuity of metal overlayers on the photovoltages in metal-coated semiconductor-liquid junction photoelectrochemical cells for efficient solar energy conversion. J Phys Chem 92(8):2316–2324. doi:10.1021/j100319a043

O'Reagan B, Grätzel M (1991) A low-cost, high-efficiency solar cell based on dye-sensitized colloidal $TiO_2$ films. Nature 353:737

Osterloh FE, Parkinson BA (2011) Recent developments in solar water-splitting photocatalysis. MRS Bull 36(01):17–22. doi:10.1557/mrs.2010.5

Peharz G, Dimroth F, Wittstadt U (2007) Solar hydrogen production by water splitting with a conversion efficiency of 18%. Int J Hydrog Energy 32(15):3248–3252, doi:http://dx.doi.org/10.1016/j.ijhydene.2007.04.036

Peter LM (2011) Towards sustainable photovoltaics: the search for new materials. Phil Trans R Soc A 369:1840–1856. doi:10.1098/rsta.2010.0348

Peter LM, Upul Wijayantha KG (2014) Photoelectrochemical water splitting at semiconductor electrodes: fundamental problems and new perspectives. ChemPhysChem 15(10):1983–1995. doi:10.1002/cphc.201402024

Prévot MS, Sivula K (2013) Photoelectrochemical tandem cells for solar water splitting. J Phys Chem C 117(35):17879–17893. doi:10.1021/jp405291g

Rajeshwar K, McConnell R, Licht S (eds) (2008) Solar hydrogen generation. Springer, New York

Reckers P, Dimamay M, Klett J, Trost S, Zilberberg K, Riedl T, Parkinson BA, Brötz J, Jaegermann W, Mayer T (2015) Deep and shallow $TiO_2$ gap states on cleaved anatase single crystal (101) surfaces, nanocrystalline anatase films, and ALD titania ante and post annealing. J Phys Chem C 119(18):9890–9898. doi:10.1021/acs.jpcc.5b01264

Reichmann J (1980) The current-voltage characteristics of semiconductor-electrolyte junction photo-voltaic cells. Appl Phys Lett 36:574–577

Reinert F, Hüfner S (2005) Photoemission spectroscopy—from early days to recent applications. New J Phys 7:97

Rhoderick EH, Williams RH (1988) Metal-semiconductor contacts, vol 19, 2nd edn, Monographs in electrical and electronic engineering. Clarendon, Oxford

Rhodin TN, Gadzuk JW (1979) In: Rhodin TN, Ertl G (eds) The nature of the surface chemical bond. North-Holland, Amsterdam, pp 113–273

Romm JJ (2005) The hype about hydrogen. Island Press, Washington

Rosenbluth ML, Lewis NS (1986) Kinetic studies of carrier transport and recombination at the n-silicon methanol interface. J Am Chem Soc 108(16):4689–4695. doi:10.1021/ja00276a001

Rosenbluth ML, Lewis NS (1989) "Ideal" behavior of the open circuit voltage of semiconductor/liquid junctions. J Phys Chem 93(9):3735–3740. doi:10.1021/j100346a072

Rossi RC, Tan MX, Lewis NS (2000) Size-dependent electrical behavior of spatially inhomogeneous barrier height regions on silicon. Appl Phys Lett 77(17):2698–2700, doi:http://dx.doi.org/10.1063/1.1319534

Schafranek R, Payan S, Maglione M, Klein A (2008) Barrier heights at (Ba, Sr)$TiO_3$/Pt interfaces studied by photoemission. Phys Rev B 77:195310

Schefold J, Kühne HM (1991) Charge transfer and recombination kinetics at photoelectrodes: a quantitative evaluation of impedance measurements. J Electroanal Chem 300(1–2):211–233, doi:http://dx.doi.org/10.1016/0022-0728(91)85396-7

Schellenberger A, Schlaf R, Pettenkofer C, Jaegermann W (1992) Synchrotron-induced surface-photovoltage saturation at intercalated Na/$WSe_2$ interfaces. Phys Rev B 45(7):3538–3545

Schlaf R, Klein A, Pettenkofer C, Jaegermann W (1993) Laterally inhomogeneous surface potential distribution and photovoltage at clustered In/$WSe_2$ (0001) interfaces. Phys Rev B 48(19):14242–14252

Schlaf R, Murata H, Kafafi ZH (2001) Work function measurements on indium tin oxide films. J Electron Spectrosc Relat Phenom 120:149–154

Schmickler W (2010) Interfacial electrochemistry, 2nd edn. Oxford University Press, London

Schulmeyer T, Hunger R, Klein A, Jaegermann W, Niki S (2004) Photoemission study and band alignment of the CdS/CuIn$Se_2$(001) heterojunction. Appl Phys Lett 84(16):3067–3069

Schwanitz K, Mankel E, Hunger R, Mayer T, Jaegermann W (2007a) Photoelectron spectroscopy at the solid-liquid interface of dye-sensitized solar cells: unique experiments with the solid-liquid interface analysis system SoLiAS at BESSY. Chimia 61(12):796–800

Schwanitz K, Weiler U, Hunger R, Mayer T, Jaegermann W (2007b) Synchrotron-induced photoelectron spectroscopy of the dye-sensitized nanocrystalline $TiO_2$/electrolyte interface: band gap states and their interaction with dye and solvent molecules. J Phys Chem C 111 (2):849–854. doi:10.1021/jp064689r

Seger B, Pedersen T, Laursen AB, Vesborg PCK, Hansen O, Chorkendorff I (2013a) Using $TiO_2$ as a conductive protective layer for photocathodic $H_2$ evolution. J Am Chem Soc 135 (3):1057–1064. doi:10.1021/ja309523t

Seger B, Tilley DS, Pedersen T, Vesborg PCK, Hansen O, Grätzel M, Chorkendorff I (2013b) Silicon protected with atomic layer deposited $TiO_2$: durability studies of photocathodic $H_2$ evolution. RSC Adv 3(48):25902–25907. doi:10.1039/c3ra45966g

Sharma BL (ed) (1984) Metal semiconductor Schottky barrier junctions and their applications. Plenum Press, New York

Shirley DA, (1972) High-Resolution X-Ray Photoemission Spectrum of the Valence Bands of Gold, Phys. Rev. B 5, 4709. Published 15 June 1972

Shockley W, Queisser HJ (1961) Detailed balance limit of efficiency of p-n junction solar cells. J Appl Phys 32(3):510–519, doi:http://dx.doi.org/10.1063/1.1736034

Siegbahn K, Nordling C, Fahlman A, Nordberg R, Hamrin K, Hedman J, Johansson G, Bergmark T, Karlsson S-E, Lindgren I, Lindberg B (1967) ESCA atomic molecular and solid state structure studied by means of electron spectroscopy. Almqvist and Wiksells, Uppsala

Siegbahn H, Svensson S, Lundholm M (1981) A new method for esca studies of liquid-phase samples. J Electron Spectrosc Relat Phenom 24(2):205–213

Sinn C, Meissner D, Memming R (1990) Charge transfer processes at $WSe_2$ electrodes with pH-controlled stability. J Electrochem Soc 137:168–172

Snaith HJ (2013) Perovskites: the emergence of a new era for low-cost. High-efficiency solar cells. J Phys Chem Lett 4(21):3623–3630. doi:10.1021/jz4020162

Spicer WE, Kendelewicz T, Newman N, Chin KK, Lindau I (1986) The mechanisms of Schottky barrier pinning in III–V semiconductors: Criteria developed from microscopic (atomic level) and macroscopic experiments. Surf Sci 168(1–3):240–259, doi:http://dx.doi.org/10.1016/0039-6028(86)90855-1

Spicer WE, Kendelewicz T, Newman N, Cao R, McCants C, Miyano K, Lindau I, Liliental-Weber Z, Weber ER (1988) The advanced unified defect model and its applications. Appl Surf Sci 33–34:1009–1029, http://dx.doi.org/10.1016/0169-4332(88)90411-4

Styring S (2012) Artificial photosynthesis for solar fuels. Faraday Discuss 155:357–376

Sun Y-Y, Agiorgousis ML, Zhang P, Zhang S (2015) Chalcogenide perovskites for photovoltaics. Nano Lett 15(1):581–585. doi:10.1021/nl504046x

Sze SM (1981) Physics of semiconductor devices, 2nd edn. Wiley, New York

Sze SM (1985) Semiconductor devices. Wiley, New York

Tanuma S, Powell CJ, Penn DR (1991) Calculations of electron mean free path. Surf Interface Anal 17:911 & 927

Tersoff J (1984a) Schottky barrier heights and the continuum of gap states. Phys Rev Lett 52 (6):465–468

Tersoff J (1984b) Theory of semiconductor heterojunction: the role of quantum dipoles. Phys Rev B 30(8):4874–4877

Thiel PA, Madey TE (1987) The interaction of water with solid surfaces: fundamental aspects. Surf Sci Rep 7(6–8):211–385, doi:http://dx.doi.org/10.1016/0167-5729(87)90001-X

Tougaard S (1986) Background removal in X-Ray photoelectron spectroscopy: relative importance of intrinsic and extrinsic processes. Phys Rev B 34(10):6779–6783
Urbain F, Smirnov V, Becker J-P, Rau U, Finger F, Ziegler J, Kaiser B, Jaegermann W (2014a) a-Si:H/μc-Si:H tandem junction based photocathodes with high open-circuit voltage for efficient hydrogen production. J Mater Res 29(22):2605–2614. doi:10.1557/jmr.2014.308
Urbain F, Wilken K, Smirnov V, Astakhov O, Lambertz A, Becker J-P, Rau U, Ziegler J, Kaiser B, Jaegermann W, Finger F (2014b) Development of thin film amorphous silicon tandem junction based photocathodes providing high open-circuit voltages for hydrogen production. Int J Photoenergy 2014, 249317. doi:10.1155/2014/249317
Urbain F, Smirnov V, Becker J-P, Lambertz A, Yang F, Ziegler J, Kaiser B, Jaegermann W, Rau U, Finger F (2015) Approaching 10 % efficiency for solar hydrogen generation based on triple and quadruple junction thin film silicon solar cells. Energy Environmental Science, submitted
Valdes A, Brillet J, Gratzel M, Gudmundsdottir H, Hansen HA, Jonsson H, Klupfel P, Kroes G-J, Le Formal F, Man IC, Martins RS, Norskov JK, Rossmeisl J, Sivula K, Vojvodic A, Zach M (2012) Solar hydrogen production with semiconductor metal oxides: new directions in experiment and theory. Phys Chem Chem Phys 14(1):49–70
van de Krol R, Grätzel M (eds) (2012) Photoelectrochemical hydrogen production. Springer, New York
Vitomirov IM, Raisanen A, Finnefrock AC, Viturro RE, Brillson LJ, Kirchner PD, Pettit GD, Woodall JM (1992) Geometric ordering, surface chemistry, band bending, and work function at decapped GaAs(100) surfaces. Phys Rev B 46(20):13293–13302
von Oertzen A, Rotermund HH, Jakubith S, Ertl G (1991) The scanning photoemission microscope: a novel tool in surface science. Ultramicroscopy 36(1–3):107–116, doi:http://dx.doi.org/10.1016/0304-3991(91)90142-S
Wagner CD, Naumkin AV, Kraut-Vass A, Allison JW, Powell CJ, Rumble Jr JR (2003) NIST X-ray photoelectron spectroscopy database. http://srdata.nist.gov/xps/
Walter MG, Warren EL, McKone JR, Boettcher SW, Mi QX, Santori EA, Lewis NS (2010) Solar water splitting cells. Chem Rev 110(11):6446–6473. doi:10.1021/Cr1002326
Warren EL, McKone JR, Atwater HA, Gray HB, Lewis NS (2012) Hydrogen-evolution characteristics of Ni-Mo-coated, radial junction, n + p-silicon microwire array photocathodes. Energy Environ Sci 5(11):9653–9661. doi:10.1039/C2EE23192A
Warren EL, Atwater HA, Lewis NS (2014) Silicon microwire arrays for solar energy-conversion applications. J Phys Chem C 118(2):747–759. doi:10.1021/jp406280x
Weber R, Winter B, Schmidt PM, Widdra W, Hertel IV, Dittmar M, Faubel M (2004) Photoemission from aqueous alkali-metal-iodide salt solutions using EUV synchrotron radiation. J Phys Chem B 108(15):4729–4736
Weiss W, Ranke W (2002) Surface chemistry and catalysis on well-defined epitaxial iron-oxide layers. Prog Surf Sci 70:1–151
Wertheim GK (1989) Electronic structure of metal clusters. Zeitschrift für Physik D 12:319–326
Würfel P (2009) Physics of solar cells: from basic principles to advanced concepts, 2nd edn. Wiley, Weinheim
Yu ET, McCaldin JO, McGill TC (1992) Band offsets in semiconductor heterojunctions. Solid State Phys 46:1–146
Zangwill A (1988) Physics at surfaces. Cambridge University Press, Cambridge
Zhang JZ (2011) Metal oxide nanomaterials for solar hydrogen generation from photoelectrochemical water splitting. MRS Bull 36(1):48–55, doi:http://dx.doi.org/10.1557/mrs.2010.9
Ziegler J, Kaiser B, Jaegermann W, Urbain F, Becker J-P, Smirnov V, Finger F (2014) Photoelectrochemical and photovoltaic characteristics of amorphous-silicon-based tandem cells as photocathodes for water splitting. ChemPhysChem 15(18):4026–4031. doi:10.1002/cphc.201402552

# Chapter 6
# Analysis of Photoelectrochemical Systems by Impedance Spectroscopy

**Juan Bisquert, Sixto Giménez, Luca Bertoluzzi, and Isaac Herraiz-Cardona**

## 6.1 Introduction

The development of efficient semiconductor structures for photoelectrochemical (PEC) fuel production from sunlight relies on the combination of fundamental semiconductor and catalytic properties as well as the coupling of different kinetic steps, to facilitate the forward flux of electronic/ionic processes, from charge generation to the electrochemical reaction at interfaces, that finally must be coupled in a cell structure that combines anodic and cathodic reactions. The aim of frequency-based characterization tools is to extract information about the internal kinetic and charge storage steps in the semiconductor electrodes and/or buried junctions, in order to discover the mechanistic operation of the system and to assess causes for efficiency or failure of it. Frequency methods operate at a given value of steady state of the system and introduce the small perturbation of a variable like voltage or light intensity. The resulting output provides a transfer function or impedance that constitutes a spectroscopy with high resolution of kinetic processes in a wide window of frequencies. Methods such as Impedance Spectroscopy (IS) are part of the standard tool kit in fields like electrochemistry and have been long used in photoelectrochemistry (Tench and Gerischer 1977). Light-modulated techniques are less widespread but also provide important insights (Peter 1990). This chapter will focus on IS applied to the understanding of PEC operation in relation to the efficient operation of semiconductor electrodes.

The foundations of IS applied to dye-sensitized solar cells have been previously described and will be taken as a basis here (Bisquert and Fabregat-Santiago 2010). A very large number of theoretical and experimental results of IS have been reported in the literature, and we do not provide a comprehensive review but focus on a number of main results and methods. We will address recent applications

J. Bisquert (✉) • S. Giménez • L. Bertoluzzi • I. Herraiz-Cardona
Institute of Advanced Materials (INAM), Universitat Jaume I, Castelló 12006, Spain
e-mail: bisquert@uji.es; sjulia@uji.es; bertoluz@uji.es; iherraiz@uji.es



S. Giménez, J. Bisquert (eds.), *Photoelectrochemical Solar Fuel Production*,
DOI 10.1007/978-3-319-29641-8_6

of several important and well-established concepts, such as the crucial role of surface states in electron or hole transfer to acceptors in solution, and the band edge shift by charging of the Helmholtz layer (Freund and Morrison 1968; Kelly and Memming 1982; Allongue and Cachet 1984; Li and Peter 1985; Vanmaekelbergh 1997). The chapter will focus on the recent developments of IS models and measurements for the analysis of the performance of the dominant types of semiconductors for water splitting, especially the set of wide bandgap metal oxides $TiO_2$, $Fe_2O_3$, $BiVO_4$, $WO_3$, and their surface modifications. The chapter is devoted to the main question of the operation of photoelectrodes; therefore, a main theme treated here is which knowledge can be gained from the frequency scan about the steady-state operation in fuel production regime. We will also describe some recent advances in the modeling and understanding of PEC operation (Bertoluzzi et al. 2016). Fundamental properties of the semiconductor photoelectrodes have been discussed in Chap. 1 by L. M. Peter. Further ideas about the application of IS methods in PEC systems will be introduced in Sect. 6.2. Then we will treat a number of models of a semiconductor electrode that can be adapted to different situations. The different types of models that we use in this chapter are summarized in Fig. 6.1. We further discuss the types of structures that will be the main object of analysis in this chapter.

Figure 6.1a shows a simple semiconductor layer where light absorption generates electrons and holes. In this model, the internal operation of the semiconductor layer is rather simple and one can focus on the properties of charge transfer at the surface. The bands are flat, indicating a small thickness that does not allow band bending. Charge transport to the interfaces occurs by diffusion and the electrode requires charge selectivity at the interfaces for PEC operation to occur. At the left side of Fig. 6.1a, the redox reaction of holes transfer is favored over the electron injection by kinetic or energetic reasons. At the right side of Fig. 6.1a, the selective contact must extract only electrons toward the complementary electrode in the cell. This simple model is very useful for the development of basic impedance models that illustrate the competition between recombination and charge extraction (Bertoluzzi and Bisquert 2012).

Figure 6.1b shows a one-dimensional model that contains a depletion region (band bending) at the semiconductor/electrolyte interface. This is the classical model for a PEC semiconductor electrode and its whole operation is illustrated in more details in Fig. 6.2a, through the energy diagram of an n-type semiconductor. This model provides a more realistic approach of the carrier dynamics inside the semiconductor layer, and such traffic of carriers can be coupled to the charge transfer and surface recombination events using numerical simulation, as discussed in Sect. 6.3. In Fig. 6.2a, the left contact equilibrates with redox potential of the relevant redox electrolyte and forms a depletion layer that will assist charge collection as indicated in Fig. 6.2b. The total charge collection length is the addition of the depletion layer $w$ and one diffusion length of the minority carrier, $L_p$. The optical absorption coefficient $\alpha$ determines the light penetration length $L_\alpha = \alpha^{-1}$ that should be shorter than charge collection length for optimal photon utilization.

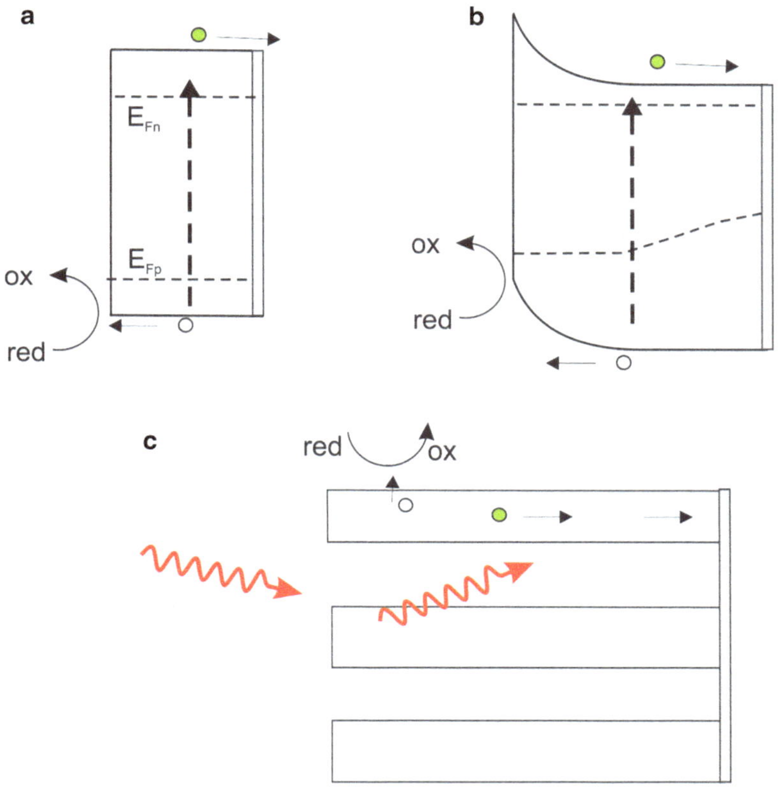


**Fig. 6.1** Scheme of the main semiconductor PEC structures for charge generation toward electrochemical reaction. (**a**) Flat nanolayer without internal field distribution. (**b**) Thick semiconductor film with depletion layer at the contact with electrolyte. (**c**) Semiconductor nanostructure permeated with electrolyte and promoting light scattering

The significance of surface states is first discussed in Sect. 6.5. The charge transfer by competition of the direct transfer from conduction band and indirect transfer via surface states, indicated in Fig. 6.2a, will be discussed in Sect. 6.5.2 Then we discuss the effect of surface modification and their characterization by IS in Sects. 6.7 and 6.8.

Finally, Fig. 6.1c shows a morphology representative of a variety of nanoscale porous films that have been widely explored. One evident advantage is the shorter distance of the photogenerated charges to the semiconductor/solution interface, which decreases the needed diffusion length, provided that the complementary carrier can be collected within the central region of nanowires. This model for a nanostructured electrode will be discussed in Sect. 6.8.

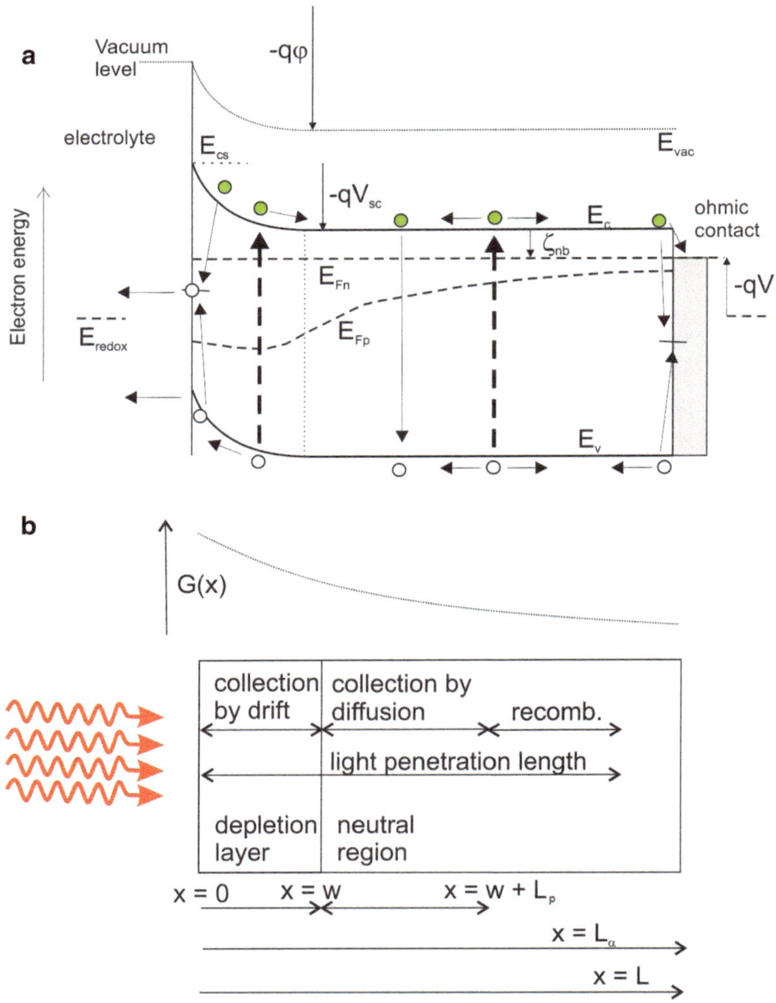


**Fig. 6.2** (**a**) Standard representation of energy diagram of a semiconductor electrode. The photogenerated electrons and holes can be extracted at the contacts or recombined in the bulk or at interface. The majority carrier electrons have a flat Fermi level at moderate illumination level. The left contact realizes the useful photoelectrochemical reaction by photogenerated holes from the valence band. Surface states at the semiconductor surface may also contribute to interfacial charge transfer, but they are also a source of recombination by trapping majority carriers. The right contact is an ohmic contact that extracts majority carriers toward a complementary electrode in the PEC. This contact may also be a locus of surface recombination. (**b**) The utilization of photogenerated carriers at the reaction interface requires extraction in depletion region and along one diffusion length of the minority carrier. Carriers generated further away are lost by recombination. On *top* of the diagram is shown the generation profile

## 6.2 Steady-State Operation of PEC Systems

Let us describe summarily the operation of a PEC semiconductor electrode under variation of voltage and illumination as described in Fig. 6.3. In the dark, we can distinguish the voltage-induced current by either majority or minority carrier injection to electrolyte (possibly by different electrochemical reactions). In an n-type semiconductor electrode, negative current increases at cathodic bias due to injection of majority carriers to the electrolyte. In the dark, the concentration of minority carriers (holes) is rather small; hence, an anodic current only occurs at strong positive bias by hole injection from the substrate, Fig. 6.3a (point 1).

The generation of electron–hole pairs produces a separation of Fermi levels, Fig. 6.3b, and causes an anodic photocurrent as indicated in Fig. 6.3a. Ideally, the photocurrent is added to the dark current that consists of a separate oxidation reaction, as in the dashed line of Fig. 6.3a (point 3). However, strong recombination combined with slow charge transfer may delay the onset of photocurrent to more anodic voltage than the flat band potential, as shown by the dark line in Fig. 6.3a (point 2). This fundamental physical operation of semiconductor electrodes has been well illustrated in the simple model by Peter that compares the initial photocurrent after onset illumination with the stable photocurrent that is decreased by the recombining holes, see Sect. 1.3.2. This point will be quantitatively addressed from the perspective of IS in Sect. 6.3. We observe that increased recombination delays the onset of photocurrent to the point 2 in the diagram. The photovoltage, defined in Fig. 1.16, as the point where the current of an illuminated semiconductor electrode is zero, is decreased. There is also a reduction of fill factor, and third feature is a decrease of the steady-state photocurrent that can be fully recovered at more positive potential, when all recombination is suppressed, see point 3 in Fig. 6.3a.

Let us note the significance of improving the three following characteristics of a semiconductor electrode: photovoltage, photocurrent, and fill factor. In the operation of a practical tandem cell, as that shown by Sivula in Fig. 12.2, the operation point for unassisted fuel production is the intersection of anode and cathode current voltage curves, as depicted in Fig. 6.3c. It is therefore very important to achieve a control of the physical, chemical, and material properties that govern the steady-state current voltage curve of a semiconductor applied to fuel production from visible light, in order to enhance the current at the intersection point, that will ultimately determine the conversion efficiency as noted by Smith in Chap. 4. One example of enhancing the PEC operation of a photoanode is shown in Fig. 6.4 (Zhong and Gamelin 2010). The current voltage curves of hematite ($Fe_2O_3$) undergo a major improvement by deposition of a cobalt–phosphate (Co–Pi) water oxidation catalyst. In Fig. 6.5a, we show the current density–voltage curves of a $Fe_2O_3$ hematite electrode for water oxidation and $[Fe(CN)_6]^{4-}$ oxidation. The latter is more facile as it is a one-hole transfer reaction by direct transfer from the valence band while the oxygen evolution reaction is a four-hole reaction (see mechanistic details in Chap. 2) that involves the trapping of holes at surface states as discussed later (Klahr et al. 2012a, b, c). One major goal of IS is to contribute to the understanding of the specific processes that cause these significant changes in current voltage curve.

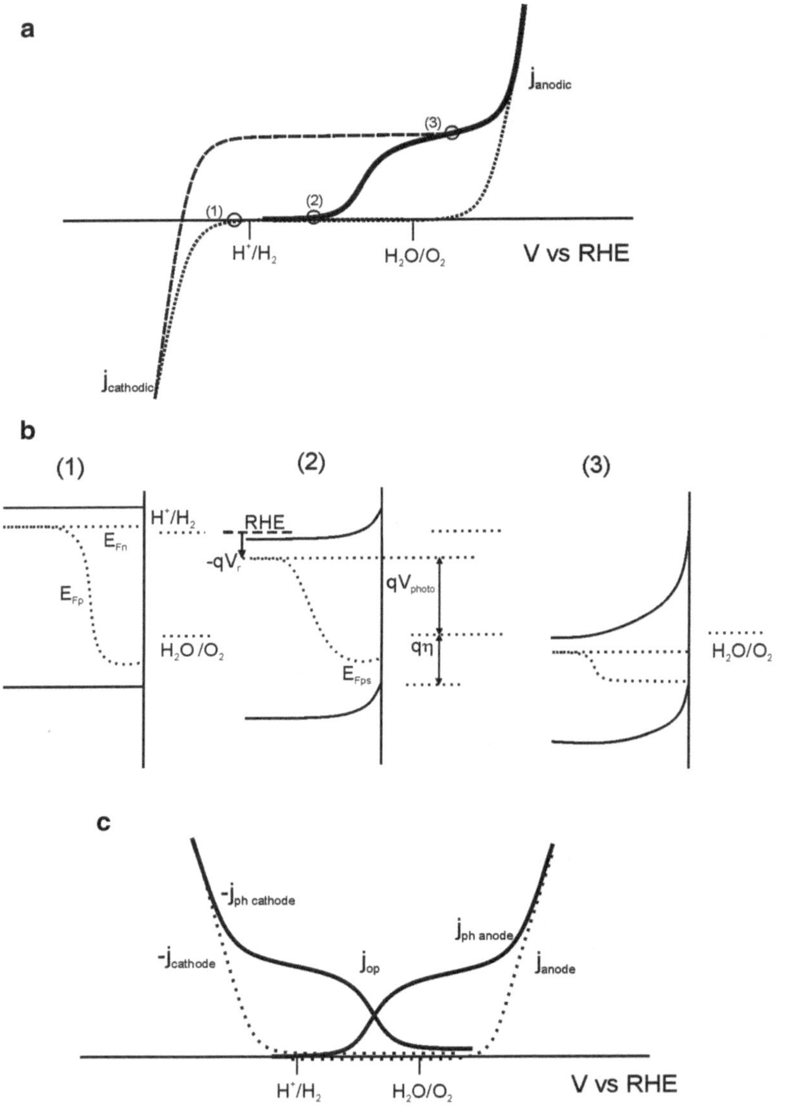


**Fig. 6.3** (**a**) Current density–voltage curves resulting from the PEC operation of a semiconductor photoanode. *Dotted line* is dark current density–voltage curve for the oxygen evolution reaction. *Dashed line* is the idealized curve under illumination. The *solid line curve* is a realistic curve including normal features of semiconductor/electrolyte junction as recombination of electrons and holes. (**b**) Representation of energy diagram of semiconductor/water interface at different voltages vs. RHE. The points *1*, *2*, *3* correspond to the different electrode potentials indicated in (**a**). (**c**) Combination of anode and cathode in a tandem PEC showing the operation point at the intersection between the curves

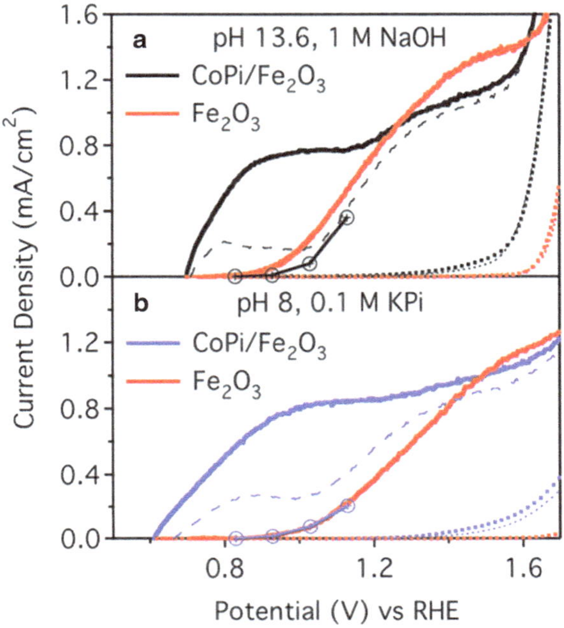


**Fig. 6.4** Dark current (*dotted*) and photocurrent (*solid* and *dashed*) densities of $Fe_2O_3$ photoanodes before and after 30 min of Co–Pi deposition, measured in pH 13.6 NaOH (**a**) and pH 8 KPi (**b**) at 50 (*thick line*) and 10 mV/s (*dashed line*). The α-$Fe_2O_3$ data (*red curves*) were collected at 10 mV/s. The *circles* denote steady-state photocurrent densities after 200 s of continuous illumination under 1 sun, AM 1.5 simulated sunlight. Reproduced with permission from Zhong and Gamelin (2010)

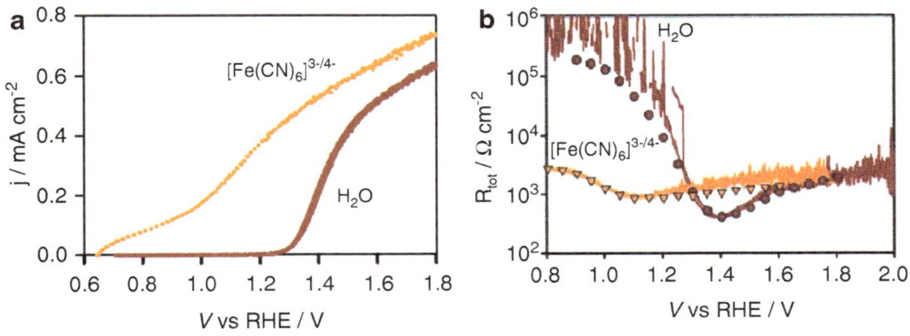


**Fig. 6.5** (**a**) *jV* curve of a hematite electrode in contact with a $H_2O$ and $[Fe(CN)^6]^{3-/4-}$ electrolyte under 1 sun illumination. (**b**) Total (dc) resistance calculated from d*V*/d*j* of the *jV* curves (*lines*) and from IS data (*symbols*) for $H_2O$ and $[Fe(CN)^6]^{3-/4-}$ electrolytes. Reproduced with permission from Klahr et al. (2012a)

## 6.3 Resistances and Capacitances in PEC Systems

Basically, IS provides a decomposition of the electrical response of the sample into resistances and capacitances, and the rationale for this technique is the following. IS consists of the measurement of the ac electrical current $\hat{I}(\omega)$ at a certain angular frequency $\omega$, when ac voltage $\widehat{V}(\omega)$ is applied to the system, or vice versa, measurement of $\widehat{V}(\omega)$ at applied $\hat{I}(\omega)$. The impedance is:

$$Z(\omega) = \frac{\widehat{V}(\omega)}{\hat{I}(\omega)} \tag{6.1}$$

The symbol $\widehat{x}$ over a quantity $x$ indicates that $\widehat{x}$ has the following properties. It is the complex amplitude of a sinusoidal (ac) perturbation of $x$ and it is a small perturbation. The "smallness" of $\widehat{x}$ is required in order to obtain in Eq. (6.1) the *linear* impedance, i.e., $\hat{I}(\omega)$ is linear with respect to $\widehat{V}(\omega)$, or vice versa, so that $Z(\omega)$ is independent of the amplitude of the perturbation. More generally, the relationship between input and output is a *transfer function*. Another example of this method which resorts to transfer functions is Intensity Modulated Photocurrent Spectroscopy (IMPS), which determines the photocurrent response to modulated illumination of a semiconductor electrode, see for instance Li and Peter (1985).

Provided that the PEC operation is stable, the analysis of the frequency domain response can be realized at each point of the steady-state curve as that shown in Fig. 6.3a, that is the truly stationary value of current for each voltage in the required condition of illumination, electrolyte, etc. The important property of IS is that we obtain a spectroscopy at each point by virtue of the frequency variation. Examples shown in Fig. 6.6 correspond to the conditions of Fig. 6.5. It is observed that IS provides different features consisting either of one or two arcs. The interpretation of these features is the key to reveal specific kinetic and charge storage properties in the water splitting process. The information contained in the experimental measurements of impedance spectroscopy and represented in the complex plane impedance plots must be analyzed in terms of a suitable equivalent circuit (EC), in order to extract the capacitances and resistances, which are related to the basic electronic processes of charge accumulation, recombination, and charge transfer.

The resistances are associated to different processes of carrier flux, either inside semiconductor materials or at interfaces. These phenomena can be included in a physicochemical model based on transport and conservation equations, as discussed later, and solved for a small perturbation condition to give the impedance model. In the analytical model, in-phase components provide resistances and out-of-phase components of the current provide capacitances. The capacitances represent a charge storage mechanism and they can be divided into two main kinds, as further discussed in the textbook by Bisquert (2014). Dielectric capacitances are associated to an internal electrical field that is produced by spatial charge separation. Chemical capacitance is due to the variation of chemical potential, or carrier concentration, in one type of electronic state. The identification of the capacitances, as well as the resistances, usually proceeds by adapting the data gathered at different voltages to a single EC. The progression of capacitances and resistances along the voltage variations gives valuable information about the meaning of each element and the overall behavior of the system.

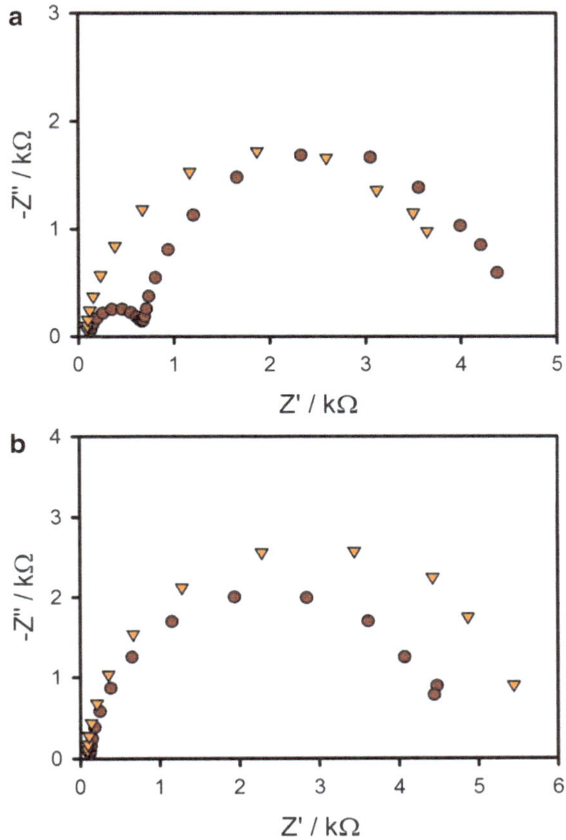


**Fig. 6.6** Complex plane impedance plots for IS data measured under 1 sun illumination for $H_2O$ (*circles*) and $[Fe(CN)_6]^{3-/4-}$ (*triangles*) electrolytes at (**a**) 1.3 V vs. RHE and (**b**) 1.6 V vs. RHE. Reproduced with permission from Klahr et al. (2012a)

In practice, the photoelectrode response is composed of different charge separation and charge injection steps. The analysis of some of these processes by IS has been well described in previous publications for related systems (Bisquert 2002, 2008; Bisquert and Fabregat-Santiago 2010; Fabregat-Santiago et al. 2011). Some of these processes are shown in Fig. 6.2a, and we can observe that we have diffusion along the bulk material, trapping and detrapping at the surface, interfacial charge transfer, etc.. . . Correspondingly, we will obtain transport resistance in the bulk and interfacial resistances associated to trapping in surface and charge transfer. Especially in the case of slow electrocatalytic reaction like the multistep oxygen evolution reaction, the characterization of the interfacial charge transfer is a challenging topic as discussed in Chap. 2 by Doyle. One particularly important aspect of IS of PEC system is an understanding of the *charge transfer resistance* $R_{ct}$ that is associated in a general sense to the interfacial charge transfer steps of the electrochemical reaction. Monitoring the charge transfer resistance can provide information about the charge transfer rate for electrochemical reactions. This charge transfer resistance must be deconvoluted from the total resistance of the system provided that the adequate equivalent circuit conveying the relevant information of the operating physicochemical processes is used. This deconvolution process is possible by virtue of the capacitive response of the electrodes.

In one simple example, let us consider a photoelectrode in contact with an electrolyte solution. If the traffic of carriers from the back contact to the solution consists of two serial processes $R_{tot} = R_1 + R_{ct}$, $R_1$ accounting for charge transport, and $R_{ct}$ for interfacial charge transfer, the capacitance allows separation of these additive resistances. Considering the classical Randles circuit (as the one shown in Fig. 6.13c), the surface capacitance is associated with the Helmholtz layer at the electrode/solution interface, $C_H$, and only affects the parallel resistance $R_{ct}$.

## 6.4 Total Resistance $R_{dc}$ and the Connection to $jV$ Curves

As mentioned, one quantity of central interest in the analysis of PEC systems is the total or dc resistance $R_{dc}$. This resistance can be obtained by removing all capacitors in the EC, and it holds a special relationship to the $jV$ curve. Let us choose a certain point of bias voltage $V_0$ with the associated current density $j_0$. At this point, a small displacement of voltage $\widehat{V}(\omega = 0)$ implies a change of current $\hat{j}\,(\omega = 0)$. The value $\omega = 0$ in parenthesis indicates that the displacement is arbitrarily slow, i.e., $\widehat{V}(0)$ and $\hat{j}(0)$ attain a value that is independent of time, i.e., the steady-state value. Since the resistance in IS is related to change of electrical current with respect to change of voltage, the resistance is associated to a derivative of a carrier flux with respect to a generalized voltage. For an electrode with macroscopic area $A$, the quotient of the small quantities gives

$$Z(0) = \frac{\widehat{V}(0)}{A\hat{j}\,(0)} \tag{6.2}$$

Therefore, we obtain a resistance $Z(0) = R_{dc}$ given by

$$R_{dc} = \left(A\frac{\partial j}{\partial V}\right)^{-1} \tag{6.3}$$

The dc resistance of the photoelectrode $R_{dc}$ (per unit area) is the reciprocal of the slope of the $jV$ curve. Figure 6.5b compares the dc resistance measured by IS with the effective resistance obtained as a reciprocal derivative indicated in Eq. (6.3). It is clearly observed that both approaches are in good agreement, which establishes the connection between the two types of measurements.

Let us further discuss the meaning of $R_{dc}$ features observed in water splitting PEC systems. We adopt the simple model of Reichman (1980) to describe the effect of internal recombination on $jV$ curves, which was qualitatively discussed in Fig. 6.3a. This model uses the standard assumption that the voltage $V$ applied to the semiconductor electrode modifies the size of space charge region. The model introduces a kinetic limitation to the extraction of minority carriers to the fuel production reaction. The hole extraction current density is given by the expression

$$j = qS_p(p_s - p_{s0}) \tag{6.4}$$

Here, $q$ is the elementary charge, $p_s$ is the concentration of holes at the surface, and parameter $S_p$ is similar to a hole surface recombination velocity. The current is found by a combination of unity collection efficiency in the SCR, and by the solution of diffusion–recombination equation in the neutral region. Neglecting recombination in SCR, the current as function of voltage has the form

$$j = \frac{j_{ph} - j_0 e^{-qV/k_B T}}{1 + \frac{j_0}{j_{s0}} e^{-qV/k_B T}} \tag{6.5}$$

Here, $j_{ph}$ is the photocurrent given by Gärtner's model, $j_0$ is the reverse saturation current, and $j_{s0} = -qS_p p_{s0}$ is the charge transfer current in equilibrium. The resistance is given by the form

$$R_{ct}^{-1} = \frac{j_0 q}{k_B T}\left(1 + \frac{j_{ph}}{j_{s0}}\right)\frac{e^{-qV/k_B T}}{\left(1 + \frac{j_0}{j_{s0}} e^{-qV/k_B T}\right)^2} \tag{6.6}$$

In Fig. 6.7a, we show the current density–voltage curve for different values of the extraction parameter $S_p$. First we note that if the extraction velocity is infinite, then the $jV$ curve is similar to a diode curve

$$j = j_{ph} - j_0 e^{-qV/k_B T} \tag{6.7}$$

This behavior corresponds to the dashed curve in Fig. 6.3a and the plain line in Fig. 6.7a. In this idealized model, the total resistance is an exponential function with no minimum feature, as shown in Fig. 6.7b. When the transfer of holes has some limitation, recombination in the neutral region is favored near the flat band potential, and the photovoltage is displaced to lower values. In this case, a minimum of the resistance is observed. Consequently, the presence of a minimum in the total resistance indicates sluggish carrier extraction.

We remark that in the IS analysis of the model of Reichman (1980), the limitation to charge transfer, observed in the $jV$ curve, is a combination of two factors: sluggish charge transfer and internal recombination. However, the model in Eq. (6.5) is a simplification with respect to recombination in the space charge region and other features. In order to establish the properties of the charge transfer resistance in a more general setting, we have developed extensive simulations of the semiconductor electrode model system indicated in Fig. 6.2, as described in the paper (Shi et al. 2016). The system is analyzed considering photogenerated

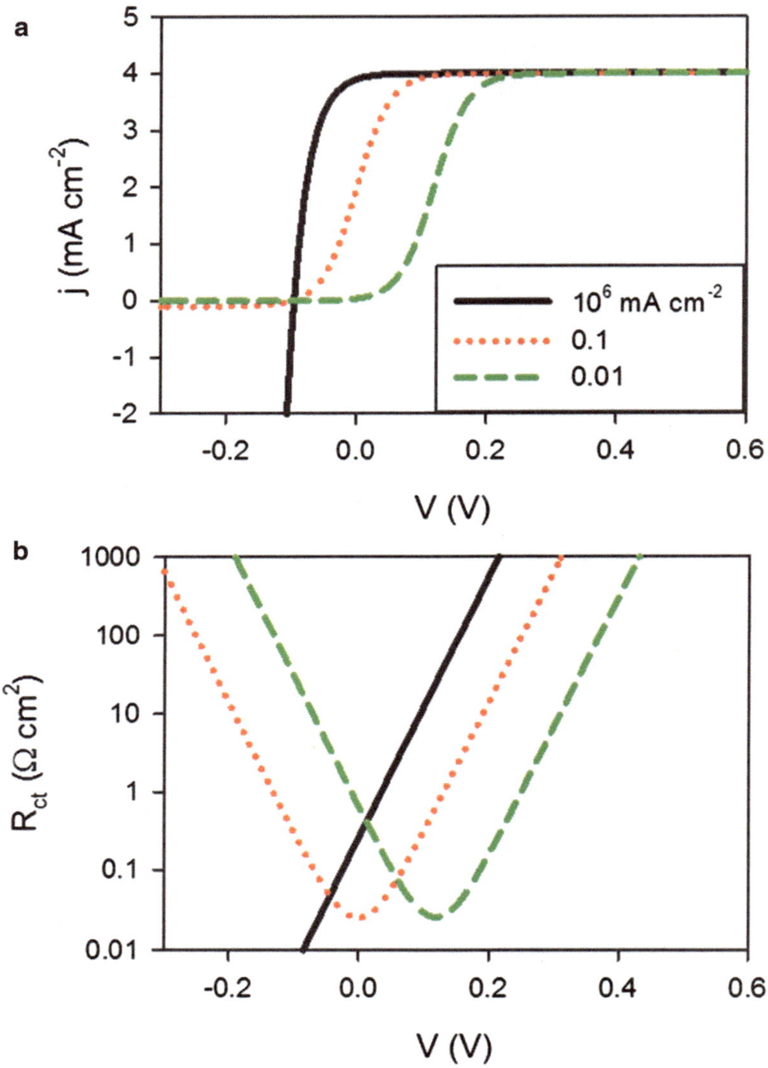


**Fig. 6.7** Current voltage curve of a photoanode (**a**) and charge transfer resistance (**b**) in the model of Reichman with limitation to extraction of the minority carrier at the semiconductor/electrolyte interface. The photocurrent is $j_{ph} = 4$ mA/cm$^2$, the reverse saturation current is $j_0 = 0.1$ mA/cm$^2$, and the hole charge transfer current $j_{s0}$ is indicated in the *inset panel*

electrons and holes with densities $n$ and $p$ per unit volume at a rate $G = \alpha\Phi_0 \exp(-\alpha x)$, being $\alpha$ the optical absorption coefficient, and $\Phi_0$ the photon flux, transport by drift and diffusion, and the formation of electrical fields including band bending at SCR according to Poisson equation. We also incorporate different rates or charge extraction at either contact. Recombination can occur either in bulk, according to the expression,

$$U_r = B(np - n_0 p_0) \tag{6.8}$$

or at the surface, as indicated in Eq. (6.4) for minority carriers. The corresponding expression for majority carrier electrons at the back contact is:

$$j = qS_n(n_s - n_{s0}) \tag{6.9}$$

Several $jV$ curves obtained in this modeling are shown in Fig. 6.8a, according to different rates of surface or bulk recombination. Again we note how the resistance of the system reaches a minimum, Fig. 6.8b, which corresponds to the inflection point in the $jV$ curve according to Eq. (6.3). The effect of increasing the surface recombination, via the parameter $S_n$, is to displace the minimum of the resistance toward cathodic potentials while for slower bulk recombination, tuned via the parameter $B$, this minimum is decreased and is associated to enhanced charge transfer kinetics.

Here, we should remark that $R_{dc}$ is not just a charge transfer (interfacial) quantity but it is influenced by recombination and diffusion events that occur far in the bulk. However, if those elements are not coupled to significant capacitive elements, then the total resistance couples to a single capacitance and the impedance gives a single arc feature as shown in Fig. 6.6b. The situation is different when we observe

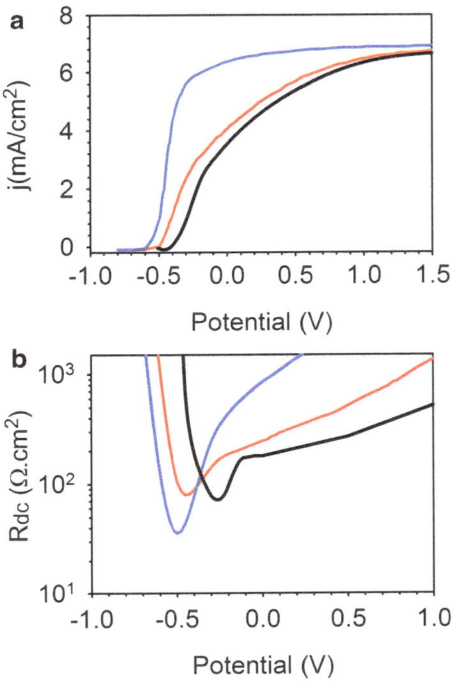


**Fig. 6.8** (**a**) $jV$ curves and (**b**) Charge transfer resistance for different values of the bulk recombination ($B$) and surface recombination at the electron selective contact ($S_n$). The values of the recombination rates are $B = 10^{-8}$ cm$^3$/s and $S_n = 10^{-2}$ cm/s (*blue line*), $B = 10^{-7}$ cm$^3$/s and $S_n = 10^{-2}$ cm/s (*red line*) and $B = 10^{-7}$ cm$^3$/s and $S_n = 10^{-4}$ cm/s (*black line*). Reproduced with permission from Shi et al. (2016)

transmission line features as commented later on in Sect. 6.7, as then the ac measurement resolves the variation of recombination and diffusion resistances across the sample thickness (Bisquert 2002; Wang et al. 2006).

## 6.5 Charge Transfer via Surface States

### *6.5.1 Properties of Surface States*

At the surface of a crystalline semiconductor material, the periodic crystal symmetry is broken and consequently, there exist electronic states within the bandgap, which are termed surface states. These surface states can be intrinsic to the semiconductor surface or extrinsic when their existence depends on the environmental conditions of the semiconductor (presence of adsorbates or electrolytes in contact with the semiconductor surface). In the context of semiconductor electrochemistry, surface states play a key role on the kinetics of interfacial reactions at illuminated electrodes. Particularly, the importance of surface states for the kinetics of electrochemical reactions involving minority carriers has been highlighted from the early studies on III–V semiconductors (Kelly and Memming 1982) and $Fe_2O_3$ (Dareedwards et al. 1983). More recent results in relation to IS have been summarized by Bertoluzzi et al. (2016). Such states can act as recombination centers for minority carriers generated by light; and consequently, surface recombination competes effectively with charge transfer from the semiconductor bands. Charge localization in surface states also leads to considerable changes in the Helmholtz potential that result from trapping, accounting for the magnitude of the overpotentials with respect to the flat band potential ($V_{\mathrm{fb}}$) required for the onset of photocurrent. This topic will be discussed in Sect. 6.6.

Impedance spectroscopy allows probing the density of surface states by monitoring the surface state chemical capacitance ($C_{\mu}^{\mathrm{ss}}$) as a function of the applied bias (electron Fermi level). In dark condition, majority carriers are in equilibrium with the transport band (conduction band for an n-type material) and this capacitance is termed the equilibrium chemical capacitance of surface states, $C_{\mu\mathrm{eq}}^{\mathrm{ss}}$, defined as (Bisquert 2014):

$$\begin{aligned} C_{\mu\mathrm{eq}}^{\mathrm{ss}} &= Aq^2 N_{\mathrm{ss}} \frac{\partial f_{\mathrm{ss}}}{\partial E_{\mathrm{Fn}}} \\ &= A\frac{q^2 N_{\mathrm{ss}}}{k_{\mathrm{B}}T} f_{\mathrm{ss}}(1-f_{\mathrm{ss}}) \end{aligned} \tag{6.10}$$

$N_{\mathrm{ss}}$ is the density of surface states, $f_{\mathrm{ss}}$ the fractional occupancy of the states, $k_{\mathrm{B}}$ the Boltzmann constant, and $E_{\mathrm{Fn}}$ the electron Fermi level. In the presence of a wide distribution of surface states such as the Gaussian distribution, the surface state capacitance is related to the density of surface states (DOS), $g_{\mathrm{ss}}(E)$, as

$$C_{\mu eq}^{ss} = A_s q g_{ss}(E_{Fn}) \tag{6.11}$$

As an illustrative example, Fig. 6.16, discussed later on, shows the DOS of hematite ($\alpha$-$Fe_2O_3$) in contact with an aqueous solution under illumination at two different pH conditions.

### *6.5.2 Theory of Charge Transfer via Surface States and Valence or Conduction Band*

Under illumination of a photoanode, the valence band obtains abundant excess of minority carriers, holes, which can be transferred to the species in solution by a direct charge transfer mechanism in order to launch the fuel production reaction. But in the presence of surface states, these localized bandgap states may capture hole carriers and consequently present another favorable pathway for indirect charge transfer as well as a new undesirable recombination pathway, as indicated in Fig. 6.2a (Salvador and Gutierrez 1984; Salvador and Gutiérrez 1984). The solar to chemical energy conversion efficiency strongly relies on the magnitude of both processes and it is therefore primordial to differentiate between each one of them. In particular, efficient solar fuel production requires reduction of the effect of surface state-assisted recombination at low applied bias.

In order to understand the effect of the competing surface state-assisted recombination and indirect charge transfer processes on water oxidation, we consider a first simple kinetic model that neglects any electrostatic influence such as the presence of an electric field. Such model is depicted in Fig. 6.9a (Bertoluzzi and Bisquert 2012). At low bias, direct charge transfer of holes from the valence band can be neglected and only three processes occur at this interface: trapping/detrapping of holes, charge transfer of holes, and trapping/detrapping of electrons. Note that the detrapping kinetic constant is directly proportional to the trapping constant through the detailed balance principle (Bisquert 2010). Hence, it is equivalent to refer to either trapping or detrapping kinetics indistinctly.

From this simple picture, it is clear that surface state-assisted recombination and hole transfer to the solution have a common factor, which is hole trapping/detrapping. In fact, Fig. 6.10 shows that the effect of trapping/detrapping does not influence notably photoelectrochemical performances of a photoanode at low applied bias. On the contrary, it is the ratio between the detrapping of electrons and the charge transfer kinetic constant ($\varepsilon_n/k_s$) which determines the relative importance of both competing processes and the onset voltage of the photooxidation reaction. In particular, the lesser the ratio $\varepsilon_n/k_s$, the smaller the onset voltage. It is therefore necessary to probe the evolution of this quantity when resorting to surface treatments such as passivating layers or catalysts.

In order to obtain a more detailed view of charge transfer processes with electrostatic features, in steady-state conditions, we have used the general

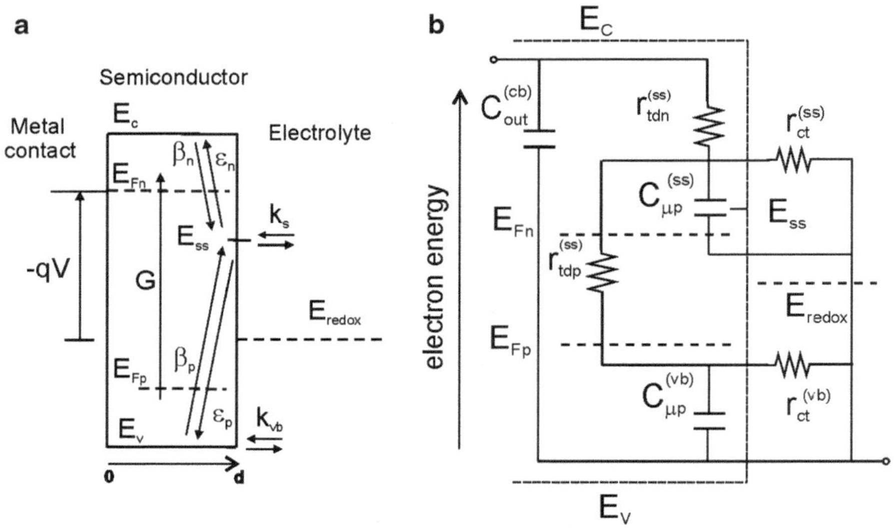


**Fig. 6.9** (**a**) Scheme of the kinetics of the processes occurring at the semiconductor/solution interface: generation of electrons and holes at a rate $G$, trapping of electrons from the conduction band (kinetic constant $\beta_n$) and detrapping ($\varepsilon_n$). Trapping of holes from the valence band ($\beta_p$) and detrapping ($\varepsilon_p$). Charge transfer of holes from the surface states ($k_s$) and from the valence band ($k_{vb}$). (**b**) Equivalent circuit obtained from a small ac perturbation. $C_{out}^{cb}$ is the semiconductor capacitance, $C_{\mu p}^{ss}$ is the traps chemical capacitance, $C_{\mu p}^{vb}$ the valence band chemical capacitance, $r_{tdn}^{ss}$ the trapping/detrapping resistance of electrons from the conduction band, $r_{tdp}^{ss}$ the trapping/detrapping resistance of holes from the valence band, $r_{ct}^{ss}$ is the charge transfer resistance from the traps and $r_{ct}^{vb}$ holes transfer resistance from the valence band. Reproduced with permission from Bertoluzzi and Bisquert (2012)

simulation procedure described in Sect. 6.4 to obtain some insight into the competition of charge transfer pathways (Bertoluzzi et al. 2016). In Fig. 6.11, we present simulations of current voltage characteristics and the corresponding evolution of the hole density in the valence band at the semiconductor/electrolyte interface. In plot 1 of Fig. 6.11a, one can appreciate the effect of direct charge transfer of holes from the valence band on the photoelectrochemical performance of a photoanode. The first feature that can be observed is a very small photocurrent at low applied bias, which arises from strong electron–hole recombination. Once the onset voltage is reached, the space charge region sustains a sufficient electric field in order to counterbalance recombination so that charge separation is favored and the anodic photocurrent increases, see Fig. 6.3b. Finally, as the applied voltage increases, the anodic photocurrent reaches saturation, whose value can be approximated within the framework of the model of Reichman (1980), as discussed earlier. This approach has been reported and discussed in several publications, and for more information on this topic, the reader is referred to a detailed review by Cendula et al. (2014).

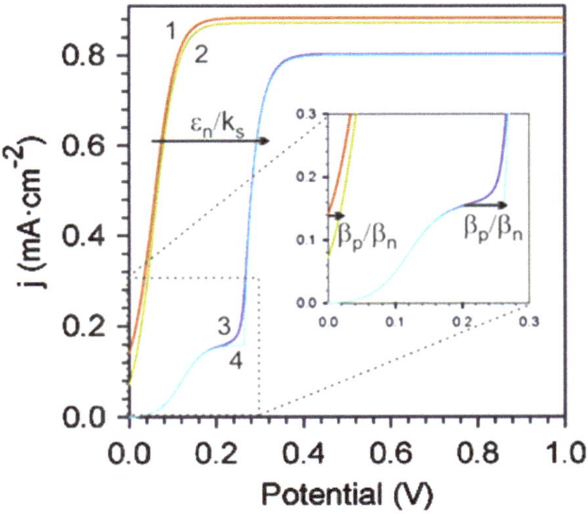


**Fig. 6.10** *jV* curves for different cases of the kinetic model including the electron trapping/detrapping, hole transfer from the surface states, and hole trapping/detrapping. Model parameters are $d = 10\,\text{nm}$, $G = \alpha\phi_0\exp(-\alpha x)$, $\alpha^{-1} = 100$ nm, $\phi_0 = 5 \times 10^{16}\,\text{cm}^{-2}\,\text{s}^{-1}$, $k_BT = 26\,\text{meV}$, $E_c = 0\,\text{eV}$, $E_{ss} = -0.12\,\text{eV}$, $N_c = 10^{20}\,\text{cm}^{-3}$, $N_{ss} = 10^{21}\,\text{cm}^{-3}$, and $k_{vb} = 10^{-1}\,\text{s}^{-1}$, *(1)* $\varepsilon_n = 10^{-1}k_s = 1\,\text{s}^{-1}$, $\beta_p = 10^{-3}\beta_n = 10^{-21}\,\text{cm}^3/\text{s}$, *(2)* $\varepsilon_n = 10^{-1}k_s = 1\,\text{s}^{-1}$, $\beta_p = 10^3\beta_n = 10^{-15}\,\text{cm}^3\,\text{s}^{-1}$, *(3)* $\varepsilon_n = 10^3k_s = 10^3\ \text{s}^{-1}$, $\beta_p = 10^{-3}\beta_n = 10^{-18}\,\text{cm}^3/\text{s}$, *(4)* $\varepsilon_n = 10^3k_s = 10^3\,\text{s}^{-1}$, $\beta_p = 10^3\beta_n = 10^{-12}\,\text{cm}^3/\text{s}$. Reproduced with permission from Bertoluzzi and Bisquert (2012)

For indirect charge transfer (plot 2 of Fig. 6.11a), the photogenerated holes are trapped in surface states and can subsequently oxidize water or recombine with trapped electrons. When a positive bias of sufficient magnitude is applied, surface states are filled with holes and hole transfer reaches a maximum; this is when anodic current reaches a first plateau, as indicated by the gray dashed lines of Fig. 6.11a (plot 3). After saturating surface states with holes, the population of holes in the valence band starts increasing. The photoanodic current subsequently keeps on increasing and saturates once the surface state-assisted recombination has been totally eliminated by the withdrawal of electrons from the surface. In principle, if surface state-assisted recombination is much smaller than photogenerated hole transfer, the maximum achievable anodic current can be higher than the photogenerated current with surface states compared to the situation without surface states, as observed in Fig. 6.11a. Once again, the discrimination between both competing processes is fundamental in the characterization of water splitting devices and determining the ratio $\varepsilon_n/k_s$, previously mentioned appears to be primordial in order to optimize the photocurrent of these devices.

The ratio $\varepsilon_n/k_s$ can be obtained by impedance spectroscopy. For a quantitative calculation of impedance spectroscopy expression and equivalent circuit involving the direct and indirect charge transfer pathways, the model of Fig. 6.9a has been utilized (Bertoluzzi and Bisquert 2012). This model provides the full equivalent circuit of

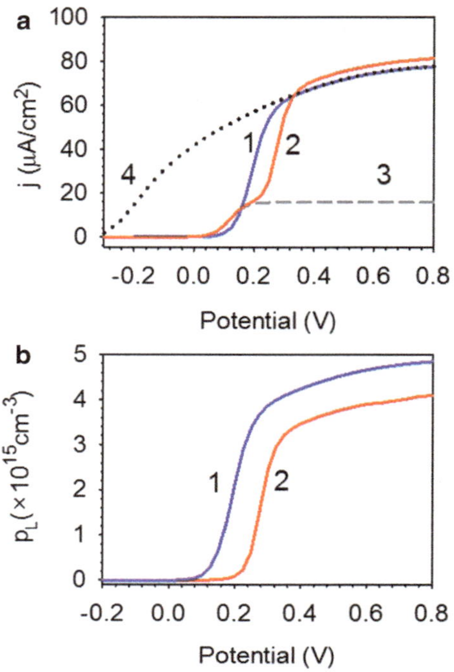


**Fig. 6.11** (**a**) Full drift-diffusion simulations of the current voltage plots when holes are directly transferred from the valence band (*1*) and when direct hole transfer competes with indirect charge transfer from the surface states (*2*). The current associated to hole transfer from the traps, in the framework of this latter model, is represented in *gray dashed lines* (*3*). We also indicate the *jV* curve for a perfect hole selective contact without surface states (*4*). (**b**) Corresponding hole density in the valence band at the semiconductor/electrolyte interface as a function of the applied voltage. Reproduced with permission from Bertoluzzi et al. (2016)

Fig. 6.9b that contains a large number of features. It is important to remark that this general equivalent circuit for a photoanode contains three types of capacitances:

1. The outer capacitance termed $C_{\mathrm{out}}^{\mathrm{cb}}$ describes the coupling of majority carriers (electrons in this case) to the external current collector. This capacitance directly responds to applied bias voltage. In a more general spatial model as that of Fig. 6.2, this outer capacitance must include the electrostatic capacitances, namely, the semiconductor capacitance due to the depletion layer $C^{\mathrm{sc}}$ and the Helmholtz capacitance.
2. The capacitance of surface states has been already discussed in Sect. 6.5.1. However, it is important to stress that this capacitance, which we note $C_{\mu\mathrm{p}}^{\mathrm{ss}}$, is different from the one measured in dark and presented in Sect. 6.5.1 ($C_{\mu\mathrm{eq}}^{\mathrm{ss}}$). In fact, in this case, the variation of the occupation probability of surface states, $f_{\mathrm{ss}}$, is not only affected by the trapping of majority carriers but also by the photogenerated minorities. The resulting expression for $C_{\mu\mathrm{p}}^{\mathrm{ss}}$ is consequently more complex (Bertoluzzi and Bisquert 2012).

3. The valence band holes capacitance $C_{\mu p}^{vb}$, explicitly depends on the chemical capacitance of the traps (Bertoluzzi and Bisquert 2012). This is due to the fact that holes in the valence band are not connected to any of the external Fermi levels, neither the Fermi level of the metal contact nor the redox level in solution. Therefore, charge in capacitor $C_{\mu p}^{vb}$ cannot be directly modulated by external bias. This latter statement has important consequences in terms of IS measurements. In fact, one necessary condition to observe the valence band capacitance is that it must be connected indirectly, via recombination, to the Fermi level of electrons. In absence of recombination, such capacitance could not be measured.

In summary, the EC in Fig. 6.9b models the kinetics of both, electrons and holes surface state-assisted recombination and charge transfer at the interface semiconductor/electrolyte in dynamic regime. This circuit explicitly includes the *three* chemical capacitances that are associated to the separate modes of carrier storage in this system: the conduction band, the valence band, and the surface states. In consequence, this model predicts the observation of three arcs in the complex impedance plot, as shown in Fig. 6.12.

However, the wide bandgap semiconductors usually used for solar fuel production do not allow for the observation of three semicircles in the complex plane. Instead, a maximum of two semicircles is generally observed, as shown in Fig. 6.6, and no more

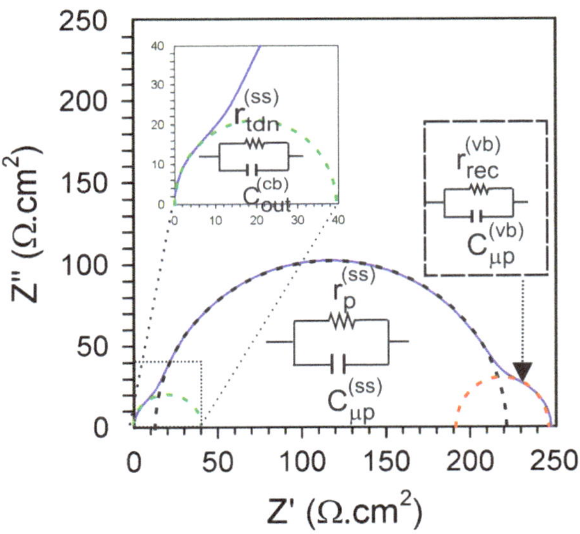


**Fig. 6.12** Illustration of the impedance in the complex plane presenting three semicircles. The first semicircle at low frequency corresponds to the valence band charge transfer, the second one to the surface states charge transfer, and the last one at higher frequencies corresponds to the electrons trapping/detrapping. The parameters used for this simulation are: $d = 10\,\text{nm}$, $\alpha^{-1} = 100\,\text{nm}$, $\phi_0 = 5 \times 10^{16}\,\text{cm}^{-2}\,\text{s}^{-1}$, $k_B T = 26\,\text{meV}$, $E_c = 0\,\text{eV}$, $E_{ss} = 0\,\text{eV}$, $N_c = 10^{20}\,\text{cm}^{-3}$, $N_{ss} = 2 \times 10^{21}\,\text{cm}^{-3}$, $k_{vb} = 0.7\,\text{s}^{-1}$, $\varepsilon_n = 10 k_s = 100\,\text{s}^{-1}$, $\beta_p = 2 \times 10^{-3} \beta_n = 2 \times 10^{-21}\,\text{cm}^3/\text{s}$. Reproduced with permission from Bertoluzzi and Bisquert (2012)

than two resistance and capacitance values can consequently be extracted from IS measurements. The extraction of these latter elements and their adequate physical interpretation relies on the proper choice of EC. In the following, we discuss a number of approaches that have been found useful in the analysis of experimental data.

Figure 6.13 shows different convenient simplifications of the kinetic model and EC of Fig. 6.9 as well as the relevant resistances which can be extracted from the IS data. Figure 6.13a is for the direct charge transfer mechanism without surface states. From this EC, we can extract values for two capacitances and resistances, which are associated to the conduction and valence band carriers. In particular, the low frequency capacitance is associated to hole storage in the valence band. Secondly, the EC of Fig. 6.13b corresponds to the classical EC for the trapping/detrapping and charge transfer of the majority carrier (Hens 1999; Bisquert 2010). This circuit allows extracting values for two resistances and two capacitances, which are associated with the conduction band and surface states. In this circuit, the low frequency capacitance is associated to hole storage in surface states. The EC of Fig. 6.13b has been employed to model the indirect hole transfer for water oxidation with $Fe_2O_3$ (Klahr et al. 2012a, b, c). Note that in this configuration, it is unfeasible to probe the hole population and consequently extract the recombination resistance associated to electron–hole recombination in the surface states. For the latter purpose, other techniques are needed, such as IMPS (Ponomarev and Peter 1995) where light intensity modulations allow to directly vary and probe the hole concentration in the valence band. However, it is worth noticing that the ratio between the electron trapping/detrapping resistance $R_{\mathrm{tdn}}^{\mathrm{ss}}$ and the charge transfer resistance from the surface states $R_{\mathrm{ct}}^{\mathrm{ss}}$ is directly related to the ratio $\varepsilon_{\mathrm{n}}/k_{\mathrm{s}}$ previously mentioned. Indeed, it can be shown that:

$$\frac{R_{\mathrm{ct}}^{\mathrm{ss}}}{R_{\mathrm{tdn}}^{\mathrm{ss}}} = \frac{\varepsilon_{\mathrm{n}} + \beta_{\mathrm{n}} n}{k_{\mathrm{s}}} \tag{6.12}$$

At reverse bias, $n \rightarrow 0$ and Eq. (6.12) reads:

$$\frac{R_{\mathrm{ct}}^{\mathrm{ss}}}{R_{\mathrm{tdn}}^{\mathrm{ss}}} = \frac{\varepsilon_{\mathrm{n}}}{k_{\mathrm{s}}} \tag{6.13}$$

Monitoring the charge transfer and trapping/detrapping resistances by IS is therefore very useful for the optimization of the onset photovoltage of the photoanodic reaction and the maximum achievable photoanodic current.

Finally, the simple Randles circuit (Fig. 6.13c) has been also employed to model the carrier dynamics of $Fe_2O_3$ photoanodes in the dark and under illumination when a hole scavenger is present in the solution, since only one arc was present in the measured Nyquist plots, and there was no evidence of the presence of the surface state affecting the carrier dynamics of the system. Note, however, that in this case the extracted charge transfer resistance from the valence band is the series combination of the recombination and charge transfer resistances of Fig. 6.13a.

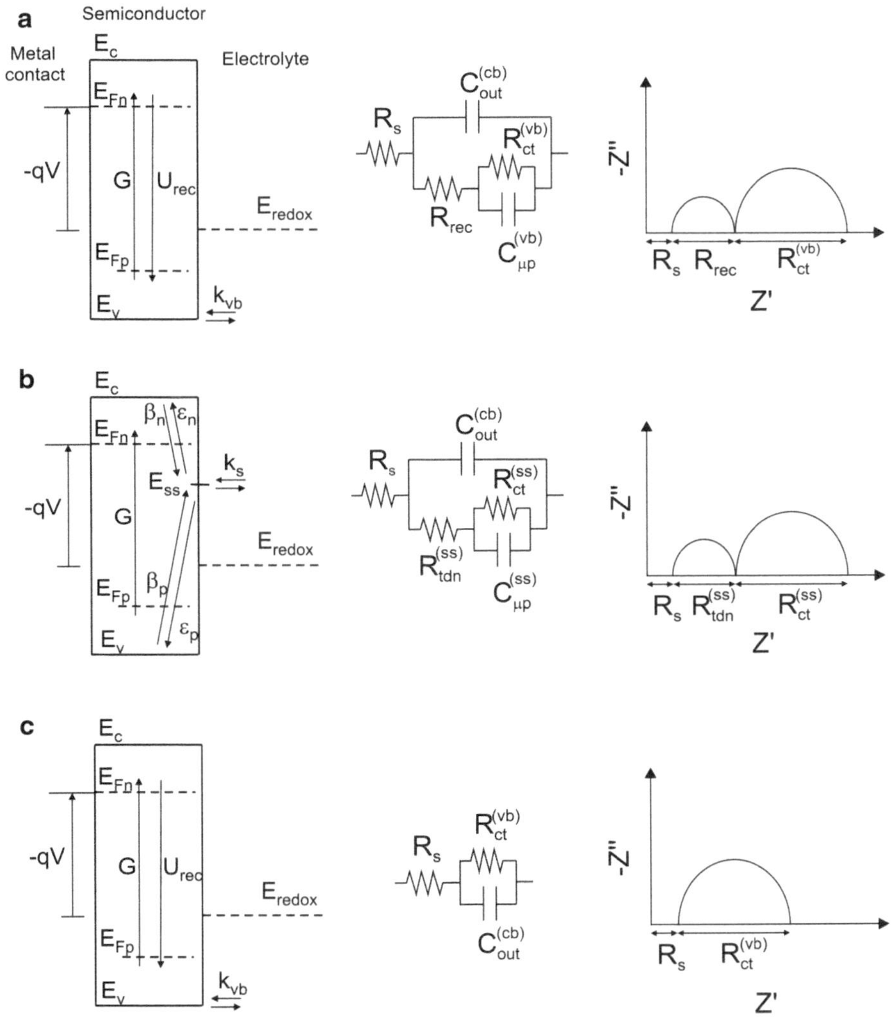


**Fig. 6.13** Main simplifications of the kinetic model of Fig. 6.9 available in the literature for the interpretation of experimental data, the associated equivalent circuit (EC) obtained from a small ac perturbation, and the corresponding complex plane IS spectra for (**a**) the direct hole transfer model, (**b**) the indirect hole transfer model, and (**c**) the reduced direct charge transfer model. In these circuits, $C_{\mathrm{out}}^{(\mathrm{cb})}$ is the semiconductor capacitance, $C_{\mu\mathrm{p}}^{(\mathrm{ss})}$ is the surface state chemical capacitance, $R_{\mathrm{rec}}$ is the band to band recombination resistance, $R_{\mathrm{ct}}^{(\mathrm{vb})}$ is the hole transfer resistance from the valence band, $R_{\mathrm{tdn}}^{(\mathrm{ss})}$ is the trapping/detrapping resistance of electrons from the conduction band, and $R_{\mathrm{ct}}^{(\mathrm{ss})}$ is the hole transfer resistance from the surface states. Note that in the case (**c**), the total charge transfer resistance is the serial combination of both resistances involved in case (**a**) ($R_{\mathrm{rec}}$ and $R_{\mathrm{ct}}^{(\mathrm{vb})}$)

### 6.5.3 Types of Analysis of IS Results

It should be pointed out that in the literature, two main approaches are considered for the interpretation of IS spectra. One relies on the full numerical resolution of transport and charge transfer equations. The other one consists in proposing a simplified analytical treatment of the impedance, which can be interpreted in terms of EC. The advantage of the first approach is that it allows taking into account more features and is certainly more accurate than the analytical approach. However, as any experimental technique, IS interpretation is limited by the amount of information one can extract from the experimental data. The equivalent circuit is therefore an excellent tool to discriminate between the main dominant features that can be experimentally unveiled. For instance, the model of Fig. 6.9b presents a high number of features and experimental data impose a reduction of this circuit in order to treat the available data, as suggested in Fig. 6.13.

Nonetheless, the EC approach presents two main issues. Firstly, we should recall that IS is a *two contact measurement* so that the information of different internal loci in the sample may be lumped into a single EC element. Secondly, the expression of a given impedance model may be interpreted in terms of several equivalent circuit representations. In order to tackle both issues, it is therefore necessary to perform further experiments and verify the physical meaning of the extracted parameters. One possibility consists in plotting the voltage variations of the capacitances and resistances obtained with a given EC.

In view of these considerations, let us further comment on the EC models discussed above. The ECs presented in Fig. 6.13a, b are identical but the electrical components of those circuits differ by their physical nature. In particular, the low frequency capacitances are of high interest and give valuable information on the origin (surface states or valence band) of the photogenerated holes involved in the water photo-oxidation. The distinction between both types of low frequency capacitances can be operated via a voltage dependence study. In fact, the valence band chemical capacitance follows the variations of the hole population in the valence band and reaches saturation at anodic voltages, as suggested in Fig. 6.14 (*blue line*). On the contrary, surface states, filled with electrons at cathodic applied bias, are progressively filled with holes when applying anodic bias. This transition of the surface state occupation leads to a peak behavior of the surface state capacitance with voltage as shown in Fig. 6.14 (*red line*). Therefore, low frequency capacitance measurements appear to be a valuable tool in order to identify the origin of the photogenerated carriers involved in solar fuel production.

### 6.5.4 Experimental Results

Recent studies on the photo-oxidation of water with $Fe_2O_3$ have clearly shown the presence of surface states, which are intrinsic to the semiconductor–water interface and determine the photocurrent onset, approximately 0.5 V more positive

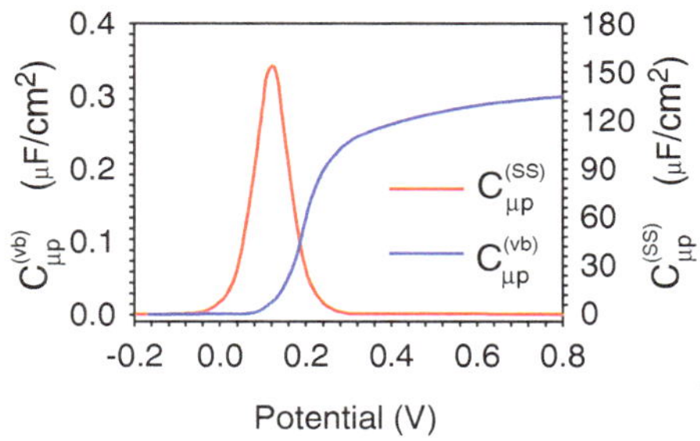


**Fig. 6.14** Low frequency capacitances for the direct charge transfer model of Fig. 6.11a (1) and the indirect charge transfer model of Fig. 6.11a (2). The former one is associated to hole storage in the valence band while the latter one is related to hole accumulation in surface states. The steady-state electrical magnitudes and parameters involved in the calculations of these capacitances are the ones obtained from the full drift-diffusion simulations of Fig. 6.11. Reproduced with permission from Bertoluzzi et al. (2016)

than the flat band potential (Klahr et al. 2012a, b, c). Indeed, when an efficient hole scavenger is present in the electrolyte, these surface states are not detectable by impedance spectroscopy and the onset of the photocurrent is cathodically shifted to the flat band potential. Figure 6.5 illustrates this situation showing in panel (a) the *jV* curves for a $Fe_2O_3$ photoanode under illumination in contact with an aqueous solution (pH 6.9) and with a $[Fe(CN)_6]^{3-/4-}$ redox couple. Panel (b) shows the total resistance obtained from impedance spectroscopy (symbols) and by derivation of the *jV* curve (lines), consistently indicating that charge transfer to the solution is more favorable in the presence of the hole scavenger at applied bias below 1.3 V vs. RHE. The characteristic complex plane impedance plots obtained for both systems under illumination at two different applied biases are shown in Fig. 6.6. At lower bias (1.3 V vs. RHE, Fig. 6.6a), the development of one or two arcs is related to the presence or absence of the hole scavenger in the solution, respectively. The high frequency arc in water photo-oxidation is related to the space charge capacitance in the semiconductor material and the low frequency arc conveys information of the capacitance of the surface state. At higher applied bias (1.6 V vs. RHE, Fig. 6.6b), the impedance spectra for water oxidation is reduced to one single semicircle associated to the space charge capacitance, as explained in Sect. 6.5.2.

Fitting the experimental data of Fig. 6.6 to the circuit depicted in Fig. 6.13b, a surface state capacitance, $C_{\mu p}^{ss}$, can be extracted, which follows a characteristic Gaussian-like dependence with applied voltage. The maximum of this capacitance is located close to the photocurrent onset suggesting that accumulation of holes at the surface state is needed before a steady-state photocurrent for water oxidation can be sustained (Fig. 6.15). On the other hand, the minimum of charge transfer resistance coincides with the inflection point of the *jV* curve. The density of states obtained by Eq. (6.11) can be mapped as shown in Fig. 6.16 for two different pH conditions and exhibits a maximum at approximately 1.23 V vs. RHE, suggesting that multistep water oxidation intermediates are strongly connected to these surface states.

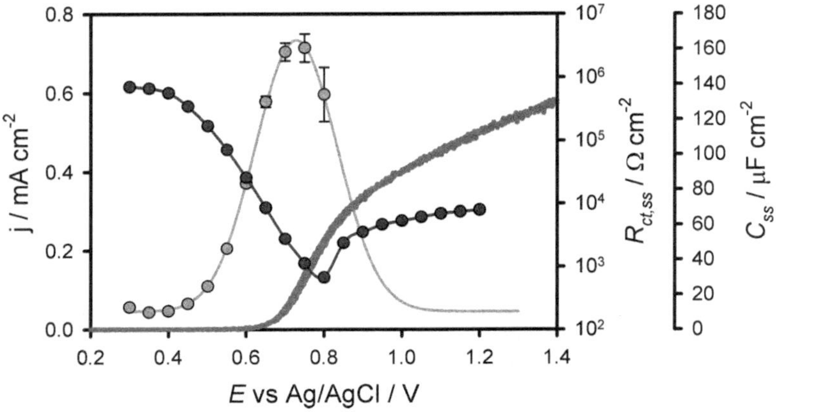


**Fig. 6.15** *jV* curve (*dark gray solid line*), $C_{ss}$ (*gray circles*) and $R_{ct,ss}$ (*black circles*) values obtained for a 60 nm hematite electrode under 1 sun illumination and pH 6.9. Adapted with permission from Klahr et al. (2012c)

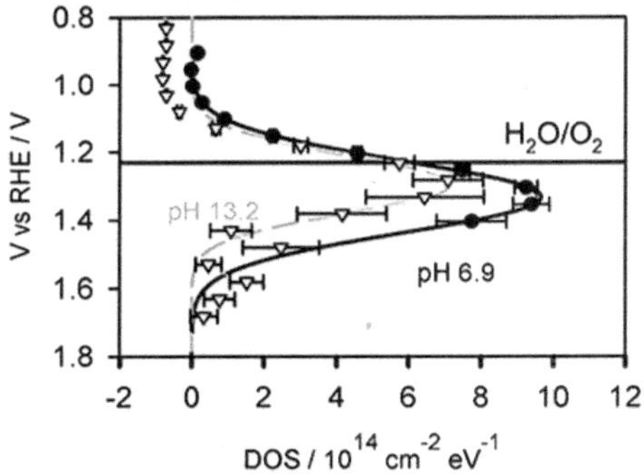


**Fig. 6.16** DOS as a function of RHE reference potential for pH 6.9 (*black circles* and *line*) and pH 13.2 (*triangles* and *gray lines*) under 1 sun illumination. Reproduced with permission from Klahr et al. (2012c)

An important feature of the surface state capacitance measured by IS and other complementary techniques such as cyclic voltammetry is that it is strongly dependent on the nature of the semiconductor/electrolyte interface and the deposition technique. For instance, the distribution of hematite surface states changes drastically in presence of water, methanol, or acetonitrile (Klahr et al. 2015). In addition, the capacitive peak associated to these states shifts anodically with pH (Iandolo and Hellman 2014). In this latter reference, it was demonstrated by DFT that the capacitive response of surface states originates from two types of surface states which differ by their terminations: –O and –OH. The former type of states is close to the middle of the gap and induces recombination while the latter ones are close to the valence band and give rise to hole transfer at higher bias. The formation mechanism of these –O termination was recently proposed by means of DFT + U calculations (Yatom et al. 2015). In particular, they showed that this midgap state

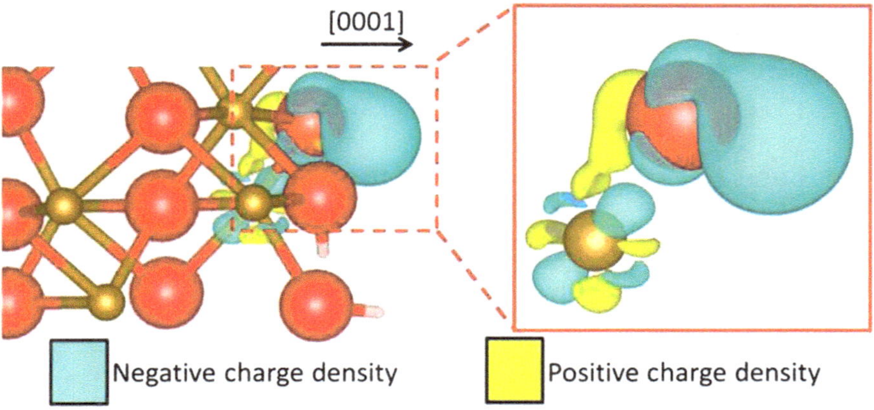


**Fig. 6.17** Charge density difference showing a surface state on pure $Fe_2O_3(0001)$. Calculated by subtracting the electron density of the *OH intermediate from the *O intermediate at the fixed ionic positions of the latter. *Red, gold,* and *white spheres* denote O, Fe, and H atoms, respectively. Negative and positive charge density iso-surfaces ($\pm 0.01[e/Bohr^3]$) are in *blue* and *yellow*, respectively, where negative being absence of electrons. Reproduced with permission from Yatom et al. (2015)

finds its origin in the hybridization of the d and p orbitals of Fe and O, respectively, as depicted by Fig. 6.17. Concomitantly, Klahr and Hamann (2014) demonstrated the existence of both types of states by impedance spectroscopy and cyclic voltammetry for hematite in aqueous electrolyte. However, the same group showed that the existence of both capacitive features is strongly dependent on the deposition method. In particular, they demonstrated that using atomic layer deposition (ALD) and annealing hematite at 800 °C allows removing recombination centers (Zandi and Hamann 2014). This work echoed an earlier study by Chou et al. (2013) who observed, via low frequency capacitance measurements, that preparation of $\alpha$-$Fe_2O_3$ electrodes by Fe oxidation is more favorable to surface state-assisted charge transfer than by anodic electrodeposition.

Similarly, Doyle and Lyons (2013) revealed through IS measurements that hydrous iron oxide is featured by a clear surface state capacitive peak associated to Fe=O intermediates at high pH (concentration of NaOH superior to 1 mol/L). Instead, for lower pH, the measured capacitance was found to be similar to the valence band capacitance displayed in Fig. 6.14 (plot 1). In this case, the EC of Fig. 6.13a must be employed for the fitting of IS data. By comparing surface state capacitance and X-ray spectroscopies of hematite samples under different plasma treatments, Braun and coworkers (Hu et al. 2016) show that iron surface states induce higher reactivity toward water oxidation than oxygen surface states.

The model of charge transfer through surface states has also been successfully employed to explain the photo-oxidation of water on other n-type semiconductor materials such as $CuWO_4$ (Pyper et al. 2013) and $TiO_2$ nanotubes (Cachet and Sutter 2015). In contrast with the behavior of $Fe_2O_3$, the presence of Fermi level pinning in the dark suggests that these midgap states are permanent and not photogenerated. Similar approach has been followed to understand carrier dynamics on $FeS_2$ (Caban-Acevedo et al. 2014).

## 6.6 Band Edge Movement

The modification of surface dipoles at the semiconductor–liquid interface originated by different factors leads to a shift of the semiconductor band edges with respect to the stable electrolyte energy level such as the Fermi energy of the redox couple (redox potential). This effect, termed *band edge movement*, or more simply *band shift*, can be monitored by IS. When a space charge region is developed in the semiconductor material in contact with the solution, the change of the flat band potential ($V_{fb}$) is a clear evidence of band shift. From Eq. (2.29), we recall that the Mott–Schottky plot (if it is reasonably straight) reveals the flat band potential and the doping density of a planar semiconductor electrode. Therefore, electrodes in which the surface conditions have produced a change of the flat band potential must show the Mott–Schottky signature horizontally displaced, and no other changes, since the doping level is obtained by the edge of the space charged region that is inside the semiconductor layer.

As an example, Fig. 6.18a shows the *jV* curves of $Fe_2O_3$ photoanodes under illumination at 100 mW/cm$^2$ in aqueous electrolyte, a mixture of water and methanol and pure methanol. There exists a clear anodic shift of the photocurrent when pure methanol is employed, which can be at least partially explained by the anodic band shift illustrated in Fig. 6.18b by the change in flat band potential. The origin of this shift is related to the difference in surface dipoles formed at the semiconductor/solution interface. On the other hand, the slopes of the Mott–Schottky plots shown in Fig. 6.18b are practically identical indicating that the donor density is not altered. The extracted doping density, $N_D$, was calculated to be $5.3 \times 10^{18}$ cm$^{-3}$, in good agreement with values previously reported (Klahr et al. 2012a, b, c). In order to account for the band shift, it is more informative to plot the *jV* curves, correcting the applied voltage to the flat band potential, $V$ vs. $V_{fb}$, as shown in Fig. 6.18c. The modulation of flat band potential can also be achieved with thin ferroelectric layer in the semiconductor surface (Yang et al. 2015).

As mentioned above, charge localization in surface states also leads to considerable changes in the Helmholtz potential that result from trapping. These changes in the potential distribution lead to Fermi level pinning, or band unpinning, as discussed in Sect. 1.2.8. Under illumination, the trapping of minority carriers at surface states produces charging of surface states and the Helmholtz layer, and hence band unpinning as first described by Kelly and Memming (1982). In Fig. 6.19, we show the mechanism of changes of Mott–Schottky plot that are observed when the semiconductor surface undergoes trapping of minority

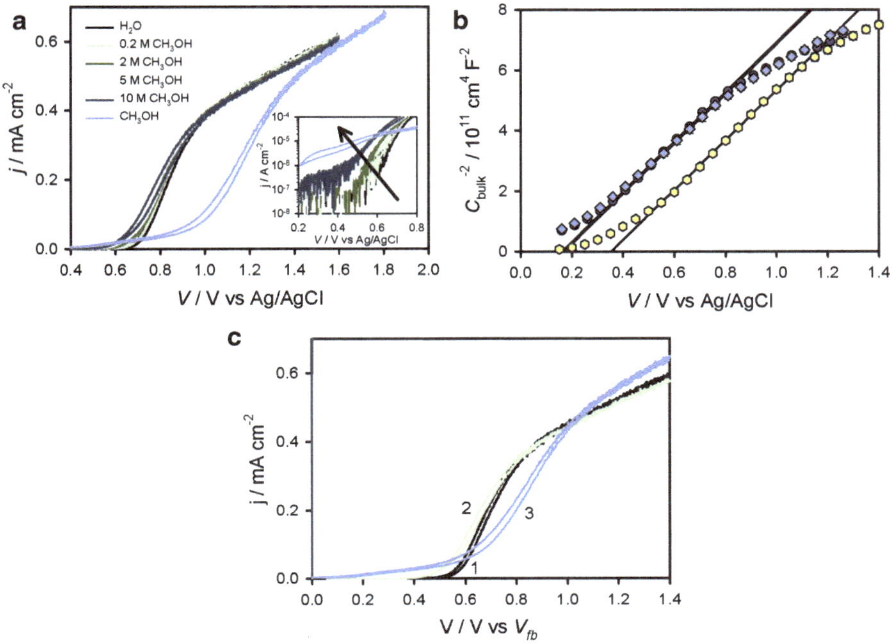


**Fig. 6.18** (**a**) *jV* curve of a hematite electrode in contact with a $H_2O$ and methanol electrolyte under 1 sun illumination. (**b**) Mott–Schottky plot of $C_{bulk}$ measured in the dark for $H_2O$ (*circles*), 5 M $CH_3OH$ (*diamonds*) and $CH_3OH$ electrolytes (*hexagons*). (**c**) *jV* curves of a hematite electrode under 1 sun illumination in contact with $H_2O$ (*1*), 5 M $CH_3OH$ (*2*) and $CH_3OH$ (*3*) plotted vs. $V_{fb}$. Adapted with permission from Klahr et al. (2015)

photogenerated carriers. Considering water oxidation with $Fe_2O_3$, Fig. 6.20 clearly illustrates Fermi level pinning associated to the presence of light-activated surface states. The Mott–Schottky plot of the space charge capacitance under illumination levels off at potentials where the surface state capacitance is measured.

## 6.7 Modification of Charge Transfer Rate

In the context of PEC solar fuels production, the characterization of the charge transfer rate is particularly relevant to monitor the effect of surface modifications of photoelectrodes, for example the deposition of co-catalysts (Klahr et al. 2012b; Badia-Bou et al. 2013; Riha et al. 2013), passivation layers (Le Formal et al. 2011; Steier et al. 2014), and the formation of homo or heterojunctions at the surface to displace energy levels (Li et al. 2012; Lin et al. 2012). IS is a widely used tool to assess changes in the photoelectrode charge transfer rate after the above-mentioned treatments. Nevertheless, many times the impedance study is exclusively based on the comparison of the complex impedance plots obtained from different samples at

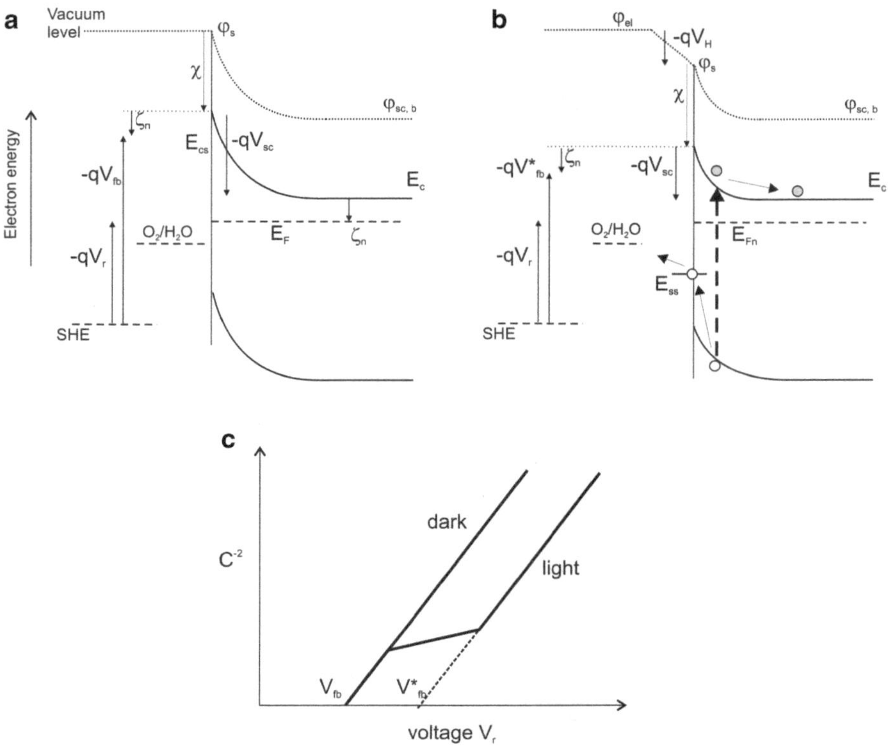


**Fig. 6.19** Energy diagram of a semiconductor/electrolyte junction, in which the semiconductor has a surface state that captures both minority and majority carriers, and transfers carriers to the electrolyte. Under dark (**a**), the surface states cannot be populated with holes and there is no Fermi level pinning effect. Under illumination (**b**), the surface states cannot be populated with holes and there is no Fermi level pinning effect. The Mott–Schottky plot with different regions of capacitance is shown in (**c**). At voltage close to flat band, dark and light curves overlap because the recombination does not allow maintaining holes in the ss. The maximum shift of the flat band potential under illumination is related to the density of surface states and the value of the Helmholtz capacitance

a single applied bias, and the reported information does not go further than an $R_{dc}$ value which could be obtained faster by dc methods. In order to fully exploit the power of this technique, the evolution of the EC components extracted from fitting the IS response with applied potential must be examined. The information provided by the charge transfer resistances coupled to that of the capacitances provides valuable insights on the basic mechanisms responsible for the functional enhancement observed.

Figure 6.21 shows the *jV* curves together with values of film capacitance and charge transfer resistance for a $Fe_2O_3$ photoelectrode coated with Co–Pi layers by photo-assisted electrodeposition with different thickness values (controlled by the charge passed through the electrodes). The capacitance of the films (Fig. 6.21a)

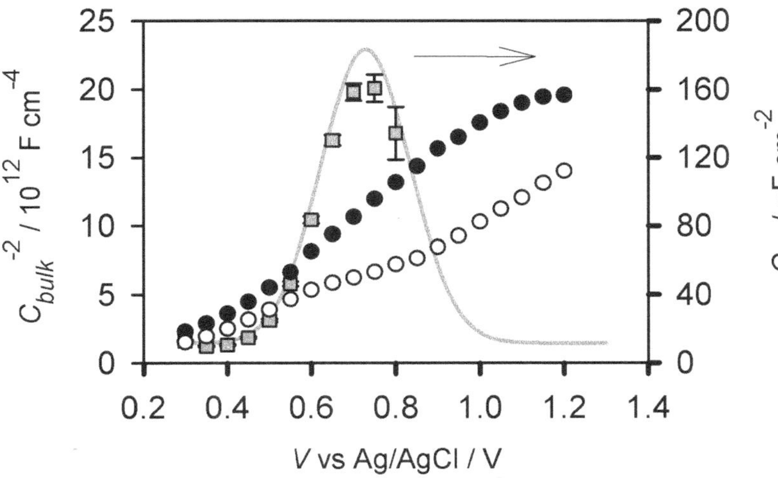


**Fig. 6.20** Mott–Schottky plots at pH 6.9 in the dark (*black circles*) and under 1 sun (*open circles*). A plot of the trap state capacitance, $C_{SS}$, (*gray squares*) is superimposed to show the Fermi level pinning. Reproduced with permission from Klahr et al. (2012c)

increases with the Co–Pi thickness, indicating that the "catalyst" controls the capacitive behavior of the electrodes. This increase of capacitance is correlated to a concomitant decrease of the charge transfer resistance (Fig. 6.21b), responsible for the faster water oxidation kinetics, which leads to the improved performance displayed in the *jV* curves (Fig. 6.21c) upon Co–Pi addition (Klahr et al. 2012b), see also Fig. 6.4. According to this picture, the Co–Pi catalyst can be seen as an efficient hole collector, which stores the photogenerated holes from the hematite electrode. This charge separation reduces recombination at the surface of $Fe_2O_3$, which results in lower photocurrent onset potentials and hence improved water oxidation efficiency. A similar mechanism has been demonstrated for $IrO_x$ catalyst on $Fe_2O_3$ (Badia-Bou et al. 2013). The evolution of film capacitance, charge transfer resistance, and photocurrent with applied voltage follows the same behavior observed for Co–Pi, validating the same interpretation provided for this catalyst. Upon deposition of either Co–Pi or $IrO_x$ on $Fe_2O_3$, any effects related to band shift were discarded, as derived from Mott–Schottky plots of the space charge capacitance. Cobalt-based materials have also been studied as catalysts for hydrogen evolution reaction. Recently, a heterostructure composed of p-Si microwires decorated with $CoSe_2$ nanorods was proposed as a promising noble metal-free photocathode candidate for solar fuel production (Basu et al. 2015). A typical volcano plot was recorded between the obtained photocurrent and the catalyst loading, which was thoroughly analyzed by optical and IS measurements. From optical absorption, it was ruled out an effect of light inhibition as responsible for the drop in photocurrent of the $CoSe_2$-richest sample. Impedance complex plane plots resolved a separated contribution of each interface of the heterostructure, i.e. Si/$CoSe_2$ and $CoSe_2$/electrolyte. From the study of the extracted resistances, it was predicted that an increased $CoSe_2$ loading on

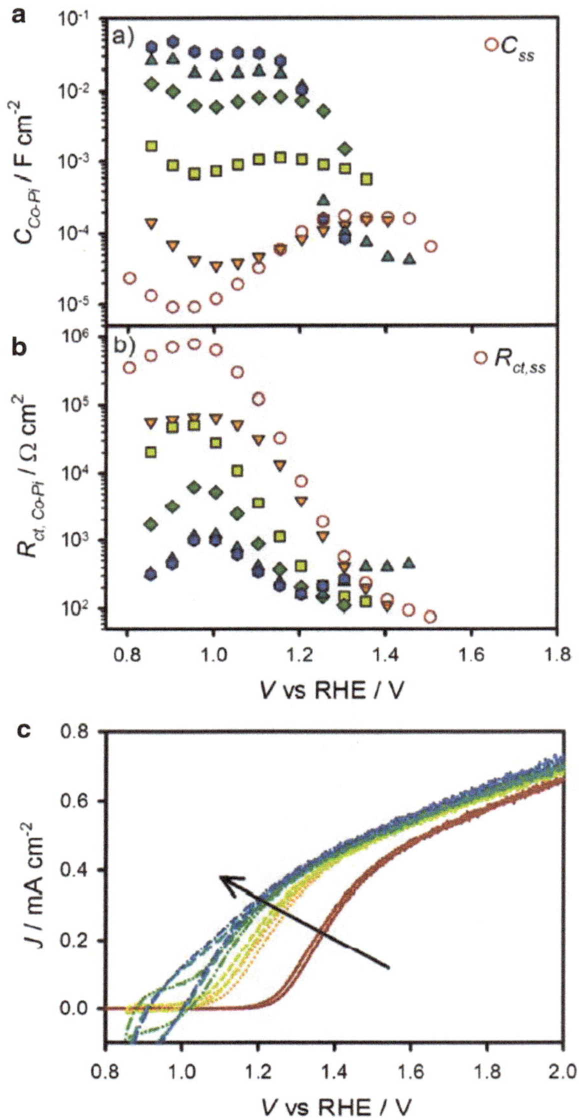


**Fig. 6.21** (**a**) Capacitance of $Fe_2O_3$/Co–Pi films ($C_{Co-Pi}$) and (**b**) Charge transfer resistance ($R_{ct,Co-Pi}$) fitted from impedance response of hematite electrodes with 1 (*triangles pointing down*), 2 (*squares*), 15 (*diamonds*), 45 (*triangles pointing up*), and 90 mC/cm$^2$ (*hexagons*) Co–Pi deposited. Bare hematite fitting parameters of (**a**) $C_{ss}$, (**b**) $R_{ct,ss}$ are shown for comparison (*red open circles*). (**c**) Corresponding *jV* curves obtained by cyclic voltammetry. Reproduced with permission from Klahr et al. (2012b)

the microwired semiconductor acts as charge recombination centres which restricts the charge transfer, playing against the reported catalytic effect of the semimetallic nanorods.

On the other hand, the deposition of $Ga_2O_3$ overlayers on $Fe_2O_3$ photoelectrodes, Fig. 6.22, also yields a favorable cathodic shift of the photocurrent onset, which is associated to the passivation of surface states at the $Fe_2O_3$ interface (Steier et al. 2014). This surface passivation has been identified by the decrease of the peak capacitance associated to the density of surface states. Additionally, a cathodic shift of the density of surface states was observed, probably indicating a change in the water oxidation mechanism. It is interesting to note that $Ga_2O_3$ does not exhibit any electrocatalytic behavior when deposited on top of FTO substrates, and does not induce any band shift when deposited on top of $Fe_2O_3$. This idea of surface state passivation of $Ga_2O_3$ coated $Fe_2O_3$ electrodes seems to be valid for the 13-group oxide overlayers with the corundum structure, which have been reported to release lattice strain of ultrathin hematite layers and decrease the density of surface states (Hisatomi et al. 2011). Photoluminescence experiments carried out on $Fe_2O_3/Al_2O_3$ systems corroborate this idea of surface states passivation, as showed in Fig. 6.23 (Le Formal et al. 2011). Another study has provided integral characterization of the passivation of Ti-doped hematite (Monllor-Satoca et al. 2015).

Studies with $WO_3/BiVO_4$ heterostructures have also shown that the charge transfer resistance is directly related to the catalytic behavior as well as the functional performance of the electrodes (Hong et al. 2011). Recently, Shi et al. (2015) thoroughly demonstrated by a systematic optoelectronic and photoelectrochemical study that the improved performance in the heterojunction was connected to a synergism between the adequate optical properties of $BiVO_4$ and the excellent charge transfer properties of $WO_3$, which controls both the transport and capacitive response in the composite.

## 6.8 Transport and Reaction in Nanostructures

In nanostructured semiconductor materials, charge transport and charge transfer (or reaction) are often coupled and impedance spectroscopy allows deconvolution of these phenomena by the use of complex physical models, generally including a transmission line element. A classical example is found in nanostructured $TiO_2$ films, which are the base of photoelectrochemical solar cells like (dye or quantum dot) sensitized solar cells (Wang et al. 2006; Raga et al. 2012). When these nanostructures are integrated as photoelectrodes for solar fuels production, impedance spectroscopy provides a powerful tool for the analysis of charge transport, charge transfer, and charge accumulation mechanisms, which are key factors to evaluate the functional performance of the device. Figure 6.24a represents the general transmission line equivalent circuit. This representation takes into account electron transport ($r_{tr}$), along the electron transport level. The transversal element $\zeta_m$ in Fig. 6.24a depends on charge accumulation and charge transfer, and the lower, resistanceless rail indicates fast transport in the electrolyte. Now considering the process of water oxidation, we take into account the presence of both electrons and holes in the semiconductor material, which is a situation absent in a sensitized solar cell. In this case, a more complex model is needed, and we suggested the circuit of Fig. 6.9b, which has been previously developed for the presence of electrons and holes in related systems. This equivalent circuit, that constitutes the

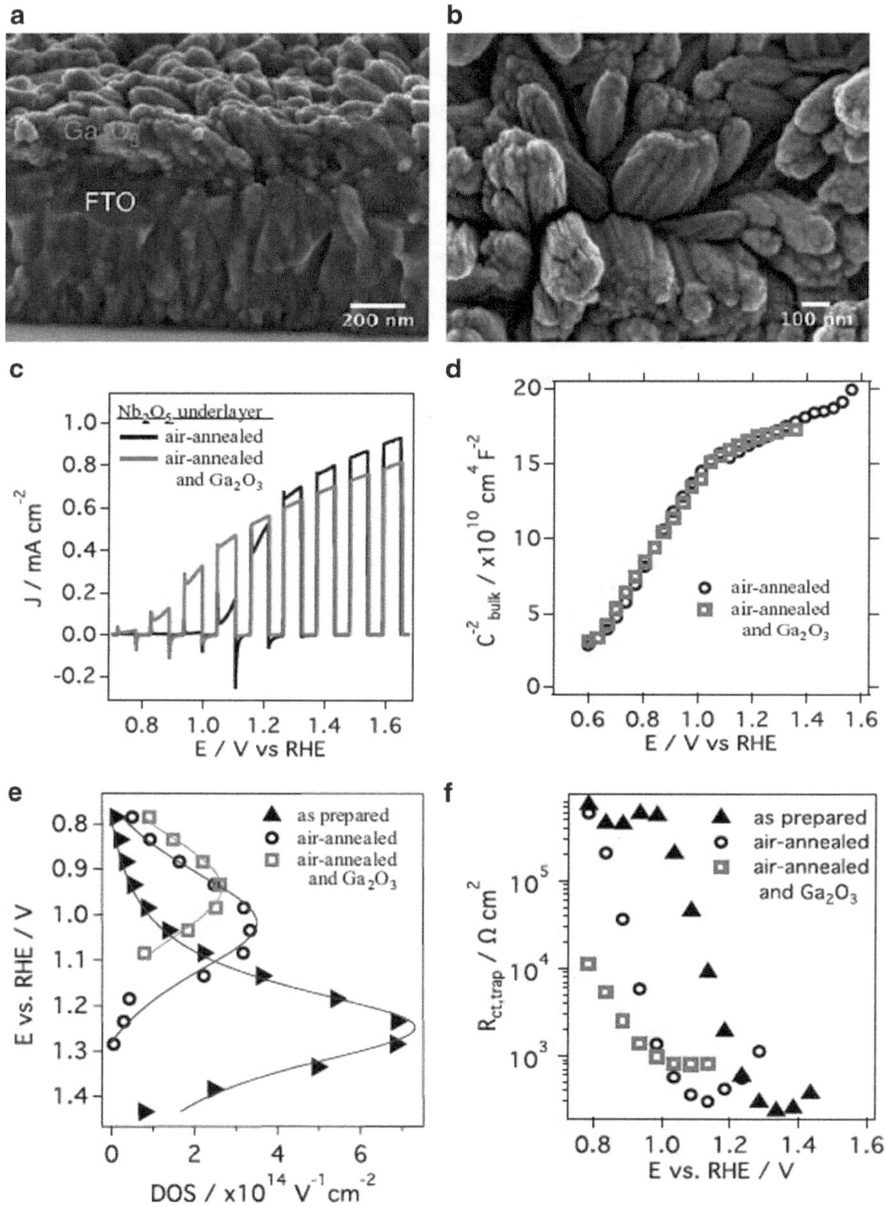


**Fig. 6.22** Characteristics of hematite photoanodes upon addition of a $Ga_2O_3$ overlayer on a Nb-doped hematite photoanode. Shown are (**a**) cross-sectional and (**b**) top-view SEM images of the $Ga_2O_3$ structure on a (barely visible) ultrathin (approximately 16 nm) hematite layer deposited on FTO glass. (**c**) Light chopping experiments of air-annealed Nb-doped thin (approximately 30 nm) hematite photoanodes before (*black*) and after $Ga_2O_3$ CBD (*gray*). (**d**) Mott–Schottky plots of the samples shown in (**c**) measured in the dark. (**e**) Applied potential versus density of surface states (DOS) extracted from $C_{ss}$. Shown are air-annealed samples from (**c**) before (*black hollow circles*) and after $Ga_2O_3$ deposition (*gray hollow squares*). In addition, a nonannealed sample is

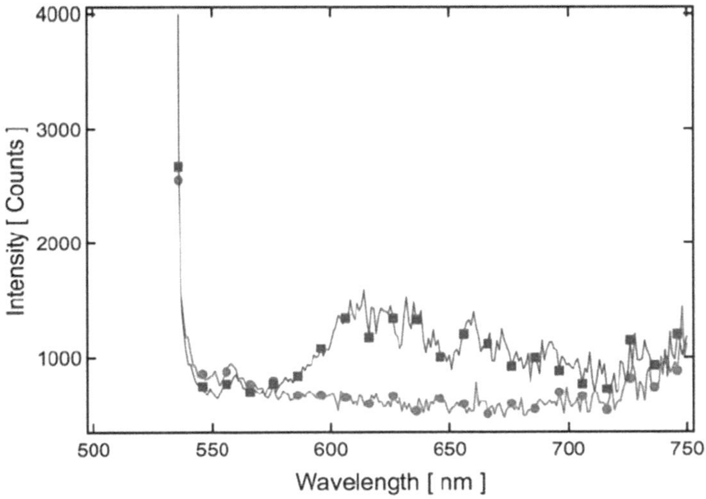


**Fig. 6.23** Photoluminescence emission spectra (excitation wavelength 520 nm) of a hematite cauliflower-type nanostructure photoanode before (*circles*) and after three ALD cycles of $Al_2O_3$ on its surface (*squares*). Reproduced with permission from Le Formal et al. (2011)

element $\zeta_m$ in Fig. 6.24a, is composed of separate chemical capacitances of electrons and holes ($C_{\mu p}^{(cb)}$ and $C_{\mu p}^{(vb)}$) and charge transfer to the solution through the conduction and valence bands ($r_{ct}^{(cb)}$ and $r_{ct}^{(vb)}$) together with electron and hole trapping ($r_{trn}^{(ss)}$ and $r_{trp}^{(ss)}$) and charge transfer through the surface state ($r_{ct}^{(ss)}$). A simplified equivalent circuit only accounting for the transport, chemical capacitance, and charge transfer of electrons from the conduction band is represented in Fig. 6.24b. If the system shows easy hole transport in the valence band, the model should be extended with a channel for hole transport, included in the equivalent circuit, as indicated in Fig. 6.24c. In this situation, the model shows the upper rail for electron transport, the lower one for hole transport, and the central one for the electrolyte. In Fig. 6.24c, the rectangular box does not represent an impedance element but indicates a more complex connection, depending on the specific charge storage, recombination, surface states features, and other specific properties that link electrons, holes, and their charge transfer to the solution. There are some illustrative examples in the literature describing the carrier dynamics of nanostructured photoanodes of several metal oxides by these advanced models, particularly

**Fig. 6.22** (continued) shown for comparison (*black filled triangles*). A Lorentzian fit for each sample guides the eye but also indicates the shift in maximum of the DOS at a certain potential. (**f**) Charge transfer resistances from surface states, $R_{ct,trap}$, versus applied potential for samples from (**e**) show a minimized resistance in presence of a $Ga_2O_3$ overlayer. Measurements shown in (**c**), (**e**), and (**f**) were carried out with simulated AM 1.5G (100 mW/cm$^2$) light. The electrolyte was 1 M NaOH (pH 13.6). Adapted with permission from Steier et al. (2014)

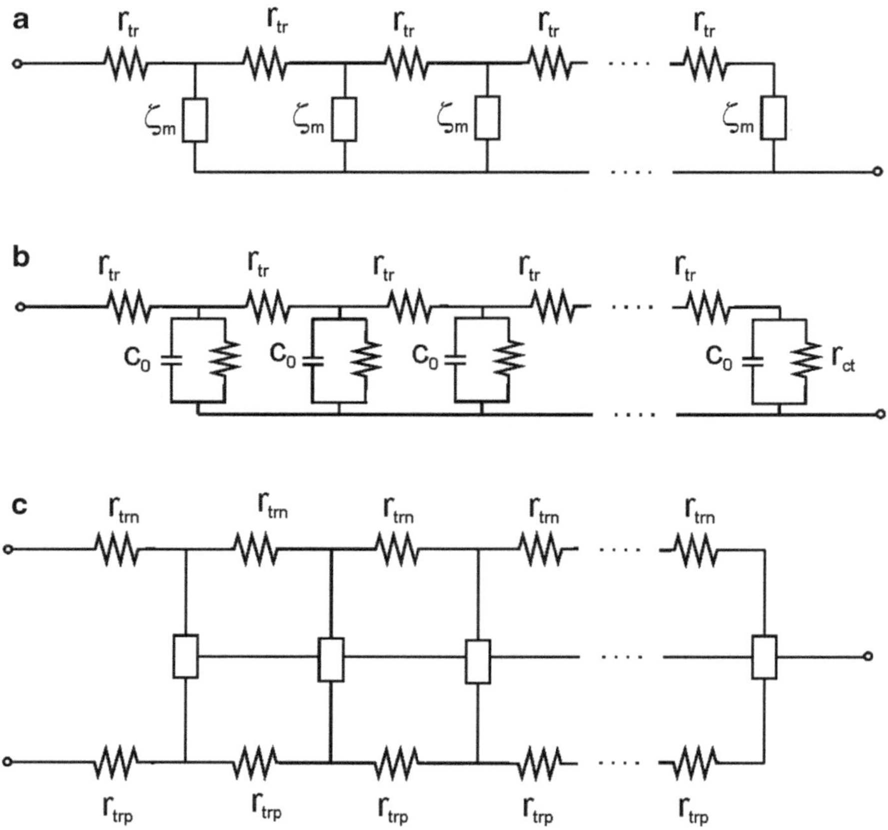


**Fig. 6.24** (**a**) Two channel transmission line equivalent circuit, with electron transport resistance and the transversal element $\zeta_m$ corresponds to the equivalent circuit represented in Fig. 6.10b. (**b**) Simplified equivalent circuit used in the present study including transport, chemical capacitance, and charge transfer of electrons. (**c**) Equivalent circuit including electron and hole transport and the central electrolyte rail. The *box* does not represent an impedance element but indicates a more complex connection, depending on the kinetic processes of electrons and holes

by the model showed as Fig. 6.24b: $TiO_2$ (Fabregat-Santiago et al. 2005; Gimenez et al. 2012; Rodenas et al. 2013; Trevisan et al. 2013), ZnO (Martinson et al. 2009) $WO_3$ (Balandeh et al. 2015) and $Fe_2O_3$ (Cummings et al. 2012).

### 6.8.1 *Transport and Charge Transfer*

In order to present a representative example of application of the transmission line model established in Fig. 6.24b, Fig. 6.25a shows the complex plane impedance plots for films of $TiO_2$ nanoparticles with different thickness. The marked area in

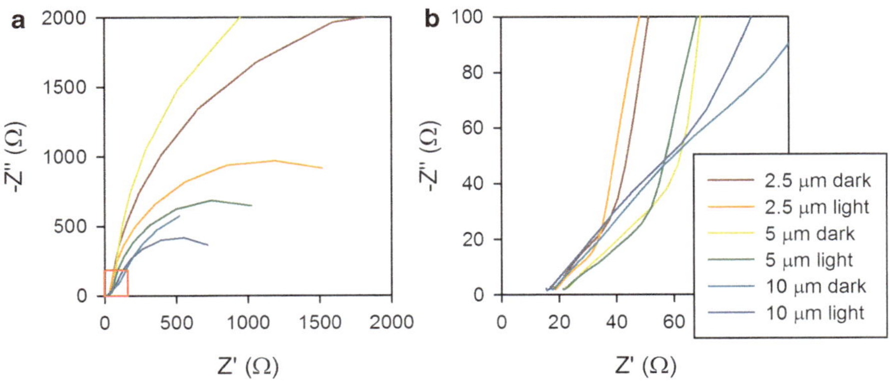


**Fig. 6.25** (**a**) EIS spectra obtained for nanoparticulated films of $TiO_2$ at −0.95 V vs. Ag/AgCl in the dark and under illumination at 100 mW/cm$^2$. (**b**) Magnification of the *squared area* in (**a**) to illustrate the 45° line related to transport resistance

Fig. 6.25a is showed as Fig. 6.25b. In this figure, the high frequency region clearly shows the 45° line characteristic of electron diffusion. It is clear that the length of this line increases with film thickness, evidencing that transport limitations are more important in thick films. From the model fitting, the chemical capacitance for electrons, $C_\mu$, the charge transfer resistance, $R_{ct}$, and the resistivity ($\rho_{TiO2}$) of the $TiO_2$ architectures could be extracted. As previously mentioned, $C_\mu$ monitors the electronic density of states at the Fermi level and probes the distribution of trap states below the conduction band. $\rho_{TiO2}$ is the reciprocal conductivity and $R_{ct}$ is related to the transfer of electrons at the $TiO_2$/solution interface (Bisquert 2002; Bisquert et al. 2004). These three quantities describe the main electronic mechanisms for the operation of the device. Figure 6.26 shows an example of the obtained results for the fitting of films of $TiO_2$ hollow nanowires (HNWs) and nanoparticles (NPs) sensitized with CdSe quantum dots by two different methods (Chemical Bath Deposition, CBD and Successive Ionic Layer Adsorption and Reaction, SILAR): the chemical capacitance exhibits the expected exponential dependence with potential, reflecting the exponential tail of the density of states of $TiO_2$ below the conduction band. To compare both transport and charge transfer resistances between the different specimens with the same density of electrons, the chemical capacitances of all samples were shifted to make them overlap and the corrected potential is termed "equivalent conduction band potential" ($V_{ecb}$). In spite of the one-dimensional nature of the hollow nanowires, the resistivity exhibits very similar values compared to nanoparticles. This could be explained by the polycrystalline nature of the nanowires since grain boundaries are preferential sites for recombination and can slow down charge transport through the material, decreasing the wire conductivity.

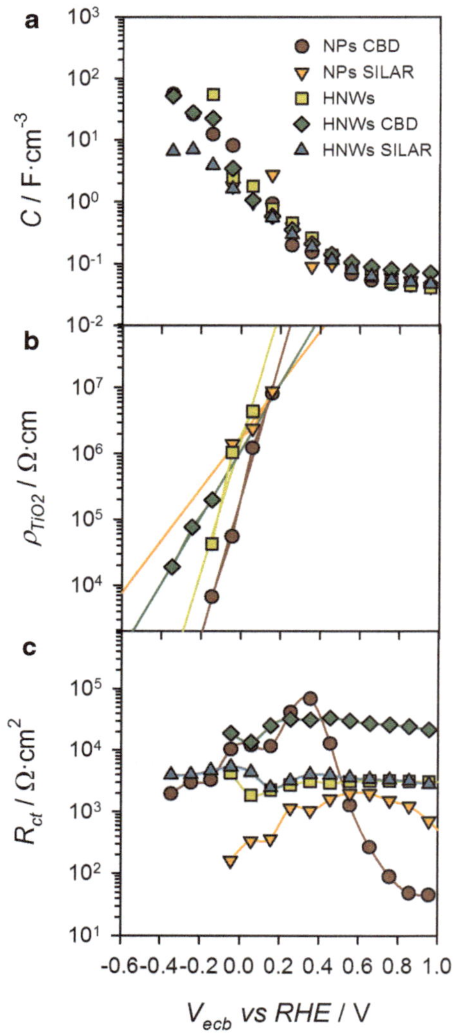


**Fig. 6.26** Parameters extracted after fitting the impedance spectroscopy spectra of $TiO_2$ nanoparticles (NPs) and hollow nanowires (HNWs) and NPs sensitized by different methods, chemical bath deposition (CBD) and successive ionic layer adsorption and reaction (SILAR), using the model presented in Fig. 9.22c. (**a**) Chemical capacitance ($C_\mu$), (**b**) $TiO_2$ resistivity ($\rho_{TiO2}$) and (**c**) recombination resistance ($R_{ct}$) as a function of potential (**a**), and versus the equivalent conduction band potential ($V_{ecb}$) (**b** and **c**). All impedance measurements were carried out in the dark. The lines in (**b**) are linear fittings of the experimental data. In (**c**) lines are drawn as eye guides. Reproduced with permission from Rodenas et al. (2013)

### 6.8.2 *Determination of Band Edge Shift by Displacement of the Chemical Capacitance*

When we consider a nanostructured semiconductor electrode, if the characteristic dimension is lower than the Debye length, the bands of the material cannot bend at the surface due to the small available size. The band edge displacement by surface modification appears as an overall displacement of the electronic energy levels of the semiconductor that can be monitored by the measurement of the chemical capacitance ($C_\mu$) that provides a powerful tool to understand band shift effects.

There are several examples in the literature for $TiO_2$ nanostructured electrodes in the field of photovoltaic and photoelectrochemical devices, as described in the previous section. A recent illustrative example shows that the deposition of carbon nitride ($C_3N_4$) on nanostructured $TiO_2$ scaffolds leads to the development of a large surface dipole (Xu et al. 2015). The $TiO_2/C_3N_4$ hybrid films were prepared by an in situ vapor-transport growth mode, using cyanuric acid–melamine (CM) or cyanuric acid–melamine–barbituric acid (CMB) supramolecular complexes as the precursor (labeled here as $TiO_2$/CM and $TiO_2$/CMB, respectively). Figure 6.27a shows the cyclic voltammetry curves obtained in 0.1 M $Na_2S$ solution for $C_3N_4$-modified $TiO_2$ substrates and comparison with reference $TiO_2$ electrodes. The voltammograms reflect the capacitive effects associated to the increasing density of states in the cathodic bias direction, typically reported on $TiO_2$ films (Bisquert 2014). Furthermore, it is also identified a cathodic peak at −0.6 V vs. Ag/AgCl for bare $TiO_2$, which is related to electron transfer through a localized bandgap state in $TiO_2$. For both $C_3N_4$-modified $TiO_2$ substrates, this peak is clearly shifted to more positive potentials. In order to confirm that this movement is related to band shift, IS measurements were carried out. The obtained complex plane impedance plots were fitted by using the transmission line model described in Fig. 6.24b, accounting for the coupling of transport and recombination with the electrolyte (Bisquert 2002; Wang et al. 2006). Figure 6.27b shows the evolution of the extracted chemical capacitance for $TiO_2$, $C_\mu$, with the applied reference electrode potential, and displays the exponential density of states of $TiO_2$ below the conduction band (Bisquert 2014). The slope of the $C_\mu$ versus the potential, in the semilogarithmic scale, is similar for all the electrodes, which indicates that the deposition of $C_3N_4$ does not significantly modify the DOS of $TiO_2$. However, a clear anodic shift of about 200 mV is observed for $C_\mu$ of the $C_3N_4$ sensitized $TiO_2$ electrodes, related to the downward displacement of the $TiO_2$ conduction band. This is consistent with a strong dipole formation after $C_3N_4$ deposition.

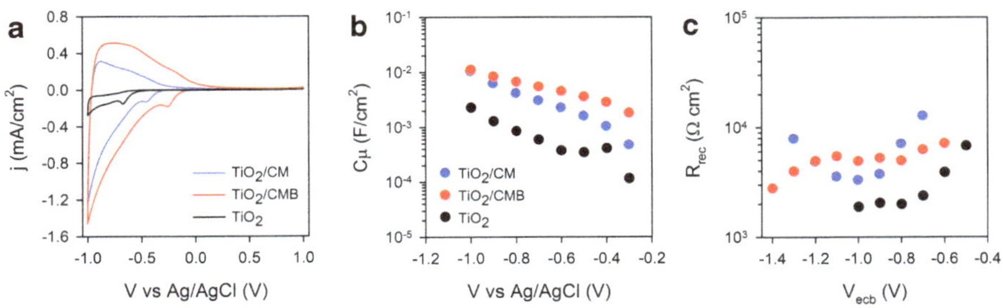


**Fig. 6.27** (**a**) Cyclic voltammetry, (**b**) chemical capacitance ($C_\mu$), and (**c**) recombination resistance ($R_{rec}$) of pristine $TiO_2$, $TiO_2$/CM, and $TiO_2$/CMB substrates obtained in the dark. The measurement was carried out in a three-electrode system, Pt as the counter and Ag/AgCl as the reference electrode and 0.1 M $Na_2S$ solution as the electrolyte. Reproduced with permission from Xu et al. (2015)

## 6.9 Conclusions

Our analysis of the transport, recombination, and charge transfer processes at the semiconductor/electrolyte system indicates that this is a problem of great complexity due to the combination of different carriers and sequential processes that determine the overall performance. One principal problem is the distinction between direct transfer from the extended states of valence band versus the hole transfer from surface states at an illuminated photoanode. We have described a variety of models that provide criteria for the separation of the different processes according to the interpretation of capacitances in the system. Similarly, IS can provide a great deal of information on the operation of catalyzed surfaces and the identification of band edge shifts. Finally, nanostructured porous electrodes are characterized by significant variation of carrier density and conductivity, and these features can be established using the distributed transmission line models.

## References


Allongue P, Cachet H (1984) I-V curve and surface state capacitance at illuminated semiconductor/liquid contacts. J Electroanal Chem 176:369–375

Badia-Bou L, Mas-Marza E, Rodenas P, Barea EM, Fabregat-Santiago F, Gimenez S, Peris E, Bisquert J (2013) Water oxidation at hematite photoelectrodes with an iridium-based catalyst. J Phys Chem C 117:3826–3833

Balandeh M, Mezzetti A, Tacca A, Leonardi S, Marra G, Divitini G, Ducati C, Meda L, Di Fonzo F (2015) Quasi-1D hyperbranched $WO_3$ nanostructures for low-voltage photoelectrochemical water splitting. J Mater Chem A 3:6110–6117

Basu M, Zhang Z-W, Chen C-J, Chen P-T, Yang K-C, Ma C-G, Lin CC, Hu S-F, Liu R-S (2015) Heterostructure of Si and $CoSe_2$: a promising photocathode based on a non-noble metal catalyst for photoelectrochemical hydrogen evolution. Angew Chem Int Ed 54:6211–6216

Bertoluzzi L, Bisquert J (2012) Equivalent circuit of electrons and holes in thin semiconductor films for photoelectrochemical water splitting applications. J Phys Chem Lett 3:2517–2522

Bertoluzzi L, Lopez-Varo P, Jimenez Tejada JA, Bisquert J (2016) Charge transfer processes at the semiconductor/electrolyte interface for solar fuel production: insight from impedance spectroscopy. J Mater Chem A 4:2873–2879

Bisquert J (2002) Theory of the impedance of electron diffusion and recombination in a thin layer. J Phys Chem B 106:325–333

Bisquert J (2008) Beyond the quasi-static approximation: impedance and capacitance of an exponential distribution of traps. Phys Rev B 77:235203

Bisquert J (2010) Theory of the impedance of charge transfer via surface states in dye-sensitized solar cells. J Electroanal Chem 646:43–51

Bisquert J (2014) Nanostructured energy devices: equilibrium concepts and kinetics. CRC Press, Boca Raton

Bisquert J, Fabregat-Santiago F (2010) In: Kalyanasundaram K (ed) Dye-sensitized solar cells. CRC Press, Boca Raton

Bisquert J, Cahen D, Hodes G, Ruhle S, Zaban A (2004) Physical chemical principles of photovoltaic conversion with nanoparticulate, mesoporous dye-sensitized solar cells. J Phys Chem B 108:8106–8118

Caban-Acevedo M, Kaiser NS, English CR, Liang D, Thompson BJ, Chen H-E, Czech KJ, Wright JC, Hamers RJ, Jin S (2014) Ionization of high-density deep donor defect states explains the low photovoltage of iron pyrite single crystals. J Am Chem Soc 136:17163–17179

Cachet H, Sutter EMM (2015) Kinetics of water oxidation at $TiO_2$ nanotube arrays at different pH domains investigated by electrochemical and light-modulated impedance spectroscopy. J Phys Chem C 119:25548–25558

Cendula P, Tilley SD, Gimenez S, Bisquert J, Schmid M, Grätzel M, Schumacher JO (2014) Calculation of the energy band diagram of a photoelectrochemical water splitting cell. J Phys Chem C 118:29599–29607

Chou J-C, Lin S-A, Lee C-Y, Gan J-Y (2013) Effect of bulk doping and surface-trapped states on water splitting with hematite photoanodes. J Mater Chem A 1:5908–5914

Cummings CY, Marken F, Peter LM, Wijayantha KGU, Tahir AA (2012) New insights into water splitting at mesoporous alpha-$Fe_2O_3$ films: a study by modulated transmittance and impedance spectroscopies. J Am Chem Soc 134:1228–1234

Dareedwards MP, Goodenough JB, Hamnett A, Trevellick PR (1983) Electrochemistry and photoelectrochemistry of Iron(III) oxide. J Chem Soc Farad Trans I 79:2027–2041

Doyle RL, Lyons MEG (2013) An electrochemical impedance study of the oxygen evolution reaction at hydrous iron oxide in base. Phys Chem Chem Phys 15:5224–5237

Fabregat-Santiago F, Bisquert J, Garcia-Belmonte G, Boschloo G, Hagfeldt A (2005) Influence of electrolyte in transport and recombination in dye-sensitized solar cells studied by impedance spectroscopy. Sol Energy Mater Sol Cells 87:117–131

Fabregat-Santiago F, Garcia-Belmonte G, Mora-Seró I, Bisquert J (2011) Characterization of nanostructured hybrid and organic solar cells by impedance spectroscopy. Phys Chem Chem Phys 13:9083–9118

Freund T, Morrison SR (1968) Mechanisms of cathodic processes on semiconductor Zinc Oxide. Surf Sci 9:119

Gimenez S, Dunn HK, Rodenas P, Fabregat-Santiago F, Miralles SG, Barea EM, Trevisan R, Guerrero A, Bisquert J (2012) Carrier density and interfacial kinetics of mesoporous $TiO_2$ in aqueous electrolyte determined by impedance spectroscopy. J Electroanal Chem 668:119–125

Hens Z (1999) The electrochemical impedance on one-equivalent electrode processes at dark semiconductor redox electrodes involving charge transfer through surface states. 1. Theory. J Phys Chem B 103:122–129

Hisatomi T, Le Formal F, Cornuz M, Brillet J, Tetreault N, Sivula K, Graetzel M (2011) Cathodic shift in onset potential of solar oxygen evolution on hematite by 13-group oxide overlayers. Energy Environ Sci 4:2512–2515

Hong SJ, Lee S, Jang JS, Lee JS (2011) Heterojunction $BiVO_4/WO_3$ electrodes for enhanced photoactivity of water oxidation. Energy Environ Sci 4:1781–1787

Hu Y, Boudoire F, Hermann-Geppert I, Bogdanoff P, Tsekouras G, Mun BS, Fortunato G, Graetzel M, Braun A (2016) Molecular origin and electrochemical influence of capacitive surface states on iron oxide photoanodes. J Phys Chem C. doi:10.1021/acs.jpcc.5b08013

Iandolo B, Hellman A (2014) The role of surface states in the oxygen evolution reaction on hematite. Angew Chem Int Ed 53:13404–13408

Kelly JJ, Memming R (1982) The influence of surface recombination and trapping on the cathodic photocurrent at p-type III-V electrodes. J Electrochem Soc 129:730–738

Klahr B, Hamann T (2014) Water oxidation on hematite photoelectrodes: insight into the nature of surface states through in situ spectroelectrochemistry. J Phys Chem C 118:10393–10399

Klahr B, Gimenez S, Fabregat-Santiago F, Bisquert J, Hamann TW (2012a) Electrochemical and photoelectrochemical investigation of water oxidation with hematite electrodes. Energy Environ Sci 5:7626–7636

Klahr B, Gimenez S, Fabregat-Santiago F, Bisquert J, Hamann TW (2012b) Photoelectrochemical and impedance spectroscopic investigation of water oxidation with "Co-Pi"-coated hematite electrodes. J Am Chem Soc 134:16693–16700

Klahr B, Gimenez S, Fabregat-Santiago F, Hamann T, Bisquert J (2012c) Water oxidation at hematite photoelectrodes: the role of surface states. J Am Chem Soc 134:4294–4302
Klahr B, Gimenez S, Zandi O, Fabregat-Santiago F, Hamann T (2015) Competitive photoelectrochemical methanol and water oxidation with hematite electrodes. ACS Appl Mater Interfaces 7:7653–7660
Le Formal F, Tetreault N, Cornuz M, Moehl T, Graetzel M, Sivula K (2011) Passivating surface states on water splitting hematite photoanodes with alumina overlayers. Chem Sci 2:737–743
Li J, Peter LM (1985) Surface recombination at semiconductor electrodes. Part III. Steady-state and intensity modulated photocurrents response. J Electroanal Chem 193:27–47
Li J, Meng F, Suri S, Ding W, Huang F, Wu N (2012) Photoelectrochemical performance enhanced by a nickel oxide-hematite p-n junction photoanode. Chem Commun 48:8213–8215
Lin Y, Xu Y, Mayer MT, Simpson ZI, McMahon G, Zhou S, Wang D (2012) Growth of p-type hematite by atomic layer deposition and its utilization for improved solar water splitting. J Am Chem Soc 134:5508–5511
Martinson ABF, Goes MS, Fabregat-Santiago F, Bisquert J, Pellin MJ, Hupp JT (2009) Electron transport in dye-sensitized solar cells based on ZnO nanotubes: evidence for highly efficient charge collection and exceptionally rapid dynamics. J Phys Chem A 113:4015–4021
Monllor-Satoca D, Bartsch M, Fabrega C, Genc A, Reinhard S, Andreu T, Arbiol J, Niederberger M, Morante JR (2015) What do you do, titanium? Insight into the role of titanium oxide as a water oxidation promoter in hematite-based photoanodes. Energy Environ Sci 8:3242–3254
Peter LM (1990) Dynamic aspects of semiconductor photoelectrochemistry. Chem Rev 90:753–769
Ponomarev EA, Peter LM (1995) A comparison of intensity modulated photocurrent spectroscopy and photoelectrochemical impedance spectroscopy in a study of photoelectrochemical hydrogen evolution at p-InP. J Electroanal Chem 397:45–52
Pyper KJ, Yourey JE, Bartlett BM (2013) Reactivity of $CuWO_4$ in photoelectrochemical water oxidation is dictated by a midgap electronic state. J Phys Chem C 117:24726–24732
Raga SR, Barea EM, Fabregat-Santiago F (2012) Analysis of the origin of open circuit voltage in dye solar cells. J Phys Chem Lett 3:1629–1634
Reichman J (1980) The current-voltage characteristics of semiconductor-electrolyte junction photovoltaic cells. Appl Phys Lett 36:574–577
Riha SC, Klahr BM, Tyo EC, Seifert S, Vajda S, Pellin MJ, Hamann TW, Martinson ABF (2013) Atomic layer deposition of a submonolayer catalyst for the enhanced photoelectrochemical performance of water oxidation with hematite. ACS Nano 7:2396–2405
Rodenas P, Song T, Sudhagar P, Marzari G, Han H, Badia-Bou L, Gimenez S, Fabregat-Santiago F, Mora-Sero I, Bisquert J, Paik U, Kang YS (2013) Quantum dot based heterostructures for unassisted photoelectrochemical hydrogen generation. Adv Energy Mater 3:176–182
Salvador P, Gutierrez C (1984) The nature of surface states involved in the photo- and electroluminescence spectra of n-titanium dioxide electrodes. J Phys Chem 88:3696–3698
Salvador P, Gutiérrez C (1984) Mechanisms of charge transfer at the semiconductor-electrolyte interface. J Electrochem Soc 131:326–336
Shi X, Herraiz-Cardona I, Bertoluzzi L, Lopez-Varo P, Bisquert J, Park JH, Gimenez S (2016) Understanding the synergistic effect of $WO_3$-$BiVO_4$ heterostructures by impedance spectroscopy. Phys Chem Chem Phys 18:9255–9261. doi:10.1039/C5CP07905E
Steier L, Herraiz-Cardona I, Gimenez S, Fabregat-Santiago F, Bisquert J, Tilley SD, Graetzel M (2014) Understanding the role of underlayers and overlayers in thin film hematite photoanodes. Adv Funct Mater 24:7681–7688
Tench DM, Gerischer H (1977) The phototransition in ZnO at 380 nm studied by anodic photocurrents. J Electrochem Soc 124:1612–1618
Trevisan R, Rodenas P, Gonzalez-Pedro V, Sima C, Sanchez RS, Barea EM, Mora-Sero I, Fabregat-Santiago F, Gimenez S (2013) Harnessing infrared photons for photoelectrochemical hydrogen generation. A PbS quantum dot based "quasi-artificial leaf". J Phys Chem Lett 4:141–146

Vanmaekelbergh D (1997) Direct and surface state mediated electron transfer at semiconductor/ electrolyte junctions. II. A comparison of the interfacial admittance. Electrochim Acta 42:1135–1141

Wang Q, Ito S, Grätzel M, Fabregat-Santiago F, Mora-Seró I, Bisquert J, Bessho T, Imai H (2006) Characteristics of high efficiency dye-sensitized solar cells. J Phys Chem B 110:19406–19411

Xu J, Herraiz-Cardona I, Yang X, Gimenez S, Antonietti M, Shalom M (2015) The complex role of carbon nitride as a sensitizer in photoelectrochemical cells. Adv Opt Mater 3(8):1052–1058

Yang W, Yu Y, Starr MB, Yin X, Li Z, Kvit A, Wang S, Zhao P, Wang X (2015) Ferroelectric polarization-enhanced photoelectrochemical water splitting in $TiO_2$–$BaTiO_3$ core–shell nanowire photoanodes. Nano Lett 15:7574–7580

Yatom N, Neufeld O, Caspary Toroker M (2015) Toward settling the debate on the role of $Fe_2O_3$ surface states for water splitting. J Phys Chem C 119:24789–24795

Zandi O, Hamann TW (2014) Enhanced water splitting efficiency through selective surface state removal. J Phys Chem Lett 5:1522–1526

Zhong DK, Gamelin DR (2010) Photoelectrochemical water oxidation by cobalt catalyst ("Co–Pi")/α-$Fe_2O_3$ composite photoanodes: oxygen evolution and resolution of a kinetic bottleneck. J Am Chem Soc 132:4202–4207

# Chapter 7
# Advanced Photoelectrochemical Characterization: Principles and Applications of Dual-Working-Electrode Photoelectrochemistry

**Fuding Lin and Shannon W. Boettcher**

## 7.1 Background

Efficient photoelectrochemical (PEC) energy conversion systems such as water-splitting photoelectrodes and dye-sensitized solar cells require optimal integration of multiple components. The ability to directly probe the dynamic state of individual components is desirable because such information provides input to accurately model and understand the functionality of the integrated device. The fundamental challenge is that typical photoelectrochemical techniques rely on analyzing the current response obtained by sweeping the voltage of a single-working electrode which only sets the "boundary condition" for the multicomponent system. For example, in a dye-sensitized solar cell, the working electrode sets the potential at the back contact, but the electrostatic potential and electronic carrier (hole and electron) distributions throughout the porous semiconductor and electrolyte are not readily observable. In a water-splitting photoelectrode, the potential of the semiconductor back ohmic contact is controlled, but illumination generates a surface potential to drive the photochemical reaction that is not easily quantified. Further, a second electrocatalytic phase on the semiconductor surface can store charge carriers, and the steady-state potential of this phase is unknown under operating conditions. Current-voltage curves measured in such scenarios therefore "hide" the complicated internal dynamics of the multiple components, and many underlying mechanisms can give rise to similar current-voltage behavior.

One strategy to provide additional information is to use secondary electrodes to sense the potential of the various other components of the system (Wilson et al. 1977; Pinson 1977; Nakato and Tsubomura 1982; Zaban et al. 2003; Lobato et al. 2006). For instance, following the initial investigation of photoelectrolysis on $TiO_2$ (Fujishima and Honda 1972), it was suggested that an n-GaP photoanode

F. Lin • S.W. Boettcher (✉)
Department of Chemistry and Biochemistry, University of Oregon, Eugene, OR 97403, USA
e-mail: swb@uoregon.edu



S. Giménez, J. Bisquert (eds.), *Photoelectrochemical Solar Fuel Production*,
DOI 10.1007/978-3-319-29641-8_7

coated with a thin electrolyte-permeable gold film overlayer could generate enough photovoltage to split water without corrosion (Nakato et al. 1976). However, direct measurement of the gold film potential during PEC operation later revealed that sustained water oxidation at the gold surface was unlikely because the potential of the gold film was not sufficiently anodic at steady state (Harris et al. 1977; Pinson 1977). In the field of dye-sensitized solar cell research, in situ measurement of the electron quasi-Fermi level at the top of the electrolyte-permeated nanocrystalline $TiO_2$ electrode (i.e., furthest from the charge-collecting back ohmic contact) using a passivated Ti metal electrode enabled direct verification of a diffusive transport model for electrons in the $TiO_2$ as well as better estimates of important material properties such as the conduction-band density of states and the electron diffusion coefficient (Lobato et al. 2006).

In this chapter we focus on the challenges of uncovering the mechanistic role of the electrocatalyst (EC) in composite photoelectrodes and new insights obtained through a dual-working-electrode PEC experimental technique which allows direct access *or control* of the dynamic state of the EC layer.

Although numerous studies have shown that integrating electrocatalysts (ECs) onto light-absorbing semiconductors (SCs) can improve the PEC performance of composite photoelectrodes (Berglund et al. 2011; Zhong et al. 2011a; Seabold and Choi 2012; Abdi et al. 2013; Kim and Choi 2014), the actual functional role of the EC layer has been a subject of active debate (Gamelin 2012; Sivula 2013; Peter and Wijayantha 2014). One question that is important from a device-engineering perspective is whether the EC layer does one or both of the following: (a) improves the reaction kinetics of the water-splitting process, i.e., functions as a catalyst; (b) improves the SC|EC interface by reducing surface recombination losses, shifting band edges of the SC relative to the solution redox energy level $E_{redox}$ (which is equal to $q\mathcal{E}_{O_2/OH^-}$ when, e.g., $O_2$ is bubbled through a basic test electrolyte), and/or increasing the band bending of the SC.

Over the past few years, several experimental techniques have been employed by different groups of researchers to characterize the performance of PEC photoelectrodes and understand the role of the EC layer in those composite devices, particularly for water-oxidizing photoanodes which will be the focus of this chapter (Barroso et al. 2011; Le Formal et al. 2011, 2012; Wijayantha et al. 2011; Klahr et al. 2012; Peter et al. 2012). Although performance-enhancing ECs have been found for all oxide photoanode SCs that absorb visible light (e.g., $BiVO_4$ and $\alpha$-$Fe_2O_3$), the mechanism of improvement is not well understood. Several studies have found that cobalt-oxyhydroxide-based ECs electrodeposited from phosphate buffer (Co–Pi) reduce the OER onset potential of $\alpha$-$Fe_2O_3$ photoanodes, but hypotheses on the role of Co–Pi differ (Zhong and Gamelin 2010; Barroso et al. 2011, 2012; Le Formal et al. 2011; Zhong et al. 2011b; Klahr et al. 2012; Pendlebury et al. 2012).

Zhong and Gamelin used linear sweep voltammetry and transient photocurrent decay measurements to study Co–Pi/$\alpha$-$Fe_2O_3$ composite photoanodes and found that the PEC performance was enhanced by sparse deposition of Co–Pi

(Zhong and Gamelin 2010; Zhong et al. 2011b). They suggest that Co–Pi suppresses electron–hole recombination by increasing the OER kinetics, but does not otherwise alter the characteristics of the electrode/electrolyte interface. A separate study by Riha et al. using impedance spectroscopy on the effect of a "Co $(OH)_2/Co_3O_4$" EC layer deposited on $\alpha$-$Fe_2O_3$ via atomic layer deposition (ALD) found evidence of reduced charge-transfer resistance, which suggests that the EC layer indeed improved OER kinetics (Riha et al. 2013). Durrant and coworkers employed transient absorption spectroscopy to study hematite electrodes modified with Co–Pi and $Ga_2O_3$. They concluded that Co–Pi forms a Schottky-type junction with $\alpha$-$Fe_2O_3$ that increases band bending and thus improves charge separation and photoelectrode performance (Barroso et al. 2011, 2012). Because catalytically inactive $Ga_2O_3$ yielded similar results, it was argued that Co–Pi did not improve the reaction kinetics. A further interpretation of these results by Gamelin suggests that photogenerated holes might bypass the Co–Pi and directly react with water from the $\alpha$-$Fe_2O_3$ surface (Gamelin 2012).

Meanwhile, Hamann and coworkers used steady-state and transient photoelectrochemical measurements as well as impedance spectroscopy to study planar thin-film hematite electrodes coated with Co–Pi catalyst and found evidence that the catalyst undergoes redox cycling while the electrode is scanned between ORR and OER potentials under illumination (indicating the catalyst is porous and permeable to electrolyte), but no evidence of increased band bending was found (Klahr et al. 2012). They suggest that photogenerated holes from hematite were efficiently collected and stored in the Co–Pi layer leading to reduced recombination and hence larger cathodic shift of $V_{oc}$. Le Formal et al. used transient photocurrent and photovoltage techniques to study the effect of adding a $Al_2O_3$ layer on $\alpha$-$Fe_2O_3$ and/or treating the hematite surface with $Co(NO_3)_2$ solution. They found that even after the SC is passivated with $Al_2O_3$, the adsorption of $Co^{2+}$ further enhances the photoanode performance (Le Formal et al. 2012). They concluded that the Co catalyst promotes charge separation by extracting photogenerated holes from the SC.

The published results obtained through different experimental techniques and interpreted by different researchers thus far have not been able to offer a consistent answer to the question of how the EC layer modifies kinetics and/or interface energetics. The difficulty of unambiguously identifying the role of the EC is in part due to the absence of direct information regarding the state of the catalyst, such as its potential, and the current passing through it. The importance of such information becomes clear upon analyzing the energy band diagram of a typical catalyzed photoanode, as shown in Fig. 7.1. Both $E_{sem}$ and $q\mathcal{E}_{O_2/OH^-}$ are known during a PEC voltammetry sweep, since the first is set by the working electrode ohmic contact and the second can be calculated for the electrolyte solution used. If the change of the EC energy $E_{cat}$ could be monitored in situ, then both the overpotential ($\eta$) that is driving the OER and the photovoltage ($V_{ph}$) generated by the SC|EC junction can be determined at any value of $E_{sem}$. By monitoring the potential of different ECs on the same SC, one could in principle determine whether a particular EC improves the photovoltage (energetics) or reduces the overpotential (kinetics) of the composite photoelectrode. Furthermore, because the catalyst is (likely)

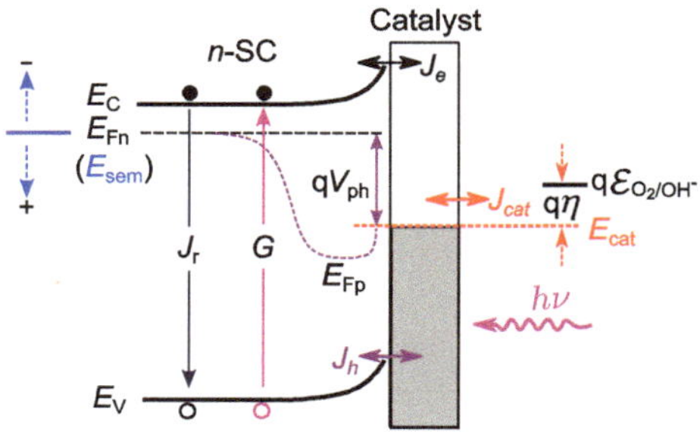


**Fig. 7.1** Energy diagram of a typical catalyzed photoanode. $\mathcal{E}_{O_2/OH^-}$ is the reversible oxygen evolution reaction potential, $E_{cat}$ is the Fermi level of the catalyst layer, $E_c$ and $E_v$ are the conduction and valence band edges, respectively. $E_{Fn}$ ($E_{sem}$) is the Fermi level of electrons in the n-type semiconductor, while $E_{Fp}$ is the quasi-Fermi level of holes under illumination. $J_r$ represents the current due to bulk recombination and $G$ represents the generation of carriers due to light excitation. Other quantities are defined as shown in the schematic. Italicized $E$s are used to represent relative energies and $\mathcal{E}$ indicates a redox potential

charged by holes from the semiconductor, its potential provides a lower-bound estimate of the quasi-Fermi level of the photogenerated holes at the semiconductor surface. In the limit where the barrier height is large and the forward electron current (i.e., "surface" recombination current) is small, the value of $E_{cat} - E_{sem}$ thus provides information on the bulk recombination-limited quasi-Fermi level splitting within the SC.

To directly access the state of the EC layer, we modified the conventional single-working-electrode photoelectrochemistry (SWE PEC) setup to include a second working electrode which is in direct contact with the EC. This experiment setup is therefore referred to as dual-working-electrode (DWE) PEC to distinguish it from the conventional single-working-electrode setup (Lin and Boettcher 2014). Although other techniques such as in situ X-ray or optical absorption of catalysts on the photoelectrode (Braun et al. 2012; Klahr and Hamann 2014; Friebel et al. 2015) might also provide information on the catalyst potential with proper calibration, only the DWE PEC method will be the focus of this chapter.

## 7.2 Single-Working-Electrode Photoelectrochemistry

The PEC behavior of composite photoelectrodes is typically characterized by experimental techniques that are based on the conventional three-electrode electrochemical cell setup, which is schematically shown in Fig. 7.2 for a catalyzed n-type semiconductor photoanode. We refer to this experimental configuration here as SWE PEC. In such a configuration, the working electrode of a potentiostat controls (or monitors) the Fermi level of the majority carrier ($E_{Fn}$) through an ohmic contact to the backside of the SC. A reference electrode is used so that $E_{Fn}$ can be compared to a known energy level such as that of a standard hydrogen

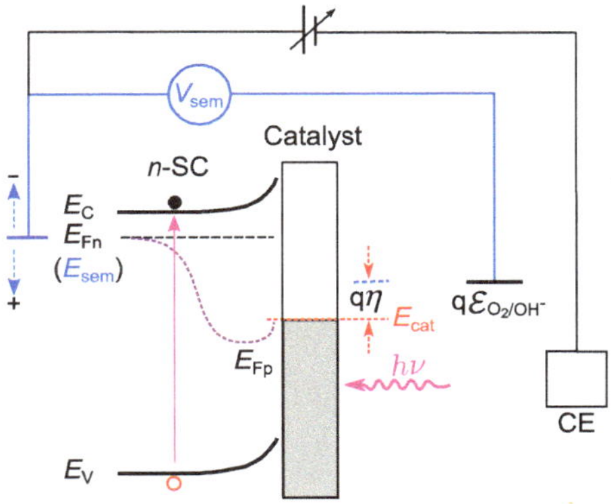


**Fig. 7.2** Energy diagram and wiring schematic of traditional SWE PEC setup. The *n*-SC potential $V_{sem}$ is controlled relative to $\mathcal{E}_{O_2/OH^-}$ by the working electrode (WE) of a potentiostat (using a reference electrode such as the standard calomel electrode). An inert material, such as Pt, is used for the counter electrode (CE)

electrode (SHE) or the reversible oxygen evolution reaction energy $q\mathcal{E}_{O_2/OH^-}$ in a particular solution. In addition to the working and reference electrodes, an inert counter electrode (CE) is used to complete the current loop so that the potentiostat can drive whatever current necessary to maintain a particular $E_{Fn}$ value.

To characterize the photoelectrochemical performance of a composite photoanode, light is introduced from either the front (EC) side or the back (SC) side depending on the optical absorption characteristics of the EC and the substrate. $V_{sem}$ is then modulated (swept, stepped, or oscillated) and the resulting changes in the current flowing through the ohmic contact ($J_{sem}$) recorded for analysis. In addition to modulating $V_{sem}$, the intensity of the light can also be varied to gain further insight.

The basic processes occurring in a catalyzed n-type photoelectrode are shown in Fig. 7.1. The n-type semiconductor, catalyst layer, and solution are all characterized by Fermi levels that equilibrate in the dark ($E_{sem}$, $E_{cat}$, and $E_{redox}$, respectively). Under illumination the concentration of minority holes increases and thus the hole quasi-Fermi level $E_{Fp}$ drops down from the electron level $E_{Fn}$. At steady state during photo-driven oxygen evolution, $E_{cat}$ must be driven lower (more positive) than $E_{redox}$ ($= q\mathcal{E}_{O_2/OH^-}$) such that there is a net forward hole current $J_{cat}$ from catalyst to solution. The degree to which $E_{Fp}$ separates from $E_{Fn}$ at the SC surface is governed by the relative forward and reverse rates of electron and hole transport at the SC|EC interface in addition to the rates of recombination ($J_r$) and generation ($G$).

Although much can be learned from conventional SWE PEC measurements using the back ohmic contact, it is not possible to determine how the individual charge transfer, catalysis, and recombination steps occurring at or near the SC|EC and EC|Sol interfaces affect the overall shape of the measured *J–V* curves. This difficulty is the result of several factors. First, it is not possible to determine which portion of the total applied potential (i.e., $qV_{app} = E_{sem} - E_{redox}$) drops at the SC|EC interface and which portion drops at the EC|Sol interface because $E_{cat}$ is not known.

Further, the current measured is the sum of the net electron and hole current and it is not possible to distinguish whether the holes or electrons flow into the catalyst or directly into the solution (e.g., in the case where the catalyst has incomplete coverage or is porous/electrolyte permeable).

These ambiguities have contributed to the variety of conflicting views introduced in the previous section. Attempts to follow the electronic processes with transient absorption spectroscopies have provided new insight (Huang et al. 2012), for example, into surface recombination rates (Barroso et al. 2011, 2012; Le Formal et al. 2014), but the resulting data is also difficult to interpret because it is challenging to connect the observed transient absorption signals resulting from a high-intensity ultrafast laser excitation to specific physical processes occurring at 1-sun. Methods based on impedance are useful (Klahr et al. 2012) but rely on fitting to equivalent circuits to account for all the electrochemical subsystems, which may lead to ambiguity in the data interpretation. New methods that complement existing ones are needed to investigate the interfacial phenomena described above.

## 7.3 Dual-Working-Electrode Photoelectrochemistry

In order to learn more about the charge-transfer process at the SC|EC|Sol interfaces, e.g., the potential of the EC layer $E_{cat}$ during operation, a logical approach is to make a direct electrical contact to the EC layer. In this case the EC potential can be either passively monitored or actively controlled so that both the SC|EC interface and the EC|Sol interface can be independently probed. A PEC setup with the additional EC contact essentially has two working electrodes that are independent of each other. This experimental setup is therefore referred to as "dual-working-electrode photoelectrochemistry" or DWE PEC for convenience. In this section and the following, we discuss how DWE PEC is implemented and what information can be obtained through this technique.

### *7.3.1 Dual-Working-Electrode Setup in Electrochemistry*

The idea of using two working electrodes in the same electrochemical cell is old. For example, Gerischer et al. as well as Kenkel and Bard have used DWE setup for generation and detection of eletroactive species in flow cells (Gerischer et al. 1965; Kenkel and Bard 1974). A DWE configuration that allows potential monitoring of an additional electrode has been employed by several researchers to investigate the function of metal overlayers and apparent "surface" quasi-Fermi level of GaP-based photoanodes (though in this case the addition of the metal layer likely modifies the actual surface potential of the semiconductor) (Pinson 1977; Wilson et al. 1977). Lin and Lonergan have used a DWE setup to monitor the in situ conductivity changes of electroactive polymers while the material undergoes oxidation and reduction cycle (Lin and Lonergan 2006). The initial conception of the DWE PEC experiment has been directly influenced by these examples. More recently we employed the same technique to study the impact of oxidation state on the conductivity of metal-oxide-based OER catalysts (Trotochaud et al. 2014; Burke et al. 2015).

## 7.3.2 *Energy Diagram of the DWE PEC Setup*

Figure 7.3 shows the energy band diagram and wiring schematic of the DWE PEC experimental setup. It is assumed that the Fermi level of the EC equilibrates with that of the metallic (e.g., Au) contact, as commonly assumed when electrocatalysts are deposited on conductive substrates. This assumption only holds, however, if the catalyst layer is of sufficient electrical conductivity. For simplicity a "metal-like" density of states (DOS) is used for the EC in Fig. 7.3, but any type of EC can be represented with a change in the shape of the DOS.

With appropriate choice of contact material and necessary equipment, the additional contact to the EC allows direct control or monitoring of the EC potential as well as the measurement of current flowing to and from the EC.

In the context of studying photoelectrodes for water splitting, the material choice for the EC contact has a few constraints. First, the contact metal needs to be stable in the electrolyte under working conditions. Second, the contact materials must be much less active than the EC in catalyzing the reaction under investigation so that it has negligible effect on reaction kinetics. In other words, the contact metal must equilibrate with EC rather than the solution. Au is a good choice for this because it has slow OER kinetics relative to typically used OER catalysts. Third, the contact needs to be porous so that intimate contact between the EC and electrolyte is preserved. We use thin Au layers, which have these properties, as will be shown later.

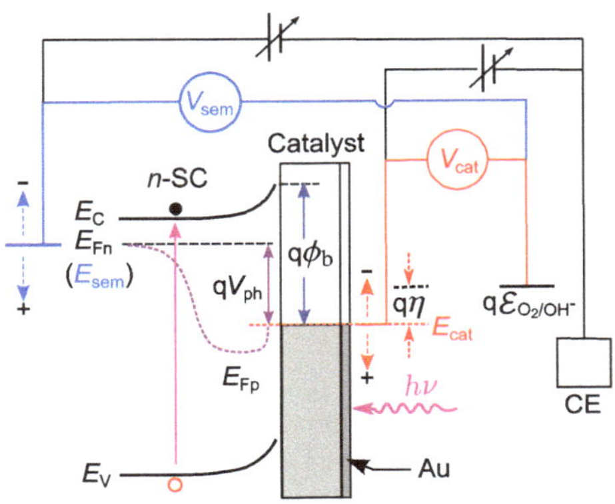


**Fig. 7.3** Energy diagram and wiring schematic of DWE PEC setup. The *n*-SC potential $V_{sem}$ and the EC potential $V_{cat}$ can be independently controlled or monitored relative to $\mathcal{E}_{O_2/OH^-}$ by a bi-potentiostat (using a single reference electrode) or an equivalent set of instruments

### 7.3.3 Instrumentation and Samples for DWE PEC

The monitoring and/or controlling of the second working electrode in a DWE PEC setup require additional consideration with regard to the instrumentation. Depending on the goal of the experiment, two different instrument setups could be employed.

When one of the two WEs is used only for potential sensing, and there is no need to either control its potential or measure the current flowing through it, the instrument setup is the simplest. In such case one can use a conventional single-channel potentiostat to control one WE and use a high-input-impedance (preferably $>10\ G\Omega$) voltmeter to monitor the potential of the other WE relative to the common reference electrode.

Alternatively, modern potentiostats usually have a built-in voltmeter that can be used for monitoring the potential of a second WE. The input impedance of such built-in voltmeter is usually sufficiently high, but it is always a good practice to verify it before use (in cases where the catalyst layer is not significantly conductive and the input impedance of the voltmeter not sufficiently high, we have found the measured catalyst potential to be inaccurate).

When independent control of both WEs in the DWE PEC setup is desirable, a bi-potentiostat could be used. In this setup, both WEs are controlled against the common reference and counter electrode. In principle one could also use two single-channel potentiostats for the same purpose, but it is then necessary to make sure no detrimental ground loops are formed between the two instruments and that the feedback loops do not interfere with each other.

Samples for the DWE PEC experiment can be constructed in three or four steps, depending on how the EC layer is deposited. Figure 7.4 illustrates the construction process of a sample with electrodeposited EC. If spin coating and thermal annealing is required for the deposition of EC (such as $IrO_x$), it is more convenient to deposit the EC before the back ohmic contact is made (Fig. 7.4a). More details of sample fabrication, testing, and experiment setup can be found in Lin and Boettcher (2014) and in the Appendix at the end of this chapter.

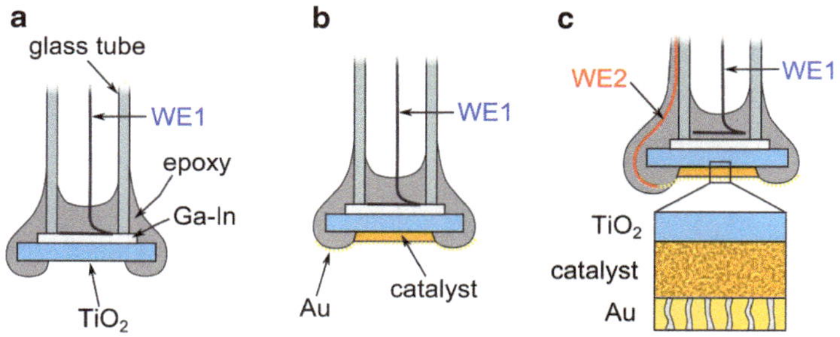


**Fig. 7.4** Schematic of DWE sample fabrication. (**a**) Ohmic contact is made to the backside of the SC substrate with appropriate doping level, and the active area on the front substrate surface is defined with epoxy. (**b**) The EC layer is electrodeposited on the front surface of the SC, followed by thermal evaporation of a thin (10–15 nm) Au film. (**c**) A second wire contact is made to the Au film, and epoxy is used to isolate the wire from the electrolyte

### 7.3.4 *Information Accessible Through DWE PEC*

With the additional contact to the EC layer as shown in Fig. 7.3, direct measurement of the EC potential ($V_{cat}$) is straightforward. Since $V_{cat}$ is equivalent to the reaction overpotential ($\eta = V_{cat} - E_{redox}/q$) when referenced to the solution energy $E_{redox}$, monitoring $V_{cat}$ allows one to quantify the loss of a PEC device due to kinetics.

Furthermore, the knowledge of $V_{cat}$ also allows one to calculate the true photovoltage ($V_{ph}$) generated by the SC|EC junction. One can then monitor how $V_{ph}$ changes as $V_{sem}$ is swept between forward and reverse biases relative to $E_{redox}$, which might shed light on how changes in the width of the space charge region affects electronic charge carrier recombination and separation in the SC. Alternatively, one can use the potentiostat to control the potential of the EC and use a high-input-impedance voltmeter to monitor $V_{sem}$ while the photoanode is illuminated. This allows direct measurement of the open-circuit $V_{ph}$ as a function of the EC charge state.

Using a bi-potentiostat, the flow of current into and out of the EC can be investigated. Because the two independent working electrodes share the same counter electrode, the net current that flows through the electrolyte can be obtained by comparing the current detected by each working electrode. In the following sections, examples of how to utilize this setup will be discussed.

## 7.4 Applications of DWE PEC

In this section we will illustrate how the DWE PEC technique was implemented in the investigation of a composite SC|EC photoanode system based on n-type single-crystal $TiO_2$ as a model SC.

### 7.4.1 *The "Adaptive" SC|EC Junction Hypothesis*

In recent years, much work concerning the improvement of SC photoanode performance through the incorporation of a suitable EC layer on the SC surface has been published (Sun et al. 2010; Zhong et al. 2011a; Barroso et al. 2011, 2012; Le Formal et al. 2011; Hisatomi et al. 2011; Ye et al. 2011; Klahr et al. 2012; Seabold and Choi 2012; Pendlebury et al. 2012). It was further noted that the best OER catalysts, as benchmarked by conventional "dark" electrolysis measurements on conductive substrates, did not necessarily lead to superior composite SC|EC performance compared to less active catalysts (Ye et al. 2011). This was inconsistent with a simple model that considers reducing the kinetic barrier of OER as the only role of the EC layer. Furthermore, the PEC performance of some composite SC|EC photoelectrodes seemed sensitive to how the EC material is prepared and deposited (Tilley et al. 2010; Badia-Bou et al. 2013; Spurgeon et al. 2014).

The difficulty in explaining the performance of SC|EC photoanodes with the assumption that the EC layer only functions as a OER catalyst naturally led to a competing hypothesis, which proposed that the EC layer improves the composite photoanode performance mainly by altering the energetics near the surface, e.g., by suppressing recombination and/or increasing band bending, and therefore the EC may be a "spectator" rather than a catalyst (Gamelin 2012). Since both hypotheses seemed to be supported by at least some of the published experimental observations, it was also proposed that the EC layer may do both, i.e., modify the OER kinetics as well as the interfacial energetics (Sivula 2013).

We further noticed that the most effective EC layers on oxide photoanodes appeared to have a porous ion-permeable structure and were able to undergo redox charging and discharging with potential cycling. These ECs included materials such as Co–Pi, which is a disordered composite of CoOOH nanosheets, phosphate anions, and electrolyte (Risch et al. 2009; McAlpin et al. 2010; Gerken et al. 2011; Du et al. 2012; Klingan et al. 2014). We hypothesized that the ability for the EC to permeate with electrolyte and change oxidation state throughout the bulk of the film may give rise to a previously undiscussed type of "adaptive" SC|EC junction which resembles a Schottky junction with tunable barrier height. Such adaptive junctions had been previously observed in other contexts (Lonergan 1997; Boettcher et al. 2007). In this adaptive-junction picture, the EC layer modifies not only the rate of electron transfer to and from the solution but also the equilibrium and nonequilibrium interface energetics.

Since the SC|EC junction is where light excitation leads to the separation of electron and hole quasi-Fermi levels ($E_{Fn}$ and $E_{Fp}$), which enables the oxidation of water at a reduced $V_{sem}$ compared to electrolysis in the dark, understanding how its properties can be dynamically modified by the state of the EC layer is important for further optimization of photoelectrode design.

Prior to discussing further the properties of the hypothesized "adaptive" junction and how it can be measured using the DWE PEC technique, we review our basic qualitative picture of the various interface possibilities.

In the absence of an EC, the SC forms a liquid junction with the aqueous electrolyte. The difference between the isolated SC Fermi level $E_{Fn}$ and the solution energy, $E_{redox}$, determines the equilibrium band bending and thus the upper limit of the attainable photovoltage $V_{ph}$ (Walter et al. 2010). With the EC layer, however, the nature of the SC|EC junction depends on the morphology of the EC material.

If the EC layer is dense and not entirely permeable to ions, such as Pt and crystalline $IrO_2$, the SC|EC junction can be expected to function like a SC|metal Schottky contact under illumination (Rhoderick and Williams 1988). In this scenario the SC|EC junction is "buried" by the solution and the flat-band potential (which is the maximum $V_{ph}$ achievable at open circuit) is defined by the equilibrium band bending in the SC, i.e., ideally the difference between the initial Fermi levels of the SC and the EC. The overpotential needed to drive the OER reaction, which is determined by the activity of the EC in the dark, accumulates only at the EC surface as a Helmholtz potential drop ($V_H$) as shown in Fig. 7.5b. Therefore, the effect of a better OER catalyst can be adequately modeled by a smaller resistance connected to a solar cell in series (Winkler et al. 2013; Mills et al. 2014).

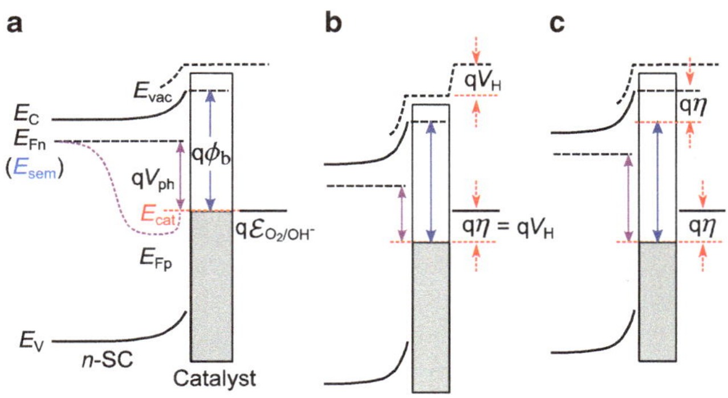


**Fig. 7.5** Energy band diagram of a catalyzed photoanode under constant illumination. (**a**) At open circuit; (**b**) at constant current (OER overpotential $= \eta$) when the EC layer is dense, i.e., the "buried junction" scenario; (**c**) at constant current (OER overpotential $= \eta$) when the EC layer is ion permeable, i.e., the "adaptive-junction" scenario

On the other hand, if the EC layer is *electrolyte/ion permeable*, then no Helmholtz potential drop can develop across the EC|Sol interface because any electric field beyond the SC surface will be screened by the electrolyte. Upon reduction or oxidation of the catalyst layer, the Fermi level of the EC changes in situ, which leads to a corresponding change of the *effective* SC|EC junction barrier height and hence the maximum $V_{ph}$ as shown in Fig. 7.5c. In other words, the maximum $V_{ph}$ of the SC|EC junction increases when the EC is held at a more oxidizing potential. Since a less active OER catalyst needs a larger overpotential to drive the same magnitude of OER current density, the *effective* barrier height ($\phi_b$) of a SC|EC junction with an inferior EC will increase further compared to that of a more active EC. Since an increase in $\phi_b$ leads to a more efficient SC|EC junction that can partially compensate for the larger $\eta$, it appears as if the SC|EC junction "adapts" to the activity of EC to maintain its overall performance. However, there is clearly a limit to this adaptive mechanism: the *effective* barrier height can only increase up to the point where the EC energy level (Fermi level) approaches the SC valence band edge (or, more precisely, the maximum hole quasi-Fermi level limited by the bulk semiconductor recombination processes). Therefore this "adaptive" mechanism is most easily observed in SC|EC systems where the SC valence band edge is substantially more positive than $E_{redox}$, which is the case for most oxide photoelectrodes under study (Peter and Wijayantha 2014).

To test the adaptive-junction hypothesis described above, the DWE PEC experiment setup was used to characterize the in situ properties of the SC|EC junction, such as $V_{cat}$ and $V_{ph}$ at open circuit ($V_{oc}$). Single-crystal $TiO_2$ was chosen as the model SC because of its deep valence band edge relative to $\mathcal{E}_{O_2/OH^-}$, its commercial availability, and the ability to easily control and characterize its carrier concentration by annealing in reducing environments. In order to contrast the difference between ion-permeable and dense ECs, electrodeposited $Ni(OH)_2$ was used to represent ion-permeable EC, and thermally deposited $IrO_x$ was used to represent dense, ion-impermeable EC. We knew that electrodeposited $Ni(OH)_2$ was ion permeable because every Ni atom in the film is electrochemically active as can be demonstrated by evaluating the integrated charge in the $Ni^{2+/3+}$ redox wave prior to the onset of OER. It is also the fastest known water oxidation catalyst in base after incorporation of Fe impurities (Trotochaud et al. 2012, 2014). In contrast, the thermally deposited nanocrystalline $IrO_x$ exhibited cyclic voltammograms with a shape and redox capacity that is indicative of only surface redox activity even after repeated cycling (Lin and Boettcher 2014). Because the electrochemistry cell and electrolyte were not specially cleaned, all the Ni-based catalysts discussed in this chapter have incorporated Fe impurities that allow for OER catalysis to occur at relatively low overpotentials.

### 7.4.2 Tracking the Path of Photogenerated Holes

Because $TiO_2$ has a deep valence band edge relative to $\mathcal{E}_{O_2/OH^-}$, it is plausible that the photogenerated holes might directly oxidize water without passing through the catalyst if the catalyst layer is permeable to electrolyte. The path of the holes should depend sensitively on the rates of the competing processes, in this case transfer to the catalyst versus direct transfer to solution. Because transfer to the catalyst involves one-electron oxidation of a transition metal cation, the kinetics are likely faster than direct oxidation of water from the semiconductor surface, which is a four-electron process. However, direct evidence in support of this hypothesis is important.

Using the DWE PEC technique, it is possible to experimentally follow the photogenerated hole current and see which path the holes follow. To track the path of photogenerated hole current, we used the $TiO_2|Ni(OH)_2$ photoanode and held $V_{sem}$ at 0 V vs. $\mathcal{E}_{O_2/OH^-}$ (equivalent to short circuit) using WE1 while sweeping $V_{cat}$ using WE2 and recording $J_{sem}$ and $J_{cat}$ both with and without illumination. As shown in Fig. 7.6, no current flows at WE1 in the dark, as is expected due to the SC|EC interface barrier and the cyclic voltammogram (CV) for the $Ni(OH)_2$ EC measured at WE2, is identical to what is observed on an Au substrate due to the reduction and oxidation of EC ($J_{cat} = J_{redox}$). Under illumination, a positive photocurrent flows at WE1 and the $Ni(OH)_2$ CV measured at WE2 is shifted negative relative to the dark CV by the same magnitude as the photocurrent measured at WE1.

As illustrated by the energy schematic in Fig. 7.6, when the ohmic contact to the back of the semiconductor collects more electrons than holes, the net current measured by the potentiostat at WE1 ($J_{sem}$) is positive. Therefore the positive $J_{sem}$ under illumination is attributed to net electron collection, which must be balanced by concurrent light-induced hole transfer ($J_{ph}$) across the semiconductor junction from left to right in Fig. 7.6. $J_{sem}$ remains largely constant until the optical

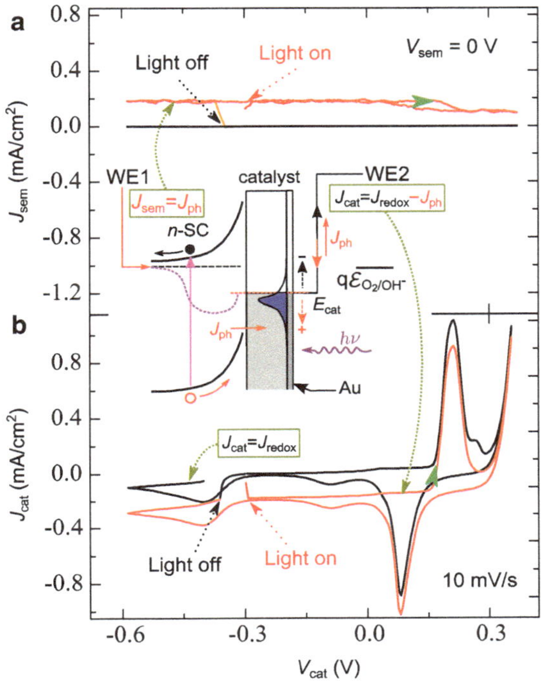


**Fig. 7.6** The effect of photocurrent on EC oxidation and reduction. (**a**) The illuminated (light on) and dark (light off) curves show $J_{sem}$ measured at the $TiO_2$ (WE1) during the scan. (**b**) Voltammetry of the $Ni(OH)_2$ EC was collected through the porous Au contact (WE2) while $V_{sem}$ was held at 0 V vs. $\mathcal{E}_{O_2/OH^-}$ in the dark and under 1-sun illumination. The *arrows* indicate the scan direction. The sign of the current is positive when the working electrode collects more electrons than holes; therefore, a positive $J_{sem}$ (equal to the photogenerated current $J_{ph}$ under the test conditions here) is measured when the SC|EC junction is illuminated. $J_{redox}$ is the current due to the reduction (−) and oxidation (+) of the EC through WE2. Figure adapted from Lin and Boettcher (2014). Copyrighted by Nature Publishing Group

absorption of the EC increases upon oxidation at ~0.2 V vs. $\mathcal{E}_{O_2/OH^-}$ (Trotochaud et al. 2013). At the semiconductor surface, the holes can transfer to either the catalyst or directly to the solution. If photogenerated holes react directly with the solution, then the current from WE1 will not affect the current measured by WE2 (as the current loop will be completed by the counter electrode). On the other hand, if the holes are transferred to the EC, this would cause a negative current shift in the catalyst voltammetry measured at WE2 (as illustrated in Fig. 7.6) because WE2 would need to supply additional electrons to compensate for the injected holes from the semiconductor as it controls the EC potential.

In Fig. 7.6b a negative current "shift" in the voltammetry of the $Ni(OH)_2$ through WE2 was indeed observed, and the magnitude of the shift is exactly equal to the positive current measured at WE1, i.e., $J_{cat} = J_{redox} - J_{ph}$. Moving $E_{cat}$ away from $\mathcal{E}_{O_2/OH^-}$ in either direction does not seem to change the magnitude of photocurrent received by the EC until light absorption by the oxidized EC becomes significant. The result of this experiment shows that, for $TiO_2$, hole transfer proceeds preferentially into the EC instead of directly oxidizing water over the $E_{cat}$ range studied. It would be useful to repeat this experiment under conditions where large OER currents are flowing from EC to solution. For sufficiently positive $E_{cat}$, hole transfer to the EC from the semiconductor should stop as the EC Fermi level $E_{cat}$ becomes lower than the surface hole quasi-Fermi level $E_{Fp}$ in the semiconductor, and thus the rate of direct hole transfer from SC to solution would become dominant.

### 7.4.3 *In Situ Measurement of the EC Potential*

Conventional SWE linear sweep voltammetry through the semiconductor ohmic contact is traditionally used to characterize the photoelectrode properties and performance. The Au contact to the EC allows for direct measurement of the EC potential in situ while such conventional photoelectrochemical experiments are performed. The experiment was done on both $TiO_2|IrO_x$ and $TiO_2|Ni(OH)_2$ samples by stepping $V_{sem}$ through a range of potentials while recording the change of $V_{cat}$. The value of $V_{cat}$ was recorded after holding $V_{sem}$ for 180 s to approximate the steady-state value. One striking observation from Fig. 7.7 is that the change of $V_{cat}$ near the photocurrent onset potential is much more abrupt in the case of $Ni(OH)_2$ relative to $IrO_x$—the potential of $Ni(OH)_2$ transitions almost stepwise from ≈0.21 to ≈−0.34 V when $V_{sem}$ varies only from −1.1 to −1.0 V, while potential of $IrO_x$ changes linearly from 0.28 to −0.30 V as $V_{sem}$ is moved between 0.1 and −0.5 V.

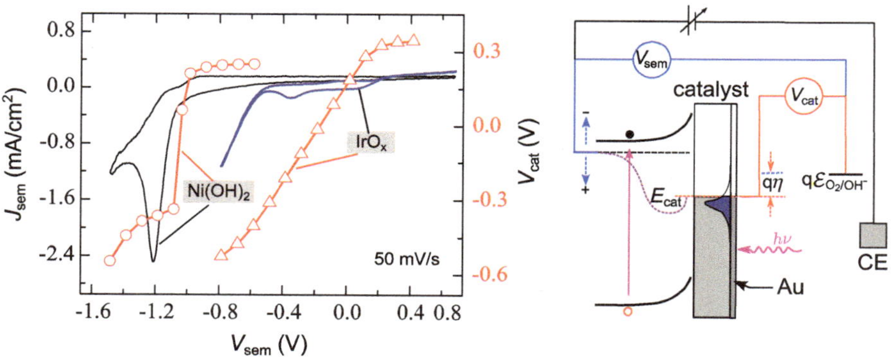


**Fig. 7.7** In situ catalyst potential measurements. The EC potential ($V_{cat}$) was measured as a function of the $TiO_2$ potential ($V_{sem}$) under constant illumination for samples coated with different catalysts. To mimic steady state, each $V_{cat}$ data point was collected after holding $V_{sem}$ for 180 s. Voltammograms collected via the $TiO_2$ electrode (WE1) under illumination are shown for reference. The corresponding energy diagram of the experimental setup is shown on the *right*. Figure adapted from Lin and Boettcher (2014). Copyrighted by Nature Publishing Group

The different dependence of $V_{cat}$ on $V_{sem}$ between $Ni(OH)_2$-coated and $IrO_x$-coated photoanodes can be explained by how the current across the SC|EC interface is coupled to the current across the EC|Sol interface. In both cases, the current across the SC|EC interface is composed of electron and hole components. For an n-type SC junction, the total current can be written as

$$J = -J_s\left(e^{qV_{jxn}/kT} - 1\right) + J_{sc} \tag{7.1}$$

where $J_s$ is the saturation current density of the junction SC|EC, $V_{jxn} = V_{cat} - V_{sem}$ is the voltage drop across the SC|EC junction, $J_{sc}$ is the short-circuit photocurrent, $T$ is the absolute temperature, and $k$ is the Boltzmann constant. When $J_{sc}$ is larger than the electron current $J_s\left(e^{qV_{jxn}/kT} - 1\right)$, the rate of hole transfer exceeds that of electron transfer and a net hole current flows into the EC. At steady state, this current must match the water oxidation current passing through the EC|Sol interface. Therefore for both samples, $V_{cat}$ plateaus after $V_{sem}$ moves sufficiently positive because the OER current across the EC|Sol interface has reached its maximum, which is determined by the constant (hole) photocurrent crossing from SC to EC.

As $V_{sem}$ is stepped more negative for the $IrO_x$-coated photoanode, $V_{cat}$ changes linearly. This is because as charge is added to a metal catalyst layer, that charge must sit at the surface of the metal (by Gauss's law) and therefore result in an electrostatic potential drop between the EC and solution. It is this electrostatic potential drop at the EC surface that provides the overpotential for driving the electrochemical reaction such that the net SC|EC and EC|Sol currents are matched. Any potential dropped at the EC|Sol interface also shifts the band edges of the SC by the same amount.

The situation is different for the $Ni(OH)_2$-coated $TiO_2$. Because $Ni(OH)_2$ is electrolyte permeable, holes can accumulate on the catalyst without resulting in an *electrostatic* potential drop normal to the SC surface. The charge is completely screened by electrolyte ions or, for example, by an electrochemical reaction $Ni(OH)_2 + OH^- \rightarrow NiOOH + e^- + H_2O$. Therefore while the *chemical potential* of the catalyst changes, this change is not coupled to a change in the band-edge position of the SC, but instead to a change in the bulk "filling" of the catalyst electronic density of states. Therefore, as $V_{sem}$ is stepped more negative, $V_{cat}$ can freely move from a potential where it is driving oxygen evolution to a potential where it is driving oxygen reduction with only small changes in $V_{sem}$.

The observed changes of $V_{cat}$ for both ion-permeable and dense ECs as a function of $V_{sem}$ under illumination agree well with our numerical simulation results (Fig. 7.8) based on a model that includes the effect of a dynamically changing EC electrochemical potential (Fermi level) under working conditions due to electrolyte permeability (Mills et al. 2014).

As explained earlier, the rapid change of $V_{cat}$ near $V_{sem} = V_{oc}$ in Figs. 7.7 and 7.8b was a result of the EC layer undergoing bulk reduction and oxidation. It is worth noting that when a thin (10–30 nm) and porous gold film was deposited on n-$TiO_2$, a similar rapid change of the Au film potential has been observed by Nakato

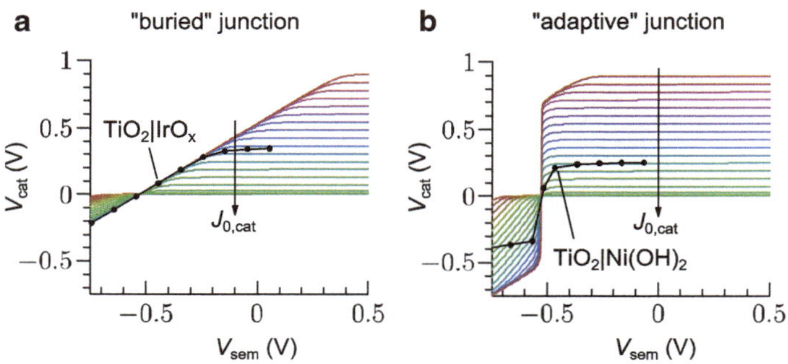


**Fig. 7.8** Comparison of simulated and experimental dependence of $V_{cat}$ on the voltage applied to the semiconductor back contact $V_{sem}$. *Lines* are results from numerical simulation. The *arrows* indicate increasing values of catalyst activity $J_{o,cat}$. The *black points* show experimental values of $V_{cat}$ for (**a**) nanocrystalline $IrO_x$ and (**b**) ion-permeable $Ni(OH)_2$/NiOOH. This figure is adapted from Mills et al. (2014). Copyrighted by the American Physical Society

et al., with accompanying signature of Au oxidation and reduction during linear sweep voltammetry using the n-$TiO_2$ as working electrode (Nakato and Tsubomura 1982). Upon saturation of the photocurrent, the Au potential was also found to plateau near the onset potential of OER presumably due to the pinning of hole quasi-Fermi level by oxidation of water either from the SC surface or from the Au surface. In this case, it appears that the thin porous Au film behaves like an ion-permeable EC of low activity. In a separate experiment by Pinson, a thin and porous gold film was evaporated onto an n-GaP semiconductor with thin surface oxide to measure the SC surface hole quasi-Fermi level using a typical PEC setup (Pinson 1977). It was found that the potential difference between n-GaP and the Au film under illumination was larger than the equilibrium barrier height of a Schottky contact between n-GaP and dense Au.

Both experiments suggest that photogenerated holes from either n-$TiO_2$ or n-GaP were able to anodically shift the Au film potential by "charging" it until the rate of the hole transfer process from the Au (e.g., water oxidation) is sufficiently fast to match the rate of hole transfer from the semiconductor and prevent further hole accumulation in the Au at steady state. In other words, the Schottky junction between the SC and porous Au film could also become a type of "adaptive" junction when the Au film is thin. This is likely due in both cases to the presence of an electrolyte-permeable hydrated oxide layer underneath the thin Au that allows the charge on the gold layer to be effectively screened from the charge in the semiconductor. In other words, the metal Au layer probably did not form a direct and intimate contact with the semiconductor.

Although it was suggested that these surface Au layers allow for direct measurement of the surface hole quasi-Fermi level in the semiconductor (Pinson 1977), we caution that this interpretation is only technically valid in the case where (a) there is negligible forward electron transfer to the Au at open circuit (i.e., in the limit where the barrier height is large and the photovoltage limited by bulk recombination and not the onset of forward majority electron transfer) and (b) where the kinetics for hole transfer from the Au to the solution are negligible compared to exchange with the semiconductor. It is unlikely both conditions were met in the cases described above.

It is also useful to contrast the experiments where Au is directly contacting the semiconductor surface with those that we reported where Au is used to sense/ control the potential of an EC layer. In the former, the Au will unavoidably affect the semiconductor surface and junction properties by affecting the rates of hole and electron transport. In the case where the Au is used only as a contact to the EC, this is not the case. Because the Au is electrolyte permeable and has slower OER kinetics than the underlying EC layer, it is not expected to affect the SC junction properties. The Au is only a passive probe electrode that does not modulate any of the underlying carrier exchange processes.

### *7.4.4 Open-Circuit Photovoltage Measurement*

In the previous subsection, we discussed how ion-permeable ECs are able to freely move their chemical potential by accumulating electronic charge that is screened by electrolyte ions, while dense (metal) catalysts store charge at their surface resulting in a change in the Helmholtz potential $V_H$. This behavior can also be analyzed in the context of a so-called adaptive junction.

In a classic solid-state Schottky junction, the open-circuit photovoltage is affected by changing the Fermi level of the metal contact or, equivalently, the barrier height $\phi_b$. Ideal Schottky theory defines the interface barrier $\phi_b$ between a metal and an n-type SC as the difference between the metal Fermi level ($E_F^m$) and the conduction-band edge of the SC ($E_C$):

$$q\phi_b = E_F^m - E_C \tag{7.2}$$

The effects of varying $\phi_b$ on the photovoltage output of the Schottky junction can be evaluated by solving the illuminated diode equation (Eq. 7.1) for the junction's open-circuit voltage $V_{oc}$. Assuming $J_{ph}/J_s \gg 1$ and the standard expression for $J_s$, we obtain:

$$-V_{oc} = \frac{kT}{q}\ln\left(\frac{J_{ph}}{J_s}\right) = \frac{kT}{q}\ln\frac{J_{ph}}{A^*T^2} + \phi_b \tag{7.3a}$$

$$J_s = A^*T^2 e^{q\phi_b/kT} \tag{7.3b}$$

where $A^*$ is the effective Richardson constant. Equation (7.3a) shows that the magnitude of $V_{oc}$ changes linearly with $\phi_b$ for an ideal diode under illumination that yields a constant $J_{ph}$.

Treating $E_{cat}$ as equivalent to $E_F^m$, a change in $E_{cat}$ should therefore lead to a corresponding change in the *effective* $\phi_b$ for ion-permeable $TiO_2|Ni(OH)_2$ junctions, which should be reflected by changes in $V_{oc}$ across the junction. The term *effective* barrier height is used here to indicate that this $\phi_b$ is not an equilibrium parameter because $E_{cat}$ is fixed at $E_{redox}$ in equilibrium.

The predicted effect of varying $E_{cat}$ can be experimentally tested by holding $V_{cat}$ at various potentials under illumination and recording $V_{sem}$ at open circuit to determine if $V_{oc} = V_{sem} - V_{cat}$ indeed changes with $V_{cat}$. Results of such an experiment conducted on both $Ni(OH)_2$-coated and $IrO_x$-coated $TiO_2$ photoanodes are shown in Fig. 7.9.

As can be seen from Fig. 7.9a, $V_{sem}$ is linearly dependent on $V_{cat}$ for the $TiO_2|IrO_x$ photoanode, leading to a constant $V_{oc}$. This result is consistent with a constant $\phi_b$ for all $V_{cat}$ because $IrO_x$ is ion impermeable and therefore $E_{cat}$ cannot change

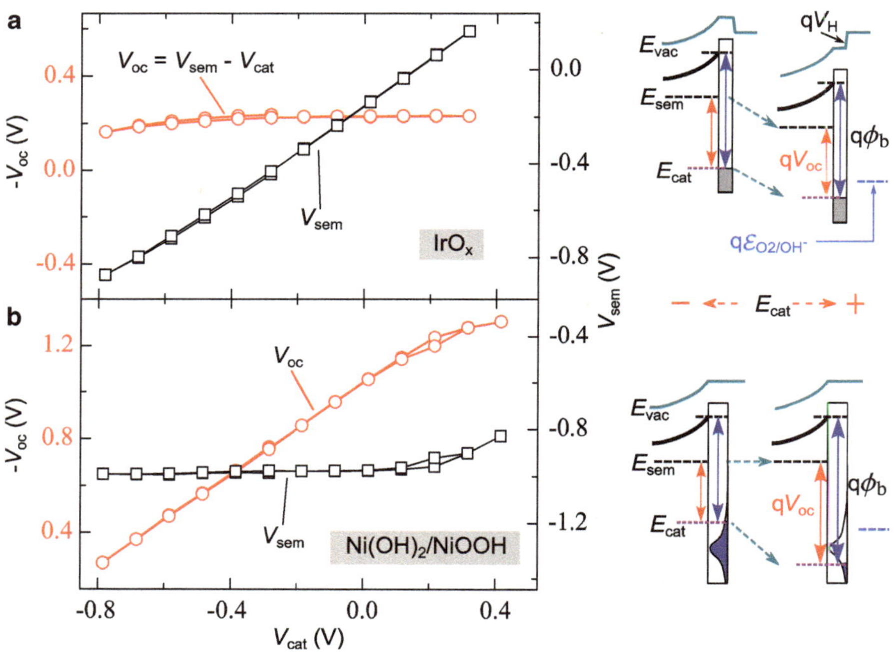


**Fig. 7.9** Open-circuit SC|EC junction photovoltage. The steady-state open-circuit SC potential $V_{sem}$ was measured as a function of the EC potential to calculate the $V_{oc}$ across the SC/EC junction. (**a**) $IrO_x$-coated $TiO_2$; (**b**) $Ni(OH)_2$-coated $TiO_2$. The band energy diagrams on the *right side* illustrate how the different electrostatic potential drops at the EC|solution interface ($V_H$) for ion-permeable and dense ECs affect the SC|EC $V_{oc}$. For dense EC in (**a**), changes to $E_{cat}$ lead to electrostatic potential drops at the EC|Sol interface and thus a constant SC|EC junction $V_{oc}$. For ion-permeable EC in (**b**), an electrostatic potential drop cannot develop within the catalyst film because it is completely screened by electrolyte ions. Therefore changes to $E_{cat}$ directly affect the relative energetics of the SC|EC interface and lead to the increase of $V_{oc}$. Figure adapted from Lin and Boettcher (2014). Copyrighted by Nature Publishing Group

relative to $E_C$ as $V_{sem}$ varies. As illustrated by the energy diagram next to Fig. 7.9a, the change in $V_{sem}$ instead leads to a change in $V_H$ across the EC|Sol interface, leaving the SC|EC junction unaffected. The small $V_{oc}$ of ~0.2 V indicates that $\phi_b$ is small and explains the poor photoelectrode performance—the $IrO_x|TiO_2$ junction is a very poor photovoltaic element.

In contrast, $V_{sem}$ is nearly independent of $V_{cat}$ for the $TiO_2|Ni(OH)_2$ photoanode (Fig. 7.9b), resulting in an increasing $V_{oc}$ as $V_{cat}$ moves toward more oxidizing potentials. This is consistent with an unpinned SC|EC junction where $E_{cat}$ is free to move independent of the SC band edges as illustrated by the energy diagram next to Fig. 7.9b. The $V_{oc}$ changes almost linearly with $V_{cat}$ as predicted by Eq. (7.3a). The deviation from linearity for $V_{cat} > 0.2$ V vs. $\mathcal{E}_{O_2/OH^-}$ is likely due to optical absorption in NiOOH (Trotochaud et al. 2013) that reduces $J_{ph}$ and hence $V_{oc}$. The increased $V_{oc}$ at more oxidizing $V_{cat}$ for the $TiO_2|Ni(OH)_2$ junction versus the constant $V_{oc}$ regardless of $V_{cat}$ for the $TiO_2|IrO_x$ junction found in Fig. 7.9 directly lead to the differing $V_{cat} - V_{sem}$ relationship observed in Fig. 7.8.

### 7.4.5 *Dark Junction* J–V *Behavior as a Function of* $V_{cat}$

In solid-state semiconductor device physics, the barrier height $\phi_b$ is a defining parameter of a Schottky junction. It is therefore expected that changes in $\phi_b$ will manifest themselves in all aspects of junction properties. One of the most important aspects is the dark *J–V* curve across the Schottky junction. To see how the junction *J–V* characteristics are affected by $\phi_b$, we rearrange Eq. (7.1) under the condition of strong forward bias in the dark, such that

$$\ln|J| \approx \ln\left(J_s e^{qV_{jxn}/kT}\right) = \frac{qV_{jxn}}{kT} + \ln\left(A^* T^2\right) - \frac{q\phi_b}{kT} \qquad (7.4)$$

Equation (7.4) predicts a shift of the $\ln|J|$ ($V_{jxn}$) curve toward larger forward voltage as $\phi_b$ increases. This prediction is also experimentally verifiable through the DWE setup. The *J–V* curve at different $E_{cat}$ (i.e., different $\phi_b$) can be collected by holding the $V_{cat}$ at different potentials relative to $\mathcal{E}_{O_2/OH^-}$ in the dark and sweeping $V_{sem}$ relative to $V_{cat}$ (or $\mathcal{E}_{O_2/OH^-}$). The relationship between ln $|J|$ and $V_{jxn}$ can then be analyzed. This measurement is therefore equivalent to measuring a standard diode *J–V* curve, but it is done in situ for different levels of charge on the EC layer. We conducted the experiment on both $Ni(OH)_2$-coated and $IrO_x$-coated $TiO_2$ photoanodes and present the results in Fig. 7.10. During the experiment, current through both the $TiO_2$ (WE1) and the EC (WE2) were recorded and verified to be identical in magnitude and opposite in sign, indicating that the electron current measured when sweeping $V_{sem}$ flows predominantly between SC and EC, and any component attributable to OER or ORR is negligible under these conditions.

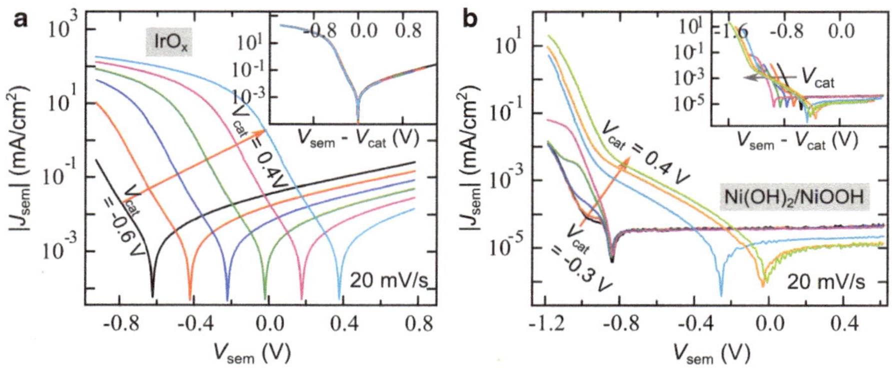


**Fig. 7.10** *J–V* curves of the SC|EC junction. The data across the SC|EC interface were collected by sweeping $V_{sem}$ while holding $V_{cat}$ constant. Panel (**a**) shows results for the $IrO_x$-coated $TiO_2$ and (**b**) the results for the $Ni(OH)_2$-coated $TiO_2$. Each curve corresponds to a different value of $V_{cat}$. All potentials presented are relative to $\mathcal{E}_{O_2/OH^-}$. The *inset* shows the same set of data plotted against the true potential drop across the SC|EC junction, $-V_{jxn} = V_{sem} - V_{cat}$. The increments in $V_{cat}$ between curves in (**a**) are 0.2 V and those in (**b**) are 0.1 V. This figure is adapted from Lin and Boettcher (2014). Copyrighted by Nature Publishing Group

In order to better relate results in Fig. 7.10 to Eq. (7.4), the *J–V* curves were also plotted (insets) against the voltage drop across the junction using $-V_{jxn} = V_{sem} - V_{cat}$. It is immediately clear from Fig. 7.10a that $\phi_b$ of the $TiO_2|IrO_x$ junction did not change with $V_{cat}$ because all *J–V* curves collapsed into one when plotting against $V_{jxn}$. The results in Fig. 7.10b, however, are more complicated.

On the one hand, the *J–V* curves for $TiO_2|Ni(OH)_2$ junction clearly depend on $V_{cat}$ as can be seen from the relative shifts when plotted against $V_{jxn}$. The cathodic shift of the forward *J–V* curves and the smaller reverse saturation currents for more positive $V_{cat}$ are both consistent with $\phi_b$ changing with $V_{cat}$ for the ion-permeable Ni $(OH)_2$ EC. On the other hand, when $V_{cat}$ moves more positive than 0.1 V vs. $\mathcal{E}_{O_2/OH^-}$, multi-exponential behavior is observed on the semi-log $J–V_{jxn}$ plots, suggesting multiple mechanisms for electron transport across the SC|EC interface and Eq. (7.4) only qualitatively explains the data. The fact that $Ni(OH)_2$ becomes resistive when reduced may have contributed to the complexity of data in Fig. 7.10b.

### 7.4.6 *Summary of DWE PEC Results*

The results obtained through DWE PEC experiments performed in this section provided direct information about the SC|EC junction that are otherwise inaccessible through other existing experimental techniques. The findings supported the hypothesis that ion permeability of the catalyst is an important factor that needs to be considered during the optimization of SC|EC junctions for solar water splitting.

With the role of ion permeability in mind, it is less surprising that porous catalysts such as Co–Pi and FeOOH have been shown to be effective in enhancing the PEC performance of photoanodes based on α-$Fe_2O_3$ and $BiVO_4$ (Kanan and Nocera 2008; Zhong and Gamelin 2010; Zhong et al. 2011a; Barroso et al. 2011, 2012; Le Formal et al. 2011; Hisatomi et al. 2011; Seabold and Choi 2012; Klahr et al. 2012; Pendlebury et al. 2012). The finding that Co–Pi outperforms $IrO_x$ when deposited on W:$BiVO_4$ (Ye et al. 2011), even though $IrO_x$ is alone a superior EC, is thus likely explained by the formation of low-barrier-height buried junction for $IrO_x$ deposited by drop casting, thermal annealing, and an adaptive junction for Co–Pi. If $IrO_x$ is deposited as hydrated nano-particulate oxide layer, the performance is substantially better than dense $IrO_x$ (Tilley et al. 2010; Badia-Bou et al. 2013; Spurgeon et al. 2014; Lin et al. 2015), thus indicating further that the electrolyte permeability of the catalyst layer plays an important role.

## 7.5 Outlook

The picture presented of the EC|SC interface remains a simple "first-order" description of the interface. It is certain that even the ion-permeable catalyst overlayers could have more complicated effects depending on their electronic densities of states, how they interact with surface states, and whether or not they modify the surface of the semiconductor in a way that shifts the band-edge positions relative to the solution potential. Understanding these effects is an important direction for the field of semiconductor photoelectrochemistry.

Further, the DWE PEC technique itself, while powerful, has significant limitations, including the following: (1) The sample preparation described above can be time consuming and relies on serial steps that are each difficult. (2) The technique in its current form is best suited for a homogenous SC surface and EC layer. Due to the macroscopic scale of the EC contact, microscopic spatial variations of either the SC or EC cannot be revealed. (3) Since the technique requires a persisting contact to the EC, it is not applicable to catalysts that undergo dynamic dissolution and redeposition. Some of these limitations might be addressed by using a local probe to measure the EC potential in situ. Further technique development is, however, needed. Such developments are likely to enable new insights into a broad array of (photo)electrochemical systems where there are multiple components interacting and the dynamic operational behavior is difficult to measure by other means.

**Acknowledgments** This material is based upon work supported by the US Department of Energy Office of Science, Office of Basic Energy Sciences Solar Photochemistry program under Award Numbers DE-SC000846 and DE-SC0014279. S.W.B. also acknowledges support from the Research Corporation for Science Advancement as a Cottrell Scholar.

## Appendix: Experimental Details

### *Au Thin-Film Contact*

In our study, Au was chosen as the contact material because OER kinetics on Au are much slower than that of many catalysts, e.g., $Ni(OH)_2$ (with Fe impurities) and $IrO_x$, deposited on an Au substrate. A comparison of OER activities among Au, $IrO_x$, and $Ni(OH)_2$ in 0.1 M KOH solution is shown in Fig. 7.11.

During the sample fabrication process, controlling the thickness of the Au layer is very important because the film needs to be thick enough to form a continuous conducting network but not too thick to hinder the free movement of ions into and out of the EC film (Doron-Mor et al. 2004). We use Au layers ~10–15 nm thick because the high surface tension of Au causes the film to "de-wet" from the surface and form a network of Au islands that is electrolyte permeable but electrically conductive. SEM images of a typical $Ni(OH)_2$ film (on an Au substrate) before and after the thin Au film deposition by thermal evaporation are shown in Fig. 7.12.

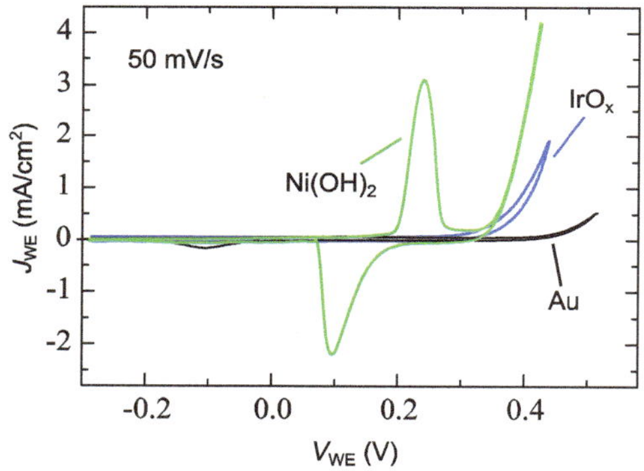


**Fig. 7.11** OER activity of $Ni(OH)_2$ (with Fe impurities) and $IrO_x$ on an Au substrate in 0.1 M KOH. The working electrode potential is referenced to $\mathcal{E}_{O_2/OH^-}$ and all curves were collected at 50 mV/s. The OER activity of bare Au substrate is also shown for comparison

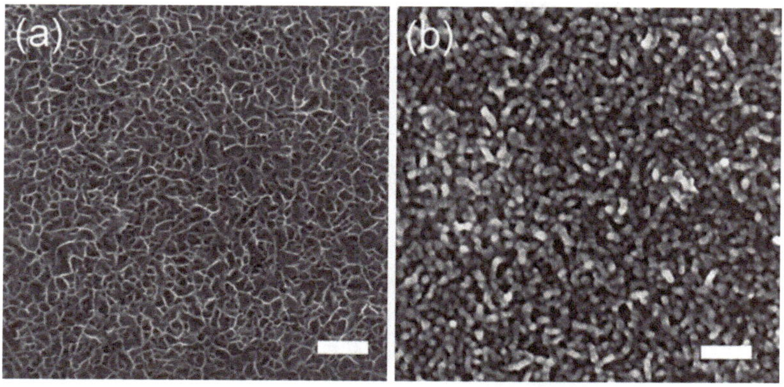


**Fig. 7.12** SEM images of $Ni(OH)_2$ EC electrodeposited on Au substrate (**a**) before and (**b**) after Au film deposition. The scale bar is 200 nm in both images

Because the Au film evaporated on the epoxy surface surrounding the EC serves as the connecting area for WE2, it is desirable to make the epoxy surface as smooth as possible. It was found that curing the epoxy (Loctite Hysol 1C) in an oven at ~90 °C produced smoother surfaces compared to curing the epoxy at room temperature.

## *Sample Viability Tests*

After samples for the DWE PEC experiment were fabricated, tests must be performed to ensure the quality of the device. One potential problem is that the Au contact film may prevent the electrolyte from permeating into the underlying EC film. Another potential problem is the Au film may form direct contact with the $TiO_2$ substrate if the EC film coverage is incomplete as a result of uneven deposition or due to the development of cracks during the drying process. If the EC film volume increases too much between the dry and wet states, it is also possible that the Au film near the edge of the epoxy surface become discontinuous which will make the EC film electrically inaccessible through WE2.

### **Test Cell and Measurement Setup**

Photoanode samples were tested in a three-neck cell with an optically flat bottom for introducing light from a solar simulator at 1-sun intensity at the SC surface. The cell was constructed from fused silica to minimize parasitic adsorption of UV photons. Typically about 15 mL of 0.1 M KOH electrolyte was used, and the solution was stirred using a magnetic stir bar and sparged with pure $O_2$ to maintain a stable $\mathcal{E}_{O_2/OH^-}$ during PEC experiments. A Pt coil was used as the counter, and the reference was either Hg/HgO filled with 0.1 M KOH or a saturated calomel electrode.

### **Impact of the Au Contact**

Deposition of an Au film on the EC layer could have a few adverse effects; it will reduce the light intensity that reaches the SC|EC junction and obscure part of the EC surface. At increased thickness, the Au film could also impede the permeation of electrolyte into the porous $Ni(OH)_2$ film. The effect of optical absorption/reflection by the Au film was tested first and shown in Fig. 7.13 for both ECs used. In both cases, the light intensity was reduced more than 50 % by the Au film as evident from the reduced photocurrent. Furthermore, the redox capacity of the $IrO_x$ catalyst was reduced more significantly than that of the $Ni(OH)_2$ catalyst,

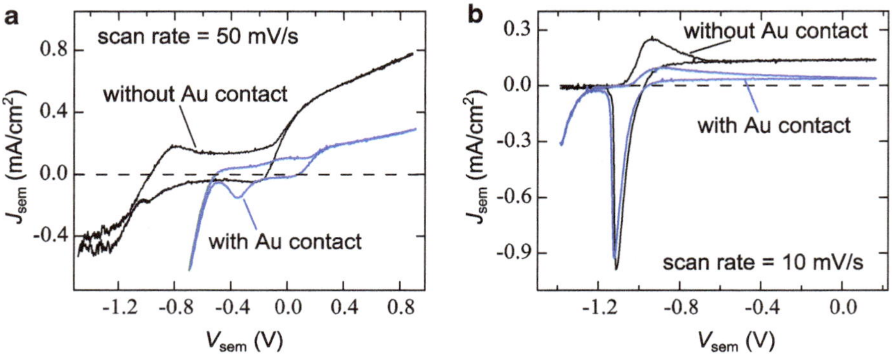


**Fig. 7.13** Cyclic voltammograms of $TiO_2$|EC samples at 1-sun illumination before and after Au film contact deposition. (**a**) $TiO_2$|$IrO_x$ sample at scan rate of 50 mV/s. (**b**) $TiO_2$|$Ni(OH)_2$ sample at scan rate of 10 mV/s. Figure adapted from Lin and Boettcher (2014). Copyrighted by Nature Publishing Group

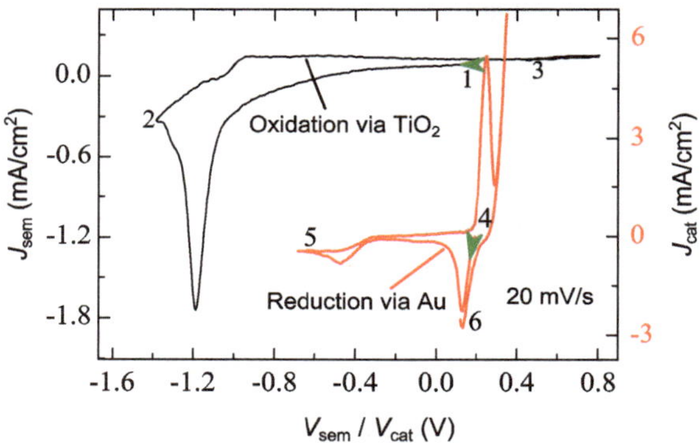


**Fig. 7.14** Oxidation and reduction of the EC on a DWE sample. The $Ni(OH)_2$ EC was oxidized via the $TiO_2$ electrode under 1-sun illumination by sweeping $V_{sem}$ negative (from points *1* to *2*) then positive (from points *2* to *3*) and reduced again via the Au film in contact with the EC (by sweeping $V_{cat}$ through points *4–5*). The scan rate was 20 mV/s. The starting point of each scan is indicated by an *arrow*. Integration of the oxidation wave on the Au, the reduction wave on the Au, and the reduction wave on the $TiO_2$ all yielded ~13 mC/cm$^2$ after background correction, indicating that the same $Ni(OH)_2$/NiOOH species are oxidized/reduced by both $TiO_2$ and Au electrodes. Figure adapted from Lin and Boettcher (2014). Copyrighted by Nature Publishing Group

presumably due to surface activity of the $IrO_x$ film versus bulk activity of the $Ni(OH)_2$ film.

To verify that the Au contact on the $Ni(OH)_2$ surface does not hinder the free flow of ions into/out of the porous $Ni(OH)_2$ film, oxidation and reduction of the EC were carried out through both WE1 ($TiO_2$) and WE2 (Au). A typical cyclic voltammetry (CV) of a $Ni(OH)_2$-coated $TiO_2$ photoanode with a top Au contact is shown in Fig. 7.14.

The observation that oxidation and reduction of the $Ni(OH)_2$ catalyst can be accomplished via both WE1 and WE2 without distortion of the reduction wave indicates that the thin Au film remains permeable to ions in the electrolyte. The cathodic shift of the NiOOH reduction wave measured via the $TiO_2$ (WE1) relative to that via Au (WE2) is due to the photovoltage $V_{ph}$ generated by the $TiO_2|Ni(OH)_2/NiOOH$ junction. The CV of $Ni(OH)_2$ via WE2 is identical to those obtained by depositing $Ni(OH)_2$ directly on Au substrate, which confirms that the electrical contact between WE2 and the EC film is intact.

Similar tests can be performed on any EC that shows either a bulk redox wave or surface capacitive charging and discharging behavior.

### Verifying the Independence of the Two WEs

After verifying the Au film permeability and WE2 contact integrity, the DWE sample was further tested for any possible direct Au–$TiO_2$ electrical short. Current-voltage curves through the SC|EC junction (between WE1 and WE2) were collected while the sample is in solution for all DWE samples, and any sample that exhibited high reverse current density (such as the green curve in Fig. 7.15, cracked EC) were excluded.

The control samples with an Au film deposited directly on the $TiO_2$ substrate (red curve in Fig. 7.15) show high reverse current due to the relatively low Schottky barrier between Au and $TiO_2$. Based on the near symmetric *J–V* curve of $TiO_2|Au$ junction, we hypothesize that high reverse current seen in some of the samples resulted from cracks in the EC film. Examination of samples with electrodeposited $Ni(OH)_2$ films revealed that some EC films developed cracks during the drying process (Fig. 7.16b), which probably caused direct Au–$TiO_2$ contact to form upon deposition of the Au layer.

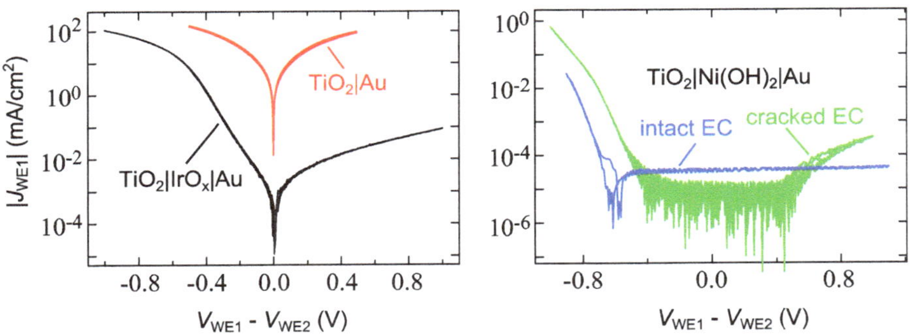


**Fig. 7.15** Representative *J–V* curves of $TiO_2|EC|Au$ and control $TiO_2|Au$ samples. The control $TiO_2|Au$ sample without catalyst showed larger, symmetric *J–V* curve, while the $IrO_x$-coated sample showed rectifying *J–V* curve. Samples with possibly cracked $Ni(OH)_2$ catalyst film showed increased reverse current, while sample with intact EC layer exhibited low reverse current. Figure adapted from Lin and Boettcher (2014). Copyrighted by Nature Publishing Group

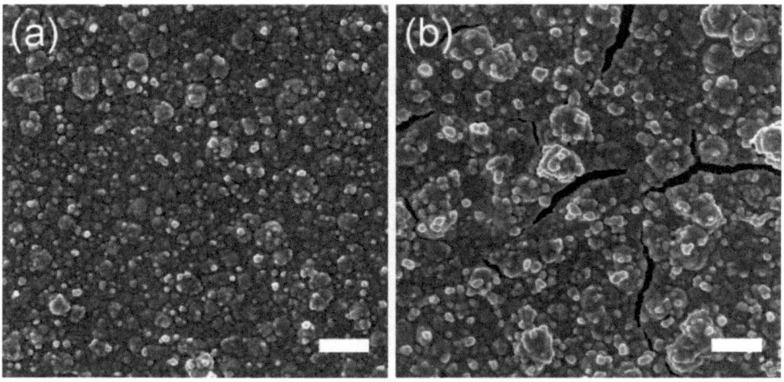


**Fig. 7.16** SEM images of electrodeposited $Ni(OH)_2$ catalyst on $TiO_2$ substrate. (**a**) Sample with continuous EC film. (**b**) Sample with cracked EC film that could lead to direct Au–$TiO_2$ contact. The scale bar is 500 nm in both images

# References


Abdi FF, Firet N, van de Krol R (2013) Efficient $BiVO_4$ thin film photoanodes modified with cobalt phosphate catalyst and W-doping. ChemCatChem 5(2):490–496. doi:10.1002/cctc.201200472

Badia-Bou L, Mas-Marza E, Rodenas P, Barea EM, Fabregat-Santiago F, Gimenez S, Peris E, Bisquert J (2013) Water oxidation at hematite photoelectrodes with an iridium-based catalyst. J Phys Chem C 117(8):3826–3833. doi:10.1021/Jp311983n

Barroso M, Cowan AJ, Pendlebury SR, Gratzel M, Klug DR, Durrant JR (2011) The role of cobalt phosphate in enhancing the photocatalytic activity of α-$Fe_2O_3$ toward water oxidation. J Am Chem Soc 133(38):14868–14871. doi:10.1021/Ja205325v

Barroso M, Mesa CA, Pendlebury SR, Cowan AJ, Hisatomi T, Sivula K, Gratzel M, Klug DR, Durrant JR (2012) Dynamics of photogenerated holes in surface modified α-$Fe_2O_3$ photoanodes for solar water splitting. Proc Natl Acad Sci U S A 109(39):15640–15645. doi:10.1073/pnas.1118326109

Berglund SP, Flaherty DW, Hahn NT, Bard AJ, Mullins CB (2011) Photoelectrochemical oxidation of water using nanostructured $BiVO_4$ films. J Phys Chem C 115(9):3794–3802. doi:10.1021/Jp1109459

Boettcher SW, Strandwitz NC, Schierhorn M, Lock N, Lonergan MC, Stucky GD (2007) Tunable electronic interfaces between bulk semiconductors and ligand-stabilized nanoparticle assemblies. Nat Mater 6(8):592–596. doi:10.1038/Nmat1943

Braun A, Sivula K, Bora DK, Zhu JF, Zhang L, Gratzel M, Guo JH, Constable EC (2012) Direct observation of two electron holes in a hematite photoanode during photoelectrochemical water splitting. J Phys Chem C 116(32):16870–16875. doi:10.1021/Jp304254k

Burke MS, Kast MG, Trotochaud L, Smith AM, Boettcher SW (2015) Cobalt–iron (oxy)hydroxide oxygen evolution electrocatalysts: the role of structure and composition on activity, stability, and mechanism. J Am Chem Soc. doi:10.1021/jacs.5b00281

Doron-Mor I, Barkay Z, Filip-Granit N, Vaskevich A, Rubinstein I (2004) Ultrathin gold island films on silanized glass. Morphology and optical properties. Chem Mater 16(18):3476–3483. doi:10.1021/Cm049605a

Du PW, Kokhan O, Chapman KW, Chupas PJ, Tiede DM (2012) Elucidating the domain structure of the cobalt oxide water splitting catalyst by X-ray pair distribution function analysis. J Am Chem Soc 134(27):11096–11099. doi:10.1021/Ja303826a

Friebel D, Louie MW, Bajdich M, Sanwald KE, Cai Y, Wise AM, Cheng M-J, Sokaras D, Weng T-C, Alonso-Mori R, Davis RC, Bargar JR, Nørskov JK, Nilsson A, Bell AT (2015) Identification of highly active Fe sites in (Ni, Fe)OOH for electrocatalytic water splitting. J Am Chem Soc 137(3):1305–1313. doi:10.1021/ja511559d

Fujishima A, Honda K (1972) Electrochemical photolysis of water at a semiconductor electrode. Nature 238(5358):37–38. doi:10.1038/238037a0

Gamelin DR (2012) Water splitting: catalyst or spectator? Nat Chem 4(12):965–967. doi:10.1038/nchem.1514

Gerischer H, Mattes I, Braun R (1965) Elektrolyse Im Stromungskanal—Ein Verfahren Zur Untersuchung Von Reaktions-Und Zwischenprodukten. J Electroanal Chem 10 (5–6):553–567. doi:10.1016/0022-0728(65)80055-9

Gerken JB, McAlpin JG, Chen JYC, Rigsby ML, Casey WH, Britt RD, Stahl SS (2011) Electrochemical water oxidation with cobalt-based electrocatalysts from pH 0–14: the thermodynamic basis for catalyst structure, stability, and activity. J Am Chem Soc 133(36):14431–14442. doi:10.1021/Ja205647m

Harris LA, Gerstner ME, Wilson RH (1977) Role of metal overlayers on gallium-phosphide photoelectrodes. J Electrochem Soc 124(10):1511–1516. doi:10.1149/1.2133103

Hisatomi T, Le Formal F, Cornuz M, Brillet J, Tetreault N, Sivula K, Gratzel M (2011) Cathodic shift in onset potential of solar oxygen evolution on hematite by 13-group oxide overlayers. Energy Environ Sci 4(7):2512–2515. doi:10.1039/C1ee01194d

Huang ZQ, Lin YJ, Xiang X, Rodriguez-Cordoba W, McDonald KJ, Hagen KS, Choi KS, Brunschwig BS, Musaev DG, Hill CL, Wang DW, Lian TQ (2012) In situ probe of photocarrier dynamics in water-splitting hematite ($\alpha$-$Fe_2O_3$) electrodes. Energy Environ Sci 5 (10):8923–8926. doi:10.1039/C2ee22681b

Kanan MW, Nocera DG (2008) In situ formation of an oxygen-evolving catalyst in neutral water containing phosphate and $Co^{2+}$. Science 321(5892):1072–1075. doi:10.1126/science.1162018

Kenkel JV, Bard AJ (1974) Dual working electrode coulometric flow cell. J Electroanal Chem 54 (1):47–54. doi:10.1016/0368-1874(74)85095-1

Kim TW, Choi KS (2014) Nanoporous $BiVO_4$ photoanodes with dual-layer oxygen evolution catalysts for solar water splitting. Science 343(6174):990–994. doi:10.1126/science.1246913

Klahr B, Hamann T (2014) Water oxidation on hematite photoelectrodes: insight into the nature of surface states through in situ spectroelectrochemistry. J Phys Chem C 118(19):10393–10399. doi:10.1021/Jp500543z

Klahr B, Gimenez S, Fabregat-Santiago F, Bisquert J, Hamann TW (2012) Photoelectrochemical and impedance spectroscopic investigation of water oxidation with "Co-Pi"-coated hematite electrodes. J Am Chem Soc 134(40):16693–16700. doi:10.1021/Ja306427f

Klingan K, Ringleb F, Zaharieva I, Heidkamp J, Chernev P, Gonzalez-Flores D, Risch M, Fischer A, Dau H (2014) Water oxidation by amorphous cobalt-based oxides: volume activity and proton transfer to electrolyte bases. ChemSusChem 7(5):1301–1310. doi:10.1002/cssc.201301019

Le Formal F, Tetreault N, Cornuz M, Moehl T, Gratzel M, Sivula K (2011) Passivating surface states on water splitting hematite photoanodes with alumina overlayers. Chem Sci 2 (4):737–743. doi:10.1039/C0sc00578a

Le Formal F, Sivula K, Gratzel M (2012) The transient photocurrent and photovoltage behavior of a hematite photoanode under working conditions and the influence of surface treatments. J Phys Chem C 116(51):26707–26720. doi:10.1021/Jp308591k

Le Formal F, Pendlebury SR, Cornuz M, Tilley SD, Gratzel M, Durrant JR (2014) Back electron-hole recombination in hematite photoanodes for water splitting. J Am Chem Soc 136 (6):2564–2574. doi:10.1021/Ja412058x

Lin F, Boettcher SW (2014) Adaptive semiconductor/electrocatalyst junctions in water-splitting photoanodes. Nat Mater 13(1):81–86. doi:10.1038/Nmat3811

Lin F, Lonergan MC (2006) Gate electrode processes in an electrolyte-gated transistor: non-Faradaically versus Faradaically coupled conductivity modulation of a polyacetylene ionomer. Appl Phys Lett 88(13):133507. doi:10.1063/1.2190077

Lin F, Bachman BF, Boettcher SW (2015) Impact of electrocatalyst activity and ion permeability on water-splitting photoanodes. J Phys Chem Lett 6(13):2427–2433. doi:10.1021/acs.jpclett.5b00904

Lobato K, Peter LM, Wurfel U (2006) Direct measurement of the internal electron quasi-Fermi level in dye sensitized solar cells using a titanium secondary electrode. J Phys Chem B 110 (33):16201–16204

Lonergan MC (1997) A tunable diode based on an inorganic semiconductor vertical bar conjugated polymer interface. Science 278(5346):2103–2106. doi:10.1126/science.278.5346.2103

McAlpin JG, Surendranath Y, Dinca M, Stich TA, Stoian SA, Casey WH, Nocera DG, Britt RD (2010) EPR evidence for Co(IV) species produced during water oxidation at neutral pH. J Am Chem Soc 132(20):6882–6883. doi:10.1021/Ja1013344

Mills TJ, Lin F, Boettcher SW (2014) Theory and simulations of electrocatalyst-coated semiconductor electrodes for solar water splitting. Phys Rev Lett 112(14):148304. doi:10.1103/Physrevlett.112.148304

Nakato Y, Tsubomura H (1982) The photo-electrochemical behavior of an n-$TiO_2$ electrode coated with a thin metal-film, as revealed by measurements of the potential of the metal-film. Isr J Chem 22(2):180–183

Nakato Y, Abe K, Tsubomura H (1976) New photovoltaic effect observed for metal-coated semiconductor electrodes and its utilization for photolysis of water. Phys Chem Chem Phys 80(10):1002–1007

Pendlebury SR, Cowan AJ, Barroso M, Sivula K, Ye JH, Gratzel M, Klug DR, Tang JW, Durrant JR (2012) Correlating long-lived photogenerated hole populations with photocurrent densities in hematite water oxidation photoanodes. Energy Environ Sci 5(4):6304–6312. doi:10.1039/C1ee02567h

Peter LM, Wijayantha KGU (2014) Photoelectrochemical water splitting at semiconductor electrodes: fundamental problems and new perspectives. ChemPhysChem 15(10):1983–1995. doi:10.1002/cphc.201402024

Peter LM, Wijayantha KGU, Tahir AA (2012) Kinetics of light-driven oxygen evolution at $\alpha$-$Fe_2O_3$ electrodes. Faraday Discuss 155:309–322. doi:10.1039/C1fd00079a

Pinson WE (1977) Quasi-Fermi level measurement in an illuminated GaP photoelectrolysis cell. Nature 269(5626):316–318. doi:10.1038/269316a0

Rhoderick EH, Williams RH (1988) Metal-semiconductor contacts, 2nd edn. Oxford University Press, New York

Riha SC, Klahr BM, Tyo EC, Seifert S, Vajda S, Pellin MJ, Hamann TW, Martinson ABF (2013) Atomic layer deposition of a submonolayer catalyst for the enhanced photoelectrochemical performance of water oxidation with hematite. ACS Nano 7(3):2396–2405. doi:10.1021/Nn305639z

Risch M, Khare V, Zaharieva I, Gerencser L, Chernev P, Dau H (2009) Cobalt-oxo core of a water-oxidizing catalyst film. J Am Chem Soc 131(20):6936–6937. doi:10.1021/Ja902121f

Seabold JA, Choi KS (2012) Efficient and stable photo-oxidation of water by a bismuth vanadate photoanode coupled with an iron oxyhydroxide oxygen evolution catalyst. J Am Chem Soc 134 (4):2186–2192. doi:10.1021/Ja209001d

Sivula K (2013) Metal oxide photoelectrodes for solar fuel production, surface traps, and catalysis. J Phys Chem Lett 4(10):1624–1633. doi:10.1021/Jz4002983

Spurgeon JM, Velazquez JM, McDowell MT (2014) Improving $O_2$ production of $WO_3$ photoanodes with $IrO_2$ in acidic aqueous electrolyte. Phys Chem Chem Phys 16 (8):3623–3631. doi:10.1039/C3cp55527e

Sun JW, Zhong DK, Gamelin DR (2010) Composite photoanodes for photoelectrochemical solar water splitting. Energy Environ Sci 3(9):1252–1261. doi:10.1039/C0ee00030b

Tilley SD, Cornuz M, Sivula K, Gratzel M (2010) Light-induced water splitting with hematite: improved nanostructure and iridium oxide catalysis. Angew Chem Int Ed 49(36):6405–6408. doi:10.1002/anie.201003110

Trotochaud L, Ranney JK, Williams KN, Boettcher SW (2012) Solution-cast metal oxide thin film electrocatalysts for oxygen evolution. J Am Chem Soc 134(41):17253–17261. doi:10.1021/Ja307507a

Trotochaud L, Mills TJ, Boettcher SW (2013) An optocatalytic model for semiconductor-catalyst water-splitting photoelectrodes based on *in situ* optical measurements on operational catalysts. J Phys Chem Lett 4(6):931–935. doi:10.1021/Jz4002604

Trotochaud L, Young SL, Ranney JK, Boettcher SW (2014) Nickel-iron oxyhydroxide oxygen-evolution electrocatalysts: the role of intentional and incidental iron incorporation. J Am Chem Soc 136(18):6744–6753. doi:10.1021/Ja502379c

Walter M, Warren E, McKone J, Boettcher SW, Mi QX, Santori L, Lewis NS (2010) Solar water splitting cells. Chem Rev 110(10):6446–6473

Wijayantha KGU, Saremi-Yarahmadi S, Peter LM (2011) Kinetics of oxygen evolution at $\alpha$-$Fe_2O_3$ photoanodes: a study by photoelectrochemical impedance spectroscopy. Phys Chem Chem Phys 13(12):5264–5270. doi:10.1039/C0cp02408b

Wilson RH, Harris LA, Gerstner ME (1977) Potential of gold overlayer on n-GaP photoelectrodes. J Electrochem Soc 124(8):1233–1234. doi:10.1149/1.2133535

Winkler MT, Cox CR, Nocera DG, Buonassisi T (2013) Modeling integrated photovoltaic-electrochemical devices using steady-state equivalent circuits. Proc Natl Acad Sci U S A 110(12):E1076–E1082. doi:10.1073/pnas.1301532110

Ye H, Park HS, Bard AJ (2011) Screening of electrocatalysts for photoelectrochemical water oxidation on W-Doped $BiVO_4$ photocatalysts by scanning electrochemical microscopy. J Phys Chem C 115(25):12464–12470. doi:10.1021/Jp200852c

Zaban A, Zhang J, Diamant Y, Melemed O, Bisquert J (2003) Internal reference electrode in dye sensitized solar cells for three-electrode electrochemical characterizations. J Phys Chem B 107 (25):6022–6025. doi:10.1021/jp034554a

Zhong DK, Gamelin DR (2010) Photoelectrochemical water oxidation by cobalt catalyst ("Co-Pi")/$\alpha$-$Fe_2O_3$ composite photoanodes: oxygen evolution and resolution of a kinetic bottleneck. J Am Chem Soc 132(12):4202–4207. doi:10.1021/Ja908730h

Zhong DK, Choi S, Gamelin DR (2011a) Near-complete suppression of surface recombination in solar photoelectrolysis by "Co-Pi" catalyst-modified W:$BiVO_4$. J Am Chem Soc 133 (45):18370–18377. doi:10.1021/Ja207348x

Zhong DK, Cornuz M, Sivula K, Graetzel M, Gamelin DR (2011b) Photo-assisted electrodeposition of cobalt-phosphate (Co-Pi) catalyst on hematite photoanodes for solar water oxidation. Energy Environ Sci 4(5):1759–1764. doi:10.1039/C1ee01034d

# Part III
# Materials and Devices

# Chapter 8
# Multinary Metal Oxide Photoelectrodes

**Fatwa F. Abdi, Sean P. Berglund, and Roel van de Krol**

## 8.1 Introduction

Sunlight is by far the largest sustainable source of energy, and there seems little doubt that it will play a major role in any conceivable future energy scenario (Lewis and Nocera 2006). One of the main challenges for the use of solar energy on a TW scale is its intermittent nature. Accommodating the day-night cycle, clouds, and seasonal variations requires intermediate storage on a truly massive scale. An attractive pathway to achieve this is by directly converting abundant and cheap resources, such as water and $CO_2$, into hydrogen or hydrocarbons using sunlight. These "solar fuels" have up to 100 times higher energy and power densities than the best batteries, can be easily transported, and can be stored on a virtually infinite time scale.

Photoelectrochemical (PEC) water splitting (Walter et al. 2010) is one of the most widely studied routes toward the production of chemical fuels with sunlight, and for good reasons; it works at room temperature, it can be applied on virtually any scale (from $cm^2$ to $km^2$), it works under both direct and diffuse sunlight, and the electrochemical half reactions allow easy separation of the reaction products. The most difficult step in this process is the four-electron water oxidation half-reaction. While far from trivial, it is still much easier than the (photo-)electrochemical reduction of $CO_2$ to methanol or methane, which requires six or even eight consecutive electron transfer steps, respectively. Once hydrogen is produced, it can be used in the reverse water-gas-shift reaction to reduce $CO_2$ to form syngas ($CO + H_2$) (Haije and Geerlings 2011). Fischer-Tropsch processes can then be used to convert the syngas into whatever liquid hydrocarbon fuel is desired, ranging from

F.F. Abdi • S.P. Berglund • R. van de Krol (✉)
Helmholtz-Zentrum Berlin für Materialien und Energie GmbH,
Institute for Solar Fuels, Hahn-Meitner-Platz 1, 14109 Berlin, Germany
e-mail: roel.vandekrol@helmholtz-berlin.de



S. Giménez, J. Bisquert (eds.), *Photoelectrochemical Solar Fuel Production*,
DOI 10.1007/978-3-319-29641-8_8

methanol and formic acid to more complex fuels such as synthetic diesel and gasoline (Herron et al. 2015). Hydrogen can of course also be used directly as a fuel, but since the safe and efficient storage of $H_2$ still poses a challenge, this may not be practical for certain applications.

Coupling a photovoltaic (PV) solar cell to an electrolyzer is the most straightforward approach toward light-induced electrochemical water splitting. The required components are technologically mature, fully developed, and commercially available, including the highly efficient (>95 %) dc-dc converters that convert the voltage from the PV cell to the optimal voltage for driving the electrolyzer. One of the concerns, however, is cost. Several studies have estimated the cost of hydrogen made by photoelectrochemical/photocatalytic systems (Pinaud et al. 2013; van de Krol and Grätzel 2012) and by PV/electrolyzer systems (Newman et al. 2012; Trainham et al. 2012). The numbers show a large spread (~3–10 \$/kg $H_2$ for PEC, ~4–8 \$/kg $H_2$ for PV/electrolysis), which is not surprising when considering the large uncertainties involved, and the fact that PEC water splitting is still near the bottom of the technological readiness level scale (TRL 1–2) (Miller 2013). Nevertheless, common sense would suggest that a single, integrated PEC device would be cheaper than a system consisting of two or three separate devices (solar cell, electrolyzer, dc-dc converter). Another crucially important consideration is long-term stability. More and more studies now also report the stability of autonomous PEC water splitting systems, but these are still measured in hours or at best a few days. Stability is also an issue for PV/electrolysis systems based on alkaline electrolyzers. The Fe/Ni electrodes quickly corrode when they are not operating, which means that they require auxiliary power at night. PEM electrolyzers do not suffer from electrode corrosion, but have the disadvantage that the use of noble metals cannot be avoided. This brings us to another important advantage of direct, integrated PEC water splitting: since the illuminated area and the catalytic surface are the same, the current densities (10–20 mA/cm$^2$) are up to ~2 orders of magnitude lower than those in commercial electrolyzers (1–2 A/cm$^2$). This strongly reduces the demands on the catalysts and is likely to enable the use of low-cost earth-abundant materials such as Fe, Ni, or Co.

To drive the water splitting reaction, the photoelectrodes in a PEC cell should deliver at least ~1.6 V (1.23 V + overpotentials) and as high as possible photocurrents. A solar-to-hydrogen (STH) efficiency of 10 % is often considered to be the threshold for practical applications, which corresponds to a photocurrent density of 8.1 mA/cm$^2$. Similar to photovoltaic solar cells, the highest efficiencies can be obtained by connecting two or more absorbers in a tandem cell configuration. The optimum bandgap values for a double-junction absorber are ~1.8 and ~1.2 eV (Seitz et al. 2014). With a bandgap of 1.12 eV, silicon is a promising candidate as a bottom absorber. The choice of the top absorber is much less obvious. The exciting new class of inorganic/organic halide perovskites has the right bandgap, but they are notoriously unstable in water. The same is true for II–VI and III–V semiconductors such as CdSe, $Ga_{1-x}Al_xAs$, and $In_{1-x}Al_xN$.

Metal oxide semiconductors are an intriguing materials class that may provide the answer. Many members of this class show semiconducting behavior, and they generally show much better stabilities than III–V, II–VI, or elemental semiconductors. They can (and must) be processed in oxygen-containing environments, and there are many thousands of candidates to choose from; chances are that one or two of them have the properties needed to make efficient and practical photoelectrodes. We just need to find them. Having said that, metal oxide semiconductors often have poor electronic properties and are usually only stable in aqueous solutions over a limited pH range. In the remainder of this chapter, an overview will be given of the properties of metal oxide semiconductors for photoelectrochemical water splitting applications. Several important challenges and bottlenecks will be discussed, and various strategies to overcome them will be outlined. The emphasis will be on ternary metal oxides, which are composed of at least three different atomic elements. These "complex" oxides have received increasing attention in the past few years, and some of them now show higher efficiencies than their simple binary counterparts.

## 8.2 Limitations of Binary Metal Oxides

Metal oxides have been studied for more than 40 years as photoelectrodes, ever since the seminal report of Fujishima and Honda on photoelectrochemical water splitting using $TiO_2$ (Fujishima and Honda 1972). Particular emphasis has been placed on binary metal oxides, i.e., oxides consisting of a single metal cation and an oxygen anion. However, these efforts have met limited success, and after decades of research, there are less than a handful of binary metal oxides that show photocurrents larger than 3 mA/cm$^2$ (Solarska et al. 2010; Tilley et al. 2010, 2014). Moreover, it is arguable whether even these oxides can be improved further.

Some of the limitations of these binary metal oxides are:

- *The bandgap is too large.*
  The large bandgap of many binary metal oxides limits the maximum number of solar photons that can be absorbed, and thus the maximum photocurrent and efficiency that can be achieved. For example, $TiO_2$ and $WO_3$ have bandgap energies of 3.0 (rutile phase) and 2.7 eV, respectively. This means that the maximum STH efficiencies are limited to ~1.5 and 5 %, assuming that all AM1.5 photons with energy larger than the bandgap are absorbed and contribute to the photocurrent. These efficiency values are definitely too low for realizing a practical solar water splitting device.
- *The material is unstable in aqueous solutions.*
  While the chemical stability of oxides is one of the main selling points of their use as photoelectrodes for solar water splitting, most oxides are actually thermodynamically unstable in aqueous solutions. Examples of unstable oxides are ZnO, $In_2O_3$, and $Cu_2O$. While efforts are currently ongoing in developing surface protection layers to improve their stability (especially in the case of $Cu_2O$) (Azevedo et al. 2014; Paracchino et al. 2011, 2012), it remains to be

proven whether these efforts will lead to true long-term stability (lifetime >5 years). It should be noted that there are also some stable oxides, such as $TiO_2$, $Fe_2O_3$, and $WO_3$.

- *The carrier lifetime is too short.*
  Short carrier lifetimes due to extensive recombination and/or poor transport properties is another performance killer of the known binary metal oxides. A particularly notorious example is hematite ($\alpha$-$Fe_2O_3$). While the bandgap (2.0 eV) and the stability of the material are excellent, it suffers from having a very short carrier lifetime (in the range of several picoseconds) (Cherepy et al. 1998). As a result, the photocurrents achieved so far (~4 $mA/cm^2$) are still far away from the theoretical maximum of 12 $mA/cm^2$, and it is unclear how much further it can be improved.

Although efforts are still ongoing in the development of binary metal oxides, the problems mentioned above really demonstrate the restrictions of binary metal oxides. At the same time, it provides a strong motivation for quickly expanding our material database, by exploring more complex "multinary" metal oxides. More than 8000 and 700,000 combinations are possible with ternary and quaternary compounds, respectively, which clearly opens up a realm of possibilities in finding an ideal semiconducting material for solar water splitting.

## 8.3 n-Type Multinary Metal Oxides

### *8.3.1 Bismuth Vanadate ($BiVO_4$)*

#### 8.3.1.1 Crystal Structure

A promising multinary oxide that has received quite a bit of attention in the last ~5 years is bismuth vanadate ($BiVO_4$). There are three polymorphs of $BiVO_4$: pucherite, dreyerite, and clinobisvanite. Pucherite is the form of $BiVO_4$ found in nature, named after the location where it was found (Pucher Shaft, Wolfgang Maaβen mine field, Schneeberg, Germany). It has an orthorhombic crystal structure (space group *Pnca* with $a = 5.332$ Å, $b = 5.06$ Å, and $c = 12.02$ Å) (Bierlein and Sleight 1975). $BiVO_4$ prepared in the laboratory normally does not have the pucherite structure. Dreyerite has a tetragonal (zircon-type) crystal structure (space group $I4_1/amd$ with $a = b = 7.303$ Å and $c = 6.584$ Å) (Dreyer and Tillmanns 1981), while clinobisvanite has a monoclinic (scheelite-type) crystal structure (space group $C_2/c$ with $a = 7.247$ Å, $b = 11.697$ Å, $c = 5.09$ Å, and $\beta = 134.226°$) (Sleight et al. 1979; Zhao et al. 2010). Figure 8.1 shows these different crystal structures for $BiVO_4$. A scheelite-type tetragonal crystal structure also exists, which is a slight modification of the monoclinic clinobisvanite $BiVO_4$ in terms of the atomic positions of Bi, V, and O (space group $I4_1/a$ with $a = b = 5.147$ Å and $c = 11.7216$ Å) (Sleight et al. 1979).

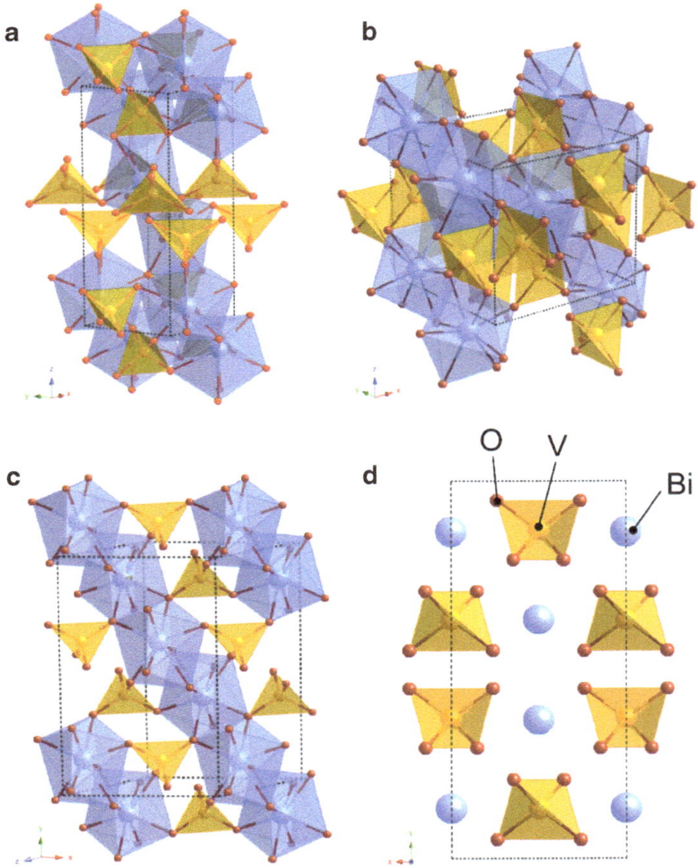


**Fig. 8.1** Crystal structures of different $BiVO_4$ polymorphs: (**a**) pucherite (orthorhombic), (**b**) dreyerite (tetragonal zircon), (**c**) clinobisvanite (monoclinic scheelite). The side view (*c*-axis) of the clinobisvanite structure is shown in (**d**)

Every V ion in all $BiVO_4$ polymorphs is tetrahedrally coordinated by four O atoms, and each Bi ion is dodecahedrally coordinated by eight O atoms. In the scheelite structure, each $BiO_8$ dodecahedral unit is surrounded by eight different $VO_4$ units. Each oxygen atom is coordinated to two Bi centers and one V center. This is true for both monoclinic and tetragonal scheelite structure. However, in the monoclinic scheelite, more distortion in the local environments of Bi and V ions is observed. As a result, the fourfold symmetry is lost. In a tetragonal scheelite, all four V-O bond lengths are equal (1.73 Å), and only two different Bi-O bond lengths exist (2.4 and 2.47 Å) (Sleight et al. 1979). On the other hand, there are two different V-O bond lengths (1.69 and 1.76 Å) and four different Bi-O bond lengths (2.35, 2.37, 2.52, and 2.63 Å) in a monoclinic scheelite structure (Sleight et al. 1979). This distortion in the monoclinic scheelite is claimed to be responsible for a higher photocatalytic activity when compared to the tetragonal scheelite structure, as shown by Tokunaga et al. (2001). This distortion has been suggested to enhance the local polarization, which improves the electron-hole separation and, thereby, the photocatalytic activity.

In the zircon structure, each $BiO_8$ dodecahedral unit is surrounded only by six $VO_4$ tetrahedral units. Two $VO_4$ units provide two O atoms, while four units provide one O atom to Bi. Finally, each $BiO_8$ unit is surrounded by seven $VO_4$ units in the orthorhombic structure, since one $VO_4$ unit gives two O atoms to Bi. The zircon-type structure is normally obtained from low-temperature synthesis, such as aqueous precipitation at room temperature, while the monoclinic scheelite structure is obtained by high-temperature preparation methods (Bhattacharya et al. 1997; Kudo et al. 1999). The zircon-type structure can be phase-transformed irreversibly to monoclinic scheelite at temperatures between 400 and 500 °C (Kudo et al. 1999; Roth and Waring 1963). The tetragonal scheelite phase is obtained at higher temperatures than the monoclinic scheelite phase, and the (reversible) transition occurs at 255 °C (Kudo et al. 1999).

#### 8.3.1.2 Electronic Structure and Optical Properties

The electronic structure of $BiVO_4$ has a strong correlation to the photocatalytic activity. An early study by Kudo et al. showed that the monoclinic scheelite $BiVO_4$ has a superior photocatalytic activity as compared to the zircon-type $BiVO_4$ (Kudo et al. 1999). They proposed that this is mainly caused by the enhanced photon absorption in the scheelite $BiVO_4$, as a result of smaller bandgap (2.4 and 2.9 eV for scheelite- and zircon-type $BiVO_4$, respectively). The optical transition in zircon-type $BiVO_4$ occurs between O 2p orbitals and empty V 3d orbitals, whereas the transition in scheelite-type $BiVO_4$ occurs between the Bi 6s or hybrid Bi 6s-O 2p orbitals and empty V 3d orbitals. This difference is illustrated in Fig. 8.2.

Further in-depth analysis based on density functional theory (DFT) calculations and X-ray spectroscopy studies was performed in the last decade (Cooper et al. 2014, 2015; Ding et al. 2013b, 2014; Payne et al. 2011; Wadnerkar and English 2013; Walsh et al. 2009; Zhao et al. 2010, 2011). It was found that

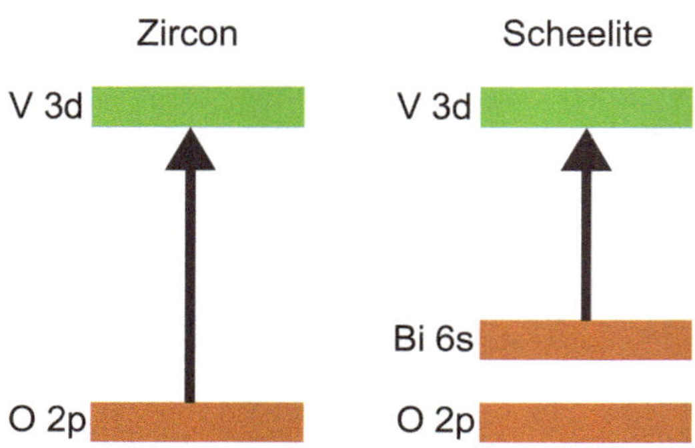


**Fig. 8.2** Schematic band structure of zircon- and scheelite-type $BiVO_4$

zircon-type $BiVO_4$ indeed has different arrangements of the density of states, as compared to the scheelite-type $BiVO_4$. In zircon-type $BiVO_4$, there is a strong interaction between Bi and O atoms, as indicated by the shorter Bi-O bonds. The valence band maximum is mainly composed of O 2p orbitals, which are located at quite low energies, causing a poor photoresponse in the visible light regime (Ding et al. 2013b). The analysis also confirmed that the scheelite-type $BiVO_4$ has a smaller bandgap. However, these studies indicate that, despite the presence of Bi 6s levels above the O 2p orbitals, the optical transition under visible light illumination still occurs between the filled O $2p_\pi$ and empty V 3d orbitals. The energy difference between these levels is, however, decreased as a result of the crystal distortion, especially in monoclinic scheelite $BiVO_4$, which pushes the O 2p states upward and reduces the bandgap (Zhao et al. 2011). This agrees with the lack of wave function overlap between the valence and conduction band edges and the large spatial distance between Bi and V atoms (~3.61–3.85 Å). Therefore, an electronic transition from the occupied Bi 6s to the empty V 3d states, as suggested earlier (Kudo et al. 1999), is rather unlikely.

The effective carrier masses in monoclinic scheelite $BiVO_4$ are also calculated to be lower than in the other $BiVO_4$ polymorphs. The lowest effective mass of electrons ($m_e^*$) at the bottom of the conduction band is ~0.9 $m_0$ ($m_0$, the rest mass of electron), while the lowest effective mass found for holes ($m_h^*$) at the top of the valence band is ~0.7 $m_0$ (Zhao et al. 2011). Zircon-type $BiVO_4$ has effective masses of electrons and holes of ~17.3 and 1.2 $m_0$, respectively (Ding et al. 2013b). These values are also smaller than those of most other oxide semiconductors for photocatalysis; for example, anatase $TiO_2$ has effective masses of electrons and holes of ~10 and 0.8 $m_0$, respectively. Lighter carriers are desired since the probability to reach the interfaces is increased. This is, therefore, likely to contribute to the superior photocatalytic activity of monoclinic scheelite $BiVO_4$.

The nature of optical transition (i.e., the fundamental bandgap) of monoclinic scheelite $BiVO_4$ was a topic of controversy in the literature. Various computational as well as experimental studies have reported monoclinic $BiVO_4$ to have either an indirect (Ding et al. 2013b; Ma and Wang 2014; Zhao et al. 2011) or a direct bandgap (Stoughton et al. 2013; Walsh et al. 2009). However, recent studies which combine advanced spectroscopic techniques (XAS, RIXS, XPS, UV-Vis, spectroscopic ellipsometry) and DFT calculations unambiguously confirm the indirect nature of $BiVO_4$ bandgap (Cooper et al. 2014, 2015). The direct transition occurs at only ~200 meV higher energies, which explains the initial controversy. It also explains the unusually high absorption coefficient ($10^7$ $m^{-1}$) for an indirect semiconductor.

#### 8.3.1.3 Photoelectrochemical Properties

$BiVO_4$ has been studied as a photocatalytic material as early as 1998, when Kudo et al. showed $O_2$ evolution under visible light irradiation in the presence of $Ag^+$ as an electron scavenger (Kudo et al. 1998). Efforts in these early years, however, focused on the application of $BiVO_4$ as a heterogeneous (powder-based) photocatalyst (Kohtani et al. 2002; Kudo et al. 1998, 1999; Long et al. 2006; Zhou et al. 2006). Later on, efforts shifted to thin-film photoelectrodes made by, e.g., metal organic decomposition (Galembeck and Alves 2000; Sayama et al. 2006). The reported photocurrents, however, remained modest (several hundreds of $\mu A/cm^2$) until a few years ago. This was mainly caused by the fact that the factors that limited the performance of $BiVO_4$ were not well understood. Rapid progress was made after several systematic studies were undertaken, and photocurrents close to the theoretical maximum have already been reported (Pihosh et al. 2015; Shi et al. 2015).

One of the main limitations of $BiVO_4$ has been attributed to the slow transfer of photogenerated holes from the $BiVO_4$ to the electrolyte, as shown by various studies involving the use of hole scavengers, such as hydrogen peroxides (Abdi et al. 2013a; Abdi and van de Krol 2012; Zhong et al. 2011), sulfites (Berglund et al. 2011; Kim and Choi 2014; Seabold and Choi 2012), methanol (Hong et al. 2011; Sayama et al. 2006), and carbonates (Saito et al. 2012; Sayama et al. 2010). Due to the sluggish water oxidation kinetics, holes accumulate at the surface and end up in recombination processes. To tackle this, many efforts have focused on modifying the surface of $BiVO_4$ with oxygen evolution catalysts (OECs). Various OECs have been investigated and reported, such as cobalt phosphate (Co-Pi), cobalt and nickel borate (Co-Bi and Ni-Bi), $CoO_x$, $Co_3O_4$, FeOOH, NiOOH, dual-layer Fe/NiOOH, as well as precious metal-based $RhO_2$ and Pt (Abdi and van de Krol 2012; Choi et al. 2013; Ding et al. 2013a; Kim and Choi 2014; Long et al. 2008; Luo et al. 2011; Seabold and Choi 2012; Ye et al. 2011; Zhong et al. 2011, 2015). The OECs generally lead to increased photocurrents, a cathodic shift of the onset potential, and improved stability. Charge transfer efficiencies close to 100 % have been reported after OEC deposition, which implies that slow water oxidation kinetics are no longer an issue in properly modified $BiVO_4$ photoanodes.

It is important to note that the effectiveness of the $BiVO_4$/OEC combination depends on the particular $BiVO_4$ synthesis technique. The photocurrents of spray-deposited $BiVO_4$ always increase after depositing Co-Pi (Abdi et al. 2013a, b). In contrast, when $BiVO_4$ is electrochemically deposited (Seabold and Choi 2012), no enhancement in photocurrent is observed upon Co-Pi deposition. A recent study in the authors' own lab with a sputtered $BiVO_4$ even shows decreased photocurrents when Co-Pi was deposited on the $BiVO_4$ surface. One hypothesis to explain these discrepancies is that these different synthesis techniques result in $BiVO_4$ with different surface properties (surface termination, oxidation state, etc.), which influence the coupling with the OECs. Various studies to elucidate the causes and the general understanding of the photoanode/OEC interface are ongoing in our lab and other labs.

The second limitation of $BiVO_4$ is its poor charge transport properties. Charge carriers are transported via a small polaron hopping conduction mechanism (Rettie et al. 2013), resulting in rather low charge carrier mobilities. A mobility of ~0.2 $cm^2/V/s$ is reported for single-crystal $BiVO_4$ (Rettie et al. 2013), with one order of magnitude smaller values being reported for polycrystalline films (Abdi et al. 2013c). This is a few orders of magnitude smaller compared to other commonly studied oxides, such as ZnO (200 $cm^2/V/s$) and $WO_3$ (10 $cm^2/V/s$). As a result of this low mobility, charge carriers recombine in the bulk before they even reach the surface or the interface, and the charge separation efficiency becomes very low.

A general way to improve conduction in a semiconductor is through doping. In $BiVO_4$, doping is expected to increase the charge carrier concentration, and therefore compensate the relatively low charge carrier mobility. Two elements stand out as highly effective dopants in $BiVO_4$: Mo and W (Abdi et al. 2013a, b; Pilli et al. 2011). A few studies have been done on other dopants (e.g., Ti, Cr, Fe), but only Mo and W resulted in significant performance improvements (Park et al. 2011; Parmar et al. 2012). Mo and W have six valence electrons and thus act as donor-type dopants when substituting for $V^{5+}$ in the $BiVO_4$ lattice. Typical dopant concentrations introduced during the synthesis process are in the range of 1–10 %, but the actual increase in free carrier concentration is normally much lower.

Oxoanion doping by replacing a small fraction of $VO_4^{3-}$ with $PO_4^{3-}$ has also been reported (Jo et al. 2012). Despite being a charge-neutral substituent for $V^{5+}$, the incorporation of $P^{5+}$ caused significant improvement of photocurrent as compared to pure $BiVO_4$. As a possible explanation, it was reported that the presence of P decreases the charge transfer resistance of $BiVO_4$. In addition, the lattice strain due to the different V-O and P-O bond lengths induces an internal electric field, which is beneficial for charge carrier separation.

Doping, however, can also have a negative influence on the overall performance of the photoelectrode. For example, an increasing dopant concentration will decrease the width of the depletion layer, thus narrowing the region of the material in which the internal electric field ensures efficient charge separation. It is therefore essential to maintain the balance between the favorable and unfavorable implications of doping. This is also one of the reasons why there seems to always be an optimum dopant concentration at which the photoelectrochemical performance peaks. One elegant way to circumvent this trade-off is to form an $n^+$-n homojunction (Abdi et al. 2013b). It takes advantage of the fact that two $BiVO_4$ layers with different carrier concentrations (undoped and W-doped $BiVO_4$, Fig. 8.3a) have a mismatch in their Fermi energy level. When these layers contact each other, their Fermi levels will equilibrate, and an internal electric field will be created at the interface (Fig. 8.3a). This enhances the charge separation efficiency at the interface of this homojunction. To fully leverage this effect, we fabricated a $BiVO_4$ film with a ten-step W dopant gradient (Fig. 8.3b) and found that this increases the carrier separation efficiency from ~38 % to more than 60 % at 1.23 V vs. RHE (Abdi et al. 2013b). The gradient dopant concept is a generally applicable strategy to increase the charge separation efficiency in highly doped semiconductors and illustrates the extent to which doping can be used to tune the materials properties.

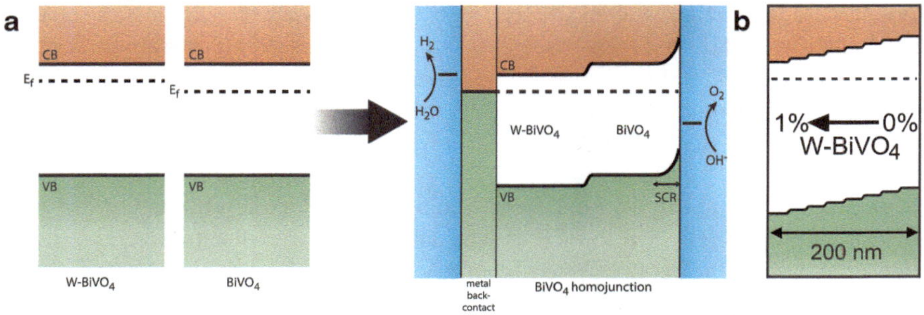


**Fig. 8.3** (**a**) The formation of a $BiVO_4$ homojunction based on the relative difference of Fermi level between W-doped $BiVO_4$ (W-$BiVO_4$) and undoped $BiVO_4$. The valence band (VB), conduction band (CB), the Fermi level energy ($E_F$), and the space charge region (SCR) are identified. (**b**) Band diagram schematic of a ten-step gradient-doped W:$BiVO_4$. Adapted with permission from Abdi et al. (2013b)

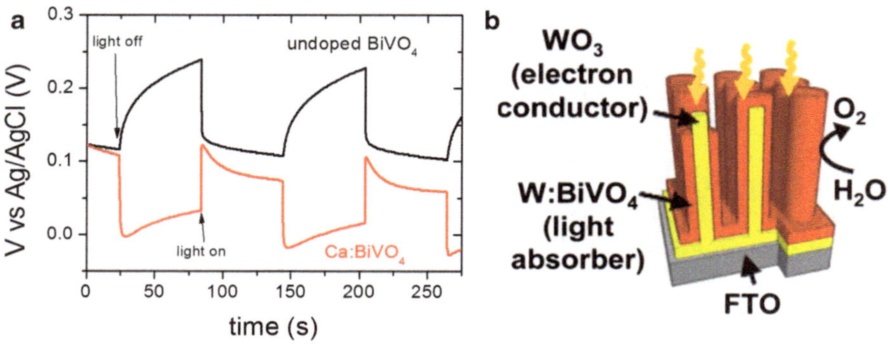


**Fig. 8.4** (**a**) Open-circuit potential ($V_{oc}$) of undoped (*black*) and 1 % Ca-doped (*red*) $BiVO_4$ under chopped AM1.5 illumination. (**b**) Schematic diagram of a guest-host nanostructured photoelectrode consisting of $WO_3$ nanowires (host) covered with W-doped $BiVO_4$ (W:$BiVO_4$, guest). Such structures can be prepared by flame vapor deposition and drop-casting (Reprinted with permission from Rao et al. (2014). Copyright 2014 American Chemical Society)

In contrast to the many reports on donor-type dopants as described above, no experimental study has been reported on acceptor-type dopants in $BiVO_4$. DFT calculations suggest that incorporating some of group II elements, such as Ca and Sr, on Bi sites would form very shallow acceptors that have low formation energies (Yin et al. 2011). A preliminary study in authors' lab shows that Ca in $BiVO_4$ may indeed act as an acceptor-type dopant. Figure 8.4a shows an open-circuit potential ($V_{oc}$) measurement of bare and Ca-doped $BiVO_4$ under chopped AM1.5 illumination. While the bare $BiVO_4$ shows a typical n-type behavior, i.e., the OCP becomes more negative under illumination, the Ca-doped $BiVO_4$ shows the exact opposite, indicating p-type behavior. Further efforts are ongoing in analyzing this type of $BiVO_4$.

Other efforts in improving the charge transport properties of $BiVO_4$ involve nanostructuring. Although the carrier mobility of $BiVO_4$ is modest, it has a relatively long carrier lifetime of ~40 ns, resulting in a carrier diffusion length of 70–100 nm (Abdi et al. 2013c; Rettie et al. 2013). As long as the feature sizes are smaller than this diffusion length, most of the charge carriers can arrive at the surface or interface before they recombine, thus ensuring high carrier separation efficiencies. $BiVO_4$ nanostructures have been fabricated with several synthesis techniques. The first is through electrochemical deposition of nanoporous $BiVO_4$ (Kim and Choi 2014). With a particle size in the range of ~70–80 nm, carrier separation efficiencies in excess of 90 % at 1.23 V vs. RHE have been achieved with this type of $BiVO_4$. A second—and more common—example of creating $BiVO_4$ nanostructures is through a guest-host approach in combination with $WO_3$ nanowires or nanoporous films (Pihosh et al. 2015; Rao et al. 2014; Shi et al. 2014, 2015), as depicted in Fig. 8.4b. $WO_3$ is normally chosen as the host material due to its excellent transport properties and suitable band alignment with $BiVO_4$. Photo-excited electrons from the $BiVO_4$ layer are transferred to the $WO_3$ nanostructure, where they can then be efficiently transported toward the back contact. $WO_3$ nanostructured films were deposited via oblique angle (or glancing angle) reactive deposition, flame synthesis, or simple drop-casting techniques. A thin conformal layer of $BiVO_4$ (~50 nm) was then deposited onto these nanostructured $WO_3$ layers by solution-based synthesis techniques, such as dip coating or electrodeposition. Since the thickness of the $BiVO_4$ was kept below the carrier diffusion length, a carrier separation efficiency close to ~1 could be achieved for these composite photoelectrodes. In addition to the high carrier separation efficiencies, these nanostructuring efforts also ensured a high optical absorption, despite the small thickness of the $BiVO_4$ layer.

At the time of writing, the highest AM1.5 photocurrent ever reported for $BiVO_4$ is 6.7 $mA/cm^2$ at 1.23 V vs. RHE, as reported by Pihosh et al. (2015). They achieved this by employing a $BiVO_4/WO_3$ guest-host nanostructure modified with a Co-Pi OEC. The onset potential of $BiVO_4$ is typically around 0.2–0.3 V vs. RHE (Abdi et al. 2013b; Kim and Choi 2014; Pihosh et al. 2015), which is more negative than other reported metal oxide photoanodes and close to the flatband potential of ~0 V vs. RHE (Sayama et al. 2006). The modest onset potential means that only a small bias voltage needs to be provided to achieve overall water splitting, which makes $BiVO_4$ a particularly well-suited candidate material for use in tandem cell devices.

#### 8.3.1.4 Future Outlook

In the past few years, $BiVO_4$ has emerged as one of the highest performing metal oxide photoanode materials. The highest reported AM1.5 photocurrent of 6.7 $mA/cm^2$ at 1.23 V vs. RHE is within 10 % of the theoretical maximum photocurrent of 7.5 $mA/cm^2$ (Pihosh et al. 2015). One of the remaining issues is the stability. $BiVO_4$ is stable in the pH range of 3–11 (Abdi et al. 2012), and as a result, almost all studies were carried out in solutions with a near-neutral pH. However, the buildup of pH gradients is difficult to avoid in neutral solutions, and this will eventually lead to a decrease in the photocurrent (Jin et al. 2014). Further developments of $BiVO_4$ should therefore be directed toward stabilization of $BiVO_4$, so that it can be used in either highly acidic or alkaline environments. The first efforts on developing surface protection layers for $BiVO_4$ have recently been reported (Lichterman et al. 2013; McDowell et al. 2014). A second challenge is to decrease the bandgap of the material, in order to be able to reach solar-to-hydrogen efficiencies above 10 %. Possible strategies are anion doping (e.g., substitution of oxygen with nitrogen) or replacing Bi with an element that pushes up the VB maximum.

## *8.3.2 Copper Tungstate ($CuWO_4$)*

### 8.3.2.1 Crystal Structure

Copper tungstate, $CuWO_4$, has a triclinic crystal structure (space group $P\overline{1}$ with $a = 4.7026$ Å, $b = 5.8389$ Å, $c = 4.8784$ Å, $\alpha = 91.677°$, $\beta = 92.469°$, and $\gamma = 82.805°$) (Kihlborg and Gebert 1970). The structure is similar to monoclinic wolframite (e.g., $NiWO_4$, $ZnWO_4$), but the cell symmetry is reduced to triclinic due to the Jahn-Teller distortion of $Cu^{2+}$ (Khyzhun et al. 2005, 2009; Kihlborg and Gebert 1970; Ruiz-Fuertes et al. 2008; Yourey and Bartlett 2011). This causes the angle $\alpha$ and $\gamma$ to deviate from 90°. Figure 8.5 shows the crystal structure of $CuWO_4$. The four positions of the oxygen atoms are nonequivalent, and both Cu and W ions are coordinated with six O atoms (octahedral). The structure can be described as a framework of oxygen atoms in an approximately hexagonal close-packing with the cations occupying half of the octahedral sites (Kihlborg and Gebert 1970). Cu and W form alternating sequence of layers between the oxygen sheets; infinite zigzag

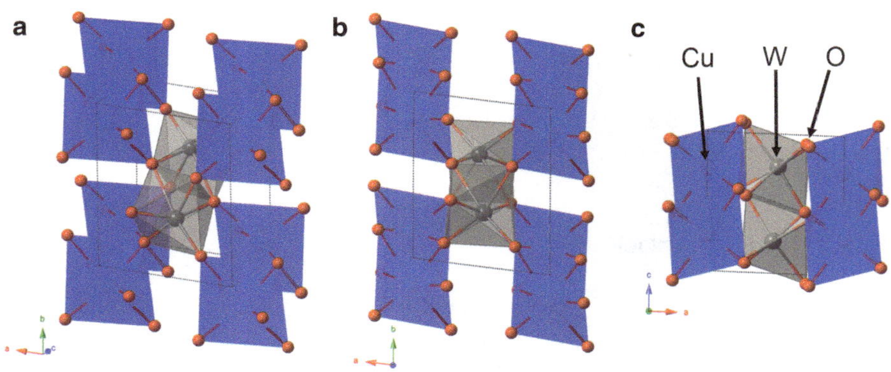


**Fig. 8.5** (**a**) Crystal structure of $CuWO_4$, which consists of $CuO_6$ and $WO_6$ octahedra. Views along the *c*-axis and *b*-axis are shown in (**b**) and (**c**), respectively

chains are formed by edge-sharing of alternating $CuO_6$ and $WO_6$ octahedra (Kihlborg and Gebert 1970; Yourey and Bartlett 2011). Antiferromagnetic behavior has been reported in $CuWO_4$, due to the existence of two kinds of connections between the $CuO_6$ octahedra in the structure of $CuWO_4$. No other crystal phase of $CuWO_4$ has been reported.

#### 8.3.2.2 Electronic Structure and Optical Properties

The electronic structure of $CuWO_4$ has been studied via first principle and DFT calculations (Jain et al. 2013b; Khyzhun et al. 2009), as well as X-ray spectroscopy experiments (Khyzhun et al. 2005, 2009). The conduction band of $CuWO_4$ mainly consists of W 5d orbitals, similar to $WO_3$, with some contributions from Cu 3d and O 2p orbitals. The valence band, however, consists of Cu 3d and O 2p orbitals (Bharati et al. 1980, 1983; Khyzhun et al. 2009). The hybridization causes the valence band maximum to raise and the bandgap to be reduced as compared to $WO_3$ where the valence band maximum mainly consists of O 2p orbitals. Although initial studies on single-crystal $CuWO_4$ reported conflicting bandgap values of 1.5 and 3.5 eV (Arora et al. 1989; Bharati et al. 1983), more recent studies have consistently reported an indirect bandgap value of ~2.2–2.4 eV (Bohra and Smith 2015; Chang et al. 2011; Gaillard et al. 2013; Yourey and Bartlett 2011). The conduction mechanism in $CuWO_4$ consists of two processes: hopping of charge carriers in the narrow 3d band at low temperatures (below 350 °C) and normal band-like conduction in the 2p band at higher temperature (Bharati et al. 1980). A hole mobility ($\mu_h$) of 10 $cm^2/V$ s has been reported for a single-crystal $CuWO_4$ (Bharati et al. 1980). Amorphous $CuWO_4$ has been reported to show p-type conductivity, but the crystalline phase is an n-type semiconductor (Chang et al. 2011).

#### 8.3.2.3 Photoelectrochemical Properties

Early reports on the photoelectrochemical properties of $CuWO_4$ date back ~30 years, when a single-crystal $CuWO_4$ was reported to be photoelectrochemically active in an acidic electrolyte (pH = 1) (Doumerc et al. 1984). The idea of using $CuWO_4$ for water oxidation was revisited recently (within the last 5 years), when thin films of $CuWO_4$ were prepared via different synthesis techniques: electrodeposition, spray deposition, and reactive co-sputtering (Bohra and Smith 2015; Chen et al. 2010a; Pandey et al. 2005; Yourey and Bartlett 2011). The flatband potential of $CuWO_4$ is ~0.4 V vs. RHE (Doumerc et al. 1984; Yourey and Bartlett 2011), which means that a modest amount of applied bias needs to be applied before the photocurrent onset can be observed. The AM1.5 photocurrent densities reported for these thin films are in the range of several hundreds of $\mu A/cm^2$. This is rather modest, considering that the theoretical maximum based on the bandgap of $CuWO_4$

is ~10 mA/cm$^2$. Several problems that explain this discrepancy have been discussed in the literature.

First, as experienced by most metal oxides, the charge carrier transfer between the surface and the electrolyte is inefficient. This is evidenced by an increase of the photocurrent when methanol (Yourey and Bartlett 2011) or hydrogen peroxide (Bohra and Smith 2015) is added to the electrolyte as hole scavengers. Surprisingly, no reports exist yet of $CuWO_4$ modified with oxygen evolution co-catalysts. The reason behind this may be related to the surface electronic structure of $CuWO_4$, which shows midgap electronic states (possibly Cu 3d related) that are involved in the charge transfer to the electrolyte (Pyper et al. 2013). Hole capture by such states can be very fast, thus preventing hole transfer to the OEC. Indeed, it is reported that standard co-catalyst loading strategies have been unsuccessful in improving the photocurrent of $CuWO_4$, unlike the case in $BiVO_4$ or $Fe_2O_3$ (Pyper et al. 2013). Further understanding of the charge transfer processes is needed to address this limitation.

However, even in the presence of hole scavengers, the AM1.5 photocurrents are still 50–100 times smaller than the theoretical maximum. The main limitation of $CuWO_4$ seems to be related to the bulk resistance, i.e., charge carrier transport within the material. An impedance analysis of $CuWO_4$ revealed that the bulk resistance is in the order of 3 k$\Omega$/cm$^2$, more than one order of magnitude higher than what was reported for better-performing metal oxide PEC materials, such as $WO_3$ (approximately 200 $\Omega$/cm$^2$) (Chang et al. 2011). An interesting way to improve the conductivity was proposed by Gaillard et al. by forming a nanocomposite of $CuWO_4$ and multiwall carbon nanotubes (MWCNT) (Gaillard et al. 2013). A TEM image of the nanocomposite is shown in Fig. 8.6. By ensuring that the $CuWO_4$ nanoparticles are dispersed on the MWCNT, carrier transport can

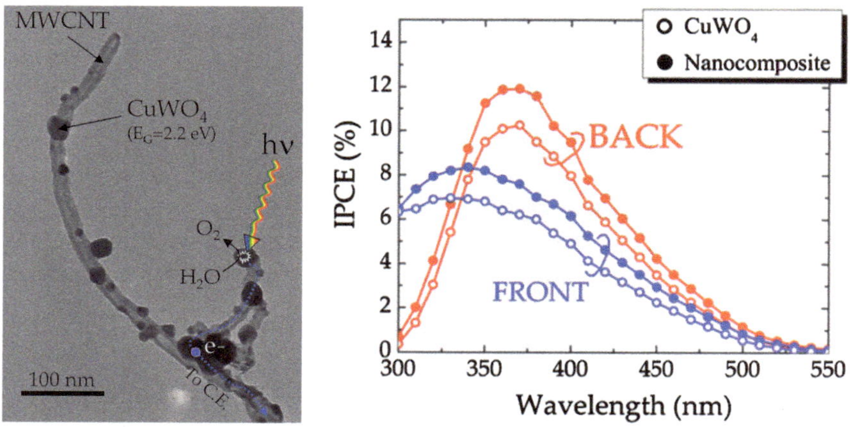


**Fig. 8.6** (*Left*) Transmission electron micrograph (TEM) of $CuWO_4$-MWCNT nanocomposite, which highlights that the $CuWO_4$ nanoparticles are firmly anchored to the MWCNT. (*Right*) This modification resulted in an improved incident photon-to-current conversion efficiency (IPCE) (Reprinted from Gaillard et al. (2013), Copyright 2013 Elsevier)

be accelerated. A lower bulk resistance was indeed measured, although the effect is not that significant (bulk resistance is still ~2 kΩ/cm$^2$). As a result, only modest improvements in the photocurrent and quantum efficiency (Fig. 8.6) were reported.

Another approach to improve the carrier transport in n-type $CuWO_4$ is by introducing donor-type dopants (Bohra and Smith 2015). In this case, $Fe^{3+}$ was chosen as a donor due to the relatively similar ionic radii compared to $Cu^{2+}$ and a good compatibility with $CuWO_4$ crystal structure. As a result of Fe-doping, the carrier separation efficiency and the photocurrent increased by 50 % at 1.23 V vs. RHE. However, the carrier separation efficiency is still lower than 8 %, which means that more than 92 % of the charge carriers still recombine before they reach the surface or interface.

The chemical stability of $CuWO_4$ has also been studied (Yourey et al. 2013). It was found that the stability is dependent on the pH and the anion of the electrolyte. In potassium phosphate, it is stable in pH 3 and 5 but not in pH 7. However, the stability in pH 7 potassium borate is excellent over the course of 12 h; the reason behind this interesting observation is still unknown. In all these cases, the Faradaic efficiency remains very close to 100 %.

#### 8.3.2.4 Future Outlook

As mentioned above, the main problem with $CuWO_4$ is its poor conductivity. The material is simply too resistive, and novel strategies are needed to improve the conductivity of $CuWO_4$ before the photoelectrochemical performance can be increased. Doping, which is typically the solution for such problem, has been shown to have limited success, although there has been only one report on this. This may be partially related to the fact that the carrier concentration in an undoped film is already very high (values of ~$5 \times 10^{19}$ up to $10^{21}$ cm$^{-3}$ have been reported) (Doumerc et al. 1984; Yourey and Bartlett 2011). The depletion layer is therefore expected to be very thin, and the majority of the photoexcited carriers are transported by diffusion, which is presumably poor in bulk $CuWO_4$. To solve this, designing a nanostructured $CuWO_4$ film that has geometrical features smaller than the diffusion length of the charge carriers may be a suitable strategy. Alternatively, introducing an internal electric field by forming homo- or heterojunction within the film may also be considered.

### 8.3.3 *Iron Tungstate ($Fe_2WO_6$)*

#### 8.3.3.1 Crystal Structure

$Fe_2WO_6$ has an orthorhombic crystal structure (space group *Pbcn*, $a = 4.576$ Å, $b = 16.766$ Å, $c = 4.967$ Å) (Harrison et al. 1985; Leiva et al. 1982; Senegas and Galy 1974; Sieber et al. 1983; Thomas and Ropital 1984; Walczak et al. 1992).

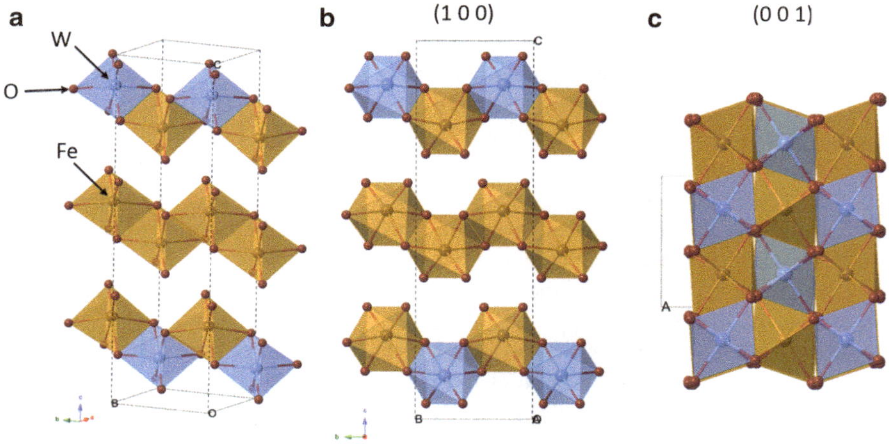


**Fig. 8.7** (**a**) Crystal structure of $Fe_2WO_6$, which consists of $FeO_6$ and $WO_6$ octahedra. Views along the (100) and (001) planes are shown in (**b**) and (**c**), respectively

Both Fe and W ions are octahedrally coordinated by oxygen. These octahedra share their edges and form zigzag chains along the *c*-axis. Figure 8.7 shows the structure of $Fe_2WO_6$ under different view axes. Two types of chains exist: one containing only Fe atoms and another with both Fe and W atoms.

#### 8.3.3.2 Electronic Structure and Optical Properties

$Fe_2WO_6$ is an n-type semiconductor (Khader et al. 1998; Leiva et al. 1982), although p-type conductivity has also been reported (Bharati and Singh 1981). The conduction band consists of Fe 3d and W 5d orbitals, while the valence band is a mixture of O 2p and Fe 3d orbitals (Bharati and Singh 1981; Jain et al. 2013b). The optical bandgap has been reported to be between 1.5 and 1.7 eV, and the transition is indirect in nature (Bharati and Singh 1981; Khader et al. 1998). A drift mobility of $4.7 \times 10^{-4}$ cm$^2$/V s was estimated at temperature of 400 K, and the small polaron conduction mechanism dominates (Bharati and Singh 1981). It should be noted that this material is still relatively unexplored; only one report exists on its electrical properties.

#### 8.3.3.3 Photoelectrochemical Properties

Only ~30 publications on $Fe_2WO_6$ have been found, and more than two thirds of these reports are more than 10 years old. Only two of these are related to the photoelectrochemical or photocatalytic properties (Harrison et al. 1985; Khader et al. 1998). Nevertheless, one single property of $Fe_2WO_6$ stands out that made it attractive for solar water splitting: its bandgap of ~1.5–1.7 eV. This means that the material can potentially achieve more than 20 % STH efficiency. This has prompted us to study this material in our own lab.

The photocurrents reported so far are modest, in the range of tens to hundreds of $\mu A/cm^2$ (Abdi et al. in preparation; Harrison et al. 1985; Khader et al. 1998). The photocurrent onset potential is very positive, about 1.0 V vs. RHE. The flatband potential based on Mott-Schottky analysis is reported to be ~0.6–0.7 V vs. RHE, and the carrier concentration is in the range of $10^{19}$–$10^{20}$ $cm^{-3}$. Similar to $CuWO_4$, charge transfer efficiency to the electrolyte is found to be limited, but the main problem lies with the bulk properties of the $Fe_2WO_6$ itself. We found that annealing at high temperature (800 °C) reduces the carrier density, but the photocurrents remain low. With a charge carrier separation efficiency of only ~5 %, the majority of the photogenerated carriers recombine within the bulk of the material. This may be related to the poor carrier transport properties, caused by the small polaron hopping mechanism in this material (Bharati and Singh 1981).

#### 8.3.3.4 Future Outlook

Other than the particularly attractive bandgap, $Fe_2WO_6$ suffers from many challenges. First, the bandgap is indirect, which causes the absorption coefficient to be very low ($<10^4$ $cm^{-1}$). Second, bulk recombination due to the poor transport properties severely limits the performance. While this may be resolved to some extent by a nanostructuring strategy, the main drawback of the material lies in the relatively high flatband potential of ~0.6–0.7 V vs. RHE. This makes it very challenging to implement this material in a tandem solar water splitting device. Unless these challenges can be overcome, it is very unlikely for $Fe_2WO_6$ to be a suitable photoelectrode material. Instead, efforts should probably be directed toward other members of the extensive and largely unexplored family of iron tungstate compounds.

## 8.4 p-Type Multinary Metal Oxides

Compared to the typical n-type metal oxides that are frequently employed as photoanodes, p-type metal oxides have received relatively little attention for use as photocathodes. Historically, photocathode research has focused on non-oxide, single-crystal semiconductors such as p-Si, p-GaP, p-InP, p-$GaInP_2$, $WS_2$, and $WSe_2$ (Aharon-Shalom and Heller 1982; McKone et al. 2013; Nakato et al. 1976; Peter and Wijayantha 2014; Walter et al. 2010). More recently, chalcopyrite thin films have been studied including $CuGaSe_2$ (CGS) and $Cu(In_{1-x}Ga_x)Se_2$ (CIGS) (Azarpira et al. 2015; Marsen et al. 2008; Yokoyama et al. 2010). The only p-type metal oxide that has been extensively tested as a photocathode material is $Cu_2O$, but its development has been limited because it is extremely unstable when illuminated in aqueous electrolytes and undergoes rapid photodecomposition (Gerischer 1977;

Peter and Wijayantha 2014; Walter et al. 2010). Recent advancements in the application of protection layers have helped to overcome this problem. For example, the research group of Michael Grätzel was able to enhance the stability of $Cu_2O$ by coating it with conformal layers of Al-doped ZnO and $TiO_2$ using atomic layer deposition (ALD) (Morales-Guio et al. 2015; Paracchino et al. 2011). Such strategies open up the possibility of using other p-type metal oxides that may otherwise be unstable. In addition, certain multinary p-type metal oxides appear to be much more stable than binary oxides like $Cu_2O$, even when the complex oxides contain a significant amount of copper themselves. Some of the recently identified copper-based p-type metal oxides that can be used as photocathodes include $CuBi_2O_4$, $CuNbO_3$, $CuNb_3O_8$, $Cu_3Ta_7O_{19}$, $Cu_5Ta_{11}O_{30}$, and $CuFeO_2$ (Arai et al. 2007; Berglund et al. 2013; Fuoco et al. 2012; Hahn et al. 2012; Joshi and Maggard 2012; Joshi et al. 2011; Prévot et al. 2015; Read et al. 2012). Numerous spinel-type materials containing iron and cobalt have also been identified as potential photocathode materials including $CaFe_2O_4$, $CoFe_2O_4$, $CuFe_2O_4$, $ZnFe_2O_4$, $Co_{3-x-y}Al_xFe_yO_4$ (Helaili et al. 2015; Ida et al. 2010; Walsh et al. 2007; Woodhouse and Parkinson 2008).

### *8.4.1 Copper Bismuth Oxide ($CuBi_2O_4$)*

Copper bismuth oxide, $CuBi_2O_4$, was first reported to be active as a photocathode in 2007 through a combinatorial chemistry study by the group of Sayama (Arai et al. 2007). Since then there have been only a few reports on $CuBi_2O_4$ as a photoelectrochemical material, so many of the photophysical and photoelectrochemical properties are not yet fully understood. $CuBi_2O_4$ (Kusachiite) is a p-type metal oxide with a tetragonal crystal structure (space group *P4/ncc*, $a = b =$ 8.511 Å, $c = 5.823$ Å, $Z = 4$) (Henmi 1995). Each bismuth atom is surrounded by a polyhedron of six oxygen atoms, and each copper atom is bonded to four oxygen atoms as shown in Fig. 8.8. $CuBi_2O_4$ has an optical absorption onset of 1.6–1.8 eV, although it is not yet clear whether the band gap is direct or indirect (Arai et al. 2007; Berglund et al. 2013; Hahn et al. 2012). The density of states and band structure have not yet been reported either. However, it is known that $CuBi_2O_4$ has a flatband potential around 1.30–1.45 V vs. RHE and a photocurrent onset potential more positive than 1 V vs. RHE, which is exceptional for a p-type oxide (Berglund et al. 2013; Hahn et al. 2012). The combination of a nearly ideal bandgap and the positive photocurrent onset potential makes $CuBi_2O_4$ an excellent candidate for the top absorber (larger bandgap material) in a dual-absorber PEC device (Seger et al. 2014; Seitz et al. 2014; Walter et al. 2010).

To date the main shortcoming of $CuBi_2O_4$ has been its low photoconversion efficiency. With a bandgap energy of 1.6–1.8 eV, $CuBi_2O_4$ has a theoretical maximum photocurrent density of at least 19.7 mA/cm$^2$ (Chen et al. 2010b). Yet the highest reported photocurrent density is only about 0.1 mA/cm$^2$ at 0.6 V vs. RHE in deaerated 0.3 M $K_2SO_4$ and 0.1 M phosphate buffer (for $CuBi_2O_4$ with

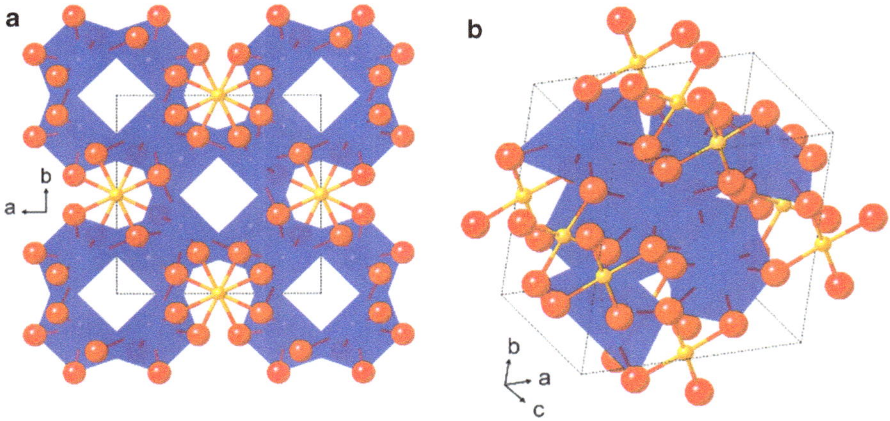


**Fig. 8.8** $CuBi_2O_4$ crystal structure as viewed (**a**) along the *c*-axis and (**b**) at an arbitrary angle. Bismuth atoms are represented the polyhedrons each of which is surrounded by six oxygen atoms. Copper is represented by the atoms that are bonded to four oxygen atoms

Pt on the surface as an electrocatalyst) (Park et al. 2014). One major limitation for $CuBi_2O_4$ as a photocathode is poor reaction kinetics for the proton reduction reaction. This is the reason that electrocatalysts such as Pt or even Ag need to be used with $CuBi_2O_4$. It was shown that Ag incorporated into $CuBi_2O_4$, remained in zero valence state at potentials more negative than 1.0 V vs. RHE, and appeared to improve reaction kinetics (Berglund et al. 2013). Nonetheless, the photocurrent densities for $CuBi_2O_4$ with an electrocatalyst are still two orders of magnitude lower than the theoretical maximum. This points to the other major limitation of charge transport within the bulk $CuBi_2O_4$ material or at the $CuBi_2O_4$-substrate interface.

Work is currently being done in the authors' lab to further understand these charge transport limitations (Berglund et al. in preparation). Time-resolved microwave conductivity (TRMC) measurements have been performed on $CuBi_2O_4$ films synthesized by drop-casting. The measurements show a mobility of $1 \times 10^{-3}$ $cm^2/V/s$ and two different lifetimes of 35 and 1087 ns, corresponding to an initial fast decay followed by a slower decay. Using the longer lifetime, the carrier diffusion length of $CuBi_2O_4$ is estimated to be approximately 50 nm. Compared to other metal oxides such as $BiVO_4$ and $Fe_2O_3$, this is a reasonable diffusion length, suggesting that, similar those other oxides, $CuBi_2O_4$ can be significantly improved by doping and/or nanostructuring (Abdi et al. 2013c; Kay et al. 2006; Kim and Choi 2014).

### *8.4.2 Delafossite ($CuFeO_2$)*

Delafossite, $CuFeO_2$, has a trigonal crystal structure (space group $R\,3m$, $a = 3.04$ Å, $c = 17.12$ Å, $Z = 3$). It has an indirect optical band gap of 1.47–1.55 eV and a conduction band edge near −0.4 V vs. RHE, so it should be capable of reducing water without a bias (Prévot et al. 2015; Read et al. 2012). In addition, the flatband potential is fairly positive at about 1 V vs. RHE, resulting in a photocurrent onset potential of about 0.9 V vs. RHE (Gu et al. 2013; Parfenov and Nazipov 2002). This makes it another attractive candidate for a dual-absorber PEC device. Additionally, $CuFeO_2$ appears to be stable under illumination in aqueous electrolytes, which is quite rare for a p-type metal oxide.

The main challenge with using $CuFeO_2$ as a photocathode material is that it typically requires high synthesis temperatures (above 700 °C) and produces relatively low photoconversion efficiencies (Parfenov and Nazipov 2002; Prévot et al. 2015; Read et al. 2012). Recently, it was shown that the photocurrent can be increased by thermal intercalation of oxygen into layered $CuFeO_2$ films to increase the acceptor density. In oxygen-saturated electrolyte, photocurrent densities of 1.5 mA/cm$^2$ at 0.35 V vs. RHE were achieved (Prévot et al. 2015). This is a promising improvement, but it is still well below the maximum theoretical value of 29 mA/cm$^2$ (based on bandgap energy of 1.5 eV), so further strategies in doping and nanostructuring will need to be explored for this material.

### 8.4.3 *Iron Spinels ($AFe_2O_4$) and Cobalt Spinels ($CoB_2O_4$)*

Spinels are a class of metal oxides that generally have the formula $AB_2O_4$ with oxidation states of $A^{2+}$, $B^{3+}$, and $O^{2-}$. In some instances, A and B are the same element, as in $Fe_3O_4$. Recently, iron-based spinels ($AFe_2O_4$) and cobalt-based spinels ($CoB_2O_4$) have been found to be photoactive for reduction reactions. One of the first reports of this phenomenon was the combinatorial synthesis study conducted by Woodhouse and Parkinson (Woodhouse et al. 2005; Woodhouse and Parkinson 2008). Ultimately, they discovered that the optimal photocurrent density for $Co_{3-x-y}Al_xFe_yO_4$ occurred for $x$ and $y$ values of 0.18 and 0.30, respectively. Later, the same research group discovered that $Co_{3-x}M_xO_4$ [M = (Al, Ga, In)] also exhibited good photoactivity, particularly for $x = 0.4$ (Newhouse and Parkinson 2015). The thickness, morphology, and surface structure of these materials has not yet been fully optimized, but the combinatorial studies offer a good starting point for further investigation of these new materials.

A communication by the group of Ishihara also showed that the iron spinel $CaFe_2O_4$ is a promising photocathode material. $CaFe_2O_4$ is a p-type material with a bandgap of 1.9 eV, a conduction band edge at −0.6 V vs. NHE, and valence band edge at 1.3 V vs. NHE (Ida et al. 2010). The $CaFe_2O_4$ photocathode prepared by calcination of $CaFe_2O_4$ particles onto a Pt substrate showed photocurrent densities on the order of 1 mA/cm$^2$ at −0.8 V vs. Ag/AgCl (in 0.1 M NaOH), and $H_2$ was produced when the photocathode was paired with a $TiO_2$ photoanode.

## 8.5 Tandem Devices with Multinary Oxides

Up till now, few reports exist on unassisted solar water splitting devices with multinary oxides. Except for very large-bandgap materials, such as $SrTiO_3$ and $NaTaO_3$ (Kato and Kudo 2001; Mavroides et al. 1976), most of these reports employ tandem device configurations, where more than one absorbers are used for the overall solar water splitting. By doing so, the stringent requirements for solar water splitting materials can be somewhat relaxed, and it is not necessary for a single material to fulfill all of them, especially in terms of its band edge position. In a typical tandem device, two (or more) absorbers are used in a stacked configuration (often referred to as D4 configuration since a dual-absorber device must absorb a total of four photons, two for each absorber, in order to create one molecule of $H_2$) (Sivula 2013). Upon AM1.5 illumination, the shorter wavelength is absorbed by the front absorber, and the longer wavelength that is transmitted is used by the consecutive absorbers.

Within the material class of multinary metal oxides, $BiVO_4$ is the only material with which tandem devices have been constructed. The reason of this is simply because $BiVO_4$ is one of the most developed and highest performing multinary oxides. However, as discussed previously, $BiVO_4$ has some distinct advantages over other oxides, most notably the favorable onset potential. As a result, $BiVO_4$-based tandem devices triumph even the efficiency of tandem devices based on more well-studied binary metal oxides, such as $Fe_2O_3$ and $WO_3$ (Brillet et al. 2012).

The first example of $BiVO_4$-based tandem device is where it was combined with a thin-film amorphous silicon solar cell (Fig. 8.9a) (Abdi et al. 2013b). The thin-film silicon solar cell used the transmitted red part of the solar spectrum to provide the bias needed by $BiVO_4$. Here, both single- and double-junction amorphous silicon solar cells were used in combination with a Co-Pi-catalyzed gradient-doped W:$BiVO_4$, and STH efficiencies of 3.6 and 4.9 % were reported for the respective combination. Later on, the efficiency was further improved to 5.2 % by replacing the amorphous solar cell with a micromorph (microcrystalline/amorphous tandem) solar cell (Han et al. 2014).

A STH efficiency of 5.7 % was then reported with a combination of a guest-host mesoporous $WO_3$/$BiVO_4$ and a dye-sensitized solar cell (DSSC) based on porphyrin dye and cobalt electrolyte (Fig. 8.9b) (Shi et al. 2015). The DSSC delivered a relatively high open-circuit potential of almost 1.0 V, which allows higher operating point of the tandem device. The use of bipolar FTO substrates in this tandem device also needs to be highlighted, which allows the device to be monolithic. The combination of $BiVO_4$ with the lead-halide perovskite has also been reported, pointing out the possibility of fabricating an all-solution-processed tandem device (Fig. 8.9c) (Chen et al. 2015). However, the STH efficiency of this device was reported to be relatively low at 2.5 %, even though the lead-halide perovskite can deliver high photocurrent and high open-circuit potential (~1.1 V). The low efficiency is mainly caused by the suboptimal performance of the $BiVO_4$ in this report.

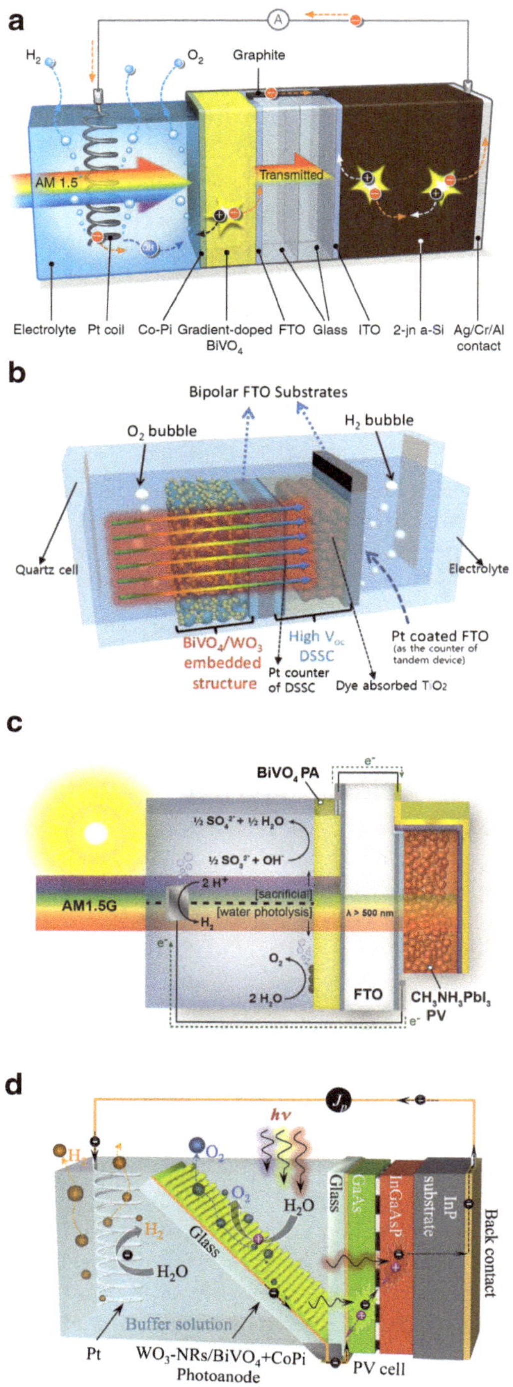


**Fig. 8.9** Various schematic structures of tandem device based on multinary oxides. In all cases, $BiVO_4$-based photoanodes were combined with (**a**) thin-film amorphous silicon solar cell (Reprinted from Abdi et al. (2013b)), (**b**) dye-sensitized solar cell (Reprinted from Shi et al. (2015), Copyright 2015 Elsevier), (**c**) lead-halide perovskite solar cell (Reprinted from Chen et al. (2015). Copyright 2015 American Chemical Society), and (**d**) GaAs/InGaAsP solar cell (Reprinted from Pihosh et al. (2015). Copyright 2015, Rights Managed by Nature Publishing Group)

$BiVO_4$-based photoelectrodes have also been combined with a multi-junction III–V solar cell (double-junction GaAs/InGaAsP) to form a tandem solar water splitting device (Pihosh et al. 2015). However, in this case, the $BiVO_4$ photoanode was positioned in 45° angle with respect to the solar cell, as depicted in Fig. 8.9d. Since a reflective substrate (ITO/Pt/ITO) was used for the $BiVO_4$ photoelectrode, the red part of the solar spectrum was fully reflected toward the solar cell and used to provide the necessary bias. Here, the $BiVO_4$ was deposited on $WO_3$ nanorods and catalyzed with Co-Pi, resulting in a record AM1.5 photocurrent of 6.7 $mA/cm^2$ at 1.23 V vs. RHE. Due to this and the similarly high performance of the III–V-based solar cell, a remarkable STH efficiency of 8.1 % was achieved, which is currently the highest reported efficiency for a metal oxide-based solar water splitting device.

Finally, a tandem solar water splitting device based on the combination of $BiVO_4$ photoanode and $Cu_2O$ photocathode has also been reported (Bornoz et al. 2014). While the STH efficiency is rather low (~0.5 %), this represents the first demonstration of an all-oxide tandem device. The efficiency is limited by the overlapping absorption regions ($BiVO_4$ and $Cu_2O$ have bandgap energies of 2.4 and 2.0 eV, respectively) and poor overlap of the photocurrent-voltage curves. Should alternative promising metal oxides be found with a better match, this type of tandem device may well be worth revisiting.

## 8.6 Challenges for Multinary Oxides

As outlined in the beginning of this chapter, the sheer number of multinary metal oxides makes this an appealing class of materials in the search for new, chemically stable semiconducting light absorbers. Compared to the simple binary oxides, however, there are several additional challenges that have to be tackled.

An obvious—but nontrivial—challenge is to identify promising candidates for initial studies. With many tens of thousands of candidates, some form of preselection is essential. Various approaches have been developed to identify or rule out promising candidates. Examples are the identification of Mott insulators as unsuitable candidates (Huda et al. 2011), high-throughput DFT calculations of the bandgap (Castelli et al. 2012), band positions, and electrochemical stability as selection criteria for large sets of materials in existing databases (Castelli et al. 2015; Singh et al. 2015) and the use of genetic algorithms to direct such searches (Ivano et al. 2013; Jain et al. 2013a). While these methods show promise, computational materials science is not yet able to truly *predict* new materials that show the desired properties. Experimental studies on large numbers of candidate materials are, therefore, still indispensable. Several high-throughput approaches have been developed in recent years to accelerate the search for new materials (Anderson et al. 2014; Sliozberg et al. 2015; Woodhouse and Parkinson 2009), and some of these will be discussed in detail in Chap. 10 of this book.

Once a promising multinary oxide has been identified, the next challenge is to synthesize it in a controlled and reproducible fashion. For most multinary oxides, this is much harder than the synthesis of simple binary oxides. The reasons for this will be discussed below.

## *8.6.1 Crystallinity and Stoichiometry*

To get a fully crystalline phase, most multinary metal oxides must be synthesized at high temperatures in order to let diffusion do its work and get all the elements at the designated lattice positions. The more complex the composition (in terms of the number of elements) and the lower the diffusivity of the individual ions, the more difficult this becomes and the higher the required temperature. At the same time, there are clear limits to the allowable temperature window. Glass substrates start to soften above ~500 °C, and the electrical conductivity of most transparent conducting oxides rapidly drops when exposed to temperatures above 400–500 °C. Indium vanadate, $InVO_4$, is an example of a ternary metal oxide photoanode material which pushes the temperature window of the substrate. The desired orthorhombic phase can be obtained when spray-depositing a precursor solution on FTO glass (F-doped $SnO_2$) at 550 °C, but traces of the unfavorable monoclinic phase are difficult to avoid (van de Krol et al. 2011). Indeed, phase-pure orthorhombic $InVO_4$ powders can only be obtained at temperatures above 680 °C (Touboul et al. 1995). Recent work in our lab on spray-deposited $Fe_2WO_6$ films on fused silica shows that temperatures as high as 800 °C are needed to obtain a crystalline phase. When doing the same deposition on FTO glass, a temperature of 500 °C results in partial crystallization of the film, but the defect density is still very high ($N_D > 10^{20}$ cm$^{-3}$). In contrast, high-quality $BiVO_4$ films can be obtained with spray deposition at the relatively modest temperature of 450 °C (Abdi et al. 2013b).

A further challenge for the synthesis of multinary oxides is the accurate control over the stoichiometry. The ratio of oxygen-to-metal ions is often easily controlled by firing the sample in air for a few hours. However, the ratio between the different metal ions is fixed during the synthesis process and cannot be easily corrected afterward. The effects can be devastating. A deviation of 0.1 % (e.g., $A_{0.9995}B_{1.0005}O_{4-\delta}$) corresponds to a defect concentration of ~$10^{19}$ cm$^{-3}$. Depending on the type of defect, such amounts can completely kill the photoactivity of the material. Accurate control over the metal-to-metal ratio is not always possible, and the suitability of the material as a photoabsorber is largely determined by its tolerance against the formation of killer defects.

The term "defect-tolerance" is often used in this context, although its exact meaning is seldom defined. Certain phases accommodate a deviation in stoichiometry by the formation of point defects, i.e., either interstitial B ions or A-vacancies in the case of $A_{0.9995}B_{1.0005}O_{4-\delta}$. In phases where the formation energies of such defects are too high, deviations in stoichiometry are accommodated by the segregation of the binary components at the grain boundaries (i.e., segregation of $B_xO_y$ for an overall composition of $A_{0.999}B_{1.0005}O_{4-\delta}$). These compounds behave as line compounds in the phase diagram. When phase segregation is preferred over the formation of point defects *and* does not adversely affect the photoactivity, the material is said to be defect tolerant.

Identifying deviations from stoichiometry can be a major challenge. Segregated binary oxides can be present as finely dispersed nanocrystalline or amorphous phases, making them invisible to standard X-ray diffraction (XRD) methods. The concentration of ionic point defects is often too low for direct detection by, e.g., electron microscopy or spectroscopy. Indirect detection by, e.g., electrical conductivity measurements is sometimes possible, but such methods do not provide any information on the chemical nature of the defect species.

### 8.6.2 High-Temperature Stability

Selective loss of one of the elements at higher temperatures can also cause deviations from the ideal stoichiometry. When heating $BiVO_4$ to temperatures above ~600 °C, part of the bismuth evaporates, leaving a Bi-deficient phase. This is due to the high vapor pressure of bismuth ($10^{-3}$ mbar at 580 °C) and the instability of $Bi_2O_3$, which decomposes into Bi and $O_2$ at higher temperatures (Béa et al. 2005). The loss of Bi from $BiVO_4$ results in pink-colored samples, as illustrated in Fig. 8.10, and a poor photoresponse.

Similar as for binary metal oxides, high-temperature treatments can lead to undesired phase transformations. The classic example for binary oxides is the conversion of the metastable anatase $TiO_2$ phase into rutile, which has a slightly smaller bandgap but generally poorer photoactivity. Ternary oxides tend to be somewhat less prone to such transformations, especially when there is a large

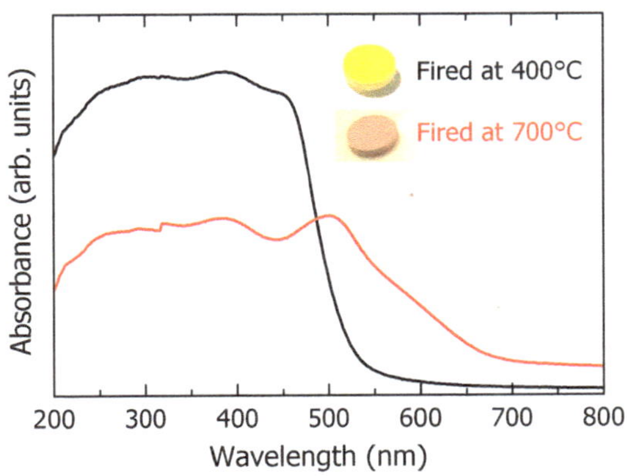


**Fig. 8.10** Photographs and diffuse reflectance spectra of $BiVO_4$ pellets fired in air at 400 and 700 °C. The X-ray diffractograms and Raman spectra shows negligible differences, suggesting that the pink color of the 700 °C sample is caused by the presence of an optically active defect state in the bandgap of $BiVO_4$

size difference between the different cations. This is exemplified by the perovskite structure ($ABO_3$), which is one of the very few crystal structures that can accommodate large size differences between the A- and B-site cations.

### 8.6.3 Doping and Alloying

It is important to realize that the incorporation of dopants in multinary metal oxides (e.g., to tune the electrical conductivity) is slightly more complicated than the doping of simply binary metal oxide (van de Krol and Grätzel 2012). Doping by straightforward dissolution of a binary oxide into a ternary oxide is energetically unfavorable, because the conservation of the number of cation sites requires the formation of highly charged metal vacancies. The formation of these negatively charged species introduces complications, as illustrated by the doping of $BiVO_4$ with $WO_3$ (written in Kröger-Vink notation (Kröger 1964)):

$$WO_3 + \tfrac{1}{2}O_2(g) + 2e' \xrightarrow{BiVO_4} V_{Bi}''' + W_V^{\bullet} + 4O_O^x$$

The consumption of free electrons is needed to compensate the charge of the Bi vacancies and effectively *decreases* the conductivity of n-type $BiVO_4$, despite W being a donor-type dopant when it substitutes for vanadium. To avoid these complications, one needs to co-dope the material with $Bi_2O_3$:

$$2WO_3 + Bi_2O_3 \xrightarrow{2BiVO_4} 2Bi_{Bi}^x + 2W_V^{\bullet} + 8O_O^x + \tfrac{1}{2}O_2(g) + 2e'$$

In a similar manner, one can write the defect chemical reactions for the acceptor-doping of $BiVO_4$ with CaO, and the donor-doping of $CuWO_4$ with $Fe_2O_3$ that were mentioned earlier in this chapter:

$$2CaO + V_2O_5 + \tfrac{1}{2}O_2(g) \xrightarrow{2BiVO_4} 2Ca_{Bi}' + 2V_V^x + 8O_O^x + 2h^{\bullet}$$

$$Fe_2O_3 + 2WO_3 \xrightarrow{2CuWO_4} 2W_W^x + 2Fe_{Cu}^{\bullet} + 8O_O^x + \tfrac{1}{2}O_2(g) + 2e'$$

When the dopant concentrations exceed 1–2 %, the term "alloying" is more appropriate. An example of this is the preparation of metal oxynitride semiconductors, which can also be viewed as multinary compounds. In oxynitrides, the partial replacement of oxygen ions by nitrogen ions decreases the bandgap of the material due to the higher energy of the N 2p orbitals compared to the O 2p orbitals. The idea is to find a compromise between the chemical stability of metal oxides and the narrower bandgap of metal nitrides. An example is the conversion of $Ta_2O_5$, a chemically stable semiconductor with a 3.9 eV bandgap, into β-TaON, which is less stable but also has a much more favorable bandgap of 2.4 eV (Chun et al. 2003; Hara et al. 2003). TaON photoanodes modified with $IrO_x$ and $CoO_x$ have shown photocurrents above 3 $mA/cm^2$ and 2 $mA/cm^2$ at 1.2 V vs. RHE in sulfate-containing solutions, respectively (Abe et al. 2010; Higashi et al. 2012). The oxynitride phase is obtained by exposing the metal oxide to gaseous ammonia at high temperatures, typically 800 °C. It turns out that this "ammonolysis" process is very sensitive to local process conditions and, therefore, reactor design. As a result, many labs have struggled to obtain high-quality, phase-pure oxynitride absorbers.

To achieve better control over the ammonolysis process, we have developed an optical fiber-based technique that uses a CCD spectrometer to monitor the optical transmission of thin-film samples during the high-temperature ammonolysis treatment (Dabirian et al. 2012; Dabirian and van de Krol 2015). Here, we make use of the fact that the bandgap directly depends on the amount of nitrogen in the material. As illustrated in Fig. 8.11a, b, the kinetics of the conversion depends strongly on the temperature, but even 575 °C is sufficient to fully convert a 160 nm $Ta_2O_5$ film to $Ta_3N_5$. XRD measurements confirmed the presence of $Ta_3N_5$ as the final product. The yellow band indicates an absorptance of ~60 % and serves as a rough indication of the absorption edge. After the conversion, it is found at a wavelength slightly below 600 nm, which indeed corresponds to the 2.1 eV bandgap of $Ta_3N_5$. These experiments suggest that the conversion proceeds directly from $Ta_2O_5$ to $Ta_3N_5$, without formation of the desired β-TaON phase. This was confirmed by the (ex situ) XRD measurements on partially ammonolyzed and quenched samples, shown in Fig. 8.11c. More recent experiments have shown that the desired β-TaON phase can be obtained by controlling not only the $NH_3$ flow but also the partial oxygen pressure (De Respinis et al. 2015).

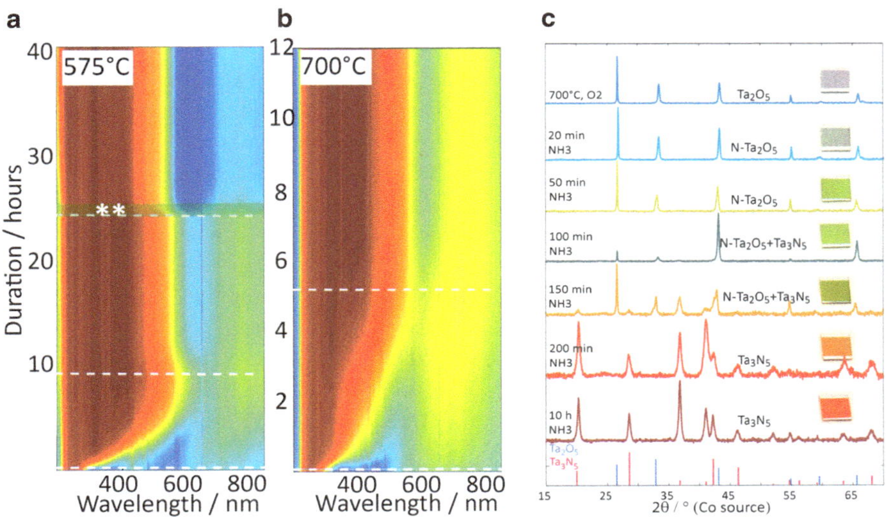


**Fig. 8.11** Optical absorptance spectra of $Ta_2O_5$ thin films on quartz, recorded during ammonolysis at (**a**) 575 °C and (**b**) 700 °C in a 25 mm diameter quartz tube under a 50 ml/min flow of dry $NH_3$. (**c**) X-ray diffractograms and photographs of partially converted films, taken out of the furnace after different ammonolysis times (Adapted with permission from Dabirian and van de Krol (2015). Copyright (2015) American Chemical Society)

## 8.7 Conclusions and Future Outlook

As previously discussed, a large research effort has been placed on using binary metal oxides ($TiO_2$, $WO_3$, $Fe_2O_3$, $Cu_2O$, ZnO, etc.) to make efficient photoelectrodes for photoelectrochemical water splitting. However, the success rate has been low due to the fundamental limitations of these materials related to bandgap energy, stability, and charge transport. To date, complex metal oxides (multinary oxides) have received much less attention, but they may offer many advantages over binary oxides including narrower bandgap energies, greater stability, and longer carrier lifetimes. Numerous multinary metal oxides have been mentioned in this chapter. Of the n-type oxides, $BiVO_4$ has been the best model for optimization of photoconversion efficiency, while $Fe_2WO_6$ demonstrates how inherent limitations such as a poor flatband position can make optimization unlikely. Several of the p-type oxides show promising attributes, including $CuBi_2O_4$ and $CuFeO_2$, but they have not yet been fully characterized and understood. Key properties for these materials are listed in Table 8.1 below. Researchers should focus on identifying the intrinsic strengths and limitations of PEC materials in addition to searching for new multinary metal oxides with the best photophysical and photoelectrochemical properties for water splitting. This will be a challenge. As described in Chap. 10, Combinatorial Synthesis, there are hundreds of thousands of possible ternary and quaternary metal oxide combinations. Combinatorial syntheses, or high-throughput screening, will be an important tool in identifying new complex metal oxides. It has already been used to successfully identify some of the materials mentioned above including $CuBi_2O_4$ and the iron and cobalt-based spinels. One drawback with using combinatorial approaches is that the screening method typically only measures a single property such as photocurrent density, $H_2$ production, open-circuit potential, or photovoltage. In addition morphology and nanostructure are often uncontrolled during combinatorial synthesis. Therefore, combinatorial synthesis and screening process will need to be improved so that promising candidates are not missed. Theory should also play a large role in identifying promising complex metal oxides to focus on. Recent advancements in DFT have been used to accurately predict the band structures of semiconductor material and the activity of electrocatalyst materials (Greeley et al. 2006; Yin et al. 2010). However, it is still challenging to simultaneously predict all of the material properties that are relevant to photoelectrochemical water splitting, including light absorption, charge transport, and reaction kinetics. It will be necessary to refine theory further so that more properties of complex metal oxides can be predicted.

**Table 8.1** Key properties for selected multinary metal oxides

| Material | Bandgap energy (eV) | Light penetration depth, $\alpha^{-1}$ (nm) | Flatband potential (V vs. RHE) | Mobility ($cm^2 V^{-1} s^{-1}$) | Carrier diffusion length (nm) | References |
|---|---|---|---|---|---|---|
| n-$BiVO_4$ | 2.4[a]<br>2.9[b] | ~100 nm<br>($\lambda = 450$ nm) | ~0 | ~0.2<br>(electrons) | 70–100 | Abdi et al. (2013c), Kudo et al. (1999), Rettie et al. (2013), Sayama et al. (2006), Stoughton et al. (2013) |
| n-$CuWO_4$ | 2.2–2.4 | | ~0.4 | 10 ($\mu_h$) | | Bharati et al. (1980), Bohra and Smith (2015), Chang et al. (2011), Doumerc et al. (1984), Gaillard et al. (2013), Yourey and Bartlett (2011)) |
| n-$Fe_2WO_6$ | 1.5–1.7 | >1000 nm | 0.6–0.7 | 4.7[c] | | Abdi et al. (in preparation), Bharati and Singh (1981), Khader et al. (1998) |
| p-$CuBi_2O_4$ | 1.6–1.8 | ~300 nm<br>($\lambda = 550$ nm) | 1.3–1.45 | $\sim 10^{-3}$ | ~50 | Arai et al. (2007), Berglund et al. (in preparation), Hahn et al. (2012) |
| p-$CuFeO_2$ | 1.47–1.55 | ~80 nm<br>($\lambda = 550$ nm) | ~1 | 0.1 ($\mu_h$)<br>$10^{-6}$ ($\mu_e$) | | Benko and Koffyberg (1987), Prévot et al. (2015), Read et al. (2012) |

[a]Scheelite
[b]Zircon
[c]Drift $\mu$ at 400 K

# References


Abdi FF, van de Krol R (2012) Nature and light dependence of bulk recombination in Co-Pi-catalyzed $BiVO_4$ photoanodes. J Phys Chem C 116(17):9398

Abdi FF, Firet N, Dabirian A, van de Krol R (2012) Spray-deposited Co-Pi catalyzed $BiVO_4$: a low-cost route towards highly efficient photoanodes. MRS Online Proc Lib 1446

Abdi FF, Firet N, van de Krol R (2013a) Efficient $BiVO_4$ thin film photoanodes modified with cobalt phosphate catalyst and W-doping. ChemCatChem 5(2):490–496

Abdi FF, Han L, Smets AHM, Zeman M, Dam B, van de Krol R (2013b) Efficient solar water splitting by enhanced charge separation in a bismuth vanadate-silicon tandem photoelectrode. Nat Commun 4:2195

Abdi FF, Savenije TJ, May MM, Dam B, van de Krol R (2013c) The origin of slow carrier transport in $BiVO_4$ thin film photoanodes: a time-resolved microwave conductivity study. J Phys Chem Lett 4(16):2752

Abdi FF, Chemseddine A, Berglund SP, van de Krol R (in preparation) Assessing the suitability of iron tungstate ($Fe_2WO_6$) photoelectrode for water oxidation

Abe R, Higashi M, Domen K (2010) Facile fabrication of an efficient oxynitride TaON photoanode for overall water splitting into $H_2$ and $O_2$ under visible light irradiation. J Am Chem Soc 132 (34):11828

Aharon-Shalom E, Heller A (1982) Efficient p-InP(Rh-H alloy) and p-InP(Re-H alloy) hydrogen evolving photocathodes. J Electrochem Soc 129(12):2865

Anderson AY, Bouhadana Y, Barad H-N, Kupfer B, Rosh-Hodesh E, Aviv H, Tischler YR, Rühle S, Zaban A (2014) Quantum efficiency and bandgap analysis for combinatorial photovoltaics: sorting activity of Cu–O compounds in all-oxide device libraries. ACS Comb Sci 16(2):53

Arai T, Konishi Y, Iwasaki Y, Sugihara H, Sayama K (2007) High-throughput screening using porous photoelectrode for the development of visible-light-responsive semiconductors. J Comb Chem 9(4):574

Arora S, Mathew T, Batra N (1989) Optical characterization of $CuWO_4$ single crystals. J Phys Chem Solids 50(7):665

Azarpira A, Lublow M, Steigert A, Bogdanoff P, Greiner D, Kaufmann CA, Krüger M, Gernert U, van de Krol R, Fischer A, Schedel-Niedrig T (2015) Efficient and stable $TiO_2$:Pt–Cu(In,Ga)$Se_2$ composite photoelectrodes for visible light driven hydrogen evolution. Adv Energy Mater 5(12):1402148

Azevedo J, Steier L, Dias P, Stefik M, Sousa C, Araújo J, Mendes A, Graetzel M, Tilley S (2014) On the stability enhancement of cuprous oxide water splitting photocathodes by low temperature steam annealing. Energy Environ Sci 7(12):4044

Béa H, Bibes M, Barthélémy A, Bouzehouane K, Jacquet E, Khodan A, Contour J-P, Fusil S, Wyczisk F, Forget A, Lebeugle D, Colson D, Viret M (2005) Influence of parasitic phases on the properties of $BiFeO_3$ epitaxial thin films. Appl Phys Lett 87(7):072508

Benko FA, Koffyberg FP (1987) Opto-electronic properties of p- and n-type delafossite, $CuFeO_2$. J Phys Chem Solids 48(5):431

Berglund SP, Flaherty DW, Hahn NT, Bard AJ, Mullins CB (2011) Photoelectrochemical oxidation of water using nanostructured $BiVO_4$ films. J Phys Chem C 115(9):3794

Berglund SP, Lee HC, Nunez PD, Bard AJ, Mullins CB (2013) Screening of transition and post-transition metals to incorporate into copper oxide and copper bismuth oxide for photoelectrochemical hydrogen evolution. Phys Chem Chem Phys 15(13):4554

Berglund SP, Abdi FF, van de Krol R (in preparation) Evaluation of $CuBi_2O_4$ as a photocathode material for photoelectrochemical water splitting

Bharati R, Singh R (1981) The electrical properties of $Fe_2WO_6$. J Mater Sci 16(2):511

Bharati R, Shanker R, Singh R (1980) Electrical transport properties of $CuWO_4$. Pramana 14(6):449

Bharati R, Singh R, Yadava Y (1983) Electrical conduction in copper tungstate. J Mater Sci Lett 2(10):623
Bhattacharya AK, Mallick KK, Hartridge A (1997) Phase transition in $BiVO_4$. Mater Lett 30(1):7
Bierlein JD, Sleight AW (1975) Ferroelasticity in $BiVO_4$. Solid State Commun 16(1):69
Bohra D, Smith WA (2015) Improved charge separation via Fe-doping of copper tungstate photoanodes. Phys Chem Chem Phys 17(15):9857
Bornoz P, Abdi FF, Tilley SD, Dam B, Van De Krol R, Graetzel M, Sivula K (2014) A bismuth vanadate–cuprous oxide tandem cell for overall solar water splitting. J Phys Chem C 118 (30):16959
Brillet J, Yum JH, Cornuz M, Hisatomi T, Solarska R, Augustynski J, Graetzel M, Sivula K (2012) Highly efficient water splitting by a dual-absorber tandem cell. Nat Photon 6(12):824
Castelli IE, Olsen T, Datta S, Landis DD, Dahl S, Thygesen KS, Jacobsen KW (2012) Computational screening of perovskite metal oxides for optimal solar light capture. Energy Environ Sci 5(2):5814
Castelli IE, Hüser F, Pandey M, Li H, Thygesen KS, Seger B, Jain A, Persson KA, Ceder G, Jacobsen KW (2015) New light-harvesting materials using accurate and efficient bandgap calculations. Adv Energy Mater 5(2):1400915
Chang Y, Braun A, Deangelis A, Kaneshiro J, Gaillard N (2011) Effect of thermal treatment on the crystallographic, surface energetics, and photoelectrochemical properties of reactively cosputtered copper tungstate for water splitting. J Phys Chem C 115(51):25490
Chen L, Shet S, Tang H, Ahn K-S, Wang H, Yan Y, Turner J, Al-Jassim M (2010a) Amorphous copper tungsten oxide with tunable band gaps. J Appl Phys 108(4):043502
Chen Z, Jaramillo TF, Deutsch TG, Kleiman-Shwarsctein A, Forman AJ, Gaillard N, Garland R, Takanabe K, Heske C, Sunkara M, McFarland EW, Domen K, Miller EL, Turner JA, Dinh HN (2010b) Accelerating materials development for photoelectrochemical hydrogen production: standards for methods, definitions, and reporting protocols. J Mater Res 25(01):3
Chen Y-S, Manser JS, Kamat PV (2015) All solution-processed lead halide perovskite-$BiVO_4$ tandem assembly for photolytic solar fuels production. J Am Chem Soc 137(2):974
Cherepy NJ, Liston DB, Lovejoy JA, Deng HM, Zhang JZ (1998) Ultrafast studies of photoexcited electron dynamics in gamma- and alpha-$Fe_2O_3$ semiconductor nanoparticles. J Phys Chem B 102(5):770
Choi SK, Choi W, Park H (2013) Solar water oxidation using nickel-borate coupled $BiVO_4$ photoelectrodes. Phys Chem Chem Phys 15(17):6499–6507
Chun W-J, Ishikawa A, Fujisawa H, Takata T, Kondo JN, Hara M, Kawai M, Matsumoto Y, Domen K (2003) Conduction and valence band positions of $Ta_2O_5$, TaON, and $Ta_3N_5$ by UPS and electrochemical methods. J Phys Chem B 107(8):1798
Cooper JK, Gul S, Toma FM, Chen L, Glans P-A, Guo J, Ager JW, Yano J, Sharp ID (2014) Electronic structure of monoclinic $BiVO_4$. Chem Mater 26(18):5365
Cooper JK, Gul S, Toma FM, Chen L, Liu Y-S, Guo J, Ager JW, Yano J, Sharp ID (2015) Indirect bandgap and optical properties of monoclinic bismuth vanadate. J Phys Chem C 119(6):2969
Dabirian A, van de Krol R (2015) High-temperature ammonolysis of thin film $Ta_2O_5$ photoanodes: evolution of structural, optical, and photoelectrochemical properties. Chem Mater 27(3):708
Dabirian A, Van't Spijker H, van de Krol R (2012) Wet ammonia synthesis of semiconducting N:$Ta_2O_5$, $Ta_3N_5$ and β-TaON films for photoanode applications. Energy Proc 22:15
de Respinis M, Fravventura M, Abdi FF, Schreuders H, Savenije TJ, Smith WA, Dam B, van de Krol R (2015) Oxynitrogenography: the controlled synthesis of high quality single phase tantalum oxynitride photoanodes. Chem Mater 27(20):7091–7099
Ding C, Shi J, Wang D, Wang Z, Wang N, Liu G, Xiong F, Li C (2013a) Visible light driven overall water splitting using cocatalyst/$BiVO_4$ photoanode with minimized bias. Phys Chem Chem Phys 15(13):4589
Ding K, Chen B, Fang Z, Zhang Y (2013b) Density functional theory study on the electronic and optical properties of three crystalline phases of $BiVO_4$. Theor Chem Acc 132(5):1

Ding K, Chen B, Fang Z, Zhang Y, Chen Z (2014) Why the photocatalytic activity of Mo-doped $BiVO_4$ is enhanced: a comprehensive density functional study. Phys Chem Chem Phys 16 (26):13465

Doumerc JP, Hejtmanek J, Chaminade JP, Pouchard M, Krussanova M (1984) A photoelectrochemical study of $CuWO_4$ single crystals. Phys Stat Sol A 82(1):285

Dreyer G, Tillmanns E (1981) Dreyerite—natural, tetragonal bismuth vanadate from Hirschhorn-Pfalz. Neues JB Miner Monat 4:151

Fujishima A, Honda K (1972) Electrochemical photolysis of water at a semiconductor electrode. Nature 238:37

Fuoco L, Joshi UA, Maggard PA (2012) Preparation and photoelectrochemical properties of p-type $Cu_5Ta_{11}O30$ and $Cu_3Ta_7O_{19}$ semiconducting polycrystalline films. J Phys Chem C 116(19):10490

Gaillard N, Chang Y, DeAngelis A, Higgins S, Braun A (2013) A nanocomposite photoelectrode made of 2.2 eV band gap copper tungstate ($CuWO_4$) and multi-wall carbon nanotubes for solar-assisted water splitting. Int J Hydrogen Energy 38(8):3166

Galembeck A, Alves O (2000) $BiVO_4$ thin film preparation by metalorganic decomposition. Thin Solid Films 365(1):90

Gerischer H (1977) On the stability of semiconductor electrodes against photodecomposition. J Electroanal Chem 82(1–2):133

Greeley J, Jaramillo TF, Bonde J, Chorkendorff I, Norskov JK (2006) Computational high-throughput screening of electrocatalytic materials for hydrogen evolution. Nat Mater 5(11):909

Gu J, Wuttig A, Krizan JW, Hu Y, Detweiler ZM, Cava RJ, Bocarsly AB (2013) Mg-Doped $CuFeO_2$ photocathodes for photoelectrochemical reduction of carbon dioxide. J Phys Chem C 117(24):12415

Hahn NT, Holmberg VC, Korgel BA, Mullins CB (2012) Electrochemical synthesis and characterization of p-$CuBi_2O_4$ thin film photocathodes. J Phys Chem C 116(10):6459

Haije W, Geerlings H (2011) Efficient production of solar fuel using existing large scale production technologies. Environ Sci Technol 45(20):8609

Han L, Abdi FF, van de Krol R, Liu R, Huang Z, Lewerenz HJ, Dam B, Zeman M, Smets AH (2014) Efficient water-splitting device based on a bismuth vanadate photoanode and thin-film silicon solar cells. ChemSusChem 7(10):2832

Hara M, Hitoki G, Takata T, Kondo JN, Kobayashi H, Domen K (2003) TaON and $Ta_3N_5$ as new visible light driven photocatalysts. Catal Today 78(1–4):555

Harrison W, Chowdhry U, Machiels C, Sleight A, Cheetham A (1985) Preparation of ferric tungstate and its catalytic behavior toward methanol. J Solid State Chem 60(1):101

Helaili N, Mitran G, Popescu I, Bachari K, Marcu I-C, Boudjemaa A (2015) Photoelectrochemical properties of $AFe_2O_4$ (A = Co, Cu, Zn) ferrospinels for water photo-reduction. J Electroanal Chem 742:47

Henmi C (1995) Kusachiite, $CuBi_2O_4$, a new mineral from Fuka, Okayama Prefecture. Japan Miner Mag 59(3):545

Herron JA, Kim J, Upadhye AA, Huber GW, Maravelias CT (2015) A general framework for the assessment of solar fuel technologies. Energy Environ Sci 8(1):126

Higashi M, Domen K, Abe R (2012) Highly stable water splitting on oxynitride TaON photoanode system under visible light irradiation. J Am Chem Soc 134(16):6968

Hong SJ, Lee S, Jang JS, Lee JS (2011) Heterojunction $BiVO_4/WO_3$ electrodes for enhanced photoactivity of water oxidation. Energy Environ Sci 4(5):1781

Huda MN, Al-Jassim MM, Turner JA (2011) Mott insulators: an early selection criterion for materials for photoelectrochemical $H_2$ production. IRESR 3(5):053101

Ida S, Yamada K, Matsunaga T, Hagiwara H, Matsumoto Y, Ishihara T (2010) Preparation of p-Type $CaFe_2O_4$ photocathodes for producing hydrogen from water. J Am Chem Soc 132 (49):17343

Ivano EC, Juan María G-L, Falco H, Kristian ST, Karsten WJ (2013) Stability and bandgaps of layered perovskites for one- and two-photon water splitting. New J Phys 15(10):105026

Jain A, Castelli I, Hautier G, Bailey D, Jacobsen K (2013a) Performance of genetic algorithms in search for water splitting perovskites. J Mater Sci 48(19):6519

Jain A, Ong SP, Hautier G, Chen W, Richards WD, Dacek S, Cholia S, Gunter D, Skinner D, Ceder G, Persson KA (2013b) Commentary: The materials project: a materials genome approach to accelerating materials innovation. APL Mater 1(1):011002

Jin J, Walczak K, Singh MR, Karp C, Lewis NS, Xiang C (2014) An experimental and modeling/simulation-based evaluation of the efficiency and operational performance characteristics of an integrated, membrane-free, neutral pH solar-driven water-splitting system. Energy Environ Sci 7(10):3371

Jo WJ, Jang JW, Kong K, Kang HJ, Kim JY, Jun H, Parmar KPS, Lee JS (2012) Phosphate doping into monoclinic $BiVO_4$ for enhanced photoelectrochemical water oxidation activity. Angew Chem Int Ed 51(13):3147

Joshi UA, Maggard PA (2012) $CuNb_3O_8$: a p-type semiconducting metal oxide photoelectrode. J Phys Chem Lett 3(11):1577

Joshi UA, Palasyuk AM, Maggard PA (2011) Photoelectrochemical investigation and electronic structure of a p-type $CuNbO_3$ photocathode. J Phys Chem C 115(27):13534

Kato H, Kudo A (2001) Water splitting into $H_2$ and $O_2$ on alkali tantalate photocatalysts $ATaO_3$ (A = Li, Na, and K). J Phys Chem B 105(19):4285

Kay A, Cesar I, Grätzel M (2006) New benchmark for water photooxidation by nanostructured α-Fe2O3 films. J Am Chem Soc 128(49):15714

Khader MM, Saleh MM, El-Naggar EM (1998) Photoelectrochemical characteristics of ferric tungstate. J Solid State Electrochem 2(3):170

Khyzhun OY, Strunskus T, Cramm S, Solonin YM (2005) Electronic structure of $CuWO_4$: XPS, XES and NEXAFS studies. J Alloys Compd 389(1):14

Khyzhun OY, Bekenev V, Solonin YM (2009) First-principles calculations and X-ray spectroscopy studies of the electronic structure of $CuWO_4$. J Alloys Compd 480(2):184

Kihlborg L, Gebert E (1970) $CuWO_4$, a distorted wolframite-type structure. Acta Cryst B Struct Sci 26(7):1020

Kim TW, Choi K-S (2014) Nanoporous $BiVO_4$ photoanodes with dual-layer oxygen evolution catalysts for solar water splitting. Science 343(6174):990

Kohtani S, Makino S, Kudo A, Tokumura K, Ishigaki Y, Matsunaga T, Nikaido O, Hayakawa K, Nakagaki R (2002) Photocatalytic degradation of 4-n-nonylphenol under irradiation from solar simulator: comparison between $BiVO_4$ and $TiO_2$ photocatalysts. Chem Lett 7:660

Kröger FA (1964) The chemistry of imperfect crystals. North-Holland Publishing Co., Amsterdam

Kudo A, Ueda K, Kato H, Mikami I (1998) Photocatalytic $O_2$ evolution under visible light irradiation on $BiVO_4$ in aqueous $AgNO_3$ solution. Catal Lett 53(3–4):229

Kudo A, Omori K, Kato H (1999) A novel aqueous process for preparation of crystal form-controlled and highly crystalline $BiVO_4$ powder from layered vanadates at room temperature and its photocatalytic and photophysical properties. J Am Chem Soc 121(49):11459

Leiva H, Dwight K, Wold A (1982) Preparation and characterization of conducting iron tungstates. J Solid State Chem 42(1):41

Lewis NS, Nocera DG (2006) Powering the planet: chemical challenges in solar energy utilization. Proc Natl Acad Sci U S A 103(43):15729

Lichterman MF, Shaner MR, Handler SG, Brunschwig BS, Gray HB, Lewis NS, Spurgeon JM (2013) Enhanced stability and activity for water oxidation in alkaline media with bismuth vanadate photoelectrodes modified with a cobalt oxide catalytic layer produced by atomic layer deposition. J Phys Chem Lett 4(23):4188

Long M, Cai W, Cai J, Zhou B, Chai X, Wu Y (2006) Efficient photocatalytic degradation of phenol over $Co_3O_4/BiVO_4$ composite under visible light irradiation. J Phys Chem B 110 (41):20211

Long MC, Cai WM, Kisch H (2008) Visible light induced photoelectrochemical properties of n-$BiVO_4$ and n-$BiVO_4$/p-$CO_3O_4$. J Phys Chem C 112(2):548

Luo WJ, Yang ZS, Li ZS, Zhang JY, Liu JG, Zhao ZY, Wang ZQ, Yan SC, Yu T, Zou ZG (2011) Solar hydrogen generation from seawater with a modified $BiVO_4$ photoanode. Energy Environ Sci 4(10):4046

Ma J, Wang L-W (2014) The role of the isolated 6s states in $BiVO_4$ on the electronic and atomic structures. Appl Phys Lett 105(17):172102

Marsen B, Cole B, Miller EL (2008) Photoelectrolysis of water using thin copper gallium diselenide electrodes. Sol Energy Mater Sol Cells 92(9):1054

Mavroides JG, Kafalas JA, Kolesar DF (1976) Photoelectrolysis of water in cells with $SrTiO_3$ anodes. Appl Phys Lett 28(5):241

McDowell MT, Lichterman MF, Spurgeon JM, Hu S, Sharp ID, Brunschwig BS, Lewis NS (2014) Improved stability of polycrystalline bismuth vanadate photoanodes by use of dual-layer thin $TiO_2$/Ni coatings. J Phys Chem C 118(34):19618

McKone JR, Pieterick AP, Gray HB, Lewis NS (2013) Hydrogen evolution from Pt/Ru-coated p-type $WSe_2$ photocathodes. J Am Chem Soc 135(1):223

Miller EL (2013) Task 26: Advanced materials for waterphotolysis. Final Report IEA Hydrogen Implementing Agreement, Washington, DC, p 14

Morales-Guio CG, Liardet L, Mayer MT, Tilley SD, Grätzel M, Hu X (2015) Photoelectrochemical hydrogen production in alkaline solutions using $Cu_2O$ coated with earth-abundant hydrogen evolution catalysts. Angew Chem Int Ed 54(2):664

Nakato Y, Tonomura S, Tsubomura H (1976) The catalytic effect of electrodeposited metals on the photo-reduction of water at p-type semiconductors. Ber Bunsenges Phys Chem 80(12):1289

Newhouse PF, Parkinson BA (2015) Combinatorial optimization of spinel $Co_{3-x}M_xO_4$ M = (Al, Ga, In) alloyed thin films prepared by ink jet printing: photoelectrochemical, optical, and structural properties. J Mater Chem A 3(11):5901

Newman J, Hoertz PG, Bonino CA, Trainham JA (2012) Review: An economic perspective on liquid solar fuels. J Electrochem Soc 159(10):A1722

Pandey PK, Bhave N, Kharat R (2005) Spray deposition process of polycrystalline thin films of $CuWO_4$ and study on its photovoltaic electrochemical properties. Mater Lett 59(24):3149

Paracchino A, Laporte V, Sivula K, Grätzel M, Thimsen E (2011) Highly active oxide photocathode for photoelectrochemical water reduction. Nat Mater 10(6):456

Paracchino A, Mathews N, Hisatomi T, Stefik M, Tilley SD, Grätzel M (2012) Ultrathin films on copper(I) oxide water splitting photocathodes: a study on performance and stability. Energy Environ Sci 5(9):8673

Parfenov VV, Nazipov RA (2002) Effect of synthesis temperature on the transport properties of copper ferrites. Inorg Mater 38(1):78

Park HS, Kweon KE, Ye H, Paek E, Hwang GS, Bard AJ (2011) Factors in the metal doping of $BiVO_4$ for improved photoelectrocatalytic activity as studied by scanning electrochemical microscopy and first-principles density-functional calculation. J Phys Chem C 115(36):17870

Park HS, Lee C-Y, Reisner E (2014) Photoelectrochemical reduction of aqueous protons with a $CuO|CuBi_2O_4$ heterojunction under visible light irradiation. Phys Chem Chem Phys 16 (41):22462

Parmar KPS, Kang HJ, Bist A, Dua P, Jang JS, Lee JS (2012) Photocatalytic and photoelectrochemical water oxidation over metal-doped monoclinic $BiVO_4$ photoanodes. ChemSusChem 5(10):1926

Payne D, Robinson M, Egdell R, Walsh A, McNulty J, Smith K, Piper L (2011) The nature of electron lone pairs in $BiVO_4$. Appl Phys Lett 98(21):212110

Peter LM, Upul Wijayantha KG (2014) Photoelectrochemical water splitting at semiconductor electrodes: fundamental problems and new perspectives. ChemPhysChem 15(10):1983

Pihosh Y, Turkevych I, Mawatari K, Uemura J, Kazoe Y, Kosar S, Makita K, Sugaya T, Matsui T, Fujita D (2015) Photocatalytic generation of hydrogen by core-shell $WO_3/BiVO_4$ nanorods with ultimate water splitting efficiency. Sci Rep 5:11141

Pilli SK, Furtak TE, Brown LD, Deutsch TG, Turner JA, Herring AM (2011) Cobalt-phosphate (Co-Pi) catalyst modified Mo-doped $BiVO_4$ photoelectrodes for solar water oxidation. Energy Environ Sci 4:5028

Pinaud BA, Benck JD, Seitz LC, Forman AJ, Chen Z, Deutsch TG, James BD, Baum KN, Baum GN, Ardo S, Wang H, Miller E, Jaramillo TF (2013) Technical and economic feasibility of centralized facilities for solar hydrogen production via photocatalysis and photoelectrochemistry. Energy Environ Sci 6(7):1983

Prévot MS, Guijarro N, Sivula K (2015) Enhancing the performance of a robust sol–gel-processed p-type delafossite $CuFeO_2$ photocathode for solar water reduction. ChemSusChem 8(8):1359

Pyper KJ, Yourey JE, Bartlett BM (2013) Reactivity of $CuWO_4$ in photoelectrochemical water oxidation is dictated by a midgap electronic state. J Phys Chem C 117(47):24726

Rao PM, Cai L, Liu C, Cho IS, Lee CH, Weisse JM, Yang P, Zheng X (2014) Simultaneously efficient light absorption and charge separation in $WO_3/BiVO_4$ Core/shell nanowire photoanode for photoelectrochemical water oxidation. Nano Lett 14(2):1099

Read CG, Park Y, Choi K-S (2012) Electrochemical synthesis of p-type $CuFeO_2$ electrodes for use in a photoelectrochemical cell. J Phys Chem Lett 3(14):1872

Rettie AJE, Lee HC, Marshall LG, Lin JF, Capan C, Lindemuth J, McCloy JS, Zhou J, Bard AJ, Mullins CB (2013) Combined charge carrier transport and photoelectrochemical characterization of $BiVO_4$ single crystals: intrinsic behavior of a complex metal oxide. J Am Chem Soc 135 (30):11389

Roth RS, Waring JL (1963) Synthesis and stability of bismutotantalite, stibiotantalite and chemically similar $ABO_4$ compounds. Am Mineral 48(11–2):1348

Ruiz-Fuertes J, Errandonea D, Segura A, Manjón F, Zhu Z, Tu C (2008) Growth, characterization, and high-pressure optical studies of $CuWO_4$. High Pressure Res 28(4):565

Saito R, Miseki Y, Sayama K (2012) Highly efficient photoelectrochemical water splitting using a thin film photoanode of $BiVO_4/SnO_2/WO_3$ multi-composite in a carbonate electrolyte. Chem Commun 48(32):3833

Sayama K, Nomura A, Arai T, Sugita T, Abe R, Yanagida M, Oi T, Iwasaki Y, Abe Y, Sugihara H (2006) Photoelectrochemical decomposition of water into $H_2$ and $O_2$ on porous $BiVO_4$ thin-film electrodes under visible light and significant effect of Ag ion treatment. J Phys Chem B 110(23):11352

Sayama K, Wang N, Miseki Y, Kusama H, Onozawa-Komatsuzaki N, Sugihara H (2010) Effect of carbonate ions on the photooxidation of water over porous $BiVO_4$ film photoelectrode under visible light. Chem Lett 39(1):17

Seabold JA, Choi KS (2012) Efficient and stable photo-oxidation of water by a bismuth vanadate photoanode coupled with an iron oxyhydroxide oxygen evolution catalyst. J Am Chem Soc 134 (4):2186

Seger B, Castelli IE, Vesborg PCK, Jacobsen KW, Hansen O, Chorkendorff I (2014) 2-Photon tandem device for water splitting: comparing photocathode first versus photoanode first designs. Energy Environ Sci 7(8):2397

Seitz LC, Chen Z, Forman AJ, Pinaud BA, Benck JD, Jaramillo TF (2014) Modeling practical performance limits of photoelectrochemical water splitting based on the current state of materials research. ChemSusChem 7(5):1372

Senegas J, Galy J (1974) L'oxyde double $Fe_2WO_6$ I Structure cristalline et filiation structurale. J Solid State Chem 10(1):5

Shi X, Choi IY, Zhang K, Kwon J, Kim DY, Lee JK, Oh SH, Kim JK, Park JH (2014) Efficient photoelectrochemical hydrogen production from bismuth vanadate-decorated tungsten trioxide helix nanostructures. Nat Commun 5:4775

Shi X, Zhang K, Shin K, Ma M, Kwon J, Choi IT, Kim JK, Kim HK, Wang DH, Park JH (2015) Unassisted photoelectrochemical water splitting beyond 5.7% solar-to-hydrogen conversion efficiency by a wireless monolithic photoanode/dye-sensitised solar cell tandem device. Nano Energy 13:182

Sieber K, Leiva H, Kourtakis K, Kershaw R, Dwight K, Wold A (1983) Preparation and properties of substituted iron tungstates. J Solid State Chem 47(3):361

Singh AK, Mathew K, Zhuang HL, Hennig RG (2015) Computational screening of 2D materials for photocatalysis. J Phys Chem Lett 6(6):1087

Sivula K (2013) Solar-to-chemical energy conversion with photoelectrochemical tandem cells. Chimia 67(3):155

Sleight AW, Chen HY, Ferretti A, Cox DE (1979) Crystal-growth and structure of $BiVO_4$. Mater Res Bull 14(12):1571

Sliozberg K, Schäfer D, Erichsen T, Meyer R, Khare C, Ludwig A, Schuhmann W (2015) High-throughput screening of thin-film semiconductor material libraries I: System development and case study for T-W-O. ChemSusChem 8(7):1270

Solarska R, Krolikowska A, Augustynski J (2010) Silver nanoparticle induced photocurrent enhancement at $WO_3$ photoanodes. Angew Chem Int Ed 49(43):7980

Stoughton S, Showak M, Mao Q, Koirala P, Hillsberry D, Sallis S, Kourkoutis L, Nguyen K, Piper L, Tenne D (2013) Adsorption-controlled growth of $BiVO_4$ by molecular-beam epitaxy. APL Mater 1(4):042112

Thomas G, Ropital F (1984) Influence des gaz sur la synthese du tungstate de fer $Fe_2WO_6$ II. Etude des mecanismes solide-solide. Mater Chem Phys 11(6):563

Tilley SD, Cornuz M, Sivula K, Grätzel M (2010) Light-induced water splitting with hematite: improved nanostructure and iridium oxide catalysis. Angew Chem Int Ed 49(36):6405

Tilley SD, Schreier M, Azevedo J, Stefik M, Graetzel M (2014) Ruthenium oxide hydrogen evolution catalysis on composite cuprous oxide water-splitting photocathodes. Adv Funct Mater 24(3):303

Tokunaga S, Kato H, Kudo A (2001) Selective preparation of monoclinic and tetragonal $BiVO_4$ with scheelite structure and their photocatalytic properties. Chem Mater 13(12):4624

Touboul M, Melghit K, Bénard P, Louër D (1995) Crystal structure of a metastable form of indium orthovanadate, $InVO_4$-I. J Solid State Chem 118(1):93

Trainham JA, Newman J, Bonino CA, Hoertz PG, Akunuri N (2012) Whither solar fuels? Curr Opin Chem Eng 1(3):204

van de Krol R, Grätzel M (eds) (2012) Photoelectrochemical hydrogen production. Springer, New York

van de Krol R, Ségalini J, Enache CS (2011) Influence of point defects on the performance of $InVO_4$ photoanodes. J photon Energy 1(1):016001

Wadnerkar N, English NJ (2013) Density functional theory investigations of bismuth vanadate: effect of hybrid functionals. Comput Mater Sci 74:33

Walczak J, Rychiowska-Himmel I, Tabero P (1992) Iron (III) tungstate and its modifications. J Mater Sci 27(13):3680

Walsh A, Wei S-H, Yan Y, Al-Jassim MM, Turner JA, Woodhouse M, Parkinson BA (2007) Structural, magnetic, and electronic properties of the Co-Fe-Al oxide spinel system: density-functional theory calculations. Phys Rev B 76(16):165119

Walsh A, Yan Y, Huda MN, Al-Jassim MM, Wei SH (2009) Band edge electronic structure of $BiVO_4$: elucidating the role of the Bi s and V d orbitals. Chem Mater 21(3):547

Walter MG, Warren EL, McKone JR, Boettcher SW, Mi Q, Santori EA, Lewis NS (2010) Solar water splitting cells. Chem Rev 110(11):6446

Woodhouse M, Parkinson BA (2008) Combinatorial discovery and optimization of a complex oxide with water photoelectrolysis activity. Chem Mater 20(7):2495

Woodhouse M, Parkinson BA (2009) Combinatorial approaches for the identification and optimization of oxide semiconductors for efficient solar photoelectrolysis. Chem Soc Rev 38(1):197

Woodhouse M, Herman GS, Parkinson BA (2005) Combinatorial approach to identification of catalysts for the photoelectrolysis of water. Chem Mater 17(17):4318

Ye H, Park HS, Bard AJ (2011) Screening of electrocatalysts for photoelectrochemical water oxidation on W-doped $BiVO_4$ photocatalysts by scanning electrochemical microscopy. J Phys Chem C 115(25):12464

Yin W-J, Tang H, Wei S-H, Al-Jassim MM, Turner J, Yan Y (2010) Band structure engineering of semiconductors for enhanced photoelectrochemical water splitting: the case of $TiO_2$. Phys Rev B 82(4):045106
Yin WJ, Wei SH, Al-Jassim MM, Turner J, Yan Y (2011) Doping properties of monoclinic $BiVO_4$ studied by first-principles density-functional theory. Phys Rev B 83(15):155102
Yokoyama D, Minegishi T, Maeda K, Katayama M, Kubota J, Yamada A, Konagai M, Domen K (2010) Photoelectrochemical water splitting using a Cu(In, Ga)Se2 thin film. Electrochem Commun 12(6):851
Yourey JE, Bartlett BM (2011) Electrochemical deposition and photoelectrochemistry of $CuWO_4$, a promising photoanode for water oxidation. J Mater Chem 21(21):7651
Yourey JE, Pyper KJ, Kurtz JB, Bartlett BM (2013) Chemical stability of $CuWO_4$ for photoelectrochemical water oxidation. J Phys Chem C 117(17):8708
Zhao ZY, Luo WJ, Li ZS, Zou ZG (2010) Density functional theory study of doping effects in monoclinic clinobisvanite $BiVO_4$. Phys Lett A 374(48):4919
Zhao ZY, Li ZS, Zou ZG (2011) Electronic structure and optical properties of monoclinic clinobisvanite $BiVO_4$. Phys Chem Chem Phys 13(10):4746
Zhong DK, Choi S, Gamelin DR (2011) Near-complete suppression of surface recombination in solar photoelectrolysis by "Co-Pi" catalyst-modified W:$BiVO_4$. J Am Chem Soc 133 (45):18370
Zhong M, Hisatomi T, Kuang Y, Zhao J, Liu M, Iwase A, Jia Q, Nishiyama H, Minegishi T, Nakabayashi M (2015) Surface modification of $CoO_x$ loaded $BiVO_4$ photoanodes with ultrathin p-type NiO layers for improved solar water oxidation. J Am Chem Soc 137(15):5053
Zhou L, Wang W, Liu S, Zhang L, Xu H, Zhu W (2006) A sonochemical route to visible-light-driven high-activity $BiVO_4$ photocatalyst. J Mol Catal A Chem 252(1):120

# Chapter 9
# Non-Oxide Materials (Nitrides, Chalcogenides, and Arsenides)

**Katsushi Fujii**

## 9.1 General Perspective

### *9.1.1 General Semiconductor Properties*

Non-oxide semiconductor materials have wide bandgap and flatband potential varieties because their anions and cations can be selected. Controllable anion especially provides different properties to non-oxide materials compared with oxide materials, that is, the valence band edge potential can be controlled. They also basically have good crystallinity and doping controllability. This is because many single crystal growth techniques exist for group IV, III–V, and II–VI semiconductor materials due to the need for high-quality single crystals in the electronic industry. These characteristics provide the bases for a wide variety of studies on non-oxide photoelectrochemistry. On the other hand, the surface of non-oxide semiconductors is difficult to stabilize in acidic and basic aqueous solutions because these semiconductors basically show chemical etching properties in aqueous solutions containing oxidation reagents. As a result, n-type materials are easy to corrode not only under photoelectrochemical-oxidized conditions, which are usually strong oxidizing circumstances. Even gallium nitride (GaN), which is famous for its chemical stability, shows anodic corrosion in the photoelectrochemical reactions of n-type materials. In order to prevent corrosion, protection layers and photoelectrochemical catalysts are also widely studied. The conditions for preventing corrosion are much easier for p-type materials because the condition is a reducing environment. Even under these conditions, some non-oxide materials corrode in electrolyte.

K. Fujii (✉)
Global Solar Plus Initiative, The University of Tokyo, 4-6-1 Komaba, Meguro-ku, Tokyo, Japan
e-mail: katsushi.fujii@riken.jp



S. Giménez, J. Bisquert (eds.), *Photoelectrochemical Solar Fuel Production*,
DOI 10.1007/978-3-319-29641-8_9

Especially in the monolithic non-oxide semiconductor applications of photoelectrochemical reactions, it is vital to know the basic properties of these non-oxide semiconductors. They are thus well studied for many device applications like transistor and diodes (Sze and Ng 2006). The electrochemical properties of non-oxide semiconductors based on the results of semiconductor engineering evaluation results are effective for real applications of electrochemistry. The electrochemical properties of semiconductor materials are well explained especially in Chaps. 5–8 of the book written by Memming (2008, 2015) and elsewhere (Nozik and Memming 1996; Mills and Le Hunte 1997). The photoelectrochemical properties of non-oxide materials are also well explained in the other books (Chap. 7 (Grimes et al. 2007), Chap. 7 (Rajeshwar et al. 2008)). Thus, after the brief summary of general but important electrochemical characteristics of semiconductors to start the discussion on non-oxide semiconductors, the individual properties of semiconductor materials for photoelectrochemical reactions of band edge potentials, photocurrent density, and stability of semiconductors will mainly be discussed here. In particular, the important properties of semiconductor materials for photoelectrochemical reactions will be focused, such as bandgap width, direct and indirect bandgap, conduction type of n- or p-type, and stability in electrolyte.

### 9.1.2 Band Edge and Flatband Potentials

The band edge potential is important for knowing the electrochemical reaction ability of a semiconductor. The relationship between the relative potential position of the band edge potential and redox potential of the reactant in electrolyte determines the possibility of the reaction progress. When the conduction band potential is much lower than that of the redox potential (the electron energy of conduction band is higher than that for the redox potential), the reduction of the reactant can occur with electron transfer from the conduction band to the reactant. The valence band is vice versa. The relationship is shown in Fig. 9.1. It should be noted that this estimation does not include the activation energy of the electron transfer. This energy is often expressed as the reaction overpotential of the electrochemical reaction.

Molecules and ions absorbed into the surface play important roles on the determination of the band edge potential of a semiconductor in electrolyte because the potential is defined by the charge balance at the interface of the semiconductor and electrolyte. In an aqueous electrolyte, since the main molecules in the electrolyte are water, the $H^+$ and $OH^-$ ions have major effects on the band edge potential position, that is, the pH of the electrolyte affects even non-oxide semiconductors. When the molecule and ion adsorption on the semiconductor surface is stable and just changes along the pH of the electrolyte, the band edge position changes according to the Nernstian relationship of water, that is, −0.059 V/pH. When the

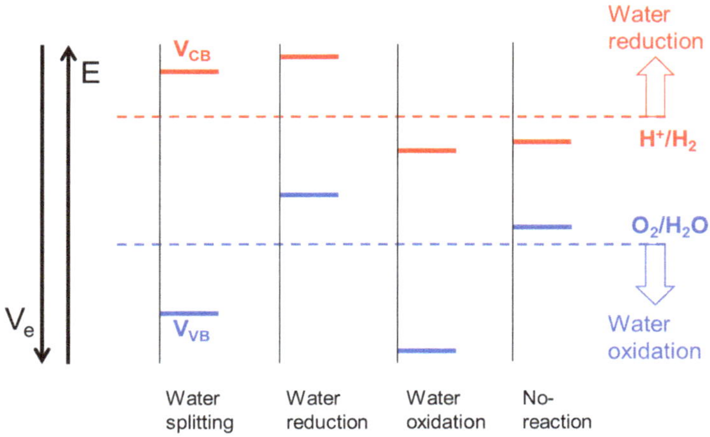


**Fig. 9.1** Relationship between conduction band edge potential ($V_{CB}$) and water-reduced redox potential ($H^+/H_2$) and valence band edge potential ($V_{VB}$) and water oxidation redox potential ($O_2/H_2O$). The conditions from left to right are water splitting, reduction, oxidation, and no reaction. The positive direction of potential is downward and that for electron energy is upward

conditions of adsorbed molecules or ions on the semiconductor surface change due to change in the pH, the Nernstian relationship is also not established.

The band edge potential can be obtained theoretically (Van de Walle and Neugebauer 2003), but experimental evaluation is required to know the precise potential position. The Mott–Schottky plot (the plot of inverse square of capacitance versus applied voltage, $1/C^2 - V$) is the most useful method to know the flatband potential (the explored point for $1/C^2 \rightarrow 0$) (Memming 2008, 2015). It should be noted that the position of the flatband potential is the Fermi level of semiconductor. Thus, the conduction band edge potentials for n-type materials are calculated from

$$\Phi_c = \Phi_{FB} - \frac{kT}{q} \ln(N_c/n), \tag{9.1}$$

where $k$ is the Boltzmann constant, $T$ is the absolute temperature, $q$ is the elemental charge, $n$ is the net carrier concentration, and $N_c$ is the conduction band effective density of states. $N_c$ is expressed as

$$N_c = 2\left(\frac{2\pi m_e^* kT}{h^2}\right)^{3/2}, \tag{9.2}$$

where $m_e$* is the electron effective mass and $h$ is the Planck's constant. For p-type semiconductors, the valence band edge potential can be identically calculated. The other band edge potential is calculated from the obtained band potential and bandgap value.

Intrinsic materials have other interesting properties regarding the relationship between the plot of log(C) and applied bias. The log(C) shows the minimum at the bias of the flatband condition (Memming 2008, 2015). This is because the carrier densities of the electron and hole are the same. This interesting phenomenon is, however, observed only for germanium (Ge), which can be made by very pure elemental materials.

Another method of band edge potential evaluation is evaluating the turn-on bias for the photocurrent versus bias characteristics. This technique is often used for non-oxide materials because obtaining the Mott–Schottky plot is sometimes difficult due to the conductivity. For this evaluation, the turn-on bias shows the position of the flatband potential. This is much easier to observe compared to the Mott–Schottky plot, but the evaluated value contains overpotential. The overpotential contains the activation energy of the reaction and some losses for the reactions. Since the activation energy can be reduced by using electron–hole scavenger, the overpotential can be reduced in the condition with the electron–hole scavenger.

### *9.1.3 Photocurrent*

The current versus voltage characteristics show that for diode-like in the dark, the current is observed when the majority carrier can flow under the applied bias. Under illumination, the photocurrent for minority carrier can be also observed, and the maximum photocurrent with the applied bias is limited by the light illumination intensity because the origin of the minority carrier is generated electron–hole pair by light in the semiconductor. Since the minority carrier is defined by the carrier type of the semiconductor, n- or p-type defines the photocurrent. That is, the minority carrier for the n-type semiconductor is a hole; thus, the photocurrent originates from the hole. On the other hand, the minority carrier for the p-type semiconductor is an electron; thus, the photocurrent originates from the electron. This affects the photoelectrochemical reaction at the interface of the semiconductor and electrolyte. The photoelectrochemical reaction for the n-type semiconductor is reactant oxidation, and the photochemical reaction for the p-type semiconductor is reactant reduction at the interface of the semiconductor and electrolyte. The schematic diagram is shown in Fig. 9.2. As a result, the control of the n- or p-type semiconductor is important for designing the photoelectrochemical reaction. Especially for n-type materials, the surface oxidation of the electrode material is a serious problem because the elements of non-oxide semiconductors are not fully oxidized. This means that the minority carrier of the hole can oxidize the semiconductor itself.

The amount of the photocurrent is basically determined by the photoabsorption, carrier separation, and carrier reaction abilities of the semiconductor. The photoabsorption is defined by the bandgap and its direct or indirect properties. The absorption coefficient of an indirect semiconductor is about 100 times smaller than that of direct semiconductor materials. Thus, a much thicker photoabsorption region is required for indirect semiconductors compared to direct ones, although the

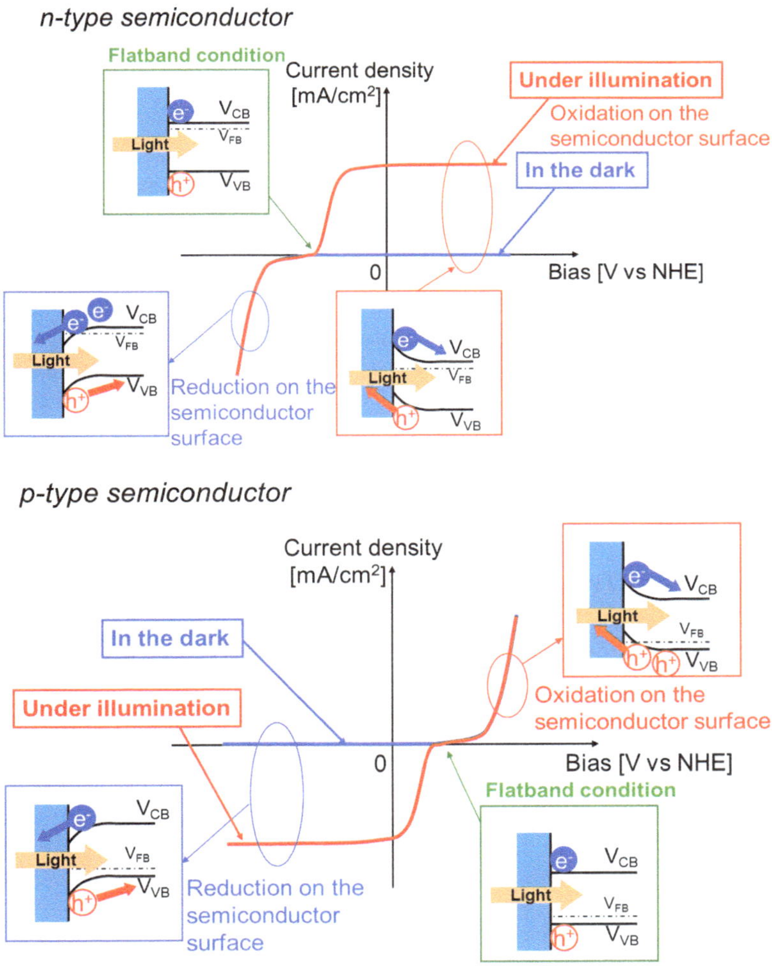


**Fig. 9.2** Schematic photocurrent–voltage characteristics for n-type (*upper*) and p-type (*lower*) semiconductors both in the dark and under illumination. $V_{CB}$, $V_{VB}$, and $V_{FB}$ are conduction band edge potential, valence band edge potential, and Fermi level, respectively

real carrier separation process is complicated due to the recombination process (Ono et al. 2007). Since the thickness of the depression layer is inversely proportional to the square of the donor (for n-type) or acceptor (for p-type) concentration, the thickness is wider when the donor or acceptor concentration is lower. The reaction between the carrier and a reactant in an electrolyte is the most difficult part of the carrier transfer process and is not well understood, but it is probably related to the molecular absorption on the semiconductor surface and its carrier transfer process.

### *9.1.4 Stability and Corrosion*

The corrosion of a semiconductor during electrochemical or photoelectrochemical reaction is another key issue. The thermodynamical estimation is summarized as follows (Gerischer 1977). When the standard electrode potential for the cathodic dissolution of the electrode material is more negative than the conduction band edge, the electrode should be catholically stable. In the same way, when the standard electrode potential for the anodic dissolution of the electrode material is more positive than the valence band edge, the electrode should be anodically stable. This means that an n-type semiconductor may be anodically stable even if the anodic dissolution potential is more negative than the valence band edge when a redox potential for the reactant in an electrolyte is lower than that for the anodic dissolution potential. The schematic relationship is shown in Fig. 9.3. The calculation results for recent semiconductors have also been reported (Chen and Wang 2012). The calculation indicates that almost all of non-oxide semiconductors are not stable for water oxidation with oxygen evolution. Thus, semiconductor protection is required for non-oxide semiconductors used especially for the anode.

A famous protection layer for the anodic electrode of n-type semiconductors is titanium dioxide ($TiO_2$). $TiO_2$ was used for metal corrosion protection as photoanode (Park et al. 2002). This protection was also applied for semiconductor photoanode. Recently, a study was conducted to observe corrosion protection as well as improvement of properties (Lin et al. 2015).

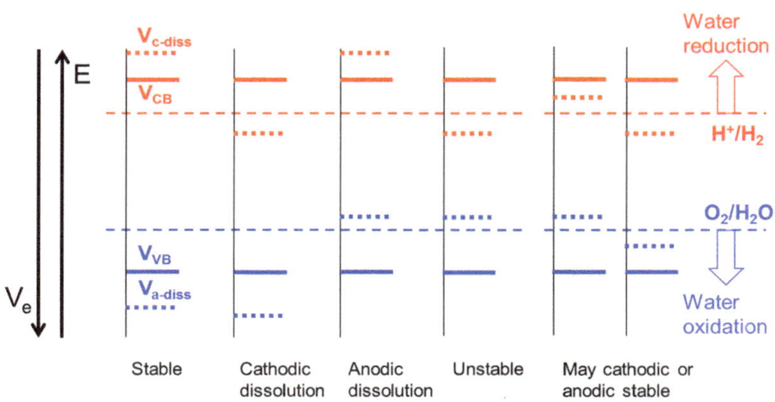


**Fig. 9.3** Relationship among conduction band potential ($V_{CB}$), standard redox potential for cathodic dissolution ($V_{c\text{-diss}}$), water reduction redox potential ($H^+/H_2$), valence band potential ($V_{VB}$), standard redox potential for anodic dissolution ($V_{a\text{-diss}}$), and water oxidation redox potential ($O_2/H_2O$). The conditions from left to right are stable, cathodic dissolution, anodic dissolution, and unstable. The two conditions of the right show the condition may be cathodic or anodic stable in unstable case. The positive direction of potential is downward and that for electron energy is upward as indicated by the *arrows*

## 9.2 Elemental Semiconductors (Si, Ge, and C (Diamond))

### *9.2.1 Semiconductor Properties*

Germanium (Ge), silicon (Si), and diamond (C) are known as elemental semiconductors and are usually used in the single crystal form except for poly- or amorphous-Si solar cells. Since they are composed of a single element, relatively pure materials can be obtained except for the diamond. The indirect bandgaps are 0.67 eV for Ge, 1.12 eV for Si, and 5.5 eV for diamond, respectively. The control of these n- and p-type semiconductors is relatively easy, and both n- and p-type versions can be obtained. The major properties are summarized in Table 9.1.

Except for the diamond, Ge and Si are most often used for semiconductor devices, especially Si. Since the bandgap of Ge is small and chemically not stable, Ge is rarely used for photoelectrochemical reactions. On the other hand, since Si has a relatively wide bandgap and it is relatively stable for chemical reaction, it is used for electrochemical reactions especially for p-type photocathodic electrode. The diamond is also rarely used for photoelectrochemical reactions due to its indirect and wide bandgap.

### *9.2.2 Flatband and Band Edge Potentials*

The flatband and band edge potentials are summarized in Table 9.1.

*Ge and Si*
The flatband potential for nearly intrinsic Ge electrodes was estimated using the plot of log (C) versus bias and surface conductivity versus bias (Harten and Memming 1962; Brattain and Boddy 1962). Minimum points were observed for

**Table 9.1** Basic semiconductor and electrochemical properties of C (diamond), Si, and Ge (details are discussed in the text)

| | Bandgap | | Structure | | Dopant | |
|---|---|---|---|---|---|---|
| | d/ind | (eV) | | (nm) | P-type | N-type |
| C (dia) | ind | 5.5 | dia | 0.3567 | B | P, As |
| Si | ind | 1.12 | dia | 0.5431 | B | P, As |
| Ge | ind | 0.67 | dia | 0.5658 | B | P, As |
| | Band edge potential [*V* vs. SHE] | | | | Slope [*V* vs. SHE/pH] | |
| | Conduction band | | Valence band | | | |
| C (dia) | −3.9[a] | | +1.6[a] | | – | |
| Si | −0.25[b] | | – | | – | |
| Ge | (inv.)[c] | | (inv.)[c] | | −0.059 | |

Bandgap: *d* direct, *ind* indirect; structure: *dia* diamond
[a]In 0.5 mol/L $H_2SO_4$ aqueous solution
[b]Flatband for n-type Si in 10 mol/L HF
[c]Inversion layer is observed (depend on pH) instead of the depletion layer

both cases. The pH dependence was −0.059 V/pH above pH 4 (Boddy and Brattain 1963). However, the pH dependence of flatband potential after anodic or cathodic pretreatment was different (Memming 2008, 2015); thus, the surface of Ge is nearly perfect and the flatband potential differs from adsorbed ions like $H^+$ or $OH^-$.

The flatband potential of Si was also evaluated using capacitance and surface conductivity method (Memming 2008, 2015). The experiments were performed on slightly n- and p-doped materials because the resistivity of intrinsic Si is too high for measuring the capacitance and surface conductivity. The measurement value did not fit the theoretical line, whereas the measurement of Ge did (Memming and Schwandt 1966b). This means that depletion layer instead of the inversion layer was observed over the whole applied bias for Si. This depletion layer is commonly observed in n- or p-type semiconductors. Re-plotted data shows that the flatband potential of n-type Si in 10 mol/L HF aqueous solution is about −0.25 V versus standard calomel electrode (SCE). This suggests that Si has the ability to reduce water but not oxidize water considering the bandgap of 1.12 eV.

*Diamond*

Diamond has interesting electrochemical properties such as high hardness, high thermal conductivity, and high carrier mobility (Pleskov 2002; Kraft 2007). However, it is not that useful for photoelectrochemical reactions due to its large bandgap. The overpotential of undoped diamond is extremely large; thus, the water oxidation (oxygen evolution) and reduction (hydrogen evolution) are in the zero current region of the overpotential (this is called "potential window" or "ideal polarizability region").

This potential window is useful for investigating nonaqueous material oxidization or reduction in aqueous solution. However, it is not worth investigating for water splitting because the overpotential prevents the progress of water oxidation and/or reduction.

The B-doped p-type diamond shows the flatband potential of the range of 0 to +1.5 V versus standard hydrogen electrode (SHE). The flatband potential for surface-oxidized p-type diamond was estimated to be +1.0 to +1.1 V versus standard calomel electrode (SCE) in 0.5 mol/L $H_2SO_4$, whereas that for as-grown sample approaches 0 V versus SCE (Pleskov 2002). Considering with the B trap level of 0.37 eV above the valence band edge and the bandgap of 5.5 eV, the valence band edge is estimated to be around +1.6 V versus SHE and conduction band potential is estimated to be around −3.9 V versus SHE in 0.5 mol/L $H_2SO_4$ aqueous solution. The band edge potential shows that diamond can split water without bias, but the absorbed light wavelength by the bandgap of 5.5 eV is in the deep ultraviolet (UV) region. It is therefore not useful for photoelectrochemical reaction. An interesting property of the B-doped diamond is electrochemical $CO_2$ reduction for forming formaldehyde (Nakata et al. 2014). $H_2$ reduction from water was reported to be suppressed in experiments.

### 9.2.3 Photocurrent–Voltage Characteristics

*Ge and Si*

Although the photocurrent properties of Ge have been well studied in the past (Memming 2008, 2015), the photoelectrochemical reactions of Ge are not of much interest at present probably due to its small bandgap and its chemical instability.

Oxidation of Si was observed, and the properties were ideal in that the photocurrent increased proportionally to the hole density at the surface (Memming 2008; Memming and Schwandt 1966a; Memming 2015). That is, the Tafel slope was 60 mV/decade. This shows that the process is almost reaction limited. The bandgap of Si is 1.12 eV; thus, Si itself cannot perform water splitting. However, since the conduction band edge potential is higher than that of water reduction (hydrogen generation) redox potential and the p-type Si can easily be obtained, Si is often used as the photocathode of the complex photoelectrochemical reaction systems.

*Diamond*

The photoelectrochemical response of p-type diamond was reported as ideal (Pleskov 2002). That is, the photocurrent was observed in the reverse-biased region.

### 9.2.4 Stability Improvements by Catalysts and Protection Layers

*Si*

Among the elemental semiconductors, Si is the most popular material for photoelectrochemical reactions, especially p-type Si for photocathodes. For p-type Si protection for chemical corrosion, polymer coating and photoelectrochemical platinum (Pt) coating of the Si surface with and without polymer were carried out (Abruna and Bard 1981; Dominey et al. 1982). These techniques are known to significantly improve photo-assisted hydrogen generation. Recently p-type Si is mainly used for photocathodes with a nanowire structure because the conduction band edge of Si is suitable for the water reduction and $H_2$ generation (Dasgupta et al. 2013). The Pt nanoparticles are often used for the catalyst in order to improve the reduced reaction of water (Wang et al. 2013).

For n-type Si, although it is not useful for complete water splitting because its valence band edge is lower than that of the water oxidation (oxygen evolution) redox potential, various methods were proposed to protect the corrosion, that is, growing a metal-doped semiconducting $SiO_x$ layer deposition on the n-type Si surface (Contractor and Bockris 1984), forming a metal silicide layer on the n-type Si surface (Nakato et al. 1982), and depositing a thin layer of manganese oxide ($MnO_x$) or analogous oxide on the n-type Si surface (Kainthla et al. 1986). The dispersion of 5 to 50 nm Pt particles on the n-type Si surface improves the stability of the Si in the aqueous solution and significantly increases the open-circuit photovoltage ($V_{OC}$) and quantum conversion efficiency (Nakato et al. 1988;

Wünsch et al. 2002). This can be explained by the fact that minority carrier accumulation is controlled by Pt particles, and the carrier is thought to accumulate in the Pt particles.

## 9.3 III–V Semiconductors (Phosphide and Arsenide)

### 9.3.1 Semiconductor Properties

The III–V semiconductors are the most popular compound semiconductors and are often used for semiconductor devices. Since their purpose is for fabricating electronic devices, a single crystal is a common structure used for these semiconductor materials. The main elements of these compound materials are aluminum (Al), gallium (Ga), and indium (In) for group III and phosphor (P) and arsenic (As) for group V. The Al compounds are indirect semiconductors and the bandgap of AlAs is 2.12 eV and AlP is 2.45 eV. AlAs and AlP are not commonly used except for mixtures, $Al_xGa_{1-x}P$, for example, because AlAs is easy to oxidize and AlP is easy to decompose in air. The GaAs is one of the basic materials for the group III–V compound semiconductor material, and the bandgap is 1.43 eV (direct). GaP is also a common material, but it is an indirect material with a bandgap of 2.26 eV. The In compounds have a direct bandgap, and the value of InAs is 0.36 eV and that of InP is 1.35 eV. InP is also a useful material for semiconductor devices. The major properties are summarized in Table 9.2.

The intermixture of these compound is also well known, and $(Al_xGa_{1-x})As$, $(Ga_xIn_{1-x})$ As, $(Ga_xIn_{1-x})$ P, $(Al_yGa_{1-y})_{0.5}In_{0.5}P$, and $GaP_zAs_{1-z}$ are common materials for electronic and optoelectronic devices. These compounds are relatively easy to chemically etch in the presence of an oxidized reagent except for GaP and $Ga_{0.5}In_{0.5}P$. The n- and p-type control of these semiconductors is relatively easy except for AlAs and AlP which tend to become an n-type material. Generally, the surfaces of these materials are 100 and 111.

Although the bandgaps of these III–V semiconductors are suitable for visible light-absorbed photoelectrochemical reactions, a protection layer for preventing chemical decomposition must be formed on the photoelectrochemical electrodes due to its chemical properties.

### 9.3.2 Flatband and Band Edge Potentials

The flatband and band edge potentials are summarized in Table 9.2.

*GaAs and GaP*

The flatband potential for n- and p-type GaP was measured in 0.5 mol/L $H_2SO_4$. It was around −0.9 V vs. standard hydrogen electrode (SHE) for n-type GaP and

**Table 9.2** Basic semiconductor and electrochemical properties of III–V compound material (details are discussed in the text)

| | Bandgap | | Structure | | Dopant | |
|---|---|---|---|---|---|---|
| | d/ind | (eV) | | (nm) | P-type | N-type |
| AlP | ind | 2.45 | zb | 0.5464 | Zn, Mg | Si, Te |
| GaP | ind | 2.26 | zb | 0.5451 | Zn, Mg | Si, Te |
| InP | d | 1.35 | zb | 0.586 | Zn, Mg | Si, Te |
| AlAs | ind | 2.12 | zb | 0.5660 | Zn, Mg | Si, Te |
| GaAs | d | 1.43 | zb | 0.5653 | Zn, Mg | Si, Te |
| InAs | d | 0.36 | zb | 0.605 | Zn, Mg | Si, Te |
| | Band edge potential [*V* vs. SHE] | | | | Slope [*V* vs. SHE/pH] | |
| | Conduction band | | Valence band | | | |
| AlP | – | | – | | – | |
| GaP | −1.05 | | +1.20 | | −0.06 | |
| InP | −1.3 (pH 14) | | +0.05 (pH 14) | | – | |
| | −0.70 (pH 7) | | +0.63 (pH 7) | | | |
| | −0.62 (pH 0) | | +0.65 (pH 0) | | | |
| AlAs | – | | – | | – | |
| GaAs | −0.90 | | +0.50 | | −0.055 | |
| InAs | – | | – | | – | |

Bandgap: *d* direct, *ind* indirect; structure: *zb* zinc blend

+1.2 V vs. SHE for p-type GaP (Memming 1969). The potential of the n-type is close to the conduction band edge and that of the p-type is close to the valence band edge. This indicates that the valence band edge potential is slightly insufficient for oxidizing water (oxygen evolution), but the electrodes in the conduction band edge can reduce water (hydrogen evolution).

The flatband potential of the n-type GaAs was $-1.5 \pm 0.2$ V vs. SHE, and the conduction band edge of the n-type GaP was estimated to be −2.2 eV vs. SHE at pH 14 (Ellis et al. 1977a). Since the flatband potential is the Fermi level when bias is applied so that the semiconductor band near the semiconductor electrolyte interface does not bend and the Fermi level is close to the conduction band edge position for n-type GaAs and GaP, the flatband potential is nearly equal to the conduction band edge position for n-type GaAs and GaP, and the conduction band edge potentials of the n-type GaAs and GaP were estimated to be −1.5 V vs. SHE and −2.2 V vs. SHE for pH 14, respectively. The valence band edge potentials of the n-type and p-type GaAs and GaP were estimated to be −0.1 V vs. SHE and +0.05 V vs. SHE for pH 14, respectively. The flatband potential of GaAs and GaP showed no specific relations with the existence of the redox of $X^{2-}/X_n{}^{2-}$ ($X =$ S, Se, Te) at pH 14 (Ellis et al. 1977a). This indicates that the GaAs and GaP have no interaction with the X-ions in the electrolyte. Considering the water reduction (hydrogen evolution) and water oxidation (oxygen evolution) redox potentials of −0.826 V vs. SHE and +0.404 V vs. SHE at pH 14, respectively, the result also shows that both GaAs and GaP can reduce water but cannot oxidize water.

From the Mott–Schottky plot of various reports, the conduction and valence band potentials for GaP are summarized (Gomes and Cardon 1982) as follows:

$$V_{\mathrm{CB}} = -1.05 - 0.06\,\mathrm{pH}\ (V\ \mathrm{vs.\ SHE}), \tag{9.3}$$

$$V_{\mathrm{VB}} = +1.20 - 0.06\mathrm{pH}\ (V\ \mathrm{vs.\ SHE}). \tag{9.4}$$

Similarly, the conduction and valence band edge potentials of GaAs are

$$V_{\mathrm{CB}} = -0.90 - 0.055\,\mathrm{pH}\ (V\ \mathrm{vs.\ SHE}), \tag{9.5}$$

$$V_{\mathrm{VB}} = +0.50 - 0.055\mathrm{pH}\ (V\ \mathrm{vs.\ SHE}). \tag{9.6}$$

These results also show that these GaAs and GaP can reduce water but cannot oxidize water. Similar results were reported later (Goossens and Schoonman 1992). The effects of the changes in the surfaces of (111) Ga-face and P-face of the n-type GaP and electrolytes on the flatband potential were also reported (Mukherjee et al. 2010).

*InP*

The flatband potential of the n-type InP was $-1.3 \pm 0.1$ V vs. SHE for pH 14 and $-0.7 \pm 0.1$ V vs. SHE for pH 7 from the Mott–Schottky plot (Ellis et al. 1977b). Since the flatband potential is the Fermi level when when bias is applied so that the semiconductor band near the semiconductor electrolyte interface does not bend and the Fermi level is close to the conduction band edge position for n-type InP, the flatband potential is nearly equal to the conduction band edge position for n-type InP, the conduction band edge potentials of n-type InP are estimated to be $-1.3$ V vs. SHE for pH 14 and $-0.7$ V vs. SHE for pH 7, and the valence band edge potentials are +0.05 V vs. SHE for pH 14 and +0.65 V vs. SHE for pH 7. The flatband potential of InP showed a negative shift with the existence of the redox of $X^{2-}/X_n{}^{2-}$ ($X =$ S, Se, Te) at pH 14 (Ellis et al. 1977b). This indicates that the InP interacts with the X-ions in the electrolyte. Considering the water reduction (hydrogen evolution) and water oxidation (oxygen evolution) redox potentials at pH 14 of $-0.826$ V vs. SHE and +0.404 V vs. SHE, respectively, InP can reduce water but cannot oxidize water.

The precise evaluation of InP flatband and band edge potentials in 1.0 mol/L HCl aqueous solution was performed (Tubbesing et al. 1986). The flatband potentials of the n- and p-type were $-0.62$ and +0.65 V vs. SHE, respectively. These values were obtained in a 1.0 mol/L HCl aqueous solution; thus, this is close to the band edge potential at standard condition. In other words, the conduction band edge potential is around $-0.62$ V vs. SHE at pH 0, and the valence band edge potential is around +0.65 V vs. SHE at pH 0. The flatband potential shifted about $-0.2$ V for the n-type InP and about +0.3 V for the p-type GaP with light illumination compared with those without light illumination. Even the flatband shifts with light illumination. This finding also confirms that these InPs can reduce water but cannot oxidize water.

### 9.3.3 Photo-corrosion and Stability Improvements by Catalysts and Protection Layers

*General for III–V*

Photo-corrosion is a common characteristic of GaAs and GaP (Frese et al. 1980, 1981b, a). The corrosion is affected by many properties of the III–V compound material (Allongue and Blonkowski 1991). It is pH dependent, and it sensitively appears in the measured results of the Tafel plot (plot of logarithm of current density vs. overpotential). It is also reflected by the type of n- or p-type conductivity, which reflects the rest potential. The doping density and origin of holes also affect corrosion. This affects the change in the potential distribution at the interface and the hole capture in surface bonds. Since many semiconductor properties affect the corrosion of the III–V compound semiconductor material, protection against corrosion is important for this material.

*GaAs*

The n-type GaAs undergoes anodic corrosion under illumination in aqueous solution (Frese et al. 1980; Allongue and Blonkowski 1991; Menezes and Miller 1983; Khader 1996). The anodic corrosion of GaAs is described as follows:

$$\mathrm{GaAs(s)} \rightarrow \mathrm{Ga^{3+}(in\ell)} + \mathrm{As(surf)} + 3e^-, \tag{9.7}$$

$$\begin{aligned} &2\mathrm{GaAs(s)} + 6\mathrm{H_2O}(\ell) \rightarrow \\ &\mathrm{Ga_2O_3(surf)} + \mathrm{As_2O_3(s)} + 12\mathrm{H^+(in\ell)} + 12e^-, \end{aligned} \tag{9.8}$$

where s is solid, $\ell$ is liquid, surf is on solid surface, and in$\ell$ is ion in liquid. Since the surface enriches the element, As, due to the anodic corrosion, the turn-on potential shifts to negative and the photocurrent density increases. The stability of element As on a GaAs surface depends on the electrolyte pH and the applied potential. When the As is oxidized to stable $As_2O_3$ at low pH with the potential of −0.45 V vs. SHE, for example, the photocurrent density decreases due to the oxide layer. The oxide on the surface affects some photoelectrochemical properties, such as consumption of photo-generated electron due to the introduction of new surface states, less light transmittance due to the scattering and absorption of light by the oxide, and anion adsorption from the electrolyte which retards the movements of the anion. Hence, a higher negative potential would be required for the displacement of the adsorbed anions. As a result, the presence of oxide shifts the turn-on potential of the photocurrent to more negative values.

The p-type GaAs can be used for hydrogen evolution photocathode due to its conduction band edge potential. From the photocurrent density versus bias characterization of p-type GaAs in 0.05 mol/L $H_2SO_4$ aqueous solution, the cathodic photocurrent generation with the turn-on potential of −0.3 to −1.3 V vs. standard calomel electrode (SCE) is observed (Khader et al. 1993). Since the expected turn-on bias is around +0.3 V vs. SCE from the flatband potential position, the hydrogen evolution from the p-type GaAs electrode has poor electrochemical catalytic activity. The p-type GaAs electrode surface modification with etching shows the

0.1 V positive shift of the turn-on potential of photocurrent and 15 % increase in the photocurrent in a 0.5 mol/L $H_2SO_4$ aqueous solution (Khader et al. 1996). The surface returns to its original state gradually during the cyclic-voltammetry measurement (this is probably due to the removal of the oxide layer from the GaAs surface), and/or the surface turns to being As rich with efficient hydrogen evolution after the etching. The coating with 5 to 10 monolayer Pt also showed photocurrent improvement. The coating with $WO_3$ on p-type GaAs surface showed a similar positive shift for the turn-on bias as a result of the improvement in the photocurrent density (Yoon et al. 1996). The p- and n-type GaAs electrodes were coated electrochemically with thin films of Pt, Ni, Cu, Au, Cd, and Pb (Khader and Saleh 1999). When metal was coated on n-type GaAs, the turn-on potential shifted. A negative direction shift was observed in the order of Pt, Ni, and Cu, and a positive direction shift was observed in the order of Cd and Pb. For metal on p-GaAs, the photocurrent showed an opposite behavior to those of n-GaAs. That is, the turn-on potential of the photocurrent shifts toward the positive direction in the order of Pt, Ni, and Cu and in the negative direction in the order of Cd and Pb. The results are explained by the Schottky barrier model for n-type GaAs and the electrocatalytic model for p-type GaAs.

*InP*

For the photoelectrochemical process of InP, surface oxidization and photocorrosion processes exist like for the other non-oxide materials (Ellis et al. 1977b). Different turn-on potentials of the photocurrent for the p-type InP with and without redox systems existing in the electrolyte were observed (Tubbesing et al. 1986). According to the relative potentials on both sides of the interface, reduction of the redox system occurs via the conduction band. The valence band process did not occur at the p-type InP surface because the hole transfer seems to be blocked by the oxide layer.

The Pt–Rh–Ru island-loaded p-type InP with a Pt counterelectrode in a 1 mol/L HCl plus 2 mol/L KCl aqueous solution showed a 20 $mA/cm^2$ photocurrent density with +0.15 V vs. SCE-applied bias under 84.7 $mW/cm^2$ sunlight illumination (Heller and Vadimsky 1981). The anode voltages for Pt were +0.98 V vs. SCE at the threshold and +1.12 V vs. SCE at 20 $mA/cm^2$. This shows that InP has high performance for the photocathode. The particle loading of Pt, Rh, and Ru on p-type InP enhances the photocurrent density (Heller et al. 1982).

### 9.3.4 Ternary Materials

*$Ga_{0.5}In_{0.5}P$*

The bandgap of the $Ga_{0.5}In_{0.5}P$ is 1.82 to 1.91 eV (direct). The bandgap changes with the ordering or disordering of the In and Ga atoms at the lattice site. The flatband potentials were −1.2 V vs. SCE for the n-type and +0.4 V vs. SCE for p-type in aqueous electrolyte of pH 2.1 (Kocha et al. 1994). The pH dependence of the flatband was −60 mV/pH. Thus, $Ga_{0.5}In_{0.5}P$ can reduce water but cannot oxidize water.

## 9.4 Nitrides (GaN, AlN, and InN)

### 9.4.1 Semiconductor Properties

III-nitride semiconductor materials are direct bandgap materials, and have wide bandgap variations, that is, 6.2 eV for aluminum nitride (AlN), 3.4 eV for gallium nitride (GaN), and 0.65 eV for indium nitride (InN). For the electrochemical studies, GaN is most widely used due to its high-crystal quality and chemical stability. The bandgap of GaN is wider than the visible light range; thus, the study is limited to mechanism analysis. The electrochemical usage of $In_xGa_{1-x}N$ is often attempted to achieve visible light absorption, but the alloyed compound has the weakness of low stability during the electrochemical reaction process. The major properties are summarized in Table 9.3

The III-nitride semiconductor material, especially GaN, shows high resistance in chemical etching compared with other major semiconductors like the elemental group IV, III–V, and II–VI compound materials. The etching for their device processes of other semiconductors is mainly by chemical etching, but reactive ion etching is used for GaN due to its high chemical stability. Generally, AlN also shows high resistance for chemical etching, but that for InN is not high.

Single III-nitride crystal layers are usually grown on a single crystal of sapphire substrates because a GaN bulk crystal is expensive. As the sapphire substrate is an insulator, the contact electrode must be formed on the same surface as the interface between the electrolyte and semiconductor when III-nitrides are used as the photoelectrochemical electrode. The single crystal layer is usually grown by

**Table 9.3** Basic semiconductor and electrochemical properties of III-nitride materials (details are discussed in the text)

| | Bandgap | | Structure | | Dopant | |
|---|---|---|---|---|---|---|
| | d/ind | (eV) | | (nm) | P-type | N-type |
| AlN | d | 6.2 | wz | a: 0.3111<br>c: 0.4978 | Mg | Si |
| GaN | d | 3.4 | wz | a: 0.3189<br>c: 0.5185 | Mg | Si |
| InN | d | 0.65 | wz | a: 0.3544<br>c: 0.5718 | Mg | Si |
| | Band edge potential [*V* vs. SHE] | | | | Slope [*V* vs. SHE/pH] | |
| | Conduction band | | Valence band | | | |
| AlN | (More negative than GaN) | | – | | – | |
| GaN | −0.53[a] | | +2.89 | | −0.055 | |
| InN | (More positive than GaN) | | – | | – | |

Bandgap: *d* direct; structure: *wz* wurtzite (the structure of zinc blende is also observed but usually wurtzite)

[a]The potential of Ag/AgCl of +0.199 V vs. SHE is used

metal-organic chemical vapor deposition (MOVPE) or molecular beam epitaxy (MBE). Relatively high-crystal quality n-type material is grown on sapphire substrate by silicon (Si) doping, but the p-type material with high-crystal quality is difficult to grow due to its deep acceptor level and high volatility of the dopant material, particularly magnesium (Mg). Therefore, a thick p-type single III-nitride crystal is difficult to grow.

Many of the semiconductor characteristics of III-nitrides are different from those of the group IV elemental semiconductor material and III–V and II–VI compound semiconductor materials. This is probably due to the hexagonal structures of III-nitride semiconductor materials, although the structures of most of the other semiconductor materials are cubic (Nakamura and Chichibu 2000). The photoelectrochemical properties of III-nitrides have been well evaluated because evaluation is much easier compared to the other non-oxide semiconductors due to its relatively high stability even in acidic and basic solutions. Since crystal growth is not easy, (0001) III-face single crystals on insulating sapphire substrate are used for the photoelectrochemical evaluations except for the discussion on surface orientations in this section.

### 9.4.2 *Flatband and Band Edge Potentials*

The flatband and band edge potentials are summarized in Table 9.3.

*GaN*

The first report of GaN photoelectrochemical properties was the pH dependence of flatband potentials by Kocha et al. (1995). They found the relation from the Mott–Schottky plot based on impedance measurements to be

$$V_{\mathrm{FB}} = -0.75 - 0.55\,\mathrm{pH}\ (V\ \mathrm{vs.\ SCE}) \tag{9.9}$$

where $V_{\mathrm{FB}}$ is flatband potential in $V$ vs. standard calomel electrode ($V$ vs. SCE) unit. The slope is close to the Nernstian relationship of water of −0.059 V/pH. This Nernstian response is linked to the ratio change of the $H^+$ and $OH^-$ molecule in the Helmholtz layer existing between an electrode and an aqueous electrolyte following the pH value. Since the response is often observed for oxide semiconductors, the GaN surface Helmholtz layer probably consists mainly of oxides and hydroxides of Ga. Using the bandgap of GaN as 3.4 eV, the conduction band edge ($V_{\mathrm{CB}}$) of GaN is −0.49 V vs. standard hydrogen electrode (SHE), and the valence band edge ($V_{\mathrm{VB}}$) is +2.91 V vs. SHE. The $E_{\mathrm{CB}}$ (conduction band edge energy) is $qV_{\mathrm{FB}} \approx 0.05$ (eV) since the used GaN crystal is of the n-type-doping type. The number shows that the GaN is capable of water splitting because the conduction band edge potential is lower than the water reduction and the valence band edge potential is higher than water oxidation for whole pH region.

A similar relationship was reported by the other groups (Huygens et al. 2000). The flatband potential pH dependence was also reported as

$$V_{\mathrm{FB}} = -0.78 - 0.060\mathrm{pH}\ (V\ \mathrm{vs.\ SCE}). \tag{9.10}$$

This relationship was obtained from n-type GaN by Hungens et al. (2000) using the Mott–Schottky plot. The next report was provided by Beach et al. (2003). The flatband potential was reported to be −0.86 V vs. SCE at pH 2, and the slope was −0.047. The reported slope is moderate compared with the others. From the comparison with the onset potential of the photocurrent and the flatband potential, they reported that the onset potential was about a 0.278 V positive shift at pH 0 with almost a similar slope for n-type GaN. The onset potential of p-type GaN was also reported to be a 0.276 V negative shift at pH 0 from an n-type flatband potential, and the slope was −0.063 pH which is much steeper than that for n-type GaN. The flatband potential obtained from the Mott–Schottky plot for p-type GaN was reported as +2.41 V vs. silver chloride electrode (Ag/AgCl) at pH 0.4 (Fujii and Ohkawa 2005). The slope and the band edge potential calculated from the flatband potential are almost coincidental to those for the n-type GaN as shown in Fig. 9.4. The conduction and valence band potentials were obtained from n-type GaN as follows (Fujii and Ohkawa 2006):

$$V_{\mathrm{CB}} = -0.73 - 0.055\mathrm{pH}\ (V\ \mathrm{vs.\ Ag/AgCl}), \tag{9.11}$$

$$V_{\mathrm{FB}} = +2.69 - 0.055\mathrm{pH}\ (V\ \mathrm{vs.\ Ag/AgCl}). \tag{9.12}$$

*$In_xGa_{1-x}N$ and $Al_yGa_{1-y}N$*

The visible light absorption is expected for $In_xGa_{1-x}N$ due to the narrowing of the bandgap. The bandgap becomes narrower with increasing In content. The change of the band edge potential with the composition, especially for the conduction band edge, which is close to the hydrogen generation redox potential (0.0 V vs. SHE at pH 0) for GaN, is very important for discussing complete water splitting using a monolithic semiconductor. The evaluation of the band edge potentials of

**Fig. 9.4** Conduction and valence band potentials for p-type (*closed marks*) and n-type (*open marks*) GaN (After ref. (Fujii and Ohkawa 2005))

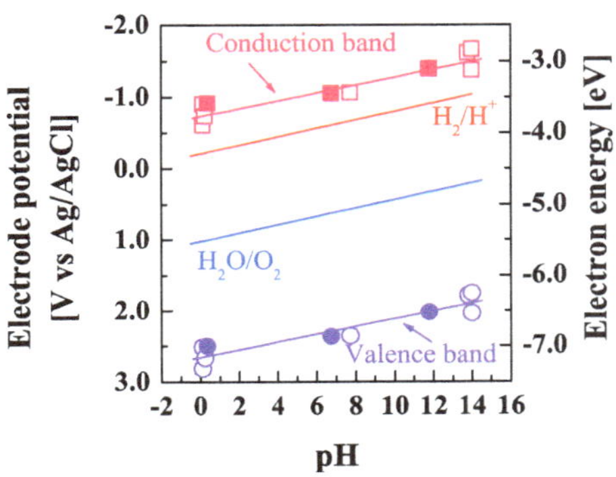

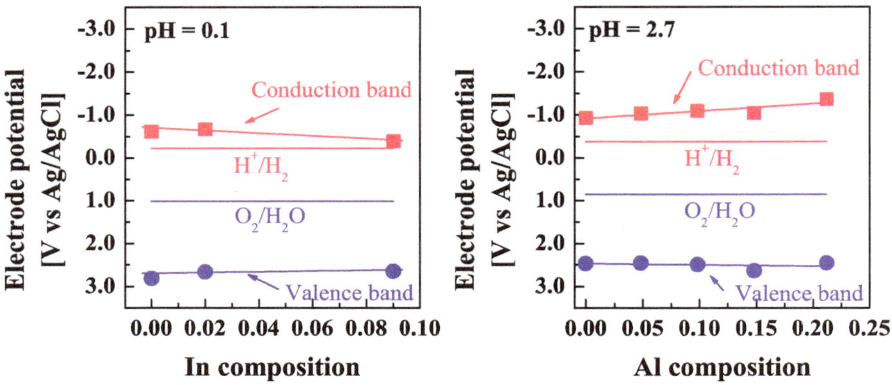


**Fig. 9.5** Conduction and valence band potentials of $In_xGa_{1-x}N$ ($0 \le x \le 0.09$, *left*) and $Al_yGa_{1-y}N$ ($0 \le y \le 0.21$, *right*) (After ref. (Fujii et al. 2007c))

$In_xGa_{1-x}N$ ($0 \le x \le 0.09$) and $Al_yGa_{1-y}N$ ($0 \le y \le 0.21$) shows that the conduction band edge decreases with increasing In composition and increasing Al composition as shown in Fig. 9.5 (Fujii et al. 2007c). The valence band edge does not move much within the experimental range. From the estimation of the slope from the In composition, the conduction band edge crosses the hydrogen generation redox potential at around $x = 0.2$, at the bandgap of around 2.8 eV (442 nm in wavelength). This means that the band edge potentials of $In_xGa_{1-x}N$ have the ability of water splitting with blue-region visible light.

### 9.4.3 *Photocurrent–Voltage Characteristics*

*GaN*

The first investigation on photocurrent–voltage characteristics was performed with n-type GaN in a 1.0 mol/L KOH aqueous solution (Fujii et al. 2005). It was difficult to observe hydrogen generation without bias; thus, +1.0 V vs. counter electrode (CE) was applied to the working electrode (WE) of photo-illuminated n-type GaN. The photocurrent reached at 1.0 mA/cm$^2$ at first, but it decreased drastically over time. Almost half of the photocurrent was used for the photo-anodic corrosion of GaN WE from the analysis of the total current flow and hydrogen generation for 300 min. The GaN corrosion equation is proposed as follows:

$$2\mathrm{GaN(s)} \rightarrow 2\mathrm{Ga}^{3+}(\mathrm{in}\ell) + \mathrm{N_2(g)} + 6\mathrm{e}^-. \quad (9.13)$$

where s is solid, g is gas, and in$\ell$ is ion in liquid. The detailed stability of GaN is discussed in the next section.

From the band edge potentials, GaN is capable of water splitting without bias; however, it has been very difficult to observe water splitting. Thus, the electrolyte and carrier concentration dependence for water splitting was evaluated (Ono et al. 2007; Fujii and Ohkawa 2006; Fujii et al. 2007a). The key properties for the high photocurrent density were the kind of electrolyte and the carrier concentration of GaN. The electrolyte of NaOH aqueous solution and the carrier concentration of GaN with around $1 \times 10^{17}$ cm$^{-3}$ showed the best performance. Hydrogen generation from CE without bias was observed using the n-type GaN with $1.7 \times 10^{17}$ cm$^{-3}$ carrier concentration dipped into 1.0 mol/L HCl aqueous solution. The oxidation reaction occurring at the photo-illuminated GaN WE was not $O_2$ evolution but $Cl_2$ evolution when the HCl aqueous solution was used (Ono et al. 2007). Hydrogen generation in the NaOH aqueous solution was observed after using HCl aqueous solution.

*$In_xGa_{1-x}N$ and $Al_yGa_{1-y}N$*

The photocurrents for n-type $Al_yGa_{1-y}N$ were reported to be much better than that for GaN at the zero bias because the onset potential of photocurrent moved to the negative direction and the photocurrent reached the plateau (Fujii et al. 2007c). The bandgap becomes wider with increasing Al content. This means that the light absorbed by $Al_yGa_{1-y}N$ has much higher energy compared to GaN, with increasing Al composition. Thus, these properties are not of much interest in general because it is opposite to the absorption of visible light. The photocurrent for the n-type $In_xGa_{1-x}N$ at zero bias was much lower than that for GaN, although the light absorption is larger than that for GaN. This is probably because the conduction band edge potential is close to the hydrogen evolution redox potential, which means that it is difficult to overcome the overpotential (Fujii et al. 2007c), or the crystal quality of $In_xGa_{1-x}N$ is not good compared with that for GaN due to the limitation of the crystal growth condition.

Efforts to improve the photocurrent are still continuing due to the possibility of visible light-absorbed hydrogen generation by water splitting. The incident photon conversion efficiency (IPCE) of $x = 0.20$ of the n-type $In_xGa_{1-x}N$ ($0 \leq x \leq 0.20$) was about 9 % from 400 to 430 nm in wavelength under +0.8 V vs. SCE-applied bias using the HBr aqueous solution. The thickness of $In_xGa_{1-x}N$ was 60 nm which was not thick enough to absorb the incident light (Luo et al. 2008). Stability in the 1.0 mol/L HBr aqueous solution (which oxidation reaction was $Br^-$ oxidation) was much better than the 1.0 mol/L HCl or 0.5 mol/L $H_2SO_4$ aqueous solutions. Samples with thickness of 0.2 μm of the n-type $In_xGa_{1-x}N$ ($x = 0.2$ and 0.4, carrier concentration of $2 \times 10^{17}$ cm$^{-3}$) were also evaluated regarding photoelectrochemical properties using the 1.0 mol/L HCl aqueous solution (which oxidation reaction was $Cl^-$ oxidation) (Li et al. 2008). The photocurrent density for $x = 0.4$ $In_xGa_{1-x}N$ was much better with the applied bias of +0.1 to +1.0 V vs. CE and under white light illumination than compared to $x = 0.2$ $In_xGa_{1-x}N$, even the crystal quality of $x = 0.4$ $In_xGa_{1-x}N$ lower than that of $x = 0.2$. The hydrogen generation was observed only for the $x = 0.4$ $In_xGa_{1-x}N$ during the experiments. These results showed that the overlap of the white light illumination and the light absorption by the band plays an important role.

*P-Type III-Nitrides*

The p-type GaN is also an interesting material because the reaction on the surface of GaN is water reduction, and it is expected to eliminate the anodic corrosion. The photocurrent of the p-type GaN is poor and is almost 1/10 of the n-type GaN at zero bias at the beginning (Fujii and Ohkawa 2005).

Attempts to use the p-type III-nitride is continuing especially for the $In_xGa_{1-x}N$ material. The photoelectrochemical properties of p-type $In_xGa_{1-x}N$ ($0 \leq x \leq 0.22$) were evaluated, and the photocurrent density of $In_xGa_{1-x}N$ was found to be much better compared to GaN (Aryal et al. 2010). The stability was also better than that for n-type materials at −0.7 V vs. CE applied bias. The photocurrent density was stable over 24 h except for the first drop with applied bias of −1.2 V vs. CE. Continuous hydrogen bubble generation and no surface corrosion were observed.

## 9.4.4 Stability Improvements by Catalysts and Protection Layers

*GaN*

Even if GaN is stable in acidic and basic chemical solutions, the anodic corrosion of n-type GaN cannot be eliminated. The first systematic evaluation of anodic corrosion of n-type GaN was probably the study on the dependence of aqueous solutions for water splitting with +1.0 V vs. CE (Fujii et al. 2007a). From the evaluation of total current and GaN weight loss, GaN corrosion in basic aqueous solutions was smaller compared to the acidic ones even when the corrosion was enhanced by the bias application. The surface morphologies after 180-min electrochemical reactions were different with the solutions. This also supports the theory that the corrosion mechanisms of acidic and basic solutions are different, which relates to the difference of stable Ga-ions in low ($Ga^{3+}$)- and high-pH ($GaO_3^{3-}$) aqueous solutions.

Thus, the detailed evaluation of photo-anodic corrosion was performed. The time dependence of n-type GaN photocurrent density in 1.0 mol/L NaOH without bias showed a "w-shaped" behavior as shown in Fig. 9.6 (Sato et al. 2009). Even after 1336-min reaction, the surface was still GaN and no sapphire substrate was observed. This means that the photoelectrochemical ability of GaN changes after the reaction. The phenomenon was explained on the basis that the surface defect formation occurred firstly, followed by the disappearance of the surface defect, and the surface orientation change was the final stage from the scanning probe microscope (SPM) and photoluminescence (PL) observation. The light intensity and applied bias dependences were also observed (Koike et al. 2010). The time dependence of photocurrent density was enhanced with increasing illuminated light intensity, but the dependence was completely changed when the applied bias was greater or equal to 0.4 V vs. CE. The dependence with the applied bias lower or equal 0.2 V vs. CE was similar to the zero bias case, and the photoelectrochemical reaction was enhanced with increasing bias similar to the light intensity. However,

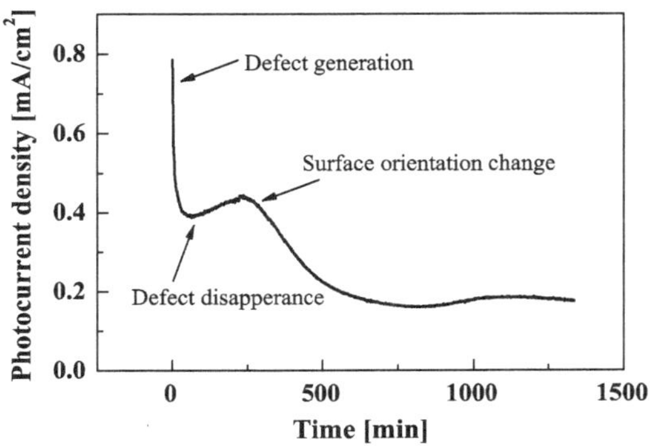


**Fig. 9.6** Time dependence of n-type GaN photocurrent density in 1.0 mol/L NaOH without bias (After ref. (Sato et al. 2009))

the photocurrent density increased drastically over the initial value within 200 min after the evaluation started and approached to almost 0 mA/cm$^2$ after 400 min when the applied bias was grater and equal to 0.4 V vs. CE. This probably indicates that the anodic photocorrosion mechanism changes with applied bias and the mechanism has some threshold energy.

The improvement of photocurrent density time dependence of the n-type GaN with changing GaN properties was also reported (Fujii et al. 2011a, 2012). The changes of the crystal quality, which was evaluated by the value of full-width half maximum (FWHM) of $(10\bar{1}2)$ X-ray rocking curve (XRC), did not affect the time dependence, although the photocurrent density increased with decreasing FWHM of XRC, which means that the crystal quality improved (dislocation density decreased) (Fujii et al. 2011a). The photocurrent density time dependence improved with decreasing concentration of Si, which was the dopant material as a donor for GaN (Fujii et al. 2011a). Since the increasing impurity in a semiconductor is often believed to increase its defects, the results show that some point defect density is related to the photocurrent degradation and dislocations do not relate to the degradation. Some excitons bounded by neutral donors related with stacking imperfection or polytype have been observed after the reactions from precise PL observation, but the detailed origin of the degradation is still obscure (Fujii et al. 2012).

The surface stability with different carrier concentrations of Si-doped n-type GaN was compared by varying the aqueous solutions of the 0.5 mol/L $H_2SO_4$ acidic solution and 1.0 mol/L NaOH basic solution (Koike et al. 2014). Even when the optimized GaN was used, the stability of the NaOH aqueous solution was the best. This shows that the complete suppression of the n-type GaN photo-anodic corrosion is difficult by only changing the GaN properties.

*Catalysts on the GaN Surface*

From the previous section, it is clear that the suppression of a photo-anodic corrosion of n-type GaN requires some other approaches apart from changing the GaN properties. The loading of an $NiO_x$ catalyst on the GaN surface was proposed following the co-catalyst approach of a water-splitting photocatalyst (Hayashi et al. 2012). The n-type GaN with 1 % coverage of about 10 μm diameter $NiO_x$

particle showed good performance not only for the stability but also for the photocurrent density. They reported that the energy conversion efficiency for the $NiO_x$-loaded GaN was almost 10 times larger compared to a reference sample without $NiO_x$ and was 1.4 %. Stable reaction was observed to approach 110 h in the 1.0 mol/L NaOH aqueous solution under 100 mW/cm$^2$ Xe-lamp illumination without bias. Smooth surface was observed after the reaction, although the surface of the reference n-type GaN without $NiO_x$ loading changes to rough with the reaction only after 10 h. During the reaction, not only hydrogen generation from the Pt counterelectrode but also oxygen generation from the $NiO_x$-loaded n-type GaN working electrode was observed. The generation ratio of $H_2$–$O_2$ was not 2:1, and the hydrogen generation was much larger than the oxygen generation.

$NiO_x$-loaded samples were made using simple diluted MOD materials (Kang et al. 2014; Kim et al. 2014). Improvements in the photocurrent density and stability were also observed. The optimized concentration of $NiO_x$ was also reported (Kim et al. 2014). The $NiO_x$ loading for the combination of a 100 nm undoped GaN layer on n-type GaN with carrier concentration of $10^{18}$–$10^{19}$ cm$^{-3}$ was also effective (Kang et al. 2014). The photocurrent density for $NiO_x$-loaded GaN was almost 3.5 times larger than that without $NiO_x$ load when the carrier concentration was optimized. The photocurrent density improvement with $NiO_x$ loading was explained by the change in the band edge energy change. The $NiO_x$ is loaded as small particles; thus, the explanation of detailed mechanisms still requires precise evaluations.

### 9.4.5 Other Properties

*Surface Orientation Dependence*

Electrochemical properties are expected to change with the surface orientation of the single crystal because the molecule absorbed on the surface is different. The absorbed molecule defines some electrochemical properties, like flatband potentials. The photoelectrochemical properties of nonpolar $(11\bar{2}0)$ and semipolar $(11\bar{2}2)$ surfaces were compared with usual (0001) Ga-polar surface of n-type GaN (Fujii et al. 2007b). The flatband potentials were −0.83 V vs. Ag/AgCl for $(11\bar{2}0)$, −0.70 V vs. Ag/AgCl for (0001), and −0.53 V vs. Ag/AgCl for $(11\bar{2}2)$. Since the most negative flatband potential was obtained from the surface of the $(11\bar{2}0)$ sample, the driving force for water reduction was expected to be the highest. (The valence band potential for water oxidation is far from the redox potential; thus, the driving force is believed to be enough.) However, the photocurrent density of $(11\bar{2}0)$ sample was almost 0 mA/cm$^2$ even under the applied bias over +1.5 V vs. Ag/AgCl. The highest photocurrent density was obtained with the sample with (0001) surface. This result shows that the important properties of photocurrent improvement is not only the flatband potential but probably also the adsorbed materials on the surface and the overpotential of the reaction.

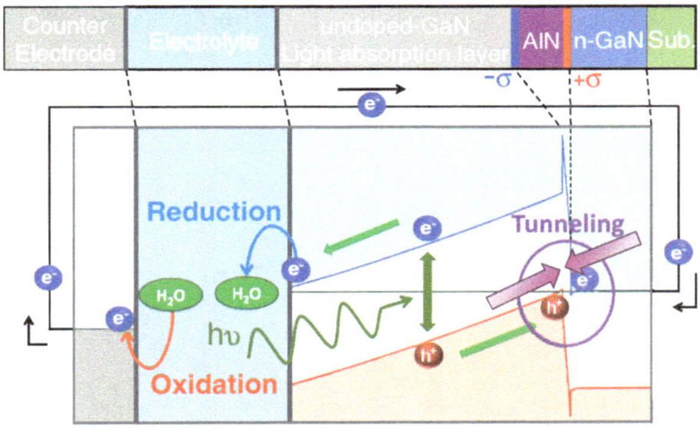


**Fig. 9.7** Schematic working electrode structure for polarization electric field application (After ref. (Nakamura et al. 2014))

*Polarization-Effect Application*

The nanostructure formation for III-nitrides is relatively easy because the crystal growth technique was developed for the electric device application. Thus, the $In_xGa_{1-x}N$/GaN quantum structure could be applied for the photoelectrochemical devices in order to improve its light absorption (Fujii et al. 2011b). Using this quantum structure technique, an undoped GaN/AlN/n-type GaN structure was proposed in order to obtain a p-type semiconductor-like photoelectrochemical response without using a p-type semiconductor as shown in Fig. 9.7 (Nakamura et al. 2014). This structure is an example of a polarization electric field application; thus, the key is the existence of a very thin 4 nm AlN layer and (0001) stack direction. The photoelectrochemical reaction was of p-type performance and was much better than the photoelectrochemical response of p-type GaN (Fujii and Ohkawa 2005).

## 9.5 Chalcogenides (Sulfide, Selenide, and Telluride) and Others

### *9.5.1 Semiconductor Properties*

The II–VI compound semiconductor material is made of zinc (Zn) or cadmium (Cd) for group II and sulfur (S), selenium (Se), or tellurium (Te) for group VI, basically. Iconicity is relatively strong compared with the III–V compound semiconductor material or group IV semiconductors, which have completely covalent characteristics. The bandgaps are 3.6 eV (ZnS), 2.7 eV (ZnSe), 2.25 eV (ZnTe), 2.42 eV (CdS), 1.73 eV (CdSe), and 1.49 eV (CdTe). The listed semiconductors have a direct bandgap and basically n-type. It is difficult to grow the crystal with the p-type II–VI semiconductor. The major properties are summarized in Table 9.4.

The single crystal is the common structure for II–VI semiconductors due to its application of electronic devices, and the chemical stability for II–VI

**Table 9.4** Basic semiconductor and electrochemical properties of chalcogenides (II–VI material) (details are discussed in the text)

| | Bandgap | | Structure | | Dopant | |
|---|---|---|---|---|---|---|
| | d/ind | (eV) | | (nm) | P-type | N-type |
| ZnS | d | 3.6 | zb | 0.541 | – | – |
| ZnSe | d | 2.7 | zb | 0.567 | – | Al |
| ZnTe | d | 2.25 | zb | 0.610 | – | Al |
| CdS | d | 2.42 | wz | a: 0.4135 | – | – |
| | | | | c: 0.6749 | | |
| CdSe | d | 1.73 | wz | a: 0.430 | P | – |
| | | | | c: 0.702 | | |
| CdTe | d | 1.49 | zb | 0.648 | P | Al |
| | Band edge potential [*V* vs. SHE] | | | | Slope [*V* vs. SHE/pH] | |
| | Conduction band | | Valence band | | | |
| ZnS | – | | – | | – | |
| ZnSe | – | | – | | – | |
| ZnTe | – | | – | | – | |
| CdS | −0.55 ~ −0.75[a] | | +1.87 ~ +1.67[a] | | – | |
| CdSe | – | | – | | – | |
| Cdte | – | | – | | – | |

Bandgap: *d* direct, *ind* indirect; structure: *zb* zinc blende, *wz* wurtzite
Dopant: Li is also used for p-type; Cl is also used for n-type
[a]Flatband for n-type CdS is regarded as the conduction band potential. Conflicting results are observed due to its strong interaction with ions. Flatband potential decreases with the sulfur the sulfur concentration: (in here, [S] means sulfur concentration)

semiconductors in aqueous solutions is relatively weak due to its strong ionic properties. Thus, protection for corrosion is required for II–VI semiconductors similar to the other III–V compound semiconductor and elemental semiconductor materials.

The layer structure semiconductor of $WSe_2$ and chalcopyrite of II–III–$VI_2$ ($Cu(Ga_xIn_{1-x})(S_ySe_{1-y})_2$) are also commented here. These are basically polycrystalline film materials.

### 9.5.2 Flatband and Band Edge Potentials

The flatband and band edge potentials are summarized in Table 9.4.

*Chalcogenides*

The flatband potential of CdX ($X = $ S, Se, Te) was reported to show a negative shift of 0.5 V with the existence of the redox of $X^{2-}/X_n^{2-}$ ($X = $ S, Se, Te) at pH 14 (Ellis et al. 1977a). This indicates that the CdX strongly interacts with the X-ions in the electrolyte.

The flatband potential for n-type CdS is complicated due to its strong interaction with the existence ions in an electrolyte (Gomes and Cardon 1982). Thus, conflicting results for flatband potential were summarized in the reference as follows. The flatband potentials for n-type CdS (donor concentration is between $10^{17}$ to $10^{18}$ $cm^{-3}$) was approximately −0.55 V and −0.75 V vs. SHE at pH = 4.8 and 9.2, respectively. A conflicting report also exists, that is, a pH-independent value of a flatband potential with −0.64 V vs. SHE in different electrolytes for the (0001) face of CdS (donor concentration is $7 \times 10^{17}$ $cm^{-3}$). In addition, an increase of −0.06 V per factor ten in sulfide concentration is observed (e.g., $Na_2S$ concentration) at constant pH, and this relationship was only observed on the (0001) Cd surface and not on the $(000\overline{1})$ S surface. The opposite effect of $Cd^{2+}$ ion was also observed. The flatband potential shifts over −0.055 V/pH in sulfide solutions were observed again in a basic aqueous solution, which coincides with the former case but conflicts with the latter case. These complicated experimental results for CdS demonstrate that the orientation and the condition of the surface as well as the composition of the electrolyte may have strong influence on the structure of surface adsorption and hence the value of the flatband potential. These effects were not remarkable for oxide or III–V semiconductor materials, possibly because the surface of CdS is primarily covered with ionized hydroxide groups.

*Layered Semiconductor of $WSe_2$*

Interesting properties were reported for layered transition metal compound of $WSe_2$. The flatband potential for the layer structure-dominant $WSe_2$ was found to be pH independent (Lewerenz et al. 1984; McEvoy et al. 1985). Since the pH dependence of flatband potential was observed when there are many steps present from one layer to the other, the pH independence can be explained by the fact that the metal atom was covered by the Se at the layered surface (Lewerenz et al. 1984). The conduction and valence band edge potentials were reported to be −0.4 and +0.75 V vs. standard calomel electrode (SCE) in the dark and +0.25 and +1.4 V vs. SCE under illumination, respectively (McEvoy et al. 1985). It was explained that the band edges at the semiconductor surface become unpinned (probably due to the changes in the interaction of the absorbed molecule) under illumination and are displaced downward. This property was also observed for $MoSe_2$ (McEvoy et al. 1985).

### 9.5.3 Photo-corrosion and Stability Improvements by Catalysts and Protection Layers

*Chalcogenides*

The bandgap and flatband potentials of n-type CdS are 2.4 eV and around −0.65 eV vs. SHE, respectively. Thus, CdS is one of the most promising semiconductors for photoelectrochemical water splitting by monolithic material using visible light. However, the CdS is not stable in aqueous solutions and undergoes self-anodic dissolution under light illumination leading to the formation of S (Inoue et al. 1977; Wilson and Park 1982). The reaction is as follows:

$$\begin{aligned} &\mathrm{CdS(s)} \rightarrow \mathrm{Cd^{2+}(in\ell)} + \mathrm{S(in\ell)} + 2\mathrm{e}^- \\ &E^0(\mathrm{vs.\,RHE, pH7}) = +0.32[V]. \end{aligned} \tag{9.14}$$

where s is solid and in$\ell$ is ion or atom in liquid. This dissolution can be suppressed by adding reducing reagents such as $S^{2-}$, $SO_3{}^{2-}$, or $S_2O_3{}^{2-}$ by the following reactions:

$$2\mathrm{S^{2-}(in\ell)} \rightarrow \mathrm{S_2^{2-}(in\ell)} + 2\mathrm{e}^-, \tag{9.15}$$

$$\mathrm{SO_3^{2-}(in\ell)} + 2\mathrm{OH^-(in\ell)} \rightarrow \mathrm{SO_4^{2-}(in\ell)} + \mathrm{H_2O(\ell)} + 2\mathrm{e}^-, \tag{9.16}$$

$$2\mathrm{SO_3^{2-}(in\ell)} \rightarrow \mathrm{S_2O_6^{2-}(in\ell)} + 2\mathrm{e}^-, \tag{9.17}$$

$$2\mathrm{S_2^{2-}(in\ell)} + \mathrm{SO_3^{2-}(in\ell)} \rightarrow \mathrm{S_2O_3^{2-}(in\ell)} + \mathrm{S^{2-}(in\ell)}. \tag{9.18}$$

where $\ell$ is liquid and in$\ell$ is ion in liquid.

Polymer film incorporation and photoinduced platinization of CdS have been reported to be effective for protecting the surface from the photodecomposition in an electrolyte (Frank and Honda 1982). Nearly 68 % of the photo-generated holes are used for oxygen evolution, while the remaining 32 % are used to produce $Cd^{2+}$ by polymer coating, whereas almost 99 % holes are used for $Cd^{2+}$ formation without coating.

The photoanodes of n-type CdSe also suffer from photocorrosion (Ellis et al. 1976; Ellis et al. 1977c):

$$\mathrm{CdSe(s)} \rightarrow \mathrm{Cd^{2+}(in\ell)} + \mathrm{Se(in\ell)} + 2\mathrm{e}^-. \tag{9.19}$$

where s is solid and in$\ell$ is ion or atom in liquid. This dissolution can be also suppressed by adding reducing reagents such as $Se^{2-}$, $SeO_3{}^{2-}$, and $Se_2O_3{}^{2-}$ by the following reactions.

The n-type CdTe has also photocorrosive nature; however, solar hydrogen generation was achieved with surface modification (p. 446 of Grimes et al. (2007)).

Interestingly, chalcogenide semiconductors can be used as the sensitizers of $TiO_2$ photoanodes, which provide high photocurrents in the presence of a hole scavenger. The $TiO_2$ hollow structure loaded with CdS/CdSe nanoparticles was used as the photoanode of water splitting (Rodenas et al. 2013). The PbS/CdS-loaded $TiO_2$ photoanode is reported to respond to infrared (IR) light (Trevisan et al. 2013).

*Chalcopyrite*

Semiconducting chalcopyrite materials of n-type $Cu(Ga_xIn_{1-x})(S_ySe_{1-y})_2$ are well-known solar cell materials. A thin film of these materials is formed by an electro-deposition process, which has a bandgap of approximately 1.6 to 2.0 eV.

Photoelectrochemical hydrogen generation was observed in 0.05 mol/L $Na_2SO_3$ in pH 10 buffer (Leisch et al. 2004). The flatband potential was shifted with the S composition. A similar work has also been performed using p-type $Cu(Ga_xIn_{1-x})Se_2$ thin-film photoelectrode in 0.5 mol/L $H_2SO_4$ (Valderrama et al. 2005). The film has the flatband potential of +0.26 V vs. SCE in 0.5 mol/L $H_2SO_4$, which means that the film can reduce water (hydrogen evolution).

## 9.6 Tandem Structures

Tandem structures for non-oxide semiconductor materials are used for improving water-splitting efficiency. The first attempted trial of this tandem structure was reported by Khaselev and Turner (Khaselev and Turner 1998). The structure is similar to the tandem-type solar cells and was p-GaAs/n-GaAs/p-$Ga_{0.52}In_{0.48}P$. The top layer of the photocathode interfacing with 3 mol/L $H_2SO_4$ aqueous solution was p-$Ga_{0.52}In_{0.48}P$. The surface was coated with a platinum (Pt) catalyst, and a tungsten lamp was used for the illumination with about 11 suns. The conversion efficiency of light-to-hydrogen energy was reported to be 12.4 %.

The structure described in the next tandem structure report that was presented after Khaselev and Turner was using n-Si/p-Si/n-$Al_{0.15}Ga_{0.85}As$/p-$Al_{0.15}Ga_{0.85}As$ tandem structure with a Pt black as a cathode and $RuO_2$ as an anode in 1 mol/L $HClO_4$ under AM1.5G simulated light (Licht et al. 2000). The uniqueness of this structure is that the photo-illuminated area is not in an electrolyte, and only conducting electrodes are dipped in electrolyte, which is similar to a structure of electrochemical cell. The conversion efficiency from the light to hydrogen was 18.3 %. This structure showed a new developing direction, which is a structure with a photo-illuminated area being out of electrolyte, to perform water splitting by solar light.

The next report on semiconductor materials dipped into electrolyte was about 10 years after the second report. The structure is well known as "artificial leaf," which consists of a three-series junctions of "n/p-amorphous silicon (Si) solar cell" connected with NiMoZn-loaded Ni metal as the cathode (Reece et al. 2011). In one of the device configurations, the cathode was wired and located in front and at the back of the anode for a second configuration. The surface of the n-type Si of the anode was covered by $(InSn)O_x$ (ITO) and was deposited by Co-oxygen evolving complex (Co-OEC) on the top. The efficiency was 4.7 % for the wired configuration and 2.5 % for backside cathode-type water splitting.

The oxide–non-oxide hybrid system of tungsten (W) gradient-doped $BiVO_4$ connected with 2-tandem amorphous-Si p/n-junction solar cells was also reported (Abdi et al. 2013). The catalyst of cobalt phosphate (Co–Pi) on the $BiVO_4$ surface was used for the anode, and the Pt cathode was wire connected to the Si solar cell. The electrolyte was a 0.1 mol/L $KH_2PO_4$ buffer aqueous solution of pH 7.3. The solar-to-hydrogen conversion efficiency was obtained to be 4.9 % under AM1.5 irradiation. Further details of this study are given in Chap. 8.

The efficiency jumped up to 10.5 % from the next report (Verlage et al. 2015). The cathode was the tandem structure of $In_{0.48}Ga_{0.52}P$ and GaAs. The $In_{0.48}Ga_{0.52}P$ surface was coated by $TiO_2$ and loaded by Ni in order to improve hydrogen generation. The anode was NiMo metal electrode. The cell was separated to the anode and cathode region and connected by an anion exchange membrane (AEM). The electrolytes were the same for both the anode and cathode and were 1.0 mol/L KOH. Hydrogen and oxygen generation was clearly observed, and it was stable over 40 h.

The tandem structure of n/p-$Ga_xIn_{1-x}P$ (bandgap, 1.78 eV) and n/p-$Ga_yIn_{1-y}As$ (bandgap, 1.26 eV) on Ge substrate was created as a photoelectrochemical electrode (May et al. 2015). This has the same structure of the three-tandem-type solar cells. The cathode and the anode were $AlInPO_z$ oxidized on the semiconductor surface and $RuO_x$, respectively. The solar-to-hydrogen conversion efficiency was over 14 % in 1 mol/L $HCLO_4$ aqueous solution under AM1.5G irradiation.

The tandem structure with a photo-illuminated area being not in the electrolyte showed also progress after the first report of 2000 (Licht et al. 2000). Water splitting was triggered by the electrochemical cell with a Pt cathode and Pt anode operated by the three-series connection of $CuIn_xGa_{1-x}Se_2$ (CIGS) solar cells without light concentration (Jacobsson et al. 2013). The solar-to-hydrogen energy conversion efficiency was 11 % with the Pt electrode and over 10 % with a Mo current transporter and a Pt catalyst under AM1.5G in 3 mol/L $H_2SO_4$.

A concentrated solar cell configuration with a tandem structure was also used. The photo-illuminated structure was the combination of n/p-$In_{0.35}Ga_{0.65}P$ on n/p-$Ga_{0.83}In_{0.17}As$ tandem solar cell (Peharz et al. 2007). The solar cell was directly mounted onto an electrochemical cell of Pt cathode and Ir anode with a polymer exchange membrane (PEM). Since this structure uses the electrochemical cell, the electrolytic solution was PEM, and the reactant was just pure water. The solar-to-hydrogen energy conversion efficiency was reported to be 18.1 % by enthalpy change in 500 times concentrated sunlight. This can be converted to 15.1 % using Gibbs energy which is commonly used.

The $In_xGa_{1-x}P/Ga_yIn_{1-y}As/Ge$ triple junction solar cell and PEM-type electrochemical cell connected by wire were also used for water splitting (Fujii et al. 2013). The solar-to-hydrogen energy conversion efficiency was 15.3 % under ten times concentrated AM1.5G light. The efficiency improved using high-performance solar cell and system optimization. The efficiency was 24.4 % in sunlight (Nakamura et al. 2015).

A similar solar cell structure with high-performance solar cell under 100 suns using a Ni electrode placed inside a 1 mol/L borate buffer (pH 9.2) with 0.23 mol/L $Na_2SO_4$ was also reported to have the light-to-hydrogen energy conversion efficiency of 22.4 % (Bonke et al. 2015).

## 9.7 Concluding Remarks

This section discussed the photoelectrochemical properties of non-oxide semiconductors, such as group IV element semiconductor materials of Si, Ge, and C; group III–V compound semiconductor materials consisting of arsenide, phosphide, and nitride; and group II–VI compound semiconductor material consisting of chalcogenide and chalcopyrite. Non-oxide semiconductors have a wide range of bandgap and band edge potentials. In some of the materials, the conduction band edge potential is more negative than that of the hydrogen generation redox potential, and the valence band edge potential is more positive than that of the oxygen generation redox potential, which means that the material can split water. Since these properties are very useful, photoelectrochemical electrodes are often applied in non-oxide semiconductors. However, almost all non-oxide semiconductors corrode chemically in an acidic or a basic aqueous solution without any bias application. Even chemically stable materials like GaP, $Ga_{0.48}In_{0.52}P$, and GaN corrode photoelectrochemically since the element is not the most oxidized state. Protection of the semiconductor surface is thus indispensable to use photoelectrodes for these non-oxide semiconductors. Oxide protection layers and oxide catalysts were proposed for photoanode, and metals were proposed for photocathode, but the detailed mechanisms are unknown.

## References

Abdi FF, Han L, Smets AHM, Zeman M, Dam B, van de Krol R (2013) Efficient solar water splitting by enhanced charge separation in a bismuth vanadate-silicon tandem photoelectrode. Nat Commun 4:2195. doi:10.1038/ncomms3195

Abruna HD, Bard AJ (1981) Semiconductor electrodes. 40. Photoassisted hydrogen evolution at poly(benzyl viologen)-coated p-type silicon electrodes. J Am Chem Soc 103(23):6898–6901. doi:10.1021/ja00413a021

Allongue P, Blonkowski S (1991) Corrosion of III-V compounds; a comparative study of GaAs and InP: Part I. Electrochemical characterization based on Tafel plot measurements. J Electroanal Chem Interf Electrochem 316(1–2):57–77. doi:10.1016/0022-0728(91)87036-4, http://dx.doi.org

Aryal K, Pantha BN, Li J, Lin JY, Jiang HX (2010) Hydrogen generation by solar water splitting using p-InGaN photoelectrochemical cells. Appl Phys Lett 96(5):052110. doi:10.1063/1.3304786, http://dx.doi.org

Beach JD, Collins RT, Turner JA (2003) Band-edge potentials of n-type and p-type GaN. J Electrochem Soc 150(7):A899–A904. doi:10.1149/1.1577542

Boddy PJ, Brattain WH (1963) The distribution of potential at the germanium aqueous electrolyte interface. J Electrochem Soc 110(6):570–576. doi:10.1149/1.2425816

Bonke SA, Wiechen M, MacFarlane DR, Spiccia L (2015) Renewable fuels from concentrated solar power: towards practical artificial photosynthesis. Energ Environ Sci 8(9):2791–2796. doi:10.1039/C5EE02214B

Brattain WH, Boddy PJ (1962) The interface between germanium and a purified neutral electrolyte. J Electrochem Soc 109(7):574–582. doi:10.1149/1.2425500

Chen S, Wang L-W (2012) Thermodynamic oxidation and reduction potentials of photocatalytic semiconductors in aqueous solution. Chem Mater 24(18):3659–3666. doi:10.1021/cm302533s

Contractor AQ, Bockris JOM (1984) Investigation of a protective conducting silica film on n-silicon. Electrochim Acta 29(10):1427–1434. doi:10.1016/0013-4686(84)87022-X, http://dx.doi.org

Dasgupta NP, Liu C, Andrews S, Prinz FB, Yang P (2013) Atomic layer deposition of platinum catalysts on nanowire surfaces for photoelectrochemical water reduction. J Am Chem Soc 135 (35):12932–12935. doi:10.1021/ja405680p

Dominey RN, Lewis NS, Bruce JA, Bookbinder DC, Wrighton MS (1982) Improvement of photoelectrochemical hydrogen generation by surface modification of p-type silicon semiconductor photocathodes. J Am Chem Soc 104(2):467–482. doi:10.1021/ja00366a016

Ellis AB, Kaiser SW, Wrighton MS (1976) Optical to electrical energy conversion. Characterization of cadmium sulfide and cadmium selenide based photoelectrochemical cells. J Am Chem Soc 98(22):6855–6866. doi:10.1021/ja00438a016

Ellis AB, Bolts JM, Kaiser SW, Wrighton MS (1977a) Study of n-type gallium arsenide- and gallium phosphide-based photoelectrochemical cells. Stabilization by kinetic control and conversion of optical energy to electricity. J Am Chem Soc 99(9):2848–2854. doi:10.1021/ja00451a002

Ellis AB, Bolts JM, Wrighton MS (1977b) Characterization of n-type semiconducting indium phosphide photoelectrodes: stabilization to photoanodic dissolution in aqueous solutions of telluride and ditelluride ions. J Electrochem Soc 124(10):1603–1607. doi:10.1149/1.2133118

Ellis AB, Kaiser SW, Bolts JM, Wrighton MS (1977c) Study of n-type semiconducting cadmium chalcogenide-based photoelectrochemical cells employing polychalcogenide electrolytes. J Am Chem Soc 99(9):2839–2848. doi:10.1021/ja00451a001

Frank AJ, Honda K (1982) Visible-light-induced water cleavage and stabilization of n-type cadmium sulfide to photocorrosion with surface-attached polypyrrole-catalyst coating. J Phys Chem 86(11):1933–1935. doi:10.1021/j100208a005

Frese KW, Madou MJ, Morrison SR (1980) Investigation of photoelectrochemical corrosion of semiconductors. J Phys Chem 84(24):3172–3178. doi:10.1021/j100461a008

Frese KW, Madou MJ, Morrison SR (1981a) Investigation of photoelectrochemical corrosion of semiconductors: II. Kinetic analysis of corrosion-competition reactions on "n-GaAS". J Electrochem Soc 128(7):1527–1531. doi:10.1149/1.2127676

Frese KW, Madou MJ, Morrison SR (1981b) Investigation of photoelectrochemical corrosion of semiconductors: III. Effects of metal layer on stability of GaAs. J Electrochem Soc 128 (9):1939–1943. doi:10.1149/1.2127770

Fujii K, Ohkawa K (2005) Photoelectrochemical properties of p-type GaN in comparison with n-type GaN. Jpn J Appl Phys 44(7L):L909

Fujii K, Ohkawa K (2006) Bias-assisted H2 Gas generation in HCl and KOH solutions using n-type GaN photoelectrode. J Electrochem Soc 153(3):A468–A471. doi:10.1149/1.2161572

Fujii K, Karasawa T, Ohkawa K (2005) Hydrogen gas generation by splitting aqueous water using n-type GaN photoelectrode with anodic oxidation. Jpn J Appl Phys 44(4L):L543

Fujii K, Ito T, Ono M, Iwaki Y, Yao T, Ohkawa K (2007a) Investigation of surface morphology of n-type GaN after photoelectrochemical reaction in various solutions for H2 gas generation. Phys Status Solidi C 4(7):2650–2653. doi:10.1002/pssc.200674917

Fujii K, Iwaki Y, Masui H, Baker TJ, Iza M, Sato H, Kaeding J, Yao T, Speck JS, DenBaars SP, Nakamura S, Ohkawa K (2007b) Photoelectrochemical properties of nonpolar and semipolar GaN. Jpn J Appl Phys 46(10R):6573

Fujii K, Ono M, Ito T, Iwaki Y, Hirako A, Ohkawa K (2007c) Band-edge energies and photoelectrochemical properties of n-type Al x Ga1 − x N and in y Ga1 − y N alloys. J Electrochem Soc 154(2):B175–B179. doi:10.1149/1.2402104

Fujii K, Koike K, Atsumi M, Goto T, Itoh T, Yao T (2011a) Time dependence of water-reducing photocurrent with change of the characteristics of n-type GaN photo-illuminated working electrodes. Phys Status Solidi C 8(7-8):2457–2459. doi:10.1002/pssc.201000937

Fujii K, Nakamura S, Yokojima S, Goto T, Yao T, Sugiyama M, Nakano Y (2011b) Photoelectrochemical properties of in x Ga1–x N/GaN multiquantum well structures in depletion layers. J Phys Chem C 115(50):25165–25169

Fujii K, Koike K, Atsumi M, Goto T, Itoh T, Yao T (2012) Photoluminescence changes in n-type GaN samples after photoelectrochemical treatment. Phys Status Solidi C 9(3-4):715–718. doi:10.1002/pssc.201100310

Fujii K, Nakamura S, Sugiyama M, Watanabe K, Bagheri B, Nakano Y (2013) Characteristics of hydrogen generation from water splitting by polymer electrolyte electrochemical cell directly connected with concentrated photovoltaic cell. Int J Hydrog Energy 38(34):14424–14432. doi:10.1016/j.ijhydene.2013.07.010, http://dx.doi.org

Gerischer H (1977) On the stability of semiconductor electrodes against photodecomposition. J Electroanal Chem Interf Electrochem 82(1–2):133–143. doi:10.1016/S0022-0728(77)80253-2, http://dx.doi.org

Gomes WP, Cardon F (1982) Electron energy levels in semiconductor electrochemistry. Prog Surf Sci 12(2):155–215. doi:10.1016/0079-6816(82)90002-8, http://dx.doi.org

Goossens A, Schoonman J (1992) An impedance study of boron phosphide semiconductor electrodes. J Electrochem Soc 139(3):893–900. doi:10.1149/1.2069321

Grimes CA, Varghese OK, Ranjan S (2007) Light, water, hydrogen—the solar generation of hydrogen by water photoelectrolysis -. Springer, New York

Harten HU, Memming R (1962) Potential distribution at the germanium electrolyte interface. Phys Lett 3(2):95–96. doi:10.1016/0031-9163(62)90020-3, http://dx.doi.org

Hayashi T, Deura M, Ohkawa K (2012) High stability and efficiency of GaN photocatalyst for hydrogen generation from water. Jpn J Appl Phys 51:112601. doi:10.1143/jjap.51.112601

Heller A, Vadimsky RG (1981) Efficient solar to chemical conversion: 12% efficient photoassisted electrolysis in the [*p*-type InP(Ru)]/HCl-KCl/Pt(Rh) cell. Phys Rev Lett 46(17):1153–1156

Heller A, Aharon-Shalom E, Bonner WA, Miller B (1982) Hydrogen-evolving semiconductor photocathodes: nature of the junction and function of the platinum group metal catalyst. J Am Chem Soc 104(25):6942–6948. doi:10.1021/ja00389a010

Huygens IM, Strubbe K, Gomes WP (2000) Electrochemistry and photoetching of n-GaN. J Electrochem Soc 147(5):1797–1802. doi:10.1149/1.1393436

Inoue T, Watanabe T, Fujishima A, Ki H, Kohayakawa K (1977) Suppression of surface dissolution of CdS photoanode by reducing agents. J Electrochem Soc 124(5):719–722. doi:10.1149/1.2133392

Jacobsson TJ, Fjallstrom V, Sahlberg M, Edoff M, Edvinsson T (2013) A monolithic device for solar water splitting based on series interconnected thin film absorbers reaching over 10% solar-to-hydrogen efficiency. Energ Environ Sci 6(12):3676–3683. doi:10.1039/C3EE42519C

Kainthla RC, Zelenay B, Bockris JOM (1986) Protection of n-Si photoanode against photocorrosion in photoelectrochemical cell for water electrolysis. J Electrochem Soc 133 (2):248–253. doi:10.1149/1.2108556

Kang J-H, Kim SH, Ebaid M, Lee JK, Ryu S-W (2014) Efficient photoelectrochemical water splitting by a doping-controlled GaN photoanode coated with NiO cocatalyst. Acta Mater 79:188–193. doi:10.1016/j.actamat.2014.07.032, http://dx.doi.org

Khader MM (1996) Surface arsenic enrichment of n-GaAs photoanodes in concentrated acidic chloride solutions. Langmuir 12(4):1056–1060. doi:10.1021/la940895r

Khader MM, Saleh MM (1999) Comparative study between the photoelectrochemical behaviors of metal-loaded n- and p-GaAs. Thin Solid Films 349(1–2):165–170. doi:10.1016/S0040-6090(99)00224-2, http://dx.doi.org

Khader MM, Nasser SA, Hannout MM, El-Dessouki MS (1993) Photoelectrochemical dissociation of water at copper-doped p-GaAs electrodes. Int J Hydrog Energy 18(11):921–924. doi:10.1016/0360-3199(93)90062-F, http://dx.doi.org

Khader MM, Hannout MM, El-Dessouki MS (1996) Catalytic effects for hydrogen photogeneration due to metallic deposition on P-GaAs. Int J Hydrog Energy 21(7):547–553. doi:10.1016/0360-3199(95)00118-2, http://dx.doi.org

Khaselev O, Turner JA (1998) A monolithic photovoltaic-photoelectrochemical device for hydrogen production via water splitting. Science 280(5362):425–427. doi:10.1126/science.280.5362.425

Kim SH, Ebaid M, Kang J-H, Ryu S-W (2014) Improved efficiency and stability of GaN photoanode in photoelectrochemical water splitting by NiO cocatalyst. Appl Surf Sci 305:638–641. doi:10.1016/j.apsusc.2014.03.151

Kocha SS, Turner JA, Nozik AJ (1994) Study of the Schottky barrier and determination of the energetic positions of band edges at the n- and p-type gallium indium phosphide electrode | electrolyte interface. J Electroanal Chem 367(1–2):27–30. doi:10.1016/0022-0728(93)03020-P, http://dx.doi.org

Kocha SS, Peterson MW, Arent DJ, Redwing JM, Tischler MA, Turner JA (1995) Electrochemical investigation of the gallium nitride-aqueous electrolyte interface. J Electrochem Soc 142(12): L238–L240. doi:10.1149/1.2048511

Koike K, Sato K, Fujii K, Goto T, Yao T (2010) Time variation of GaN photoelectrochemical reactions affected by light intensity and applied bias. Phys Status Solidi C 7(7-8):2221–2223. doi:10.1002/pssc.200983450

Koike K, Nakamura A, Sugiyama M, Nakano Y, Fujii K (2014) Surface stability of n-type GaN depending on carrier concentration and electrolytes under photoelectrochemical reactions. Phys Status Solidi C 11(3-4):821–823. doi:10.1002/pssc.201300466

Kraft A (2007) Doped diamond: a compact review on a new, versatile electrode material. Int J Electrochem Sci 2(5):355–385

Leisch JE, Bhattacharya RN, Teeter G, Turner JA (2004) Preparation and characterization of Cu (In, Ga)(Se, S)2 thin films from electrodeposited precursors for hydrogen production. Sol Energy Mater Sol Cells 81(2):249–259. doi:10.1016/j.solmat.2003.11.006, http://dx.doi.org

Lewerenz HJ, Gerischer H, Lübke M (1984) Photoelectrochemistry of WSe2 electrodes: comparison of stepped and smooth surfaces. J Electrochem Soc 131(1):100–104. doi:10.1149/1.2115467

Li J, Lin JY, Jiang HX (2008) Direct hydrogen gas generation by using InGaN epilayers as working electrodes. Appl Phys Lett 93(16):162107. doi:10.1063/1.3006332, http://dx.doi.org

Licht S, Wang B, Mukerji S, Soga T, Umeno M, Tributsch H (2000) Efficient solar water splitting, exemplified by RuO2-catalyzed AlGaAs/Si photoelectrolysis. J Phys Chem B 104 (38):8920–8924. doi:10.1021/jp002083b

Lin Y, Kapadia R, Yang J, Zheng M, Chen K, Hettick M, Yin X, Battaglia C, Sharp ID, Ager JW, Javey A (2015) Role of TiO2 surface passivation on improving the performance of p-InP photocathodes. J Phys Chem C 119(5):2308–2313. doi:10.1021/jp5107313

Luo W, Liu B, Li Z, Xie Z, Chen D, Zou Z, Zhang R (2008) Stable response to visible light of InGaN photoelectrodes. Appl Phys Lett 92(26):262110. doi:10.1063/1.2955828, http://dx.doi.org

May MM, Lewerenz H-J, Lackner D, Dimroth F, Hannappel T (2015) Efficient direct solar-to-hydrogen conversion by in situ interface transformation of a tandem structure. Nat Commun 6:8286. doi:10.1038/ncomms9286

McEvoy AJ, Etman M, Hemming M (1985) Interface charging and intercalations effects on d-band transition metal diselenide photoelectrodes. J Electroanal Chem Interf Electrochem 190 (1–2):225–241. doi:10.1016/0022-0728(85)80089-9, http://dx.doi.org

Memming R (1969) Mechanism of the electrochemical reduction of persulfates and hydrogen peroxide. J Electrochem Soc 116(6):785–790. doi:10.1149/1.2412052

Memming R (2008) Semiconductor electrochemistry. John Wiley & Sons, Hoboken, NJ

Memming R (2015) Semiconductor electrochemistry, 2nd edn. John Wiley & Sons Inc, Hoboken, NJ

Memming R, Schwandt G (1966a) Anodic dissolution of silicon in hydrofluoric acid solutions. Surf Sci 4(2):109–124. doi:10.1016/0039-6028(66)90071-9, http://dx.doi.org

Memming R, Schwandt G (1966b) Potential distribution and formation of surface states at the silicon-electrolyte interface. Surf Sci 5(1):97–110. doi:10.1016/0039-6028(66)90052-5, http://dx.doi.org

Menezes S, Miller B (1983) Surface and redox reactions at GaAs in various electrolytes. J Electrochem Soc 130(2):517–523. doi:10.1149/1.2119742
Mills A, Le Hunte S (1997) An overview of semiconductor photocatalysis. J Photochem Photobiol A Chem 108(1):1–35. doi:10.1016/S1010-6030(97)00118-4, http://dx.doi.org
Mukherjee J, Erickson B, Maldonado S (2010) Physicochemical and electrochemical properties of etched GaP(111)a and GaP(111)B surfaces. J Electrochem Soc 157(4):H487–H495. doi:10.1149/1.3314305
Nakamura S, Chichibu SF (2000) Introduction to nitride semiconductor blue lasers and light emitting diodes. CRC Press, Boca Raton
Nakamura A, Fujii K, Sugiyama M, Nakano Y (2014) A nitride based polarization-engineered photocathode for water splitting without a p-type semiconductor. Phys Chem Chem Phys 16 (29):15326–15330. doi:10.1039/C4CP01599A
Nakamura A, Ota Y, Koike K, Hidaka Y, Nishioka K, Sugiyama M, Fujii K (2015) A 24.4% solar to hydrogen energy conversion efficiency by combining concentrator photovoltaic modules and electrochemical cells. Appl Phys Express 8(10):107101. doi:10.7567/apex.8.107101
Nakata K, Ozaki T, Terashima C, Fujishima A, Einaga Y (2014) High-yield electrochemical production of formaldehyde from $CO_2$ and seawater. Angew Chem Int Ed 53(3):871–874. doi:10.1002/anie.201308657
Nakato Y, Tsumura A, Tsubomura H (1982) Efficient photo-electrochemical conversion of solar-energy with normal-type silicon semiconductor electrodes surface-doped with IIIA-group elements. Chem Lett 7:1071–1074. doi:10.1246/cl.1982.1071
Nakato Y, Ueda K, Yano H, Tsubomura H (1988) Effect of microscopic discontinuity of metal overlayers on the photovoltages in metal-coated semiconductor-liquid junction photoelectrochemical cells for efficient solar energy conversion. J Phys Chem 92(8):2316–2324. doi:10.1021/j100319a043
Nozik AJ, Memming R (1996) Physical chemistry of semiconductor–liquid interfaces. J Phys Chem 100(31):13061–13078. doi:10.1021/jp953720e
Ono M, Fujii K, Ito T, Iwaki Y, Hirako A, Yao T, Ohkawa K (2007) Photoelectrochemical reaction and H2 generation at zero bias optimized by carrier concentration of n-type GaN. J Chem Phys 126(5):054708. doi:10.1063/1.2432116, http://dx.doi.org
Park H, Kim KY, Choi W (2002) Photoelectrochemical approach for metal corrosion prevention using a semiconductor photoanode. J Phys Chem B 106(18):4775–4781. doi:10.1021/jp025519r
Peharz G, Dimroth F, Wittstadt U (2007) Solar hydrogen production by water splitting with a conversion efficiency of 18%. Int J Hydrog Energy 32(15):3248–3252. doi:10.1016/j.ijhydene.2007.04.036, http://dx.doi.org
Pleskov YV (2002) Electrochemistry of diamond: a review. Russ J Electrochem 38 (12):1275–1291. doi:10.1023/a:1021651920042
Rajeshwar K, McConnell RD, Licht S (2008) Solar hydrogen generation. Springer, New York
Reece SY, Hamel JA, Sung K, Jarvi TD, Esswein AJ, Pijpers JJH, Nocera DG (2011) Wireless solar water splitting using silicon-based semiconductors and earth-abundant catalysts. Science 334(6056):645–648. doi:10.1126/science.1209816
Rodenas P, Song T, Sudhagar P, Marzari G, Han H, Badia-Bou L, Gimenez S, Fabregat-Santiago F, Mora-Sero I, Bisquert J, Paik U, Kang YS (2013) Quantum dot based heterostructures for unassisted photoelectrochemical hydrogen generation. Adv Energy Mater 3(2):176–182. doi:10.1002/aenm.201200255
Sato K, Fujii K, Koike K, Goto T, Yao T (2009) Anomalous time variation of photocurrent in GaN during photoelectrochemical reaction for H2 gas generation in NaOH aqueous solution. Phys Status Solidi C 6(S2):S635–S638. doi:10.1002/pssc.200880814
Sze SM, Ng KK (2006) Physics of semiconductor devices. John Wiley & Sons, Hoboken, NJ
Trevisan R, Rodenas P, Gonzalez-Pedro V, Sima C, Sanchez RS, Barea EM, Mora-Sero I, Fabregat-Santiago F, Gimenez S (2013) Harnessing infrared photons for photoelectrochemical

hydrogen generation. A PbS quantum dot based "quasi-artificial leaf". J Phys Chem Lett 4 (1):141–146. doi:10.1021/jz301890m

Tubbesing K, Meissner D, Memming R, Kastening B (1986) On the kinetics of electron transfer reactions at illuminated InP electrodes. J Electroanal Chem Interf Electrochem 214 (1–2):685–698. doi:10.1016/0022-0728(86)80132-2, http://dx.doi.org

Valderrama RC, Sebastian PJ, Pantoja Enriquez J, Gamboa SA (2005) Photoelectrochemical characterization of CIGS thin films for hydrogen production. Sol Energy Mater Sol Cells 88 (2):145–155. doi:10.1016/j.solmat.2004.10.011, http://dx.doi.org

Van de Walle CG, Neugebauer J (2003) Universal alignment of hydrogen levels in semiconductors, insulators and solutions. Nature 423(6940):626–628

Verlage E, Hu S, Liu R, Jones RJR, Sun K, Xiang C, Lewis NS, Atwater HA (2015) A monolithically integrated, intrinsically safe, 10% efficient, solar-driven water-splitting system based on active, stable earth-abundant electrocatalysts in conjunction with tandem III-V light absorbers protected by amorphous TiO2 films. Energ Environ Sci 8(11):3166–3172. doi:10.1039/C5EE01786F

Wang Y, Wang T, Da P, Xu M, Wu H, Zheng G (2013) Silicon nanowires for biosensing, energy storage, and conversion. Adv Mater 25(37):5177–5195. doi:10.1002/adma.201301943

Wilson JR, Park SM (1982) Photoanodic dissolution of n—CdS studied by rotating ring-disk electrodes. J Electrochem Soc 129(1):149–154. doi:10.1149/1.2123739

Wünsch F, Nakato Y, Tributsch H (2002) Minority carrier accumulation and interfacial kinetics in nanosized Pt-dotted silicon electrolyte interfaces studied by microwave techniques. J Phys Chem B 106(44):11526–11530. doi:10.1021/jp021016+

Yoon KH, Lee JW, Cho YS, Kang DH (1996) Photoeffects in WO3/GaAs electrode. J Appl Phys 80(12):6813–6818. doi:10.1063/1.363810, http://dx.doi.org

# Chapter 10
# Combinatorial Synthesis and Screening of Oxide Materials for Photoelectrochemical Energy Conversion

**Katarzyna Skorupska and Bruce A. Parkinson**

## 10.1 The Combinatorial Approach for Developing New Materials

The concept of a combinatorial approach for discovery of new compounds was mentioned for the first time by Hannak (1970) where he questioned the efficiency of the serial approach, studying only one composition at a time, to the search for new materials. Since the more widespread adoption of combinatorial approaches, a few valuable general reviews of its applications in materials science have appeared including a global analysis of the state of the art by Potyrailo et al. (2011) and a review by Green et al. (2013). A review of combinatorial approaches to electrochemistry was published by Muster et al. (2011).

The combinatorial approach is based on a high-throughput synthesis and rapid screening of large numbers or so-called libraries of materials aiming at discovery and optimization of a new material with superior properties. The concept has had a very broad impact on medical and material science. Its first large-scale practical application was for drug discovery in the pharmaceutical industry (Terrett et al. 1995).

The concept of combinatorial electrochemistry for simultaneous screening of many electrode compositions was for the first demonstrated by the group of Mallouk (Reddington et al. 1998). Looking for a new material with desired

K. Skorupska (✉)
Department of Chemistry, University of Wyoming, Laramie, WY 82071, USA

Center for Photoconversion & Catalysis, University of Wyoming, Laramie, WY 82071, USA
e-mail: katarzynaskoruspka@gmail.com

B.A. Parkinson
Center for Photoconversion & Catalysis, University of Wyoming, Laramie, WY 82071, USA
e-mail: bparkin1@uwyo.edu



S. Giménez, J. Bisquert (eds.), *Photoelectrochemical Solar Fuel Production*,
DOI 10.1007/978-3-319-29641-8_10

properties can be like searching for a needle in the haystack and so a high-throughput combinatorial approach can shorten the time and as a consequence lower the costs of discovering and optimizing new materials.

The combinatorial high-throughput (CHT) search for a material is focused on screening for desirable properties for a particular application. Below, the most studied materials are grouped by their applications and properties.

By properties (here as examples):

1. Electronic materials (gate dielectrics).
2. Magnetic materials (magnetic recording).
3. Optical materials (phosphors, light emitting (LED), light absorbing (PV)).
4. Catalytic materials (heterogeneous and homogeneous catalysts).

By application:

1. Sensing materials.
2. Polymer-based industrial coatings.
3. Passivation.
4. Pharmacology—drug/gen delivery biomaterials.
5. Inorganic/organic surfaces for cell systems (implants materials).
6. *Energy related materials*
   - Thermoelectrics.
   - Battery materials.
   - Hydrogen storage materials.
   - Photoactive materials (absorber, transparent conducting oxides (TCOs).
   - Photovoltaics, organic PV, all-oxide PV (Rühle et al. 2012, 2014).
   - *Solar water splitting.*

This last group of materials is the subject of this book where the combinatorial approach plays a particularly important role in materials discovery and development for photoelectrochemical (PEC) solar water splitting where there are no known materials that possess all the needed properties to efficiently and cost effectively split water with sunlight.

## 10.2 Materials for Photoelectrochemical Water Splitting

The concept of PEC water splitting was introduced in the 1970s using n-$TiO_2$ (Fujishima and Honda 1972) and was recognized as a very promising, potentially cheap path to a clean and environmentally friendly renewable energy; only recently has substantial attention and funding been devoted to this subject (McKone et al. 2013; Walter et al. 2010). Despite early work using conventional materials, such as p-CdTe, p-GaP, n-$TiO_2$, n-$SrTiO_3$ (Ohashi et al. 1977), p-InP (Heller et al. 1982), and p-$GaInP_2$ (Khaselev and Turner 1998) up to now, there is no operating device or even a stable material or materials that could be used for construction of such a device. Indeed one of the authors was quoted in 1982 during an interview for Chem. Eng. News saying "Right now, there is a lot of basic research that still needs to be done to understand even some of the fundamentals in the field" and even in the three decades since that message has not resulted in a breakthrough in the field.

Although the concept of PEC water splitting seems to be straightforward, the criteria that have to be fulfilled by the needed materials are daunting:

- They must have an appropriate band gap (absorption of visible light with a single absorber at ~2.0 eV and dual absorbers at ~1.0 and ~1.8 eV)
- The $E_{CB}$, $E_{VB}$ energy levels must be aligned to the potentials for water oxidation and proton reduction (taking into account needed overpotentials) (Lewerenz 2010; Weber and Dignam 1984).
- They must be stable under illumination in either basic or acidic electrolytes (Hernández-Pagán et al. 2012).
- They must be inexpensive, contain abundant elements, nontoxic, inexpensive to process and have >10 % solar to hydrogen efficiency (Pinaud et al. 2013).

Although there are well-known issues with metal oxides, such as short carrier life times and diffusion lengths (Pendlebury et al. 2014), it is believed that the criteria presented above can be fulfilled by an appropriate metal oxide semiconductor (Walter et al. 2010) especially if nanostructuring of the photoelectrode can be done to minimize the path for photogenerated carriers to reach the electrolyte interface. Oxides have the best chance of being stable for many years under illumination in an electrolyte solution. After all most stable rocks are oxides. The case for oxide semiconductors with acceptable electronic properties is given some support by the surprising properties discovered and exploited in metal oxide compounds such as high temperature superconductivity and giant magnetoresistance. Theory also suggests that the properties of a promising material can be altered in the rational way (Abdi et al. 2013; Akimov et al. 2013; Castelli et al. 2012a; Vojvodic and Nørskov 2011). In the past, stabilization layers have been applied to conventional materials in unsuccessful attempts to stabilize them while illuminated in electrolytes. However, a recent breakthrough has revived this approach (Hu et al. 2014) and resulted in a bifurcation of the field between stabilization of well-known PV materials, that have been shown to be quite unstable under illumination in solution, and discovering and developing new materials. We concentrate on the latter in this chapter.

The mixed metal oxides are a group of materials that could lead to the "Holy Grail" for PEC $H_2O$ splitting. The bad news is the huge number of possible compositions to be produced and screened. Taking into account the number of candidate metals in the periodic table ($N \sim 50$) the possible combinations if only 1 stoichiometry is considered is equal to $N!/3!(N-3)!$ for a ternary and $N!/4!(N-4)!$ for a quaternary resulting in >19,000 and >220,000, respectively (Woodhouse and Parkinson 2008a). If the more likely multiple stoichiometric ratios are considered the numbers become very much larger. Therefore, a combinatorial approach becomes necessary.

The original work employing combinatorial techniques to develop materials capable of photoelectrochemical water splitting was independently initiated in the McFarland and Parkinson laboratories. Others, realizing the potential of this approach, also joined in. The biggest scale up occurred in 2010 when the Joint Center for Artificial Photosynthesis (JCAP), a US Department of Energy (DOE) Energy Innovation Hub, was established as the world largest research program dedicated to the development of an artificial solar-fuel generation technology. JCAP devoted a substantial portion of their effort to the establishment of a high-throughput group to discover and develop new materials for PEC water splitting. Related efforts throughout the world are being pursued in the Bard group at the University of Texas at Austin, the Stahl group at the University of Wisconsin, Madison in the USA, the Schuhmann and Ludwig group at the Ruhr-University Bochum in Germany, the Hassel group at the Johannes Kepler University of Linz in Austria and Sayama group at the National Institute of Advanced Industrial Science and Technology (AIST) in Japan and the Zaban group in Israel. The rest of this chapter is devoted to reviewing the details of their various approaches.

## 10.3 The Preparation and Characterization of Libraries

Combinatorial high-throughput materials development requires a sequence of characteristic steps. First is the design of the combinatorial experiment protocols. The next step is to develop of a parallel synthesis technique to deposit the materials on substrates and then screening the materials in the library for their performance of the desired properties.

The last, and often underappreciated step, is the analysis, automation and archiving of the hopefully large amounts of data to minimize false negatives and false positives and to provide easy access for data mining to search for trends in the data set.

The following subsections follow these steps and present the current literature status.

## 10.4 Library Design

Probably the first step of a combinatorial experimental design is to design the layout of the pattern or patterns for preparing a materials library. Library patterns can be divided into two groups: discrete compositions and gradient compositions.

In most cases the library consists of an array of patterns with a desired shape and size having defined areas of compositions containing the desired mixture of elements. The use of gradient patterns, where the concentration of an element changes gradually within the pattern, is a convenient method to produce and screen all stoichiometries within a composition range. Composition regions showing promise

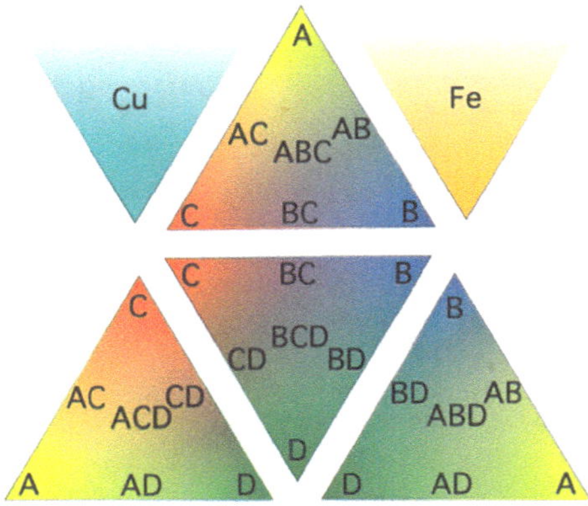


**Fig. 10.1** The faces of a tetrahedral 4-component phase diagram where three elements are mixed at the time

can then be prepared using discrete patterns to more precisely determine compositions. One example of a gradient pattern can be seen in Fig. 10.1 where the four triangular faces of a tetrahedral 4-component phase diagram, corresponding to ternary phase diagrams, for mixing four metals three at a time is shown. The interior of the three dimensional phase diagram can also be explored by taking small triangular slices through the interior of the tetrahedron that will all be smaller than the triangular faces (Reddington et al. 1998). A major drawback of this method is the risk that the prepared library will not exactly follow the gradient design within the pattern due to inhomogeneous mixing of elements during the deposition process or due to physical limitations of the deposition process to prepare a controlled continuous gradient. For example spray pyrolysis or sputtering can produce gradients that are very dependent on the geometry of the deposition system whereas in ink-jet printing any gradient that you can draw can be printed; however, it is limited by the placement and mixing of the small droplets. Controlled gradients can be obtained from sputtering by using a movable shutter that masks parts of the substrate during some period of the sputtering process.

There are many possible ways to prepare discrete libraries. One design is shown in Fig. 10.2 where three different materials may be mixed over a controlled range of stoichiometries.

## 10.5 Substrates

The substrates used for combinatorial library deposition have to be relatively cheap and easy to obtain and prepare. The commonly utilized substrate is glass coated with a thin layer (~300–400 nm) of a transparent conducting oxide (TCO), allowing for either front or back illumination, both options in a device. Fluorine-doped tin oxide (FTO), and less often indium tin oxide (ITO), are used as transparent high band gap oxides having band gaps between 3 and 4 eV, making them transparent in

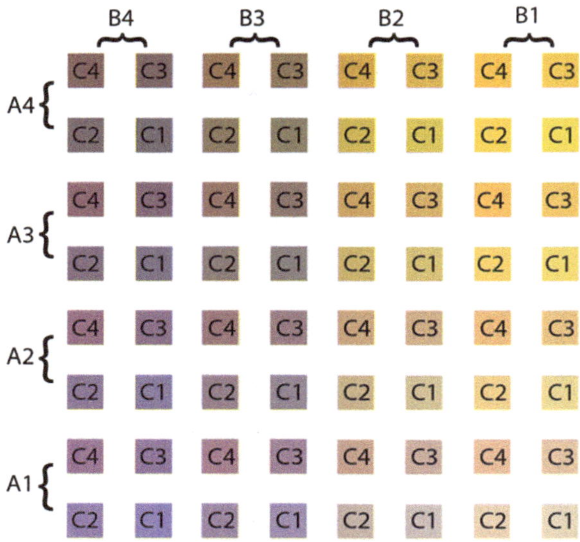


**Fig. 10.2** Discrete library pattern consisting of squares with homogenous composition of three mixed A, B, and C elements where *double rows* of different amounts of elements A and B are deposited followed by four different element C amounts deposited within the areas containing identical A and B amounts

the visible range ($T > 75$ % for 400–900 nm) (Hsu and Chen 2012); however, there is some absorption in the near infrared (NIR) and infrared (IR) due to free carrier absorption (Minami 2008). Indium tin oxide is more expensive than FTO and the diffusion of indium into the absorber can influence its properties (Benck et al. 2014). The diffusion of tin into the absorbers could also be a problem but no detailed studies of the extent of this problem have been reported although it could be argued that since FTO is probably the preferred substrate in an eventual device it is not bad to discover and optimize materials using it.

The surface pretreatment of TCOs before the library deposition is cleaning with a detergent followed by an ultrasonic bath in DI water and an organic solvent such as methanol, ethanol, or acetone. Prior to deposition the TCO surface can be chemically treated to terminate the surface with functional groups to make it more hydrophilic or hydrophobic leading to a more homogeneous film depending on the deposition method. For instance to increase the hydrophilicity of the surface, rinsing in methanol (Woodhouse and Parkinson 2008b) or soaking in KOH solution in isopropanol was used (Jiang et al. 2013). Soaking the substrate in a solution of dichlorodimethylsilane in n-hexane was used to prepare hydrophobic FTO surfaces (Berglund et al. 2013).

Other substrates have been used for used for library preparation such as titanium foil (Baeck et al. 2002; Minguzzi et al. 2008) or Si/$SiO_2$/Pt (Sliozberg et al. 2015b; Stepanovich et al. 2012; Khare et al. 2013), molybdenum on glass and single crystal silicon wafers; however, these are more expensive than FTO glass.

## 10.6 Methods for Preparing Libraries

Library preparation usually employs well-known deposition methods modified in a manner to be scalable and high throughput for quickly producing many libraries. The selection of the most suitable method results from the tradeoff between the quality of the obtained films, the quantity of libraries that need to be produced, the speed of the deposition process and the cost. The selection of methods may also vary depending if the goal is to discover new materials and properties, where the parameter space may be very large, or to optimize the properties of a known material or materials where a more finite set of compositions or additives may be needed. In either case methods of library preparation can be divided into two groups, preparation in ambient air and vacuum and/or controlled pressure deposition.

### *10.6.1 Preparation Methods in Ambient Air*

An automated pipet system with x–y–z positioning was utilized for preparation of a large variety of libraries: (a) Cu:M:Bi, M:Cu, M—21 transition and post-transition metals (Berglund et al. 2013), (b) Bi:M, M = Zn, Sr, Ca, Al, Y, Mn, Fe, Ni, Mg, Cu and Fe:N, N = Ti, Nb, V (Arai et al. 2007), (c) Fe:Ti:M, M—various metal elements (Kusama et al. 2010), (d) $Sn_{1-x}Ir_xO_2$ (Minguzzi et al. 2008).

An array dispenser (Berglund et al. 2013) is a system that allows one to prepare an array of patterns on a conductive substrate by dispensing a series of drops with a controlled volume. Different compositions can be obtained by dispensing a different volume of each of the four precursor solutions available at the print head onto different regions of the substrate. A piezoelectric microdispensing device and robotic positioner are often used.

Another example of a system based on liquid dispensing is the Automated Semiconductor Synthesis System (Arai et al. 2007; Kusama et al. 2010). Here an automated liquid handling platform is used to mix the precursors solutions before dispensing them onto a conductive substrate. In this way metal ions are mixed at the atomic level making the ratio of elements more defined. Use of organic solvents can also allow the synthesis of porous films. Another report describes an automated library synthesis based on a Zinsser pipetting robot connected with *a* sol–gel synthesis (Seyler et al. 2007; Minguzzi et al. 2008).

An automated electrochemical deposition device, designed for the high-throughput synthesis and screening was utilized for the preparation of $W_nO_mM_x$ materials where M = Ni, Co, Cu, Zn, Pt, Ru, Rh, Pd, Ag (Baeck et al. 2002), $Zn_{1-x}Co_xO$ (Jaramillo et al. 2004a), and ZnO mixtures with other metals (Jaramillo et al. 2004b). In this method the deposition of the library on a conductive substrate is performed in separated electrochemical cells using a system of counter and reference electrodes connected to an x–y–z scanner capable moving to submerge into each cell. A Teflon block perforated with a multitude of holes acts as series of electrochemical cells where each cell is sealed from the FTO substrate by an O-ring and can be filled with a prescribed electrolyte composition and in the presence of a reference and counter electrode a rapid serial electrodeposition is performed at room temperature. The number of semiconductors that can be synthesized using this method is limited by those that can be electrodeposited and the consideration that electrodeposition from solution can require quite different conditions for each element.

Ink-jet printing is an extremely fast, inexpensive method with the potential to produce a large number of libraries in a very short time. The idea of using ink-jet printing of overlapping precursors patterns onto a conductive substrate in order to identify the materials suitable for PEC water splitting was brought up by the Parkinson (Woodhouse et al. 2005) and later by the Lewis group (Katz 2008; Katz et al. 2009). The former used ink-jet printing for the discovery and optimization of photoelectrodes for instance p-type CoAlFe (Woodhouse and Parkinson 2008b), n-type $BiVO_4$ (Jiang et al. 2013), and p-type AlCrFe-ox (Rowley et al. 2014) and also electrocatalyst materials (Seley et al. 2013). The Lewis group adapted the ink-jet printing method for the High-Throughput Experimentation (HTE) program for the Joint Center for Artificial Photosynthesis (JCAP) that was established in the year 2010. Currently they claim to be able to print tens of thousands of new alloys per hour (Liu et al. 2012) but have mainly published work on electrocatalytic materials.

Ink-jet printing can be performed on any substrate; however, FTO is most commonly used. The printing of a precursor is conducted using cartridges with a fluid reservoir and a jetting module containing a set of nozzles. Ink-jetting cartridges use two main methods for dispensing the droplets, piezoelectric and thermal. The ink formulations that contain the metal oxide precursor (Table 10.2) must fulfill requirements for either of these technologies in order to produce reproducible and stable droplets. It depends on the following properties of the printing fluid: viscosity, surface tension, volatility, and density, the values of which can vary depending on the ink-jetting technology used in the print head. Additives to improve the performance of the ink-jetting include glycols and/or polymer based additives and in the case of thermal printing ammonium nitrate. Printers may be able to print from only one head at a time or may have multiple print heads that can greatly improve throughput.

One drawback of ink-jet printing is the unknown molecular level mixing of the metal precursors on the substrate that can lead to uncontrolled distribution of the final oxide compositions. Also when using nitrate salts as precursors the decomposition goes through several steps during pyrolysis first driving off the water and solvents, then melting of the nitrate mixtures and then decomposition of the nitrates to oxides with pyrolysis of any organic compounds and finally formation and sintering of the mixed oxides that may be single or multiple phases. Hence the printed films can be thin and inhomogeneous. Despite those issues, and taking into account that ink-jet printing is fast, easy to use, relatively cheap, and versatile, it can be a good choice for initial discovery of promising new materials for PEC water splitting.

### 10.6.2 Preparation Under Reduced Pressure

Methods operating under reduced pressure are capable of producing libraries of very high quality films to screen for PEC water splitting activity. Their drawback is the cost of the equipment and the maintenance of the high vacuum deposition facilities. Two physical vapor deposition methods were reported for the preparation of metal oxides thin film libraries using combinatorial approach. Magnetron sputtering has been very successfully used by the collaboration between the groups of Schuhmann and Ludwig to produce and study libraries of absorbers for PEC water splitting such as n-type $WO_3$:$TiO_2$ (Khare et al. 2013), n-type $WO_3$ (Stepanovich et al. 2012), p- and n-types AlCrFe (Sliozberg et al. 2015b), Ti-W (Sliozberg et al. 2015a), and Fe-W (Meyer et al. 2015).

The simultaneous co-sputtering of more than one element is performed in high vacuum using metallic sputtering targets as precursors. By ion bombardment sputtered material is ejected from the target and deposited onto a substrate. Magnetron sputtering uses magnetic fields to confine the plasma in front of the target to intensify the ion bombardment. The sputtering gas is often Ar and $O_2$ is used as a reactive gas to produce oxides. Only a little heat is radiated from the water-cooled targets so that FTO covered glass substrates can be used and their temperature can be independently controlled. The alignment of the targets and the substrate can be used to control the thickness and composition gradients of the prepared library (Fig. 10.3). In addition moving shutters are used for gradients that are not naturally produced by the geometry of the chamber. In this case the most uniform area of the

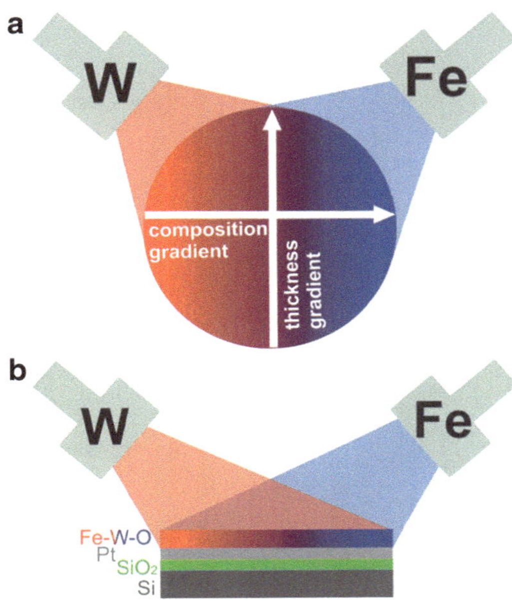


**Fig. 10.3** Schematic of the combinatorial sputtering setup for combined composition and thickness gradients on a material library in *top view* (**a**) and *side view* (**b**) (Meyer et al. 2015)

sputtered beam is used. Sputtering can produce very smooth, dense and uniform layers but often post-annealing is still used to promote reaction and grain growth.

Pulsed laser deposition (PLD) was used to prepare a library of transition metal-containing oxides with the perovskite structure (Kim et al. 2012) for all-oxide photovoltaic applications (Rühle et al. 2012, 2014; Anderson et al. 2014). This method is also applicable to the synthesis of transition metal oxides for PEC water splitting. Pulsed laser deposition uses high power laser pulses impinging on a target that absorbs the energy and results in the ablation of material that deposits on a heated substrate that can be FTO. The thin film deposition process can be performed in high vacuum or in the presence of background gas like oxygen as commonly used for deposition of metal oxides. A composition gradient can be formed during a single deposition procedure by using shutters, masks, and dual laser beams or by a sequential deposition of submonolayers and offsetting the position of the target and substrate.

Simultaneous thermal co-evaporation was used to study $WO_3$-NiO libraries (Kollender et al. 2014). The thermal evaporation is performed in high vacuum where the heated source materials are evaporated and condense on a substrate such as FTO. In this case the desired compositional gradient was obtained by individual control of the flux of each evaporated material using quartz crystal thickness monitors placed above each thermal source. Thermal co-evaporation is generally faster than sputtering and can therefore more easily produce thicker films.

Table 10.1 summarizes precursors utilized for the deposition of various elements. The selection of precursor is dictated by the deposition method.

**Table 10.1** Precursors and the deposition methods used for different elements in the combinatorial approach

| Element | Precursor | Method | References |
|---|---|---|---|
| Ag | $AgNO_3$ | Dispensing | Berglund et al. (2013) |
| Al | Nitrate | Ink-jet printing | Woodhouse et al. (2005); Woodhouse and Parkinson (2008b); Rowley et al. (2014) |
| | | Dispensing | Berglund et al. (2013) |
| | Metallic Al | Magnetron sputtering | Stepanovich et al. (2012) |
| Bi | Nitrate | Ink-jet printing | Jiang et al. (2013) |
| | | Dispensing | Berglund et al. (2013) |
| Cd | $Cd(NO_3)_2$ | Dispensing | Berglund et al. (2013) |
| Ce | Cerium (III) nitrate | Ink-jet printing | Haber et al. (2014a) |
| Co | Nitrate | Electrochemical deposition | Jaramillo et al. (2004a); Gregoire et al. (2013b) |
| | | Ink-jet printing | Woodhouse et al. (2005); Woodhouse and Parkinson (2008b); Xiang et al. (2013) |
| | | Dispensing | Berglund et al. (2013) |
| | Chloride | Electrochemical deposition | Baeck et al. (2002) |

(continued)

**Table 10.1** (continued)

| Element | Precursor | Method | References |
|---|---|---|---|
| Cr | $Co(NO_3)_2$ | Ink-jet printing | Jiang et al. (2013); Rowley et al. (2014) |
| | | Dispensing | Berglund et al. (2013) |
| Cs | Nitrate | Ink-jet printing | Woodhouse et al. (2005) |
| Cu | Chloride | Electrochemical deposition | (Baeck et al. 2002) |
| | Nitrate | Ink-jet printing | Woodhouse et al. (2005); Jiang et al. (2013) |
| | | Dispensing | Berglund et al. (2013) |
| Fe | $Fe(NO_3)_3$ | Ink-jet printing | Woodhouse et al. (2005); Woodhouse and Parkinson (2008b); Jiang et al. (2013); Rowley et al. (2014); Xiang et al. (2013) |
| | | Dispensing | Berglund et al. (2013) |
| | $FeCl_3$ | Ink-jet printing | Gregoire et al. (2013b); Xiang et al. (2014); Haber et al. (2014a) |
| | Metallic Fe | Magnetron sputtering | Stepanovich et al. (2012); Sliozberg et al. (2015b) |
| Ga | $Ga(NO_3)_3$ | Dispensing | Berglund et al. (2013) |
| | | Ink-jet printing | Newhouse and Parkinson (2015) |
| In | $In(NO_3)_3$ | Dispensing | Berglund et al. (2013) |
| | | Ink-jet printing | Newhouse and Parkinson (2015) |
| Ir | $(NH_4)_2IrCl_6$ | Ink-jet printing | Seley et al. (2013) |
| | $IrCl_3$ | Sol–gel | Minguzzi et al. (2008) |
| La | $La(NO_3)_2$ | Dispensing | Berglund et al. (2013) |
| Mg | Nitrate | Ink-jet printing | Jiang et al. (2013) |
| Mn | Nitrate | Ink-jet printing | Jiang et al. (2013) |
| | | Dispensing | Berglund et al. (2013) |
| Mo | Nitrate | Ink-jet printing | Jiang et al. (2013) |
| | $(NH_4)_6Mo_7O_{24}$ | Dispensing | Berglund et al. (2013) |
| Nb | $NbCl_5$ | Dispensing | Berglund et al. (2013) |
| Nd | Nitrate | Ink-jet printing | Woodhouse et al. (2005) |
| Ni | Chloride | Electrochemical deposition | Baeck et al. (2002); Haber et al. (2014a) |
| | $Ni(NO_3)_2$ | Ink-jet printing | Gregoire et al. (2013b); Xiang et al. (2013) |
| | | Dispensing | Berglund et al. (2013) |
| Pd | Nitrate | Ink-jet printing | (Seley et al. 2013) |
| Pt | Chloride | Electrochemical deposition | Baeck et al. (2002) |
| | Chloroplatinic acid | Ink-jet printing | Seley et al. (2013) |
| Rh | Chloride | Ink-jet printing | Seley et al. (2013) |
| Ru | Chloride | Electrochemical deposition | (Baeck et al. 2002) |
| | | Ink-jet printing | Seley et al. (2013) |

(continued)

**Table 10.1** (continued)

| Element | Precursor | Method | References |
|---|---|---|---|
| Sn | $SnCl_2$ | Ink-jet printing | Xiang et al. (2014) |
| | $SnCl_4$ | Dispensing | Berglund et al. (2013); Minguzzi et al. (2008) |
| Ta | $TaCl_5$ | Dispensing | Berglund et al. (2013) |
| Ti | $TiCl_4$—chloride | Ink-jet printing | Gregoire et al. (2013b) |
| | | Dispensing | Berglund et al. (2013) |
| | $Ti(OBu)_4$ | Ink-jet printing | Xiang et al. (2014) |
| | Metallic Ti | Magnetron sputtering | Khare et al. (2013) |
| V | $VCl_3$ | Ink-jet printing | Jiang et al. (2013) |
| | | Dispensing | Berglund et al. (2013) |
| W | Tungsten peroxo | Electrochemical deposition | Baeck et al. (2002) |
| | Ammonium metatungstate hydrate, $(NH_4)_{10}W_{12}O_{41}$ | Ink-jet printing | Jiang et al. (2013) |
| | | Dispensing | Berglund et al. (2013) |
| | Metallic W | Magnetron sputtering | Stepanovich et al. (2012); Khare et al. (2013); Sliozberg et al. (2015b) |
| Y | $Y(NO_3)_3$ | Dispensing | Berglund et al. (2013) |
| Zn | Chloride | Electrochemical deposition | Jaramillo et al. (2004a) |
| | $Zn(NO)_3$ | Ink-jet printing | Xiang et al. (2014) |
| | | Dispensing | Berglund et al. (2013) |
| Zr | $ZrCl_4$ | Dispensing | Berglund et al. (2013) |

## 10.7 Pyrolysis and Annealing

The usual final step in a oxide library preparation is a heat treatment that leads to calcination of the precursors and formation of oxide phases. The limiting factor for TCO substrates is the softening of the glass that starts at about 550 °C. This temperature may not be sufficient for either producing high temperature phases or to adequately produce good grain sizes from some materials. In addition the FTO on glass begins to lose some of its conductivity at high temperatures. Since the use of FTO as a glass substrate may not be predicated on producing the material in a thin film on this substrate other substrates have been used to achieve higher temperature phases. More expensive FTO or ITO on boron-aluminosilicate glasses can be an alternative if higher temperatures are required (Fluegel et al. 2008). Degenerately doped crystal Si wafers are another example that can be used to at least 1000 °C; however, pyrolysis in air then becomes problematic. Pyrolysis and annealing in an atmosphere with low or no oxygen can then be used. A hydrogen containing atmosphere can be used to obtain lower metal oxidation states in the oxide material without oxidizing the Si substrate as was the case for some CuNb oxide phases.

Hydrogen containing annealing atmospheres are not compatible with FTO substrates since they will reduce the Sn(IV) to Sn(II), destroying the conductivity (Skorupska et al. 2015).

## 10.8 Screening Libraries

The construction of a screening system depends on the specific properties to be measured. PEC water splitting searches are mainly focused on finding optimal semiconducting candidates for a photoanode, a photocathode as well as electrocatalysts for oxygen or hydrogen evolution. The reported screening systems can then be grouped by the properties that they are designed to measure. Experimental parameters for PEC can be the photocurrent response, the photovoltage, the full dark and photocurrent voltage curves, the photocurrent spectra or the optical properties, that would give an indication of the band gap, or the products formed under illumination ($O_2$ and/or $H_2$). Some systems are capable of measuring many of these properties during one screening pass on the library. Screening for many properties will slow the throughput but if only one property is screened more follow-up studies are usually needed to evaluate the actual usefulness of the material. Therefore, the high-throughput conundrum can then be stated as "you can learn more and more about fewer and fewer materials until you know everything about none of them or you can learn less and less about more and more materials until you know nothing about all of them" (Parkinson 2015).

### *10.8.1 Measuring (Photo)electrochemical Properties*

A completed library needs to have its properties determined usually as a function of position across the library since this is related to the different compositions. The most relevant parameters for photoelectrochemical screens are the band gap and the production of photocurrent and photovoltage. The band gap can be estimated by UV–Vis spectroscopy on the dry library as is done at JCAP (Gregoire et al. 2013b) or it can be done with photocurrent spectroscopy in an electrochemical cell as is done in Bochum (Sliozberg et al. 2015a). Electrochemical cells used for screening can be categorized into miniature electrochemical cells that collect data when moved from point to point across an array of different compositions and those where a whole library is immersed into an electrolyte. These variations are discussed in more detail below.

#### 10.8.1.1 Screening with the Whole Library Immersed into an Electrochemical Cell

*Laser or light rastering*. One method is where the complete library is dipped into an electrochemical cell with a counter-electrode of comparable size and geometry to the library plate (working electrode) and an optional reference electrode. The large size of the counter electrode ensures a homogeneous distribution of the electrical field and ion flow because it is placed parallel to the library (Newhouse and Parkinson 2015). The use of a transparent substrate such as FTO allows either back or front illumination of the library. The photocurrent is generated only on the illuminated area of the sample and so is a property of only that composition. The illumination usually uses a laser of an appropriate wavelength (Woodhouse et al. 2005) that is rastered over the sample either by galvanometer mirrors or by moving the laser or the cell. A doubled YAG laser with a wavelength of 532 nm is convenient since they are rather low cost per milliwatt of power and will detect photocurrent on any material with a band gap less than about 2.3 eV. A material with a band gap greater that this value will not be useful in a practical water splitting system (Osterloh and Parkinson 2011). Alternatively a full spectrum lamp can be used (Katz et al. 2009) that can be either white light (Katz 2008) or passed through a monochromator. Either the beam is rastered or the library is translated across the beam generation to generate a false color photocurrent map of the photocurrent response as a function of position/composition. An example of such a map is shown in Fig. 10.14 left. The light source can be used directly or chopped with a chopper or intensity modulated by changing the power to the light source. Chopped or modulated light allows the photocurrent to be amplified by lock-in amplifier (Fig. 10.4) to discriminate it from DC dark currents that may be significant on a large area electrode.

Another variation of library screening is an Array Scanner (Fig. 10.5) that uses an optical fiber (400 μm diameter, illuminated by a 100 W xenon lamp) that is immersed into an electrolyte and kept at about 150 μm above the working electrode (Berglund et al. 2013). This system consists of a Teflon electrochemical cell with a three-electrode arrangement to record both dark and photocurrent under a constant potential applied to the working electrode.

If the whole library is immersed in an electrolyte and rastered with a light source it can be usually scanned much faster and is much simpler when compared to systems where the microelectrode or an small electrochemical cell has to be moved and approached directly to each sample area; however, it cannot accurately measure the important photovoltage since the voltage is a global property of the whole electrode. Local (photo)electrochemical screening techniques are discussed next.

*Scanning Electrochemical Microscopy or SECM*. It is very useful to directly detect the products of the photoelectrochemical reaction to unambiguously determine that the photocurrent is generating the desired product, either hydrogen or oxygen. A microelectrode system operating very close to the surface, such as in a scanning electrochemical microscope (SECM), can be used to directly measure the presence

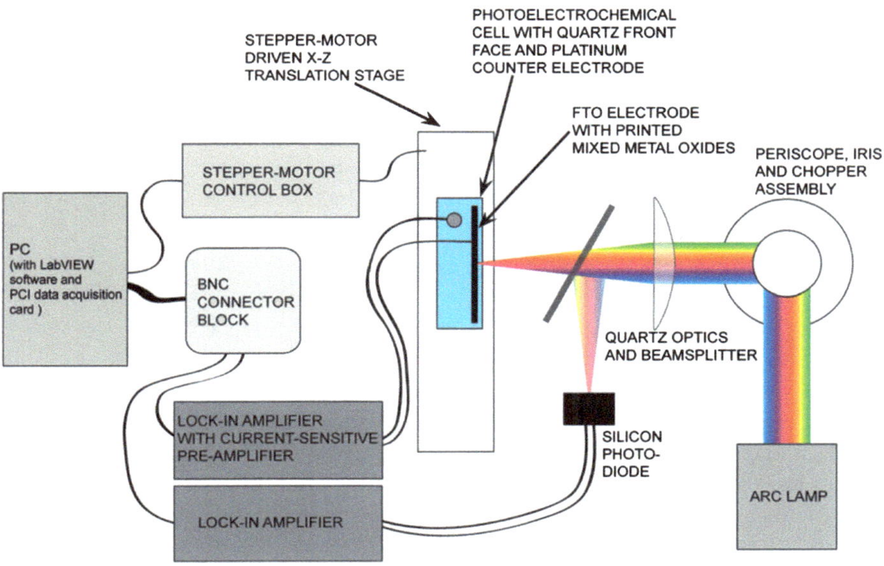


**Fig. 10.4** Scheme of the scanning photoelectrochemical cell with light source (in this case ARC lamp) illuminating each printed material, the plate with a library is mounted to a X–Z translation stage (Katz 2008)

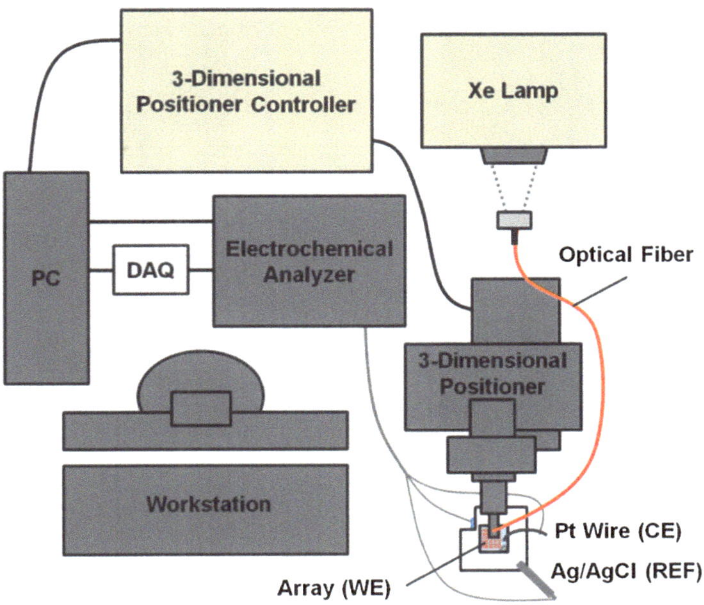


**Fig. 10.5** Scheme of array scanner (Berglund et al. 2013)

of hydrogen or oxygen generation. The Bard group has developed such a system (Minguzzi et al. 2008) for evaluation of the activity of various material compositions, specifically for screening oxygen evolution electrocatalysts. The system consists of a conductive tip that is scanned over an array of different compositions and kept at a constant potential operating as an amperometric sensor by reducing any photogenerated oxygen. An optical fiber can be used for a local illumination source near the microelectrode if photoelectrochemical oxygen or hydrogen evolution is to be detected.

#### 10.8.1.2 Screening Using Independent Electrochemical Cells

*Multiplexed Arrays*. Another approach somewhere between using an independent electrochemical cell or immersing the entire library consists of using a multiplexed combinatorial array in close proximity with an oppositely arranged set of counter electrodes addressed by four 32-channel multiplexers (Xiang et al. 2014). The photocurrent measurement was performed for each area by aligning the illumination source near the desired sample and connecting the corresponding Pt counter electrode (Fig. 10.6). The advantage of this system is that its geometry, with a dedicated counter electrode for each single sample within the library, allows high-throughput photoabsorber screening and eliminates the influence of a shared electrolyte and electrical contacts. Dark and photocurrent voltage curves could be measured for each sample spot with very little crosstalk from the other spots.

An automated high-throughput photoelectrochemical screening system (Jaramillo et al. 2004a) allows the measurement of each sample individually and

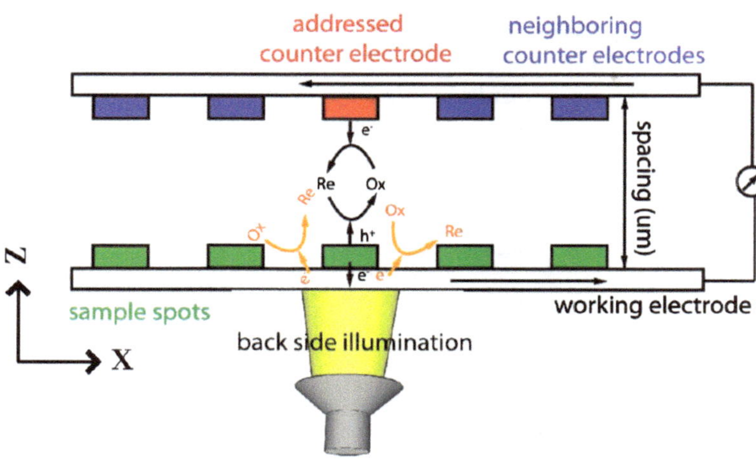


**Fig. 10.6** Schematic illustration of the quantum yield screening system where the *rectangles* in the *lower row* represent the ink-jet printed material library and the *rectangles* in the *upper row* represent the Pt counter electrodes showing the electrically addressed one in the center. The photocurrent shunt pathways that occur at the nearby FTO region are indicated by the *lower curved arrows* (Xiang et al. 2014)

**Fig. 10.7** Image of a high-throughput screening photocurrent probe (Jaramillo et al. 2004b)

consists of a probe with a platinum counter and an Ag/AgCl reference electrode mounted into an electrolyte container with a hole at the bottom. The probe with an inert O-ring at the bottom steps from sample to sample across the library. As the probe is moved down on each sample the O-ring seals the electrochemical cell to the substrate that can be filled with an electrolyte (Fig. 10.7). The photoelectrochemical measurements were performed under chopped illumination from a xenon lamp delivered to the sample by an optical fiber. This assembly has the advantage of individually measuring each sample within a library but moving the quite complicated cell structure to each sample and sealing it slows down the screening process.

#### 10.8.1.3 Scanning Droplet Cells for Photoelectrochemical Screening

There have been several other designs for screening systems that move a small electrochemical cell over a library to individually measure each material or area. The apparatus shown in Fig. 10.8 uses a capillary shaped electrochemical cell containing the reference, counter electrodes and the light fiber (Kollender et al. 2013a, 2014). The silicon gasket seals the cell and a force sensor ensures the wetting of the measurement spot and the proper force on the gasket. The whole electrode arrangement is controlled by a X–Y–Z translation stage.

Another similar concept has been implemented by JCAP for the high-throughput PEC scanning of libraries with a capillary tubing system delivering an individual drop of electrolyte at each compositional area on the library with a counter and an optional reference electrode (Fig. 10.9) (Gregoire et al. 2013a). This construction

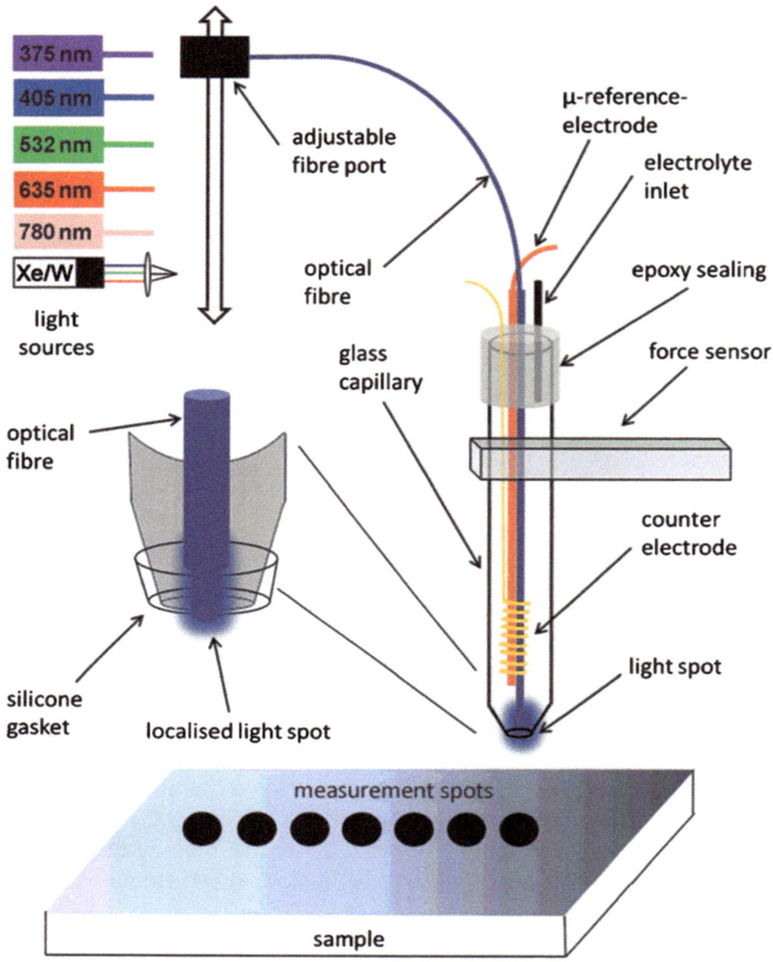


**Fig. 10.8** Scheme of the photoelectrochemical scanning droplet cell (Kollender et al. 2013a)

allows for rapid rastering of the cell over a planar working electrode containing the library.

Probably the most sophisticated optical scanning droplet cell (OSDC) was constructed by the combinatorial group from Bochum (Sliozberg et al. 2015a). The system operates in three-electrode configuration with a Xe lamp connected to a monochromator with an optical fiber to illuminate the interrogated area. A force sensor detects the contact of the drop to the library surface and, if the hydrophobicity or hydrophilicity does not vary too much across the sample, a reproducible wetting area is obtained for each measurement providing high reproducibility of the measurements (Fig. 10.10). This system is capable of measuring complete dark and light current voltage curves, intensity dependence of both photocurrent and voltage

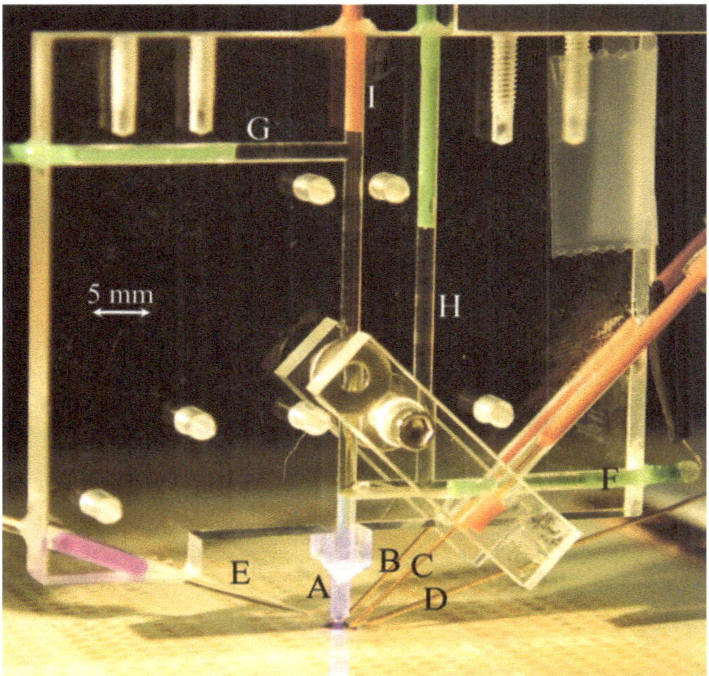


**Fig. 10.9** A photograph of the scanning droplet cell suspended over a composition library. The drop of solution at the bottom of the cell contacts a 2.5 mm-diameter region of the working electrode containing a single 1 mm square sample. Nine ports (labeled *A* through *I*) are used for the solution flow and insertion of electrodes and fiber optic. The 385 nm light from an internal fiber optic illuminates the working electrode and produces a visible purple glow in the port A PTFE tube (Gregoire et al. 2013a)

as well as the spectral response of each area for band gap determination. A full suite of such measurements can take some time making this system perhaps more suited to optimization of promising materials rather than discovery of new materials. This system is combined with high-throughput SEM and X-ray diffraction instruments that will provide morphological and structural data for the libraries to a database where one can access and assess the material with all this data available.

A Parallel Electrochemical Treatment System complements the high-throughput systems at JCAP in order to determine if currents detected during initial scans are due to corrosion currents (Shinde et al. 2015). The experiments are performed at a constant current applied to the library (Fig. 10.11) at an anodic potential of approximately 600 mV overpotential for the OER. The UV–Vis measurements done before screening and after the treatment system can be compared to determine changes in color or optical absorption that would indicate that a material is corroding or changing its composition.

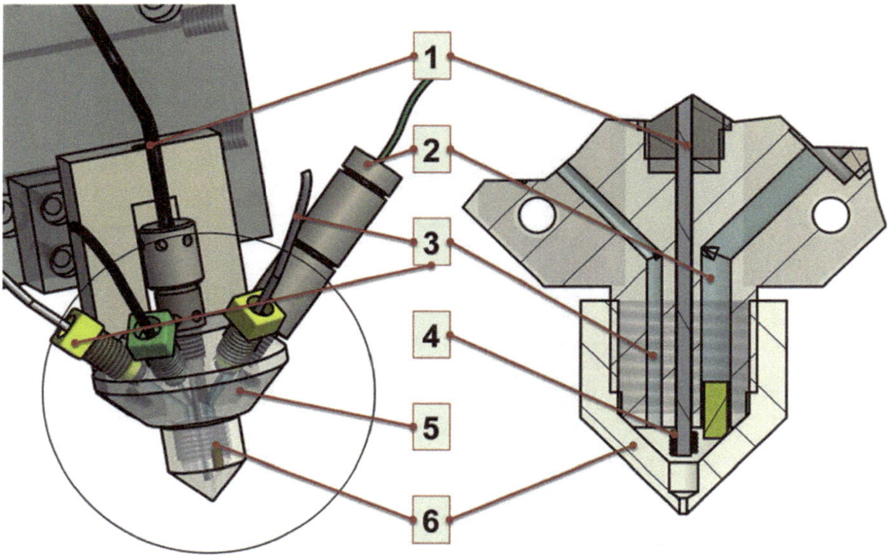


**Fig. 10.10** Scheme of the OSDC: Optical fiber *1*, integrated double-junction Ag/AgCl/3 M KCl reference electrode *2*, electrolyte inlet and outlet *3*, and a Pt-wire counter electrode *4* incorporated in a PMMA body *5* with a PTFE capillary tip *6* (Sliozberg et al. 2015a)

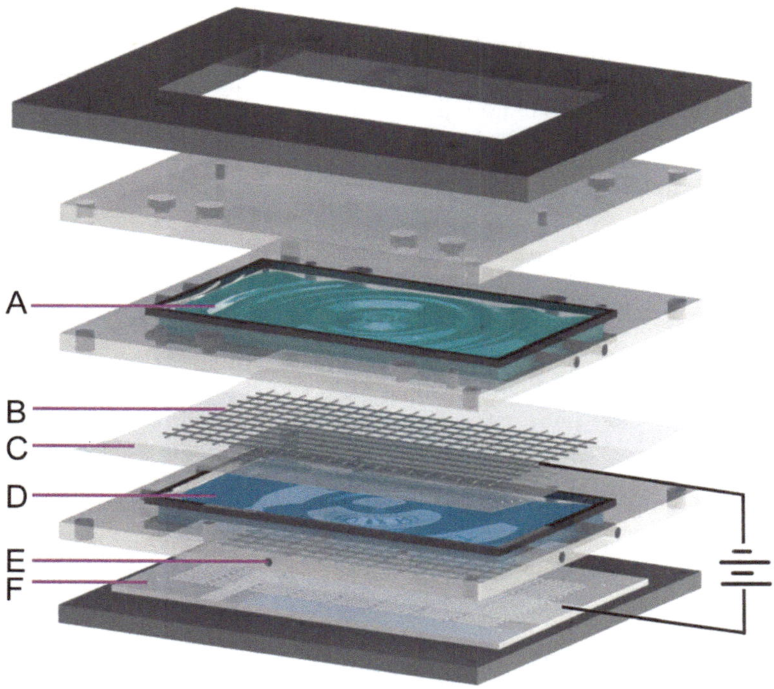


**Fig. 10.11** An illustration of the parallel electrochemical treatment system: *A* catholyte, *B* planar counter electrode, *C* Nafion membrane, *D* anolyte, *E* reference electrode port, and *F* working electrode (library substrate) (Shinde et al. 2015)

### 10.8.2 Detecting the Changes in Optical Properties Due to Evolution of the Products

This group of screening devices uses a detector that changes its optical properties in the presence of the products. It is a very useful method of qualitative libraries screening. It can be especially convenient for non-semiconducting electrocatalyst studies where scanning with light is not an option.

A Pd-coated $WO_3$ sensor (Jaramillo et al. 2001) detects the reflectance changes of an electrocatalyst/$WO_3$ photocathode system in the presence of $H_2$, allowing to find the optimal electrocatalyst material for the HER. The concept is based on the chemo-optical properties of $WO_3$ changing its reflectance in the presence of $H_2$. Apparently the system is not universal and cannot work on other than $WO_3$ photocathodes; however, it can still be helpful in screening for a suitable HER electrocatalyst.

An electrochemical cell (Gerken et al. 2012) is configured to detect the quenched fluorescence by produced dissolved oxygen diffusing from the individual catalysts that quenches the red fluorescence of a fluorescent screen that is captured by a digital camera producing a signal proportional to the amount of $O_2$ evolved. The screen is a mesh covered with a fluorescent paint that acts as the oxygen detector (Fig. 10.12). It is important to mention that the presented screening method is also pH independent.

### 10.8.3 Detection of Products

Here the setups for more direct detection of products are presented.

A bubble screening cell (Xiang et al. 2013) was used for a high-throughput scanning of a electrocatalyst activity by scientists from JCAP where bubble evolution, indicating catalytic activity, is captured by a camera during the screening (Fig. 10.13). A big advantage of the reported technique is that it is pH independent and can be employed for screening both HER, OER catalysts.

Gas chromatography can be used for $H_2$ detection (Berglund et al. 2013) and is a convenient and sensitive method although definitely not a high-throughput method. Here a syringe was used to collect a gas amount from single sealed samples within a library for injection into a gas chromatograph. It is difficult to imagine a scanning of large libraries using this approach, unless an automated injection system is developed, but for small numbers of samples it can be quite helpful.

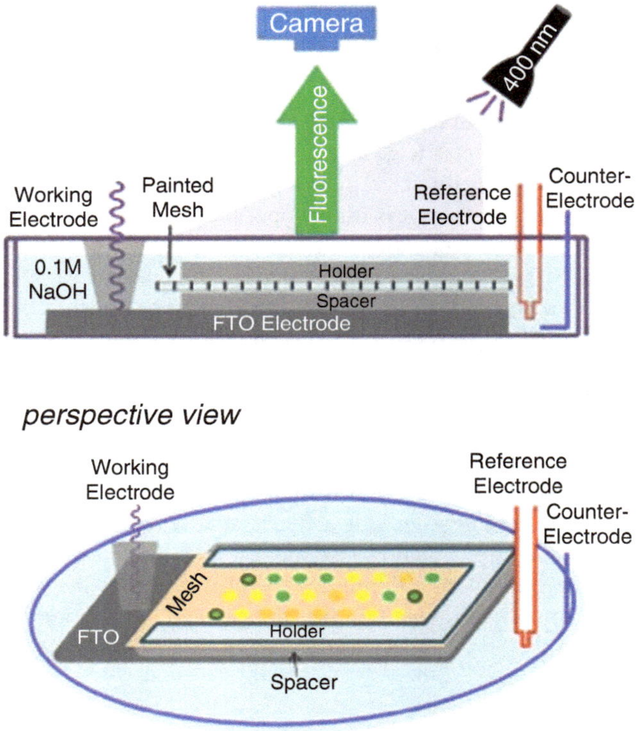


**Fig. 10.12** Schematic drawing (side and perspective views) of the electrochemical screening apparatus showing the positions of the electrode array, $O_2$-sensing mesh, light source, and camera (Gerken et al. 2012)

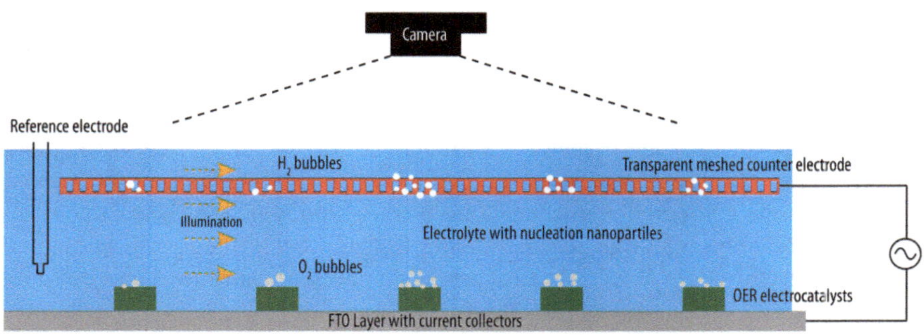


**Fig. 10.13** Cross-sectional schematic illustration of the bubble screening setup where the *rectangles* at the *bottom* represent the testing material library, and the counter electrode, above, consists of a transparent mesh (Xiang et al. 2013)

## 10.9 Data Processing

Data from screening libraries are usually stored in a database that can consist of a number of individually measured parameters collected for each composition. All laboratories to this point have had their own database structure and information and the preservation and archiving of this data is so far up to each group. JCAP has promised to make their database accessible and searchable by outside users but other labs have not yet done this nor is there any agreed upon format for the content and organization of such a database. Standards may emerge from the Materials Genome Initiative (https://www.whitehouse.gov/mgi), a multi agency initiative designed to create new resources and infrastructure to support the discovery and deployment of advanced materials.

## 10.10 Combinatorial Studies of Mixed Metal Oxide Photoelectrodes and Electrocatalysts

A device for solar water splitting requires a light absorber, most likely a hydrogen and/or oxygen evolution electrocatalyst, an electrolyte and a means to keep the gaseous products separate and conduct ions between the anode and cathode compartments (McKone et al. 2013) such as a separator or a membrane. Combinatorial approaches emphasize the discovery and optimization of semiconducting light absorbers for the photoanode and photocathode and the so-called co-catalysts to enable hydrogen or oxygen evolution.

## 10.11 Libraries Prepared with Combinatorial Methods

Table 10.2 presents many of the material libraries that were prepared and evaluated with combinatorial methods. The materials are organized by their applications, which are (1) p-type semiconductors or n-type semiconductors light absorbers for the hydrogen evolution reaction (HER) or for the oxygen evolution reaction (OER) respectively and (2) electrocatalysts for either HER or OER. Up until now the combinatorial approach has not produced the Holy Grail of new materials for PEC water splitting but has produced a few promising leads and the techniques have been used order to improve the properties of already known materials.

**Table 10.2** Photoelectrodes and electrocatalyst studied with the combinatorial approach—reported in literature

| Material | Acting as | References |
|---|---|---|
| Cu:M:Bi, M:Cu<br>M = Ag, Cd, Zn, Al, Ga, In, La, Sc, Y, Sn, Ti, Zr, Nb, Ta, V, Cr, Mo, W, Mn, Co, Fe, Ni | Photocathode HER | Berglund et al. (2013) |
| Bi:M, M = Zn, Sr, Ca, Al, Y, Mn, Fe, Ni, Mg, Cu | Photocathode HER | Arai et al. (2007) |
| $Zn_{1-x}Co_xO$ | Photocathode HER | Jaramillo et al. (2004a) |
| Fe:Al:Co, Cs:Al:Co | Photocathode HER | Woodhouse et al. (2005) |
| Al:Co:Fe | Photocathode HER | Woodhouse and Parkinson (2008b) |
| $Co_{3-x}M_xO_4$, M = Al, Ga, In | Photocathode HER | Newhouse and Parkinson (2015) |
| Al:Cr:Fe | Photocathode HER | Rowley et al. (2014); Sliozberg et al. (2015c) |
| Bi:V:M<br>M = Fe, Cu, Mn, Cr, Mg, W, Mo | Photoanode OER | Jiang et al. (2013) |
| Ti FeSnZn | Photoanode OER | Xiang et al. (2014) |
| Fe:Ti:M, M = Al, B, Ba, Bi, Ca, Ce, Co, Cr, Cu, Dy, In, La, Mg, Mn, Nb, Ni, Pb, Sb, Si, Sm, Sn, Sr, Ta, Y, Zn | Photoanode OER | Kusama et al. (2010) |
| Fe:Cs:Nb | Photoanode OER | Woodhouse et al. (2005) |
| Fe:M, M = Ti, Nb, V | Photoanode OER | Arai et al. (2007) |
| Fe:W | Photoanode OER | Sliozberg et al. (2015b); Kollender et al. (2013b) |
| $W_nO_mM_x$<br>M = Ni, Co, Cu, Zn, Pt, Ru, Rh, Pd, Ag | Photocathode HER | Baeck et al. (2002) |
| $WO_3$:NiO | Photoanode OER | Kollender et al. (2014) |
| Nanoporous $WO_3$ films | Photoanode OER | Stepanovich et al. (2012) |
| $WO_3/TiO_2$ nanostructures | Photoanode OER | Khare et al. (2013) |
| Ru:Rh:Pt, Pt:Ru:Ir, Ru:Rh:Pd, Ir:Pt:Rh, Ir:Rh:Ru, Ir:Pt:Pd, Ir:Ru:Pd<br>Ir:Ru:M, M = Al, Nb, Mn, Sn | OER electrocatalyst | Seley et al. (2013) |
| Ru:M, M = Pd, Ir, Cu, Co, Re, Cr, Ni | OER electrocatalyst | Forgie et al. (2010) |
| $Sn_{1-x}Ir_xO_2$, x = 1 to 0 | OER electrocatalyst | Minguzzi et al. (2008) |
| Co:Fe:Ni | OER electrocatalyst | Xiang et al. (2013) |
| Ti:Co:Fe:Ni | OER electrocatalyst | Gregoire et al. (2013b) |
| Ce:Co:Fe:Ni and Ce:La:Co:Ni | OER electrocatalyst | (Haber et al. 2014a, b) |
| Co:Mn:Ta:Sb | | Shinde et al. (2015) |
| Ni:Ms, Co:Ms, Ni:Fe:M, M = Fe, Al, Ga, Cr | OER electrocatalyst | Gerken et al. (2014) |
| Ni:Fe:Al | OER electrocatalyst | Chen et al. (2014) |

### 10.11.1 Photocathodes for the HER

An extensively studied earth abundant metal oxide for a HER photocathode is p-type $Cu_2O$, with a direct band gap of 2 eV and photocurrents at solar intensities in the range of $mA/cm^2$ (Paracchino et al. 2011, 2012). CuO also has an attractive bandgap ($E_g = 1.35$–$1.7$ eV) with a good onset potential (~0.9 V vs. RHE) making it an appealing photocathode material for PEC water splitting (Hardee and Bard 1977). However, cuprous and cupric oxides are not stable in aqueous solution due to rapid photoreductive decomposition ($CuO \rightarrow Cu_2O \rightarrow Cu$) (Gerischer 1977). In order to study the stability of copper oxides the screening of 22 transition and post-transition metals added to CuO was investigated with a combinatorial approach (Berglund et al. 2013). The best results were obtained by adding Cd, Zn, Sn, and Co. Unfortunately, the incorporation of the additional element did not improve the stability of the material in aqueous solutions but enhancement of the photocurrent in a nonaqueous electrolyte was observed.

Combinatorial studies indicated that p-type $CuBi_2O_4$ was a candidate for HER with visible light (Arai et al. 2007). Later its favorable band alignment and stability in high pH solutions was reported (Hahn et al. 2012). However the low photoconversion efficiency of the material suggests that its electronic properties need to be improved. This was the aim of the combinatorial studies performed in the Bard group where they screened Bi and Cu mixed with 22 transition and post-transition metals. The best results were obtained for Ag, Cd, and Zn the later showing four times higher photocurrent compared to $CuBi_2O_4$ while maintaining its stability.

High band gap semiconductors such as ZnO have a long history of consideration as a photoelectrode (Micka and Gerischer 1972); however, their lack of light absorption in the visible range would preclude any efficient solar energy conversion (Inoue et al. 1979). Following the reported improvement of ZnO light absorption properties when doped with Co (Hirano and Kozuka 2003). Jaramillo et al. developed a combinatorial technique in order to optimize the $Zn_{1-x}Co_xO$ composition for solar hydrogen production (Jaramillo et al. 2004a). It was found that a ~2–5 % concentration of cobalt increases the photocurrent originating from visible light illumination by a factor of four but this improvement was still too low for a practical system.

The compositions that have been revealed by combinatorial studies in the Parkinson group are Co-rich regions of Co-Al-Fe, Co-Al-Cs (Woodhouse et al. 2005) and $Co_{3-x}(Al,Ga,In)_xO_4$ (Newhouse and Parkinson 2015) and Cr-rich regions of the Al-Cr-Fe system (Rowley et al. 2014) using ink-jet printing. It is worth mentioning that the first indication of the activity of the Al-Cr-Fe system comes from the distributed combinatorial outreach effort called the Solar Hydrogen Activity research Kit (SHArK) program (for more details Chap. 11).

More in depth studies of the Co-Al-Fe composition showed that the optimal composition, which is ~84 % Co, 10 % Fe, and 6 % Al has an indirect band gap of 1.5 eV with the onset of hydrogen evolution 0.9 V positive of the thermodynamic value with photocurrents limited by slow kinetics for HER since no hydrogen evolution catalysts were employed. Follow-up combinatorial studies of the p-type Al-Cr-Fe-oxide by two groups (Rowley et al. 2014; Sliozberg et al. 2015c) showed the material has a band gap below 2 eV and photocurrent onset at about 1 V vs. RHE; however, it remains to be determined if the material is stable at hydrogen evolving potentials.

## *10.11.2 Photoanodes for the OER*

Metal oxide photoanodes that have been modified using combinatorial methods include mixtures based on $\alpha$-$Fe_2O_3$, $WO_3$, and $BiVO_4$. These materials are often considered as candidates for a photoanode but all of them have significant flaws that will impede their practical application. In addition to their band gaps being too large, $WO_3$ has indirect high band gap material of 2.7 eV, $\alpha$-$Fe_2O_3$ has failed to deliver any significant photovoltage considering its 2.2 eV band gap and $BiVO_4$ with a band gap of 2.4 eV is unstable in aqueous electrolytes.

Tungsten oxide has a conduction band position below the $H^+/H_2$ redox Fermi level and so cannot operate as a single water splitting photoelectrode; nevertheless, $WO_3$ can photooxidize water to $O_2$ without addition of an electrocatalyst (Valdés and Kroes 2009). The efficiency of light absorption and separation of photogenerated electron–hole pairs can be improved by texturing the photoelectrode (Zhang et al. 2015). This approach was used in combinatorial studies of $WO_3$, as well as for the mixtures of $WO_3$ with $TiO_2$, where the nanostructuring the film increases the photocurrent (Stepanovich et al. 2012; Khare et al. 2013). Studies where tungsten oxide is mixed with various metals (Pt, Ru, Rh, Pd, Au), mixed with NiO metal oxides and prepared as metal–metal composites (Co, Cu, Zn) have shown that Ni is the most effective PEC dopant (Baeck et al. 2002). The combinatorial studies of the performance of tungsten oxide mixed with Fe (Sliozberg et al. 2015b; Kollender et al. 2013b) and Ni (Kollender et al. 2014) show a change in the film morphology at different ratios of the elements. The optimal amount of Fe in $WO_3$ was established to be about 20 atomic percent resulting in a photocurrent of 20–65 $\mu A\ cm^{-2}$ whereas with just 6.2 atomic percent of Ni, 2.5 $mA\ cm^{-2}$ was measured.

$Fe_2O_3$ is often considered a promising material but its short charge carrier lifetimes and slow electron-transfer kinetics are reflected in very low photovoltages (Hamann 2012). Combinatorial attempts were made to improve the properties of $Fe_2O_3$, by making ternary compositions based on Fe-Ti, Fe-Nb, and Fe-V (Arai et al. 2007; Kusama et al. 2010) as well as the quaternary compositions based on Fe-Zn-Sn-Ti (Xiang et al. 2014). Mixing with wide band gap transition metal oxides (Nb, Ti, Ta) can be beneficial due to their stability and ability to facilitate four-photon water splitting (Walter et al. 2010). In response to reports where individually Si, Al, or Fe impurities were either purposely or accidently added to $\alpha$-$Fe_2O_3$, ink-jet printing was employed to study the photoelectrochemical activity

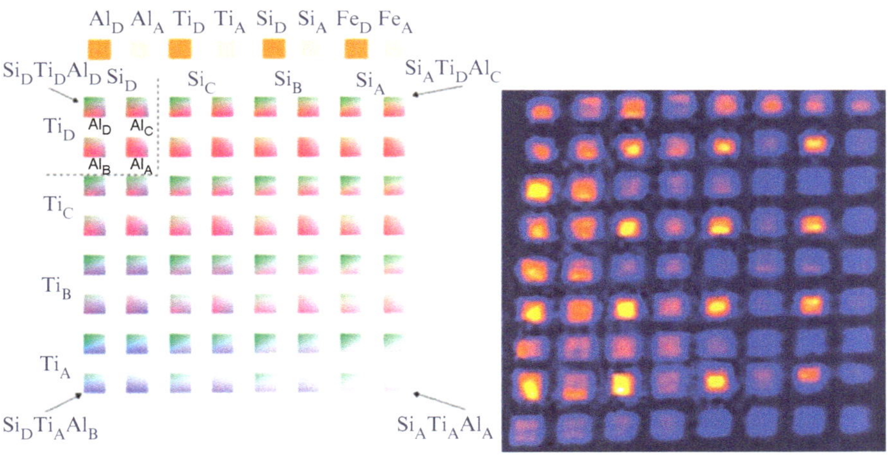


**Fig. 10.14** Combinatorial investigation of adding small amounts of Si, Al, and Ti to a uniform pattern of ink-jet printed $Fe_2O_3$ squares. The template on the *left* shows the printed pattern with the *top row* of *squares* illustrating the extremes of the individual element additions or with no addition on the *far right*. In this case $A = 0.67$ %, $B = 1.1$ % $C = 2.5$ %, and $D = 9.5$ %. The false color photocurrent map (*right*) shows that areas with all three elements present produced higher photocurrents than the extremes of the addition of just one element (*top row*) (He and Parkinson 2011)

of α-$Fe_2O_3$ with incorporated combinations of these elements (He and Parkinson 2011). When Si and Al were individually added to iron oxide at the studied levels the photoelectrolysis activity decreased, whereas low levels of Ti addition enhanced the photocurrents. Synergistic effects were observed resulting in enhanced photocurrents when multiple impurities were added to α-$Fe_2O_3$. Figure 10.14 shows the photocurrents obtained from such a library. Even though α-$Fe_2O_3$ has not been shown to be a useful material for photoelectrochemical water splitting the implications for optimization of a material once its basic composition has been identified are apparent. The example of trace sodium in the glass substrate unexpectedly substantially improving the efficiency of $CuInGaSe_2$ thin film solar cells and the long development time to commercialization of modules made with this material are examples of the time needed to deploy a new material.

Theoretical work concerning the electronic structure of $BiVO_4$ suggested that the addition of $nd^0$ and $ns^2/nd^{10}$ cations could reduce the oxide band gap and increase its conductivity (Walsh et al. 2009). Combinatorial studies showed that the addition of tungsten (W) to $BiVO_4$ substantially improved its photoresponse but only at 473 mm (Jiang et al. 2013). The tungsten doped bismuth vanadate (W-$BiVO_4$) is thus absorbing only a small portion of the visible light ($E_g > 2$ eV, direct bandgap) and is stable only in neutral pH which eliminates it from application in a real device (Hernández-Pagán et al. 2012); however, in lack of a "perfect candidate" W-$BiVO_4$ is still a very good material for the fundamental studies.

The identification of a semiconductor with band gap below 2 eV and valence band aligned to the OER potential remains a very active field of research, and through a combination of high-throughput theory and experimentation, JCAP has recently identified $Mn_2V_2O_7$ as an Earth-abundant metal oxide that meets these requirements (Yan et al. 2015). Any substantial photocurrent and long-term stability of this compound have yet to be demonstrated, but the identification of desirable band energetics in complex metal oxide systems indicates that the community is heading in promising directions with respect to photoanode development.

### 10.11.3 Electrocatalysts

The combinatorial search for electrocatalysts for PEC water splitting has been concentrated on improving electrocatalysts for the oxygen evolution reaction (OER). To the best of our knowledge there are no published combinatorial studies focused on HER electrocatalysts. The work on OER electrocatalysts has been concerned with either noble metals (Ru, Pt, Rh, Ir, Pd) or on non-noble containing mixtures containing Ni. Up until now the only OER catalysts that are very active and stable in acidic conditions are noble metal oxides, especially $IrO_2$, although there has been recent progress from high-throughput development of mixed non-noble metal oxide catalysts exhibiting moderate stability (Shinde et al. 2015).

The most active noble metal catalysts for water oxidation in acid solution are $RuO_2$ and $IrO_2$ (Trasatti 1984). $RuO_2$ is characterized by the highest catalytic activity, but its price is too high for large-scale application and its stability decreases with increases of the applied potential (Loučka 1977). The Strasser Group performed a combinatorial study of seven binary alloy systems based on ruthenium with Pd, Ir, Cu, Co, Re, Cr, and Ni (Forgie et al. 2010). The authors reported that the Ru–Co mixed-phase electrocatalysts clearly exhibited improved catalytic activities compared to the benchmark Ru catalysts. Theory predicts that Ru and Ir form thermodynamically stable mixed oxides (Rossmeisl et al. 2007). The Parkinson group studied Ir with different additions of Ru, Rh, and Pt and with non-noble metals as Al, Nb, Mn, and Sn (Seley et al. 2013). It was concluded that within the studied metal mixed oxides (Ir, Ru, Rh, Pt, Pd) the mixtures of Ir and Rh stabilizes and enhances the OER in comparison to the standard $IrO_2$ but at low current densities a composition of Ir, Sn, and Ru performed well. The Bard group performed a combinatorial survey on $IrO_2$ mixed with the non-noble element Sn to produce the mixed phases $Sn_{1-x}Ir_xO_2$ (Minguzzi et al. 2008). Ruthenium oxide was chosen due to its higher stability over $RuO_2$ and the combinatorial approach was applied to study additions of tin in order to find a cheaper OER catalyst.

Non-noble OER catalysts operate effectively only in alkaline solutions for example Ni-oxide/hydroxide electrodes are employed in commercial electrolyzers. It is believed that incorporation of different metal ions may improve the performance of Ni oxide in PEC water oxidation. In most combinatorial approaches iron was added to nickel which is known to enhance Ni oxide activity (Młynarek et al. 1984). The Stahl group also found that combinations of Co and Ni have low over-potentials for water oxidation and studied various mixtures of those two metals in combination with a third metal (Gerken et al. 2014). It was found that Cr and Fe have a beneficial effect on the mixed oxide catalysts in contrast to Mn that showed a decreased activity relative to pure Co or Ni oxides. The combinatorial studies performed by the Stahl group led to the discovery of a very promising inverse spinel $NiFeAlO_4$ that showed a higher catalytic activity relative to NiO and a binary NiFe oxide. Mixtures of Fe, Ni, and Co were intensively studied at JCAP (Xiang et al. 2013) and also in quaternary mixtures with Ti (Gregoire et al. 2013b) and with Ce (Haber et al. 2014a). Both of these material classes had compositions with overpotentials of about 400 mV for oxygen evolution at currents of tens of $mA/cm^2$. Combinatorial studies would aid the search for non-noble metal OER catalysts that are stable under acidic conditions.

One aspect of coupling absorbers with catalyst materials is that the best catalyst on a metal support may not be the best catalyst on a particular absorber surface. The well-known "catalyst support interaction" found in heterogeneous catalysis supports this assertion. For example $BiVO_4$ shows much better photooxidation of water with a layer of Pt or Co oxide electrocatalyst than with $IrO_2$ despite the fact that $IrO_2$ is considered a much better OER catalyst (Ye et al. 2011).

## 10.12 Theoretical Approaches

Due to the large number of possible candidate materials for PEC water splitting to be synthesized and screened, a theoretical approach could be helpful for suggesting promising metal oxides. Recently a comprehensive computational screening based on criteria for stability, the band gap and the position of the bands revealed 15 metal oxide compounds potentially suitable for PEC water splitting. The study included more than 5400 oxide/oxynitride compounds that may exist in the cubic perovskite structure and covered 52 metals (Castelli et al. 2012a, b).

Despite the theoretical indications of perovskite type metal oxides as promising candidates for PEC water splitting (Montoya et al. 2014; Vojvodic and Nørskov 2011), to our knowledge there is not yet any concerted effort to produce and screen the materials with a combinatorial approach to determine their usefulness. There is work, published by the group of Ross, about transition metal substituted perovskite-structured films based on $SrTiO_3$ mixed with Fe, Cr, Ni, but it does not focus on their photoelectrochemical properties and instead only cites a wide range of possible applications (Kim et al. 2012). Ideally theory could solve the difficult inverse design problem where once the desired properties are defined, such as band gap, band positions, and stability, the theory produces a material where the appropriate atoms are placed in a stable structure to produce those properties. Until theory can reach this point we are still reliant on actually producing and measuring many materials in order to find the "holy grail."

## 10.13 Distributed Searches in an Outreach Environment

In contrast to a combinatorial approach where a few people are producing and screening many compositions one could adopt a distributed approach where many people are producing and screening a few materials to search for materials needed for PEC water splitting. This is akin to the well-known SETI at Home or Protein Folding at Home distributed science programs; however, these programs only use the participant's spare computer capacity. The SHArK (Solar Hydrogen Activity research Kit) project was established by the Parkinson group in 2007 and provides undergraduate and high school students with the simple and inexpensive tools needed to produce and screen metal oxides for PEC water splitting (Parkinson 2010). The goals of the program are dual, (a) to increase the throughput of the photoelectrocatalyst screening process and (b) to enlist the participation of undergraduates and high school students where they learn about research and the science behind water splitting in a hands on process and have a real chance of discovering something new and important.

Other groups emulated this idea and developed spin off projects that currently make up the "Solar Army" (http://thesolararmy.org/).

These projects include:

- SHArK (Solar Hydrogen Activity research Kit) (Parkinson 2010; Anunson et al. 2013).
- HArPoON (Heterogeneous Anodes Rapidly Perused for Oxygen Overpotential Neutralization) (Gerken et al. 2012).
- SEAL (Solar Energy Activity Laboratory) (Winkler and Winkler 2011).

The aim of the follow-up projects is also to design easy experimental protocols accessible to the students and to equip them with the necessary materials to enable them to produce and screen materials. The SHArK and SEAL screening kits are based on detection of photocurrent to discover new semiconductors whereas the HArPoON kits are used to screen electrocatalysts for oxygen evolution by using fluorescence quenching by oxygen as a function of electrode potential.

## 10.14 Outlook

An energy economy based on hydrogen fuel is perhaps the only sustainable fuel based system that can replace fossil fuels and mitigate their effect on the climate. We are hopeful that photoelectrochemical water splitting can play an important role in the future. We cannot forget that photovoltaic cells have recently seen steep price declines and hooking them up to a conventional electrolyzer can also produce hydrogen from solar energy. The only way photoelectrochemical cells can compete is if we can produce hydrogen more cheaply than in the two devices. However, slow advancement of photoelectrochemical water splitting in the last three decades is attributable to not having suitable materials. It is clear that a combinatorial approach can speed up the discovery and optimization of materials and compositions that cannot now be predicted. Even after the discovery of suitable materials there is still a long path to their implementation in a real working device. A joint effort, combining a multitude of sophisticated characterization, fabrication, testing and measurement methods, is a promising synergistic method to discover and optimize a material that may be useful in an actual device. It is clear that unsuitable materials such as $WO_3$, $TiO_2$, $BiVO_4$, and $Fe_2O_3$ should be abandoned and work on new candidate materials should not be continued if a fatal flaw is identified as in the case for the aforementioned materials. The field should abide by one of the tenets of industrial development, that is, "fail quickly" meaning unproductive paths should be abandoned to focus efforts on more productive directions.

## References


Abdi FF, Han L, Smets AHM, Zeman M, Dam B, Krol R. van de (2013) Efficient solar water splitting by enhanced charge separation in a bismuth vanadate-silicon tandem photoelectrode. Nature Communications 4. doi: 10.1038/ncomms3195. http://www.nature.com/ncomms/2013/130729/ncomms3195/full/ncomms3195.html

Akimov AV, Neukirch AJ, Prezhdo OV (2013) Theoretical insights into photoinduced charge transfer and catalysis at oxide interfaces. Chem Rev 113(6):4496–4565. doi:10.1021/cr3004899

Anderson AY, Bouhadana Y, Barad H-N, Kupfer B, Rosh-Hodesh E, Aviv H, Tischler YR, Rühle S, Zaban A (2014) Quantum efficiency and bandgap analysis for combinatorial photovoltaics: sorting activity of Cu–O compounds in all-oxide device libraries. ACS Comb Sci 16(2):53–65. doi:10.1021/co3001583

Anunson PN, Winkler GR, Winkler JR, Parkinson BA, Schuttlefield Christus JD (2013) Involving students in a collaborative project to help discover inexpensive, stable materials for solar photoelectrolysis. J Chem Educ 90(10):1333–1340. doi:10.1021/ed300574x

Arai T, Konishi Y, Iwasaki Y, Sugihara H, Sayama K (2007) High-throughput screening using porous photoelectrode for the development of visible-light-responsive semiconductors. J Comb Chem 9(4):574–581. doi:10.1021/cc0700142

Baeck SH, Jaramillo TF, Brändli C, McFarland EW (2002) Combinatorial electrochemical synthesis and characterization of tungsten-based mixed-metal oxides. J Comb Chem 4 (6):563–568. doi:10.1021/cc020014w

Benck JD, Pinaud BA, Gorlin Y, Jaramillo TF (2014) Substrate selection for fundamental studies of electrocatalysts and photoelectrodes: inert potential windows in acidic, neutral, and basic electrolyte. PLoS One 9(10):e107942. doi:10.1371/journal.pone.0107942

Berglund SP, Lee HC, Núñez PD, Bard AJ, Mullins CB (2013) Screening of transition and post-transition metals to incorporate into copper oxide and copper bismuth oxide for photoelectrochemical hydrogen evolution. Phys Chem Chem Phys 15(13):4554–4565. doi:10.1039/C3CP50540E

Castelli IE, Landis DD, Thygesen KS, Dahl S, Chorkendorff I, Jaramillo TF, Jacobsen KW (2012a) New cubic perovskites for one- and two-photon water splitting using the computational materials repository. Energy Environ Sci 5(10):9034–9043. doi:10.1039/C2EE22341D

Castelli IE, Olsen T, Datta S, Landis DD, Dahl S, Thygesen KS, Jacobsen KW (2012b) Computational screening of perovskite metal oxides for optimal solar light capture. Energy Environ Sci 5(2):5814–5819. doi:10.1039/C1EE02717D

Chen JYC, Miller JT, Gerken JB, Stahl SS (2014) Inverse spinel $NiFeAlO_4$ as a highly active oxygen evolution electrocatalyst: promotion of activity by a redox-inert metal ion. Energy Environ Sci 7(4):1382–1386. doi:10.1039/C3EE43811B

Fluegel A, Earl DA, Varshneya AK, Seward TP (2008) Density and thermal expansion calculation of silicate glass melts from 1000°C to 1400°C. Eur J Glass Sci Technol B 49(5):245–257

Forgie R, Bugosh G, Neyerlin KC, Liu Z, Strasser P (2010) Bimetallic Ru electrocatalysts for the OER and electrolytic water splitting in acidic media. Electrochem Solid-State Lett 13(4):B36. doi:10.1149/1.3290735

Fujishima A, Honda K (1972) Electrochemical photolysis of water at a semiconductor electrode. Nature 238(5358):37–38. doi:10.1038/238037a0

Gerischer H (1977) On the stability of semiconductor electrodes against photodecomposition. J Electroanal Chem Interfacial Electrochem 82(1–2):133–143. doi:10.1016/S0022-0728(77)80253-2

Gerken JB, Chen JYC, Massé RC, Powell AB, Stahl SS (2012) Development of an $O_2$-sensitive fluorescence-quenching assay for the combinatorial discovery of electrocatalysts for water oxidation. Angew Chem Int Ed 51(27):6676–6680. doi:10.1002/anie.201201999

Gerken JB, Shaner SE, Massé RC, Porubsky NJ, Stahl SS (2014) A survey of diverse earth abundant oxygen evolution electrocatalysts showing enhanced activity from Ni–Fe oxides containing a third metal. Energy Environ Sci 7(7):2376–2382. doi:10.1039/C4EE00436A

Green ML, Takeuchi I, Hattrick-Simpers JR (2013) Applications of high throughput (combinatorial) methodologies to electronic, magnetic, optical, and energy-related materials. J Appl Phys 113(23):231101. doi:10.1063/1.4803530

Gregoire JM, Xiang C, Mitrovic S, Liu X, Marcin M, Cornell EW, Fan J, Jin J (2013a) Combined catalysis and optical screening for high throughput discovery of solar fuels catalysts. J Electrochem Soc 160(4):F337–F342. doi:10.1149/2.035304jes

Gregoire JM, Xiang C, Liu X, Marcin M, Jin J (2013b) Scanning droplet cell for high throughput electrochemical and photoelectrochemical measurements. Rev Sci Instrum 84(2):024102–024106. doi:10.1063/1.4790419

Haber JA, Cai Y, Jung S, Xiang C, Mitrovic S, Jin J, Bell AT, Gregoire JM (2014a) Discovering Ce-rich oxygen evolution catalysts, from high throughput screening to water electrolysis. Energy Environ Sci 7(2):682–688. doi:10.1039/C3EE43683G

Haber JA, Guevarra D, Jung S, Jin J, Gregoire JM (2014b) Discovery of new oxygen evolution reaction electrocatalysts by combinatorial investigation of the Ni–La–Co–Ce oxide composition space. ChemElectroChem 1(10):1613–1617. doi:10.1002/celc.201402149

Hahn NT, Holmberg VC, Korgel BA, Mullins CB (2012) Electrochemical synthesis and characterization of P-$CuBi_2O_4$ thin film photocathodes. J Phy Chem C 116(10):6459–6466. doi:10.1021/jp210130v

Hamann TW (2012) Splitting water with rust: hematite photoelectrochemistry. Dalton Trans 41(26):7830–7834. doi:10.1039/C2DT30340J

Hanak JJ (1970) The 'multiple-sample concept' in materials research: synthesis, compositional analysis and testing of entire multicomponent systems. J Mater Sci 5(11):964–971. doi:10.1007/BF00558177

Hardee KL, Bard AJ (1977) Semiconductor electrodes X. Photoelectrochemical behavior of several polycrystalline metal oxide electrodes in aqueous solutions. J Electrochem Soc 124(2):215–224. doi:10.1149/1.2133269

He J, Parkinson BA (2011) Combinatorial investigation of the effects of the incorporation of Ti, Si, and Al on the performance of $\alpha$-$Fe_2O_3$ photoanodes. ACS Comb Sci 13(4):399–404. doi:10.1021/co200024p

Heller A, Aharon-Shalom E, Bonner WA, Miller B (1982) Hydrogen-evolving semiconductor photocathodes: nature of the junction and function of the platinum group metal catalyst. J Am Chem Soc 104(25):6942–6948. doi:10.1021/ja00389a010

Hernández-Pagán EA, Vargas-Barbosa NM, Wang T, Zhao Y, Smotkin ES, Mallouk TE (2012) Resistance and polarization losses in aqueous buffer–membrane electrolytes for water-splitting photoelectrochemical cells. Energy Environ Sci 5(6):7582–7589. doi:10.1039/C2EE03422K

Hirano T, Kozuka H (2003) Photoanodic properties of ZnO thin films prepared from zinc acetate solutions containing cobalt acetate and polyvinylpyrrolidone. J Mater Sci 38(20):4203–4210. doi:10.1023/A:1026341908829

Hsu C-H, Chen D-H (2012) CdS nanoparticles sensitization of Al-doped ZnO nanorod array thin film with hydrogen treatment as an ITO/FTO-free photoanode for solar water splitting. Nanoscale Res Lett 7(1):593. doi:10.1186/1556-276X-7-593

Hu S, Shaner MR, Beardslee JA, Lichterman M, Brunschwig BS, Lewis NS (2014) Amorphous $TiO_2$ coatings stabilize Si, GaAs, and GaP photoanodes for efficient water oxidation. Science 344(6187):1005–1009. doi:10.1126/science.1251428

Inoue T, Fujishima A, Honda K (1979) Competitive redox reactions at the ZnO semiconductor photoelectrode. Bull Chem Soc Jpn 52(11):3217–3220. doi:10.1246/bcsj.52.3217

Jaramillo TF, Ivanovskaya A, McFarland EW (2001) High-throughput screening system for catalytic hydrogen-producing materials. J Comb Chem 4(1):17–22. doi:10.1021/cc010054k

Jaramillo TF, Baeck S-H, Kleiman-Shwarsctein A, Choi K-S, Stucky GD, McFarland EW (2004a) Automated electrochemical synthesis and photoelectrochemical characterization of $Zn_{1-x}Co_xO$ thin films for solar hydrogen production. J Comb Chem 7(2):264–271. doi:10.1021/cc049864x

Jaramillo TF, Baeck S-H, Kleiman-Shwarsctein A, McFarland EW (2004b) Combinatorial electrochemical synthesis and screening of mesoporous ZnO for photocatalysis. Macromol Rapid Commun 25(1):297–301. doi:10.1002/marc.200300187

Jiang C, Wang R, Parkinson BA (2013) Combinatorial approach to improve photoelectrodes based on BiVO4. ACS Comb Sci 15(12):639–645. doi:10.1021/co300119q

Katz JE (2008) Metal oxide-based photoelectrochemical cells for solar energy conversion. http://resolver.caltech.edu/CaltechETD:etd-10192007-190231

Katz JE, Gingrich TR, Santori EA, Lewis NS (2009) Combinatorial synthesis and high-throughput photopotential and photocurrent screening of mixed-metal oxides for photoelectrochemical water splitting. Energy Environ Sci 2(1):103–112. doi:10.1039/B812177J

Khare C, Sliozberg K, Meyer R, Savan A, Schuhmann W, Ludwig A (2013) Layered $WO_3/TiO_2$ nanostructures with enhanced photocurrent densities. Int J Hydrogen Energy 38(36):15954–15964. doi:10.1016/j.ijhydene.2013.09.142

Khaselev O, Turner JA (1998) A monolithic photovoltaic-photoelectrochemical device for hydrogen production via water splitting. Science 280(5362):425–427. doi:10.1126/science.280.5362.425

Kim DH, Bi L, Aimon NM, Jiang P, Dionne GF, Ross CA (2012) Combinatorial pulsed laser deposition of Fe, Cr, Mn, and Ni-substituted $SrTiO_3$ films on Si substrates. ACS Comb Sci 14 (3):179–190. doi:10.1021/co2001185

Kollender JP, Mardare AI, Hassel AW (2013a) Photoelectrochemical scanning droplet cell microscopy (PE-SDCM). ChemPhysChem 14(3):560–567. doi:10.1002/cphc.201200656

Kollender JP, Mardare AI, Hassel AW (2013b) Localized photoelectrochemistry on a tungsten oxide–iron oxide thin film material library. ACS Comb Sci 15(12):601–608. doi:10.1021/co400051g

Kollender JP, Gallistl B, Mardare AI, Hassel AW (2014) Photoelectrochemical water splitting in a tungsten oxide - nickel oxide thin film material library. Electrochim Acta 140:275–281. doi:10.1016/j.electacta.2014.04.186

Kusama H, Wang N, Miseki Y, Sayama K (2010) Combinatorial search for iron/titanium-based ternary oxides with a visible-light response. J Comb Chem 12(3):356–362. doi:10.1021/cc9001845

Lewerenz HJ (2010) Tailoring of interfaces for the photoelectrochemical conversion of solar energy. Photoelectrochemical Materials and Energy Conversion Processes: 61–181. http://onlinelibrary.wiley.com/doi/10.1002/9783527633227.ch2/summary

Liu X, Shen Y, Yang R, Zou S, Ji X, Shi L, Zhang Y, Liu D, Xiao L, Zheng X, Li S, Fan J, Stucky GD (2012) Inkjet printing assisted synthesis of multicomponent mesoporous metal oxides for ultrafast catalyst exploration. Nano Lett 12(11):5733–5739. doi:10.1021/nl302992q

Loučka T (1977) The reason for the loss of activity of titanium anodes coated with a layer of $RuO_2$ and $TiO_2$. J Appl Electrochem 7(3):211–214. doi:10.1007/BF00618987

McKone JR, Lewis NS, Gray HB (2013) Will solar-driven water-splitting devices see the light of day? Chem Mater 26(1):407–414. doi:10.1021/cm4021518

Meyer R, Sliozberg K, Khare C, Schuhmann W, Ludwig A (2015) High-throughput screening of thin-film semiconductor material libraries II: characterization of Fe-W-O libraries. ChemSusChem 8(7):1279–1285. doi:10.1002/cssc.201402918

Micka K, Gerischer H (1972) Anodic photooxidation of formic acid and methanol at a zinc oxide electrode and the influence of anions. J Electroanal Chem Interfacial Electrochem 38(2):397–402. doi:10.1016/S0022-0728(72)80350-4

Minami T (2008) Present status of transparent conducting oxide thin-film development for indium-tin-oxide (ITO) substitutes. Thin Solid Films 516(17):5822–5828. doi:10.1016/j.tsf.2007.10.063

Minguzzi A, Alpuche-Aviles MA, López JR, Rondinini S, Bard AJ (2008) Screening of oxygen evolution electrocatalysts by scanning electrochemical microscopy using a shielded tip approach. Anal Chem 80(11):4055–4064. doi:10.1021/ac8001287

Młynarek G, Paszkiewicz M, Radniecka A (1984) The effect of ferric ions on the behaviour of a nickelous hydroxide electrode. J Appl Electrochem 14(2):145–149. doi:10.1007/BF00618733

Montoya JH, Garcia-Mota M, Nørskov JK, Vojvodic A (2014) Theoretical evaluation of the surface electrochemistry of perovskites with promising photon absorption properties for solar water splitting. Phys Chem Chem Phys 17(4):2634–2640. doi:10.1039/C4CP05259E

Muster TH, Trinchi A, Markley TA, Lau D, Martin P, Bradbury A, Bendavid A, Dligatch S (2011) A review of high throughput and combinatorial electrochemistry. Electrochim Acta 56 (27):9679–9699. doi:10.1016/j.electacta.2011.09.003

Newhouse PF, Parkinson BA (2015) Combinatorial optimization of spinel $Co_3{-}xMxO_4$ M = (Al, Ga, In) alloyed thin films prepared by ink jet printing: photoelectrochemical, optical, and structural properties. J Mater Chem A 3:5901–5907. doi:10.1039/C4TA05671J

Ohashi K, Mccann J, Bockris JO (1977) Stable photoelectrochemical cells for the splitting of water. Nature 266(5603):610–611. doi:10.1038/266610a0

Osterloh FE, Parkinson BA (2011) Recent developments in solar water-splitting photocatalysis. MRS Bulletin 36(01):17–22. doi:10.1557/mrs.2010.5

Paracchino A, Laporte V, Sivula K, Grätzel M, Thimsen E (2011) Highly active oxide photocathode for photoelectrochemical water reduction. Nat Mater 10(6):456–461. doi:10.1038/nmat3017

Paracchino A, Brauer JC, Moser J-E, Thimsen E, Graetzel M (2012) Synthesis and characterization of high-photoactivity electrodeposited $Cu_2O$ solar absorber by photoelectrochemistry and ultrafast spectroscopy. J Phy Chem C 116(13):7341–7350. doi:10.1021/jp301176y

Parkinson B (2010) Distributed research: a new paradigm for undergraduate research and global problem solving. Energy Environ Sci 3(5):509–511. doi:10.1039/B919523H

Pendlebury SR, Wang X, Le Formal F, Cornuz M, Kafizas A, Tilley SD, Grätzel M, Durrant JR (2014) Ultrafast charge carrier recombination and trapping in hematite photoanodes under applied bias. J Am Chem Soc 136(28):9854–9857. doi:10.1021/ja504473e

Pinaud BA, Benck JD, Seitz LC, Forman AJ, Chen Z, Deutsch TG, James BD, Baum KN, Baum GN, Ardo S, Wang H, Miller E, Jaramillo TF (2013) Technical and economic feasibility of centralized facilities for solar hydrogen production via photocatalysis and photoelectrochemistry. Energy Environ Sci 6(7):1983–2002. doi:10.1039/C3EE40831K

Potyrailo R, Rajan K, Stoewe K, Takeuchi I, Chisholm B, Lam H (2011) Combinatorial and high-throughput screening of materials libraries: review of state of the art. ACS Comb Sci 13 (6):579–633. doi:10.1021/co200007w

Reddington E, Sapienza A, Gurau B, Viswanathan R, Sarangapani S, Smotkin ES, Mallouk TE (1998) Combinatorial electrochemistry: a highly parallel, optical screening method for discovery of better electrocatalysts. Science 280(5370):1735–1737. doi:10.1126/science.280.5370.1735

Rossmeisl J, Qu Z-W, Zhu H, Kroes G-J, Nørskov JK (2007) Electrolysis of water on oxide surfaces. J Electroanal Chem 607(1–2):83–89. doi:10.1016/j.jelechem.2006.11.008

Rowley JG, Do TD, Cleary DA, Parkinson BA (2014) Combinatorial discovery through a distributed outreach program: investigation of the photoelectrolysis activity of P-Type Fe, Cr, Al oxides. ACS Appl Mater Interfaces 6(12):9046–9052. doi:10.1021/am406045j

Rühle S, Anderson AY, Barad H-N, Kupfer B, Bouhadana Y, Rosh-Hodesh E, Zaban A (2012) All-oxide photovoltaics. J Phy Chem Lett 3(24):3755–3764. doi:10.1021/jz3017039

Rühle S, Barad HN, Bouhadana Y, Keller DA, Ginsburg A, Shimanovich K, Majhi K, Lovrincic R, Anderson AY, Zaban A (2014) Combinatorial solar cell libraries for the investigation of different metal back contacts for $TiO_2$–$Cu_2O$ hetero-junction solar cells. Phys Chem Chem Phys 16(15):7066–7073. doi:10.1039/C4CP00532E

Seley D, Ayers K, Parkinson BA (2013) Combinatorial search for improved metal oxide oxygen evolution electrocatalysts in acidic electrolytes. ACS Comb Sci 15(2):82–89. doi:10.1021/co300086h

Seyler M, Stoewe K, Maier WF (2007) New hydrogen-producing photocatalysts—a combinatorial search. Appl Catal Environ 76(1–2):146–157. doi:10.1016/j.apcatb.2007.05.020

Shinde A, Jones RJR, Guevarra D, Mitrovic S, Becerra-Stasiewicz N, Haber JA, Jin J, Gregoire JM (2015) High-throughput screening for acid-stable oxygen evolution electrocatalysts in the (Mn–Co–Ta–Sb)O X composition space. Electrocatalysis 6(2):229–236. doi:10.1007/s12678-014-0237-7

Skorupska K, Maggard PA, Eichberger R, Schwarzburg K, Shahbazi P, Zoellner B, Parkinson BA (2015) Combinatorial investigations of high temperature CuNb oxide phases for photoelectrochemical water splitting. ACS Combinatorial Science. doi:10.1021/acscombsci.5b00142. http://dx.doi.org/10.1021/acscombsci.5b00142

Sliozberg K, Schäfer D, Meyer R, Ludwig A, Schuhmann W (2015a) A combinatorial study of photoelectrochemical properties of Fe-W-O thin films. ChemPlusChem 80(1):136–140. doi:10.1002/cplu.201402277

Sliozberg K, Schäfer D, Erichsen T, Meyer R, Khare C, Ludwig A, Schuhmann W (2015b) High-throughput screening of thin-film semiconductor material libraries I: System development and case study for Ti-W-O. ChemSusChem 8(7):1270–1278. doi:10.1002/cssc.201402917

Sliozberg K, Stein HS, Khare C, Parkinson BA, Ludwig A, Schuhmann W (2015c) Fe–Cr–Al containing oxide semiconductors as potential solar water-splitting materials. ACS Appl Mater Interfaces 7(8):4883–4889. doi:10.1021/am508946e

Stepanovich A, Sliozberg K, Schuhmann W, Ludwig A (2012) Combinatorial development of nanoporous $WO_3$ thin film photoelectrodes for solar water splitting by dealloying of binary alloys. Int J Hydrogen Energy 37(16):11618–11624. doi:10.1016/j.ijhydene.2012.05.039

Terrett NK, Gardner M, Gordon DW, Kobylecki RJ, Steele J (1995) Combinatorial synthesis — the design of compound libraries and their application to drug discovery. Tetrahedron 51 (30):8135–8173. doi:10.1016/0040-4020(95)00467-M

Trasatti S (1984) Electrocatalysis in the anodic evolution of oxygen and chlorine. Electrochim Acta 29(11):1503–1512. doi:10.1016/0013-4686(84)85004-5

Valdés Á, Kroes G-J (2009) First principles study of the photo-oxidation of water on tungsten trioxide (WO3). J Chem Phys 130(11):114701. doi:10.1063/1.3088845

Vojvodic A, Nørskov JK (2011) Optimizing perovskites for the water-splitting reaction. Science 334(6061):1355–1356. doi:10.1126/science.1215081

Walsh A, Yan Y, Huda MN, Al-Jassim MM, Wei S-H (2009) Band edge electronic structure of BiVO4: elucidating the role of the Bi s and V d orbitals. Chem Mater 21(3):547–551. doi:10.1021/cm802894z

Walter MG, Warren EL, McKone JR, Boettcher SW, Mi Q, Santori EA, Lewis NS (2010) Solar water splitting cells. Chem Rev 110(11):6446–6473. doi:10.1021/cr1002326

Weber MF, Dignam MJ (1984) Efficiency of splitting water with semiconducting photoelectrodes. J Electrochem Soc 131(6):1258–1265. doi:10.1149/1.2115797

Winkler GR, Winkler JR (2011) A light emitting diode based photoelectrochemical screener for distributed combinatorial materials discovery. Rev Scientific Ins 82(11):114101. doi:10.1063/1.3657155

Woodhouse M, Parkinson BA (2008a) Combinatorial approaches for the identification and optimization of oxide semiconductors for efficient solar photoelectrolysis. Chem Soc Rev 38 (1):197–210. doi:10.1039/B719545C

Woodhouse M, Parkinson BA (2008b) Combinatorial discovery and optimization of a complex oxide with water photoelectrolysis activity. Chem Mater 20(7):2495–2502. doi:10.1021/cm703099j

Woodhouse M, Herman GS, Parkinson BA (2005) Combinatorial approach to identification of catalysts for the photoelectrolysis of water. Chem Mater 17(17):4318–4324. doi:10.1021/cm050546q

Xiang C, Suram SK, Haber JA, Guevarra DW, Soedarmadji E, Jin J, Gregoire JM (2013) High-throughput bubble screening method for combinatorial discovery of electrocatalysts for water splitting. ACS Comb Sci 16(2):47–52. doi:10.1021/co400151h

Xiang C, Haber J, Marcin M, Mitrovic S, Jin J, Gregoire JM (2014) Mapping quantum yield for (Fe–Zn–Sn–Ti)Ox photoabsorbers using a high throughput photoelectrochemical screening system. ACS Comb Sci 16(3):120–127. doi:10.1021/co400081w

Yan Q, Li G, Newhouse PF, Yu J, Persson KA, Gregoire JM, Neaton JB (2015) Mn2V2O7: an earth abundant light absorber for solar water splitting. Advanced Energy Materials: n/a–n/a. doi:10.1002/aenm.201401840

Ye H, Park HS, Bard AJ (2011) Screening of electrocatalysts for photoelectrochemical water oxidation on W-doped $BiVO_4$ photocatalysts by scanning electrochemical microscopy. J Phy Chem C 115(25):12464–12470. doi:10.1021/jp200852c

Zhang P, Gao L, Song X, Sun J (2015) Micro- and nanostructures of photoelectrodes for solar-driven water splitting. Adv Mater 27(3):562–568. doi:10.1002/adma.201402477

# Chapter 11
# Nanostructured Materials

**James E. Thorne, Yumin He, and Dunwei Wang**

## 11.1 A Brief History of Nanoscale Photoelectrochemistry

Nanoscale photoelectrochemistry is created by introducing nanostructured materials as photoelectrodes. Modern nanoscience and photoelectrochemistry are both relatively new branches of science without much history. The concept of nanoscience can go back to the famous talk "There's Plenty of Room at the Bottom" by R. Feynman in 1959, but modern nanoscience started from Dingle's discovery of quantum size effects in GaAs thin films in 1974 (Dingle et al. 1974). Later, A. Henglein studied small CdS semiconductor nanoparticles in 1982 and 1984 (Henglein 1982; Fojtik et al. 1984). Almost at the same time, in 1983, L. Brus set up a simple model for the ionization potential, electron affinity, and aqueous redox potentials of small semiconductor crystals (Brus 1983). On the other hand, Fujishima and Honda reported the electrochemical photolysis of water at $TiO_2$ electrodes in 1971, demonstrating the use of photoelectrochemical cells for artificial photosynthesis (Fujishima and Honda 1972). Later A.J. Bard's group demonstrated the photocatalytic effects of suspended semiconductor particles (Bard 1979). This was further improved by M. Grätzel's group by using colloidal $TiO_2$ nanoparticles with $RuO_2$ and Pt catalyst to achieve water decomposition under UV light (Duonghong et al. 1981). Recently, more nanoscale semiconductor particles have been synthesized to achieve bias free water splitting, such as $NiO_x$-loaded $NaTaO_3$ nanocrystals (Yokoi et al. 2011) and $NiO_x$-modified $SrTiO_3$ nanocrystals (Townsend et al. 2012).

Despite these early examples of water splitting using nanomaterials, many problems persist for bias-free nanocomposite water splitting. Many designs are

J.E. Thorne • Y. He • D. Wang (✉)
Department of Chemistry, Boston College, 2609 Beacon Street, Chestnut Hill, MA 02467, USA
e-mail: thornejb@bc.edu; heym@bc.edu; dwang@bc.edu



S. Giménez, J. Bisquert (eds.), *Photoelectrochemical Solar Fuel Production*,
DOI 10.1007/978-3-319-29641-8_11

operated with large bandgap materials that only absorb ultraviolet light resulting in low efficiencies. Compared with early colloid nanostructures, photoelectrodes made from different morphologies of nanostructured materials have made significant progress over the past 10 years. Many semiconductor materials, such as $TiO_2$, $Fe_2O_3$, $WO_3$, silicon, have shown improved performances using nanostructures. Compared with bulk materials, nanostructured photoelectrodes can provide higher surface area and a shorter pathway for charge carrier to transport between semiconductor and electrolyte. The PEC performance of many explored nanostructured materials is discussed in greater detail in Sects. 11.5 and 11.6. However, there are difficulties associated with nanoscale morphologies as well. In order to further improve the system, both the positive and negative effects caused by nanostructuring should be considered.

## 11.2 Pros and Cons of Nanostructured Photoelectrodes

### *11.2.1 Shorter Pathway for Charge Carrier Collection*

When a semiconductor absorbs a photon with energy higher than the band gap of the material, an electron–hole pair is generated. In order to drive the water oxidation and reduction reactions, charges need to reach the semiconductor–electrolyte interface (SCLI) as well as the back contact. The longest length for charges to reach the surface, $L$, is the combination of the depletion layer width $W$ and charge carrier diffusion distance $L_D$. If charges are generated too far from the surface, electrons and holes will simply recombine rather than being used for photoelectrochemistry. Recombination is common in bulk materials that have low charge carrier mobilities and lifetimes, since electron–hole generation can happen at absorption depths larger than the $L$.

Nanoscaling of electrodes can help solve the problem of recombination. In Fig. 11.1 models for n-type planar and nanostructured photoelectrodes display how nanostructuring improves charge transport. In Fig. 11.1a, the electron–hole pairs generated deep within the planar photoelectrode cannot reach the surface due to restraints of the $L$. In the case of the nanostructured material, as seen in Fig. 11.1b, no matter where the charge carriers are generated, the distance for the charge to reach the surface is never greater than the $L$ (Lin et al. 2011a, b). This increases the amount of charges reaching the semiconductor–electrolyte interface, making charge collection efficient for the nanostructured material.

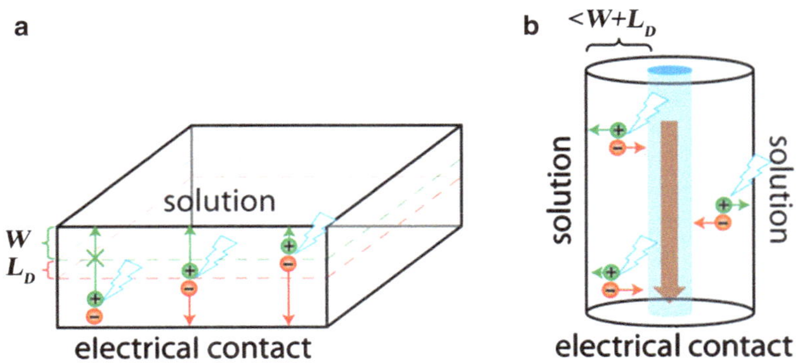


**Fig. 11.1** The schematic for the pathway of photogenerated electron–hole pairs in (**a**) bulk and (**b**) nanostructure. $W$ is the width of space charge region and $L_D$ is the charge carrier diffusion distance. Reproduced with permission from ref. Lin et al. (2011a). Copyright 2011, Elsevier

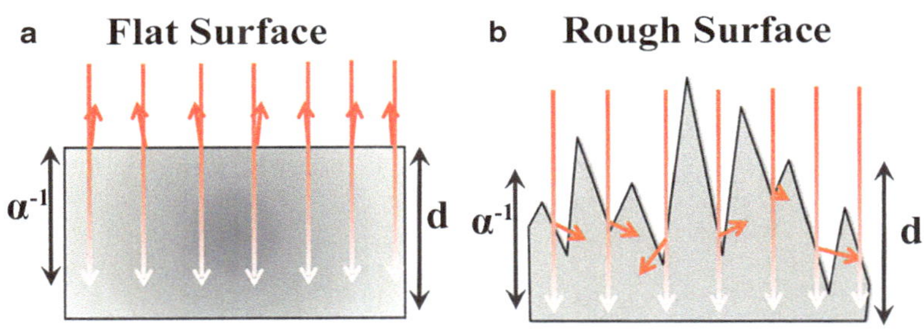


**Fig. 11.2** Light distribution in (**a**) flat and (**b**) nanostructured films. In the picture, $d$ is the film thickness and $\alpha^{-1}$ is the optical penetration depth. *Short arrows* stand for scattered or reflected light. Reproduced with permission from ref. Osterloh (2013). Copyright 2012, Royal Society of Chemistry

### *11.2.2 Higher Light Absorption*

The absorption coefficient, $\alpha$, of a semiconductor indicates how far the light of a particular wavelength can penetrate into the material before being absorbed (Osterloh 2013). The higher the value, the more readily photons are absorbed. The absorption coefficient is determined by the band gap, the type of band-to-band transition within a material, as well as the wavelength of the incident light. Although the absorption coefficient remains the same for a nanostructured material, a nanostructure can enhance light absorption by absorbing light that has been scattered as indicated in Fig. 11.2. The high aspect ratio of a nanostructured surface creates a greater chance that light will be scattered rather than being directly absorbed or reflected, as in the case of a flat surface. Large increases, due to scattering, assume that the incident light is perpendicular to the surface and the morphology is well controlled.

### 11.2.3 Increased Surface Area

The surface area of photoelectrodes can be greatly enlarged through the using of nanostructures. However, this advantage will bring both positive and negative effects on the PEC performance of the electrode. Increased surface areas allow for more charges to reach the SCLI, which enhances the current density of electrode at a fixed illumination area. On the other hand, the larger surface area will result in a greater number of charges recombining at surface defects. In order to improve the overall performance of the photoelectrode, the surface area needs to increase without proportionately increasing the surface defects.

### 11.2.4 Reduced Depletion Layer Thickness

In order to separate photogenerated electrons and holes it is essential to maximize the depletion layer, or space charge layer. The direction of the electric field within the depletion layer pushes minority carriers to the surface and majority carriers to the back contact. If a nanomaterial is made smaller than the depletion layer then it will be more difficult to separate charges. The thickness of the depletion layer can be derived from equation seen below.

$$W = \left[\frac{2\varepsilon_{S} V_{SC}}{q N_{D}}\right]^{\frac{1}{2}} \tag{11.1}$$

In the equation, $W$, is the thickness of the depletion layer, while $V_{SC}$ is the built-in voltage within the depletion layer. The static dielectric constant is $\varepsilon_{S}$, while the electric charge is $q$, and the doping density of the semiconductor $N_{D}$, which is known for a certain materials. Under zero bias the $V_{SC}$ equals the flat-band potential, which can be measured. Typical dimensions of $W$ are around 10–1000 nm. Therefore it is important to consider the dimensions of the depletion layer when nanostructuring photoelectrodes.

## 11.3 Different Morphologies of Nanostructures

The morphologies of nanostructures can have several different dimensions from 0 to 3. In this section, a brief description of each of these different structures is given. The performance of certain nanostructured materials are discussed further in Sect. 11.5.

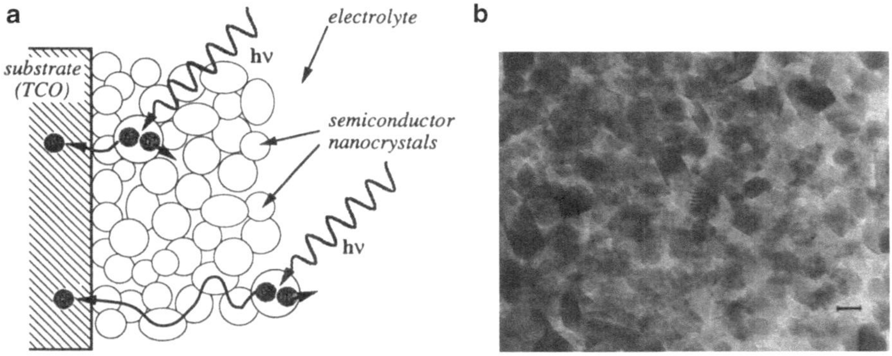


**Fig. 11.3** (**a**) Model of the charge carrier separation and charge transport in a nanocrystalline film. Reproduced with permission from ref. Hagfeldt and Graetzel (1995). Copyright 1995, American Chemical Society. (**b**) Transmission electron microscopy (TEM) image of $TiO_2$ particles in DSSC. The scale bar is 10 nm. Reproduced with permission from ref. O'Regan and Gratzel (1991). Copyright 1991, Nature Publishing Group

### *11.3.1 Zero-Dimensional Nanoparticles*

Although networks of nanoparticles such as $TiO_2$ nanoparticles have been widely used in dye-sensitized solar cells (DSSC) (O'Regan and Gratzel 1991), photoelectrodes consisting of such nanoparticles are relatively rare for PEC water splitting applications. The low charge transfer rates of nanoparticle electrodes make them less than ideal nanostructure photoelectrodes (Fig. 11.3).

### *11.3.2 One-Dimensional Nanostructures for Water Splitting*

One-dimensional (1D) nanostructures such as nanowires, nanotubes, and nanorods have received significant attention as photoelectrode materials. Compared with nanoparticles, charge transfer within 1D nanostructures do not suffer from charge transport problems as charge collection by the crystalline nanomaterial proceeds with ease. As illustrated in Sect. 11.2, the effect of a shortened pathway for charge carrier transport is an important advantage of the 1D nanostructured materials, especially for semiconductors with low charge mobilities and lifetimes.

Nanotubes are different from the other two 1D nanostructures because of their hollow structure. One of the most explored nanotube materials is $TiO_2$ synthesized by electrochemical oxidation. Grimes et al. have made significant improvements by using highly ordered $TiO_2$ nanotube arrays for photoelectrochemistry (Mor et al. 2004). By controlling a set of parameters such as anodization potential, electrolyte compositions, and anodization time and temperature, $TiO_2$ nanotubes can be synthesized with different pore sizes and wall thicknesses. While $TiO_2$

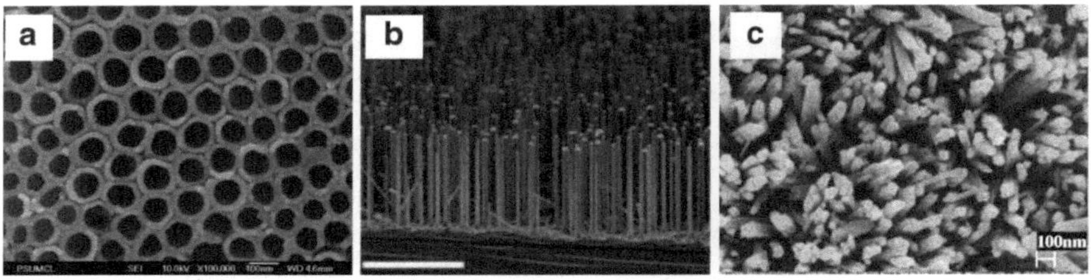


**Fig. 11.4** Scanning electron microscopy (SEM) images of different 1D nanostructured materials. (**a**) $TiO_2$ nanotubes. Reproduced with permission from ref. Grimes (2007). Copyright 2007, Royal Society of Chemistry. (**b**) Si nanowires. Reproduced with permission from ref. Yuan et al. (2009). Copyright 2009, John Wiley and Sons. (**c**) $Fe_2O_3$ nanorods. Reproduced with permission from ref. Vayssieres et al. (2001)). Copyright 2001, American Chemical Society. The scale bar in (**a**), (**b**), and (**c**) is 100 nm, 10 μm, and 100 nm, respectively

nanotube photoanodes can show 16.25 % power conversion efficiency under UV illumination (320–400 nm), the large band gap of $TiO_2$ ($E_g = 3.2$ eV) prevents it from absorbing visible light. Therefore, nanotubes of materials with smaller band gap such as $Fe_2O_3$ and $WO_3$ have also been pursued. In addition, vertically oriented Ti-Fe-O (Mor et al. 2007), Cu-Ti-O (Mor et al. 2008), and $Ta_3N_5$ (Feng et al. 2010) nanotube arrays have also been reported.

A wide range of materials have also been synthesized into nanowires and nanorods as photoelectrodes for water splitting, including Si (Peng et al. 2009; Yuan et al. 2009, 2011; Dai et al. 2013), $Fe_2O_3$ (Beermann et al. 2000; Vayssieres et al. 2001; Mao et al. 2011a, b; Mayer et al. 2012), $WO_3$ (Chakrapani et al. 2009; Rao et al. 2013), $TiO_2$ (Feng et al. 2008; Shankar et al. 2009), ZnO (Yang et al. 2009), $BiVO_4$ (Su et al. 2009), and $W_2N$ (Chakrapani et al. 2009) (Fig. 11.4).

### *11.3.3 Two/Three-Dimensional Nanostructures for Water Splitting*

Compared with 1D nanostructures, two-dimensional (2D) nanostructures are less common. The Grimes group has synthesized $WO_3$ nanoflakes on FTO substrates, which show higher photocurrent than $WO_3$ nanowires obtained through similar synthesis (Su et al. 2010). (Zhou et al. 2010) prepared ordered $BiVO_4$ nanowall films, which expose high-active facets. After the deposition of a Co-Pi catalysts, the $BiVO_4$ photoanodes yielded better absorbed photon-to-current efficiency (APCE) and a significant cathodic shift of onset potential (Zhou et al. 2012).

Three-dimensional (3D) nanostructures usually consist of porous thin films in which the sizes of pores are on the nanoscale. Macroporous or mesoporous semiconductor films such as $Fe_2O_3$ (Prakasam et al. 2006; Brillet et al. 2010), $WO_3$ (Santato et al. 2001; Watcharenwong et al. 2008; Li et al. 2010; Kim et al. 2011), $Cu_2O$ (Li et al. 2007), $Cu_2ZnSnS_4$ (Ma et al. 2011), and GaP (Price and Maldonado

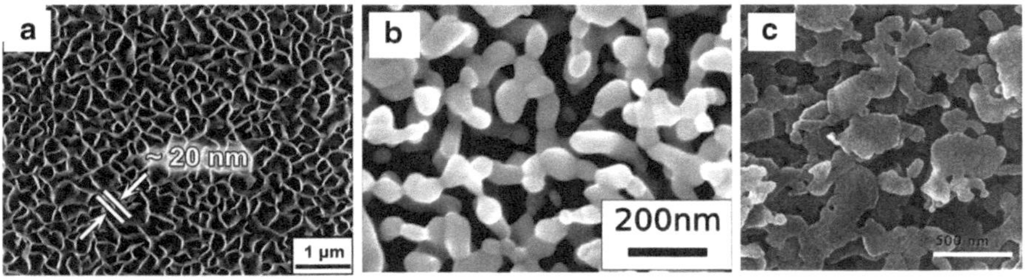


**Fig. 11.5** SEM images of different 2D nanostructured materials. (**a**) $BiVO_4$ nanowalls. Reproduced with permission from ref. Zhou et al. (2012). Copyright 2012, John Wiley and Sons. (**b**) $Fe_2O_3$ porous films. Reproduced with permission from ref. Brillet et al. (2010). Copyright 2010, American Chemical Society. (**c**) $WO_3$ porous films. Reproduced with permission from ref. Santato et al. (2001). Copyright 2001, American Chemical Society. The scale bar in (**a**), (**b**) and (**c**) is 1 μm, 200 nm and 500 nm, respectively

2009; Hagedorn et al. 2010) have been reported. Such porous structures provide a highly accessible surface and better structural support when compared with 0D and 1D nanomaterials. In addition, some 3D branched nanostructures, such as branched $TiO_2$ nanorods, have also synthesized (Cho et al. 2011). The branched nanorods outperform $TiO_2$ nanorods due to its higher surface area and better charge separation/collection (Fig. 11.5).

### *11.3.4 Heteronanostructures for Water Splitting*

In addition to different morphologies for nanoscale structures, the performance of photoelectrodes can be enhanced through the heteronanostructures, which consist of two layered nanoscale semiconductors. Typically, semiconductors with different band gaps are chosen which can lead to a synergistic effect. This effect can extend the absorption spectrum of the photoelectrodes, while also maintaining the advantages gained through nanostructuring. Different combinations include n–n junctions, p–p junctions, and p–n junctions, depending on the band gaps and band positions. Specific materials chosen for building heteronanostructures are discussed further in Sect. 11.6.

## 11.4 Synthesis Methods

There are many ways to synthesize nanostructured semiconductors based on the desired morphology. Of the different synthesis methods that have been explored, these methods are summarized and divided into solution phase methods, electrochemical anodization, gas phase methods, and template induced methods.

### *11.4.1 Solution Phase Methods*

Solution phase methods such as sol–gel, hydrothermal and solvothermal syntheses are common ways to prepare nanostructured materials. They provide good control of the product's composition, while also producing a large amount of product, compared with gas phase syntheses. Additionally, other types of solution phase synthesis, such as chemical bath deposition and chemical etching, are also reported as methods to prepare nanostructured materials.

The sol–gel method is commonly used to synthesize porous $WO_3$ thin films. This procedure involves the elution of W precursor solution by adding organic solvent such as ethanol or isopropanol. The organic solvent slows down the condensation of tungstate. Then the organic stabilizer, such as poly(ethylene glycol) (PEG) 300, glycerol, mannitol, or ethylene glycol, is added to improve the porous structure. Finally, the $WO_3$ film is prepared by spreading the synthesized solution onto substrate and annealing in oxygen to increase the crystallinity. Santato et al. (2001) used $H_2WO_4$, obtained by the acidification of $Na_2WO_4$ solution, to produce mesoporous $WO_3$ thin films. The size of the plate-like nanoparticles and the porosity were both increased by raising the annealing temperature. The optimal 2.5 μm thin film annealed under 550 °C showed about 1.7 mA/cm$^2$ of photocurrent density at 0.6 V vs Hg/$HgSO_4$ in 1 M aqueous $HClO_4$ with simulated AM1.5 solar light. Another type of W precursor, peroxy-tungstate, was used by Kim et al. (2011) to synthesize mesoporous films of $WO_3$. The highly transparent sphere-like $WO_3$ nanoparticle films could provide 3.7 mA/cm$^2$ of photocurrent density at 1.3 V vs Ag/AgCl in 1 M $H_2SO_4$ under 100 mW/cm$^2$ illumination. By controlling the W/PEG 300 ratio, nanoparticles with different diameters and porosity could be obtained.

Hydrothermal and solvothermal methods are similar, in that the product is synthesized from a high temperature solution at high vapor pressure in a stainless steel autoclave. Feng et al. (2008) reported using hydrothermal method to synthesize $TiO_2$ nanowires on FTO substrates. Toluene, $Ti^{4+}$ precursors, and a small portion of hydrochloric acid were mixed in an autoclave. Under 180 °C, the water would diffuse away from the water–toluene interface on substrate surface while $Ti^{4+}$ precursor formed a seed on the interface, which would continue growing with constant hydrolysis. This method is highly reproducible and the organic and water phases can be recycled for further use. Vertically aligned $WO_3$ nanowires and nanoflakes were also synthesized through a solvothermal method (Su et al. 2010). By varying parameters such as the amount of water, pH, the amount of oxalic acid and urea, the morphology of $WO_3$ can be selectively tuned to produce nanowires or nanoflakes.

Chemical bath deposition (CBD) is the cheapest method for the deposition of thin films and nanomaterials. In a typical CBD procedure, the substrate is immersed into a precursor solution and the reaction is performed under a certain temperature. Wang et al. reported that CdS/CdSe core–shell nanorod arrays gave a maximum photocurrent density of 5.3 mA/cm$^2$ for $SO_3{}^{2-}$ oxidation at 0.7 V vs Ag/AgCl in

0.5 M $Na_2SO_3$ under 100 mW/cm$^2$ illumination, which was much better compared with bare CdS or CdSe (Wang et al. 2013). This heteronanostructure was prepared by a CBD synthesis of CdSe on hydrothermally synthesized CdS nanorods. Another example of a chemical bath nanostructure synthesis was reported by Vayssieres et al., in which β-FeOOH nanorods were grown in a mixture of sodium nitrate and ferric chloride solution followed by annealing at temperatures higher than 390 °C to form hematite (Vayssieres et al. 2001).

Electroless etching is used as an effective way to prepare Si nanowires. The typical procedure involves two steps. The first step is to obtain a clean Si substrate while the second step is etching in $HF/AgNO_3$ solution. After the deposition of a platinum catalyst on the electroless etched Si nanowires, Peng et al. (2009) reported an apparent energy conversion efficiency of up to 8.14 % in $Br^-/Br_2$ electrolyte. Following a similar synthesis method, Yuan et al. (2009) reported an efficiency of 10.1 % in 1 M $LiClO_4$ with $Me_2Fc^{+/0}$ redox pair under 100 mW/cm$^2$ of illumination. Dai et al. (2013) used atomic layer deposition (ALD) to deposit Pt nanoparticles, as HER catalyst, on Si nanowires prepared using electroless etching, which gave better turn-on voltages than planar silicon (Fig. 11.6).

### *11.4.2 Electrochemical Anodization*

Electrochemical anodization uses a high electric field in a metal anode to form nanostructures. The metal ions will be driven toward the solution by the applied field and can form an oxide with $OH^-$. After the formed metal oxide becomes thicker, the intensity of the electric field, within the oxide layer, becomes smaller until metal oxide can no longer be formed. Different morphologies of metal oxide can be obtained such as compact oxide, random porous oxide, oriented porous oxide and oriented tubular oxide depending on the degree of self-ordering process. This process will be affected by the electrolyte, anodization voltage, temperature, and the impurities in the material (Asoh and Ono 2007).

For the formation of $TiO_2$ nanotubes, fluoride ions are essential, due to the water-soluble $TiF_6^{2-}$. Grimes (2007) synthesized $TiO_2$ nanotubes in an ethylene glycol electrolyte containing 0.25 wt% $NH_4F$ under 60 V. The specific dimensions of the $TiO_2$ nanotubes can be controlled by varying the synthesis conditions. For example, Shankar et al. (2009) pointed out that by increasing size of the electrolyte cations they could obtain longer $TiO_2$ nanotubes.

Nanostructured hematite, such as nanotubes and porous thin films, can also be synthesized using an anodization method. Smooth and ultrathin hematite nanotubes can be obtained through the anodization of Fe foil, as reported by the Misra group (Mohapatra et al. 2009). Their electrolyte consisted of 0.5 wt% ammonium fluoride in an aqueous ethylene glycol solution (3 vol.% water) and the anodization was performed under 50 V for varying amounts of time. When the anodization time was limited to 6 min, nanoporous thin films were obtained. When the anodization time was extended to 13 min, $Fe_2O_3$ nanotubes with a thickness of 5–7 nm and 3–4 μm

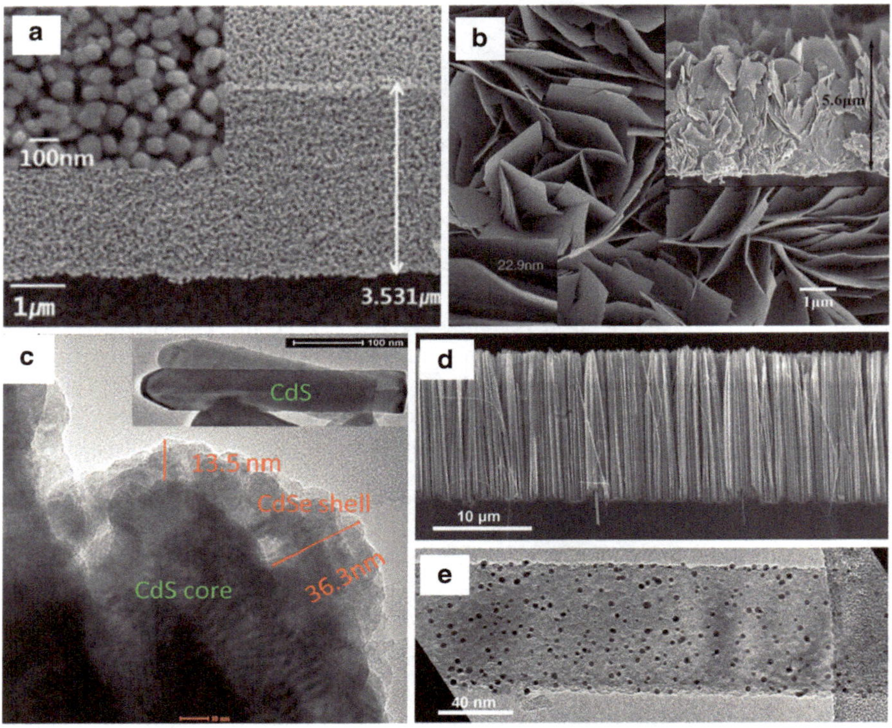


**Fig. 11.6** Electron micrograph of nanostructured materials synthesized using solution methods: (**a**) porous $WO_3$ film on FTO glass using a sol–gel method. *Inset* shows detailed morphology under higher magnification. Reproduced with permission from ref. Kim et al. (2011). Copyright 2011, Royal Society of Chemistry. (**b**) $WO_3$ nanoflakes using a hydrothermal method. *Inset* shows film cross section. Reproduced with permission from ref. Su et al. (2010). Copyright 2011, American Chemical Society. (**c**) CdS/CdSe core–shell nanorods using CBD method. The scale bar is 10 nm. *Inset* shows the morphology of bare CdS nanorod. Reproduced with permission from ref. Wang et al. (2013). Copyright 2013, American Chemical Society. (**d**) Si nanowires grown using an electroless etching decorated with Pt nanoparticles using atomic layer deposition (ALD). (**e**) TEM image of Si nanowire from (**e**) showing the distribution of ALD Pt nanoparticles. Both (**d**) and (**e**) are reproduced with permission from ref. Dai et al. (2013). Copyright 2013, John Wiley and Sons

length were produced. After annealing in $O_2$ and $H_2$ (10 % in Ar) atmosphere under 500 °C for 3 h, 1.4 mA/cm$^2$ of photocurrent was obtained at 0.5 V vs Ag/AgCl in 1 M KOH solution under AM 1.5 of illumination.

The anodization strategy can also be used for the formation of $WO_3$ nanostructures. Li et al. (2010) reported that when W foil was anodized in electrolyte without $F^-$, compact $WO_3$ films were formed. By adding NaF to the electrolyte during anodization, the barrier oxide layer was rapidly dissolved into a soluble fluoride complex, leading to the formation of a nanoporous film with a pore diameter size of 70–90 nm. After annealing at 450 °C for 3 h, the nanoporous film showed

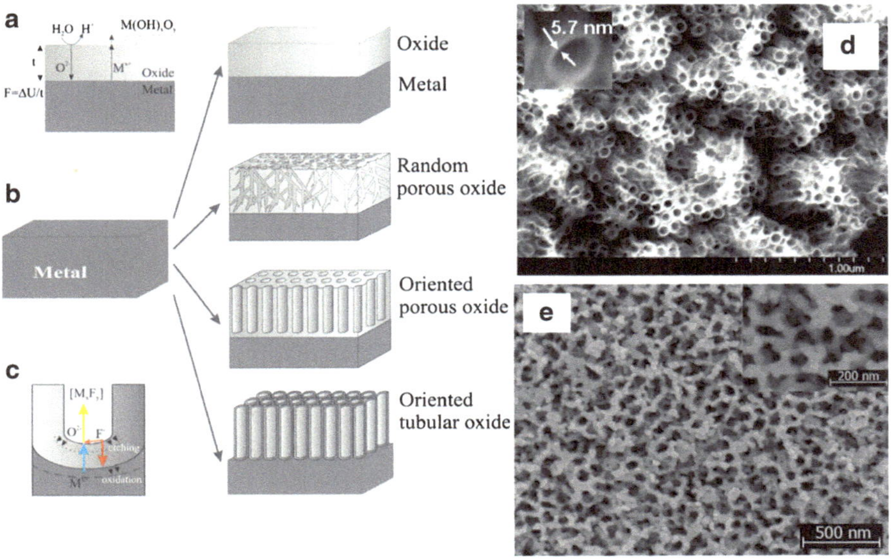


**Fig. 11.7** (**a**) The schematic of the oxide formation mechanism during anodization. (**b**) Different morphologies that can be obtained by anodization for Titanium under different conditions. (**c**) The mechanism for Titanium nanotube formation. (**a–c**) are reproduced with permission from ref. Ghicov and Schmuki (2009). Copyright 2009, Royal Society of Chemistry. (**d**) SEM image of $Fe_2O_3$ nanotubes. Reproduced with permission from ref. Mohapatra et al. (2009). Copyright 2009, American Chemical Society. (**e**) SEM image of porous $WO_3$ films grown by anodization. Reproduced with permission from ref. Li et al. (2010). Copyright 2010, Elsevier

four times higher photocurrent than when compared to the compact film. Cristino et al. (2011) prepared a wormlike $WO_3$ film by anodizing W foil in the electrolyte of NMF/$H_2O$/$NH_4F$ at 40 V for 72 h (Fig. 11.7).

In summary, the potentiostatic anodization of metal foil combined with an annealing treatment, to improve the crystallinity of the metal oxide, is an effective way to synthesize porous or tubular nanostructure materials for water splitting.

### *11.4.3 Gas Phase Methods*

Synthesis of nanostructured materials for PEC water splitting by gas phase deposition is another route that has been studied. Common syntheses include chemical vapor deposition, molecular beam epitaxy, and flame vapor deposition.

Chemical vapor deposition (CVD) has been widely used to synthesize semiconducting nanowires. For example, Si nanowires can be grown using a vapor–liquid–solid (VLS) method, which involves alloying, nucleation, and then growth. After the temperature reaches the Au-Si eutectic point, Au-Si alloy droplets will form, which then adsorbs the Si precursor, resulting in nucleation. Yuan et al. (2009)

synthesized n-type Si nanowire in CVD system by a slow Si precursor feeding rate to achieve a slow growth rate. Efficient energy conversion was achieved using the highly oriented Si nanowires, with 20 μm length and 200 nm width, in a $Me_2Fc/Me_2Fc^+$ electrolyte. Cauliflower-type α-$Fe_2O_3$ was synthesized by the Grätzel group using $Fe(CO)_5$ as precursor in atmospheric pressure CVD (APCVD) (Kay et al. 2006). The photoanode, without a catalyst, showed 2.2 mA/cm$^2$ at 1.23 V vs RHE in 1 M NaOH under 100 mW/cm$^2$ of illumination. Aerosol-assisted CVD (AACVD) has been used to synthesize $Bi_2S_3$ nanotubes reported by Tahir et al. (2010) by creating an aerosol precursor, $Bi_2[(S_2CN(C_2H_5)_2)_3]_2$. By adjusting the solvent and temperature, different morphologies could be obtained.

Another type of CVD to synthesize semiconducting nanowires is called metalorganic chemical vapor deposition (MOCVD). In this method, metalorganics and hydrides, containing the required chemical elements, are used as the precursors to synthesize semiconductors containing more than one element. Gao et al. (2014) synthesized p-type InP nanowire arrays using MOCVD, which also followed the VLS mechanism. By introducing a $MoS_3$ HER catalyst on the nanowire, the p-InP nanowire showed higher photocurrent and fill factor than planar InP.

$WO_3$ nanowires can also be synthesized using an oxygen flow over a hot tungsten filament in a CVD system. Instead of forming a liquid metal droplet, as a medium for further nucleation and growth as in VLS mechanism, $WO_3$ nanowires were formed due to a $WO_2$ cluster on the substrate as nucleation site, which enhanced the oxidation and formation of $WO_{3-x}$ nanowires (Thangala et al. 2007). Chakrapani et al. (2009) used $WO_3$ nanowires synthesized by this method as photoanode and achieved 85 % IPCE at 370 nm at 1.2 V vs SCE (Fig. 11.8).

Unlike CVD, which utilizes the chemical reaction of gas precursors, molecular beam epitaxy (Zhang et al. 2014) grows semiconducting nanowires by the physical deposition of element gases. Wang et al. prepared GaN nanowires using a radio frequency plasma-assisted MBE under nitrogen-rich conditions (Wang et al. 2011). Unlike using metal droplets as nucleation site to grow nanowires, as with CVD, the GaN nanowires were synthesized by a nucleation free method. Following a similar method, Kamimura et al. (2013) synthesized p-type (In,Ga)N nanowires using Mg as dopant and a photocathode with platinum, and these nanowires showed 5 mA/cm$^2$ of photocurrent density at −0.5 V vs NHE in 0.5 M $H_2SO_4$ under ~40 mW/cm$^2$ of illumination.

Another interesting method used to synthesize nanostructured materials by gas phase, is flame vapor deposition (FVD). Compared with the CVD method, the FVD method can provide a more rapid growth rate of the nanomaterials as well as scalability and low cost. The strategy is to use high-temperature flame to generate metal oxide vapor from metal precursor and the vapors will deposit onto a colder substrate in the downstream. Rao et al. (2013) synthesized $WO_3$ nanowires by annealing as-grown $W_{18}O_{49}$ nanowires through FVD in oxygen atmosphere. This $WO_3$ photoanode showed higher saturation photocurrent densities than the $WO_3$ nanowire synthesized by hot-wire CVD or the hydrothermal method. The flame can be also used to enhance the PEC performance of metal oxide nanowires, including $TiO_2$, hematite, and ZnO by increasing the oxygen vacancies, as indicated by the Zheng group (Cho et al. 2013).

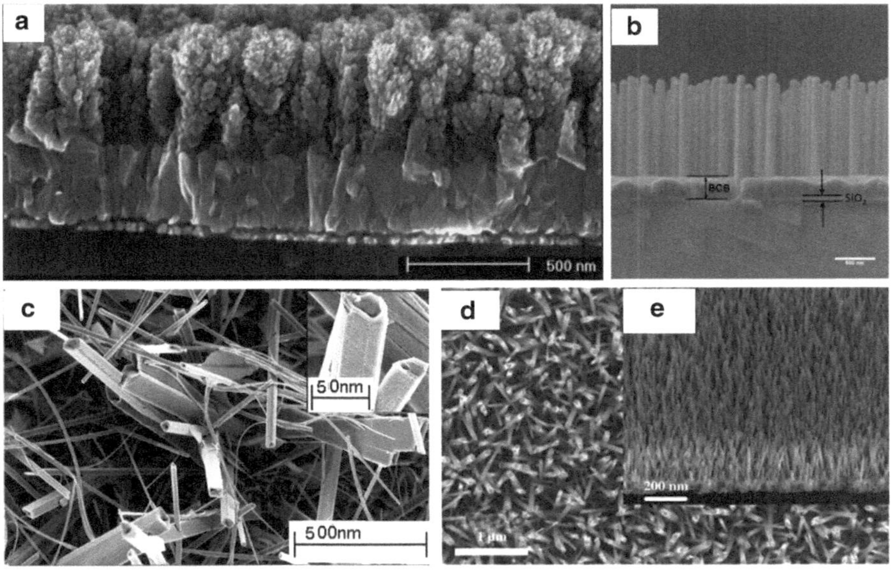


**Fig. 11.8** SEM images of nanostructured materials synthesized through CVD. (**a**) Cauliflower-like $Fe_2O_3$ synthesized by APCVD. Reproduced with permission from ref. Kay et al. (2006). Copyright 2006, American Chemical Society. (**b**) InP nanowire synthesized by MOCVD. The scale bar is 500 nm. Reproduced with permission from ref. Gao et al. (2014). Copyright 2014, American Chemical Society. (**c**) $Bi_2S_3$ nanotubes synthesized by AACVD. *Inset* shows the detailed morphology under higher magnification. Reproduced with permission from ref. Tahir et al. (2010). Copyright 2010, American Chemical Society. (**d**) $WO_3$ nanowires synthesized by hot-filament-assisted CVD. The scale bar is 1 μm. (**e**) The cross-section SEM image of $WO_3$ nanowires on FTO substrate. The scale bar is 200 nm. Reproduced with permission from ref. Thangala et al. (2007). Copyright 2007, John Wiley and Sons

### *11.4.4 Template Induced Synthesis*

Due to the intrinsic properties of semiconductor materials, some complex nanostructures are not easily grown from typical starting materials. Therefore, another helpful method to synthesize those nanostructures is the use of a nanostructured template. There are two kinds of usage for templates. One type of template serves only as the mold to synthesize nanostructure, which will be removed after the synthesis. In another condition, the template itself has the PEC performance or can serve as a beneficial substrate in the photoelectrode. Occasionally, the semiconductor material will show better performance when it is grown on a template, which is known as a heteronanostructure.

Anodized aluminum oxide (AAO) templates are used to prepare hematite nanorods, as reported by Mao et al. (2011a, b). At first, short Au nanorods were electrodeposited into the pores of AAO. Then Fe nanorods were co-electrodeposited onto Au nanorods, followed by annealing in $O_2$ for the conversion into hematite. The AAO template could be removed by soaking in a 0.5 M NaOH solution. Similar method can be used to prepare hematite nanotube by introducing polyaniline (Chakrapani et al.) nanorod before Fe deposition. When the lengths of both the nanorods and nanotubes were about 10 μm, the nanotubular hematite showed more negative onset potential than the nanorod ones. In addition, the hematite nanotubes could achieve 2.18 mA/cm$^2$ photocurrent density at 0.43 V vs Ag/AgCl under AM 1.5G light illumination in 1 M NaOH, which was higher than that of the nanorod ones. Brillet et al. (2010) prepared mesoporous hematite films with the help of a $SiO_2$ confinement scaffold. Compared with porous hematite, annealed at high temperatures, the scaffold consisted of a mesoporous layer of varying particle sizes, depending on the annealing temperature. The smaller particle sizes tended to reduce the scattering of unabsorbed photons, which improved the photocurrent significantly. Another highly ordered template used to synthesize 3D macroporous material is a polystyrene (PS) colloidal crystal template. Li et al. (2007) prepared 3D ordered macroporous $Cu_2O$ through electrochemical reduction of copper lactate in alkaline solution using the PS template. Compared with the bulk $Cu_2O$, an enhanced photocurrent was observed in the nanostructured photocathode.

Some nanostructured materials can be synthesized from their oxide or hydroxide nanostructures. CdSe nanotubes (NT) are synthesized from facile solution-based selenization of $Cd(OH)_2$ nanowires bundles (Santori et al.) as reported by Shim et al. (2009). The diameter of the nanotubes depends on the dimension of the Cd $(OH)_2$ template. The nitridation of $WO_3$ nanowires in $NH_3$ atmosphere yields $W_2N$ nanowires (Chakrapani et al. 2009). A Similar strategy has been applied to form TaON nanotubes, in which was nitridation of $Ta_2O_5$ nanotubes was performed (Banerjee et al. 2009).

In order to form heteronanostructures, the base layer is used as a nanostructured template. $TiSi_2$ nanonets are a unique 2D nanostructured material developed by the Wang group (Zhou et al. 2009). Because of its high surface area and excellent conductivity, $TiSi_2$ nanonets have been used as a substrate for other semiconductor materials. Lin et al. and Liu et al. deposited $TiO_2$ (Lin et al. 2009), $WO_3$ (Liu et al. 2011), and hematite (Lin et al. 2011a, b) on the $TiSi_2$ nanonets using ALD. The ALD procedure provides good crystallinity and uniformity of materials such as $TiSi_2$. Hwang et al. used ALD to grow $TiO_2$ on Si nanowires prepared by electroless etching (Hwang et al. 2008). The heteronanostructures showed better performance than the combination of planar Si and $TiO_2$, due to its lower reflectance and higher surface area (Fig. 11.9).

## 11.5 Explored Nanostructured Materials

This section discusses the photoelectrochemical properties of some of the materials which have been made into nanostructures. Nanostructuring has proven to be beneficial for most photoelectrodes for water splitting and in some cases, the ability

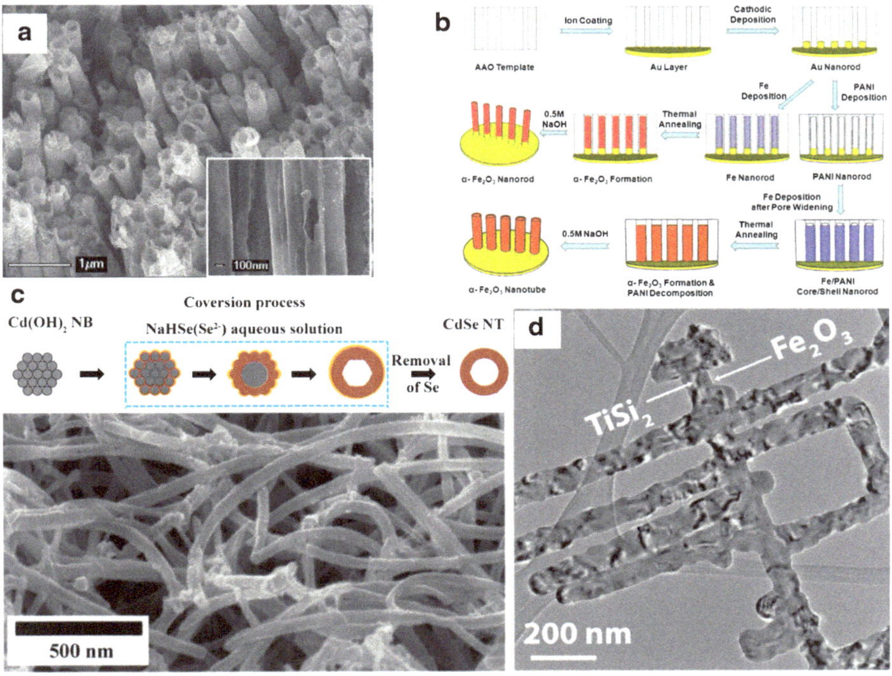


**Fig. 11.9** (**a**) SEM image of hematite nanotubes and (**b**) the schematic for the synthesis of hematite nanorods and nanotubes; Reproduced with permission from ref. Mao et al. (2011a, b). Copyright 2011, American Chemical Society. (**c**) SEM image of CdSe nanotubes and the schematic for its synthesis from *top view*; Reproduced with permission from ref. Shim et al. (2009). Copyright 2009, American Chemical Society. (**d**) TEM image of $TiSi_2$ coated with hematite. Reproduced with permission from ref. Lin et al. (2011a, b). Copyright 2011, American Chemical Society

to create nanostructures, has rejuvenated the research around the material. The different morphologies used and their corresponding performances are discussed (Fig. 11.10).

### 11.5.1 Titanium Dioxide

$TiO_2$ nanoparticles were one of the first nanostructures put to use for PEC water splitting. A mesoporous film sensitized with a ruthenium dye showed a vast improvement over the planar $TiO_2$ crystal with the main difference being the improved light collection made possible by the mesoporous $TiO_2$ nanoparticles (O'Regan and Gratzel 1991). Following this demonstration, a great deal of research has been put into the development of $TiO_2$ nanostructures and subsequently

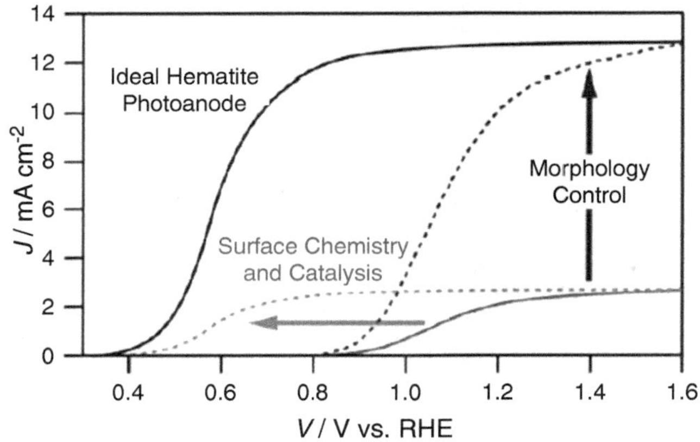


**Fig. 11.10** A J-V curve depicting the photocurrent that can be gained by creating beneficial morphologies. The *solid grey line* depicts one of the best hematite photoanodes to date, with the *dashed lines* showing performance enhancements to the onset potential and photocurrent by the addition of catalyst and surface medications and better morphology developments, respectively. 0.4 V vs RHE and ~12 mA/cm$^2$ are approximately the theoretical maximums for hematite. Reproduced from ref. Sivula et al. (2011)

procedures to grow $TiO_2$ nanowires, nanotubes, and branched rods have been developed.

Of the different structures that have been explored, nanotubes have shown a great deal of promise for water splitting applications. As Mor et al. (2006) showed, titania nanotubes grown using an anodization process can be used for water splitting and also as a support for dye molecules. The nanotubes grown at 5 °C and with dye were found to have photocurrent densities as high as 7.87 mA/cm$^2$ under 100 mW/cm$^2$ of illumination, which is competitive with similar nanoparticle designs.

To continue to improve the $TiO_2$ performances, much effort has been put into increasing the amount of solar light absorption by introducing dopants. The large bandgap of $TiO_2$, 3.2 eV, allows it to only absorb the UV portion of sunlight. Doping $TiO_2$ nanostructures creates a method to maximize the charge transportations while also possibly increasing the light absorption to include more of the visible light. The Bard group has shown that C and N doping of $TiO_2$ can lead to light absorption in the visible region which can increase the STH efficiency to 0.55 % (Park et al. 2005).

### 11.5.2 *Zinc Oxide*

While ZnO has a bandgap that is even slightly larger than $TiO_2$ efforts have still been made to explore the nanostructuring of ZnO. One of the first reports of nanostructured ZnO came from the Hagfeldt group in 1999 when they showed that ZnO nanowires that were ~7 μm in length could increase the IPCE values of ZnO up to around 80 % (Keis et al. 1999). While more recently Wolcott et al. (2009) found that a film of ZnO nanoparticles was able to deliver 175 μA/$cm^2$ of photocurrent densities at 1.6 V vs NHE at pH 7.4. Using nitrogen doping Yat Li's group was able to show a significant increase in performance of ZnO nanorods to 400 μA/$cm^2$ at 1 V vs Ag/AgCl compared with 17 μA/$cm^2$ for the undoped nanorods under the same conditions (Yang et al. 2009).

### 11.5.3 Tungsten Trioxide

With a band gap of 2.6–2.7 eV and a sufficiently anodic valence band, $WO_3$ can be used as a photoanode for water splitting. An interesting property of $WO_3$ is its stability in acidic conditions, making it an interesting candidate as a tandem cell absorber. However the photocurrents for the bulk material are low, typically less than 3 mA/$cm^2$, which is at least in part limited by the diffusion length of holes in $WO_3$, which is known to be about 150 nm (Butler 1977). Using nanostructuring the higher surface area allows for higher light absorption while the nanostructure allows photogenerated holes a higher probability to diffuse to the semiconductor–electrolyte interface. Augustynski's group demonstrated in 2001 that the performance of $WO_3$ can be improved by nanostructuring (Santato et al. 2001). In this report, it is shown that a 2.5 μm thin films of $WO_3$ nanocrystals, synthesized by annealing a tungsten acid and ethylene glycol solution in oxygen at 550 °C for 30 min, yielded higher IPCE values than when compared to planar $WO_3$, indicating that the nanosheets were indeed more effective at capturing the incident photons. It has been shown that the photocurrent density of nanoporous $WO_3$ is six times that of its bulk counterpart at 1.0 V vs RHE, as demonstrated by van de Krol et al. (de Respinis et al. 2013). The Bisquert group produced photocurrents of 3.8 mA/$cm^2$ at 1 V vs SCE in 1 M $H_2SO_4$ at 0.3 W/$cm^2$ illumination intensity (Tacca et al. 2012). While the Grimes group has shown that nanoflakes of $WO_3$ which are vertically aligned to the substrate, are able to more easily shuttle photogenerated electrons to the back contact (Su et al. 2010). This led to photocurrents densities of 2.3 mA/$cm^2$ at 1.2 V vs Ag/AgCl in 0.1 M $NaSO_4$ which was an improvement to previous studies of nanoflakes that were randomly oriented.

### 11.5.4 Iron Oxide

Iron Oxide, hematite, or α-$Fe_2O_3$ is one of the most promising materials for water splitting. The favorable band gap of 2.1 eV could lead to theoretical photocurrent densities of 12.6 mA/$cm^2$ (Sivula et al. 2011). Also, the high abundance of the material in the earth's crust, and the good stability of hematite are both intriguing properties. However, due to the very fast recombination rates $<10$ ps and short hole diffusion lengths smaller than 4 nm, iron oxide has proven to be a difficult material to optimize to its theoretical photocurrent and onset potentials (Le Formal et al. 2010). Again, nanostructuring has shown that hematite is a viable candidate for water oxidation which has led to significant attention to the material in the past decade.

One of the first examples of nanostructured hematite was demonstrated by the Grätzel group in 1987 when they showed that α-$Fe_2O_3$ nanoparticles can be used to generate hydrogen when a small bias is applied (Kiwi and Gratzel 1987). Later the Hagfelt group showed that hematite can be made in perpendicular nanowires and they were able to observe photocurrents of 160 $\mu A/cm^2$ for water oxidation (Beermann et al. 2000). As mentioned previously, the Grätzel group has shown that with silicon doping, iron oxide is capable of producing competitive photocurrent densities up to 2.3 $mA/cm^2$ without a cocatalyst added and 3.0 $mA/cm^2$ with a $IrO_2$ catalyst (Beermann et al. 2000; Kay et al. 2006).

Of the various structures that have been explored, the cauliflower like nanostructure has proven to be the best in terms of delivering photocurrent density. However, many other nanostructures have been developed for hematite, such as nanocorals, which can give photocurrent densities of up to 2.6 $mA/cm^2$ at 1.8 V vs RHE in 1 M NaOH and under 100 $mW/cm^2$ of illumination, and also nanotubes that have shown 1.41 $mA/cm^2$ of photocurrent density at 0.5 V vs Ag/AgCl under 87 $mW/cm^2$ of illumination (Mohapatra et al. 2009; Vincent et al. 2012). Yat Li's group has shown that nanowires with Sn doping can achieve photocurrents up to 1.86 $mA/cm^2$ (Ling et al. 2011). While significant progress has been made in the last decade, the photocurrents of hematite are still only ~25 % of the theoretical maximum.

### 11.5.5 Bismuth Vanadate

As seen in Fig. 11.11, $BiVO_4$ is also a very promising material for water oxidation. Bulk $BiVO_4$ was shown in 1998 to have water oxidation properties by Kudo et al. (1998). With the bandgap of 2.4–2.5 eV $BiVO_4$ has a theoretical photocurrent density of 7.6 $mA/cm^2$ (Park et al. 2013). A part of the problem to realize these efficiencies is that $BiVO_4$ has poor ICPE values. Since nanostructuring helps collect photons more efficiently with the increased area for light absorption, many different nanostructures of $BiVO_4$ have been explored in hopes to improve its water oxidation photocurrents.

The development of nanowires, nanotubes, and thin films has indeed shown that the performance of $BiVO_4$ can be improved by nanostructuring. In an effort to improve the IPCE values of $BiVO_4$, Liang et al. used a $SnO_2$ layer between FTO and films of $BiVO_4$ that were 300–500 nm in thickness (Liang et al. 2011). They found that the $SnO_2$ layer acted as a hole blocker and increased IPCE values to 36 % at 450 nm, which would correspond to a photocurrent density of 1.4 $mA/cm^2$ at 1.23 V vs RHE. A wet coating method was used by the Sayama group to deposit a thin film of $BiVO_4$ onto a $SnO_2$ hole blocking layer which was on top of a $WO_3$.

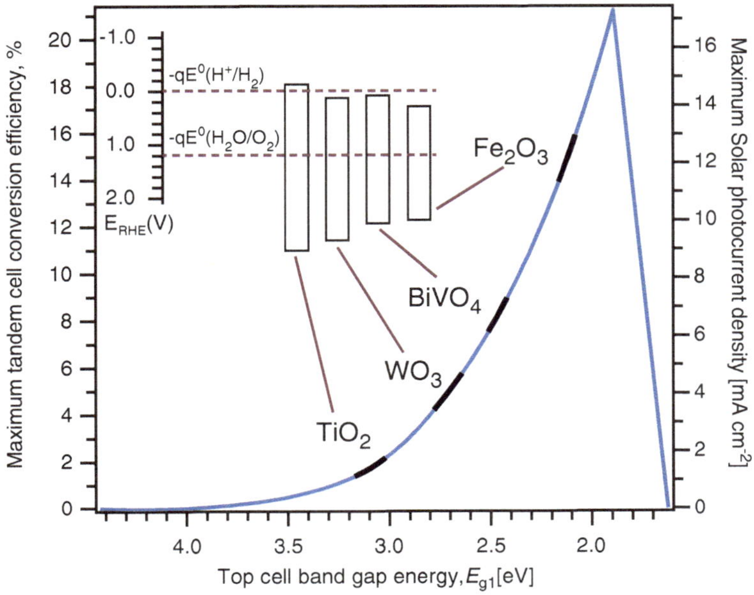


**Fig. 11.11** The maximum theoretical photocurrent densities and tandem cell efficiencies derived from four potential photoanodes for water splitting. It can be seen that the smaller bandgap of hematite is more favorable and has theoretically more potential than other popular materials. Reproduced from ref. Huang et al. (2014)

This approach led to a 3.04 mA/cm$^2$ photocurrent density at 1.23 V vs RHE in 0.1 M $KHCO_3$ (Saito et al. 2012). Choi's group has demonstrated photocurrent densities of 4.61 mA/cm$^2$ at 1.2 V vs RHE in a pH 7 potassium phosphate solution illuminated at 100 mW/cm$^2$ (Seabold and Choi 2012). Their approach was using an electrodeposition technique to deposit an amorphous layer of $BiVO_4$, between 200 and 500 nm thick. They then used a photodeposition process to put a ~150 nm thick layer of FeOOH on top of the $BiVO_4$. This gave them an increase in photocurrent performance as well as an increase in stability (Fig. 11.12).

## 11.5.6 Cuprous Oxide

Cuprous oxide, or $Cu_2O$, has a band gap that aligns for water oxidation and proton reduction, however the valence band is not very positive relative to the water oxidation potential. The conduction band on the other hand is negative enough for $Cu_2O$ to act as a photocathode and reduce protons to hydrogen gas. The bandgap of $Cu_2O$ is a favorable 2.0 eV which absorbs a major portion of the solar spectrum and gives a theoretical maximum efficiency of 14.7 mA/cm$^2$ (Paracchino et al. 2011). The main issue with $Cu_2O$ is that it is a very instable material which is easily reduced to form copper oxide (CuO) and the minority diffusion length is

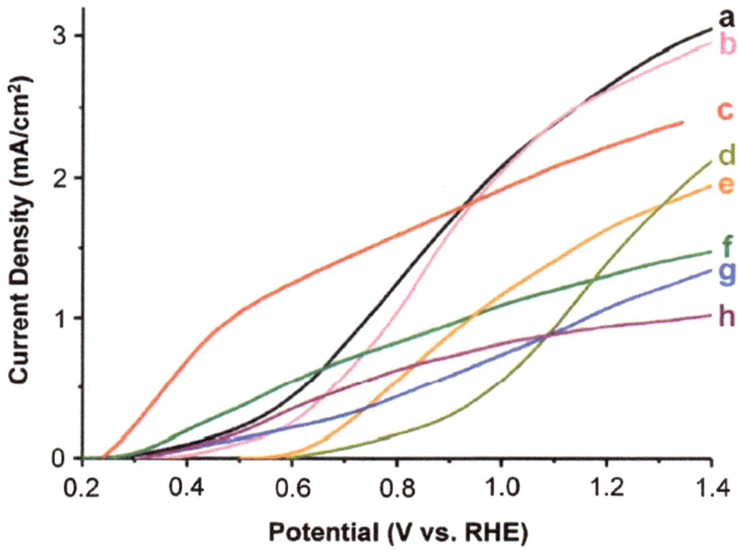


**Fig. 11.12** A JV curve showing an overlay of high performing $BiVO_4$. Curves *a* and *b* are from spin coating methods. Reproduced from ref. Park et al. (2013)

around 20–100 nm. Mostly for the latter reason some efforts have been made to improve the performance of $Cu_2O$ by creating nanostructures.

The Choi group has shown that dendritic structures of $Cu_2O$ can give 0.25 mA/cm$^2$ of photocurrent density while in 0.2 M $K_2SO_4$ and under 100 mW/cm$^2$ of illumination (Mor et al. 2008). However, their photocurrents degraded quickly. The Grätzel group showed that $Al_2O_3$ and $TiO_2$ overlayers can help create a more stable $Cu_2O$ films and measured photocurrent densities of 7.6 mA/cm$^2$ were achieved at 0 V vs RHE (Paracchino et al. 2011). More recently, Li et al. demonstrated that a protection layer of CuO helps with stability issues and the corresponding $Cu_2O$ films could give a –0.45 mA/cm$^2$ photocurrent density at 0 V vs RHE (Han et al. 2015).

### *11.5.7 Manganese Dioxide*

Manganese oxide is known as a water oxidation catalyst, but it also has a favorable bandgap of 1.8–2.7 eV depending upon the crystal structure. While manganese dioxide is not active towards water oxidation in its bulk form, some nanostructures have shown to give small water oxidation photocurrents. For instance the Jaramillo group showed that $MnO_2$ could deliver up to 50 μA/cm$^2$ of photocurrent density while at 1.45 V vs RHE (Pinaud et al. 2011). This study used a Na birnessite type $MnO_2$ films with a direct bandgap of 2.7 eV and an indirect bandgap of 2.1 eV. The Brudvig group showed in 2011 that $MnO_2$ could deliver up to 18 μA/cm$^2$ of photocurrent density in 0.1 $KNO_3$ at 1.2 V vs NHE while under 500 mW/cm$^2$ of illumination (Young et al. 2011). This was done by impregnating Nafion with $MnO_2$ nanocrystals on a FTO surface. While the photocurrent densities are low compared to other photoanodes, there has been little research surrounding $MnO_2$ as a photoanode.

### *11.5.8 Metal Nitrides*

With a bandgap of 2.1 eV and suitable band edge positions for water oxidation, $Ta_3N_5$ is also a potential photoanode for water splitting (Hitoki et al. 2002). In 2010 Grimes et al. showed that nanotubes of $Ta_3N_5$ can increase the IPCE values as they hypothesized that the recombination rate within the crystalline nanotubes is decreased (Feng et al. 2010). More recently, the Domen group has shown that vertically aligned $Ta_3N_5$ nanowires can deliver photocurrent densities of up 3.8 mA/cm$^2$ at 1.23 V vs RHE in 0.5 M $Na_2SO_4$ under 100 mW/cm$^2$ illumination which is 3.2 times the value of the bulk material (Li et al. 2013). They prepared the wires by using a template induced synthesis using porous anodic alumina mask then filling the pores with $Ta_2O_5$ then used a nitridation process at 1000 °C to convert the wires to $Ta_3N_5$.

### *11.5.9 Silicon*

Silicon, a material which has already been matured in terms of synthesis, can also be made into nanostructures, as long micron length wires with nanoscale diameters. Silicon's 1.1 eV bandgap is ideal for absorbing large amounts of the solar spectrum and its band edge positions also align well for it to be used as a photocathode for water splitting with a theoretical photocurrent of 43.6 mA/cm$^2$ (Meillaud et al. 2006). Combined with its high earth abundance this makes silicon a very interesting study for water splitting tandem cells.

To date silicon microwires are synthesized by using metal seeds, to initiate growth, then either vapor–liquid–solid deposition or an HF etch method. The Lewis group has shown that Silicon microwires with a p-n junction can perform at a high level (Santori et al. 2012). They have used a VLS deposition technique to create their buried junction on the silicon nanowires and have seen photocurrent densities of 10 mA/cm$^2$ at 0 V vs RHE in a $Me_2Fc^{+/0}$-$CH_3OH$ electrolyte under 100 mW/cm$^2$ of illumination. The Nam group found that the HF etching method could grow Si microwires that performed about 1.5 times better than the planar silicon films. Their Si microwires were about 30 μm in length and gave photocurrent densities of −10.65 mA/cm$^2$ at 0 V vs RHE with 100 mW/cm$^2$ of illumination in pH 0 $HClO_4$ (Sim et al. 2013) (Fig. 11.13).

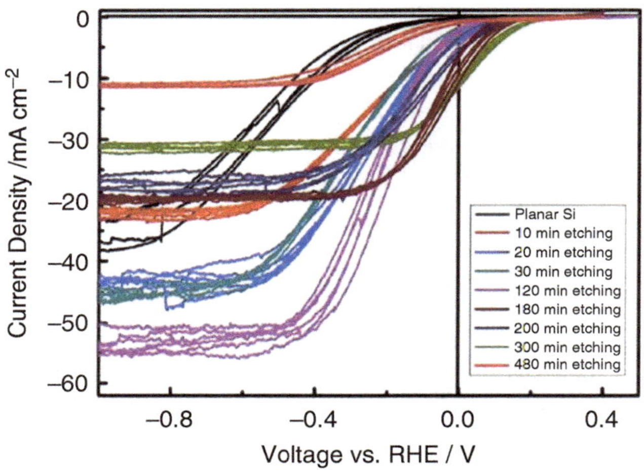


**Fig. 11.13** A JV curve under illumination showing the increase in onset potential and photocurrent by nanowires compared to a planar Si wafer. Reproduced from ref. Sim et al. (2013)

## *11.5.10 Metal Chalcogenides*

Materials such as CdS and CdSe have bulk bandgaps of 2.25 and 1.7 eV respectively, as well as suitable band edge positions for water reduction (Gratzel 2001). However, metal chalcogenides tend to suffer severe photocorrosion. Nonetheless, CdS, CdSe, PbS, PbSe have all received significant research attention due to their quantum confinement properties that lead to a tunable absorption spectra.

Creating nanoparticles of metal chalcogenides that are smaller and smaller in size eventually leads to the size of the particles to become smaller than the bohr radius. When this happens, the electron–hole pair or excitons begin to experience quantum effects which in turn allows for tuning of the band gap of metal chalcogenide materials which are commonly referred to as quantum dots. The ability to easily control the bandgap of a material has been an intriguing topic for researchers to explore for light harvesting and PEC water splitting applications.

Another interesting property of quantum dots is the ability to achieve enhanced quantum efficiencies due to multiple exciton generation, as shown by the Parkinson group in 2010. Using small band gap quantum dots, such as PbS, they demonstrated that photons, with energies around three times that of the quantum dot band gap, could generate two excitons per photon (Sambur et al. 2010). The possibility of harvesting multiple excitons per photon has the potential to surpass the Schockley–Queisser limitation and reach theoretical efficiency of 45 % (Nozik 2008). These exciting nanostructuring data have large implications for future designs of light harvesting devices such as PEC water splitting. In the case of quantum dots, these applications are typically in the form of heteronanostructures which brings us to our next topic of discussion.

## 11.6 Heteronanostructures

In many cases, creating a layered heterostructure is beneficial for the performance of water splitting photoelectrodes. Heteronanostructures have the ability to enhance the light absorption by using multiple band gaps, promote charge separation by creating energetic band offsets, increase the kinetics at the semiconductor–liquid interface by passivating trap states, and to create materials of unique energetics by the combination of different materials (Kronawitter et al. 2011). In this section some of the different heteronanostructures that have been explored and their corresponding performances are discussed (Fig. 11.14).

In many examples of heteronanostructures wide band gap materials such as $TiO_2$ and ZnO are used as the conductive support for smaller bandgap materials. An example of this is using an aluminum doped inverse opal ZnO structure then depositing a thin film of $BiVO_4$ as shown by Baumberg's group (Zhang et al. 2014). This resulted in a photocurrent densities of 1.5 mA/cm$^2$ at 1.23 V vs RHE under 100 mW/cm$^2$ of illumination in a pH 7 phosphate buffer solution. This was a 3 times increase from the performance of a thin film of $BiVO_4$, where the increase in photocurrent was correlated to higher IPCE values. Another example is the use of $TiO_2$ nanotubes as a support for $SrTiO_3$ particles that led to a drastic increase in water oxidation photocurrent densities (Zhang et al. 2014). After the hydrothermal growth of $SrTiO_3$ nanocrystals, photocurrents were improved to 1.75 mA/cm$^2$ at 0.2 V vs SCE from 1.0 mA/cm$^2$ in 0.1 M NaOH solutions under the same illumination intensity. Yat Li's group used a double sided array of ZnO nanowires with CdS on one side and CdS on the other side to show high hydrogen

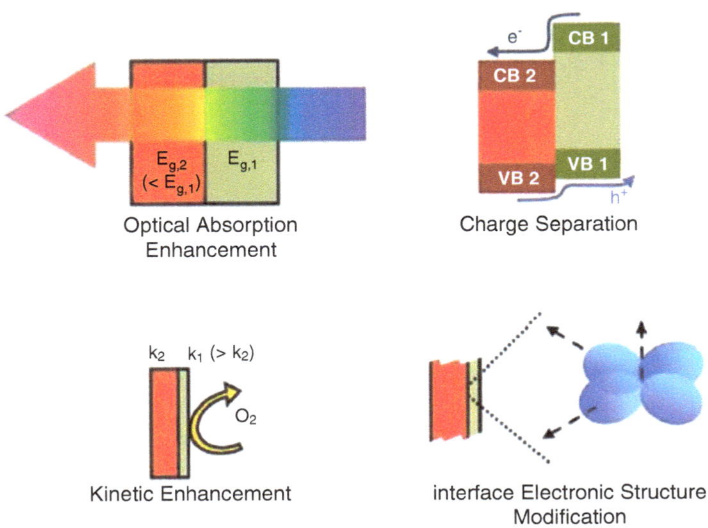


**Fig. 11.14** Examples of four different ways that heteronanostructures can lead to improved water splitting performances (Kronawitter et al. 2011)

production photocurrents of 12 mA/cm$^2$ at 0.4 V vs Ag/AgCl under 100 mW/cm$^2$ of illumination (Wang et al. 2010).

$WO_3$ has also been used as a support for a variety of heteronanostructures studies. The Grimes group created a $WO_3/BiVO_4$ photoanode by growing $WO_3$ nanowires using solvothermal technique (Su et al. 2011). They then used a spin coater to create a thin film of $BiVO_4$ on the $WO_3$ wires. The resulting structure resulted in enhanced photoconversion as the IPCE increased with each spin coat layer of $BiVO_4$. The stability of $BiVO_4$ against photocorrosion was also seen to increase compared to $BiVO_4$ without the $WO_3$ underlayer. $WO_3$ has also been used as a support for $\alpha$-$Fe_2O_3$ (Sivula et al. 2009). In this study a 60 nm layer of hematite was grown by APCVD onto a film of $WO_3$ also grown by a CVD method. The resulting photocurrent densities increased by more than 20 % from 1.41 to 1.71 mA/cm$^2$ at 1.43 V vs RHE in 1 M NaOH and 100 mW/cm$^2$ of illumination.

Recently, ALD has proven to be a very useful method for deposition of highly uniform films of light absorbing material onto a conducting substrate to create heteronanostructures. Some examples of this are $\alpha$-$Fe_2O_3$ being deposited by ALD onto silicon microwires which are grown by an HF etch (Mayer et al. 2012). This creates a dual absorbing structure where the holes from silicon are injected into hematite and the photogenerated electrons from hematite are injected into the silicon microwires. This set up resulted in improved onset potentials of 0.6 V vs RHE (one of the lowest onset potentials in the literature) and gave photocurrents of slightly less than 1 mA/cm$^2$ at 1.2 V vs RHE in 1 M NaOH all without the addition of dopants or surface catalyst. Similarly, ALD has been used to add a layer of ZnS to the surface of $Cu_2O$, which leads to improved photovoltages (Dai et al. 2014). In the same system ALD was also used to apply a layer of $TiO_2$, which is known to increase the stability of the $Cu_2O$ (Paracchino et al. 2012). The combination of these two ALD layers led to an increase in HER performance by over 30 % at 0 V vs RHE.

## 11.7 Conclusions and Outlook

Nanostructures have proven to be an effective route in order to increase the performance of water splitting photoanodes and photocathodes. Different synthesis methods to create 0D, 1D, 2D, and 3D nanostructures have provided many different pathways of research and many of the new nanostructures have indeed helped improve performance of the photoanode or photocathode. Heteronanostructures are also providing new ways to enhance the performance of water splitting materials. The combination of both strategies has spearheaded the exploration for a material which has all of the right characteristics needed for complete unbiased water splitting. However, the search for new materials is ongoing. In one effort a combinatorial approach is under way as a high throughput method designed to discover new ternary and tertiary components that are photoactive and active towards water splitting (Woodhouse and Parkinson 2009). Using metal nitrates and an inkjet printer gradients of metals are printed then are baked in air to create metal oxides. Several new compounds, such as a p-type $Co_{3-x-y}Al_xFe_yO_4$ with a bandgap of 1.7 eV, have already been discovered using this method.

The last decade has shown considerable improvement in onset potentials and photocurrent densities, and as these materials become closer to their theoretical limitations the next challenge will be the maximization of the stability of these materials and the successful incorporation of different materials into a tandem configuration. Further understanding of surface states on nanostructures and their prevention will need to be explored in order to optimize performances, and also catalysts and their incorporation will be needed.

## References

Asoh H, Ono S (2007) Electrocrystallization in nanotechnology. Wiley-VCH, Berlin, pp 138–166

Banerjee S, Mohapatra SK, Misra M (2009) Synthesis of TaON nanotube arrays by sonoelectrochemical anodization followed by nitridation: a novel catalyst for photoelectrochemical hydrogen generation from water. Chem Commun 46:7137–7139

Bard AJ (1979) Photoelectrochemistry and heterogeneous photo-catalysis at semiconductors. J Photochem 10:59–75

Beermann N, Vayssieres L, Lindquist SE, Hagfeldt A (2000) Photoelectrochemical studies of oriented nanorod thin films of hematite. J Electrochem Soc 147:2456–2461

Brillet J, Grätzel M, Sivula K (2010) Decoupling feature size and functionality in solution-processed, porous hematite electrodes for solar water splitting. Nano Lett 10:4155–4160

Brus LE (1983) A simple model for the ionization potential, electron affinity, and aqueous redox potentials of small semiconductor crystallites. J Chem Phys 79:5566–5571

Butler MA (1977) Photoelectrolysis and physical properties of the semiconducting electrode $WO_3$. J Appl Phys 48:1914–1920

Cristino V, Caramori S, Argazzi R, Meda L, Marra GL, Bignozzi CA (2011) Efficient photoelectrochemical water splitting by anodically grown $WO_3$ electrodes. Langmuir 27:7276–7284

Chakrapani V, Thangala J, Sunkara MK (2009) $WO_3$ and $W_2N$ nanowire arrays for photoelectrochemical hydrogen production. Int J Hydrogen Energ 34:9050–9059

Cho IS, Chen Z, Forman AJ, Kim DR, Rao PM, Jaramillo TF, Zheng X (2011) Branched $TiO_2$ nanorods for photoelectrochemical hydrogen production. Nano Lett 11:4978–4984

Cho IS, Logar M, Lee CH, Cai L, Prinz FB, Zheng X (2013) Rapid and controllable flame reduction of $TiO_2$ nanowires for enhanced solar water-splitting. Nano Lett 14:24–31

Dai P, Li W, Xie J, He Y, Thorne J, McMahon G, Zhan J, Wang D (2014) Forming buried junctions to enhance the photovoltage generated by cuprous oxide in aqueous solutions. Angew Chem Int Ed 53:13493–13497

Dai P, Xie J, Mayer MT, Yang X, Zhan J, Wang D (2013) Solar hydrogen generation by silicon nanowires modified with platinum nanoparticle catalysts by atomic layer deposition. Angew Chem Int Ed 52:11119–11123

de Respinis M, De Temmerman G, Tanyeli I, van de Sanden MC, Doerner RP, Baldwin MJ, van de Krol R (2013) Efficient plasma route to nanostructure materials: case study on the use of m-$WO_3$ for solar water splitting. ACS Appl Mater Interfaces 5:7621–7625

Dingle R, Wiegmann W, Henry CH (1974) Quantum states of confined carriers in very thin $Al_xGa_{1-x}As$-GaAs-$Al_xGa_{1-x}As$ heterostructures heterostructures. Phys Rev Lett 33:827–830

Duonghong D, Borgarello E, Graetzel M (1981) Dynamics of light-induced water cleavage in colloidal systems. J Am Chem Soc 103:4685–4690
Feng X, LaTempa TJ, Basham JI, Mor GK, Varghese OK, Grimes CA (2010) $Ta_3N_5$ nanotube arrays for visible light water photoelectrolysis. Nano Lett 10:948–952
Feng X, Shankar K, Varghese OK, Paulose M, Latempa TJ, Grimes CA (2008) Vertically aligned single crystal $TiO_2$ nanowire arrays grown directly on transparent conducting oxide coated glass: synthesis details and applications. Nano Lett 8:3781–3786
Fojtik A, Weller H, Koch U, Henglein A (1984) Photo-chemistry of colloidal metal sulfides 8. Photo-physics of extremely small CdS particles: Q-state CdS and magic agglomeration numbers. Berich Bunsengesellsch Phys Chem 88:969–977
Fujishima A, Honda K (1972) Electrochemical photolysis of water at a semiconductor electrode. Nature 238:37–38
Gao L, Cui Y, Wang J, Cavalli A, Standing A, Vu TTT, Verheijen MA, Haverkort JEM, Bakkers EPAM, Notten PHL (2014) Photoelectrochemical hydrogen production on InP nanowire arrays with molybdenum sulfide electrocatalysts. Nano Lett 14:3715–3719
Ghicov A, Schmuki P (2009) Self-ordering electrochemistry: a review on growth and functionality of $TiO_2$ nanotubes and other self-aligned MOx structures. Chem Commun 20:2791–2808
Gratzel M (2001) Photoelectrochemical cells. Nature 414:338–344
Grimes CA (2007) Synthesis and application of highly ordered arrays of $TiO_2$ nanotubes. J Mater Chem 17:1451–1457
Hagedorn K, Collins S, Maldonado S (2010) Preparation and photoelectrochemical activity of macroporous p-GaP(100). J Electrochem Soc 157:D588–D592
Hagfeldt A, Graetzel M (1995) Light-induced redox reactions in nanocrystalline systems. Chem Rev 95:49–68
Han J, Zong X, Zhou X, Li C (2015) $Cu_2O$/CuO photocathode with improved stability for photoelectrochemical water reduction. RSC Adv 5:10790–10794
Henglein A (1982) Photo-degradation and fluorescence of colloidal-cadmium sulfide in aqueous solution. Berich Bunsengesellsch Phys Chem 86:301–305
Hitoki G, Ishikawa A, Takata T, Kondo JN, Hara M, Domen K (2002) $Ta_3N_5$ as a novel visible light-driven photocatalyst (lambda $< 600$ nm). Chem Lett 2002:736–737
Huang ZF, Pan L, Zou JJ, Zhang X, Wang L (2014) Nanostructured bismuth vanadate-based materials for solar-energy-driven water oxidation: a review on recent progress. Nanoscale 6:14044–14063
Hwang YJ, Boukai A, Yang P (2008) High density n-Si/n-$TiO_2$ core/shell nanowire arrays with enhanced photoactivity. Nano Lett 9:410–415
Kamimura J, Bogdanoff P, Lähnemann J, Hauswald C, Geelhaar L, Fiechter S, Riechert H (2013) Photoelectrochemical properties of (In, Ga)N nanowires for water splitting investigated by in situ electrochemical mass spectroscopy. J Am Chem Soc 135:10242–10245
Kay A, Cesar I, Grätzel M (2006) New benchmark for water photooxidation by nanostructured $\alpha$-$Fe_2O_3$ films. J Am Chem Soc 128:15714–15721
Keis K, Vayssieres L, Lindquist S-E, Hagfeldt A (1999) Nanostructured ZnO electrodes for photovoltaic applications. Nanostruct Mater 12:487–490
Kim JK, Shin K, Cho SM, Lee T-W, Park JH (2011) Synthesis of transparent mesoporous tungsten trioxide films with enhanced photoelectrochemical response: application to unassisted solar water splitting. Energ Environ Sci 4:1465–1470
Kiwi J, Gratzel M (1987) Light-induced hydrogen formation and photo-uptake of oxygen in colloidal suspensions of [α]-Fe2O3. Faraday transactions 1: physical chemistry in condensed phases. J Chem Soc 83:1101–1108
Kronawitter CX, Vayssieres L, Shen S, Guo L, Wheeler DA, Zhang JZ, Antoun BR, Mao SS (2011) A perspective on solar-driven water splitting with all-oxide hetero-nanostructures. Energ Environ Sci 4:3889–3899
Kudo A, Ueda K, Kato H, Mikami I (1998) Photocatalytic $O_2$ evolution under visible light irradiation on $BiVO_4$ in aqueous $AgNO_3$ solution. Catal Lett 53:229–230

Le Formal F, Grätzel M, Sivula K (2010) Controlling photoactivity in ultrathin hematite films for solar water-splitting. Adv Funct Mater 20:1099–1107

Li W, Li J, Wang X, Luo S, Xiao J, Chen Q (2010) Visible light photoelectrochemical responsiveness of self-organized nanoporous $WO_3$ films. Electrochim Acta 56:620–625

Li X, Tao F, Jiang Y, Xu Z (2007) 3-D ordered macroporous cuprous oxide: fabrication, optical, and photoelectrochemical properties. J Colloid Interface Sci 308:460–465

Li Y, Takata T, Cha D, Takanabe K, Minegishi T, Kubota J, Domen K (2013) Vertically aligned $Ta_3N_5$ nanorod arrays for solar-driven photoelectrochemical water splitting. Adv Mater 25:125–131

Liang Y, Tsubota T, Mooij LPA, van de Krol R (2011) Highly improved quantum efficiencies for thin film $BiVO_4$ photoanodes. J Phys Chem C 115:17594–17598

Lin Y, Yuan G, Liu R, Zhou S, Sheehan SW, Wang D (2011a) Semiconductor nanostructure-based photoelectrochemical water splitting: a brief review. Chem Phys Lett 507:209–215

Lin Y, Zhou S, Liu X, Sheehan S, Wang D (2009) $TiO_2/TiSi_2$ heterostructures for high-efficiency photoelectrochemical $H_2O$ splitting. J Am Chem Soc 131:2772–2773

Lin Y, Zhou S, Sheehan SW, Wang D (2011b) Nanonet-based hematite heteronanostructures for efficient solar water splitting. J Am Chem Soc 133:2398–2401

Ling Y, Wang G, Wheeler DA, Zhang JZ, Li Y (2011) Sn-doped hematite nanostructures for photoelectrochemical water splitting. Nano Lett 11:2119–2125

Liu R, Lin Y, Chou L-Y, Sheehan SW, He W, Zhang F, Hou HJM, Wang D (2011) Water splitting by tungsten oxide prepared by atomic layer deposition and decorated with an oxygen-evolving catalyst. Angew Chem Int Ed 50:499–502

Ma GJ, Minegishi T, Yokoyama D, Kubota J, Domen K (2011) Photoelectrochemical hydrogen production on $Cu_2ZnSnS_4$/Mo-mesh thin-film electrodes prepared by electroplating. Chem Phys Lett 501:619–622

Mao A, Park N-G, Han GY, Park JH (2011a) Controlled growth of vertically oriented hematite/Pt composite nanorod arrays: use for photoelectrochemical water splitting. Nanotechnology 22:175703

Mao A, Shin K, Kim JK, Wang DH, Han GY, Park JH (2011b) Controlled synthesis of vertically aligned hematite on conducting substrate for photoelectrochemical cells: nanorods versus nanotubes. ACS Appl Mater Interfaces 3:1852–1858

Mayer MT, Du C, Wang D (2012) Hematite/Si nanowire dual-absorber system for photoelectrochemical water splitting at low applied potentials. J Am Chem Soc 134:12406–12409

Meillaud F, Shah A, Droz C, Vallat-Sauvain E, Miazza C (2006) Efficiency limits for single-junction and tandem solar cells. Sol Energ Mat Sol Cells 90:2952–2959

Mohapatra SK, John SE, Banerjee S, Misra M (2009) Water photooxidation by smooth and ultrathin $\alpha$-$Fe_2O_3$ nanotube arrays. Chem Mater 21:3048–3055

Mor GK, Prakasam HE, Varghese OK, Shankar K, Grimes CA (2007) Vertically oriented $Ti-Fe-O$ nanotube array films: toward a useful material architecture for solar spectrum water photoelectrolysis. Nano Lett 7:2356–2364

Mor GK, Shankar K, Paulose M, Varghese OK, Grimes CA (2004) Enhanced photocleavage of water using titania nanotube arrays. Nano Lett 5:191–195

Mor GK, Shankar K, Paulose M, Varghese OK, Grimes CA (2006) Use of highly-ordered $TiO_2$ nanotube arrays in dye-sensitized solar cells. Nano Lett 6:215–218

Mor GK, Varghese OK, Wilke RHT, Sharma S, Shankar K, Latempa TJ, Choi K-S, Grimes CA (2008) p-Type $Cu-Ti-O$ nanotube arrays and their use in self-biased heterojunction photoelectrochemical diodes for hydrogen generation. Nano Lett 8:1906–1911

Nozik AJ (2008) Multiple exciton generation in semiconductor quantum dots. Chem Phys Lett 457:3–11

O'Regan B, Gratzel M (1991) A low-cost, high-efficiency solar cell based on dye-sensitized colloidal $TiO_2$ films. Nature 353:737–740

Osterloh FE (2013) Inorganic nanostructures for photoelectrochemical and photocatalytic water splitting. Chem Soc Rev 42:2294–2320

Paracchino A, Laporte V, Sivula K, Gratzel M, Thimsen E (2011) Highly active oxide photocathode for photoelectrochemical water reduction. Nat Mater 10:456–461

Paracchino A, Mathews N, Hisatomi T, Stefik M, Tilley SD, Grätzel M (2012) Ultrathin films on copper(i) oxide water splitting photocathodes: a study on performance and stability. Energ Environ Sci 5:8673

Park JH, Kim S, Bard AJ (2005) Novel carbon-doped $TiO_2$ nanotube arrays with high aspect ratios for efficient solar water splitting. Nano Lett 6:24–28

Park Y, McDonald KJ, Choi KS (2013) Progress in bismuth vanadate photoanodes for use in solar water oxidation. Chem Soc Rev 42:2321–2337

Peng K-Q, Wang X, Wu X-L, Lee S-T (2009) Platinum nanoparticle decorated silicon nanowires for efficient solar energy conversion. Nano Lett 9:3704–3709

Pinaud BA, Chen Z, Abram DN, Jaramillo TF (2011) Thin films of sodium birnessite-type $MnO_2$: optical properties, electronic band structure, and solar photoelectrochemistry. J Phys Chem C 115:11830–11838

Prakasam HE, Varghese OK, Paulose M, Mor GK, Grimes CA (2006) Synthesis and photoelectrochemical properties of nanoporous iron (III) oxide by potentiostatic anodization. Nanotechnology 17:4285

Price MJ, Maldonado S (2009) Macroporous n-GaP in nonaqueous regenerative photoelectrochemical cells. J Phys Chem C 113:11988–11994

Rao PM, Cho IS, Zheng X (2013) Flame synthesis of $WO_3$ nanotubes and nanowires for efficient photoelectrochemical water-splitting. P Combust Inst 34:2187–2195

Saito R, Miseki Y, Sayama K (2012) Highly efficient photoelectrochemical water splitting using a thin film photoanode of $BiVO_4/SnO_2/WO_3$ multi-composite in a carbonate electrolyte. Chem Commun (Camb) 48:3833–3835

Sambur JB, Novet T, Parkinson BA (2010) Multiple exciton collection in a sensitized photovoltaic system. Science 330:63–66

Santato C, Odziemkowski M, Ulmann M, Augustynski J (2001) Crystallographically oriented mesoporous $WO_3$ films: synthesis, characterization, and applications. J Am Chem Soc 123:10639–10649

Santori EA, Maiolo Iii JR, Bierman MJ, Strandwitz NC, Kelzenberg MD, Brunschwig BS, Atwater HA, Lewis NS (2012) Photoanodic behavior of vapor–liquid–solid-grown, lightly doped, crystalline Si microwire arrays. Energ Environ Sci 5:6867

Seabold JA, Choi KS (2012) Efficient and stable photo-oxidation of water by a bismuth vanadate photoanode coupled with an iron oxyhydroxide oxygen evolution catalyst. J Am Chem Soc 134:2186–2192

Shankar K, Basham JI, Allam NK, Varghese OK, Mor GK, Feng X, Paulose M, Seabold JA, Choi K-S, Grimes CA (2009) Recent advances in the use of $TiO_2$ nanotube and nanowire arrays for oxidative photoelectrochemistry. J Phys Chem C 113:6327–6359

Shim H-S, Shinde VR, Kim JW, Gujar TP, Joo O-S, Kim HJ, Kim WB (2009) Diameter-tunable CdSe nanotubes from facile solution-based selenization of $Cd(OH)_2$ nanowire bundles for photoelectrochemical cells. Chem Mater 21:1875–1883

Sim U, Jeong H-Y, Yang T-Y, Nam KT (2013) Nanostructural dependence of hydrogen production in silicon photocathodes. J Mater Chem A 1:5414–5422

Sivula K, Formal FL, Grätzel M (2009) $WO_3-Fe_2O_3$ photoanodes for water splitting: a host scaffold, guest absorber approach. Chem Mater 21:2862–2867

Sivula K, Le Formal F, Gratzel M (2011) Solar water splitting: progress using hematite $\alpha$-$Fe_2\,O_3$ photoelectrodes. ChemSusChem 4:432–449

Su J, Feng X, Sloppy JD, Guo L, Grimes CA (2010) Vertically aligned $WO_3$ nanowire arrays grown directly on transparent conducting oxide coated glass: synthesis and photoelectrochemical properties. Nano Lett 11:203–208

Su J, Guo L, Bao N, Grimes CA (2011) Nanostructured $WO_3/BiVO_4$ heterojunction films for efficient photoelectrochemical water splitting. Nano Lett 11:1928–1933

Su J, Guo L, Yoriya S, Grimes CA (2009) Aqueous growth of pyramidal-shaped $BiVO_4$ nanowire arrays and structural characterization: application to photoelectrochemical water splitting. Cryst Growth Des 10:856–861

Tacca A, Meda L, Marra G, Savoini A, Caramori S, Cristino V, Bignozzi CA, Gonzalez Pedro V, Boix PP, Gimenez S, Bisquert J (2012) Photoanodes based on nanostructured $WO_3$ for water splitting. Chemphyschem 13:3025–3034

Tahir AA, Ehsan MA, Mazhar M, Wijayantha KGU, Zeller M, Hunter AD (2010) Photoelectrochemical and photoresponsive properties of Bi2S3 nanotube and nanoparticle thin films. Chem Mater 22:5084–5092

Thangala J, Vaddiraju S, Bogale R, Thurman R, Powers T, Deb B, Sunkara MK (2007) Large-scale, hot-filament-assisted synthesis of tungsten oxide and related transition metal oxide nanowires. Small 3:890–896

Townsend TK, Browning ND, Osterloh FE (2012) Nanoscale strontium titanate photocatalysts for overall water splitting. ACS Nano 6:7420–7426

Vayssieres L, Beermann N, Lindquist S-E, Hagfeldt A (2001) Controlled aqueous chemical growth of oriented three-dimensional crystalline nanorod arrays: application to iron(III) oxides. Chem Mater 13:233–235

Vincent T, Gross M, Dotan H, Rothschild A (2012) Thermally oxidized iron oxide nanoarchitectures for hydrogen production by solar-induced water splitting. Int J Hydrogen Energ 37:8102–8109

Wang D, Pierre A, Kibria MG, Cui K, Han X, Bevan KH, Guo H, Paradis S, Hakima A-R, Mi Z (2011) Wafer-level photocatalytic water splitting on GaN nanowire arrays grown by molecular beam epitaxy. Nano Lett 11:2353–2357

Wang G, Yang X, Qian F, Zhang JZ, Li Y (2010) Double-sided CdS and CdSe quantum dot co-sensitized ZnO nanowire arrays for photoelectrochemical hydrogen generation. Nano Lett 10:1088–1092

Wang M, Jiang J, Shi J, Guo L (2013) CdS/CdSe core–shell nanorod arrays: energy level alignment and enhanced photoelectrochemical performance. ACS Appl Mater Interfaces 5:4021–4025

Watcharenwong A, Chanmanee W, de Tacconi NR, Chenthamarakshan CR, Kajitvichyanukul P, Rajeshwar K (2008) Anodic growth of nanoporous $WO_3$ films: morphology, photoelectrochemical response and photocatalytic activity for methylene blue and hexavalent chrome conversion. J Electroanal Chem 612:112–120

Wolcott A, Smith WA, Kuykendall TR, Zhao Y, Zhang JZ (2009) Photoelectrochemical study of nanostructured ZnO thin films for hydrogen generation from water splitting. Adv Funct Mater 19:1849–1856

Woodhouse M, Parkinson BA (2009) Combinatorial approaches for the identification and optimization of oxide semiconductors for efficient solar photoelectrolysis. Chem Soc Rev 38:197–210

Yang X, Wolcott A, Wang G, Sobo A, Fitzmorris RC, Qian F, Zhang JZ, Li Y (2009) Nitrogen-doped ZnO nanowire arrays for photoelectrochemical water splitting. Nano Lett 9:2331–2336

Yokoi T, Sakuma J, Maeda K, Domen K, Tatsumi T, Kondo JN (2011) Preparation of a colloidal array of $NaTaO_3$ nanoparticles via a confined space synthesis route and its photocatalytic application. Phys Chem Chem Phys 13:2563–2570

Young KJ, Gao Y, Brudvig GW (2011) Photocatalytic water oxidation using manganese compounds immobilized in nafion polymer membranes. Aust J Chem 64:1221–1228

Yuan G, Aruda K, Zhou S, Levine A, Xie J, Wang D (2011) Understanding the origin of the low performance of chemically grown silicon nanowires for solar energy conversion. Angew Chem Int Ed 50:2334–2338

Yuan G, Zhao H, Liu X, Hasanali ZS, Zou Y, Levine A, Wang D (2009) Synthesis and photoelectrochemical study of vertically aligned silicon nanowire arrays. Angew Chem Int Ed 48:9680–9684

Zhang L, Reisner E, Baumberg JJ (2014) Al-doped ZnO inverse opal networks as efficient electron collectors in BiVO4 photoanodes for solar water oxidation. Energ Environ Sci 7:1402–1408

Zhou M, Bao J, Bi W, Zeng Y, Zhu R, Tao M, Xie Y (2012) Efficient water splitting via a heteroepitaxial $BiVO_4$ photoelectrode decorated with Co-Pi catalysts. ChemSusChem 5:1420–1425

Zhou M, Zhang S, Sun Y, Wu C, Wang M, Xie Y (2010) C-oriented and {010} facets exposed $BiVO_4$ nanowall films: template-free fabrication and their enhanced photoelectrochemical properties. Chem Asian J 5:2515–2523

Zhou S, Liu X, Lin Y, Wang D (2009) Rational synthesis and structural characterizations of complex $TiSi_2$ nanostructures. Chem Mater 21:1023–1027

# Chapter 12
# Advanced Device Architectures and Tandem Devices

**Kevin Sivula**

## 12.1 Introduction

Independent of the material, nanostructure, or catalyst used in a photoelectrochemical (PEC), solar-to-fuel device, one aspect remains apparent: in order to achieve the conversion of solar energy to chemical fuel, sufficient external (photo)chemical potential must be generated to overcome the thermodynamic energy difference between the reactants and the products plus the kinetic (electrochemical) overpotentials for the oxidation and reduction reactions, $\eta_{ox}$ and $\eta_{red}$, respectively. For the classic situation of PEC water splitting for solar hydrogen production this amounts to around 1.6–1.8 eV depending on the materials employed and the operating current density. In this chapter, routes to manage this requirement together with the need to maximize solar light harvesting (and thus conversion efficiency) are explored by presenting examples of advanced device architectures. Practical device design aspects are also considered. Before discussing operational PEC cells it is helpful to first examine the maximum possible solar energy conversion based on the light absorbing properties of the materials employed—namely the semiconductor band-gap energy.

K. Sivula (✉)
Institute of Chemical Sciences and Engineering, École Polytechnique Fédérale de Lausanne, 1015 Lausanne, Switzerland
e-mail: kevin.sivula@epfl.ch



S. Giménez, J. Bisquert (eds.), *Photoelectrochemical Solar Fuel Production*,
DOI 10.1007/978-3-319-29641-8_12

## 12.2 Harvesting the Solar Spectrum and Developing Photopotential

An ideal semiconductor is capable of absorbing an incident photon with an energy, $E_{photon} = h\nu = hc/\lambda$, greater than or equal to the band gap energy, $E_g$, to create an excited electron in the conduction band and an electron hole in the valence band. After a rapid thermalization these excited states differ in energy by the band gap, $E_g$. However, not all of this energy is directly available as thermodynamic free energy (i.e., available to perform work). The entropic energy loss resulting from the amount of work that must be done to extract the excited states from the semiconductor must be considered. Thus, for a single semiconductor the externally available free energy developed upon light absorption, $\Delta\mu_{ex}$, is significantly less than $E_g$. The difference between these quantities depends on the band-gap of the semiconductor and the illumination intensity. For a typical semiconductor at terrestrial solar intensities and with the assumption that each absorbed photon creates one excited electron–hole pair, the entropic loss can be approximated to be about $E_g/4$ (Bolton et al. 1981) . Thus, $\Delta\mu_{ex} \approx 0.75\ E_g$ is a reasonable assumption. For PEC water splitting, this means that a minimum $E_g$ of 2.2 eV would be required to afford unassisted simultaneous water oxidation and reduction with a single semiconductor junction (Weber and Dignam 1984). Based on the energy distribution of the standard solar spectrum, this corresponds to a maximum solar-to-hydrogen conversion efficiency, *STHCE*, of 11.6 %. Figure 12.1 shows the standard solar spectrum

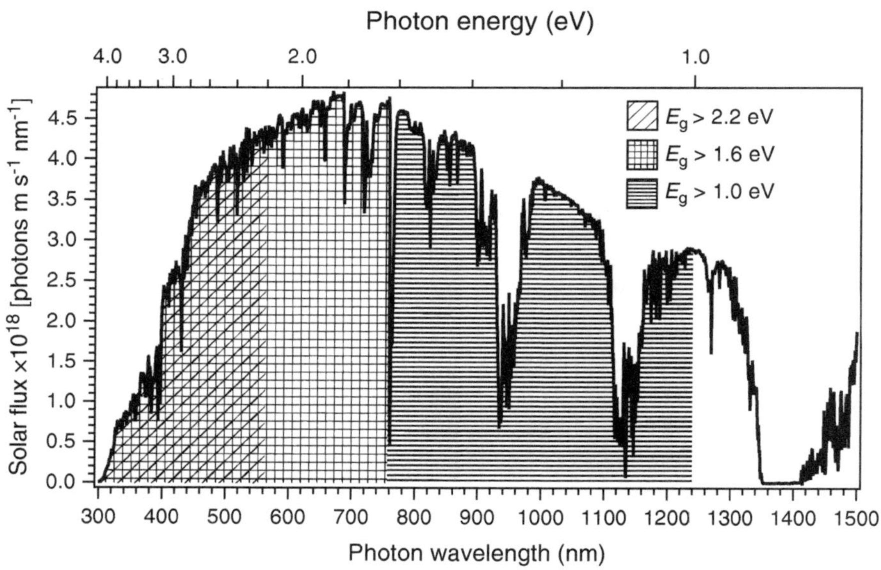


**Fig. 12.1** The AM 1.5 Global Tilt Spectra shown as solar photon flux as a function of wavelength. The portion of the solar spectra harvested by various band-gap energy semiconductors are indicated

and the portion of it harvested in this ideal single junction case (i.e., the portion with $E_{photon} > 2.2\,eV$). In reality, attaining a PEC water splitting device based on a single semiconductor junction with *STHCE* > 10 % has proven to be a very difficult task. While many promising materials exist (e.g., GaP and InP), their cost and stability in aqueous environments have to date remained critical limitations and prevented the realization of an efficient PEC water splitting device based on a single semiconductor (Walter et al. 2010).

The challenge of developing sufficient $\Delta\mu_{ex}$ for PEC fuel production while also harvesting an appreciable amount of the solar spectrum is a problem that is similar to optimizing solar-to-power energy conversion in a photovoltaic device. In analogy to the photovoltaic field, higher solar energy conversion efficiency can be achieved using multiple semiconductor junctions. Moreover, since a PEC reaction must consist of two half reactions, using two photelectrodes—a photocathode for reduction and a photoanode for oxidation—is an obvious choice. The simplest way to accomplish an overall PEC fuel production reaction with two semiconductor junctions is to place two cells side-by-side, one functioning as the photoanode and one as the photocathode. The maximum potential solar-to-fuel energy conversion of this approach for the classic situation of PEC water splitting has been estimated to be 16.6 % with both semiconductor band gaps equal to 1.4 eV—a moderate increase over the single absorber approach (Weber and Dignam 1984). However, using an integrated tandem approach (where the cells are placed one on top of the other and photons not absorbed by the first cell can be transmitted and absorbed by the second) can achieve more efficient light harvesting and thus higher *STHCE*s. Schematic examples of this approach are shown in Fig. 12.2. Figure 12.2a shows a simple photoanode/photocathode device where the photopotential is developed at the semiconductor liquid junction (due to the interfacial band-bending), while Fig. 12.2b shows a three junction cell where the photopotential is developed at buried p-n junctions in the device. In both cases the first (top) cell, with band-gap $E_{g1}$ absorbs photons from the sun with a wavelength smaller than $\lambda_1 = hc/E_{g1}$. Ignoring reflection and scattering events, photons with $\lambda > \lambda_1$ are transmitted through to the next cell with a band-gap $E_{g2} < E_{g1}$, and so on. Since the individual cells must be connected electrically in series to add the photopotential, the total photocurrent density will be limited by the cell producing the least current.

In order to assess the potential benefit of a tandem cell approach and determine the optimum band-gap semiconductors that should be employed, next a method for determining the maximum possible solar-to-fuel conversion efficiency using a tandem system will be developed. In analogy with the formalisms presented in Chap. 4, a general expression for the maximum possible solar-to-fuel conversion efficiency (*STFCE*, for an arbitrary fuel produced) for a series-connected tandem cell can be given as:

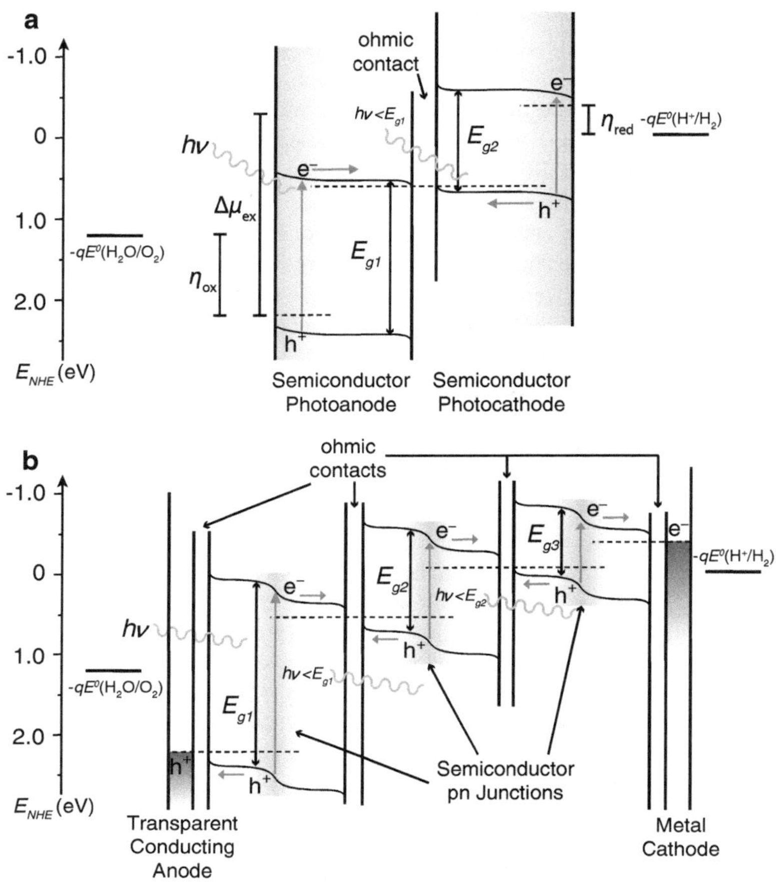


**Fig. 12.2** Electron energy schemes of two possible tandem water splitting device configurations: (**a**) shows an n-type photoanode connected to p-type photocathode and (**b**) shows a triple (buried) pn-junction. In (**a**) the externally available free energy developed along with the overpotentials (as described in the main text, are indicated

$$\mathrm{STFCE} = \frac{J_{\min}\Delta G_{\mathrm{F}}^{0}\eta_{\mathrm{F}}}{rqP_{\mathrm{S}}} \tag{12.1}$$

where $\Delta G_{\mathrm{F}}^{0}$ represents the standard free energy of the fuel produced (for hydrogen this is the lower heating value at 2.46 eV per molecule of $H_2$), $\eta_{\mathrm{F}}$ is the Faradaic efficiency for the fuel producing reaction, $r$ is the stoichiometric ratio between one molecule of the produced fuel and the number of electrons needed to produce it electrochemically ($r = 2$ for $H_2$ production from water), and $q$ is the elementary charge ($1.6022 \times 10^{-19}\ C = 1$ eV $V^{-1}$). To account for the power coming from the illumination source $P_{\mathrm{S}}$ represents the power density from the sun incident on the tandem cell. For standard conditions (AM 1.5 Global tilt) $P_S$ is 1000 W $m^{-2}$. In the case of nonstandard illumination conditions $P_S$ can be expressed by integrating the incident spectral irradiance as a function of photon wavelength, $E_S(\lambda)$ (given in W $m^{-2}$ $nm^{-1}$), accordingly so that

$$P_S = \int_{\lambda} E_S(\lambda) d\lambda \tag{12.2}$$

Finally, $J_{min}$ in Eq. (12.1) represents photocurrent density (in units A $m^{-2}$) flowing through the tandem device during operation (i.e., the electrical current travelling between the series connected cells divided by the illuminated area). It is important to note that $J_{min}$ is zero if there is not sufficient $\Delta\mu_{ex}$ generated by the tandem cell for the desired fuel production reaction (here $\Delta\mu_{ex}$ is the sum of the free energies produced by each cell, $\Delta\mu_{ex,1} + \Delta\mu_{ex,2} + \ldots + \Delta\mu_{ex,n}$ where $n$ is the number of photopotential-producing junctions). The minimally required free energy can be further defined as $\Delta\mu_{ex}^{min} = \Delta G_{H_2}^{0}/r + \eta_{ox} + \eta_{red}$. Otherwise $J_{min}$ is limited by the individual cell producing the least amount of photocurrent, as the cells are assumed be connected in series. Thus, an expression for $J_{min}$ can be explicitly written using a piecewise function and the "min" operator, which selects the minimum value from a set:

$$J_{min} = \begin{cases} 0, & |\Delta\mu_{ex} \leq \Delta\mu_{ex}^{min} \\ \min(J_1, J_1, \ldots, J_n), & |\Delta\mu_{ex} > \Delta\mu_{ex}^{min} \end{cases} \tag{12.3}$$

The individual (maximum possible) current density in the $i$th cell, $J_i$, can be expressed with respect to $E_S(\lambda)$, and the wavelength cut-offs for each semiconductor ($\lambda_i = hc/E_{g,i}$, $i = 1, 2, \ldots, n$ under the assumption that $E_{g,i} > E_{g,i+1}$) as

$$J_i = \int_{\lambda_{i-1}}^{\lambda_i} \frac{E_S(\lambda)\phi_{C,i}(\lambda)}{(hc/\lambda)} d\lambda, \quad \text{for } i = 1, 2, \ldots, n. \tag{12.4}$$

where $E_S(\lambda)$ is the previously defined spectral distribution of solar illumination, $\phi_{C,i}(\lambda)$ represents the quantum conversion efficiency for each semiconductor as a function of wavelength (i.e., the ratio of photocurrent density and absorbed photon flux), and $\lambda_0$ represents the smallest wavelength for which there is an appreciable photon flux from the sun ($\lambda_0 \approx$ ca. 300 nm for AM1.5G solar irradiation). Dividing by the factor of $hc/\lambda$ in Eq. (12.4) serves to convert the spectral output from the sun, into a photon flux (photons $m^{-2}$ $s^{-1}$). For simplicity, this method for calculating the $J_i$'s assumes that each semiconductor absorbs all photons with energy $hc/\lambda > E_g$ and transmits all photons with energy $hc/\lambda < E_g$. It is important to keep in mind that no reflection or scattering losses are included for this ideal case. Thus, calculated values for the $J_i$'s are upper-bound values. In real tandem systems, reflection and scattering losses exist, but can be addressed with device engineering approaches (e.g., antireflection coatings or nanostructuring). Also in real systems, both $\Delta\mu_{ex}$ and $\phi_C$ are complicated functions of the light intensity and the semiconductor properties as previously mentioned. Moreover, the overpotentials for the oxidation and reduction reactions, $\eta_{ox}$ and $\eta_{red}$, are dependent on the actual reactions taking place and on $J_{min}$ by the Tafel relation. Thus, additional assumptions are needed to obtain reasonable values for *STFCE*. In principle, $\phi_C$ can be made very close to unity even in real systems, so $\phi_C = 1$ can realistically be used to evaluate Eq. (12.4). If we also assume that no corrosion or undesirable side reactions occur, then $\eta_F = 1$. To move toward a reasonable estimation for the maximum possible solar-to-fuel conversion efficiency, plausible values for the remaining loss processes need to be estimated. For the case of PEC water splitting, reasonable assumptions regarding the overpotentials and the free energy can be lumped into one value for the energy loss, $U_{loss}$, defined as the sum of the loss of energy from each semiconductor (which is ca. 0.25 $E_g$) plus the overpotentials. This lumped loss parameter ($U_{loss} = 0.25 \sum_i E_{g,i} + \eta_{ox} + \eta_{red}$) has been estimated to be about 1.4 eV for the case of two semiconductor junctions (Bolton et al. 1985).

With these assumptions, and the known spectral distribution of the standardized solar spectrum, all that is needed to predict the maximum possible *STFCE* is contained in the above equations. One needs only choose the number of junctions and their band-gaps (ensuring that $\Delta\mu_{ex} > \Delta\mu_{ex}^{min}$) and find the smallest $J_i$ using Eq. (12.4), $J_{min}$. Calculation of the max *STFCE* is then straightforward with $J_{min}$ in Eq. (12.1). For the case of PEC water splitting for hydrogen production and using two semiconductor junctions in tandem, the maximum *STHCE* can be solved for using even a simple spreadsheet software and plotted as a function of the two band-gaps on a contour plot. Figure 12.3 shows this contour plot for the case of $U_{loss} = 1.4$ eV. The shapes of the contours, which represent values of $E_{g1}$ and $E_{g2}$ that result in the same value of *STHCE*, are easily understood. First, the contours are present only where the set condition $E_{g1} > E_{g2}$ is satisfied. Thus, the lower-right part of the plot is undefined. The lower-left region, where the value of *STHCE* is also undefined, represents semiconductor combinations that do not possess sufficient $\Delta\mu_{ex}$ for water splitting in a two-junction configuration given $U_{loss}$.

Calculating the upper limit of *STHCE* and optimum $E_{g1}$ and $E_{g2}$ pairing for an arbitrary system can be performed simply using this plot. For example, take $E_{g1} = 2.1$ eV (corresponding to for example $Fe_2O_3$, an oxide semiconductor commonly used as a photoanode for water splitting) as the top cell. A horizontal line drawn at $E_{g1} = 2.1$ eV indicates that the maximum *STHCE* with this top cell would be slightly more than 15 %. Looking at the range of bottom cell band-gaps that would be suitable for a 2.1 eV band gap top cell it is clear that $E_{g2}$ must be between 0.6 and 1.6 eV. A higher band gap energy for the bottom cell and it will not be able to harvest sufficient photons from those transmitted by the first cell. Then the bottom cell will be $J_{min}$ and the *STHCE* will drop off accordingly as $E_{g2}$ increases. On the other hand choosing a bottom cell with $E_{g2} < 0.6$ eV means that insufficient photopotential will be developed to drive the water splitting reactions.

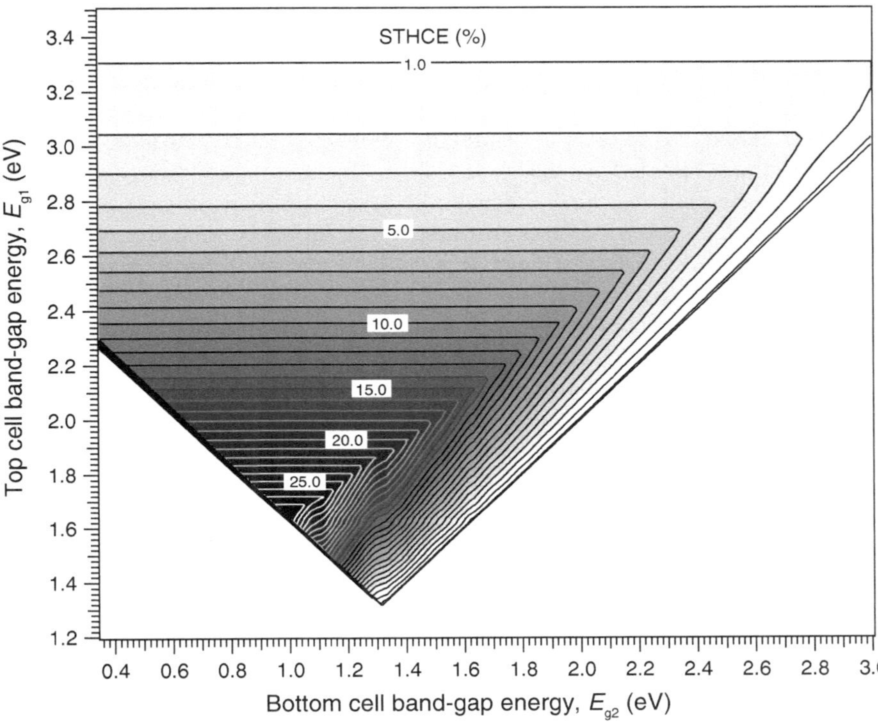


**Fig. 12.3** Contour plot showing the maximum reasonable *STHCE* for a tandem cell with two semiconductor junctions. Each semiconductor is assumed to absorb all photons with energy greater than its band-gap, and transmit photons with lower energy as described in the main text. The incident radiation is taken to be AM 1.5G (1000 W $m^{-2}$) and a total loss, $U_{loss}$, is set at 1.4 eV

The global optimum value *STHCE* using $U_{loss} = 1.4$ eV occurs at 29.9 % with $E_{g1} = 1.64$ eV and $E_{g2} = 0.99$ eV. A higher assumed value for $U_{loss}$ results in a new contour plot with a lower value for the maximum *STHCE* and new optimum band-gaps as more $\Delta\mu_{ex}$ is needed to overcome the losses. For example, with $U_{loss} = 1.6$ eV the optimum conditions give $STHCE = 27.1$ % using $E_{g1} = 1.72$ eV and $E_{g2} = 1.11$ eV, and with $U_{loss} = 2.0$ eV, $STHCE = 21.6$ % using $E_{g1} = 1.89$ eV and $E_{g2} = 1.34$ nm. The maximum predicted *STHCE* for a particular $E_{g1}$, $E_{g2}$ does not change when assuming higher values for $U_{loss}$, as long as sufficient $\Delta\mu_{ex}$ can still be generated. Indeed, with $U_{loss} = 2.0$, the maximum *STHCE* for $Fe_2O_3$ ($E_{g1} = 2.1$ eV) remains at just over 15 %. However, in this case, $E_{g2}$ must be greater than 1.1 eV to ensure that sufficient $\Delta\mu_{ex}$ is generated.

Overall the model presented in this section gives only reasonable expected maximum values for the *STFCE*, which can be of use in describing the basic capability (or limitation) of specific materials or material combinations. This is useful in determining the interest in developing specific systems. For example, $WO_3$ ($E_g = 2.6$ eV) has been extensively investigated as a photo anode material, and while Fig. 12.3 certainly suggests that a wide range of low band-gap materials could work with $WO_3$ to afford overall solar to hydrogen conversion, the *STHCE* will be limited to about 6 %. This limitation may dissuade serious investment if it can be further determined that this *STHCE* is not sufficient for practical application (see Sect. 12.6). While in this sense the above-presented model is valuable, however, it must be noted that the model does not include any additional constraints on selecting the materials with respect to the conduction and valence band energy levels, or work function—factors that add to the difficulty of finding ideal and complementary materials for a water splitting tandem cell. It should also be noted that the above formalisms are appropriate for an arbitrary number of semiconductor junctions; however, since a PEC cell naturally has two electrodes, and a two junction cell is clearly capable of considerable *STHCE* even with large assumed losses, this two-absorber tandem cell is a good compromise between device complexity and potential performance.

## 12.3 Employing thin Semiconductor Layers in Tandem Cells

One important limitation of the model described in the previous section is the assumption that each semiconductor absorbs all photons with energy $hc/\lambda > E_g$. This assumption is not suitable when considering absorber layers that have a thickness equal to or less than the photon penetration depth in the material (defined as $\alpha^{-1}$ where $\alpha = f(\lambda)$ is the absorption coefficient of the semiconductor). Indeed, it is often necessary to employ absorber layers thinner than $\alpha^{-1}$ due to poor charge carrier transport. Materials like $Fe_2O_3$ and amorphous Si:H are good examples of this and are often employed using layers too thin to absorb all photons with energy $h\nu > E_g$. Assuming that the same $\Delta\mu_{ex}$ can still be generated even with fewer absorbed photons (an assumption that becomes less appropriate as the amount of absorbed photons becomes small as $\Delta\mu_{ex}$ will become $\ll E_g$), using a thin top cell will relax the constraint on the maximum band gap possible for the bottom cell because more light will reach the bottom cell. However, the maximum *STFCE* will be still be limited by $J_{min}$ in Eq. (12.1). A modified version of Eq. (12.4) can be written to recast $J_i$ under the notion of a thin film where there is partial transmission of photons with energy greater than the band-gap (while still ignoring reflection and scattering):

$$J_i = \int_\lambda \frac{\phi_{C,i}(\lambda)}{\left(\frac{hc}{\lambda}\right)} \left[E_S(\lambda)\exp\left(-\sum_{k=0}^{k=(i-1)} \alpha_k(\lambda)d_k\right)\right][1 - \exp(-\alpha_i(\lambda)d_i)], \quad \text{for } i = 1, 2, \ldots, n. \tag{12.5}$$

Here $\alpha_i(\lambda)$ is the absorption coefficient of the $i$th semiconductor material as a function of the photon wavelength and $d_i$ is the thickness of the $i$th layer. In this case the integration can be performed over all wavelengths for each semiconductor layer considering that the band-gap will be implicitly present in the value of $\alpha_i(\lambda)$. In this case the expression for the light incident on the $i$th semiconductor is the original solar spectrum $E_S(\lambda)$ multiplied by the transmittance of all of the layers above the $i$th layer as expressed by the exponential of the sum of the thicknesses times the absorption coefficient for all layers encountered before the $i$th layer (for $k = 0$, $\alpha_o(\lambda)d_o = 0$).

To see the effect of having a semi-transparent top cell in a two-absorber tandem cell for PEC water splitting, a hypothetical semiconductor with an absorption coefficient consistent with a typical indirect band-gap semiconductor (e.g., Silicon) and $E_{g1}$ close to the optimum condition (when $U_{loss} = 1.4$ eV), i.e., $E_{g1} = 1.7$ eV, is next considered for application in solar water splitting. The absorption coefficient, $\alpha(\lambda)$, is shown in Fig. 12.4a. If a layer of the semiconductor 100 nm thick is used as the top absorber, only a small fraction of the incident solar illumination with photon energy above the band-gap is absorbed as shown in Fig. 12.4b. The remaining photons will be transmitted to the bottom cell, and, since they do not produce current in the top cell, the solar energy conversion is limited to only ca. 2 %. As the thickness of the semiconductor increases the amount of light absorbed (and the

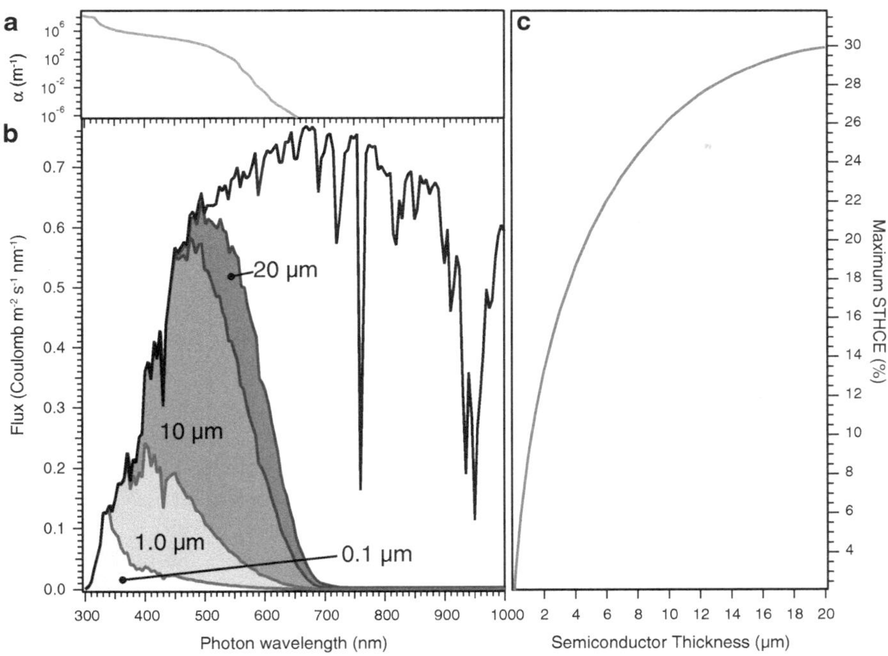


**Fig. 12.4** Effects of using semi-transparent thin films as electrodes for a PEC tandem cell. (**a**) Shows a hypothetical absorption coefficient of an indirect band-gap semiconductor ($E_g = 1.7$ eV). (**b**) Shows the fraction of the solar spectrum that material can absorb at different thicknesses as a function of photon wavelength. (**c**) Shows the maximum solar-to-hydrogen conversion efficiency of a tandem cell using this semiconductor as the top cell as a function of semiconductor thickness

maximum possible *STHCE,* accordingly) increases. A thickness of 10 μm is required to absorb enough light to represent a maximum *STHCE* of 26 % and, in this case, 20 μm is required before the maximum *STHCE* levels off at just under 30 % as shown in Fig. 12.4c. Overall this concept is discussed later in this chapter with respect to actual demonstrated tandem cells, and it is important to note again that strategies to increase light absorption without increasing the thickness of the semiconductor are very much of interest to reduce the material used and lessen charge transport distances. These strategies are not the focus of this chapter. Finally it is important to mention that the actual tandem device performance will also depend on the equilibrium energy level positions and how the overpotential depends on the photocurrent density. Insight into these aspects and a better estimation of the actual tandem cell performance can be achieved by comparing the actual $J-V$ curves for working electrodes. As described in Chap. 4 the point where the $J-V$ curves of the photocathode and photoanode intersect gives the operating current density $J_{op}$, which can be used in the place of $J_{min}$ to estimate *STFCE* in Eq. (12.1).

## 12.4 Photoelectrode Configurations and Device Design Considerations

Regardless of the nature of the semiconductor junction combination chosen to make a tandem cell for solar water splitting, consideration must be taken for how the device will be illuminated and how the evolved gases will be collected. The latter point is especially important given that the stoichiometric combination of $H_2$ and $O_2$ produced during water splitting is highly explosive when mixed. Thus, effective gas separation is preferable. Fortunately, the intrinsic configuration of a water splitting tandem cell—one electrode (or photoelectrode) for water oxidation and another side for water reduction—facilitates the separate production and collection of gases.

A simple device can be described that uses two electrolyte compartments separated by a glass frit or membrane for ionic transport and to prevent the reverse water splitting reactions from occurring. An idealized variation of this device is shown in Fig. 12.5. Here, the photoanode and photocathode are directly connected electronically (by an ohmic contact) to facilitate device fabrication. Indeed this "wireless" device geometry can be simply prepared by coating the two sides of a transparent conducting substrate (e.g., indium tin oxide) with the absorber materials, and placed as a divider in the electrolyte container. In this straightforward design gravity would be the driving force for separating the bubbles of gas forming on the electrodes surface and the remaining electrolyte. Bubble removal from the electrolyte is indeed a critical device parameter (Vogt and Balzer 2005). However, it must be pointed out that the gravitationally driven force is most effective when

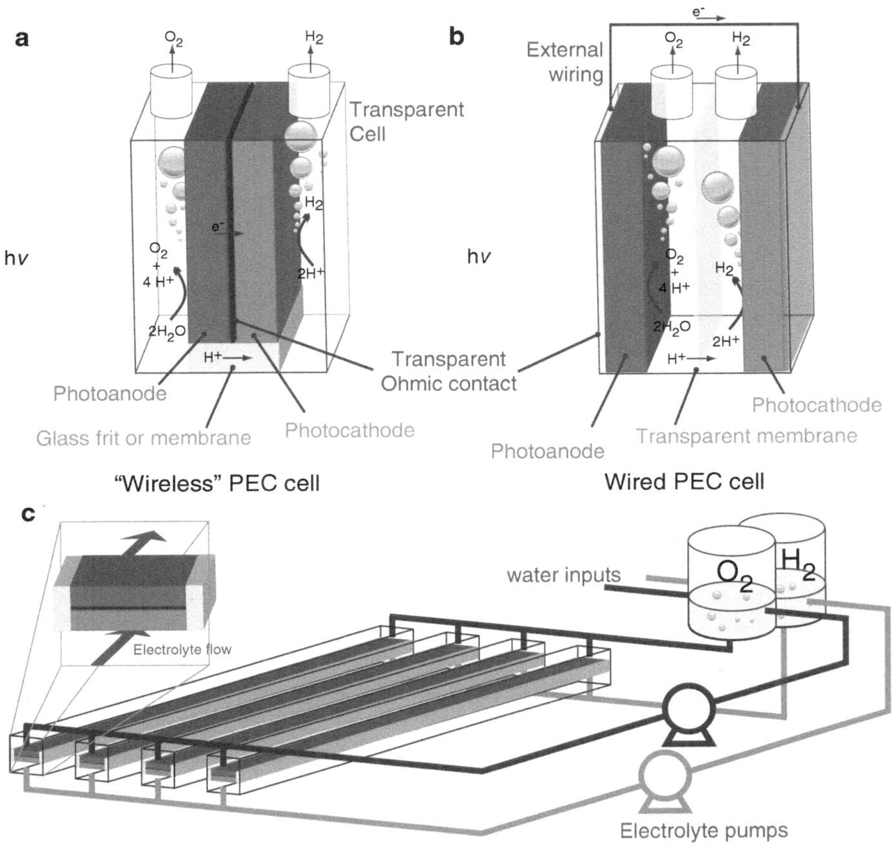


**Fig. 12.5** Tandem cell configurations for solar water splitting. (**a**) Shows a wireless cell configuration while (**b**) shows the wired configuration. Both examples assume that the band-gap of the photoanode is larger than the photocathode. (**c**) Shows a scalable trough-type flow system proposed for the large-scale implementation of the PEC cell

the photoelectrodes are vertically oriented. This orientation requires the illumination direction to be perpendicular to the gravitational force—a condition that only occurs at sunrise or sunset (without using mirrors, which would add considerable device cost). Thus, additional device considerations are necessary. For example, the device could simply be placed with the top-cell facing in the direction of the Earth's equator (i.e., facing south if installed in the northern hemisphere) and at an angle optimized to maximize insolation while still providing sufficient buoyancy force for bubble detachment. This angle depends on the force needed to remove a bubble from the surface of an electrode, which can vary widely between different systems (Vogt and Balzer 2005). At large latitudes (north or south) this titled geometry would allow for near optimal illumination during the whole day and while also permitting the collection of the gases by gravity alone. Since near the equator the illumination direction is parallel to the gravitational force, the cells should be oriented parallel to the earth's surface to maximize insolation. This leads to difficulties with bubble accumulation at the surface of the bottom cell which would reduce the performance of the tandem cell by lowering the surface area of the electrode exposed to the electrolyte.

The inherent difficulties in relying on gravity (or rather buoyancy, to be more precise) alone to separate the evolved gas bubbles from the electrode and afford the separate collection of the two gases have inspired research to consider alternative systems designs. One such design is a trough-type flow system (Miller et al. 2003) shown schematically in Fig. 12.5c. Here the electrolyte is actively pumped though the two electrolyte compartments. The flowing electrolyte adds an extra force to shear the bubbles off the surface (Eigeldinger and Vogt 2000), which are then carried along with the electrolyte to separating tanks where the gases are collected at ambient pressure. This type of system allows the tandem cell to be at any angle enabling the optimum insolation at any latitude or by using solar tracking without any concern of bubbles masking the active area of either electrode. The scalable nature of this design and its simplicity makes it a promising solution over other approaches using expensive solar tracking or solar concentration.

The above description of a "wireless" PEC cell geometry, which provides a direct electronic connection between the anode and the cathode, is a convenient device configuration, but the ionic conduction takes a comparatively long path (i.e., though one electrolyte compartment, through the membrane and then through the other electrolyte compartment). Indeed an important practical consideration in any real, large scale device is the ohmic loss in the device. A rough look at the conductivities of the media involved can give some quick insight into the magnitude of the resistive losses. In the best case the ionic conductor may have a conductivity as high as 0.2–0.5 S $cm^{-1}$, which is the case for concentrated NaOH or $H_2SO_4$. Copper, on the other hand, has a conductivity of $6 \times 10^5$ S $cm^{-1}$, and indium tin oxide, a transparent conducting oxide suitable for application in a tandem cell, has a conductivity of about 300 S $cm^{-1}$. Thus, to keep ohmic losses as small as possible, it is clear that the tandem cell should include shortest, most direct pathway of electrolyte between the two electrodes. In contrast to the "wireless" device, a "wired" device (Fig. 12.5b) possesses an electronic conductor that completes the circuit between the photoanode and the photocathode by a relatively long path, and the ionic path is as short as possible. To accomplish this and also facilitate the separate collections of gases, a gas-impermeable membrane that could conduct only ions in the electrolyte can be placed between the two electrodes in the "wired" device configuration. This creates two electrolyte compartments similar to the wireless design.

Many authors have discussed and addressed the aspects or wired and wireless devices and ohmic losses in device design. Of note is work performed by Mallouk and coworkers (2012). Here the authors analyzed the individual potential losses in electrolytic systems of buffer solutions and commercially available anion- and cation-exchange membranes at the current density of 25 mA $cm^{-2}$. The pH gradient which builds up across the membrane was found to be the most problematic source of loss in these systems. However, the authors showed that monoprotic buffers can minimize the pH gradient by diffusion of the neutral acidic or basic form of the buffer across the membrane.

As the losses across the membrane are minimized, the ohmic losses in the electrolyte become dominate. While these can be minimized as previously discussed by using the "wired" device geometry, the simplicity of the "wireless" device geometry is appealing to facilitate device fabrication. One apparent way to reduce series resistance resulting from the long path length of ionic conductance in a wireless device design is to use perforated electrodes as suggested by Nocera and coworkers (2011). Indeed by using a "wireless" tandem cell with a design like that shown in Fig. 12.6a the distance for ionic conductivity can be tuned by choosing the size and pitch of the perforations, which contain the ion transporting membrane, at the cost of reducing the active area of the electrode. Lewis and coworkers (2012) developed an extensive model of the transport in the electrolyte, and simulated this

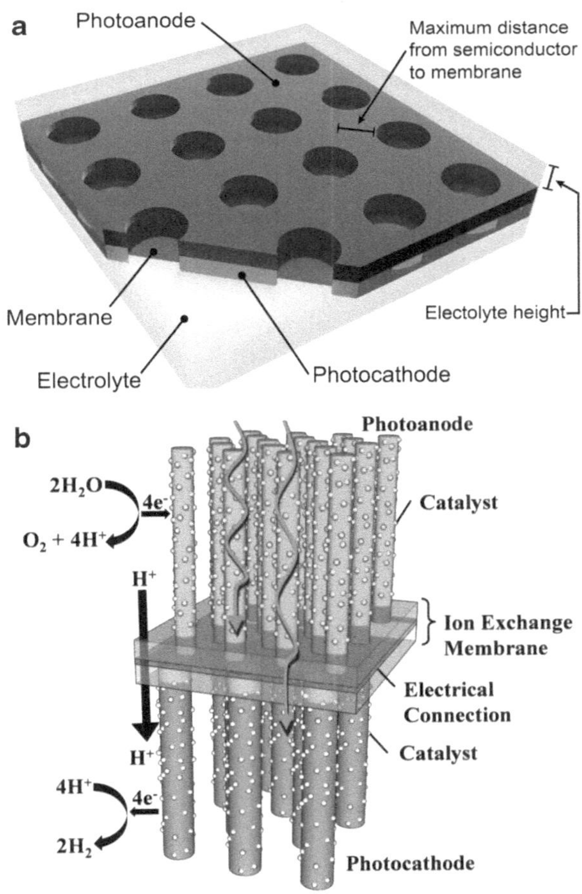


**Fig. 12.6** Proposed "wireless" tandem cell configurations to reduce Ohmic losses due to ionic transport. (**a**) Shows a perforated-type device design. (**b**) Shows a schematic of a proposed tandem cell concept employing a ion exchange membrane together with nanowire semiconductor arrays. Part (**b**) was reprinted with permission from J. M. Spurgeon et al., *Energy Environ. Sci.* 2011, *4*, 1772–1780. http://dx.doi.org/10.1039/C1EE01028J. Copyright 2011 The Royal Society of Chemistry

type of perforated cell in two dimensions. Setting the area of the semiconductor and membrane to be equal and assuming concentrated electrolytes (1 M), the size of the perforations was varied and the resulting ohmic losses were calculated and reported as voltage losses for a PEC cell under standard illumination conditions. It was found that the maximum distance from any point on the semiconductor should be no more than ca. 30 mm to keep the voltage drop due to the ionic conductivity less than 100 mV when the height of the electrolyte was 50 mm. For shorter electrolyte heights (corresponding to smaller electrolyte compartments) the voltage drop increased. With a height of 1 mm the voltage drop rose to over 300 mV. Assuming a larger ratio of semiconductor area to membrane was found to also create larger resistance losses. For the case of a 90 % semiconductor area (10 % membrane area) the maximum distance from any point on the semiconductor to a membrane perforation should only be 10–15 mm (depending on the height of the electrolyte) to keep the voltage drop below 100 mV. Overall this work showed clearly that the losses due to ionic conductivity and the membrane can be mitigated, even for the wireless device geometry. However, the size of the perforations and their pitch need to be on the millimeter to tens of millimeter scale to keep the voltage drop under 100 mV. While this would complicate device fabrication, the distances involved seem feasible.

Similar systems have been proposed to reduce ohmic losses and also facilitate the harvesting of minority carriers using nanowire or microwire electrodes protruding through an ion exchange membrane (Spurgeon et al. 2011). A schematic of this device concept is shown in Fig. 12.6b. Here catalysts are also shown integrated at the semiconductor liquid junctions to facilitate the electrochemical reactions taking place. The smallness of this system avoids ohmic problems; however, the economic feasibility of this design is not clear, especially at the production scale needed for hydrogen to become a primary energy storage vector (Newman 2013).

## 12.5 Demonstrations of Operational Tandem Cells for Solar Hydrogen Production

Regardless of economic feasibility of any device configuration, to date there are many demonstrated examples of functioning tandem cells for overall solar fuel production. Typically research has focused on solar water splitting (although $CO_2$ reduction is also feasible) and buried multijunction PV cells have been used together with direct semiconductor-liquid junction photo electrodes in many combinations to enable the direct solar to hydrogen energy conversion. As of the time of this writing the field is quickly evolving. Some examples are presented in the following paragraphs.

Devices employing two or more (buried) pn-junctions in tandem have attained the highest reported *STHCE* (Khaselev et al. 2001; Licht 2001). Champion devices use the well-known band-gap engineering of III–V semiconductor systems to optimize light harvesting and $\Delta\mu_{ex}$. In particular, Licht et al. (2000) have employed an AlGaAs/Si (pn-pn) structure using Pt-black and $RuO_2$, as reduction and oxidation catalysts, respectively and obtained solar-to-hydrogen conversion efficiencies as high as 18.3 % under simulated AM0 sunlight (135 mW $cm^{-2}$). Turner and coworkers investigated monolithic $GaAs/GaInP_2$ (pn-p, pn-pn, or pn-pn-p) systems and attained *STHCE* up to 12.5 % under concentrated AM 1.5G illumination (Gao et al. 1999; Khaselev and Turner 1998a, b; Kocha et al. 1998). The conversion efficiencies reported with these systems are impressive, but major concerns exist about the price and stability of these devices (Khaselev and Turner 1998a, b).

A more economically feasible system using buried junctions is based on amorphous silicon (a-Si). Bockris and coworkers first showed that a triple stack of n-type/intrinsic/p-type (nip) a-Si on a Ti substrate could give up to *STHCE* = 7.5 % under AM1 (100 mW $cm^{-2}$) illumination when islands of Pt and $RuO_2$ were used as a reduction and oxidation catalysts (Lin et al. 1989). In this case the band-gap of the a-Si was the same for each junction, so the layer thicknesses had to be optimized to ensure that each layer produced the same photocurrent. This optimization can be done by using the absorption coefficient and Eq. (12.5) as described in the previous section. Amorphous silicon based tandem cells have been advanced by mixing some Ge in with the Si to modify the $E_g$ of the bottom two layers. An *STHCE* of 7.8 % under AM1.5 (100 mW $cm^{-2}$) illumination was achieved using these devices for water splitting (Rocheleau et al. 1998). Various other thin-film photovoltaic technologies have recently been employed in a similar manner to demonstrate wired or wireless solar water splitting devices at similar *STHCE*s. Despite the reasonable demonstrated performance, the advantage of using these buried junction PV tandem devices over simply using photovoltaics coupled to an electrolyser is not clear, especially given the added requirement to ensure the PV cells are protected from the electrolyte.

Using a direct semiconductor electrolyte junction can reduce the complexity of the device and reduce potential fabrication costs. The difficulty in identifying a stable p-type cathode (without protection layers) makes a photoanode/PV tandem device a good compromise with device complexity and stability. Miller and coworkers (2005) have investigated combining a tungsten oxide photoanode in tandem with double junction a-Si:Ge PV and 3 % *STHCE* was obtained. Using $BiVO_4$ as a photoanode with a double junction a-Si back cell gave *STHCE* up to 4.9 % (Abdi et al. 2013).

Alternative technologies can be employed for the PV part in these classes of devices as well. Tandem devices assembled with either $WO_3$ or $Fe_2O_3$ as the photoanode and a high-voltage dye sensitized solar cell (DSSC) gave *STHCE*s of 3.10 % and 1.17 % with the $WO_3$/DSSC and $Fe_2O_3$/DSSC combinations, respectively (Brillet et al. 2012). Notably only two junctions were used in this case compared to the three junction devices based on a-Si. An optical analysis compared the predicted photocurrent from the integration of IPCE measurements and the actual *J–V* behavior measured in situ in the device. This optical analysis and the *J–V* curves for each device are shown in Fig. 12.7. In the case of the $Fe_2O_3$/DSSC tandem cell, the crossing point of the *J–V* curves, the operating current density $J_{op}$,

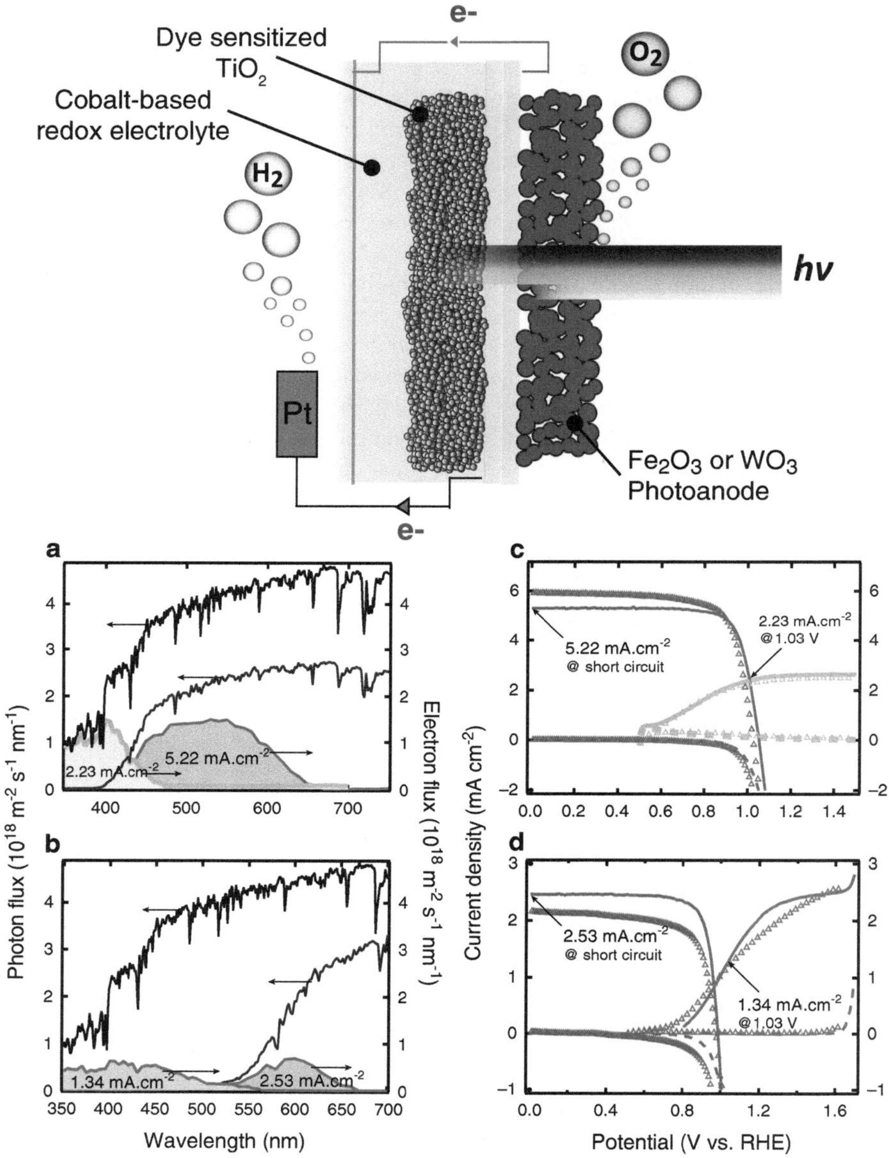


**Fig. 12.7** General scheme (*top*) of a Photoanode /DSSC tandem cell. Spectral response and *J–V* characteristics of the $WO_3$/DSSC (**a**, **c**) and $Fe_2O_3$/DSSC (**b**, **d**) tandem cells. The transmittance of the photoanode (not shown) convoluted to the AM 1.5G photon flux on the photoanode (**a**, **b**) allows the calculation of the photon flux incident to the DSSC (**a**, **b**). IPCE data (not shown) and the photon flux incident to each element is used to estimate the photocurrent density (shaded areas under the curves in **a** and **b**). *J–V* curves (**c** and **d**) of the cells are shown under AM 1.5G irradiation. The *filled lines* represent the *J–V* curves predicted from calculation and the triangles represent data from in situ device measurements. Adapted from Brillet et *a*l., *Nat. Photonics* **6**, 824–828 (2012). Copyright 2012 Macmillan Publishers Limited

is far from the plateau region of the $Fe_2O_3$ electrode photocurrent. This results in a performance far from the maximum obtainable. The limitation of this system is thus attributable to the late onset of the photocurrent in the photoanode. However, in the case of the $WO_3$/DSSC tandem cell, the photocurrent onset is not a limitation and the $J_{op}$ is very close to the plateau region of the photoanode (approx. 1000 mV). The limiting factor in this case is the low photocurrent obtainable by the photoanode due to less than ideal absorption capability of tungsten trioxide in the visible region of the solar spectrum. Despite this, the $Fe_2O_3$/DSSC device still exhibited a near unity faradaic efficiency and a good stability over an 8 h testing period. Overall this work suggests that the low *STHCE* of the $Fe_2O_3$/DSSC tandem cell offers the larger room for improvement. In particular, further reduction of the overpotential for water oxidation is expected to rapidly advance *STHCE*s.

As previously mentioned, the most straightforward and simple tandem cell for solar fuel production (arguably) is direct semiconductor junction liquid junction device with a photoanode and a photocathode as it contains the fewest material layers. This approach is much less developed, compared to approaches using PV cells or buried junctions, due to a lack of suitable photocathode materials for water reduction. Despite this a few efforts have been made to demonstrate photoanode/photocathode tandem cells. Early work by Nozik introduced the n-$TiO_2$/p-GaP system. A high internal resistance limited the *STHCE* to 0.25 % at zero bias (Nozik 1976). More recently, oxidized Cu-Ti films, primarily CuO ($E_g = 1.4$ eV) with a $Cu_4Ti_3$ impurity phase were placed in tandem with $TiO_2$ to give a working tandem device (Mor et al. 2008). A $J_{op} = 0.25$ mA $cm^{-2}$ (*STHCE* around 0.30 %) under standard illumination conditions and a reasonable stability over minutes were observed. A protected p-type $Cu_2O$ photocathode was also used with a $BiVO_4$ photoanode to give a tandem cell with $J_{op} = 1.0$ mA $cm^{-2}$ (*STHCE* around 1.2 %) but poor stability was attributed to dissolution for the catalysts used for water oxidation (Bornoz et al. 2014). Overall the development of suitable photocathode/photoanode pairs is needed to realize high *STFCE* in this type of device configuration. The above-described examples all employ the photoanode as the top cell, as the photoanode possessed the higher band-gap energy. However, as pointed by out Chorkendorf and coworkers, in some cases (e.g., if a protecting window layer needs to be employed for the photocathode) it makes sense to choose a photocathode with a larger band gap than the photoanode and use it as the top cell (Seger et al. 2014).

## 12.6 Economic Considerations

The above descriptions of demonstrated tandem cells for solar fuel production already touches on the ideas of the cost to fabricate the device. Indeed, economic considerations will ultimately be the driving force of the adoption of any PEC solar fuel production technology. The potential cost of producing the tandem device, its *STFCE* and its longevity are all important a factors to the price at which a cell can produce fuel. An important technoeconomic assessment estimated that a device lifetime of 10 years, a cost of *ca.* 150 US\$ $m^{-2}$, and a $STHCE = 10$ % are necessary to produce hydrogen by PEC water splitting at a cost comparable to that PV + electrolysis ($H_2$ at a price of about 10 US\$ $kg^{-1}$) (Pinaud et al. 2013). Moreover this assessment also showed that the price of the produced hydrogen was relatively insensitive to extending the tandem cell lifetime (8.6 US\$ $kg^{-1}$ for a 20 year lifetime) but more sensitive to reducing the cost of the PEC cell (6.9 US\$ $kg^{-1}$ for an 80 US\$ $m^{-2}$ tandem cell). While it was also found possible to reduce the cost of $H_2$ to 6.1 US\$ $kg^{-1}$ by doubling the efficiency to 20 %, this goal seems unlikely given the demonstrated performance to date. Thus, reducing device complexity and employing inexpensive fabrication methods while using stable, widely available materials are the main issues for the practical implementation of the solar water splitting tandem cell. Realizing the goal of an 80 US\$ $m^{-2}$ tandem cell seems difficult if using a device technology that requires many layers deposited at high temperature, or using high vacuum, e.g., chemical vapor deposition. Slow, conformal growth methods, like atomic layer deposition have proved very valuable in protecting unstable semiconductor materials (Sivula 2014); however, the economic viability of this technique for the preparation of large-area devices is questionable. Solution-based roll-to-roll device fabrication techniques are, however, viable to produce solar energy conversion devices in a scalable way at a price that can easily meet the economic constraints (Søndergaard et al. 2012); thus, the development of electrodes for the tandem cell should focus on materials for which this type of processing is possible, while the other criteria of device performance and stability can be also met. Solution-processed metal oxides or chalcogenides are thus promising materials for this application.

## 12.7 Summary and Conclusion

Overall, in this chapter the motivation and theoretical framework for the PEC fuel-producing tandem cell are examined. While a single junction device could theoretically be capable of a 10 % conversion efficiency for solar water splitting, a perfect material has not yet been identified. An integrated tandem approach is more flexible regarding material choice, and can develop suitable photopotential for fuel production while also harvesting an appreciable amount of the solar spectrum. The simple theoretical framework presented in this chapter shows how using two absorbers could reasonably convert up to 30 % of the sun's energy into hydrogen via water splitting (even with large assumed losses) if ideal materials can be found. Furthermore a large range of different combinations of band-gaps can potentially give rise to a device with over 10 % conversion efficiency. The design of the PEC cell was also discussed. Resistance losses due to membranes and ionic conduction are important to practical application, but can be mitigated by optimizing the geometry of the device. Many different systems have been investigated using various combinations of photovoltaic cells and photoelectrodes and in some cases the conversion efficiency have been over 10 %. However, in order to be economically competitive with simple "brute force" strategies or the production of $H_2$ from PV + electrolysis, a practical water splitting tandem cell must optimize cost, longevity and performance. A simple photoanode/photocathode tandem fabricated with stable oxide semiconductors using inexpensive (e.g., solution-based) processing would be ideal, but identifying a suitable material combination and also minimizing electrochemical overpotentials still remain challenging.

## References

Abdi FF, Han L, Smets AH, Zeman M, Dam B, van de Krol R (2013) Efficient solar water splitting by enhanced charge separation in a bismuth vanadate-silicon tandem photoelectrode. Nat Commun 4:2195. doi:10.1038/ncomms3195

Bolton JR, Haught AF, Ross RT (1981) Photochemical energy storage: an analysis of limits. In: Connolly JS (ed.) Photoelectrochemical conversion and storage of solar energy. Academic, New York, NY, pp 297–339

Bolton JR, Strickler SJ, Connolly JS (1985) Limiting and realizable efficiencies of solar photolysis of water. Nature 316:495–500

Bornoz P et al (2014) A bismuth vanadate–cuprous oxide tandem cell for overall solar water splitting. J Phys Chem C 118:16959–16966. doi:10.1021/jp500441h

Brillet J et al (2012) Highly efficient water splitting by a dual-absorber tandem cell. Nat Photon 6:824–828. doi:10.1038/NPHOTON.2012.265

Eigeldinger J, Vogt H (2000) The bubble coverage of gas-evolving electrodes in a flowing electrolyte. Electrochim Acta 45:4449–4456. doi:10.1016/s0013-4686(00)00513-2

Gao X, Kocha S, Frank AJ, Turner JA (1999) Photoelectrochemical decomposition of water using modified monolithic tandem cells. Int J Hydrogen Energ 24:319–325

Haussener S, Xiang C, Spurgeon JM, Ardo S, Lewis NS, Weber AZ (2012) Modeling, simulation, and design criteria for photoelectrochemical water-splitting systems. Energy Environ Sci 5:9922–9935. doi:10.1039/C2EE23187E

Hernandez-Pagan EA, Vargas-Barbosa NM, Wang T, Zhao Y, Smotkin ES, Mallouk TE (2012) Resistance and polarization losses in aqueous buffer-membrane electrolytes for water-splitting photoelectrochemical cells. Energy Environ Sci 5:7582–7589. doi:10.1039/C2EE03422K

Khaselev O, Turner JA (1998a) A monolithic photovoltaic-photoelectrochemical device for hydrogen production via water splitting. Science 280:425–427

Khaselev O, Turner JA (1998b) Electrochemical stability of p-GaInP[sub 2] in aqueous electrolytes toward photoelectrochemical water splitting. J Electrochem Soc 145:3335–3339. doi:10.1149/1.1838808

Khaselev O, Bansal A, Turner JA (2001) High-efficiency integrated multijunction photovoltaic/electrolysis systems for hydrogen production. Int J Hydrogen Energ 26:127–132

Kocha SS, Montgomery D, Peterson MW, Turner JA (1998) Photoelectrochemical decomposition of water utilizing monolithic tandem cells. Sol Energ Mat Sol Cells 52:389–397

Licht S (2001) Multiple band gap semiconductor/electrolyte solar energy conversion. J Phys Chem B 105:6281–6294. doi:10.1021/jp010552j

Licht S, Wang B, Mukerji S, Soga T, Umeno M, Tributsch H (2000) Efficient solar water splitting, exemplified by RuO2-catalyzed AlGaAs/Si photoelectrolysis. J Phys Chem B 104:8920–8924. doi:10.1021/jp002083b

Lin GH, Kapur M, Kainthla RC, Bockris JOM (1989) One step method to produce hydrogen by a triple stack amorphous silicon solar cell. Appl Phys Lett 55:386–387

Miller EL, Rocheleau RE, Deng XM (2003) Design considerations for a hybrid amorphous silicon/photoelectrochemical multijunction cell for hydrogen production. Int J Hydrogen Energ 28:615–623

Miller EL, Marssen B, Paluselli D, Rocheleau R (2005) Optimization of hybrid photoelectrodes for solar water-splitting. Electrochem Solid-State Lett 8:A247–A249. doi:10.1149/1.1887196

Mor GK et al (2008) p-Type Cu – Ti – O nanotube arrays and their use in self-biased heterojunction photoelectrochemical diodes for hydrogen generation. Nano Lett 8:1906–1911. doi:10.1021/nl080572y

Newman J (2013) Scaling with Ohm's Law; wired vs. wireless photoelectrochemical cells. J Electrochem Soc 160:309–311. doi:10.1149/2.020304jes

Nozik AJ (1976) p-n photoelectrolysis cells. Appl Phys Lett 29:150–153

Pinaud BA et al (2013) Technical and economic feasibility of centralized facilities for solar hydrogen production via photocatalysis and photoelectrochemistry. Energy Environ Sci 6:1983–2002. doi:10.1039/C3EE40831K

Reece SY et al (2011) Wireless solar water splitting using silicon-based semiconductors and earth-abundant catalysts. Science 334:645–648. doi:10.1126/science.1209816

Rocheleau RE, Miller EL, Misra A (1998) High-efficiency photoelectrochemical hydrogen production using multijunction amorphous silicon photoelectrodes. Energ Fuel 12:3–10

Seger B, Castelli IE, Vesborg PCK, Jacobsen KW, Hansen O, Chorkendorff I (2014) 2-Photon tandem device for water splitting: comparing photocathode first versus photoanode first designs. Energy Environ Sci 7:2397–2413. doi:10.1039/C4EE01335B

Sivula K (2014) Defects give new life to an old material: electronically leaky titania as a photoanode protection layer. ChemCatChem 6:2796–2797. doi:10.1002/cctc.201402532

Søndergaard R, Hösel M, Angmo D, Larsen-Olsen TT, Krebs FC (2012) Roll-to-roll fabrication of polymer solar cells. Mater Today 15:36–49. doi:10.1016/S1369-7021(12)70019-6

Spurgeon JM, Walter MG, Zhou J, Kohl PA, Lewis NS (2011) Electrical conductivity, ionic conductivity, optical absorption, and gas separation properties of ionically conductive polymer membranes embedded with Si microwire arrays. Energy Environ Sci 4:1772–1780. doi:10.1039/C1EE01028J

Vogt H, Balzer RJ (2005) The bubble coverage of gas-evolving electrodes in stagnant electrolytes. Electrochim Acta 50:2073–2079. doi:10.1016/j.electacta.2004.09.025

Walter MG et al (2010) Solar water splitting cells. Chem Rev 110:6446–6473. doi:10.1021/cr1002326

Weber MF, Dignam MJ (1984) Efficiency of splitting water with semiconducting photoelectrodes. J Electrochem Soc 131:1258–1265

# Chapter 13
# Dye Sensitized Photoelectrosynthesis Cells for Making Solar Fuels: From Basic Science to Prototype Devices

**Michael K. Coggins and Thomas J. Meyer**

Research in solar energy conversion and storage by using the energy of the sun to drive chemical reactions that create solar fuels was inspired by the initial experiments of Honda and Fujishima in the 1970s. They showed that UV band gap excitation of a $TiO_2$ semiconductor oxide electrode with a Pt counter electrode and small applied bias resulted in water splitting, Fig. 13.1 (Fujishima and Honda 1972) Despite the simplicity of the approach, the need for band gap excitation, and the small fraction of the solar spectrum that was used ruled out practical applications.

Significant progress has been made since on using semiconductors for water splitting and solar fuels but large scale, economically feasible device applications remain elusive. Technically formidable challenges remain arising from the aqueous stability of semiconductor materials, achieving light absorption sufficiently far into the visible, enabling efficient electron–hole pair separation, and using the electron–hole pair to drive surface catalysts of water oxidation and water or $CO_2$ reduction.

An alternate, molecular-based approach appeared shortly after the initial report by Honda and Fujishima. It was based on light absorption by the ($d\pi^6$) Ru(II) polypyridyl complex $[Ru(bpy)_3]^{2+}$ (bpy is 2,2′-bipyridine) in solution to give its $^3(d\pi^5\pi^1)$ Metal-to-Ligand Charge Transfer excited state, $[Ru(bpy)_3]^{2+*}$, with a lifetime of ~1 μm in water at room temperature. With the added oxidative quencher methylviologen ($MV^{2+}$), diffusional electron transfer quenching of the excited state, $[Ru(bpy)_3]^{2+*} + MV^{2+} \rightarrow [Ru(bpy)_3]^{3+} + MV^{+}$, occurred to give

M.K. Coggins
Department of Chemistry, University of North Carolina at Chapel Hill, Chapel Hill, NC 27516, USA

Eastman Chemical Company, Kingsport, TN 37662, USA

T.J. Meyer (✉)
Department of Chemistry, University of North Carolina at Chapel Hill, Chapel Hill, NC 27516, USA
e-mail: tjmeyer@unc.edu



S. Giménez, J. Bisquert (eds.), *Photoelectrochemical Solar Fuel Production*,
DOI 10.1007/978-3-319-29641-8_13

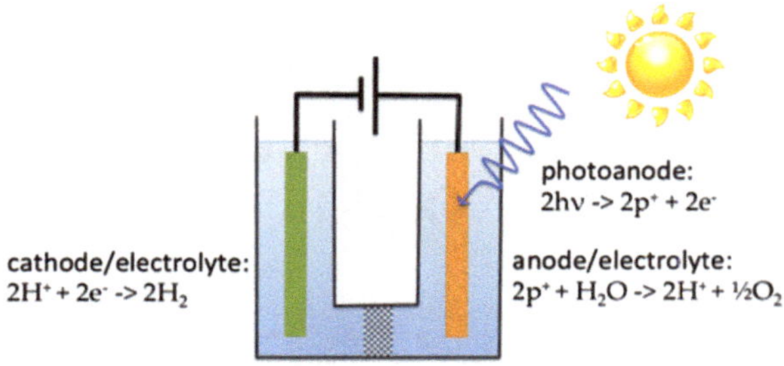


**Fig. 13.1** A diagram illustrating the Honda–Fujishima photoelectrochemical cell for water splitting

**Scheme 13.1** Oxidative quenching of $[Ru(bpy)_3]^{2+}$ by $MV^{2+}$

$$Ru(bpy)_3^{2+}\,(d\pi^6) + h\upsilon \rightarrow Ru^{III}(bpy^{-\cdot})(bpy)_2^{2+*}\,(d\pi^5)(\pi^*)^1$$
$$Ru(bpy)_3^{2+*} + MV^{2+} \rightarrow Ru(bpy)_3^{3+} + MV^{+\cdot}$$
$$Ru(bpy)_3^{3+} + MV^{+\cdot} \rightarrow Ru(bpy)_3^{2+} + MV^{2+}$$

$[Ru(bpy)_3]^{3+}$ and $MV^+$ as products, Scheme 13.1. The oxidized chromophore is thermodynamically capable of oxidizing water with $E^{o\prime}([Ru(bpy)_3]^{3+/2+}) = 1.26$ V vs. NHE and $MV^+$, with $E^{o\prime}(MV^{2+/+}) \sim -0.40$ V vs. NHE, is capable of reducing protons to $H_2$ in acidic solutions (Young et al. 1976; Bock et al. 1974; Meyer 1990).

Following the quenching step, diffusional back electron transfer occurs and the stored redox equivalents are lost as heat to the surroundings. Nonetheless, the result in Scheme 13.1 was provocative in demonstrating a chemical basis for solar water splitting. However, there were two significant issues to address before the reactivity could be exploited. The first was that there were no known molecular catalysts for water oxidation. The second was, that to overcome diffusional back electron transfer, a molecular architecture would have to be created to separate and store the transient redox equivalents where they could be used to activate catalysts for water oxidation and proton reduction, Eqs. (13.1) and (13.2).

$$2H_2O \rightarrow O_2 + 4e^- + 4H^+ \tag{13.1}$$

$$2H^+ + 2e^- \rightarrow H_2 \tag{13.2}$$

## 13.1 Water Oxidation

The first complexes investigated as water oxidation catalysts were $Ru^{II}$ polypyridyl derivatives of $[Ru(bpy)_3]^{2+}$ which had been modified by adding a water molecule to the inner coordination sphere—*cis*-$[Ru(bpy)_2(py)(OH_2)]^{2+}$ and $[Ru(tpy)(bpy)(OH_2)]^{2+}$ (tpy is 2,2′,2″-terpyridine) (Moyer and Meyer 1978; Binstead et al. 1992; Dwyer et al. 1952). They underwent stepwise one-electron oxidation, first to $Ru^{III}$-$OH_2^{2+}$ and then to $Ru^{IV}{=}O^{2+}$, by proton coupled electron transfer (PCET). The $Ru^{IV}{=}O^{2+}$ forms proved to have remarkably rich oxidative reactivities toward a variety of organic and inorganic functional groups but redox potentials were ~0.1–0.2 V too low for water oxidation. As noted below, further oxidation to $Ru^V(O)^{3+}$ was accessible for these and related complexes, but at higher potentials and, for $[Ru^V(tpy)(bpy)(O)]^{3+}$, with very slow rates.

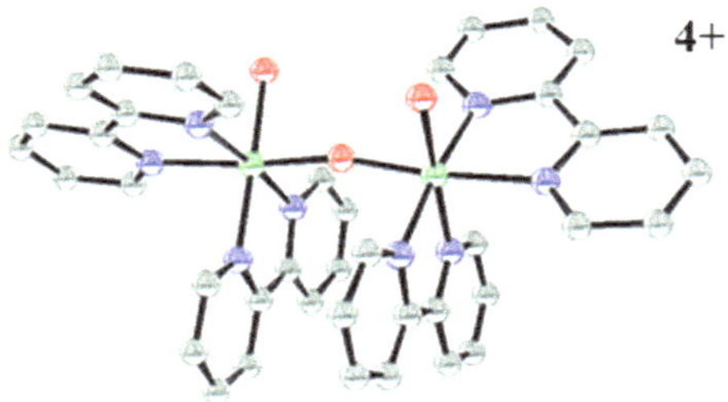


**Fig. 13.2** Structure of the Ru(III) "blue dimer," $[(bpy)_2Ru(OH_2)(\mu\text{-}O)Ru(OH_2)(bpy)_2]^{4+}$

The coordination chemistry of $Ru^{II}$ and $Os^{II}$ polypyridyl complexes was developed extensively by Dwyer and coworkers in the 1950s and 1960s (Dwyer et al. 1952). Their results on related complexes provided a basis for the synthesis of the, now famous, μ-oxo-bridged $Ru^{III}$ "blue dimer," $[(bpy)_2Ru(OH_2)(\mu\text{-}O)Ru(OH_2)(bpy)_2]^{4+}$, Fig. 13.2. The synthesis was a success. Addition of Ce(IV) in acid to solutions containing the blue dimer resulted in $O_2$ evolution (Gersten et al. 1982; Gilbert et al. 1985; Hurst et al. 1992; Hurst 2005; Liu et al. 2008).

An important consideration in the initial design of the blue dimer was the potential value of linking the two $Ru^{III}$ sites with a μ-oxo bridge. Following $4e^-/4H^+$ oxidation of $[(H_2O)Ru^{III}\text{-}O\text{-}Ru^{III}(OH_2)]^{4+}$ to $[(O)Ru^{V}\text{-}O\text{-}Ru^{V}(O))]^{4+}$, the μ-oxo bridge could provide a structural basis for intra-molecular O–O coupling. Detailed kinetic and mechanistic studies later revealed that water oxidation actually occurred by a single-site mechanism related to the one shown in Scheme 13.2 for the water oxidation catalyst, $[Ru(Mebimpy)(bpy)(OH_2)]^{2+}$ (Mebimpy is 2,6 (1-methylbenzimidazol-2-yl) (Concepcion et al. 2010a, b). For this single-site catalyst, O–O bond formation occurs by O-atom transfer to a water molecule in the surrounding solvent.

Inspired by the mechanism of catalysis for the blue dimer, a renewed search for single site water oxidation catalysts some 20 years later led to a detailed mechanistic study on Ce(IV) catalyzed water oxidation by $[Ru(tpy)(bpz)(OH_2)]^{2+}$ and $[Ru(tpy)(bpm)(OH_2)]^{2+}$ (bpz is 2,2′-bipyrizine and bpm is 2,2′-bipyrimidine (Zong and Thummel 2005; Concepcion et al. 2010a, b).

The mechanism in Scheme 13.2 illustrates the sequential reactions that occur in water oxidation by the catalyst $[Ru(Mebimpy)(bpy)(OH_2)]^{2+}$ (Chen et al. 2010a, b, c; Chen et al. 2011a, b, c; Norris et al 2013; Tamaki et al 2014). For this catalyst, the key O–O bond forming step occurs to give a hydroperoxide intermediate, $Ru^{III}\text{-}OOH^{2+}$ following $3e^-/2H^+$ oxidative activation. Further oxidation, proton loss, and $O_2$ evolution completes the water oxidation cycle.

The mechanism in Scheme 13.2 is complex but, given the demands of the overall reaction, Eq. (13.1), any water oxidation mechanism is necessarily complex. At a single-site catalyst, activation occurs by stepwise $1e^-$ oxidation through a series of oxidation states. As noted below, depending on the catalyst and conditions, rate-limiting behavior has been observed at three different stages in the overall catalytic cycle: PCET (proton coupled electron transfer) oxidation of $Ru^{III}\text{-}OH_2{}^{3+}$ or $Ru^{III}\text{-}OH^{2+}$ to $Ru^{IV}{=}O^{2+}$; the O–O forming step by $Ru^{IV}{=}O^{2+}$, or more commonly by $Ru^{V}{=}O^{3+}$; and loss of $O_2$ from the peroxo intermediate, $Ru^{IV}(OO)^{2+}$. Polypyridyl

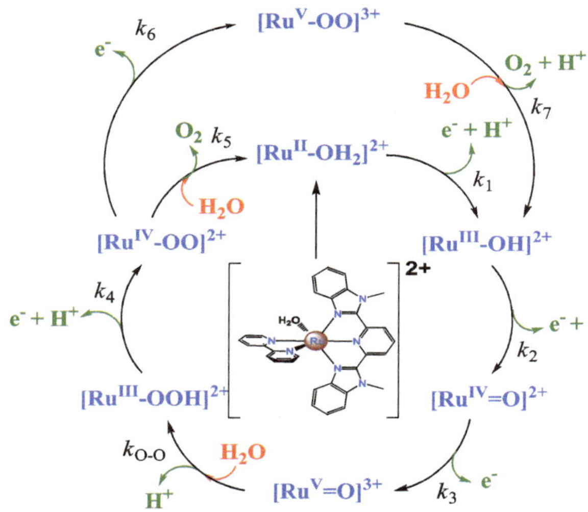


**Scheme 13.2** Single-site catalytic water oxidation cycle for water oxidation by [Ru(Mebimpy)(bpy)(OH$_2$)]$^{2+}$. Reprinted with permission from American Chemical Society and Chemical Reviews (http://pubs.acs.org/doi/abs/10.1021%2Facs.chemrev.5b00229)

complexes of Ru(II) have played an outsized role in water oxidation catalysis. These complexes are coordinatively stable in oxidation states from II to V and systematic ligand variations can be used to tune redox potentials for the couples involved in the water oxidation cycle.

PCET, with coupled electron/proton loss, is important in preventing an increase in charge build-up at the catalyst allowing multiple oxidative equivalents to accumulate at a single catalytic site or cluster with small, or even no, increase in potential (Weinbert et al. 2012; Gagliardi et al 2012) Oxo-based reactivity is triggered by $2e^-/2H^+$ oxidation to $Ru^{IV}{=}O^{2+}$. The electrophilic oxo ligand enables multiple electron pathways to occur—oxygen-atom transfer, hydride transfer, C–H bond insertion (Meyer and Huynh 2003). The oxo-based reactivity is further enhanced in $Ru^V(O)^{3+}$ which is an even more powerful and reactive oxidant.

In a mechanistic study on single-site water oxidation catalysis by [Ru(tpy)(bpm)(OH$_2$)]$^{2+}$ with Ce(IV) as the oxidant, the rate-limiting step was shown to be O–O bond formation by $Ru^V(O)^{3+}$ but, with a $H_2O/D_2O$ kinetic isotope effect (KIE) of 6.6 at room temperature (Concepcion et al. 2010a, b; Chen et al. 2011a, b, c). This led to the suggested pathway for the O–O bond forming step in Eq. (13.3). In this step, O–O bond formation is accompanied by concerted proton transfer to a second water molecule or water cluster by a mechanism described as Atom-Proton Transfer (APT). Although complex, this pathway avoids the high energy hydrogen peroxide intermediate, $Ru^{III}(OOH_2)^{3+}$. As discussed below, with added bases such as $Ac^-$, $HPO_4^{2-}$, or $PO_4^{3-}$, it can lead to significant rate enhancements if O–O bond formation is rate-limiting.

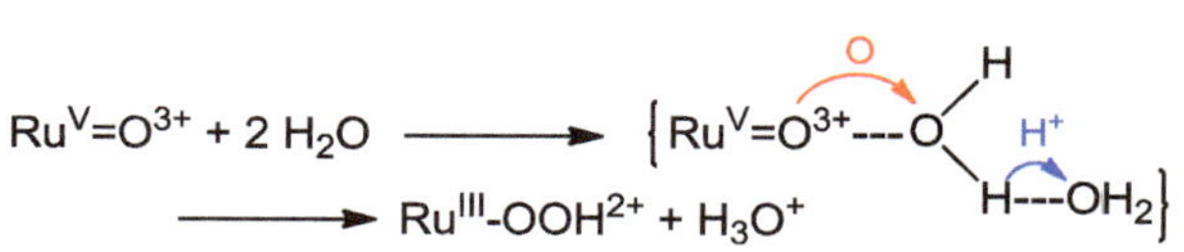


(13.3)

## 13.2 Dye-Sensitized Photoelectrosynthesis Cells (DSPECs)

A second issue in exploiting molecular excited state reactivity in water splitting was the architecture issue. Could a way be found to separate and use transient oxidative and reductive equivalents before they undergo back electron transfer? Although other approaches were explored, success was found by integrating molecular assemblies that combined a light absorbing chromophore and a catalyst for water oxidation with nanoparticle films of high band gap semiconductor oxides, notably $TiO_2$ (O'Regan and Gratzel 1991; O'Regan and Durrant 2009; Gratzel 2009; Ardo and Meyer 2009). As in a Dye-Sensitized Solar Cell (DSSC), light absorption by the chromophore and excited state injection release an electron at the conduction band potential of the oxide for transfer to a cathode where reduction occurs (O'Regan and Gratzel 1991). In a DSPEC, light absorption and injection are integrated with a catalyst for water oxidation, Fig. 13.3. The solar input is used to drive water splitting or $CO_2$ reduction rather than to create a photophotential and photocurrent as in a DSSC.

An initial report of a molecular assembly-based DSPEC, for the dehydrogenation of *iso*-propanol, was reported in 1999. In aqueous solutions, phosphonic-acid groups are typically used for surface binding to the semiconductor oxide (Song et al. 2011). They form relatively stable surface ester links although they do undergo hydrolysis from the surface as the pH is raised (Chen et al. 2009, 2010a, b, c). Other functional groups for surface attachment, including carboxylates and siloxanes, have also been used but carboxylates are readily hydrolyzed and siloxanes have proven to be synthetically challenging (Treadway et al. 1999). Successful stabilization of surface-bound phosphonate assemblies has been demonstrated by using atomic layer deposition (ALD) to add stabilizing overlayers of $Al_2O_3$ or $TiO_2$ and by the addition of a hydrophobic polymer film overlayer by a simple dipping procedure (Vannucci et al. 2013; Wee et al. 2014).

Light absorption by the assembly chromophore at a DSPEC photoanode initiates the sequence of events shown in Fig. 13.3. The example assembly in Fig. 13.3 is $[(4,4'\text{-}PO_3H_2)_2bpy)_2Ru(4\text{-Mebpy-}4'\text{-bimpy})Ru(tpy)(OH_2)]^{4+}$ ($[Ru_a^{II}\text{-}Ru_b^{II}\text{-}OH_2]^{4+}$: $(4,4'\text{-}PO_3H_2)_2bpy$ is 4,4′-bisphosphonic acid-2,2′-bipyridine; 4-Mebpy-4′-bimpy is 4-(methylbipyridin-4′-yl)-*N*-(benzimid)-*N*′-pyridine) (Norris et al. 2013; Alibabaei et al. 2013). As in a DSSC, light absorption and excited state formation are followed by injection into the conduction band, Eq. (13.4a). Following injection, intra-assembly electron transfer from the catalyst to the oxidized chromophore begins the process of catalyst activation, Eq. (13.4b). Injection is followed by electron transport through the nanoparticle film to a transparent conducting oxide collector electrode, typically fluorine-doped tin oxide (FTO), Eq. (13.5a), where it is transferred to an external cathode for proton reduction to $H_2$. As shown in Fig. 13.3, in order to maximize cell efficiency and drive $H^+/H_2O$ reduction to $H_2$ to completion, an applied bias of 0.2 V is required. At the photoanode, electron migration to FTO is in competition with back electron transfer to the assembly, Eq. (13.5b). Back electron transfer is deleterious resulting in loss of the transiently stored oxidative

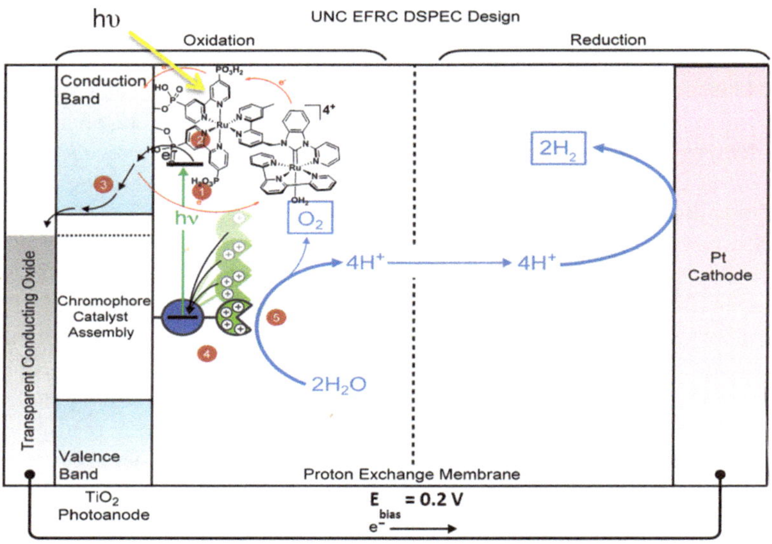


**Fig. 13.3** Schematic diagram of a dye-sensitized photoelectrosynthesis cell (DSPEC) for light-driven water oxidation and proton reduction illustrating: (1) excitation; (2) injection; (3) electron transport through the nanoparticle film; (4) electron transfer activation of the catalyst; (5) proton transfer to the cathode. A bias of 0.2 V is required to maximize cell efficiency and drive $H^+/H_2O$ reduction to $H_2$ to completion

equivalent. Core/shell oxide structures have been found to be advantageous for controlling these interfacial electron transfer dynamic (Alibabaei et al. 2014; Kim et al. 2014). It is worth noting that the design inspiration for DSPEC devices loosely comes from the natural photosynthetic apparatus, photosystem II, where molecular level, light-driven water oxidation occurs (Renger and Renger 2008).

$$\begin{aligned} &\mathrm{FTO}|\mathrm{TiO_2}|\text{-}\left[\mathrm{Ru_a}^{\mathrm{II}}\text{-}\mathrm{Ru_b}^{\mathrm{II}}\text{-}\mathrm{OH_2}\right]^{4+} \xrightarrow{h\upsilon} \mathrm{FTO}|\mathrm{TiO_2}|\text{-}\left[\mathrm{Ru_a}^{\mathrm{II}*}\text{-}\mathrm{Ru_b}^{\mathrm{II}}\text{-}\mathrm{OH_2}\right]^{4+} \\ &\quad\rightarrow \mathrm{FTO}|\mathrm{TiO_2}(e^-)|\text{-}\left[\mathrm{Ru_a}^{\mathrm{III}}\text{-}\mathrm{Ru_b}^{\mathrm{II}}\text{-}\mathrm{OH_2}\right]^{5+} \end{aligned} \tag{13.4a}$$

$$\mathrm{FTO}(e^-)|\mathrm{TiO_2}(e^-)|\text{-}\left[\mathrm{Ru_a}^{\mathrm{III}}\text{-}\mathrm{Ru_b}^{\mathrm{II}}\text{-}\mathrm{OH_2}\right]^{5+} \rightarrow \mathrm{TiO_2}(e^-)|\text{-}\left[\mathrm{Ru_a}^{\mathrm{II}}\text{-}\mathrm{Ru_b}^{\mathrm{III}}\text{-}\mathrm{OH_2}\right]^{5+} \tag{13.4b}$$

$$\mathrm{FTO}|\mathrm{TiO_2}(e^-)|\text{-}\left[\mathrm{Ru_a}^{\mathrm{II}}\text{-}\mathrm{Ru_b}^{\mathrm{III}}\text{-}\mathrm{OH_2}\right]^{5+} \rightarrow \mathrm{FTO}|\mathrm{TiO_2}|\text{-}\left[\mathrm{Ru_a}^{\mathrm{II}}\text{-}\mathrm{Ru_b}^{\mathrm{II}}\text{-}\mathrm{OH_2}\right]^{4+} \tag{13.5a}$$

$$\mathrm{FTO}|\mathrm{TiO_2}(e^-)|\text{-}\left[\mathrm{Ru_a}^{\mathrm{II}}\text{-}\mathrm{Ru_b}^{\mathrm{III}}\text{-}\mathrm{OH_2}\right]^{5+} \rightarrow \mathrm{FTO}(e^-)||\mathrm{TiO_2}|\text{-}\left[\mathrm{Ru_a}^{\mathrm{II}}\text{-}\mathrm{Ru_b}^{\mathrm{III}}\text{-}\mathrm{OH_2}\right]^{5+} \tag{13.5b}$$

## 13.3 The Elements of a DSPEC

As illustrated in Fig. 13.3, the key elements of a DSPEC photoanode for water oxidation include: (1) a chromophore or chromophores that absorb light broadly from the near UV well into the visible, ideally into the near IR, (2) a chromophore excited state or states with an orbital pathway and excited state potential sufficiently reducing to undergo rapid injection, (3) a stable, high bandgap metal-oxide semiconductor, transparent in the visible, with a conduction band potential suitable for electron injection by the chromophore excited state or states, (4) a molecular catalyst or catalysts that are activated by the oxidized chromophore and undergo rapid (< seconds) water oxidation, and (5) proton equilibration. DSPEC performance is dictated by interfacial dynamics with injection and intra-assembly electron transfer activation of the catalyst followed by a competition between intra-film electron transport and back electron transfer.

Although DSPEC solar water splitting was the initial target of DSPEC research, an ultimate goal is solar driven reduction of $CO_2$ by water, ultimately to hydrocarbons, but initially to the 2e$^-$ products CO or formate/formic acid, $HCOO^-$/HCOOH. Significant progress has been made on the electrocatalyzed reduction of $CO_2$ both to formate/formic acid and to syngas mixtures of $H_2$/CO. Efficient, robust electrocatalysts are available for both reactions (Kang et al. 2012, 2013, 2014a, b; Chen et al. 2013). As shown in Fig. 13.4, formate/formic acid is a potential target for off-grid solar applications based on DSPEC reduction of $CO_2$ with the reduced carbon fuel used in formic acid/$O_2$ fuel cells for electric power generation. Capture and reuse of the evolved $CO_2$ from the fuel cell would create a self-contained, renewable power generator with a built in capacity for energy storage.

Syngas, $H_2$/CO mixtures in different ratios, are precursors for both methanol and hydrocarbons by Fischer–Tropsch synthesis. As discussed below, we recently reported a rapid, stable catalyst for electrochemical reduction of $CO_2/HCO_3^-$ solutions in water to syngas with the $H_2$/CO ratio variable by varying either the applied potential or the concentration of $HCO_3^-$ (Kang et al. 2014a, b).

Given the thermodynamic requirement for $CO_2$ reduction to CO, with $E^{o\prime} = -0.53$ V for the $CO_2$/CO couple at pH 7 vs. NHE, reduction of $CO_2$ by water in a DSPEC requires a tandem configuration with both a photoanode for water oxidation and photocathode for $CO_2$ reduction. A tandem design is also required for

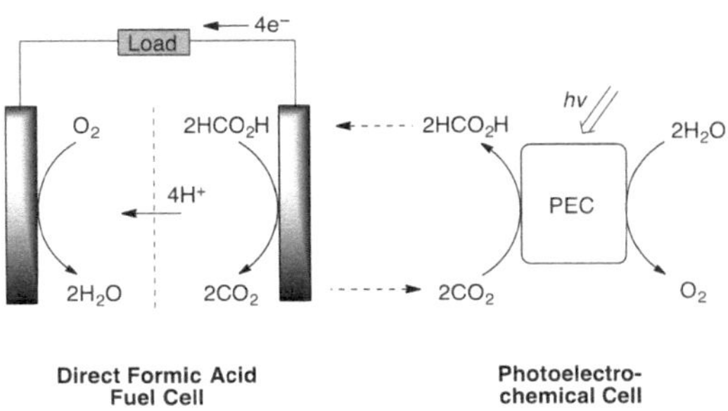


**Fig. 13.4** An integrated DSPEC-formate/formic acid fuel cell for solar energy conversion and storage. Reprinted with permission from Springer Science + Business Media

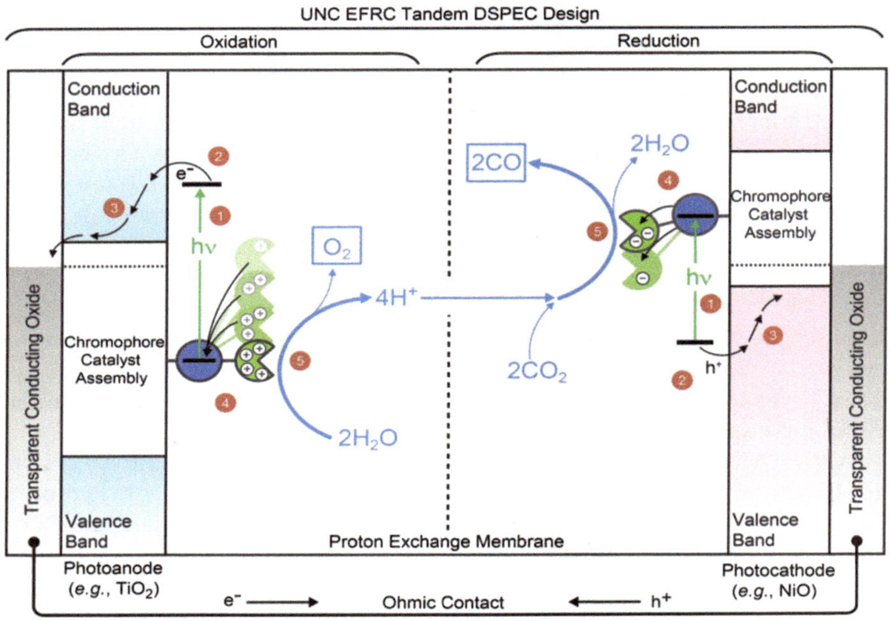


**Fig. 13.5** Illustrating a tandem DSPEC with a photoanode as in Fig. 13.3 paired with a second chromophore–catalyst based photocathode for reduction of $CO_2$ to CO. Image courtesy of James Cahoon. Reprinted with permission from American Chemical Society and Chemical Reviews (http://pubs.acs.org/doi/abs/10.1021%2Facs.chemrev.5b00229)

bias-free water splitting at $TiO_2$ to supply the 0.2 V applied external bias needed to drive $H^+/H_2O$ reduction to completion.

A DSPEC tandem cell design for $CO_2$ reduction to CO is illustrated in Fig. 13.5. The use of a tandem cell design introduces additional complexities arising from current, spectral, and potential matching of the two separate photoelectrodes. In a maximally designed cell, the solar spectrum would be split between the two photoelectrodes to maximize solar efficiency.

## 13.4 Chromophore–Catalyst Assemblies

Initially, there were no real guidelines for the design of chromophore–catalyst assemblies for DSPEC applications, only a recognition of the importance of integrating light absorption and catalysis. Insights were available from extensive studies on DSSCs in nonaqueous solvents. They highlighted the importance of spectral coverage, excited state potentials, efficient electron injection, and the manipulation of semiconductor surfaces to inhibit back electron transfer as a way

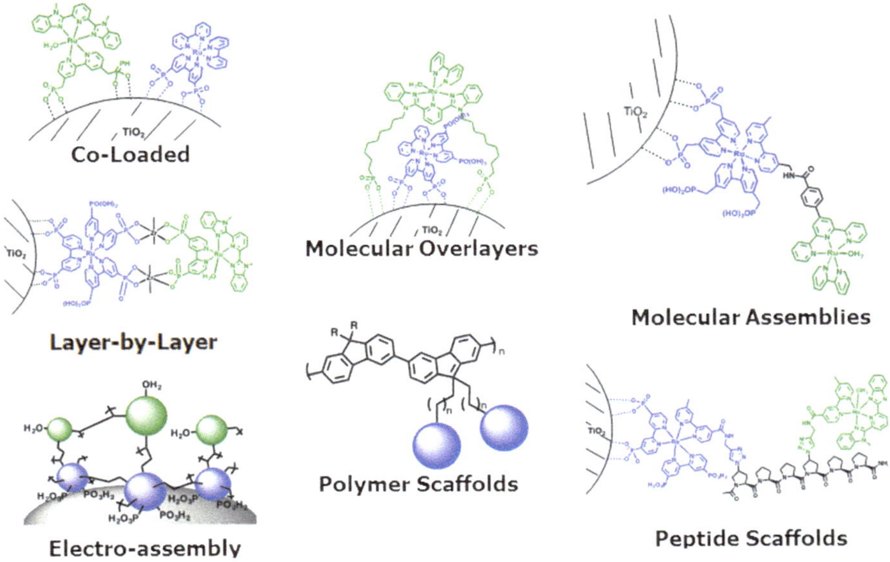


**Fig. 13.6** Chromophore–catalyst assembly strategies for DSPEC photoanodes

of maximizing efficiencies and open circuit potentials (Zhang et al. 2013a, b; Durrant et al. 2004; Clifford et al. 2011; Zhang and Cao 2011).

The types of chromophore–catalyst assemblies that have been investigated in our own work are shown in Fig. 13.6. They utilize phosphonic acid derivatives for oxide surface binding. The examples shown include examples with covalent bonds linking chromophore and catalyst with the ability to extend the spacer as a way to control rates of back electron transfer. Later transient absorption measurements revealed a fundamental limitation to this strategy. Following injection into $TiO_2$, intra-particle electron transfer to the surface, where back electron transfer occurs, is relatively slow, ~1 μs, typically masking the influence of back electron transfer from the surface to the oxidized catalyst (Ashford et al. 2012a, b; Puodziukynaite et al. 2014).

Other assembly strategies include multi-derivatized polymers by our EFRC (Energy Frontier Research Center) collaborators at the University of Florida and Georgia Institute of Technology. They offer a platform for integrating chromophores and catalysts and, with multiple chromophores, of providing a possible antenna effect for concentrating solar insolation. Transient laser flash photolysis measurements have been used to demonstrate rapid intra-strand energy and electron transfer in designed polymers. A goal in this area is to create polymer assemblies that combine the light concentrating advantages of the chlorophyll-auxiliary pigment antennae of natural photosynthesis with molecular catalysts for water oxidation.

The peptide scaffold strategy in Fig. 13.6 is an elegant approach based on the step-by-step assembly of polypeptides on polymer beads pioneered by Merrifield and coworkers (Leem et al. 2014; Merrifield 1986). Based on this well-developed synthetic approach, it is possible to control both the content and relative spatial disposition of functional groups as they are added to the growing polypeptide chain to control distances and rates of intra-assembly electron and energy transfer.

Preformed assemblies are appealing but they often require tedious, multistep syntheses and complex chromatographic separations resulting in low yields. For large scale, practical applications, the surface assembly strategies shown in Fig. 13.6 are appealing. Co-loading, in which sequential or simultaneous soaking procedures are used to load both chromophore and catalyst on the same surface, is especially so. It allows for both the chromophore–catalyst ratio and extent of surface loading to be varied systematically (Ryan et al. 2014; Song et al. 2013; Zhao et al. 2012; Gao et al. 2013) Complete or near-complete surface loading is required for rapid cross-surface electron and energy transfer to occur.

The catalyst "molecular overlayer" approach shown in Fig. 13.6 is another co-loading strategy in which the chromophore is first linked to the oxide surface, in this case, followed by a catalyst overlayer linked to the surface by long chain –$(CH_2)$– phosphonate linkers (Glasson et al. 2012). The "layer-by-layer" strategy in Fig. 13.6 is based on earlier work by Mallouk and Haga and their coworkers (Hanson et al. 2012; Lee et al. 1988; Ishida et al. 2009). It utilizes Zr(IV) as a bridge based on its high affinity for phosphonate coordination (Bettis et al. 2014; Nayak et al. 2014). In this approach a multiply phosphonic acid derivatized chromophore is initially bound to the oxide surface. As shown for $[Ru((4,4'\text{-}PO_3H_2)_2bpy))_3]^{2+}$ in Fig. 13.7, oxide surface binding leaves uncoordinated phosphonate groups oriented toward the external solution for coordination to Zr(IV). After Zr(IV) binding, a second soaking procedure is used to link a phosphonate derivatized water oxidation catalyst completing the synthesis of the assembly.

In the electro-assembly procedure in Fig. 13.6, a doubly derivatized chromophore, with both phosphonate and vinyl substituted bpy ligands, is initially bound to an oxide surface (Lapides et al. 2013; Ashford et al. 2014). Electrochemical reduction at π*(bpy) levels on the ligands induces C–C bond formation, linking

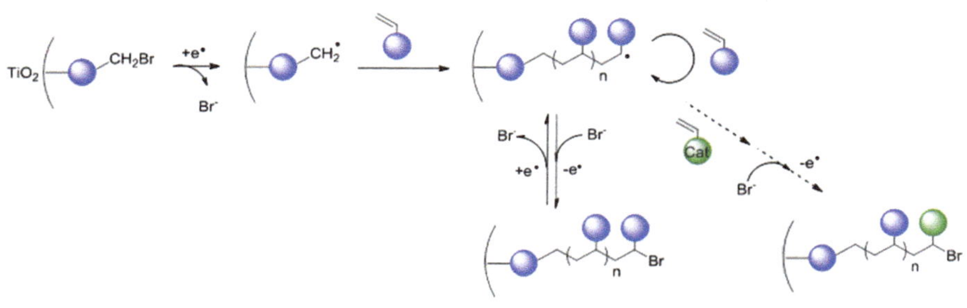


**Fig. 13.7** Stepwise electro-assembly formation of a multi-chromophore assembly in the cavities of a nanoparticle $TiO_2$ film

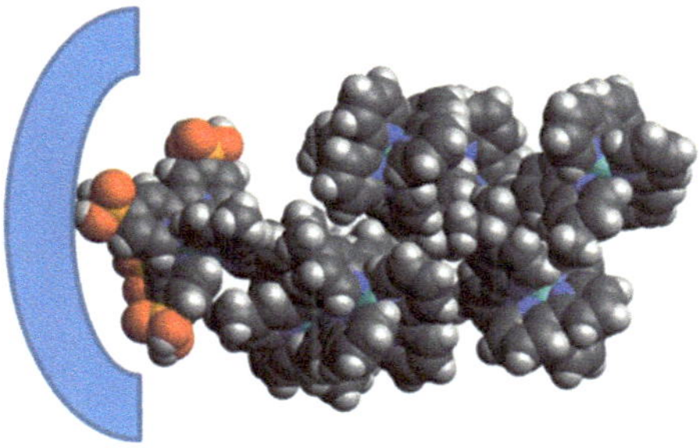

**Fig. 13.8** Illustration of a seven member chromophore assembly prepared by electro-assembly within the cavities of mesoscopic, nanoparticle films of $TiO_2$

the vinyl groups. With repeated reductive scans, a C–C linked assembly is formed on the surface with the chromophore: catalyst ratio controllable by the number of reductive scans.

In a notable extension, stepwise preparation of a multi-chromophore assembly was demonstrated within the cavities of mesoscopic, nanoparticle films of $TiO_2$ (Ashford et al. 2015). Phosphonate binding of a bpy-$CH_2$Br derivative was followed by ligand-based reduction. Reduction and $Br^-$ loss give a ligand-centered radical which couples with the vinyl-bpy derivative $[Ru(bpy)_2(4\text{-vinyl-bpy})]^{2+}$ in the external solution. Continued reductive cycling results in the step-by-step formation of the oligomer shown in Fig. 13.7. Based on the results of a molecular modeling study, up to six polypyridyl units could be added to the surface-bound phosphonate complex, completely filling the internal cavity volume of the films, Fig. 13.8.

## 13.5 Interfacial Dynamics

The results of transient absorption measurements on timescales ranging from femtoseconds to milliseconds are available for a variety of surface-bound assemblies. A broad overview of interfacial dynamic events and timescales following chromophore excitation, are shown in Fig. 13.9 with the amide-linked assembly, $[(4,4'\text{-}(PO_3H_2\text{-}CH_2)_2\text{-bpy})_2Ru_a(\text{bpy-NH-CO-py})Ru_b(\text{bpy})(OH_2)]^{4+}$, as the example (Ashford et al. 2012a, b).

Injection is kinetically complex and, if favorable, occurs rapidly. Ultrafast transient Raman and UV-visible measurements on surface-bound $TiO_2|\text{–}[(4,4'\text{-}PO_3H_2)_2\text{bpy})Ru(\text{bpy})_2]^{2+}$ and related complexes show that ultrafast injection occurs from an upper $^1$MLCT (metal-to-ligand charge transfer) state before it relaxes to the corresponding triplet, $^3$MLCT, and from excited vibrational levels of the triplet state. Slower injection occurs from thermally equilibrated $^3$MLCT states.

Intra-assembly electron transfer following injection is complete within a ns for the amide-bridged assembly shown in Fig. 13.10. In acidic solutions, the combination of rapid electron injection and intra-assembly electron transfer results in electron transfer activation of the catalyst with near unity efficiency. Even with

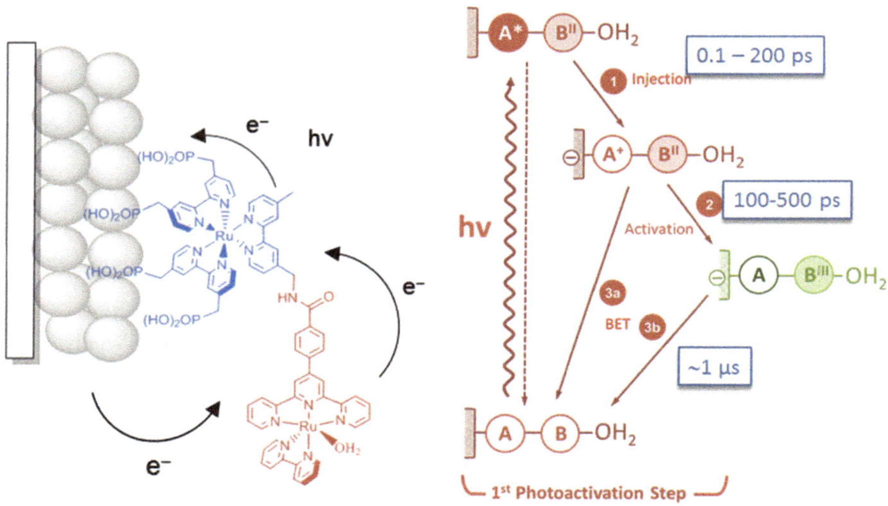


**Fig. 13.9** Interfacial electron transfer events and timescales for $TiO_2$-bound chromophore–catalyst assemblies following chromophore excitation with the amide-linked assembly $[(4,4'\text{-}(PO_3H_2\text{-}CH_2)_2\text{-bpy})_2Ru_a(\text{bpy-NH-CO-py})Ru_b(\text{bpy})(OH_2)]^{4+}$ shown as the example

**Fig. 13.10** (**a**) Proposed Ce (IV)-driven water oxidation mechanism for Ru(bda)(L)$_2$ (bda = 2,2′-bipyridine-6,6′-dicarboxylate). L is a neutral ligand with L = picoline shown in (**b**)

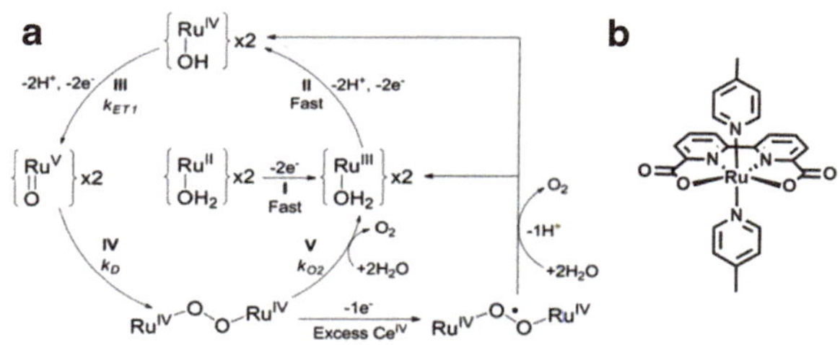


efficient 1e$^-$ activation, water oxidation requires three additional excitation-injection-electron transfer cycles to meet the overall 4e$^-$ demand for water oxidation, all in competition with back electron transfer.

For the assembly in Fig. 13.10, back electron transfer on $TiO_2$ occurs on the ~ μs timescale, largely dictated by relatively slow electron migration within the $TiO_2$ nanoparticles to the surface. Back electron transfer is in competition with electron migration through the film to the FTO electrode which occurs on the millisecond timescale. This unfavorable mismatch in rates, and the requirement for four excitation/injection/electron transfer cycles, ensure inefficient water oxidation. This is especially true for the sun as a light source with the rate of solar insolation for ambient sunlight only 1–2 s$^{-1}$. Transiently stored redox equivalents must be stored for extended periods within the assembly for multiple steps while avoiding back electron transfer at each step.

In natural photosynthesis, solar-driven water oxidation occurs at the oxygen evolving complex (OEC) in photosystem II (PSII). It is integrated with an antenna array consisting of hundreds of chlorophyll and auxiliary pigments. They absorb light broadly throughout the visible region of the spectrum and concentrate the solar input at the OEC increasing the effective rate of solar insolation to milliseconds rather than seconds. The antenna array delivers excited state energy to a terminal chlorophyll, $P_{680}$. Its singlet excited state undergoes electron transfer quenching to a neighboring pheophytin. The initial quenching event is followed by four follow up electron transfers resulting in an oxidative equivalent at the OEC and a reductive equivalent at quinone A with the two separated by ~50 Å. The long separation distance slows back electron transfer to the >ms timescale allowing for the required four oxidative equivalents needed for water oxidation to accumulate in the OEC.

## 13.6 Molecular Catalysis: Water Oxidation

Significant progress has been made in developing molecular catalysts for both $H_2O$ oxidation and $CO_2$ reduction with reviews for both in the recent literature. The focus here is on results obtained at UNC (University of North Carolina at Chapel Hill) to illustrate the key issues involved in carrying out this essential half reaction. As noted above, the first designed water oxidation catalyst was the blue Ru(III) dimer $[(bpy)_2(H_2O)Ru^{III}ORu^{III}(H_2O)(bpy)_2]^{4+}$ with mechanistic insight gained from the blue dimer leading to the design of single-site Ru(II) polypyridyl catalysts and the catalytic cycle in Scheme 13.1 (Gersten et al 1982; Gilbert et al 1985; Hurst et al 1992; Hurst 2005; Liu et al 2008). The ability of these molecular water oxidation catalysts to perform water oxidation is impressive. Electrocatalyzed water oxidation by $[Ru(Mebimpy)(bpy)(OH_2)]^{2+}$ has been shown to continue for 2 weeks through >43,000 turnovers at pH 7.0 in 0.1 M $H_2PO_4^-/HPO_4^{2-}$.

Oxidative activation of these aqua-containing catalysts occurs by PCET with proton loss at the coordinated water induced by enhanced acidity in Ru(III) and, for coordinated hydroxide, at Ru(IV). PCET, with coupled $e^-/H^+$ loss, is important in avoiding charge build-up, decreasing the difference between redox potentials for adjacent couples, allowing multiple oxidative equivalents to accumulate at a single sites or cluster at relatively moderate potentials.

The potential difference between the separate redox couples in a catalytic cycle, such as those in Scheme 13.1, are dictated by both pH and the coordination environments at the catalyst. For $[Ru(tpy)(bpm)(OH_2)]^{2+}$ and the 4,4′-bipyrazine analog, the tetrazine ligands are strongly backbonding with low lying $\pi^*$ acceptor acceptor levels inducing extensive $d\pi^6$-$\pi^*$(L-L) orbital mixing. Backbonding stabilization of Ru(II) is sufficient to increase the formal potential for the the $Ru^{III}$-$OH^{2+}/Ru^{II}$-$OH_2^{2+}$ couple above $E^{o\prime}$ for the $Ru^{IV}{=}O^{2+}/Ru^{III}$-$OH^{2+}$ couple and Ru(III) is unstable with respect to disproportionation, Eq. (13.6).

From cyclic voltammetry (CV) measurements in aqueous solution, only a single $2e^-$ wave is observed for the $Ru^{IV}{=}O^{2+}/Ru^{II}(OH_2)^{2+}$ couple.

$$2\,Ru(III) \rightarrow \left[Ru^{II}(tpy)(L\text{-}L)(OH_2)\right]^{2+} + \left[Ru^{IV}(tpy)(L\text{-}L)(O)\right]^{2+} \tag{13.6}$$

In the cycle shown in Scheme 13.2, stepwise $3e^-/2H^+$ oxidation of $d\pi^6$ $Ru^{II}$-$OH_2{}^{2+}$ to $d\pi^3$ $Ru^V(O)^{3+}$ is followed by O-atom transfer to a water molecule to give an intermediate peroxide by the APT mechanism in Eq. (13.3). Two electron activation to $Ru^{IV}{=}O^{2+}$ can also initiate water oxidation catalysis if the catalyst is sufficiently oxidizing, see below. For the steps in Scheme 13.2, rate-limiting, or a contribution to rate-limiting behavior, has been found at three stages in the overall cycle: (1) PCET oxidation of $Ru^{III}$-$OH_2{}^{3+}$ or $Ru^{III}$-$OH^{2+}$ to $Ru^{IV}{=}O^{2+}$ with kinetic inhibition arising from the requirement for proton loss and oxo formation, (2) O–O bond formation with O-atom transfer occurring from the oxo ligand of $Ru^{IV}{=}O^{2+}$ or $Ru^V(O)^{3+}$ to a water molecule in concert with loss of a proton to the surrounding solvent, Eq. (13.3), and (3) loss of $O_2$ from the $Ru^{IV}(OO)^{2+}$ intermediate which releases $O_2$ and reenters the catalytic cycle as $Ru^{II}(OH_2)^{2+}$. Further oxidation of $Ru^{IV}(OO)^{2+}$ to $Ru^V(OO)^{3+}$ results in accelerated $O_2$ loss with reentry into the catalytic cycle as $Ru^{III}(OH_2)^{3+}$ or $Ru^{III}(OH)^{2+}$ depending on the pH.

There is indirect evidence that Ru(IV)-peroxo complexes accumulate as intermediates in Ce(IV) water oxidation and direct evidence from electrochemical measurements, where pH-dependent $Ru^{IV}(O_2)^{2+}/Ru^{III}(O_2H)^{2+}$ and $Ru^{III}(O_2H)^{2+}/Ru^{II}(HOOH)^{2+}$ couples for the surface-bound catalyst $[Ru(Mebimpy)(4,4'\text{-}(PO_3H_2CH_2)_2bpy)(OH_2)]^{2+}$ on nanoITO in 0.1 M $HClO_4$ have been observed. Although Ru-peroxide intermediates appear in water oxidation cycles involving surface-bound catalysts, they do not accumulate during electrocatalytic measurements because oxidation of $Ru^{IV}(OO)^{2+}$ to $Ru^V(OO)^{3+}$ with $O_2$ loss occurs at a potential lower than $E^{o\prime}$ for the oxidation of $Ru^{IV}(O)^{2+}$ to $Ru^V(O)^{3+}$ which reinitiates water oxidation.

## 13.7 The Role of Atom-Proton Transfer in Water Oxidation Catalysis

The rate-limiting step in most catalytic cycles for water oxidation is O-atom transfer from the oxidized catalyst to a water molecule with O–O bond formation (Concepcion et al. 2010a, b). As noted in Sect. 3.6.1, based on the results of DFT calculations and the observation of a $H_2O/D_2O$ kinetic isotope effect of 6.6 for water oxidation by $[Ru^V(tpy)(bpm)(O)]^{3+}$, the rate-limiting step was proposed to involve concerted Atom-Proton Transfer (APT). In this pathway, O–O bond formation occurs in concert with proton loss to a hydrogen bonded water molecule or cluster, Eq. (13.3). APT is advantageous in forming a hydroperoxide intermediate, $Ru^{III}$-$O_2H^{2+}$, rather than the higher energy hydrogen peroxide complex, $Ru^{III}(OOH_2)^{3+}$, by direct reaction with water.

**Table 13.1** Rate constants for water oxidation by the $Ru^{IV}$ and $Ru^{V}$ forms of $[Ru(tpy)(bpm)(OH_2)]^{2+}$ with $H_2O$, $PO_4^{3-}$, and $OH^-$ as the proton acceptor base (B)

| B | $k_{Ru(IV),B}/M^{-1}s^{-1}$ | $k_{Ru(V),B}/M^{-1}s^{-1}$ |
|---|---|---|
| $H_2O$ | $k_{H_2O} \approx 0.15\ s^{-1}$ | $k_{H_2O} \approx 0.97\ s^{-1}$ |
| $PO_4^{3-}$ | 5.4 | $3.0 \times 10^2$ |
| $OH^-$ | 19 | $6.8 \times 10^2$ |

Further evidence for APT in water oxidation has come from experiments with added buffers with kinetic rate accelerations that are first order in added buffer base. The appearance of these pathways in rate laws for water oxidation have been interpreted as evidence for APT with the added buffer base—$H_2PO_4^-$, acetate, $HPO_4^{2-}$—initially hydrogen bonded to the reacting water molecule, Eq. (13.3). The extent of rate acceleration varies directly with the base strength of the buffer base with rate accelerations of up to $10^4$ observed in phosphate buffers with high concentrations of added $HPO_4^{2-}$. The appearance of buffer base effects is symptomatic of rate-limiting O–O bond formation and, in their absence, of rate-limiting behavior in another step in the catalytic cycle. In any case, interpretation of relative rates of water oxidation and comparisons between catalysts under different solution conditions need to be undertaken with great care.

The impact of APT pathways on reactivity can be significant. Even though water oxidation by $[Ru^{IV}(tpy)(bpz)(O)]^{2+}$ is thermodynamically favorable by ~0.1 eV, water oxidation in acidic solution is slow. With high concentrations of added $PO_4^{3-}$ at high pH, the rate of water oxidation is greatly enhanced as shown by the data in Table 13.1. For the $Ru^{IV}(O)^{2+}$ complex, the observed rate constants under these conditions is given by, $k_{obs} = k_{H2O} + k_{[PO4]3-} + k_{OH-}$. In this rate constant expression, the $k_{[PO4]3-}$ term arises from APT with $PO_4^{3-}$ as the acceptor base and the $k_{OH-}$ term from direct reaction between $Ru^{IV}{=}O^{2+}$ and $OH^-$.

## 13.8 BDA-Catalysts

Sun and coworkers have demonstrated impressive catalytic rates for water oxidation with complexes of the type $Ru(bda)(L)_2$ (bda = 2,2′-bipyridine-6,6′-dicarboxylate, L = pyridine or other neutral donor ligands, Fig. 13.10 (Duan et al. 2012; Gao et al. 2013). These catalysts can be systematically varied by varying the axial ligands including the addition of phosphonic acid and vinyl derivatives for oxide surface binding or electro-assembly/electropolymerization. Although the mechanistic chemistry of this class of catalysts continues to evolve, kinetic measurements in acidic solution with Ce(IV) as an oxidant have provided evidence for a pathway involving oxidation, formation of a transient seven-coordination Ru(V) oxo intermediate, and bimolecular O–O coupling, Fig. 13.10a (Duan et al. 2013).

The mechanism of electrocatalytic water oxidation by $Ru^{II}(dba)(isoq)_2$ (isoq is isoquinoline) has been investigated by cyclic voltammetry (Song et al. 2015).

**Table 13.2** Kinetic data for water oxidation by $Ru^V$(dba)$(isoq)_2(O)^+$ in the presence of added proton-acceptor bases, B, in 0.5 M ($NaClO_4$) in 4 % $CH_3CN$ at 22 ± 2 °C

| B | $pK_{a,HB}$ | $k_B$, $M^{-1}$ $s^{-1}$ |
|---|---|---|
| $H_2O$ | −1.74 | $k_{H_2O}$ ≈ 6 $s^{-1}$ |
| $OAc^-$ | 4.76 | 93 ± 14 |
| $HPO_4^{2-}$ | 7.20 | 499 ± 29 |
| $PO_4^{3-}$ | 12.35 | ≈(1.0 ± 0.1) × $10^5$ |
| $OH^-$ | 15.75 | (9.3 ± 0.6) × $10^5$ |

$pK_a$ values for the conjugate acids are also listed in the table

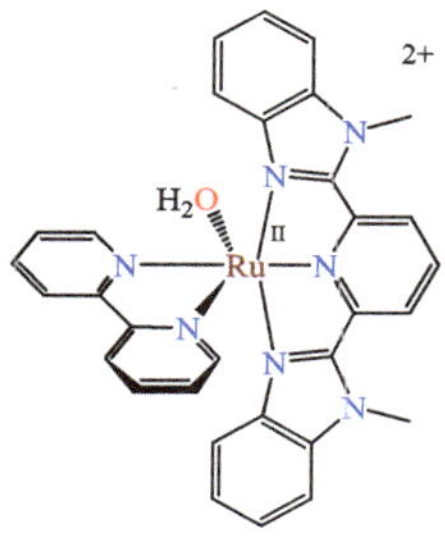


**Fig. 13.11** Single-site water oxidation catalyst [Ru(Mebimpy)(bpy)$(H_2O)]^{2+}$

With small amounts of added acetonitrile (MeCN), the complex exists as the chelate ring-opened nitrile complex [Ru($CO_2$-bpy-$CO_2H$)$(isoq)_2(NCCH_3)]^+$ with $pK_a$ ~ 2.4 for the uncoordinated carboxylate arm. Above pH 7 the complex undergoes $2e^-/2H^+$ oxidation to $Ru^{IV}$=O followed by further pH-independent oxidation to $Ru^V(O)^+$ and rate-limiting O–O bond formation. Dramatic rate enhancements are observed with added buffer bases with the effect increasing with base strength, Table 13.2. Under these conditions, water oxidation is rate-limited by APT O–O bond formation. At high pH, with the concentration of $OH^-$ kinetically significant, a pathway first-order in hydroxide appears in the rate law. In aqueous solutions 1.0 M in $PO_4^{3-}$, the half time for water oxidation is ~7 μs! Rate constant data are summarized in Table 13.2.

## 13.9 Water Oxidation in Propylene Carbonate

Studies on oxidation catalysis by [Ru(Mebimpy)(bpy)$(OH_2)]^{2+}$ (Fig. 13.11) have been extended to propylene carbonate (PC) with water added as a limiting reagent (Chen et al. 2010a, b, c). Based on the results of an electrochemical study, water oxidation in PC was shown to be first order in catalyst and first order in water up to its 8 % miscibility limit in PC. The experimental rate law is consistent with rate-limiting O-atom transfer from $Ru^V(O)^{3+}$ to $H_2O$ to give the hydrogen peroxide intermediate in Eq. (13.7). It undergoes further oxidation and $O_2$ release, Eq. (13.8). The rate of water oxidation is enhanced in PC by ~300 compared to water as the solvent with the increase attributed to the loss in solvent stabilization for $H_2O$ in PC compared to $H_2O$ in $H_2O$.

$$Ru^{V}=O^{3+} + H_2O \rightarrow Ru^{III}(HOOH)^{3+} \tag{13.7}$$

$$Ru^{III}(HOOH)^{3+} - 4e^{-} + H_2O \rightarrow Ru^{V}=O^{3+} + O_2 + 4H^{+} \tag{13.8}$$

## 13.10 First Row Transition Metal Complex Catalysts for Water Oxidation

Ru(II) polypyridyl complexes have played a dominant role in DSPEC chromophore–catalyst assemblies. The underlying synthetic chemistry is well-developed and accessible and polypyridyl complexes of Ru(II) have highly evolved photochemical and photophysical properties. In addition, these complexes are coordinatively stable in multiple oxidation states and have a long history as catalysts for water oxidation and $CO_2$ reduction. Nonetheless, Ru is a second row transition metal with limited abundance. Ultimate assembly targets are to use inexpensive organic dyes as chromophores and first row transition metal complexes—Mn, Fe, Co, Ni—as water oxidation catalysts. Investigation of both are important targets of current UNC EFRC research.

The evolution of first row water oxidation catalysts has not been without complications arising primarily from catalyst instability. Despite initial promise, initial results with complexes of Co(III) have been plagued by decomposition from ligand loss following oxidation to Co(IV), giving catalytically active oxide clusters or films (Wasylenko et al. 2011, 2013; Wang and Groves 2013). Although the oxide clusters have been applied to water oxidation productively, the loss of molecular integrity greatly hinders their use in assemblies. Complexes of Mn(II) and Ni(II) suffer similar fates, although it is notable that a ligand-stabilized $Mn_4$ oxo-bridged cluster is the reactive site in the oxygen evolving complex of photosystem II (Ashmawy et al. 1985; Limburg et al. 1999; Young et al. 2013; Wang et al. 2014). Polypyridyl complexes of Fe(II), the first row congener of Ru(II), are far more labile than complexes of Ru(II). In acidic solutions either they, or their oxidized Fe(III) forms, undergo ligand loss and decomposition to iron oxides and/or free $Fe^{3+}$ depending on pH. Further progress in this area hinges on the design and use of robust ligand systems that can withstand unlimited oxidative activation/water oxidation cycles (Ellis et al. 2010; Hong and Yamada 2012; Coggins et al. 2014a, b).

The most promising results to date have come with complexes of Cu(II). The first example, reported by Mayer and coworkers, was a $Cu^{II}$(bpy) complex, active toward catalytic water oxidation but only at high pH with $Cu(bpy)(OH)_2$ as the active form (Barnett et al. 2012). Water oxidation catalysis has also been reported for the peptide complex $[(TGG)Cu(OH_2)]^{2-}$ and for the pyridyl complex $Cu(Py_3P)$ with the structures of the TGG (tetra-anionic macrocycle triglycylglycine) and $Py_3P$ (*N,N*-bis(2-(2-pyridyl)ethyl)pyridine-2,6-dicarboxamidate) complexes shown in Fig. 13.12 (Zhang et al. 2013a, b; Coggins et al. 2014a, b). For both, water oxidation

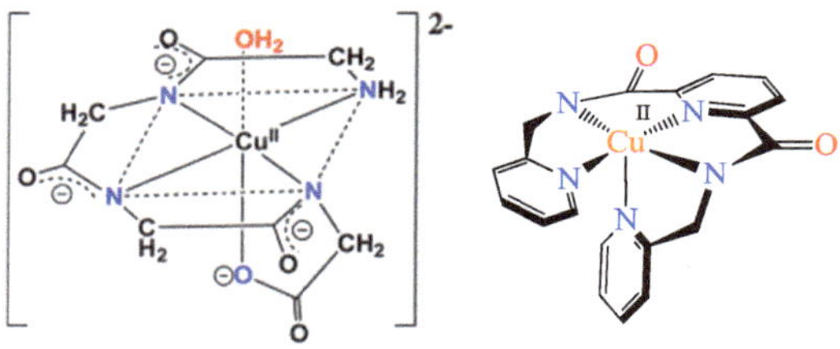

**Fig. 13.12** Water oxidation catalysts [(TGG)Cu$(OH_2)]^{2-}$ (*left*) and Cu$(Py_3P)$ (*right*)

occurs by single-site mechanisms analogous to the mechanism for [Ru(Mebimpy)(bpy)$(OH_2)]^{2+}$ in Scheme 13.2. In these reactions the results of electrochemical studies show that 1e$^-$ oxidation of Cu(II) to Cu(III) is followed by a further, pH-dependent 1e$^-$ oxidation to give a twice-oxidized intermediate—$Cu^{IV}$(O), $Cu^{III}(O^+)$, or possibly Cu(III) with a ligand-centered radical. Based on these data, the O–O bond forming step appears to be rate-limiting with an important role for APT. For Cu$(Py_3P)$ as the catalyst, rates of water oxidation at pH 8.0 in a 0.2 M phosphate buffer are sufficiently rapid to have promise for its use as the catalyst in DSPEC molecular assemblies.

The ubiquitous nature of water oxidation catalysis by Cu(II) was demonstrated by the use of simple Cu(II) salts as catalysts in aqueous solutions 1.0 M in $CO_3^{2-}$. Under these conditions, precipitation of Cu(II) as the Cu$(CO_3)$ is avoided by the formation of soluble carbonate-containing anionic coordination complexes. They act as catalysts for sustained electrocatalytic water oxidation with increasingly important contributions from a surface-bound copper oxide coating on the electrode as the electrolyses proceed.

## 13.11 Water Oxidation on Oxide Surfaces

For DSPEC and related applications, catalysis is integrated with light absorption and excited state injection on a semiconductor oxide surface. Relatively stable oxide surface binding is available with phosphonic acid derivatives in acidic solutions. Figure 13.13 illustrates the catalyst [Ru(Mebimpy)(4,4′-$(PO_3H_2CH_2)_2$bpy)$(OH_2)]^{2+}$ surface-bound to a nanoparticle film of tin-doped indium oxide (*nano*ITO). This optically transparent, conducting oxide film allows for simultaneous spectral and electrochemical monitoring of surface redox events. As shown in Fig. 13.14, based on the results of these studies, the catalytic cycle and water oxidation mechanism for the catalyst are retained on the oxide surface with a minimal impact on reactivity.

A first example of water oxidation by a surface-bound catalyst was demonstrated for the assembly [(4,4′-$(HO)_2OPCH_2)_2$bpy$)_2Ru^{II}$(bpm)$Ru^{II}$(Mebimpy)$(OH_2)]^{4+}$ on FTO (Concepcion et al. 2009). It was shown to undergo sustained water oxidation catalysis in acidic solution through 28,000 turnovers at a rate of 0.6 $s^{-1}$ without decomposition.

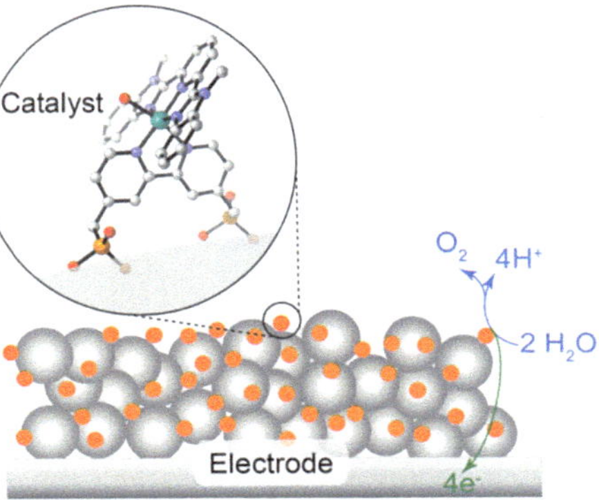


**Fig. 13.13** Illustration of the single-site water oxidation catalyst [Ru(Mebimpy)(4,4′-$(PO_3H_2CH_2)_2$bpy)$(OH_2)]^{2+}$ bound to a *nano*ITO electrode surface

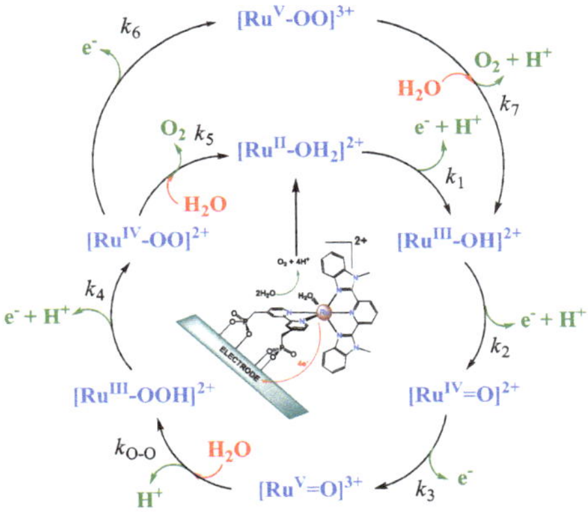


**Fig. 13.14** Catalytic cycle for water oxidation by [Ru(Mebimpy)(4,4′-$(PO_3H_2CH_2)_2$bpy)$(OH_2)]^{2+}$ surface-bound to an oxide electrode

The assembly $[(4,4'\text{-}PO_3H_2)_2bpy)_2Ru(4\text{-}Mebpy\text{-}4'\text{-}bimpy)Ru(tpy)(OH_2)]^{4+}$ ($[Ru_a{}^{II}\text{-}Ru_b{}^{II}\text{-}OH_2]^{4+}$), shown in Figs. 13.3 and 13.15, combined with oxide core/shell electrodes provided the basis for a first successful DSPEC for visible water splitting. The catalytic cycle for water oxidation by this assembly on *nano*ITO has been investigated independently and is shown in Scheme 13.3. As for related single-site Ru polypyridyl catalysts, stepwise oxidation of *nanoI*TO-$[Ru_a{}^{II}\text{-}Ru_b{}^{II}\text{-}OH_2]^{4+}$ to *nanoI*TO-$[Ru_a{}^{II}\text{-}Ru_b{}^{IV}{=}O]^{4+}$ is followed by further oxidation of the chromophore to *nanoI*TO-$[Ru_a{}^{III}\text{-}Ru_b{}^{IV}{=}O]^{5+}$ which triggers water oxidation. The appearance of water oxidation at the $3e^-/2H^+$ stage is an essential element in light-driven DSPEC water oxidation. For the catalyst itself, $3e^-/2H^+$ oxidation to $Ru^V(O)^{3+}$ leads to water oxidation but $E^{o\prime}$ for the $Ru^V(O)^{3+}/Ru^{IV}{=}O^{2+}$ couple at ~1.6 V vs. NHE is inaccessible to the -$Ru_a{}^{3+/2+}$-chromophore couple with $E^{o\prime}$ ~1.3 V.

The ability of the $3e^-$ oxidized assembly, -$[Ru_a{}^{III}\text{-}Ru_b{}^{IV}{=}O]^{5+}$, to undergo water oxidation has been attributed to a special APT pathway, Electron-Atom Proton Transfer (EAPT), in which O–O bond formation and proton loss occur in concert with single electron transfer to both –$Ru_b{}^{IV}{=}O$ and -$Ru_a{}^{III}$-.

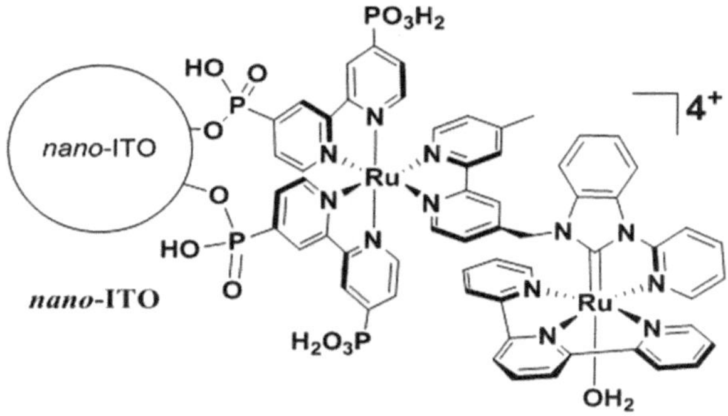


**Fig. 13.15** The assembly $[(4,4'\text{-}PO_3H_2)_2bpy)_2Ru(4\text{-}Mebpy\text{-}4'\text{-}bimpy)Ru(tpy)(OH_2)]^{4+}$ bound to a *nano*ITO electrode

$$nanoITO\text{-}[Ru_a^{II}\text{-}Ru_b^{II}\text{-}OH_2]^{4+} \xrightarrow{-e^-,\ -H^+} nanoITO\text{-}[Ru_a^{II}\text{-}Ru_b^{III}OH]^{4+}$$

$$nanoITO\text{-}[Ru_a^{II}\text{-}Ru_b^{III}\text{-}OH]^{4+} \xrightarrow{-e^-,\ -H^+} nanoITO\text{-}[Ru_a^{II}\text{-}Ru_b^{IV}{=}O]^{4+}$$

$$nanoITO\text{-}[Ru_a^{II}\text{-}Ru_b^{IV}{=}O]^{4+} \xrightarrow[-H^+]{-e^-} nanoITO\text{-}[Ru_a^{III}\text{-}Ru_b^{IV}{=}O]^{5+}$$

$$nanoITO\text{-}[Ru_a^{III}\text{-}Ru_b^{IV}{=}O]^{5+} + H_2O \xrightarrow[-H^+]{RDS} nanoITO\text{-}[Ru_a^{II}\text{-}Ru_b^{III}\text{-}OOH]^{4+}$$

$$nanoITO\text{-}[Ru_a^{II}\text{-}Ru_b^{III}\text{-}OOH]^{4+} \xrightarrow[+H_2O]{-H^+,\ -2e^-} nanoITO\text{-}[Ru_a^{II}\text{-}Ru_b^{III}\text{-}OH_2]^{5+} + O_2$$

**Scheme 13.3** Catalytic cycle for stepwise oxidative activation and water oxidation by $[(4,4'\text{-}PO_3H_2)_2bpy)_2Ru(4\text{-}Mebpy\text{-}4'\text{-}bimpy)Ru(tpy)(OH_2)]^{4+}$ on *nano*ITO

## 13.12 Molecular Catalysis: $CO_2$ Reduction

An ultimate DSPEC target is $CO_2$ reduction to $CH_4$ and higher hydrocarbons. However, with the exception of the pyridine/pyridinium catalysts developed by Bocarsly and coworkers, there are few if any reports of molecular catalysts for $CO_2$ reduction past the $2e^-$ stage to formate ($HCOO^-$) or CO (Barton et al. 2010; yang et al. 2013). The two electron reduction chemistry has been described in recent reviews. Most of the results on $CO_2$ reduction to date have been carried out in nonaqueous solvents. The emphasis in our research has been on aqueous reduction under conditions consistent with water oxidation catalysis and, ultimately, DSPEC applications.

## 13.13 $CO_2$ Reduction to Formate

Selective reduction of $CO_2$ to $HCOO^-$ by the Ir(III) pincer catalyst (POCOP)Ir$(MeCN)_2(H)$ was first investigated in MeCN with added water. The catalytic cycle in Scheme 13.4 features initial reduction of the complex to the corresponding

**Scheme 13.4** Catalytic cycle for $CO_2$ reduction to $HCOO^-$ by (POCOP)Ir(MeCN)$(H)_2$

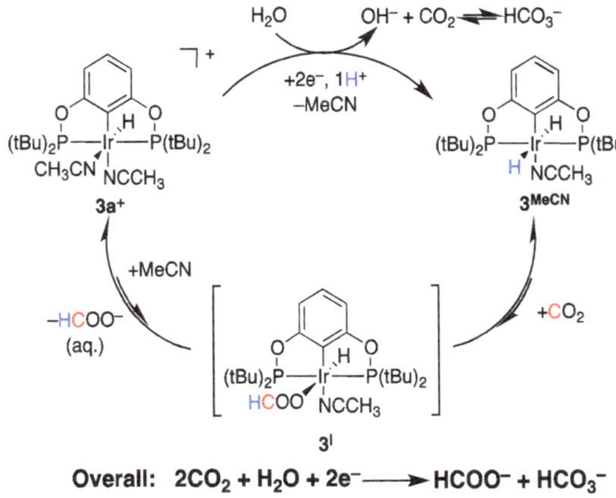


**Fig. 13.16** Structure of the water soluble catalyst, $[(POCOP)Ir(MeCN)_2(H)]^+$

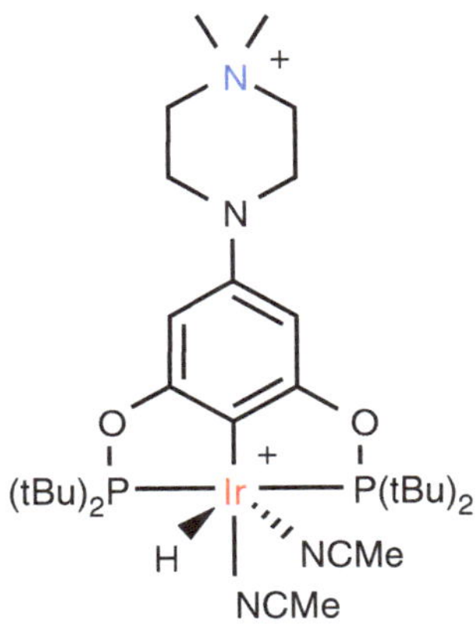


dihydride followed by $CO_2$ insertion into an Ir–H bond to give a format complex intermediate. Upon subsequent hydrolysis of the coordinated formate, the catalyst reenters the catalytic cycle. The rate-limiting step is $CO_2$ insertion into the Ir–H bond, which is relatively facile with $k = 20(2)\,M^{-1}\ s^{-1}$. It is also remarkably selective without interference from $H_2$ production in neutral solutions.

In later experiments, selectivity toward formate as a product was exploited by using water as the solvent with the solubilized aminium derivative of the complex shown in Fig. 13.16. In aqueous $CO_2$-saturated solutions with added $HCO_3^-$, selectivity toward $CO_2$ reduction is maintained with an increase in rate constant of ~3.5 due to a solvent effect on the rate-limiting insertion step. The overall rate of $CO_2$ reduction is slower in water because of the reduced solubility of $CO_2$ in water compared to MeCN.

Recently a pyrene derivative was synthesized and, as shown in Fig. 13.17, attached to a high surface area carbon-based gas diffusion electrode for flow-through applications.

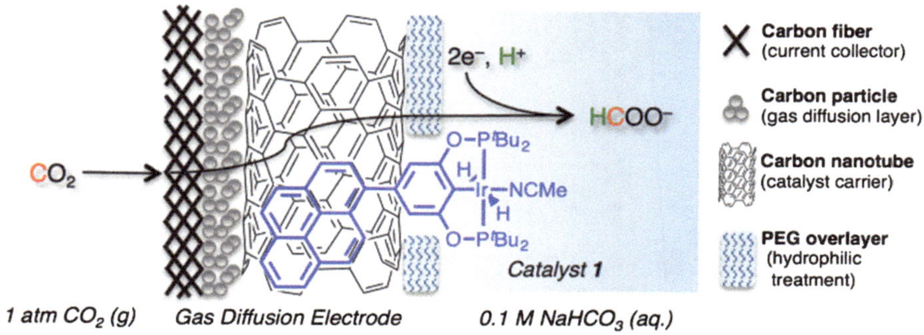


**Fig. 13.17** Pyrene-derivitized Ir(POCOP) $CO_2$ reduction catalyst immobilized on a gas diffusion electrode

## 13.14 $CO_2$ Reduction to Syngas

Electrocatalytic reduction of $CO_2$ to CO has been investigated for polypyridyl complexes of Re(I), Ru(II), and Os(II) (Machan et al. 2014; Tamaki et al. 2013; Chen et al. 2011a, b, c). In a recent mechanistic study in MeCN, it was shown that tpy-based reduction of $[Ru(tpy)(bpy)(NCCN)]^{2+}$ at −1.1 V vs. NHE is followed by a second ligand-based reduction. Addition of the second electron labilizes MeCN by dissociative ligand loss with the lower coordinate intermediate captured by $CO_2$ to give the corresponding metallocarboxylate, $[Ru(tpy)(bpy)]^0 + CO_2 \rightarrow$ [Ru(tpy)(bpy)(COO)]. The metallocarboxylate undergoes further reduction and $O^{2-}$ loss, to give the carbonyl, [Ru(tpy)(bpy)(CO)], which, following loss of CO and re-coordination of MeCN, reenters the catalytic cycle.

With added $H_2O$ or other weak acids, reduction of $[Ru(tpy)(bpy)(NCCN)]^{2+}$ results in electrocatalytic $H_2$ generation. The mechanism for $H_2$ production involves the same ligand-based reduction sequence to give twice-reduced [Ru(tpy)(bpy)(NCMe]. However, with an added proton source, the intermediate undergoes associative addition of the acid to give the hydride, $[Ru(tpy)(bpy)(H)]^+$, Eq. (13.9). Once formed, the hydride undergoes further ligand-based reduction with a second proton added to give a transient $H_2$ intermediate. Following release of $H_2$, the catalyst reenters the catalytic cycle as $[Ru(tpy)(bpy)(NCCN)]^{2+}$.

$$[Ru(tpy)(bpy)(NCCN)]^0 + HA \rightarrow [Ru(tpy)(bpy)H]^+ + A^- \qquad (13.9)$$

Reduction of HA to give $H_2$ and $CO_2$ to give CO by the nitrile complex, occurs through the common $2e^-$ reduced intermediate $[Ru(tpy)(bpy)]^0$. In MeCN with added $CO_2$ and $H_2PO_4^-$ as the added acid, a competition for the intermediate exists between the two and syngas mixtures of $H_2$ and CO are formed as products. As noted above, $H_2$:CO syngas mixtures are the feedstock for Fischer–Tropsch synthesis of methanol or hydrocarbons on the industrial scale with the products depending on the syngas ratio. For $[Ru(tpy)(bpy)(NCCH_3)]^{2+}$ as catalyst, the rate of $H_2$ production increases with the concentration of added acid providing a basis for controlling the syngas ratio over a wide range by varying the concentration of acid in $CO_2$ saturated solutions.

Electrochemical synthesis of syngas has been extended to $CO_2/HCO_3^-$ mixtures in water (Gao et al. 2013). As shown in Fig. 13.18, the syngas ratio can be varied from ~4:1 to ~1:3 by varying the electrolysis potential at a fixed concentration of $HCO_3^-$. Similarly, at −1.2 V the ratio can be varied from ~4:1 to ~1:2 by increasing the $HCO_3^-$ concentration. A schematic diagram of the electrolysis cell operating under conditions that give a syngas ratio of 2:1 is shown in Fig. 13.18.

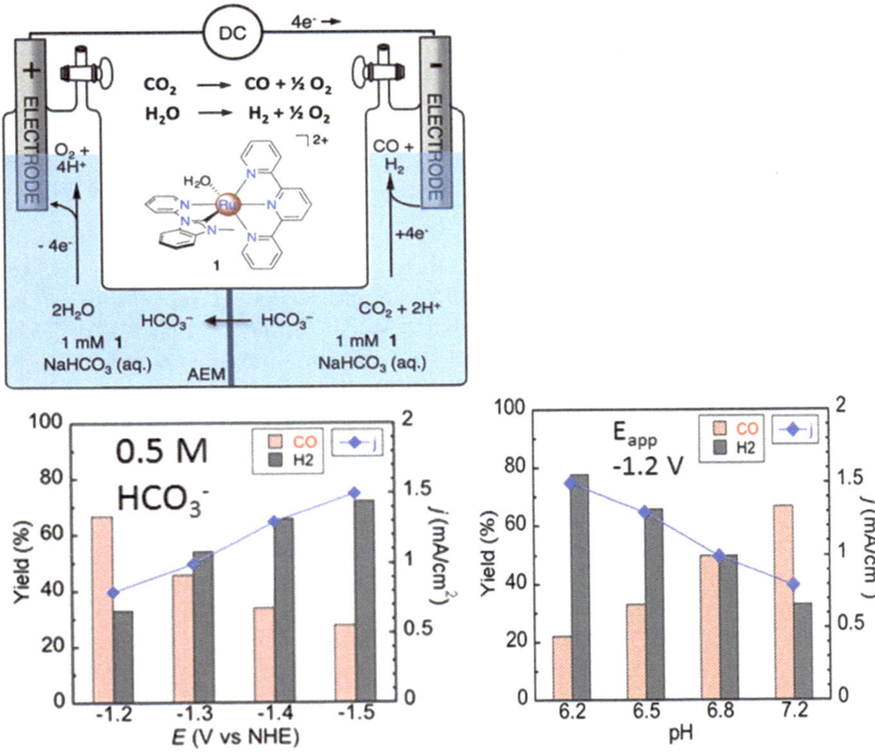


**Fig. 13.18** Electrolysis cell for water oxidation (anode) and $CO_2$ reduction (cathode) by [Ru(tpy)(Mebim-py)($H_2O$)]$^{2+}$. (*Left*) At pH 7.2 in $CO_2/HCO_3^-$ buffered solutions as a function of applied potential. The graph shows percent yield (*bars*) and current density (*blue rhombi*) of CO and $H_2$ production as a function of applied potential in $CO_2$-saturated 0.5 M $HCO_3^-$. (*Right*) Percent yield (*bars*) and current density (*blue rhombi*) for CO and $H_2$ as a function of pH at −1.2 V

## 13.15 Catalysis of $CO_2$ Splitting into CO and $O_2$

A remarkable feature in the electrolysis cell diagram in Fig. 13.17 is the use of *the same catalyst for both water oxidation and* $CO_2$ reduction in the separate anode and cathode cell compartments. The microscopic origin of the underlying catalyst reactivity in the two half reactions is completely different, but remarkably, both occur at relatively rapid rates for extended electrolysis periods. A related observation has been made with Cu as a cathode for $CO_2$ reduction and Cu(II) as the catalyst for water oxidation in aqueous carbonate solutions.

## 13.16 Stabilization of Surface Binding

A major challenge in exploiting hybrid DSPEC molecular assembly/semiconductor devices in artificial photosynthesis is stabilization of surface-bound molecules and assemblies on oxide surfaces. Although phosphonate-surface binding gives relatively stable surface phosphonate esters in acidic solution, even this surface link is unstable toward hydrolysis as the pH is increased above 5 with added buffers.

*ALD*. Progress has been made in stabilizing phosphonate surface binding by using *atomic layer deposition* (ALD) to deposit inert overlayers of $Al_2O_3$ or $TiO_2$. ALD is a gas phase, layer-by-layer conformal technique. Oxide surfaces with acidic hydroxyl groups are exposed to a reactive precursor in the gas phase—$Al(CH_3)_3$ for $Al_2O_3$,$TiCl_4$ and other precursors for $TiO_2$—followed by treatment with water which releases $CH_4$ or HCl for $TiCl_4$. This sequence of reactions creates layers or sublayers of the oxide on the surface. Sequential gas phase-surface hydrolysis cycles are used to build up multiple oxide layers on the surface.

ALD has been used to stabilize oxide surface binding. Phosphonate-derivatized chromophores, catalysts, or assemblies are first bound to the surface followed by ALD stabilization by creating added overlayers of $Al_2O_3$ or $TiO_2$ to prevent hydrolysis from the surface. A schematic diagram illustrating the $TiO_2$ surface-stabilized water oxidation catalyst $[Ru(Mebimpy)(4,4'\text{-}(PO_3H_2CH_2)_2bpy)(OH_2)]^{2+}$ is shown in Fig. 13.19. Figure 13.19 also compares current-time traces for water oxidation by ALD stabilized and unstabilized catalyst at pH 11 in a phosphate buffer. Under these conditions, the rate of water oxidation is ~$10^4\ s^{-1}$.

*Polymer Overlayers*. Stabilization of surface-bound chromophores on oxide surfaces by a poly-methylmethacrylate (PMMA) polymer film is illustrated in Fig. 13.20. In this approach, the surface-bound molecule or assembly is exposed to dichloromethane solutions containing the hydrophobic polymer. Overlayer film thicknesses can be varied reproducibly by varying the concentration of polymer in the loading solution with well-defined films formed as shown by Transmission Electron Microscopy (TEM) imaging.

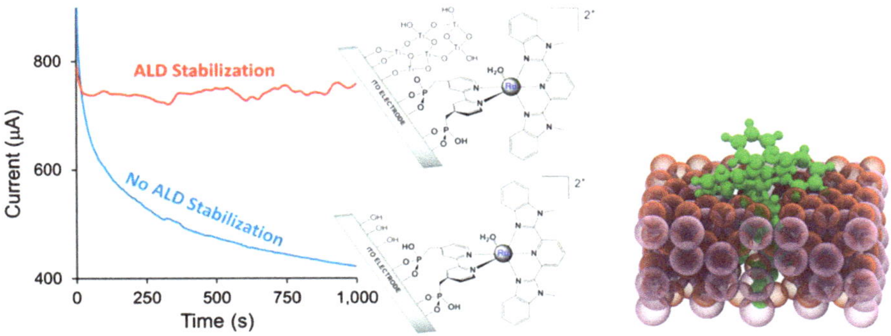


**Fig. 13.19** (*Left*) Controlled potential current-time traces for $[Ru(Mebimpy)(4,4'\text{-}((HO)_2OPCH_2)_2bpy)(OH_2)]^{2+}$ bound to *nano*ITO with (*red trace*) and without (*blue trace*) ALD $TiO_2$ overlayer stabilization. Electrolyses were performed at pH 11 in phosphate buffers at 1.6 V vs. NHE. A schematic illustration of surface stabilization by $TiO_2$ is also shown. (*Right*) Results of molecular modeling of ALD surface stabilization of $Ru((4,4'\text{-}PO_3H_2)_2bpy)(bpy)_2]^{2+}$

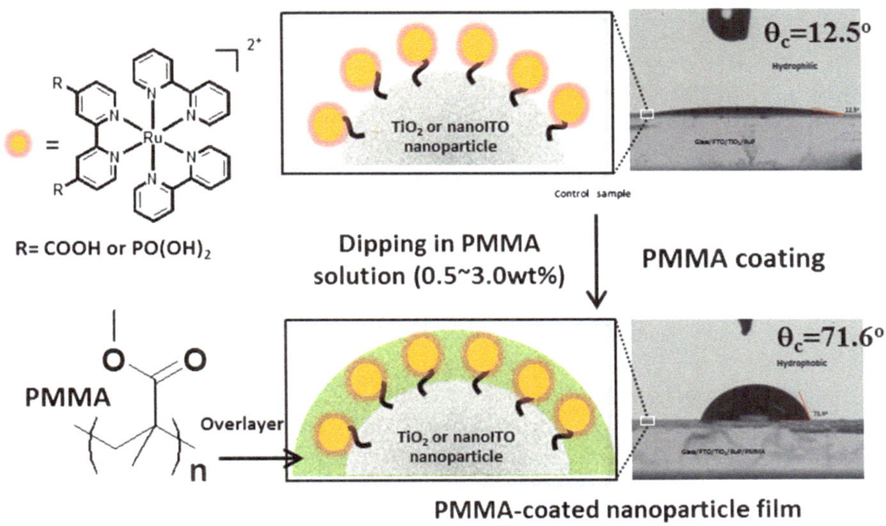


**Fig. 13.20** Cartoons and contact angle measurements illustrating PMMA stabilization of a surface-bound $[Ru(bpy)_2(4,4'\text{-}R\text{-}bpy)$ chromophore (R = $-CO_2H$ or $-PO(OH)_2$) on $TiO_2$

*Electro-Assembly*. There is an extensive literature on surface formation of electroactive films by reductive electropolymerization of vinyl-derivatized pyridyl and polypyridyl complexes. Reduction of the complexes at ligand-based π* levels induces vinyl-based C-C coupling and film formation at the electrode–solution interface. As described above, a related approach has been used to synthesize assemblies in the cavities of nanoparticle oxide films. In this “electro-assembly”

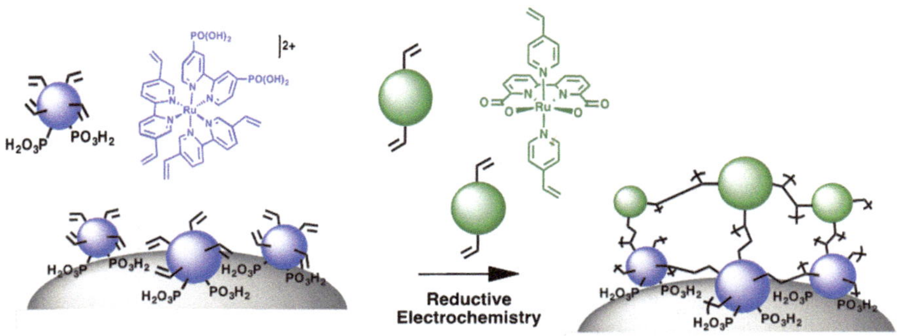


**Fig. 13.21** Illustration of electro-assembly formation by ligand-based reduction and C–C bond formation with the catalyst Ru(dba)(4-vinylpyridine)$_2$ in the external solution and [Ru(dvb)$_2$((PO$_3$H$_2$)$_2$bpy)]$^{2+}$ bound to a nanoparticle oxide electrode; dvb is 5,5′-divinyl-2,2′-bipyridine and (PO$_3$H$_2$)bpy is 2,2′-bipyridine-4,4′-diylbis(phosphonic acid)

approach, doubly derivatized, phosphonate-vinyl-2,2′-bipyridine or vinyl-pyridine complexes are first bound to the oxide surface. Voltammetric scans to negative potentials, with a vinyl-derivatized ligand water oxidation catalyst in the external solution induces C–C bond formation linking the catalyst to the chromophore and to assembly formation on the surface (Fig. 13.21).

## 13.17 DSPEC Water Splitting

Electrochemically driven water oxidation by the assembly [(4,4′-PO$_3$H$_2$)$_2$bpy)$_2$Ru(4-Mebpy-4′-bimpy)Ru(tpy)(OH$_2$)]$^{4+}$ on the surface of *nano*ITO was described in Section 3.6.11. With an applied potential sufficient to oxidize the Ru$^{III/II}$ chromophore couple and a Pt cathode, the assembly-modified electrode functions as the anode in an electrochemical cell for water splitting with $H_2O/H^+$ reduction to $H_2$ at the cathode. The same assembly bound to a nanoparticle $TiO_2$ electrode is shown in Fig. 13.15. Visible light illumination under the same conditions results in low photocurrents and highly inefficient photoelectrochemical water splitting.

Inefficient water splitting is a consequence of the unfavorable interfacial electron transfer dynamics at $TiO_2$. Timescales for the processes that dictate microscopic efficiencies and cell performance are illustrated in Fig. 13.22. The range in timescales is extraordinary. They vary from the fsec-psec timescale for injection by the chromophore excited state to the 1–2 s$^{-1}$ timescale for solar insolation to the indefinitely slow time scale for desorption in an operating device.

As noted in Fig. 13.3, for the assembly in Fig. 13.22, light absorption is followed, in sequence, by excited state injection and intra-assembly electron transfer from the catalyst to the oxidized chromophore. In appropriately designed assemblies, both are complete within a nanosecond and occur with high efficiencies Eq. (4). Following injection and catalyst oxidation, Eq. (13.10), there is a

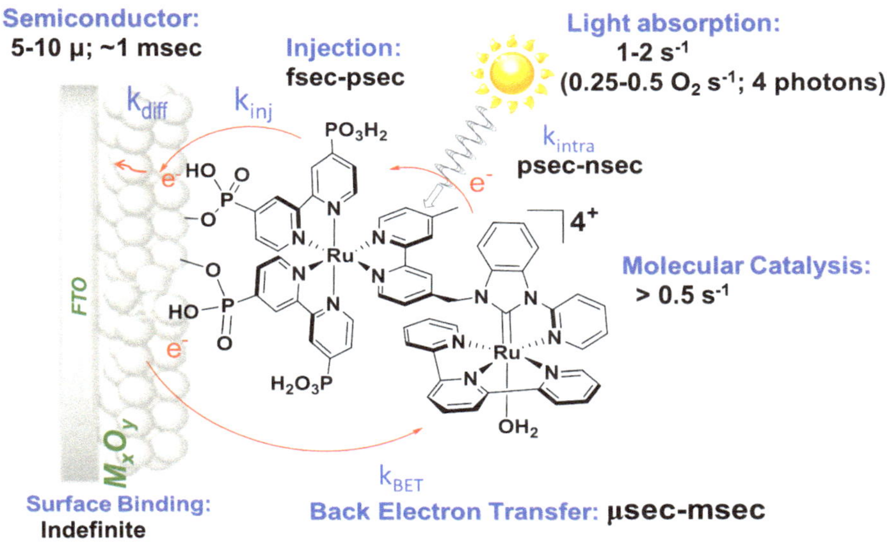


**Fig. 13.22** Timescales that determine DSPEC efficiencies and performance

competition for the injected electron between back electron transfer, Eq. (13.5a), and electron migration through the oxide film to the underlying FTO electrode for transfer to the cathode, Eq. (13.5b). As noted in a previous section, relatively rapid back electron transfer compared to electron migration necessarily leads to low overall efficiencies, especially given the requirement for four excitation/electron transfer cycles to drive water oxidation. There do appear to be mechanisms on $TiO_2$ which circumvent, in part, the unfavorable surface kinetics. In current transient absorption studies, we find evidence for a role for cross-surface electron transfer following initial $1e^-$ oxidation of the assembly which allows the $2e^-$ oxidized assembly to build up on the surface.

## 13.18 Core/Shell Electrodes

As noted in a previous section, in natural photosynthesis the effective rate of solar insolation is enhanced from 1 to 2 $s^{-1}$ into the millisecond time range by using a highly efficient "antenna" of coupled chlorophylls and auxiliary pigments. In addition, following sensitization of $^1P_{680}$*, in the Reaction Center, a sequence of five electron transfer reactions is used to create and separate the oxidative and reductive equivalents by ~50 Å greatly slowing the rate of back electron transfer.

We recently introduced oxide core/shell structures as a way to control interfacial electron transfer at $TiO_2$. They provide greatly enhanced efficiencies for DSPEC water splitting. In preparing these structures, ALD was used to prepare thin, 2–4 nm

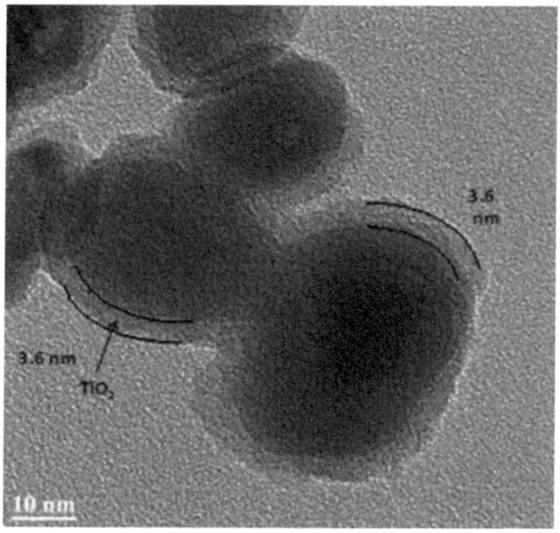


**Fig. 13.23** TEM of a *nano*ITO/$TiO_2$ core/shell electrode with a 3.6 nm $TiO_2$ shell. Reprinted with permission from the Proceedings of the National Academy of Sciences

shells of $TiO_2$ on the surfaces of nanoparticle films of *nano*ITO. Figure 13.23 shows a TEM image of a *nano*ITO/$TiO_2$ core/shell structure with 3.6 nm $TiO_2$ shell.

At FTO|*nano*ITO|$TiO_2$ core/shell electrodes, surface-derivatized by the assembly $[(4,4'\text{-}PO_3H_2)_2bpy)_2Ru(4\text{-}Mebpy\text{-}4'\text{-}bimpy)Ru(tpy)(OH_2)]^{4+}$, Fig. 13.22 ($[Ru_a{}^{II}\text{-}Ru_b{}^{II}\text{-}OH_2]^{4+}$), chromophore excitation and injection, Eq. (13.10), are followed by electron transport through the thin $TiO_2$ shell to the conducting *nano*ITO core, Eq. (13.11). Because of the thin shell of $TiO_2$, electron transport to the core occurs on the sub-nanosecond timescale rather than on the millisecond timescale. The injected electrons are transported rapidly to the FTO collector electrode allowing oxidative equivalents to accumulate at the catalyst.

$$\begin{aligned}&\text{FTO}|\text{nanoITO}|\text{TiO}_2|\left[\text{Ru}_a{}^{\text{II}}\text{-Ru}_b{}^{\text{II}}\text{-OH}_2\right]^{4+} + h\nu \rightarrow \\ &\quad \text{FTO}|\text{nanoITO}|\text{TiO}_2(\text{e}^-)| - \left[\text{Ru}_a{}^{\text{II}}\text{-Ru}_b{}^{\text{III}}\text{-OH}_2\right]^{5+}\end{aligned} \tag{13.10}$$

$$\begin{aligned}&\text{FTO}|\text{nanoITO}|\text{TiO}_2(\text{e}^-)| - \left[\text{Ru}_a{}^{\text{II}}\text{-Ru}_b{}^{\text{III}}\text{-OH}_2\right]^{5+} \rightarrow \\ &\quad \text{FTO}(\text{e}^-)|\text{nanoITO}|\text{TiO}_2| - \left[\text{Ru}_a{}^{\text{II}}\text{-Ru}_b{}^{\text{III}}\text{-OH}_2\right]^{5+}\end{aligned} \tag{13.11}$$

Back electron transfer by tunneling through the thin $TiO_2$ layer to the oxidized assembly is competitive with electron migration as shown by per photon absorbed efficiencies for water splitting with monochromatic irradiation of only 4.4 %. In more recent studies, results with the same assembly on a $SnO_2$/$TiO_2$ core/shell were used to obtain per photon efficiencies of ~20 % with monochromatic light. In the $SnO_2$/$TiO_2$ core/shell, back electron transfer is inhibited because of the conduction band offset between the oxides with $SnO_2$ ($e^-$) → $TiO_2$ electron transfer thermodynamically unfavorable by ~0.4 eV.

Figure 13.24 shows the final core/shell configuration at the photoanode for DSPEC water splitting, FTO|*nano*ITO|$TiO_2$|–$[Ru_a{}^{II}\text{-}Ru_b{}^{II}\text{-}OH_2]^{4+}$, and the

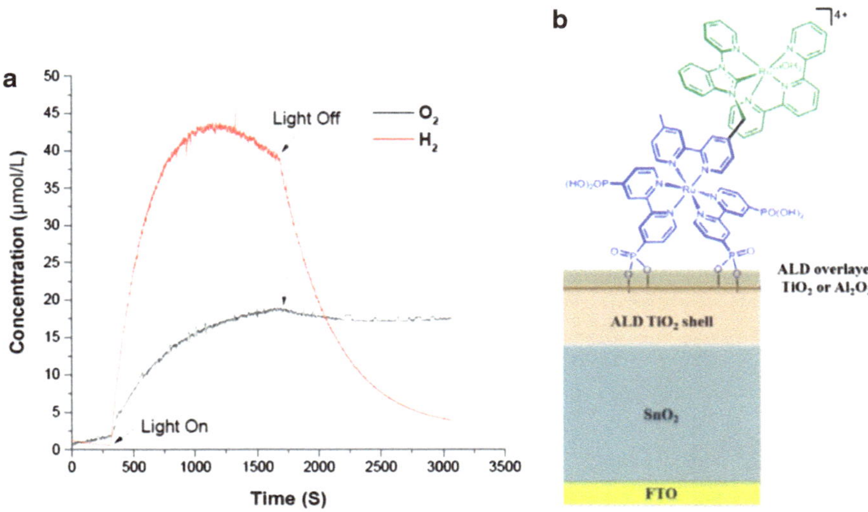


**Fig. 13.24** DSPEC water splitting. (**a**) Direct monitoring of $O_2$ (photoanode) and $H_2$ (cathode) by microelectrode sensors with 455 nm 46.2 mW/cm$^2$ LED excitation with a 600 mV applied bias in a 0.1 M $H_2PO_4^-/HPO_4^{2-}$ buffer at pH 7 with the ionic strength adjusted to 0.5 M with $NaClO_4$. (**b**) Structure of the surface-stabilized core/shell assembly

appearance of $O_2$ and $H_2$ by direct microelectrode monitoring. As shown in Fig. 13.24b, the assembly is surface-stabilized by an ALD overlayer of $TiO_2$ or $Al_2O_3$.

## 13.19 DSPEC: Design Strategies

A variety of chromophore–catalyst assembly strategies and approaches have been reported with additional examples under investigation (Fig. 13.25). Surface binding by co-surface loading by surface adsorption, phosphonate surface binding, and the use of overlayer films of ion exchange polymers for cationic catalysts have all been reported. Notable has been the work of Sun and coworkers who have used co-loading strategies with phosphonate-derivatized chromophores, such as [Ru((4,4′-$PO_3H_2)_2$bpy)(bpy)$_2]^{2+}$, in combination with phosphonate-derivatized bda catalysts.

In our current research, a number of assembly-based photoanode structures are under investigation. Vinyl-based reductive electro-assembly was illustrated in Fig. 13.22. As shown in Fig. 3.25, other examples include: (1) linear chain electro-assemblies with multiple chromophores for antenna applications, (2) layer-by-layer assemblies with Zr(IV)-phosphonate bridging and energy-tuned porphyrins, (3) ALD-stabilized assemblies with the inner chromophore ([Ru((4,4′-$PO_3H_2)_2$bpy)(bpy)$_2]^{2+}$) buried by an overlayer of $Al_2O_3$ and the catalyst [Ru(Mebimpy)(bpy)

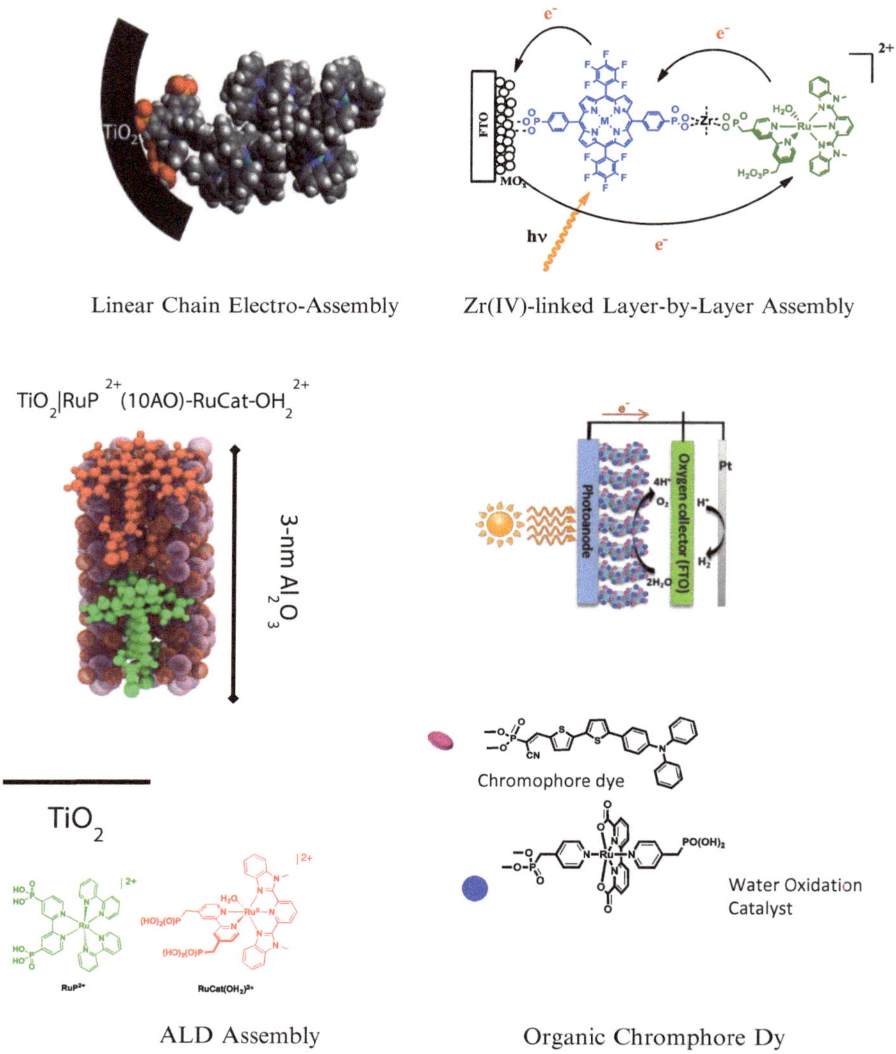


**Fig. 13.25** Chromophore–catalyst assembly structures under investigation for DSPEC photoanode applications. Reprinted with permission from the Royal Society of Chemistry

$(H_2O)]^{2+}$ added and surface-stabilized by ALD. (4) An organic dye, co-loaded with a bda water oxidation catalyst.

Experimental protocols for the evaluation of DSPEC devices are evolving based on earlier studies on semiconductor water splitting. Typically, the results of relatively short term photolyses have been reported because of the instability of the hybrid molecular-oxide interfacial structures under photoelectrochemical conditions. Based on the limited quantitative data available, photocurrents are typically rate-limited by the rate of water oxidation with only a minor influence from variations in the incident light intensity. Cell efficiencies on $TiO_2$ are greatly enhanced by the use of core/shell structures to control interfacial dynamics and avoid back electron transfer. As noted above, on $TiO_2$ an important role may be at play for cross-surface electron transfer enabling the buildup of oxidized assemblies on the oxide surface.

In order to obtain reasonable photocurrent levels, added buffer bases such as $Ac^-$ with $pK_a(HA) = 4.75$ and $HPO_4^{2-}$ with $pK_a(H_2PO_4^-) = 7.2$ are typically added to enhance the rate of water oxidation. However, as the pH is increased, base-catalyzed hydrolysis from the surface is accelerated with stabilization of surface binding by ALD or other approaches essential. Increased pH and buffer base concentrations also induce ligand-based, oxidative decomposition in the oxidized forms of the polypyridyl chromophores. The rates of ligand decomposition are also enhanced with increasing pH and buffer base concentration leading to significant, nonproductive background photocurrents. Under these conditions, quantitative evaluation of cell performance by $O_2$ measurements is essential. Contributions to photocurrents and $H_2$ evolution include contributions from chromophore decomposition as well as water splitting.

## 13.20 Looking Ahead: Challenges for the Future

Significant progress has been made in evolving the DSPEC concept past the drawing board toward laboratory-scale prototypes. Much remains to be done in validating the concept as a viable approach to solar energy conversion and storage. Key elements in future research in this area will include:

1. *Light Absorption*: Light absorption and spectral coverage are keys to maximizing cell efficiencies. Extending light absorption further into the visible must be balanced by redox potentials for the lowest excited state sufficient for electron injection into $TiO_2$ or $SnO_2$ or hole injection into NiO or another p-type semiconductor. The resulting oxidized or reduced forms of the chromophore must be sufficiently oxidizing to drive water oxidation catalysis at the photoanode or water/$CO_2$ reduction at the photocathode. For bias-free water splitting or $CO_2$ reduction in tandem cells, the solar spectrum needs to be maximally split between the photoanode and photocathode.

   The ultimate light absorbers will be inexpensive organic dyes with broad spectral coverage subject to the energy constraints of injection and water oxidation or water/$CO_2$ reduction, all in aqueous environments. As in natural photosynthesis, kinetically linked multi-chromophoric arrays could be advantageous for concentrating the solar input at single catalyst sites to enhance rates of excitation and catalyst activation in competition with back electron transfer.

2. *Interfacial Dynamics*: The combination of photocurrent and transient absorption measurements provide powerful tools for predicting and analyzing DSPEC electrode efficiencies. They will continue to be used to analyze the kinetic characteristics of interfacial structures and to monitor cell performance.
3. *Interfacial Structure*: The influence of surface structure on interfacial dynamics is becoming increasingly well-documented. In the future, important issues will include the manipulation of core/shell and other surface and near-surface structures to control interfacial electron transfer dynamics and overall DSPEC performance.
4. *Water Oxidation Catalysis*: The family of Ru(II) polypyridyl water oxidation catalysts has demonstrated promise in DSPEC photoanode applications both in rates with added buffer bases and longevity. Extensions to acidic solutions with new catalysts would enhance surface stability and aid long term performance. Significant progress has been made on Cu(II)-based water oxidation catalysis with promise in future DSPEC applications. Further developments in the coordination chemistry of other first row metals in highly stabilized ligand environments will, hopefully, lead to new families of first row catalysts.
5. *Carbon Dioxide Reduction Catalysis*: Rapid, efficient Ir, Re, and Ru-based catalysts for $CO_2$ reduction to formate or syngas mixtures are available for use in DSPEC photocathodes. They remain to be integrated into photocathode assemblies based on p-type oxides or transparent conducting oxides for tandem cell applications with the chemistry extended to first row catalysts.
6. *DSPEC Design*: A number of promising assembly strategies have emerged for photoanode applications with further investigation required to establish the most promising leads for DSPEC applications. Assemblies for $CO_2$ reduction need to be developed for photocathode applications and integrated with photoanodes for use in tandem cell configurations.
7. *Stabilization*: Stabilization strategies for long term performance based on ALD will continue to evolve with additional strategies under investigation for possible integration with new assembly structures as they emerge.

**Acknowledgments** Funding by the UNC EFRC Center for Solar Fuels, an EFRC funded by the US DOE, Office of Science, Office of Basic energy Sciences, under award number DE-SC0001011 is gratefully acknowledged.

## References

Alibabaei L, Brennaman M, Norris M, Kalanyan B, Song W, Losego M, Concepcion J, Binstead R, Parsons G, Meyer T (2013) Solar water splitting in a molecular photoelectrochemical cell. Proc Natl Acad Sci U S A 110:20008

Alibabaei L, Farnum B, Kalanyan B, Brennaman M, Losego M, Parsons G, Meyer T (2014) Atomic layer deposition of $TiO_2$ on mesoporous nanoITO: conductive core-shell photoanodes for dye-sensitized solar cells. Nano Lett 14:3255

Ardo S, Meyer G (2009) Photodriven heterogeneous charge transfer with transition-metal compounds anchored to $TiO_2$ semiconductor surfaces. Chem Soc Rev 38:115
Ashford D, Stewart D, Glasson C, Binstead R, Harrison D, Norris M, Concepcion J, Fang Z, Templeton J, Meyer T (2012a) An amide-linked chromophore-catalyst assembly for water oxidation. Inorg Chem 51:6428
Ashford D, Song W, Concepcion J, Glasson C, Brennaman M, Norris M, Fang Z, Templeton J, Meyer T (2012b) Photoinduced electron transfer in a chromophore-catalyst assembly anchored to $TiO_2$. J Am Chem Soc 134:19189
Ashford D, Lapides A, Vannucci A, Hanson K, Torelli D, Harrison D, Templeton J, Meyer T (2014) Water oxidation by an electropolymerized catalyst on derivatized mesoporous metal oxide electrodes. J Am Chem Soc 136:6578
Ashford D, Sherman B, Binstead R, Templeton J, Meyer T (2015) Electro-assembly of a chromophore-catalyst bilayer for water oxidation and photocatalytic water splitting. Angew Chem Int Ed 54:4778
Ashmawy F, McAuliffe C, Parish R, Tames J (1985) Water photolysis. Part 1. the photolysis of coordinated water in $[\{MnL\text{-}(H_2O)\}_2][ClO_4]_2$ (L=Dianion of Tetradentate $O_2N_2$-Donor Schiff Bases). A model for the manganese site in photosystem II of green plant photosynthesis. J Chem Soc Dalton Trans 1985:1391
Barnett S, Goldberg K, Mayer J (2012) A soluble copper–bipyridine water-oxidation electrocatalyst. Nat Chem 4:498
Barton C, Lakkaraju P, Rampulla D, Morris A, Abelev E, Bocarsly A (2010) Using a one-electron shuttle for the multielectron reduction of $CO_2$ to methanol: kinetic, mechanistic, and structural insights. J Am Chem Soc 132:11539
Bettis S, Hanson K, Wang L, Gish M, Concepcion J, Fang Z, Meyer T, Papanikolas J (2014) Photophysical characterization of a chromophore/water oxidation catalyst containing a layer-by-layer assembly on nanocrystalline $TiO_2$ using ultrafast spectroscopy. J Phys Chem A 118:10301
Binstead R, McGuire M, Dovletoglou A, Seok W, Roecker L, Meyer T (1992) Oxidation of hydroquinones by [(bpy)2(py)RuIV(O)]2+ and [(bpy)2(py)RuIII(OH)]2+. Proton-coupled electron transfer. J Am Chem Soc 114:173
Bock C, Meyer T, Whitten D (1974) Electron transfer quenching of the luminescent excited state of Tris(2,2′-bipyri¬dine)ruthenium(II). A flash photolysis relaxation technique for measuring the rates of very rapid electron transfer reactions. J Am Chem Soc 96:4710
Chen Z, Concepcion J, Jurss J, Meyer T (2009) Single-site, catalytic water oxidation on oxide surfaces. J Am Chem Soc 131:15580
Chen Z, Concepcion J, Hu X, Yang W, Hoertz P, Meyer T (2010a) Concerted O atom-proton transfer in the O—O bond forming step of water oxidation. Proc Natl Acad Sci U S A 107:7225
Chen Z, Concepcion J, Hull J, Hoertz P, Meyer T (2010b) Catalytic water oxidation on derivatized nanoITO. Dalton Trans 39:6950
Chen Z, Concepcion J, Luo H, Hull J, Paul A, Meyer T (2010c) Nonaqueous catalytic water oxidation. J Am Chem Soc 132:17670
Chen Z, Concepcion J, Meyer T (2011a) Rapid catalytic water oxidation by a single site, Ru carbene catalyst. Dalton Trans 40:3789
Chen Z, Vannucci A, Concepcion J, Jurss J, Meyer T (2011b) Electrocatalytic reduction of $CO_2$ to CO by polypyridyl ruthenium complexes. Proc Natl Acad Sci U S A 108:1461
Chen Z, Chen C, Weinberg D, Weinberg D, Kang P, Concepcion J, Harrison D, Brookhart M, Meyer T (2011c) Chem Commun 47:12607
Chen Z, Kang P, Zhang M-T, Stoner B, Meyer T (2013) Cu(II)/Cu(0) electrocatalyzed $CO_2$ and $H_2O$ splitting. Energ Environ Sci 6:813
Clifford J, Martinez-Ferrero E, Viterisi A, Palomares E (2011) Sensitizer molecular structure-device efficiency relationship in dye sensitized solar cells. Chem Sci Rev 40:1635
Coggins M, Zhang M-T, Vannucci A, Dares C, Meyer T (2014a) Electrocatalytic water oxidation by an amidate-ligated Fe(III)-aqua complex. J Am Chem Soc 136:5531

Coggins M, Zhang M-T, Chen Z, Song N, Meyer T (2014b) Single-site $Cu^{II}$ water oxidation electrocatalysis. Rate enhancements with $HPO_4^{2-}$ as a proton acceptor at pH8**. Angew Chem Int Ed 53:12226

Concepcion J, Jurss J, Hoertz P, Meyer T (2009) Catalytic and surface-electrocatalytic water oxidation by redox mediator–catalyst assemblies. Angew Chem Int Ed 48:9473

Concepcion J, Tsai M-K, Muckerman J, Meyer T (2010a) Mechanism of water oxidation by single-site ruthenium complex catalysts. J Am Chem Soc 132:1545

Concepcion J, Jurss J, Norris M, Chen Z, Templeton J, Meyer T (2010b) Catalytic water oxidation by single site ruthenium catalysts. Inorg Chem 49:1277

Duan L, Bozoglian F, Mandal S, Stewart B, Privalov T, Llobet A, Sun L (2012) A molecular ruthenium catalyst with water-oxidation activity comparable to that of photosystem II. Nat Chem 4:418

Durrant J, Haque S, Palomares E (2004) Towards optimisation of electron transfer processes in dye sensitised solar cells. Coord Chem Res 248:1247

Dwyer F, Gyarfas E, Rogers W, Koch J (1952) Biological activity of complex ions. Nature 170:190

Ellis W, McDaniel N, Bernhard S, Collins T (2010) Fast water oxidation using iron. J Am Chem Soc 132:10990

Fujishima A, Honda K (1972) Electrochemical photolysis of water at a semiconductor electrode. Nature 238:37

Gagliardi C, Vannucci A, Concepcion J, Chen Z, Meyer T (2012) The role of proton coupled electron transfer in water oxidation. Energ Environ Sci 5:7704

Gao Y, Ding X, Liu J, Wang L, Lu Z, Li L, Sun L (2013) Visible light driven water splitting in a molecular device with unprecedentedly high photocurrent density. J Am Chem Soc 135:4219

Gersten S, Samuels G, Meyer T (1982) Catalytic oxidation of water by an oxo-bridged ruthenium dimer. J Am Chem Soc 104:4029

Gilbert J, Egglesteon D, Murphy W, Geselowitz D, Gersten S, Hodgson D, Meyer T (1985) Structure and redox properties of the water-oxidation catalyst $[(bpy)_2(OH_2)RuORu(OH_2)(bpy)_2]_4+$. J Am Chem Soc 107:3855

Glasson C, Song W, Ashford D, Vannucci A, Chen Z, Concepcion J, Holland P, Meyer T (2012) Self-assembled bilayers on indium-tin oxide (SAB-ITO) electrodes: a design for chromophore-catalyst photoanodes. Inorg Chem 51:8637

Gratzel M (2009) Recent advances in sensitized mesoscopic solar cells. Acc Chem Res 42:1788

Hanson K, Torelli D, Vannucci A, Brennaman M, Luo H, Alibabaei L, Song W, Ashford D, Norris M, Glasson C, Concepcion J, Meyer T (2012) Self-assembled bilayer films of Ru(II) polypyridyl complexes by layer-by-layer deposition on nanostructured metal oxides. Angew Chem Int Ed 51:12782

Hong D, Yamada Y, Nagatomi T, Takai Y, Fukuzumi S (2012) Catalysis of nickel ferrite for photocatalytic water oxidation using $[Ru(bpy)_3]^{2+}$ and $S_2O_8^{2-}$. J Am Chem Soc 134:19572

Hurst J (2005) Water oxidation catalyzed by dimeric μ-oxo bridged ruthenium diimine complexes. Coord Chem Rev 249:313

Hurst J, Zhou J, Lei Y (1992) Pathways for water oxidation catalyzed by the (.mu.-oxo)bis[aquabis(bipyridine)ruthenium](4+) ion. Inorg Chem 31:1010

Ishida T, Terada K-I, Hasegawa K, Kuwahata H, Kusama K, Sato R, Nakano M, Haga M-A (2009) Self-assembled monolayer and multilayer formation using redox-active Ru complex with phosphonic acids on silicon oxide surface. Appl Surf Sci 255:8824

Kang P, Cheng C, Chen Z, Schauer C, Meyer T, Brookhart M (2012) Selective electrocatalytic reduction of $CO_2$ to formate by water-stable iridium dihydride pincer complexes. J Am Chem Soc 134:5500

Kang P, Meyer T, Brookhart M (2013) Selective electrocatalytic reduction of carbon dioxide to formate by a water-soluble iridium pincer catalyst. Chem Sci 4:3497

Kang P, Zhang S, Meyer T, Brookhart M (2014a) Rapid, selective electrocatalytic reduction of carbon dioxide to formate by an iridium pincer catalyst immobilized on carbon nanotube electrodes. Angew Chem Int Ed 53:8709

Kang P, Chen Z, Nayak A, Zhang S, Meyer T (2014b) Single catalyst electrocatalytic reduction of $CO_2$ in water to $H_2$+CO syngas mixtures with water oxidation to $O_2$. Energ Environ Sci 7:4007

Kim D, Losego M, Hanson K, Alibabaei L, Lee K, Meyer T, Parsons G (2014) Stabilizing chromophore binding on $TiO_2$ for long-term stability of dye-sensitized solar cells using multicomponent atomic layer deposition. Phys Chem Chem Phys 16:8615

Lapides A, Ashford D, Hanson K, Torelli D, Templeton J, Meyer T (2013) Stabilization of a ruthenium(II) polypyridyl dye on nanocrystalline $TiO_2$ by an electropolymerized overlayer. J Am Chem Soc 135:15440

Lee H, Kepley L, Hong H, Mallouk T (1988) Inorganic analogs of Langmuir-Blodgett films: adsorption of ordered zirconium 1,10-decanebisphosphonate multilayers on silicon surfaces. J Am Chem Soc 110:618

Leem G, Morseth Z, Puodziukynaite E, Jiang J, Fang Z, Gilligan A, Reynolds J, Papanikolas J, Schanze K (2014) Light harvesting and charge separation in a π-conjugated antenna polymer bound to $TiO_2$. J Phys Chem C 118:28535

Limburg J, Vrettos J, Liable-Sands L, Rheingold A, Crabtree R, Brudvig G (1999) A functional model for O-O bond formation by the $O_2$-evolving complex in photosystem II. Science 283:1524

Liu F, Concepcion J, Jurss J, Cardolaccia T, Templeton J, Meyer T (2008) Mechanisms of water oxidation from the blue dimer to photosystem II. Inorg Chem 47:1727

Machan C, Chabolla S, Yin J, Gilson M, Tezcan F, Kubiak D (2014) Supramolecular assembly promotes the electrocatalytic reduction of carbon dioxide by Re(I) bipyridine catalysts at a lower overpotential. J Am Chem Soc 136:14598

Merrifield B (1986) Solid physe synthesis. Science 232:341

Meyer T (1990) Intramolecular control of excited state electron and energy electron transfer. Pure Appl Chem 62:1003

Meyer T, Huynh M (2003) The remarkable reactivity of high oxidation state ruthenium and osmium polypyridyl complexes. Inorg Chem 42:8140

Moyer B, Meyer T (1978) Oxobis(2,2′-bipyridine)pyridineruthenium(IV) Ion, $[(bpy)_2(py)Ru{=}O]^2$. J Am Chem Soc 100:3601

Nayak A, Knauf R, Hanson K, Alibabaei L, Concepcion J, Ashford D, Dempsey J, Meyer T (2014) Synthesis and photophysical characterization of a porphyrin-Ru(II) polypyridyle chromophore catalyst assembly on mesoporous metal oxides. Chem Sci 5:3115

Norris M, Concepcion J, Harrison D, Binstead R, Ashford D, Fang Z, Templeton J, Meyer T (2013) Redox mediator effect on water oxidation in a ruthenium based chromophore-catalyst assembly. Am Chem Soc 135:2080

O'Regan B, Durrant J (2009) Kinetic and energetic paradigms for dye-sensitized solar cells: moving from the ideal to the real. Acc Chem Res 42:1799

O'Regan B, Gratzel M (1991) A low-cost, high-efficiency solar cell based on dye-sensitized colloidal $TiO_2$ films. Nature 353:737

Puodziukynaite E, Wang L, Schanze K, Papanikolas J, Reynolds J (2014) Poly(fluorene-*co*-thiophene)-based ionic transition-metal complex polymers for solar energy harvesting and storage applications. Polym Chem 5:2363

Renger G, Renger T (2008) Photosystem II: the machinery of photosynthetic water splitting. Photosynth Res 98:53

Ryan D, Coggins M, Concepcion J, Ashford D, Fang Z, Alibabaei L, Ma D, Meyer T, Waters M (2014) Synthesis and electrocatalytic water oxidation by electrode-bound helical peptide chromophore-catalyst assemblies. Inorg Chem 53:8120

Song W, Chen Z, Brennaman M, Concepcion J, Patrocinio A, Iha N, Meyer T (2011) Making solar fuels by artificial photosynthesis. Pure Appl Chem 83:749

Song W, Ito A, Binstead R, Hanson K, Luo H, Brennaman M, Concepcion J, Meyer T (2013) Accumulation of multiple oxidative equivalents at a single site by cross-surface electron transfer on $TiO_2$. J Am Chem Soc 135:11587

Song N, Concepcion J, Binstead R, Rudd J, Vannucci A, Dares C, Coggins M, Meyer T (2015) Base enhanced catalytic water oxidation by a carboxylate-bipyridine Ru(II) complex. Proc Natl Acad Sci U S A 112:4935

Tamaki Y, Koike K, Morimoto T, Yamazaki Y, Ishitani O (2013) Red-light-driven photocatalytic reduction of $CO_2$ using Os(II)–Re(I) supramolecular complexes. Inorg Chem 52:11902

Tamaki Y, Vannucci A, Dares C, Binstead R, Meyer T (2014) One-electron activation of water oxidation catalysis. J Am Chem Soc 136:6854

Vannucci A, Alibabaei L, Losego M, Concepcion J, Kalanyan B, Parsons G, Meyer T (2013) Crossing the divide between homogeneous and heterogeneous catalysis in water oxidation. Proc Natl Acad Sci U S A 110:20919

Wang D, Groves J (2013) Efficient water oxidation catalyzed by homogeneous cationic cobalt porphyrins with critical roles for the buffer base. Proc Natl Acad Sci U S A 110:15579

Wang D, Ghirlanda G, Allen J (2014) Water oxidation by a nickel-glycine catalyst. J Am Chem Soc 136:10198

Wasylenko D, Ganesamoorthy C, Borau-Garcia J, Berlinguette C (2011) Electrochemical evidence for catalyticwater oxidation mediated by a high-valent cobalt complex. Chem Commun 47:4249

Wasylenko D, Tatlock H, Bhandari L, Gardinier J, Berlinguette C (2013) Proton-coupled electron transfer at a $[Co\text{-}OH_x]^z$ unit in aqueous media: evidence for a concerted mechanism. Chem Sci 4:734

Wee K, Brennaman M, Alibabaei L, Farnum B, Sherman S, Lapides A, Meyer T (2014) Water oxidation by an electropolymerized catalyst on derivatized mesoporous metal oxide electrodes. J Am Chem Soc 136:13514

Weinbert D, Gagliardi C, Hull J, Murphy C, Kent C, Westlake B, Paul A, Ess D, McCafferty D, Meyer T (2012) Proton-coupled electron transfer. Chem Rev 112:4016

Yang Y, Zeitler E, Gu J, Hu Y, Bocarsly A (2013) Electrochemistry of aqueous pyridinium: exploration of a key aspect of electrocatalytic reduction of $CO_2$ to methanol. J Am Chem Soc 135:14020

Young R, Meyer T, Whitten D (1976) Electron transfer quenching of excited states of metal complexes. J Am Chem Soc 98:286

Young K, Takase M, Brudvig G (2013) An anionic N-donor ligand promotes manganese-catalyzed water oxidation. Inorg Chem 52:7615

Zhang Q, Cao C (2011) Nanostructured photoelectrodes for dye-sensitized solar cells. Nano Today 6:91

Zhang S, Yang X, Numata Y, Han L (2013a) Highly efficient dye-sensitized solar cells: progress and future challenges. Energ Environ Sci 6:1443

Zhang M-T, Chen Z, Kang P, Meyer T (2013b) Electrocatalytic water oxidation with a copper(II) polypeptide complex. J Am Chem Soc 135:2048

Zhao Y, Swiek J, Megiatoo J Jr, Sherman B, Youngblood W, Qin D, Lentz D, Moore A, Moore T, Gust D, Mallouk T (2012) Improving the efficiency of water splitting in dye-sensitized solar cells by using a biomimetic electron transfer mediator. Proc Natl Acad Sci U S A 109:15612

Zong R, Thummel R (2005) A new family of Ru complexes for water oxidation. J Am Chem Soc 127:12802

# Index





S. Giménez, J. Bisquert (eds.), *Photoelectrochemical Solar Fuel Production*,
DOI 10.1007/978-3-319-29641-8

MIX
Papier | Fördert gute Waldnutzung
FSC® C083411

Kolibri 360 GmbH
Ferdinand-Jühlke-Straße 7
99095 Erfurt, Deutschland
produktsicherheit@kolibri360.de